ESSENTIAL ALGEBRA

THE JOHNSTON/WILLIS
DEVELOPMENTAL MATHEMATICS SERIES

Essential Arithmetic, **Sixth Edition** (paperbound, 1991)
Johnston/Willis/Lazaris

Essential Algebra, **Sixth Edition** (paperbound, 1991)
Johnston/Willis/Lazaris

Developmental Mathematics, **Third Edition** (paperbound, 1991)
Johnston/Willis/Hughes

Elementary Algebra, **Third Edition** (hardbound, 1991)
Johnston/Willis/Buhr

Intermediate Algebra, **Third Edition** (hardbound, 1991)
Johnston/Willis/Buhr

Intermediate Algebra, **Fifth Edition** (paperbound, 1991)
Johnston/Willis/Lazaris

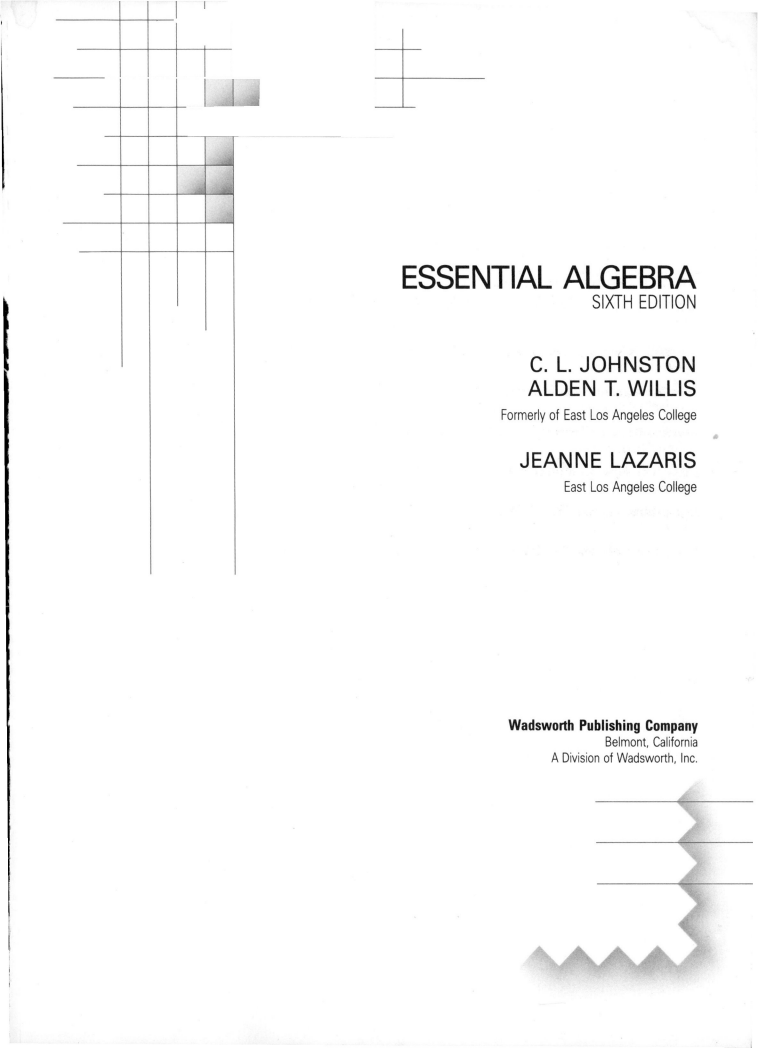

ESSENTIAL ALGEBRA
SIXTH EDITION

C. L. JOHNSTON
ALDEN T. WILLIS
Formerly of East Los Angeles College

JEANNE LAZARIS
East Los Angeles College

Wadsworth Publishing Company
Belmont, California
A Division of Wadsworth, Inc.

This book is dedicated to our students,
who inspired us to do our best
to produce a book worthy of their time.

The following problems are used with permission from J. Michael
Shaughnessy, Brian M. McCay, Carla A. Randall, Gary L. Winckler,
Application Clusters for Intermediate Algebra (Boston: Prindle, Weber
& Schmidt, 1978): Exercises 4.5, Set I, #14; Exercises 4.5, Set II, #13;
Exercises 4.6, Set I, #31, 32; Exercises 8.11, Set I, #13, 14, 15, 16;
Exercises 8.11, Set II, #15, 16.

Mathematics Editor: Anne Scanlan-Rohrer
Assistant Editor: Tamiko Verkler
Special Projects Editor: Alan Venable
Editorial Assistant: Leslie With
Production: Ruth Cottrell
Print Buyer: Randy Hurst
Designer: Julia Scannell
Copy Editor: Mary Roybal
Technical Illustration: Reese Thornton and Jonathan Peck Typographers
Compositor: Jonathan Peck Typographers
Cover: Frank Miller
Signing Representative: Kenneth King

Printed in the United States of America

2 3 4 5 6 7 8 9 10—95 94 93 92 91

Library of Congress Cataloging-in-Publication Data

Johnston, C. L. (Carol Lee), 1911–
   Essential algebra / C. L. Johnston, Alden T. Willis, Jeanne
Lazaris. —6th ed.
        p.   cm.
   Includes index.
   ISBN 0-534-14004-1
   1. Algebra.   I. Willis, Alden T.   II. Lazaris, Jeanne, 1932–
III. Title.
QA152.2.J63   1991
512—dc20   90-12802
                                                CIP

Contents

4   WORD PROBLEMS

5   EXPONENTS

6   POLYNOMIALS

10 SYSTEMS OF EQUATIONS

11 RADICALS

12 QUADRATIC EQUATIONS

APPENDIXES

A SETS 593

B BRIEF REVIEW OF ARITHMETIC 603

Preface

Essential Algebra, Sixth Edition, can be used in an elementary algebra course in any community college or four-year college, in either a lecture format or a learning laboratory setting, or it can be used for self-study. Our main goal is to prepare those students who have studied from this book for intermediate algebra and for any other course that requires the knowledge of elementary algebra.

Features of This Book
The major features of this book include the following:

1. The book uses a one-step, one-concept-at-a-time approach. The topics are divided into small sections, each with its own examples and exercises. This approach allows students to master each section before proceeding confidently to the next section.

2. Many concrete, annotated examples illustrate the general algebraic principles covered in each section. To prevent confusion, each example ends with these symbols: ———— and ■.

3. Important concepts and algorithms are enclosed in boxes for easy identification and reference.

4. The approach to solving word problems includes a detailed method for translating a word statement into an algebraic equation or inequality and a step-by-step problem-solving technique.

5. Visual aids such as shading, color, and annotations guide students through worked-out problems.

6. In special "Words of Caution," students are warned against common algebraic errors.

7. The importance of checking solutions is stressed throughout the book.

8. A review section with Set I and Set II exercise sets appears at the end of each chapter, and some chapters also have a midchapter review that includes Set I and Set II exercise sets. The Set II review exercises allow space for working problems and for answers; they can be removed from the text for grading. (Removal of these pages will not interrupt the continuity of the text.)

9. This book contains more than 6,000 exercises.
   Set I Exercises. The complete solutions for odd-numbered Set I exercises are included in the back of the book, together with the answers for all of the Set I exercises. In most cases (except in the review exercises), the even-numbered exercises are matched with the odd-numbered exercises. Thus, students can use the solutions for odd-numbered exercises as study aids for doing the even-numbered Set I exercises.
   Set II Exercises. No answers for Set II exercises are given in the text; answers to all these exercises are included in the Instructor's Manual. The odd-numbered exercises of Set II are, for the most part (except in the review exercises), matched to the odd-numbered exercises of Set I, while the even-numbered exercises of Set II are *not* so matched. Thus, while students

can use the odd-numbered exercises of Set I as study aids for doing the odd-numbered exercises of Set II, they are on their own in doing the even-numbered exercises of Set II.

10. A diagnostic test at the end of each chapter can be used for study and review or as a pretest. There are also midchapter diagnostic tests in Chapters 1 and 8. Complete solutions to all problems in these diagnostic tests, together with section references, appear in the answer section of this book.

11. A set of Cumulative Review Exercises is included at the end of each chapter except Chapter 1; the answers are in the answer section of this book.

Using This Book

Essential Algebra, Sixth Edition, can be used in three types of instructional programs: lecture, laboratory, and self-study.

The conventional lecture course. This book has been class-tested and used successfully in conventional lecture courses by the authors and by many other instructors. It is not a workbook, and therefore it contains enough material to stimulate classroom discussion. Examinations for each chapter are provided in the Instructor's Manual, and two different kinds of computer software enable instructors to create their own tests. One software program utilizes a test bank with full graphics capability, and the other is a random-access test generator. Tutorial software is available to help students who require extra assistance.

The learning laboratory class. This text has also been used successfully in many learning labs. The format of explanation, example, and exercises in each section of the book and the tutorial software make the book easy to use in laboratories. Students can use the diagnostic test at the end of each chapter as a pretest or for review and diagnostic purposes. Because several forms of each chapter test are available in the Instructor's Manual, and because test generators are available, a student who does not pass a test can review the material covered on that test and can then take a different form of the test.

Self-study. This book lends itself to self-study because each new topic is short enough for the student to master before continuing, and because more than 900 examples and over 1,500 completely solved exercises show students exactly how to proceed. Students can use the diagnostic test at the end of each chapter to determine which parts of that chapter they need to study and can thus concentrate on those areas in which they have weaknesses. The tutorial software and the random test generator, which provides answers and cross-references to the text and permits the creation of individualized work sheets, extend the usefulness of the new edition in laboratory and self-study settings.

Changes in the Sixth Edition

The sixth edition includes changes that resulted from many helpful comments from users of the first five editions as well as the authors' own classroom experience in teaching from the book. The major changes in the Sixth Edition include the following:

1. Throughout the text, we've tried to give more of an explanation of *why* we do what we do and what we're aiming toward. More details have been included in many of the examples.

2. The distributive property and the multiplicative inverse property (as they apply to the arithmetic of real numbers) and order of operations are now included in Chapter 1. A new section on word problems—"Preparation for Solving Word Problems"—has been added to Chapter 1.

3. Chapter 2 now includes several topics from Chapter 3 of the Fifth Edition: using the distributive property to simplify an algebraic expression, removing grouping symbols, and combining like terms (all with problems carefully selected so that

none of the laws of exponents are needed). Evaluating algebraic expressions and using formulas are also included in Chapter 2, and a completely new section—"Solving Word Problems by Using Formulas"—has been added.

4. Much of the material on solving equations and inequalities (now Chapter 3) has been rewritten. We now discuss solving equations by using the multiplication and division properties of equality in *one* section rather than in two separate sections.

5. A new section—"Miscellaneous Word Problems"—has been added to the chapter on solving word problems (Chapter 4). This section includes word problems for which no examples are given, and the odd- and even-numbered problems are not matched. Throughout the chapter, the emphasis is on using the general problem-solving techniques for solving word problems, with "step 1," "step 2," and so forth included in the examples.

6. Most of the material from Chapter 3 of the Fifth Edition has been moved to Chapter 5 ("Exponents"), and the section on scientific notation has been largely rewritten, with better motivation and with the inclusion of some word problems that require the use of scientific notation.

7. Chapter 6 ("Polynomials") now has a new section, "Word Expressions and Polynomials," in which we translate a word expression into a polynomial.

8. Much of the terminology in Chapter 8 (now titled "Rational Expressions") has been changed, and the definition of the *least common multiple* of two or more polynomials has been included.

9. In Chapter 9, we've given a better explanation of what the graph of an equation *is*, and more work has been included on writing the equations of lines parallel to or perpendicular to a given line.

10. We have added midchapter diagnostic tests to Chapters 1 and 8. Thus, those instructors who wish to do so can test students on the first half of each of those chapters before finishing them. The diagnostic tests at the *ends* of Chapters 1 and 8 cover the entire chapter.

Ancillaries

The following ancillaries are available with this text:

1. The Instructor's Manual contains five different tests for each chapter, two forms of three midterm examinations, five forms of the midchapter tests for Chapters 1 and 8, and two final examinations that can be easily removed and duplicated for class use. These tests are prepared with adequate space for students to work the problems. Answer keys for these tests are provided in the manual, as are the answers to the Set II exercises. The manual also contains essays to help the instructor teach developmental mathematics students. Essays cover such topics as: writing in the mathematics classroom, running a lab, cooperative learning, and more.

2. The test bank is also available in a computerized format entitled EXP-Test. EXP-Test is a fast, highly flexible computerized testing system for the IBM PC and compatibles. Instructors can edit and scramble test items or create their own tests.

3. In addition, Wadsworth offers the *Johnston/Willis/Lazaris Computerized Test Generator* (JeWeL TEST) software for Apple II and IBM PC or compatible machines. This software, written by Ron Staszkow of Ohlone College, allows instructors to produce many different forms of the same test for quizzes, work sheets, practice tests, and so on. Answers and cross-references to the text provide additional instructional support.

4. An "intelligent" tutoring software system is available for the IBM PC and compatibles. *Expert Tutor*, written by Sergei Ovchinnikov of San Francisco State University, uses a highly interactive format and sophisticated techniques to tailor lessons to the specific algebra and prealgebra learning problems of students. The result is individualized tutoring strategies with specific page references to problems, examples, and explanations in the textbook. The software has been fully revised for the new edition.

5. Nineteen videotapes, created by John Jobe of Oklahoma State University, review the most essential and difficult topics from the textbook.

To obtain additional information about these supplements, contact your Wadsworth–Brooks/Cole representative.

Acknowledgments

We wish to thank the members of the editorial staff at Wadsworth Publishing Company for their help with this edition. Special thanks go to Anne Scanlan-Rohrer, Tamiko Verkler, Leslie With, Ruth Cottrell, Mary Douglas, Alan Venable, Sally Uchizono, and Mary Roybal.

We also wish to thank our many friends for their valuable suggestions. We are deeply grateful to Gale Hughes for preparing the Instructor's Manual, and we thank the following reviewers for their helpful comments: Anthony Brunswick, Delaware Technical and Community College–Terry Campus; Karen Sue Cain, Eastern Kentucky University; Daniel Henry, Los Medanos College; Adele LeGere, Oakton Community College; Letty Macdonald, Piedmont Virginia Community College; Juanita Peterson, College of Alameda; Susan Piccione, Dundalk Community College; Douglas Robertson, University of Minnesota; Gerald Smith, Cayuga Community College; Joan Tomaszewski, SUNY–Farmingdale; and Stan Wilson, Murray State College. We'd also like to thank the following, who checked the text for mathematical accuracy: Mike Connor, Gavilan College; Letty Macdonald, Piedmont Virgina Community College; Jack Murphy, Pennsylvania College of Technology; Richard Spangler, Tacoma Community College; and Ruth Vanderkarr, Los Medanos College.

1 Operations on Real Numbers

Elementary algebra is sometimes considered "generalized arithmetic." It does deal with the six fundamental operations of arithmetic (addition, subtraction, multiplication, division, raising to powers, and extracting roots), but it includes operations on negative as well as positive numbers. In algebra, we often use letters to represent some of the numbers, and we learn to solve algebraic equations. Our major goal is to be able to use algebra to solve mathematical problems that arise in higher mathematics and science courses. Before we begin the study of algebra, we review a few basic definitions relating to sets and numbers. (See Appendix A for a more complete discussion of sets and Appendix B for a brief review of arithmetic.)

1.1 Basic Definitions

Set   A **set** is a collection of objects or things.

Example 1   Examples of sets:

a. A 48-piece set of dishes

b. A basket of birthday presents

c. A basket containing an apple, a pillow, and a cat

d. The first seven letters of our alphabet   ■

Element of a Set   The objects that make up a set are called its **elements** (or **members**). We often represent sets by listing their elements within braces { }. We never use parentheses for sets. Thus, (2, 1, 7) does *not* name a set. The order in which the elements of a set are listed is not important. A set may contain just a few elements, many elements, or no elements at all.

Example 2   Examples showing the elements of sets:

a. Set {1, 2, 3} has elements 1, 2, and 3.

b. Set {a, d, f, h, k} has elements $a, d, f, h,$ and k.

c. Set {Ben, Kay, Frank, Albert} has elements Ben, Kay, Frank, and Albert.

d. Set { } contains *no* elements. It is called the **empty set**.   ■

The Meaning of the Equal Sign   The equal sign (=) in a statement means that the expression on the left side of the equal sign *has the same value or values* as the expression on the right side of the equal sign. We say two *sets* are equal if they contain the same elements.

Naming a Set   A set is usually named by a capital letter, such as $A, N, W,$ and so on. The expression "$A = \{1, 5, 7\}$" is read "A is the set whose elements are 1, 5, and 7."

Roster Method of Representing a Set   A **roster** is a list of members of a group. When we represent a set by {3, 8, 9, 11}, we are representing the set by a roster (or list) of its members. This method of representing a set is called the **roster method**.

Important Sets of Numbers

Natural Numbers   The numbers

$$1, 2, 3, 4, 5, 6, 7, 8, 9, 10, 11, 12, \text{ and so on}$$

are called the **natural numbers** (or **counting numbers**). These were probably the first numbers invented to enable people to count their possessions. The largest natural number can never be found, because no matter how far we count there are always larger natural numbers. Since it is impossible to write all the natural numbers, it is customary to represent the set of natural numbers as follows:

{1, 2, 3, 4, . . . }
———— Read "and so on"

The three dots to the right of the number 4 indicate that the remaining numbers are to be found by counting in the same way we have begun, namely, by adding 1 to each number to find the next number. We call the set of natural numbers N; that is,

$$N = \{1, 2, 3, 4, \ldots\}$$

Number Line   Natural numbers can be represented by numbered points equally spaced along a straight line, as in Figure 1.1.1. Such a line is called a **number line**.

```
   ┬  ┬  ┬  ┬  ┬  ┬  ┬  ┬  ┬  ┬  ┬  ┬  ──▶
   0  1  2  3  4  5  6  7  8  9  10 11
```

FIGURE 1.1.1   NUMBER LINE

We put an arrowhead at the right, showing the direction in which numbers get larger; some authors put arrowheads at *both* ends of the number line. Later we will discuss other kinds of numbers, such as fractions, decimal numbers, and negative numbers, that can be placed on the number line.

The Graph of a Number   We **graph a number** by placing a dot on the number line above that number.

Whole Numbers   When 0 is included with the natural numbers, we have the set of **whole numbers**, which we call W. Figure 1.1.2 shows the graphs of the first twelve whole numbers.

$$W = \{0, 1, 2, 3, \ldots\}$$

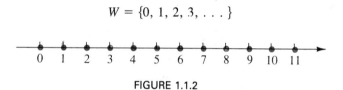

```
   ●  ●  ●  ●  ●  ●  ●  ●  ●  ●  ●  ●  ──▶
   0  1  2  3  4  5  6  7  8  9  10 11
```

FIGURE 1.1.2

Set of Digits   One important set of numbers is the set of **digits**. This set contains the numbers 0, 1, 2, 3, 4, 5, 6, 7, 8, and 9. These symbols make up our entire number system; *any* number can be written by using some combination of these digits.

Numbers are often referred to as *one-digit* numbers, *two-digit* numbers, *three-digit* numbers, and so on. Also, we sometimes wish to refer to the *first*, *second*, or *third* digit of a number; when we do this, we count from left to right.

Example 3   Examples to show the use of digits:

a. An example of a two-digit number is 35.

b. An example of a one-digit number is 7.

c. Three-digit numbers include 275.

d. The first digit of 785 is 7.

e. The third digit of 785 is 5.   ■

Fractions   A **fraction** is an indicated division; the fraction $\dfrac{a}{b}$ is equivalent to the division $a \div b$. (While $\dfrac{a}{b}$ is the preferred form, in text fractions are often denoted a/b.) We call a and b the **terms** of the fraction. In this section, we consider only those fractions in which the *numerator* a is a whole number and the *denominator* b is a natural number. *The denominator of a fraction can never equal zero.*

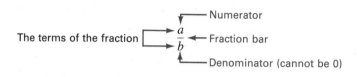

The number of the fraction
— Numerator
$\dfrac{a}{b}$
— Fraction bar
— Denominator (cannot be 0)

When the numerator of a fraction is less than the denominator, we can think of the fraction as being part of a whole. In this case, the denominator tells us how many equal parts the whole has been divided into, and the numerator tells us how many of those equal parts are being considered.

Example 4   An example of the meaning of a fraction:

The fraction $\dfrac{3}{4}$ is equivalent to the division problem $3 \div 4$. The numerator is 3 and the denominator is 4.

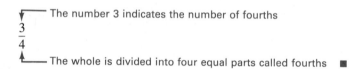

── The number 3 indicates the number of fourths
$\dfrac{3}{4}$
── The whole is divided into four equal parts called fourths ■

Mixed Numbers   A **mixed number** is made up of both a whole number part and a fraction part. There is an understood plus sign between the whole number part and the fraction part.

Example 5   Examples of mixed numbers:

a. $2\dfrac{1}{2}$ means $2 + \dfrac{1}{2}$

b. $3\dfrac{5}{8}$ means $3 + \dfrac{5}{8}$

c. $5\dfrac{1}{4}$ means $5 + \dfrac{1}{4}$

d. $12\dfrac{3}{16}$ means $12 + \dfrac{3}{16}$ ■

Decimal Fractions   A **decimal fraction** is a fraction whose denominator is 10, or 100, or 1,000, and so on.

Example 6   Examples of decimal fractions:

a. $\dfrac{4}{10} = 0.4$ Read "four tenths"

b. $\dfrac{5}{100} = 0.05$ Read "five hundredths"

c. $\dfrac{6}{1,000} = 0.006$ Read "six thousandths"

d. $\dfrac{23}{10} = 2.3$ ◀── Read "two and three tenths"

── Read "twenty-three tenths" ■

Decimal Places   The number of decimal places in a number is the number of digits written to the right of the decimal point.

Example 7   Examples of the number of decimal places in a number:

a. 75.14 has two decimal places.

b. 1.086 has three decimal places.

c. 2.5000 has four decimal places. ■

Real Numbers   Natural numbers, whole numbers, fractions, decimals, and mixed numbers are all elements of a set we call the set of **real numbers**, and they can all be represented by points on the number line. Points representing a few fractions, decimals, and mixed numbers are shown in Figure 1.1.3.

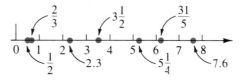

FIGURE 1.1.3

Other kinds of real numbers are introduced in Sections 1.2 and 1.10, and a chart that shows the relationships among the real numbers is given in Section 1.10. It is shown in higher-level mathematics courses that any number that can be represented by a point on the number line is a real number.

"Greater Than" and "Less Than" Symbols   The symbol $>$ is read "is greater than," and the symbol $<$ is read "is less than." These *inequality symbols* are among the symbols we can use between numbers that are *not* equal to each other. Numbers get larger as we move to the right on the number line; recall that the arrowhead at the right of the number line indicates the direction in which numbers get larger. Numbers get smaller as we move to the left.

Example 8 $5 > 2$ is read "5 is greater than 2."

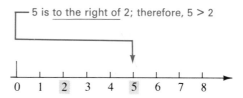

■

Example 9 $2 < 5$ is read "2 is less than 5."

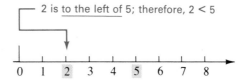

■

Notice that $5 > 2$ and $2 < 5$ give the same information, even though they are read differently. An easy way to remember the meaning of the symbol is to notice that the wide part of the symbol is next to the larger number. Some people like to think of the symbols $>$ and $<$ as arrowheads that point toward the smaller number.

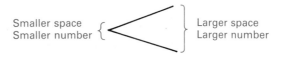

Example 10 Inequalities are written and read as follows:

a. $7 > 6$ is read "7 is greater than 6."

b. $7 > 1$ is read "7 is greater than 1."

——— c. $5 < 10$ is read "5 is less than 10." ∎

Another inequality symbol is $\neq$. A slash line drawn through a symbol puts a *not* in the meaning of the symbol.

Example 11 Examples of the use of the slash line:

a. $=$ is read "is equal to."
$\neq$ is read "is *not* equal to."

b. $<$ is read "is less than."
$\not<$ is read "is *not* less than."

c. $>$ is read "is greater than."
$\not>$ is read "is *not* greater than."

d. $4 \neq 5$ is read "4 is *not* equal to 5."

e. $3 \not< 2$ is read "3 is *not* less than 2."

——— f. $5 \not> 6$ is read "5 is *not* greater than 6." ∎

EXERCISES 1.1

Set I

1. What is the second digit of the number 159?

2. What is the fourth digit of the number 1,975?

3. What is the smallest natural number?

4. What is the smallest digit?

5. What is the largest one-digit natural number?

6. What is the largest two-digit whole number?

7. What is the smallest two-digit natural number?

8. What is the smallest three-digit whole number?

9. Which symbol, $<$ or $>$, should be used to make the statement 10 _?_ 0 true?

10. Which symbol, $<$ or $>$, should be used to make the statement 3 _?_ 15 true?

11. Which symbol, $<$ or $>$, should be used to make the statement 8 _?_ 7 true?

12. Which symbol, $<$ or $>$, should be used to make the statement 0 _?_ 1 true?

13. Is 2.3 a real number?

14. Is $\frac{7}{8}$ a real number?

15. Is 1.8 a real number?

16. Is $\frac{9}{5}$ a real number?

17. Is 1.8 a natural number?

18. Is $\frac{3}{4}$ a natural number?

19. Is 15 a digit?

20. Is 15 a natural number?

21. How many decimal places are there in the number 7.010?

22. How many decimal places are there in the number 41.0005?

Set II
1. What is the third digit of the number 3,187?

2. Is 12 a natural number?

3. What is the largest natural number?

4. What is the smallest whole number?

5. What is the largest digit?

6. What is the smallest three-digit natural number?

7. Is 58.4 a real number?

8. What is the smallest one-digit natural number?

9. Which symbol, < or >, should be used to make the statement 18 _?_ 5 true?

10. Which symbol, < or >, should be used to make the statement 8 _?_ 17 true?

11. Which symbol, < or >, should be used to make the statement 11 _?_ 6 true?

12. Write all the whole numbers < 4.

13. Is $5\frac{1}{2}$ a real number?

14. Is $5\frac{1}{2}$ a natural number?

15. Is 0 a real number?

16. Is 58.4 a natural number?

17. Is 3,628 a natural number?

18. Is $\frac{2}{5}$ a real number?

19. Is 12 a digit?

20. Is $\frac{2}{3}$ a digit?

21. How many decimal places are there in the number 50.602?

22. How many decimal places are there in the number 23.0?

1.2 Negative Numbers and Rational Numbers

In Section 1.1, we showed that whole numbers can be represented by points equally spaced along the number line. We now extend the number line to the left of zero and continue with the equally spaced points.

Numbers used to name the points to the left of zero on the number line are called **negative numbers**. Thus, the point two units to the left of zero is called −2 (read "negative two"), and the point one-half unit to the left of zero is called −1/2 (read "negative one-half"). Numbers used to name the points to the right of zero on the number line are called **positive numbers** (see Figure 1.2.1). Zero itself is neither positive nor negative. The positive and negative numbers are referred to as **signed numbers**.

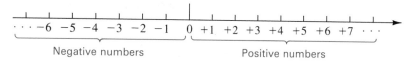

FIGURE 1.2.1

Integers   The set of **integers** includes the set of whole numbers as well as the numbers $-1, -2, -3$, and so on; it can be represented in the following way:

$$\{\ldots, -3, -2, -1, 0, +1, +2, +3, \ldots\}$$

When reading or writing positive numbers, we usually omit the word *positive* and the plus sign. Therefore, when there is no sign in front of a number, it is understood to be positive.

Example 1   Examples of reading positive and negative integers:

a. -1 is read "negative one."

b. -575 is read "negative five hundred seventy-five."

c. 25 is read "twenty-five" or "positive twenty-five."   ■

Example 2   On an unusually cold day in Minnesota, the temperature was $-40°$ F. This means that the temperature was $40°$ F below $0°$ F.   ■

Example 3   The altitudes of some unusual places on earth are as follows:

a.   Mt. Everest   29,028 ft
     This means that the peak of Mt. Everest
     is 29,028 feet *above* sea level.

b.   Mt. Whitney (California)   14,494 ft

c.   Lowest point in Death Valley (California) -282 ft
     This means that the lowest point in
     Death Valley is 282 ft *below* sea level.

d.   Dead Sea (Jordan) $-1,299$ ft

e.   Mariana Trench (Pacific Ocean) $-36,198$ ft   ■

Using Inequality Symbols with Integers   Recall that any number to the right of a given number is greater than the given number, and any number to the left of a given number is less than the given number. These facts are true for negative integers (and for other negative numbers) as well as for positive numbers.

Example 4 $-3 > -5$ is read "-3 is greater than -5."

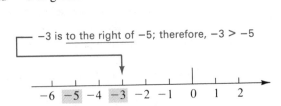

■

Example 5 $-5 < -1$ is read "-5 is less than -1."

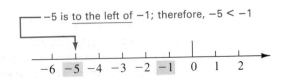

■

Example 6   Verify the following inequalities by noting whether the first number of each pair is to the right or left of the second number on the number line:

a. $6 > 4$ 6 is to the right of 4

b. $0 > -1$ 0 is to the right of -1

c. $-2 > -5$ -2 is to the right of -5

d. $-20 < -10$ -20 is to the left of -10

e. $-5 < 3$ -5 is to the left of 3   ■

Rational Numbers   A **rational number** is any number that can be expressed in the form a/b, where a and b are integers and $b \neq 0$. All integers and all fractions are rational numbers, and all rational numbers are real numbers. When a rational number is expressed in decimal form, the decimal always terminates (see Example 8) or repeats (see Example 9). See Appendix B, if necessary, for a brief review of converting fractions to decimal form.

Example 7   Examples of rational numbers:

a. $\dfrac{3}{4}$, because 3 and 4 are integers and $4 \neq 0$.

b. -5, because $-5 = -\dfrac{5}{1}$.

c. $1\dfrac{8}{9}$, because $1\dfrac{8}{9} = \dfrac{17}{9}$.

d. -3.7, because $-3.7 = -\dfrac{37}{10}$.

e. 0, because $0 = \dfrac{0}{1}$ or $\dfrac{0}{5}$ or $\dfrac{0}{23}$, and so on.   ■

Example 8   Examples of rational numbers whose decimals terminate:

a. $\dfrac{3}{4} = 0.75$ The decimal terminates

b. $\dfrac{1}{16} = 0.0625$ The decimal terminates   ■

When the decimal form of a rational number is a nonterminating, repeating decimal, it is customary to place a bar over the digit or group of digits that repeats. The bar indicates that that group of digits repeats forever. Thus, $8.\overline{1738} = 8.173817381738\ldots$, and $8.1\overline{73} = 8.1737373\ldots$.

Example 9   Examples of rational numbers whose decimals repeat:

a. $\dfrac{1}{3} = 0.333333\ldots = 0.\overline{3}$ The bar over the 3 indicates that the 3 repeats forever

b. $\dfrac{4}{33} = 0.121212\ldots = 0.\overline{12}$ The pair of digits "12" repeats forever   ■

Negative Numbers and the Real Number Line   In Section 1.1, we mentioned that natural numbers, whole numbers, fractions, decimals, and mixed numbers are all real numbers. Negative integers and negative rational numbers are also real numbers. They can be represented by points on the real-number line; a few are shown in Figure 1.2.2.

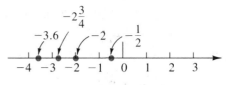

FIGURE 1.2.2

In Section 1.1, we stated that a largest natural number could never be found, because no matter how far we count, there are always larger natural numbers. Similarly, no matter how far we count along the number line to the left of 0, we never reach a smallest negative number.

Other sets of numbers that are often referred to in word problems are the following:

Consecutive Numbers   Integers that follow one another (without interruption) are called **consecutive numbers**. Thus, 5, 6, 7, and 8 are consecutive numbers, because 6 follows 5, 7 follows 6, and 8 follows 7.

Even Integers   Integers that are *exactly* divisible by 2 are called **even integers**. Therefore, 8, −6, and 0 are even integers.

Odd Integers   Integers that are *not* exactly divisible by 2 are called **odd integers**. Therefore, 1, −7, and 5 are odd integers.

EXERCISES 1.2

Set I

1. Write −75 in words.

2. Write −49 in words.

3. Use digits to write negative fifty-four.

4. Use digits to write negative one hundred nine.

5. Which is larger, −2 or −4?

6. Which is larger, 0 or −10?

7. Which is larger, −5 or −10?

8. Which is smaller, −1 or −15?

9. What is the largest negative integer?

10. Is 18 a rational number?

11. Is −3 a rational number?

12. Is 18 a real number?

13. Is −3 a real number?

14. Is $1.\overline{35}$ a rational number?

15. Is 0.74 a rational number?

16. What is the smallest negative integer?

17. Write, in consecutive order, the natural numbers < 5.

18. Write, in consecutive order, the digits > 6.

19. Write, in consecutive order, the even digits that are < 6.

20. Write, in consecutive order, the odd natural numbers that are < 9.

In Exercises 21–26, determine which symbol, $<$ or $>$, should be used to make each statement true.

21. $0 \underline{\ ?\ } -3$ **22.** $-2 \underline{\ ?\ } -6$ **23.** $-5 \underline{\ ?\ } 2$

24. $-7 \underline{\ ?\ } -4$ **25.** $-2 \underline{\ ?\ } -10$ **26.** $-8 \underline{\ ?\ } -3$

27. A scuba diver descends to a depth of sixty-two feet. Represent this number by an integer.

28. The temperature in Fairbanks, Alaska, was forty-five degrees Fahrenheit below zero. Represent this number by an integer.

Set II

1. Write -17 in words.

2. Write, in consecutive order, the negative integers > -5.

3. Use digits to write negative two hundred four.

4. Write, in consecutive order, the even natural numbers < 12.

5. Which is larger, -6 or -3?

6. Which is smaller, -5 or 0?

7. Which is larger, -8 or 5?

8. Which is smaller, -10 or 2?

9. What is the largest real number?

10. What is the smallest positive integer?

11. Is $\frac{2}{7}$ a rational number?

12. Is $-\frac{2}{15}$ a rational number?

13. Is $\frac{2}{7}$ a real number?

14. Is $0.\overline{62}$ a rational number?

15. Is 2.35 a real number?

16. Is $-\frac{2}{15}$ a real number?

17. Write, in consecutive order, the odd negative integers > -7.

18. Write, in consecutive order, the odd digits < 8.

19. Write, in consecutive order, the even digits that are > 6.

20. Write, in consecutive order, the even natural numbers that are < 15.

In Exercises 21–26, determine which symbol, $<$ or $>$, should be used to make each statement true.

21. $0 \underline{\ ?\ } -5$ **22.** $-7 \underline{\ ?\ } -3$ **23.** $3 \underline{\ ?\ } -5$

24. $-8 \underline{\ ?\ } -2$ **25.** $-6 \underline{\ ?\ } -3$ **26.** $-16 \underline{\ ?\ } 0$

27. The temperature in Fairbanks, Alaska, was six degrees Fahrenheit below zero. Represent this number by an integer.

28. Nitrogen becomes a liquid at (about) 195 degrees Celsius below zero. Represent this number by an integer.

1.3 Absolute Value; Adding Integers and Other Signed Numbers

Absolute Value

The **absolute value** of a number is the distance between that number and 0 on the number line *with no regard to direction* (see Figure 1.3.1). The symbol for the *absolute value* of a real number x is $|x|$.

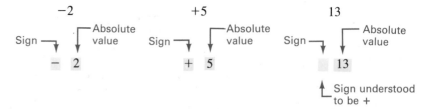

FIGURE 1.3.1 ABSOLUTE VALUE

Example 1   Examples of the absolute value of numbers:

a. $|9| = 9$ A positive number

b. $|0| = 0$ Zero

c. $|-4| = 4$ A positive number

Note that the absolute value of a number can never be negative ∎

A signed number has two distinct parts: its absolute value and its sign.

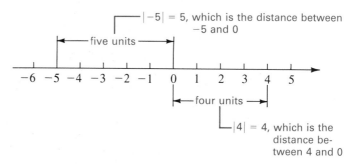

Note that the absolute value of a signed number is the number written without its sign.

Adding Integers and Other Signed Numbers

Representing Integers by Arrows   We can represent an integer by an arrow whose length represents the absolute value of the number. The arrow must point toward the right if the number is positive (see Example 2) and toward the left if the number is negative (see Example 3). The arrow that represents an integer need not start at zero (see the arrow that represents −7 in Example 4).

Example 2   Represent 4 by an arrow.

This arrow represents a movement of four units to the *right*. ∎

Example 3   Represent −5 by an arrow.

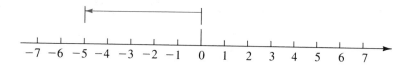

This arrow represents a movement of five units to the *left*. ■

Sum   The *answer* to an addition problem is called the **sum**.

Adding Integers on the Number Line   Before we state the rules for adding integers and other signed numbers, we will give a few examples of adding integers on the number line. We draw arrows for the numbers we're adding. We start the first arrow at zero, and we start the second arrow at the *end* of the first arrow; the arrow for the answer goes *from* zero *to* the end of the second arrow (see Examples 4–7).

Example 4   Add −7 to 5 on the number line; that is, find 5 + (−7).
Solution We begin by drawing the arrow that represents 5; then we start the arrow that represents −7 at the *end* of the first arrow.

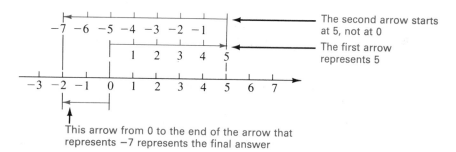

Because the arrow that represents the final answer is 2 units long, the absolute value of the answer is 2. Because the arrow that represents the final answer points to the *left*, the answer is *negative*. Therefore, 5 + (−7) = −2. ■

NOTE   When the problem from Example 4 is written as 5 + (−7), we consider the parentheses around the −7 necessary, as we don't like to see two operation symbols next to each other with no symbol between them. That is, we feel the problem should not be written as 5 + −7 or as ⁺5 + ⁻7. (Some authors and some instructors do not object to these forms.)   ☑

Example 5   Add −4 to −3 on the number line.
Solution We begin by drawing the arrow that represents −3; then we start the arrow that represents −4 at the *end* of the first arrow.

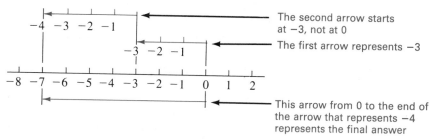

Because the arrow that represents the final answer is 7 units long, the absolute value of the answer is 7. Because the arrow that represents the final answer points to the *left*, the answer is *negative*. Therefore, −3 + (−4) = −7. ■

Example 6   Add $+8$ to -5 on the number line.

Solution We begin by drawing the arrow that represents -5; then we start the arrow that represents $+8$ at the *end* of the first arrow.

The second arrow starts at -5, not at 0

The first arrow represents -5

This arrow from 0 to the end of the arrow that represents $+8$ represents the final answer

Because the arrow that represents the final answer is 3 units long, the absolute value of the answer is 3. Because the arrow that represents the final answer points to the *right*, the answer is *positive*. Therefore, $-5 + (+8) = 3$. ∎

Example 7   On the number line, add 0 to -3.

Solution We begin by drawing an arrow that represents -3. To add 0, we then move zero units from that point.

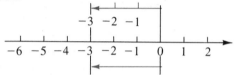

Because the arrow that represents the final answer is 3 units long and points to the left, the answer is -3. Therefore, $-3 + 0 = -3$. ∎

You should verify that we can use the method just discussed to add two positive integers on the number line; for example, verify by using arrows and the number line that $(+3) + (+2) = 5$.

The Rules for Adding Integers or Other Signed Numbers   Adding integers on the number line is easy to understand but is a very slow process. Trying to add 8,317,005 to $-17,461,037$ on the number line, for example, would be impractical. The following rules give an easier, faster method for adding integers *and* other signed numbers.

TO ADD TWO INTEGERS OR OTHER SIGNED NUMBERS

1. When the numbers have the same sign,

> **a.** Add the absolute values
> *and*
> **b.** Attach (to the left of the result from step a) the sign of the answer; *the answer has the sign of both numbers.**

2. When the numbers have different signs,

> **a.** *Subtract* the smaller absolute value from the larger absolute value
> *and*
> **b.** Attach (to the left of the result from step a) the sign of the answer; *the answer has the sign of the number with the larger absolute value.**

---

*When the sign is positive, it can often be omitted.

Notice that adding can involve finding a *difference*.
Examples 8–10 show the addition of integers using the rules given above.

Example 8
Find $(-29) + (-35)$.
Solution Because the numbers have the same sign, we use Rule 1 and begin by adding the absolute values:
Using rule 1a:

$$|-29| + |-35| = 29 + 35 = 64.$$

According to rule 1b, the answer has the sign of both numbers, which is $-$:

$$\underset{\Big\uparrow}{^-64}$$

Sign of both numbers attached to the left of the result from step a

Therefore, $(-29) + (-35) = -64$. ∎

Example 9
Find $(-9) + (+23)$.
Solution Because the numbers have different signs, we use Rule 2 and begin by deciding which number has the larger absolute value. $|-9| = 9$; $|+23| = 23$; because $23 > 9$, the number with the larger absolute value is $+23$.
Using rule 2a, we subtract the smaller absolute value from the larger absolute value:

$$|+23| - |-9| = 23 - 9 = 14$$

According to rule 2b, the answer has the sign of the number with the larger absolute value, which is $+$:

$$\underset{\Big\uparrow}{+14}$$

Sign of $+23$ attached to the left of the result from step a

Therefore, $(-9) + (+23) = +14$, or 14. ∎

In Example 8, the parentheses around the first number were not necessary; $(-29) + (-35) = -29 + (-35)$. In Example 9, the parentheses and one plus sign could have been omitted; $(-9) + (+23) = -9 + 23$.

Example 10
Find $-24 + 17$.
Solution Because the numbers have different signs, we use Rule 2 and begin by deciding which number has the larger absolute value. $|-24| = 24$; $|17| = 17$; because $24 > 17$, the number with the larger absolute value is -24.
Using rule 2a, we subtract the smaller absolute value from the larger absolute value:

$$|-24| - |17| = 24 - 17 = 7$$

According to rule 2b, the answer has the sign of the number with the larger absolute value, which is $-$:

$$\underset{\Big\uparrow}{^-7}$$

Sign of -24 attached to the left
of the result from step a

Therefore, $-24 + 17 = -7$. ∎

As you first use the rules for adding integers, you should do each problem by using the rules *and* check your results by using arrows and the number line. You should practice adding integers until you can write the correct answer quickly without writing down any intermediate steps. In Examples 11 and 12, we add signed numbers that are not integers. You may want to refer to Appendix B for a review of the operations of arithmetic.

Example 11 Find $-2\frac{1}{4} + 8\frac{1}{8}$.

Solution The least common denominator (LCD) is 8. (The LCD is discussed more fully in Chapter 8.) Since the two numbers have different signs, we use Rule 2. $\left| -2\frac{1}{4} \right| = 2\frac{1}{4}$ and $\left| 8\frac{1}{8} \right| = 8\frac{1}{8}$. The number with the larger absolute value is $8\frac{1}{8}$. To subtract the smaller absolute value from the larger, we'll set up the problem vertically.

Using rule 2a, and subtracting $2\frac{1}{4}$ from $8\frac{1}{8}$:

$$8\frac{1}{8} = \quad 7\frac{9}{8} \quad \longleftarrow \text{We rewrite } 8\frac{1}{8} \text{ as } 7\frac{9}{8} \text{ because } \frac{1}{8} \text{ is smaller than } \frac{1}{4}$$

$$-2\frac{1}{4} = -2\frac{2}{8} \qquad \frac{1}{4} = \frac{1 \cdot 2}{4 \cdot 2} = \frac{2}{8}$$

$$\rule{2cm}{0.4pt}$$

$$5\frac{7}{8} \quad \longleftarrow \text{We subtract } \frac{2}{8} \text{ from } \frac{9}{8} \text{ and 2 from 7}$$

According to rule 2b, the answer has the sign of the number with the larger absolute value, which is $+$:

$$+5\frac{7}{8}$$

$$\text{Sign of } 8\frac{1}{8} \text{ is } + \text{ and can be written or omitted}$$

Therefore, $-2\frac{1}{4} + 8\frac{1}{8} = 5\frac{7}{8}$. ∎

Example 12 Find $-2\frac{1}{2} + \left(-4\frac{1}{3} \right)$.

Solution LCD = 6. Since the numbers have the same sign, we use Rule 1 and add the absolute values. $\left| -2\frac{1}{2} \right| = 2\frac{1}{2}$ and $\left| -4\frac{1}{3} \right| = 4\frac{1}{3}$.

Using rule 1a, and adding the absolute values:

$$2\frac{1}{2} = \quad 2\frac{3}{6}$$

$$+4\frac{1}{3} = +4\frac{2}{6}$$

$$\rule{2cm}{0.4pt}$$

$$6\frac{5}{6}$$

According to rule 1b, the answer has the sign of both numbers, which is $-$:

$$-6\frac{5}{6}$$

$$\text{Sign of both numbers attached to the left of the result from step a}$$

Therefore, $-2\frac{1}{2} + \left(-4\frac{1}{3} \right) = -6\frac{5}{6}$. ∎

Example 13   At 6 A.M. the temperature in Alamosa, Colorado, was $-15°$ F. By 10 A.M., the temperature had risen $9°$ F. What was the temperature at 10 A.M.?

Solution We must *add* the rise in temperature ($9°$) to the original temperature ($-15°$), and $-15 + (9) = -6$. Therefore, the new temperature is $-6°$ F. ■

Example 14   Add $|-8| + (-3)$.

Solution We must remove the absolute-value symbols before adding:

$$|-8| + (-3) = 8 + (-3) = 5 \quad ■$$

Additive Identity (Addition Involving Zero)   Because adding zero to a number gives us a number *identical* to the one we started with (see Example 7), we call zero the **additive identity**.

THE ADDITIVE IDENTITY IS 0

If a represents any real number, then

$$a + 0 = a \quad \text{and} \quad 0 + a = a$$

NOTE   All the problems in this section are *addition problems*, even though we must sometimes subtract absolute values in order to find the sums. ☑

EXERCISES 1.3

Set I   In Exercises 1–40, find the sums.

1. $4 + 5$

2. $6 + 2$

3. $-3 + (-4)$

4. $-7 + (-1)$

5. $-6 + 5$

6. $-8 + 3$

7. $7 + (-3)$

8. $9 + (-4)$

9. $-8 + (-4)$

10. $-5 + (-6)$

11. $3 + (-9)$

12. $4 + (-8)$

13. $-5 + 0$

14. $0 + (-17)$

15. $-7 + 9$

16. $-5 + 8$

17. $-2 + (-11)$

18. $-3 + (-6)$

19. $5 + (-15)$

20. $4 + (-12)$

21. $-8 + 9$

22. $-7 + 13$

23. $-4 + 4$

24. $-9 + 9$

25. $-27 + (-13)$

26. $-42 + (-12)$

27. $-80 + 121$

28. $-69 + 134$

29. $105 + (-73)$

30. $218 + (-113)$

31. $-1\frac{1}{2} + \left(-3\frac{2}{5}\right)$

32. $-2\frac{1}{2} + \left(-5\frac{1}{4}\right)$

33. $4\frac{5}{6} + \left(-1\frac{1}{3}\right)$

34. $6\frac{3}{4} + \left(-2\frac{1}{8}\right)$

35. $5\frac{1}{4} + \left(-2\frac{1}{3}\right)$

36. $-3\frac{4}{5} + 8\frac{1}{2}$

37. $6.075 + (-3.146)$

38. $-4.745 + 93.118$

39. $5.2 + (-2.345)$

40. $-5.325 + 6.1$

In Exercises 41–44, remove the absolute value symbols.

41. $|-73|$ **42.** $|-55|$ **43.** $-|48|$ **44.** $-|26|$

In Exercises 45–50, find the sums.

45. $|-6| + (-2)$ **46.** $|-8| + (-5)$ **47.** $-8 + |-17|$

48. $-6 + |-27|$ **49.** $-9 + |23|$ **50.** $-1 + |32|$

51. At 6 A.M. the temperature in Hibbing, Minnesota, was $-35°$ F. If the temperature had risen $53°$ F by 2 P.M., what was the temperature at that time?

52. At midnight in Billings, Montana, the temperature was $-50°$ F. By noon the temperature had risen $67°$ F. What was the temperature at noon?

Set II   In Exercises 1–40, find the sums.

1. $8 + 7$ **2.** $8 + (-5)$ **3.** $-9 + (-3)$

4. $-19 + (-3)$ **5.** $-5 + 9$ **6.** $-28 + 28$

7. $-4 + 7$ **8.** $0 + (-4,162)$ **9.** $-2 + (-5)$

10. $18 + (-6)$ **11.** $6 + (-10)$ **12.** $1,468 + (-1,468)$

13. $-16 + 0$ **14.** $-\frac{2}{3} + \left(-\frac{1}{5}\right)$ **15.** $-8 + 5$

16. $-146 + (-362)$ **17.** $-8 + (-9)$ **18.** $-1.724 + 3.6$

19. $26 + (-35)$ **20.** $-4,728 + (-35)$ **21.** $-29 + 32$

22. $-67 + 28$ **23.** $-6 + 6$ **24.** $147 + (-362)$

25. $-38 + (-17)$ **26.** $-849 + (-738)$ **27.** $-132 + 261$

28. $-536 + (-2)$ **29.** $872 + (-461)$ **30.** $621 + (-1,417)$

31. $-1\frac{1}{3} + \left(-5\frac{3}{5}\right)$ **32.** $-6\frac{1}{4} + \left(-8\frac{1}{2}\right)$ **33.** $3\frac{2}{3} + \left(-2\frac{2}{9}\right)$

34. $3\frac{1}{3} + \left(-5\frac{1}{6}\right)$ **35.** $8\frac{1}{6} + \left(-3\frac{1}{3}\right)$ **36.** $-2\frac{5}{6} + 7\frac{1}{2}$

37. $-18.0164 + 2.281$ **38.** $-9.6 + (-57.356)$ **39.** $8.3 + (-3.523)$

40. $-6.127 + 9.3$

In Exercises 41–44, remove the absolute value symbols.

41. $|-81|$ **42.** $|-462|$ **43.** $-|17|$ **44.** $-|-84|$

In Exercises 45–50, find the sums.

45. $(-17) + |-4|$ **46.** $|-5| + |-2|$ **47.** $-18 + |-25|$

48. $|-12| + |-17|$ **49.** $-3 + |-3|$ **50.** $|-18| + |-25|$

51. In Fairbanks, Alaska, the temperature at 2 A.M. was $-35°$ F. By noon the temperature had risen $27°$ F. What was the temperature at noon?

52. In Duchesne, Utah, the temperature at midnight was $-18°$ F. By 10 A.M. the temperature had risen $12°$ F. What was the temperature at 10 A.M.?

1.4 Subtracting Integers and Other Signed Numbers

Subtraction is the *inverse* operation of addition; that is, subtraction "undoes" addition.

Difference   The *answer* to a subtraction problem is called the **difference**.

Additive Inverse (The Negative of a Number)   When the sum of two numbers is 0 (the additive identity), we say that the numbers are the **additive inverses** of each other or that they are the **negatives** of each other. To find the additive inverse or negative of a signed number, we simply change the sign of the number. (The additive inverse, or negative, of 0 is 0.)

ADDITIVE INVERSE

If *a* represents any real number, the additive inverse of *a* is $-a$.

$$a + (-a) = 0$$
$$-a + a = 0$$

Example 1   Examples of the additive inverse (or negative) of a *positive* number:

   a. The additive inverse (or negative) of 5 is -5.

   b. The additive inverse (or negative) of 12 is -12.   ∎

Example 2   Examples of the additive inverse (or negative) of a *negative* number:

   a. The additive inverse (or negative) of -10 is 10. We can write this as

$$-(-10) = 10$$

   b. The additive inverse (or negative) of -14 is 14. We can write this as

$$-(-14) = 14$$ ∎

DEFINITION OF SUBTRACTION

$$a - b = a + (-b)$$

In words: To subtract *b* from *a*, *add* the additive inverse of *b* to *a*.

This definition leads to the following rule for subtracting integers or other signed numbers.

TO SUBTRACT ONE INTEGER OR OTHER SIGNED NUMBER FROM ANOTHER

1. Change the subtraction symbol to an addition symbol, *and* change the sign of the number being subtracted.

2. Add the resulting signed numbers as shown in Section 1.3.

Example 3 Find $-2 - (-5)$.

Solution

Change the subtraction symbol to an addition symbol
Change the sign of the number being subtracted

$$(-2) \; - \; (\, -\, 5)$$

$$= (-2) \; + \; (\, +\, 5)$$

$$= -2 + 5 \qquad \text{Now use Rule 2 for } adding \text{ integers}$$
$$\text{(adding integers with different signs)}$$

$$= 3 \quad \blacksquare$$

Example 4 Find $13 - (-14)$.

Solution

Change the subtraction symbol to an addition symbol
Change the sign of the number being subtracted

$$13 \; - \; (\, -\, 14)$$

$$= 13 \; + \; (\, +\, 14) \qquad \text{Now use Rule 1 for } adding \text{ integers}$$
$$\text{(adding integers with the same sign)}$$

$$= 27 \quad \blacksquare$$

In Example 5, we use the fact that "subtract a from b" means "$b - a$."

Example 5 Subtract 9 from 6.

Solution Subtract 9 from 6 *means* $6 - 9$.

Change the subtraction symbol to an addition symbol
Change the sign of the number being subtracted

$$6 \; - \; (\, +\, 9)$$

$$= 6 \; + \; (\, -\, 9) \qquad \text{Now use Rule 2 for } adding \text{ integers}$$
$$\text{(adding integers with different signs)}$$

$$= -3 \quad \blacksquare$$

Example 6 What is -3 diminished by 6?

Solution "-3 diminished by 6" means to subtract 6 (the second number) from -3 (the first number).

Change the subtraction symbol to an addition symbol
Change the sign of the number being subtracted

$$-3 \; - \; (\, +\, 6)$$

$$= -3 \; + \; (\, -\, 6) \qquad \text{Now use Rule 1 for } adding \text{ integers}$$
$$\text{(adding integers with the same sign)}$$

$$= -9 \quad \blacksquare$$

Example 7 Find $-3\frac{1}{2}$ reduced by $+2\frac{1}{4}$.

Solution We must subtract $2\frac{1}{4}$ from $-3\frac{1}{2}$.

$$-3\frac{1}{2} - \left(+2\frac{1}{4}\right)$$

$$= -3\frac{2}{4} + \left(-2\frac{1}{4}\right)$$

$$= -5\frac{3}{4}$$

If the addition is set up vertically, we have

$$-3\frac{1}{2} \;\; = \;\; -3\frac{2}{4}$$
$$+\left(-2\frac{1}{4}\right) = +\left(-2\frac{1}{4}\right)$$
$$\rule{3cm}{0.4pt} \quad \rule{3cm}{0.4pt}$$
$$-5\frac{3}{4}$$

$\blacksquare$

Example 8   Find $(-4.56) - (-7.48)$.
Solution

$$(-4.56) - (-7.48) \quad \text{Or} \quad 7.48$$
$$= (-4.56) + (+7.48) \qquad \underline{-4.56}$$
$$= 2.92 \qquad\qquad\qquad +2.92 \quad \blacksquare$$

Example 9   At 4 A.M. the temperature in Missoula, Montana, was $-26°$ F. At noon, the temperature was $-4°$ F. What was the rise in temperature?
Solution We must find the *difference* in temperatures; that is, we must subtract the *lower* temperature $(-26°)$ from the *higher* temperature $(-4°)$. The problem then becomes

$$-4 - (-26)$$
$$= -4 + (+26)$$
$$= 22$$

Therefore, the rise in temperature was $22°$ F. $\blacksquare$

Subtraction Involving Zero   The rules for subtraction involving zero are derived from the rules for addition, since the subtraction $a - b$ has been defined as $a + (-b)$.

SUBTRACTION INVOLVING ZERO

If a represents any real number, then

1. $a - 0 = a$ **2.** $0 - a = 0 + (-a) = -a$

EXERCISES 1.4

Set I   **1.** Find the additive inverse of -6.   **2.** Find the additive inverse of $2/3$.

In Exercises 3–38, find the differences.

3. $-3 - (-2)$ **4.** $-4 - (-3)$ **5.** $-6 - 2$

6. $-8 - 5$ **7.** $9 - (-5)$ **8.** $7 - (-3)$

9. $2 - (-7)$ **10.** $3 - (-5)$ **11.** $-5 - (-9)$

12. $-2 - (-8)$ **13.** $-4 - 3$ **14.** $-7 - 8$

15. $6 - 11$ **16.** $5 - 9$ **17.** $-9 - (-4)$

18. $-6 - (-3)$ **19.** $4 - (-7)$ **20.** $8 - (-9)$

21. $-15 - 11$ **22.** $-24 - 16$ **23.** $0 - 7$

24. $0 - 15$ **25.** $0 - (-9)$ **26.** $0 - (-25)$

27. $16 - 0$ **28.** $10 - 0$ **29.** $156 - (-97)$

30. $284 - (-89)$ **31.** $-354 - (-286)$ **32.** $-484 - (-375)$

33. $-7 - (-2.009)$ **34.** $-16 - (-7.89)$ **35.** $-6\frac{1}{2} - \left(-3\frac{2}{3}\right)$

36. $-8\frac{1}{3} - \left(-2\frac{5}{6}\right)$ **37.** $6\frac{1}{3} - 8\frac{1}{4}$ **38.** $12\frac{1}{5} - 9\frac{1}{2}$

39. Subtract (-2) from $(+5)$.   **40.** Subtract (-10) from (-15).

41. Subtract $\left(2\frac{1}{2}\right)$ from $\left(-5\frac{1}{4}\right)$.   **42.** Subtract $\left(3\frac{1}{6}\right)$ from $\left(-7\frac{1}{3}\right)$.

43. What is -8 diminished by 5?   **44.** What is -3 diminished by 6?

45. What is 7 reduced by 12?   **46.** What is 2 reduced by 12?

47. What is 0 diminished by 4?   **48.** What is 0 diminished by 8?

49. Mr. Reyes has a balance of $473.29 in his checking account. Find his new balance after he writes a check for $238.43.

50. Ms. Johnson made a $45 deposit on a quadraphonic home music system costing $623.89. What is the balance due?

51. At 5 A.M. the temperature at Mammoth Mountain, California, was $-7°$ F. At noon the temperature was $42°$ F. What was the rise in temperature?

52. At 4 A.M. the temperature in Massena, New York, was $-5.6°$ F. At 1 P.M. the temperature was $37.5°$ F. What was the rise in temperature?

53. A scuba diver descends to a depth of 141 ft below sea level. Her buddy dives 68 ft deeper. What is her buddy's depth at the deepest point of her dive?

54. When Fred checked his pocket altimeter at the seashore on Friday afternoon, it read -150 ft. Saturday morning it read 9,650 ft when he checked it on the peak of a nearby mountain. Allowing for the obvious error in his altimeter reading, what is the correct height of that peak?

55. Mt. Everest (the highest known point on earth) has an altitude of 29,028 ft. The Mariana Trench in the Pacific Ocean (the lowest known point on earth) has an altitude of $-36,198$ ft. Find the difference in altitude of these two places.

56. An airplane is flying 75 ft above the level of the Dead Sea (elevation $-1,299$ ft). How high must it climb to clear a 2,573-ft peak by 200 ft?

Set II   **1.** Find the additive inverse of $-\frac{1}{5}$.   **2.** Find the additive inverse of 0.

In Exercises 3–38, find the differences.

3. $-8 - 5$	**4.** $7 - (-3)$	**5.** $-6 - (-8)$
6. $-4 - 6$	**7.** $2 - 7$	**8.** $9 - (-7)$
9. $-14 - 10$	**10.** $184 - (-286)$	**11.** $-473 - 389$
12. $-784 - (-528)$	**13.** $-17 - (-12)$	**14.** $0 - (-12)$
15. $18 - 367$	**16.** $83 - (-5)$	**17.** $-24 - (-15)$
18. $-63 - (-63)$	**19.** $24 - (-15)$	**20.** $17 - 23$
21. $-28 - 17$	**22.** $1 - 26$	**23.** $0 - 862$
24. $361 - 0$	**25.** $0 - (-816)$	**26.** $19 - (-26)$
27. $83 - 0$	**28.** $28 - 362$	**29.** $352 - (-89)$

30. $-352 - (-89)$ **31.** $-563 - (-825)$ **32.** $563 - (-825)$

33. $-15 - (-9.794)$ **34.** $-25 - (-5.63)$ **35.** $-4\frac{1}{3} - \left(-1\frac{3}{4}\right)$

36. $-9\frac{1}{5} - \left(-2\frac{2}{3}\right)$ **37.** $5\frac{1}{4} - 8\frac{1}{8}$ **38.** $11\frac{1}{6} - 8\frac{1}{2}$

39. Subtract (-120) from (-285).

40. Subtract (-4) from 8.

41. Subtract $\left(3\frac{1}{5}\right)$ from $\left(-5\frac{1}{10}\right)$.

42. Subtract $\left(4\frac{1}{4}\right)$ from $\left(-9\frac{1}{2}\right)$.

43. What is -5 diminished by 6?

44. What is -2 diminished by 8?

45. What is 4 reduced by 10?

46. What is 3 reduced by 8?

47. What is 0 diminished by 6?

48. What is 6 diminished by 0?

49. John has a balance of $281.42 in his checking account. Find his new balance after he writes a check for $209.57.

50. Sue has a balance of $563.24 in her checking account. Find her new balance after she writes a check for $347.87.

51. At 2 A.M. the temperature in Burlington, Vermont, was $-3°$ F. At 11 A.M. the temperature was $9°$ F. What was the rise in temperature?

52. At 5 A.M. Don's temperature was $101.8°$ F. By noon his temperature had risen to $103.2°$ F. What was the increase in his temperature?

53. At midnight the temperature in Fairbanks, Alaska, was $-5°$ F. At 10 A.M. the temperature was $24°$ F. What was the rise in temperature?

54. A dune buggy starting from the floor of Death Valley $(-282$ ft$)$ is driven to the top of a nearby mountain having an elevation of 5,782 ft. What was the change in the dune buggy's altitude?

55. A jeep starting from the shore of the Dead Sea $(-1,299$ ft$)$ is driven to the top of a nearby hill having an elevation of 723 ft. What was the change in the jeep's altitude?

56. At 2 P.M. Cindy's temperature was $103.4°$ F. By 6 P.M. her temperature had dropped to $99.9°$ F. What was the drop in her temperature?

1.5 Multiplying Integers and Other Signed Numbers

Product   The *answer* to a multiplication problem is called the **product**.

Factors   The numbers that are multiplied together to give a product are called the **factors** of that product.

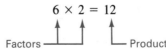

The numbers 6 and 2 are *factors* of 12; 12 is the *product* of 6 and 2. Similarly, 3 and 4 are factors of 12, because $3 \times 4 = 12$.

Symbols Used in Multiplication Multiplication may be shown in a number of different ways:

1. $3 \times 2 = 6$

2. $3 \cdot 2 = 6$ The multiplication dot "·" is written a little higher than a decimal point.

3. $3(2) = 6$ When there is no operation symbol between a number (or letter) and parentheses, the operation is understood to be multiplication.

4. $(3)(2) = 6$ The operation is understood to be multiplication.

5. ab When two expressions (other than two *numbers*) are written next to each other in this way, it is understood that they are to be multiplied.

6. $3a$ In this example, it is understood that the value of a is to be *multiplied* by 3. Thus, if a is 7, $3a = 3(7) = 21$.

Multiplication Involving a Positive Integer Multiplication by a positive integer is a short method for doing repeated addition of the same number.

Example 1 $3 \times 5 =$ the sum of three 5s
                      $= 5 + 5 + 5 = 15$ ∎

Example 2 $6 \times 2 =$ the sum of six 2s
                      $= 2 + 2 + 2 + 2 + 2 + 2 = 12$ ∎

In Examples 3 and 4, we use repeated addition for multiplying a negative integer by a positive integer.

Example 3 $3 \times (-2) =$ the sum of three negative 2s
                      $= (-2) + (-2) + (-2) = -6$ ∎

Example 4 $4 \times (-6) =$ the sum of four negative 6s
                      $= (-6) + (-6) + (-6) + (-6) = -24$ ∎

Example 5 $1 \times (-8) =$ one negative 8
                      $= -8$ ∎

We see from Examples 3, 4, and 5 that the product of two numbers that have opposite signs appears to be negative.

Example 6 $3 \times 1 = 1 + 1 + 1$
                      $= 3$ ∎

Multiplicative Identity Examples 5 and 6 illustrate the fact that multiplying any real number by 1 gives us back the *identical* number we started with. Because this is true, we call 1 the **multiplicative identity**.

THE MULTIPLICATIVE IDENTITY IS 1

If a represents any real number, then

$$a \cdot 1 = a \quad \text{and} \quad 1 \cdot a = a$$

Multiplication Involving Zero   Since multiplication is a method for doing repeated addition of the same number, multiplying a number by zero gives a product of zero.

Example 7   Examples of multiplying by zero:

a. $3 \cdot 0 = 0$ $3 \cdot 0 = 0 + 0 + 0 = 0$

b. $4 \cdot 0 = 0$ $4 \cdot 0 = 0 + 0 + 0 + 0 = 0$

c. $0 \cdot 3 = 0$ $0 \cdot 3 = 0 + 0 + 0 = 0$

d. $0 \cdot 4 = 0$ $0 \cdot 4 = 0 + 0 + 0 + 0 = 0$ ■

MULTIPLICATION INVOLVING ZERO

If a represents any real number, then

$$a \cdot 0 = 0 \qquad \text{and} \qquad 0 \cdot a = 0$$

The Product of Two Negative Numbers   *The product of two negative numbers is positive.* We will not prove this statement; however, an examination of the pattern of products as shown below may convince you that it is true.

Decreasing by 1 →

$$
\begin{aligned}
4 \,(-2) &= -8 \\
3 \,(-2) &= -6 \\
2 \,(-2) &= -4 \\
1 \,(-2) &= -2 \\
0 \,(-2) &= 0 \\
-1 \,(-2) &= 2 \\
-2 \,(-2) &= 4 \\
-3 \,(-2) &= 6
\end{aligned}
$$

← Increasing by 2

The product of two negative numbers is positive

Multiplying Integers and Other Signed Numbers   The rules for multiplying two integers or other signed numbers are summarized as follows:

TO MULTIPLY TWO INTEGERS OR OTHER SIGNED NUMBERS

Multiply the absolute values *and* attach the correct sign to the left of the product; that sign is

a. *positive* when the factors have the same sign;

b. *negative* when the factors have different signs.

Example 8   Multiply − 7(4).
Solution

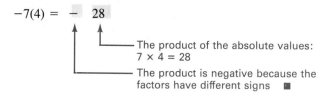

$$-7(4) = -\ 28$$

The product of the absolute values:
7 × 4 = 28

The product is negative because the
factors have different signs ∎

Example 9   Multiply 23(−11).
Solution

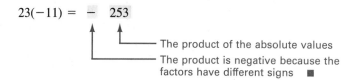

$$23(-11) = -\ 253$$

The product of the absolute values

The product is negative because the
factors have different signs ∎

Example 10   Multiply −14(−10).
Solution

$$-14(-10) = +\ 140$$

The product of the absolute values

The product is positive because the
factors have the same sign ∎

Example 11   Multiply $\left(4\frac{1}{2}\right)\left(-1\frac{1}{3}\right)$.
Solution

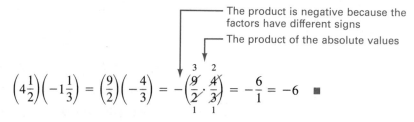

The product is negative because the
factors have different signs

The product of the absolute values

$$\left(4\frac{1}{2}\right)\left(-1\frac{1}{3}\right) = \left(\frac{9}{2}\right)\left(-\frac{4}{3}\right) = -\left(\frac{\overset{3}{\cancel{9}}}{\underset{1}{\cancel{2}}} \cdot \frac{\overset{2}{\cancel{4}}}{\underset{1}{\cancel{3}}}\right) = -\frac{6}{1} = -6 \quad \blacksquare$$

Example 12   Multiply −2.7(−4.6).
Solution

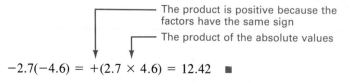

The product is positive because the
factors have the same sign

The product of the absolute values

$$-2.7(-4.6) = +(2.7 \times 4.6) = 12.42 \quad \blacksquare$$

Now that we can multiply by a negative number, we have an alternate way of finding the *additive inverse* (or *negative*) of a number. The additive inverse of any real number can be found by multiplying the number by −1.

Example 13   Examples of finding the additive inverse of a real number:

a. The additive inverse of 5 is 5(−1), or −5.

b. The additive inverse of −2 is −2(−1), or 2.   ∎

The Reciprocal of a Fraction   To find the **reciprocal** of a fraction, interchange the numerator and the denominator; the reciprocal of a positive number is always positive, and the reciprocal of a negative number is always negative. Zero has no reciprocal. If $a \neq 0$ and $b \neq 0$, then the reciprocal of $\dfrac{a}{b}$ is $\dfrac{b}{a}$. Consider the product of $\dfrac{a}{b}$ and its reciprocal:

$$\frac{\overset{1}{\cancel{a}}}{\underset{1}{\cancel{b}}} \cdot \frac{\overset{1}{\cancel{b}}}{\underset{1}{\cancel{a}}} = 1$$

In fact, the product of any nonzero fraction and its reciprocal is always 1.

Example 14 Find the reciprocals of each of the following numbers:

a. $\dfrac{5}{6}$ The reciprocal of $\dfrac{5}{6}$ is $\dfrac{6}{5}$; notice that $\left(\dfrac{5}{6}\right)\left(\dfrac{6}{5}\right) = 1$.

b. -8 The reciprocal of -8 is $-\dfrac{1}{8}$; notice that $(-8)\left(-\dfrac{1}{8}\right) = 1$.

c. $-\dfrac{5}{7}$ The reciprocal of $-\dfrac{5}{7}$ is $-\dfrac{7}{5}$; notice that $\left(-\dfrac{5}{7}\right)\left(-\dfrac{7}{5}\right) = 1$.

d. $-\dfrac{1}{2}$ The reciprocal of $-\dfrac{1}{2}$ is -2; notice that $\left(-\dfrac{1}{2}\right)(-2) = 1$. ∎

Multiplicative Inverse When the product of two numbers equals the multiplicative identity (that is, when the product is 1), we say that the numbers are **multiplicative inverses** of each other. Therefore, the reciprocal of a fraction can also be called its multiplicative inverse.

MULTIPLICATIVE INVERSE

If a represents any real number except 0, the multiplicative inverse of a is $1/a$. (There is no multiplicative inverse for 0.)

$$a\left(\frac{1}{a}\right) = 1 \quad \text{and} \quad \frac{1}{a}(a) = 1$$

EXERCISES 1.5

Set I In Exercises 1–38, find the products.

1. $3(-2)$	**2.** $4(-6)$	**3.** $-5(2)$	**4.** $-7(5)$
5. $-8(-2)$	**6.** $-6(-7)$	**7.** $8(-4)$	**8.** $9(-5)$
9. $-7(9)$	**10.** $-6(8)$	**11.** $-10(-10)$	**12.** $-9(-9)$
13. $8(-7)$	**14.** $12(-6)$	**15.** $-26(10)$	**16.** $-11(12)$
17. $-20(-10)$	**18.** $-30(-20)$	**19.** $75(-15)$	**20.** $86(-13)$
21. $-30(+5)$	**22.** $-50(+6)$	**23.** $-7(-20)$	**24.** $-9(-40)$

25. $\left(-5\frac{1}{2}\right)(0)$

26. $\left(-7\frac{2}{7}\right)(0)$

27. $\left(6\frac{3}{4}\right)\left(-8\frac{1}{4}\right)$

28. $\left(9\frac{2}{9}\right)\left(-7\frac{1}{3}\right)$

29. $\left(-3\frac{3}{5}\right)\left(-8\frac{1}{5}\right)$

30. $\left(-2\frac{5}{7}\right)\left(-7\frac{1}{5}\right)$

31. $-3.5(-1.4)$

32. $-4.7(-1.6)$

33. $2.74(-100)$

34. $3.04(-100)$

35. $\left(2\frac{1}{3}\right)\left(-3\frac{1}{2}\right)$

36. $\left(-5\frac{1}{4}\right)\left(-2\frac{3}{5}\right)$

37. $0\left(-\frac{2}{3}\right)$

38. $0\left(-\frac{7}{8}\right)$

39. Find the *additive* inverse of -12.

40. Find the *additive* inverse of $5/7$.

41. Find the reciprocal of -3.

42. Find the reciprocal of -2.

43. Are 5 and $-\frac{1}{5}$ the multiplicative inverses of each other? Why or why not?

44. Are $\frac{2}{3}$ and $-\frac{3}{2}$ the multiplicative inverses of each other? Why or why not?

Set II   In Exercises 1–38, find the products.

1. $5(-4)$

2. $-6(3)$

3. $-7(-3)$

4. $-4(8)$

5. $-9(-6)$

6. $8(+5)$

7. $-15(10)$

8. $15(-10)$

9. $-300(-10)$

10. $15(0)$

11. $-15(-15)$

12. $0(-28)$

13. $9(-12)$

14. $-23(-28)$

15. $-35(100)$

16. $-100(42)$

17. $-56(-100)$

18. $83(0)$

19. $27(-81)$

20. $-33(42)$

21. $-25(+6)$

22. $0(-3,679)$

23. $-18(-47)$

24. $-73(61)$

25. $\left(-8\frac{1}{3}\right)(0)$

26. $\left(-8\frac{1}{3}\right)\left(-2\frac{1}{5}\right)$

27. $\left(5\frac{1}{7}\right)\left(-3\frac{4}{5}\right)$

28. $\left(2\frac{1}{8}\right)\left(-2\frac{8}{9}\right)$

29. $\left(-12\frac{1}{2}\right)\left(-3\frac{1}{4}\right)$

30. $\left(-18\frac{1}{2}\right)(0)$

31. $-5.67(10)$

32. $-15.1(32)$

33. $-0.0032(-100)$

34. $6.07(0)$

35. $\left(-2\frac{1}{5}\right)\left(3\frac{3}{4}\right)$

36. $\left(8\frac{1}{3}\right)\left(-2\frac{3}{5}\right)$

37. $\left(-5\frac{3}{7}\right)(0)$

38. $0\left(-9\frac{1}{2}\right)$

39. Find the *additive* inverse of $-6/11$.

40. What is the multiplicative identity?

41. Find the reciprocal of $-\frac{3}{7}$.

42. Find the reciprocal of 5.

43. Are $\frac{3}{5}$ and $-\frac{5}{3}$ the multiplicative inverses of each other? Why or why not?

44. Show that $-\frac{7}{8}$ and $-\frac{8}{7}$ are the multiplicative inverses of each other.

1.6 Dividing Integers and Other Signed Numbers

Division is the *inverse* operation of multiplication; that is, division "undoes" multiplication. For example, $12 \div 4 = 3$ *because* $3 \times 4 = 12$. The division problem $a \div b$ is equivalent to the multiplication problem $a \cdot \dfrac{1}{b}$.

Quotient The *answer* to a division problem is called the **quotient**.

Divisor The number we are dividing *by* is called the **divisor**.

Dividend The number we are dividing *into* is called the **dividend**.

Remainder If the divisor does not divide *exactly* into the dividend, the part that is "left over" is called the **remainder**.

Division may be shown in several ways:

$$12 \div 4 = 12/4 = \frac{12}{4} = 4\overline{)12}$$ The *quotient* in this case is 3

All of these statements are equivalent to the multiplication problem $12 \cdot \dfrac{1}{4}$. That is, we can solve the division problem $a \div b$ by multiplying the dividend by the multiplicative inverse (the reciprocal) of the divisor.

Check the answer in a division problem as follows:

> (Divisor × quotient) + remainder = dividend

If the remainder is zero, then the *divisor* and the *quotient* are both *factors* of the *dividend*, and the problem could be checked as follows:

> Divisor × quotient = dividend

Example 1 Identify the quotient, divisor, dividend, and remainder for $23 \div 6$, and check the solution.

Solution

$$\begin{array}{r} \text{Quotient} \rightarrow 3 \\ \text{Divisor} \rightarrow 6\overline{)23} \leftarrow \text{Dividend} \\ \underline{18} \\ 5 \leftarrow \text{Remainder} \end{array}$$

Check (Divisor × quotient) + remainder = $(6 \times 3) + 5 = 18 + 5 = 23$, the dividend. ∎

Example 2   Identify the quotient, divisor, and dividend for $12 \div 4$, and check the solution.

Solution

$$\underset{\text{Divisor} \longrightarrow 4\overline{)12}\longleftarrow \text{Dividend}}{\overset{\lceil\!\text{Quotient}}{\overset{3}{}}}$$

The remainder is zero. (Because the remainder is zero, 4 and 3 are both *factors* of 12.)

Check Divisor $\times$ quotient $= 4 \times 3 = 12$, the dividend. ∎

Because of the inverse relation between division and multiplication, the rules for finding the sign of a quotient are the same as the ones used for finding the sign of a product.

TO DIVIDE ONE INTEGER OR OTHER SIGNED NUMBER BY ANOTHER

Divide the absolute value of the dividend by the absolute value of the divisor *and* attach the correct sign to the left of the quotient; that sign is

a. *positive* when the numbers have the same sign;

b. *negative* when the numbers have different signs.

Division Involving Zero

Division of zero by a number other than zero is possible and the quotient is always 0. That is, $0 \div 2 = \dfrac{0}{2} = 0$, or $2\overline{)0}^{\,0}$, because $0 \times 2 = 0$.

Division of a nonzero number by zero is impossible. Let's try to divide some number (not 0) by 0. For example, let's try $4 \div 0$, or $\dfrac{4}{0}$. Suppose the quotient is some unknown number we call q. Then $4 \div 0 = q$ means that we must find a number q such that $q \times 0 = 4$. But $q \times 0 = 4$ is impossible, since any number times 0 is 0. Therefore, no number q exists, and $4 \div 0$ has no answer (or is *undefined*). (Similarly, there is no answer for $1 \div 0$, or $\dfrac{1}{0}$; it is for this reason that zero has no reciprocal.)

Division of zero by zero cannot be determined. What about $0 \div 0$, or $\dfrac{0}{0}$?

Consider the following examples:

$$0\overline{)0}^{\,1} \text{ means } 0 \times 1 = 0, \text{ which is true.}$$

$$0\overline{)0}^{\,27} \text{ means } 0 \times 27 = 0, \text{ which is true.}$$

$$0\overline{)0}^{\,111} \text{ means } 0 \times 111 = 0, \text{ which is true.}$$

In other words, $0 \div 0 = 1$, 27, and 111. In fact, $0 \div 0$ can be *any* number. Therefore, we don't know what answer to write for $0 \div 0$, and we say that $0 \div 0$ *cannot be determined*.

For these reasons, we say that division by 0 is *undefined*. The important thing to remember about division involving zero is that you cannot divide *by* zero.

Division involving zero can be summarized as follows:

DIVISION INVOLVING ZERO

If a represents any real number except 0, then

1. $\dfrac{0}{a} = 0 \div a = 0$

2. $\dfrac{a}{0} = a \div 0$ is not possible.

3. $\dfrac{0}{0} = 0 \div 0$ cannot be determined

Division *by* zero is not defined.

Example 3 Solve $(-30) \div (5)$.
Solution
$$(-30) \div (5) = -6$$

The quotient of the absolute values: $30 \div 5 = 6$

The quotient is negative because the numbers have different signs ■

Example 4 Solve $(-64) \div (-8)$.
Solution
$$(-64) \div (-8) = +8$$

The quotient of the absolute values: $64 \div 8 = 8$

The quotient is positive because the numbers have the same sign ■

Example 5 Divide $\dfrac{2}{7}$ by $-\dfrac{5}{14}$.

Solution The problem is $\dfrac{2}{7} \div \left(-\dfrac{5}{14}\right)$. We must multiply the dividend $\left(\dfrac{2}{7}\right)$ by the reciprocal of the divisor.

The sign is negative because the numbers have different signs

The product of the absolute values

$$\frac{2}{7} \div \left(-\frac{5}{14}\right) = \left(\frac{2}{7}\right)\left(-\frac{14}{5}\right) = -\left(\frac{2 \cdot 14}{7 \cdot 5}\right) = -\frac{4}{5}$$ ■

Example 6 Divide $\dfrac{-42}{8}$.
Solution
$$\frac{-42}{8} = -\frac{21}{4} \text{ or } -5\frac{1}{4} \text{ or } -5.25$$ ■

31

Example 7   Solve $0 \div (-3)$.
             Solution

$$0 \div (-3) = 0 \quad \blacksquare$$

Example 8   Divide -7 by 0.
             Solution $-7 \div 0$ is not defined. (We cannot divide by zero.)   ∎

EXERCISES 1.6

Set I   Find the following quotients, or write "Not defined."

1. $-10 \div (-5)$ **2.** $-12 \div (-4)$ **3.** $-8 \div (2)$ **4.** $-6 \div (3)$

5. $\dfrac{+10}{-2}$ **6.** $\dfrac{+8}{-4}$ **7.** $\dfrac{-6}{-3}$ **8.** $\dfrac{-10}{-2}$

9. $-40 \div (8)$ **10.** $-60 \div (10)$ **11.** $16 \div (-4)$ **12.** $25 \div (-5)$

13. $-15 \div (-5)$ **14.** $-27 \div (-9)$ **15.** $\dfrac{12}{-4}$ **16.** $\dfrac{24}{-6}$

17. $\dfrac{-18}{-2}$ **18.** $\dfrac{-49}{-7}$ **19.** $\dfrac{-150}{10}$ **20.** $\dfrac{-250}{100}$

21. $36 \div (-12)$ **22.** $56 \div (-8)$ **23.** $-45 \div 15$ **24.** $-39 \div 13$

25. $\dfrac{4}{0}$ **26.** $\dfrac{8}{0}$ **27.** $0 \div 0$ **28.** $\dfrac{0}{0}$

29. $0 \div 7$ **30.** $0 \div (-12)$ **31.** $\dfrac{-1}{0}$ **32.** $\dfrac{-15}{0}$

33. $\dfrac{-15}{6}$ **34.** $\dfrac{-27}{12}$ **35.** $\dfrac{7.5}{-0.5}$ **36.** $\dfrac{1.25}{-0.25}$

37. $\dfrac{-6.3}{-0.9}$ **38.** $\dfrac{-4.8}{-0.6}$ **39.** $\dfrac{-367}{100}$ **40.** $\dfrac{-4,860}{1,000}$

41. $-\dfrac{3}{8} \div \dfrac{2}{5}$ **42.** $-\dfrac{2}{3} \div \left(-\dfrac{1}{6}\right)$

 43. $23.1616 \div (-4.136)$ **44.** $-17.0312 \div (-2.792)$

Set II   Find the following quotients, or write "Not defined."

1. $-8 \div (-2)$ **2.** $-10 \div (5)$ **3.** $12 \div (-6)$

4. $\dfrac{14}{-7}$ **5.** $\dfrac{-8}{2}$ **6.** $\dfrac{-50}{-10}$

7. $36 \div (-12)$ **8.** $-18 \div (-9)$ **9.** $-16 \div (4)$

10. $\dfrac{-15.6}{10}$ **11.** $\dfrac{13.8}{-10}$ **12.** $\dfrac{-6.3}{-0.7}$

13. $-8.4 \div (0.6)$ **14.** $-9.6 \div (-0.8)$ **15.** $18.5 \div (-3.7)$

16. $-14 \div 0$ **17.** $\dfrac{-18}{6}$ **18.** $-\left(\dfrac{0}{0}\right)$

19. $\dfrac{21}{-14}$ **20.** $\dfrac{-17}{-34}$ **21.** $81 \div (-15)$

22. $\dfrac{-51}{17}$

23. $-8 \div 16$

24. $-35 \div (-7)$

25. $\dfrac{13}{0}$

26. $0 \div (-10)$

27. $-8 \div 0$

28. $\dfrac{-12}{-2}$

29. $0 \div 9$

30. $\dfrac{72}{-18}$

31. $\dfrac{-11}{22}$

32. $\dfrac{16}{-24}$

33. $\dfrac{31}{-62}$

34. $-35 \div 0$

35. $\dfrac{-1.2}{0.24}$

36. $\dfrac{37}{-10}$

37. $\dfrac{-42}{-36}$

38. $\dfrac{-8.1}{2.7}$

39. $\dfrac{64}{-12}$

40. $\dfrac{-315}{100}$

41. $\dfrac{15}{16} \div \left(-\dfrac{3}{5} \right)$

42. $-4 \div \dfrac{1}{6}$

 43. $-15.327 \div 23.58$ 44. $5.63992 \div (-0.0572)$

1.7 Review: 1.1–1.6

Natural Numbers
1.1

$\{1, 2, 3, \ldots\}$

Whole Numbers
1.1

$\{0, 1, 2, \ldots\}$

Integers
1.2

$\{\ldots, -3, -2, -1, 0, 1, 2, 3, \ldots\}$

Digits
1.1

$\{0, 1, 2, 3, 4, 5, 6, 7, 8, 9\}$

Fractions
1.1

Fractions are numbers that can be put in the form a/b, where a is a whole number and b is a natural number.

Decimal Fractions
1.1

A **decimal fraction** is a fraction whose denominator is 10 or 100 or 1,000, and so on.

Mixed Numbers
1.1

A **mixed number** is made up of both a whole-number part and a fraction part. There is an understood plus sign between the two parts.

Equal Sign
1.1

The **equal sign** ($=$) in a statement means that the expression on the left side of the equal sign *has the same value or values* as the expression on the right side of the equal sign.

"Greater-Than" and "Less-Than" Symbols
1.1

The symbol $>$ is read "is greater than," and the symbol $<$ is read "is less than." Numbers get larger as we move to the right on the number line and get smaller as we move to the left.

Positive Numbers
1.2

All real numbers greater than 0 are positive numbers.

Negative Numbers
1.2

All real numbers less than 0 are negative numbers.

Real Numbers
1.1 and 1.2

All the numbers that can be represented by points on the number line are called real numbers.

Rational Numbers
1.2

Rational numbers are numbers that can be put in the form a/b, where a and b are integers and $b \neq 0$. (All natural numbers, whole numbers, fractions, decimals, and mixed numbers are rational numbers.) All rational numbers are *real* numbers.

The *decimal form* of any rational number is always a *terminating* or a *repeating* decimal.

Absolute Value
1.3

The **absolute value** of a number is the distance between that number and 0 on the number line with no regard to direction. It can never be negative. The absolute value of a real number x is written $|x|$.

Addition of Two Signed Numbers
1.3

1. When the numbers have the same sign,
   a. Add the absolute values *and*
   b. Attach (to the left of the result from step a) the sign of both numbers.

2. When the numbers have different signs,
   a. Subtract the smaller absolute value from the larger absolute value *and*
   b. Attach (to the left of the result from step a) the sign of the number with the larger absolute value.

Additive Identity
1.3

The **additive identity** is 0.

$$a + 0 = a \quad \text{and} \quad 0 + a = a$$

Additive Inverse
1.4

The **additive inverse** (or **negative**) of a is $-a$.

$$a + (-a) = 0 \quad \text{and} \quad (-a) + a = 0$$

Subtraction of One Signed Number from Another
1.4

1. Change the subtraction symbol to an addition symbol, *and* change the sign of the number being subtracted.

2. Add the resulting signed numbers.

$$a - b = a + (-b)$$

Subtraction is the inverse operation of addition.

Subtraction Involving 0
1.4

$$a - 0 = a$$
$$0 - a = -a$$

Multiplication of Two Signed Numbers
1.5

Multiply the absolute values *and* attach the correct sign to the left of the product of the absolute values. That sign is *positive* when the numbers have the same sign and *negative* when the numbers have different signs.

Multiplicative Identity
1.5

The **multiplicative identity** is 1.

$$a \cdot 1 = a \quad \text{and} \quad 1 \cdot a = a$$

Multiplication Involving 0
1.5

$$a \cdot 0 = 0 \quad \text{and} \quad 0 \cdot a = 0$$

Multiplicative Inverse or Reciprocal 1.5

If $a \neq 0$, the **multiplicative inverse** or **reciprocal** of a is $1/a$.

$$a\left(\frac{1}{a}\right) = 1 \quad \text{and} \quad \frac{1}{a}(a) = 1$$

Division of One Signed Number by Another 1.6

Divide the absolute value of the dividend by the absolute value of the divisor *and* attach the correct sign to the left of the quotient of the absolute values. The sign is *positive* when the numbers have the same sign and *negative* when the numbers have different signs.

Division is the inverse operation of multiplication.

Division Involving Zero 1.6

If a represents any real number except 0:

$$\frac{0}{a} = 0$$

$\frac{a}{0}$ is not possible

$\frac{0}{0}$ cannot be determined

$\Big\}$ Division *by* 0 is not defined.

Review Exercises 1.7 Set I

1. Write all the digits greater than 7.

2. Write the smallest two-digit natural number.

3. Write the smallest one-digit integer.

4. Write the largest one-digit integer less than zero.

5. Which symbol, $<$ or $>$, should be used to make the statement true?
   a. $-3 \underline{\ ?\ } 8$ b. $5 \underline{\ ?\ } -2$

6. Which symbol, $<$ or $>$, should be used to make the statement true?
   a. $7 \underline{\ ?\ } -8$ b. $-3 \underline{\ ?\ } -9$

7. What is the multiplicative identity?

8. What is the additive inverse of 4?

In Exercises 9–42, perform the indicated operations, or write "Not defined."

9. $-2 + (+3)$ **10.** $-5 + (+4)$ **11.** $-6 \div (-2)$

12. $-8 \div (-4)$ **13.** $-5 - (-3)$ **14.** $-7 - (-2)$

15. $(+5) - |-2|$ **16.** $(+8) - |-3|$ **17.** $-3(-4)$

18. $-5(-4)$ **19.** $-7 - (3)$ **20.** $-2 - (4)$

21. $4 + (-12)$ **22.** $6 + (-8)$ **23.** $8(-15)$

24. $5(-18)$ **25.** $(24) \div (-3)$ **26.** $-\frac{7}{4} \div \frac{3}{4}$

27. $9 - (-4)$ **28.** $4 - (-7)$ **29.** $\left(-2\frac{2}{3}\right)\left(2\frac{1}{2}\right)$

30. $\left(5\frac{4}{5}\right) + \left(-1\frac{1}{2}\right)$ **31.** $-6 \div (2)$ **32.** $-12 \div (4)$

33. $-10 + (-2)$ **34.** $-8 + (-3)$ **35.** $-4(6)$

36. $-5(7)$ **37.** $\dfrac{-25}{-5}$ **38.** $\dfrac{-16}{-2}$

39. $0 \div (-4)$ **40.** $(0)(-5)$

41. Subtract (-6) from (-10).   **42.** Subtract (-8) from (-5).

Review Exercises 1.7 Set II

NAME _____

1. Write all the digits greater than 5.

2. Write all the even whole numbers less than 12.

3. What is the additive identity?

4. Write all the whole numbers less than 4.

5. Write the smallest digit.

6. What is the multiplicative inverse of $-1/6$?

7. Which symbol, $<$ or $>$, should be used to make the statement true?
   a. $-5 \ \underline{\ ?\ } \ -1$ b. $-2 \ \underline{\ ?\ } \ 0$

8. Which symbol, $<$ or $>$, should be used to make the statement true?
   a. $-3 \ \underline{\ ?\ } \ -8$ b. $5 \ \underline{\ ?\ } \ -2$

In Exercises 9–42, perform the indicated operations, or write "Not defined."

9. $-5 + (+2)$ 10. $-7 - (2)$ 11. $8 \div (-4)$ 12. $-5(+2)$

13. $-8 \div (-2)$ 14. $-8 + (-2)$ 15. $|0| - |-5|$ 16. $5 - (-12)$

17. $-8(-2)$ 18. $-8 - 2$ 19. $-9 - (3)$

1. _____

2. _____

3. _____

4. _____

5. _____

6. _____

7a. _____

b. _____

8a. _____

b. _____

9. _____

10. _____

11. _____

12. _____

13. _____

14. _____

15. _____

16. _____

17. _____

18. _____

19. _____

20. $-9(-3)$ **21.** $9 + (-3)$ **22.** $16 \div (-12)$

23. $41(-1)$ **24.** $\dfrac{8}{9} \div \left(-\dfrac{1}{2}\right)$ **25.** $\dfrac{1}{0}$

26. $\dfrac{3}{-12}$ **27.** $24 \div 0$ **28.** $-72 \div (-8)$

29. $\dfrac{0}{2}$ **30.** $-15 + (-23)$ **31.** $-|-4|$

32. $(-437)(0)$ **33.** $0 \div (-15)$ **34.** $0 - 5$

35. $0(-5)$ **36.** $(-20) - 25$ **37.** $(-20)(-25)$

38. $\left(-1\dfrac{7}{8}\right)\left(-3\dfrac{1}{5}\right)$ **39.** $\dfrac{-10}{-2}$ **40.** $0 - (+7)$

41. Subtract -12 from 7.   **42.** Subtract 3 from -15.

20. _____
21. _____
22. _____
23. _____
24. _____
25. _____
26. _____
27. _____
28. _____
29. _____
30. _____
31. _____
32. _____
33. _____
34. _____
35. _____
36. _____
37. _____
38. _____
39. _____
40. _____
41. _____
42. _____

Sections 1.1–1.7 Diagnostic Test

The purpose of this test is to see how well you understand addition, subtraction, multiplication, and division of two integers or other signed numbers. If you will be tested on Sections 1.1–1.7, we recommend that you work this diagnostic test *before* your instructor tests you on this material. Allow yourself about 50 minutes to do this test.

Complete solutions for all the problems on this test, together with section references, are given in the answer section at the end of the book. We suggest that you study the sections referred to for the problems you do incorrectly.

In Problems 1–10, write "True" if the statement is always true; otherwise, write "False."

1. The smallest natural number is 0. **2.** 0 is a rational number.

3. $|-6| = -6$ **4.** $-4 > -3$

5. The additive identity element is 1. **6.** $6 > -10$

7. 18 is a digit. **8.** 3.762 is an integer.

9. $-\dfrac{1}{2}$ is a rational number. **10.** The reciprocal of -4 is 4.

11. Which symbol, $<$ or $>$, should be used to make the statement -5 __?__ -1 true?

12. Which symbol, $<$ or $>$, should be used to make the statement 2 __?__ -8 true?

13. Remove the absolute value symbols: $|-7.6|$.

14. Remove the absolute value symbols: $-|-56|$.

15. Subtract 17 from 3. **16.** Subtract -5 from -12.

17. What is 6 diminished by -2? **18.** What is the additive inverse of $\dfrac{5}{8}$?

19. What is the additive inverse of -6?

20. What is the multiplicative inverse of -6?

In Problems 21–50, perform the indicated operation, or write "Not defined."

21. $8 + (-26)$ **22.** $-13 + (-5)$ **23.** $-21 + (-5)$

24. $-\dfrac{2}{5} + \dfrac{3}{10}$ **25.** $6.16 + (-8.3)$ **26.** $-3\dfrac{1}{2} + 2\dfrac{1}{8}$

27. $-8 - (-3)$ **28.** $6 - (-12)$ **29.** $-4 - 1$

30. $8 - 37$ **31.** $-5\dfrac{2}{3} - \left(-2\dfrac{8}{9}\right)$ **32.** $3\dfrac{1}{3} - 8\dfrac{1}{6}$

33. $-2.325 - (-6.3)$ **34.** $0 - 12$ **35.** $0(-12)$

36. $-4(-1)$ **37.** $12(-6)$ **38.** $-16(0)$

39. $-\dfrac{5}{6}\left(-\dfrac{3}{10}\right)$ **40.** $-1\dfrac{1}{3}\left(1\dfrac{3}{4}\right)$ **41.** $(6.32)(-0.1)$

42. $8 \div (-2)$ **43.** $\dfrac{17}{0}$ **44.** $\dfrac{-24}{10}$

45. $-36 \div (-12)$ **46.** $\dfrac{18}{-24}$ **47.** $0 \div (-2)$

48. $\dfrac{0}{0}$ **49.** $-54 \div 9$ **50.** $\dfrac{-4.9}{-7}$

1.8 Commutative, Associative, and Distributive Properties of Real Numbers

We have already discussed several important properties of the real-number system. Namely, we've mentioned that 0 is the additive identity, that 1 is the multiplicative identity, that the additive inverse of a is $-a$ (where a represents any real number), and that if a is any real number except 0 the multiplicative inverse of a is $1/a$. The *commutative*, *associative*, and *distributive properties* are also important properties of the set of real numbers. *All* of these properties are **axioms**, that is, statements that are accepted as true without proof.

Commutative Properties

Addition Is Commutative   Reversing the order of two numbers in an addition problem does not change the sum (see Example 1). This important property is called the **commutative property of addition**.

> COMMUTATIVE PROPERTY OF ADDITION
>
> If a and b represent any real numbers, then
>
> $$a + b = b + a$$

Example 1
a. Verify that the commutative property holds for the sum of 2 and 3.
Solution $2 + 3 = 5$, and $3 + 2 = 5$. Since $2 + 3$ and $3 + 2$ both equal 5, they must equal each other. Therefore, $2 + 3 = 3 + 2$.

b. Verify that the commutative property holds for the sum of -6 and 2.
Solution $-6 + 2 = -4$, and $2 + (-6) = -4$. Since $-6 + 2$ and $2 + (-6)$ both equal -4, they must equal each other. Therefore, $-6 + 2 = 2 + (-6)$.

c. Verify that the commutative property holds for the sum of -4 and -8.
Solution $-4 + (-8) = -12$, and $-8 + (-4) = -12$. Since $-4 + (-8)$ and $-8 + (-4)$ both equal -12, they must equal each other. Therefore, $-4 + (-8) = -8 + (-4)$.   ■

Subtraction is not commutative, as Example 2 shows.

Example 2
Show that subtraction is *not* commutative by showing that $3 - 2 \neq 2 - 3$.
Solution $3 - 2 = 1$, and $2 - 3 = -1$. Since $3 - 2$ and $2 - 3$ do not both equal the same number, $3 - 2 \neq 2 - 3$. By finding one subtraction problem for which the commutative property does not hold, we've shown that subtraction is *not* commutative.   ■

Multiplication Is Commutative   Reversing the order of two numbers in a multiplication problem does not change the product (see Example 3). This important property is called the **commutative property of multiplication**.

> COMMUTATIVE PROPERTY OF MULTIPLICATION
>
> If a and b represent any real numbers, then
>
> $$a \cdot b = b \cdot a$$

Example 3   a. Verify that the commutative property holds for the product of 4 and 5.
Solution $4 \cdot 5 = 20$, and $5 \cdot 4 = 20$. Since $4 \cdot 5$ and $5 \cdot 4$ both equal 20, they must equal each other. Therefore, $4 \cdot 5 = 5 \cdot 4$.

  b. Verify that the commutative property holds for the product of -9 and 3.
Solution $-9(3) = -27$, and $3(-9) = -27$. Since $-9(3)$ and $3(-9)$ both equal -27, they must equal each other. Therefore, $-9(3) = 3(-9)$. ■

Division is not commutative, as Example 4 shows.

Example 4   Show that division is *not* commutative by showing that $10 \div 5 \neq 5 \div 10$.
Solution $10 \div 5 = 2$, and $5 \div 10 = \frac{1}{2}$. Since $10 \div 5$ and $5 \div 10$ do not both equal the same number, $10 \div 5 \neq 5 \div 10$. By finding one division problem for which the commutative property does not hold, we've shown that division is *not* commutative. ■

Associative Properties

Addition Is Associative   When we are given three numbers to add, we obtain the same answer when we add the first two numbers together first as when we add the last two numbers together first. This property is called the **associative property of addition**. We use grouping symbols such as parentheses ( ), brackets [ ], or braces { } to show which two numbers are to be added together first.

ASSOCIATIVE PROPERTY OF ADDITION

If a, b, and c represent any real numbers, then

$$(a + b) + c = a + (b + c)$$

Example 5   Verify that $(2 + 3) + 4 = 2 + (3 + 4)$.
Solution

$(2 + 3) + 4$ The *parentheses* indicate that 2 and 3 must be *added together first*

$= \quad 5 \quad + 4$

$= \quad 9$

and

$2 + (3 + 4)$ The *parentheses* indicate that 3 and 4 must be *added together first*

$= 2 + \quad 7$

$= \quad 9$

Since $(2 + 3) + 4$ and $2 + (3 + 4)$ both equal 9, they must equal each other. Therefore, $(2 + 3) + 4 = 2 + (3 + 4)$, and we've shown that the associative property of addition holds for the sum $2 + 3 + 4$. ■

Subtraction is not associative, as Example 6 will prove.

Example 6   Show that subtraction is *not* associative by showing that $(12 - 6) - 2 \neq 12 - (6 - 2)$.
Solution

$(12 - 6) - 2$	$12 - (6 - 2)$
$= \quad 6 \quad - 2$	$= 12 - \quad 4$
$= \quad 4$	$= \quad 8$

Since $(12 - 6) - 2$ and $12 - (6 - 2)$ do not both equal the same number, $(12 - 6) - 2 \neq 12 - (6 - 2)$. By finding one subtraction problem for which the associative property does not hold, we've shown that subtraction is *not* associative. ∎

Multiplication Is Associative   When we are given three numbers to multiply, we obtain the same answer when we multiply the first two numbers together first as when we multiply the last two numbers together first. This property is called the **associative property of multiplication**. Parentheses, brackets, or braces are used to show which two numbers are to be multiplied together first.

ASSOCIATIVE PROPERTY OF MULTIPLICATION

If a, b, and c represent any real numbers, then

$$(a \cdot b) \cdot c = a \cdot (b \cdot c)$$

Example 7   Verify that $(3 \cdot 4) \cdot 2 = 3 \cdot (4 \cdot 2)$.

Solution

$(3 \cdot 4) \cdot 2$	$3 \cdot (4 \cdot 2)$
$= \quad 12 \cdot 2$	$= 3 \cdot 8$
$= \quad 24$	$= 24$

Since $(3 \cdot 4) \cdot 2$ and $3 \cdot (4 \cdot 2)$ both equal 24, they must equal each other. Therefore, $(3 \cdot 4) \cdot 2 = 3 \cdot (4 \cdot 2)$, and we've shown that the associative property of multiplication holds for the product $3 \cdot 4 \cdot 2$. ∎

Example 8   Verify that $[(-6)(+2)](-5) = (-6)[(+2)(-5)]$.

Solution

$[(-6)(+2)](-5)$	$(-6)[(+2)(-5)]$
$= \quad [-12] \cdot (-5)$	$= (-6) \cdot [-10]$
$= \quad 60$	$= 60$

Since $[(-6)(+2)](-5)$ and $(-6)[(+2)(-5)]$ both equal 60, they must equal each other. Therefore, $[(-6)(+2)](-5) = (-6)[(+2)(-5)]$, and we've shown that the associative property of multiplication holds for the product $(-6)(+2)(-5)$. ∎

Division is not associative, as a single example will prove.

Example 9   Show that division is *not* associative by showing that $(16 \div 4) \div 2 \neq 16 \div (4 \div 2)$.

Solution

$(16 \div 4) \div 2$	$16 \div (4 \div 2)$
$= \quad 4 \quad \div 2$	$= 16 \div 2$
$= \quad 2$	$= 8$

Since $(16 \div 4) \div 2$ and $16 \div (4 \div 2)$ do not both equal the same number, $(16 \div 4) \div 2 \neq 16 \div (4 \div 2)$. By finding one division problem for which the associative property does not hold, we've shown that division is *not* associative. ∎

SUMMARY

1. The *commutative properties* guarantee that when we have two numbers to add or two numbers to multiply, *reversing the order* of the numbers does not change the answer.

2. The *associative properties* guarantee that when we have three numbers to add or three numbers to multiply, we obtain the same answer when we operate on the first two numbers first as when we operate on the last two numbers first.

How to Determine Whether Commutativity or Associativity Has Been Used   In commutativity, the numbers or letters actually exchange places (commute):

$$a + b = b + a \qquad c \cdot d = d \cdot c$$

The first element occupies the second place, and vice versa.

In associativity, the numbers or letters stay in their original places, but the grouping is changed:

$$a + (b + c) = (a + b) + c \qquad d \cdot (e \cdot f) = (d \cdot e) \cdot f$$

Distributive Property

Multiplication Is Distributive Over Addition and Over Subtraction   The **distributive property** is another axiom—one of the fundamental properties of real numbers that we've accept as true without proof. Whereas each of the properties of real numbers that we've discussed so far was concerned with just one operation, the distributive property is concerned with *two* operations: multiplication *and* addition or multiplication *and* subtraction.

MULTIPLICATION IS DISTRIBUTIVE OVER ADDITION AND OVER SUBTRACTION

If a, b, and c represent any real numbers,

then $\qquad\qquad a(b + c) = (ab) + (ac)$

and $\qquad\qquad a(b - c) = (ab) - (ac)$

Example 10   Verify that $5(3 + 7) = (5 \cdot 3) + (5 \cdot 7)$.

Solution

$5(3 + 7)$		$(5 \cdot 3) + (5 \cdot 7)$
$= 5 \;\;(10)$		$=\;\; 15 + 35$
$=\;\; 50$		$=\;\; 50$

Because $5(3 + 7)$ and $(5 \cdot 3) + (5 \cdot 7)$ both equal 50, they must equal each other. Therefore, $5(3 + 7) = (5 \cdot 3) + (5 \cdot 7)$, and we've verified that the distributive property holds in this example. ∎

Example 11   Verify that $8[3 + (-10)] = [8 \cdot 3] + [8 \cdot (-10)]$
Solution $8[3 + (-10)]$ $[8 \cdot 3] + [8 \cdot (-10)]$

$= 8 \ [-7]$ $= \ 24 \ + \ [-80]$

$= \ -56$ $= \ -56$

Because $8[3 + (-10)]$ and $[8 \cdot 3] + [8 \cdot (-10)]$ both equal -56, they must equal each other. Therefore, $8[3 + (-10)] = [8 \cdot 3] + [8 \cdot (-10)]$, and we've verified that the distributive property holds in this example. The problem could have been written as "Verify that $8(3 - 10) = (8 \cdot 3) - (8 \cdot 10)$." ∎

In Example 12 and Exercises 1.8, we include problems that involve the identity and inverse properties as well as the commutative, associative, and distributive properties. We also include a few problems dealing with the multiplication property of zero.

Example 12   State whether each of the following is true or false. If the statement is true, give the reason.

a. $(-7) + 5 = 5 + (-7)$ *True* because of the commutative property of addition (*order* of numbers changed)

b. $(+6)(-8) = (-8)(+6)$ *True* because of the commutative property of multiplication (*order* of numbers changed)

c. $[(-3) + 5] + (-2)$ *True*; associative property of addition
 $= (-3) + [5 + (-2)]$ (*grouping* changed)

d. $-6(3 + 5) = (-6 \cdot 3) + (-6 \cdot 5)$ *True*; multiplication is distributive over addition

e. $-8 + 0 = -8$ *True*; additive identity property (0 is the additive identity)

f. $-5 + 5 = 0$ *True*; additive inverse property (additive inverse of -5 is 5)

g. $[(7) \cdot (-4)] \cdot (2)$ *True*; associative property of multiplication
 $= (7) \cdot [(-4) \cdot (2)]$ (*grouping* changed)

h. $(+8) - (-7) = (-7) - (+8)$ *False*

i. $3(4 \cdot 5) = (3 \cdot 4) \cdot (3 \cdot 5)$ *False*

j. $a + (b + c) = (a + b) + c$ *True*; associative property of addition
                            (*grouping* changed)

k. $y \div z = z \div y$ *False*

l. $(p \cdot r) \cdot s = p \cdot (r \cdot s)$ *True*; associative property of multiplication
                            (*grouping* changed)

m. $3 \times 0 = 3$ *False*

n. $(3 + 5) + 7 = 3 + (7 + 5)$ *True*; commutative *and* associative properties of addition

o. $8 \cdot (3 \cdot 6) = (8 \cdot 6) \cdot 3$ *True*; commutative *and* associative properties of multiplication. ∎

Because the commutative and associative properties of addition and of multiplication hold for all real numbers, we can add more than two numbers *in any order*, and we can multiply more than two numbers *in any order*.

Example 13   Add $-8 + 3 + (-4) + (-6)$.

Solution $-8 + 3 + (-4) + (-6)$

$= [(-8) + (-4) + (-6)] + 3$ Collecting all the negative numbers

$= [-18] + 3$ Adding all the negative numbers first

$= -15$ ■

> In multiplying several numbers together, an *odd* number of negative factors gives a product that is *negative* (see Example 14), and an *even* number of negative factors gives a product that is *positive* (see Example 15).

Example 14   Multiply $(-2)(-5)(-3)$.

Solution $(-2)(-5)(-3)$ Three negative factors

$= (+10)(-3)$

$= -30$ Notice that an *odd* number of negative signs gives a product that is *negative* ■

Example 15   Multiply $(-2)(-5)(-3)(-4)$.

Solution $(-2)(-5)(-3)(-4)$ Four negative factors

$= (+10)(-3)(-4)$

$= (-30)(-4)$

$= +120$ Notice that an *even* number of negative signs gives a product that is *positive* ■

EXERCISES 1.8

Set I   In Exercises 1–36, state whether each of the following is true or false. If the statement is true, give the reason.

1. $7 + 5 = 5 + 7$ **2.** $9 + 4 = 4 + 9$

3. $(2 + 6) + 3 = 2 + (6 + 3)$ **4.** $(1 + 8) + 7 = 1 + (8 + 7)$

5. $6 - 2 = 2 - 6$ **6.** $4 - 7 = 7 - 4$

7. $(a \cdot b) \cdot c = a \cdot (b \cdot c)$ **8.** $(p \cdot q) \cdot r = p \cdot (q \cdot r)$

9. $8 \div 4 = 4 \div 8$ **10.** $3 \div 6 = 6 \div 3$

11. $(p)(t) = (t)(p)$ **12.** $(m)(n) = (n)(m)$

13. $(4) + (-5) = (-5) + (4)$ **14.** $(-7) + (2) = (2) + (-7)$

15. $5 + (3 + 4) = 5 + (4 + 3)$ **16.** $6 + (8 + 2) = 6 + (2 + 8)$

17. $6(2 \cdot 3) = (6 \cdot 2) \cdot (6 \cdot 3)$ **18.** $7(3 + 5) = (7 \cdot 3) + (7 \cdot 5)$

19. $e + f = f + e$ **20.** $j + k = k + j$

21. $-8 \times 1 = -8$ **22.** $11 + 0 = 11$

23. $15 + (-15) = 0$ **24.** $-4 + 4 = 0$

25. $8(5 - 2) = (8 \cdot 5) - (8 \cdot 2)$

26. $9(6 \cdot 4) = (9 \cdot 6) \cdot (9 \cdot 4)$

27. $8 \times 0 = 8$

28. $0 \times (-4) = -4$

29. $9 + (5 + 6) = (9 + 6) + 5$

30. $3 \cdot (8 \cdot 4) = (3 \cdot 4) \cdot 8$

31. $x - 4 = 4 - x$

32. $5 - y = y - 5$

33. $4(a \cdot 6) = 4a(6)$

34. $m(7 \cdot 5) = (m \cdot 7)5$

35. $H + 8 = 8 + H$

36. $4 + P = P + 4$

In Exercises 37–46, complete each statement by using the property indicated.

37. Commutative Property: $\qquad -7 + 3 = $ _____

38. Commutative Property: $\qquad 12 + (-5) = $ _____

39. Associative Property: $\qquad 4(-3 \cdot 6) = $ _____

40. Associative Property: $\qquad -8[4 \cdot (-2)] = $ _____

41. Distributive Property: $\qquad 3(5 - 8) = $ _____

42. Distributive Property: $\qquad 6(15 + 2) = $ _____

43. Commutative Property: $\qquad (-4)(3) = $ _____

44. Commutative Property: $\qquad 6(-3) = $ _____

45. Associative Property: $\qquad 4 + (-3 + 6) = $ _____

46. Associative Property: $\qquad (-3 + 4) + (-2) = $ _____

In Exercises 47–56, perform the indicated operations.

47. $5 + (-2) + 4 + (-8) + (-5)$

48. $-2 + 6 + (-8) + (-12) + 5$

49. $8 + (-3) + (-7) + (-1)$

50. $-2 + (-5) + 6 + (-11)$

51. $(-5)(-4)(-2)$

52. $(-3)(-2)(-8)$

53. $2(-5)(-9)$

54. $(-3)(4)(-2)$

55. $(-2)(-3)(-5)(-4)$

56. $(-4)(-2)(-1)(-7)$

Set II In Exercises 1–36, state whether each of the following is true or false. If the statement is true, give the reason.

1. $5 + 3 = 3 + 5$

2. $(3 + 1) + 5 = 3 + (1 + 5)$

3. $8 - 2 = 2 - 8$

4. $(x \cdot y) \cdot z = x \cdot (y \cdot z)$

5. $10 \div 2 = 2 \div 10$

6. $(3)(-2) = (-2)(3)$

7. $2 + (3 + 4) = 2 + (4 + 3)$

8. $x + y = y + x$

9. $8 + (2 + 5) = (8 + 5) + 2$

10. $a - 2 = 2 - a$

11. $(4 \cdot c)(3) = 4(3 \cdot c)$

12. $3 \div x = x \div 3$

13. $5(-2) = -2(5)$

14. $8 + (3 + 2) = 8 + (2 + 3)$

15. $12 - (6 - 9) = (12 - 6) - 9$

16. $9 \times (-3) = (-3) \times 9$

17. $8(7 \cdot 2) = (8 \cdot 7) \cdot (8 \cdot 2)$

18. $6(5 - 1) = (6 \cdot 5) - (6 \cdot 1)$

19. $3 + (4 + 7) = (3 + 7) + 4$

20. $25 \div 3 = 3 \div 25$

21. $3 + 0 = 3$

22. $0 \times (-13) = -13$

23. $\frac{1}{2} + \left(-\frac{1}{2}\right) = 0$

24. $4 \times (-4) = 0$

25. $3(9 - 4) = (3 \cdot 9) - (3 \cdot 4)$

26. $-3(2 \cdot 4) = (-3 \cdot 2) \cdot (-3 \cdot 4)$

27. $-5 \times 0 = -5$

28. $-16 \times 0 = 0$

29. $18 - 3 = 3 - 18$

30. $5 \cdot (12 \cdot 6) = (5 \cdot 6) \cdot 12$

31. $24 + 16 = 16 + 24$

32. $24 \div (12 \div 2) = (24 \div 12) \div 2$

33. $2 \times (3 \times 4) = (2 \times 3) \times 4$

34. $6 - 15 = 15 - 6$

35. $18 + (3 + 5) = (18 + 3) + 5$

36. $9 - (4 - 12) = (9 - 4) - 12$

In Exercises 37–46, complete each statement by using the property indicated.

37. Commutative Property: $\qquad 6 + (-8) = $ _____

38. Commutative Property: $\qquad -6(-4) = $ _____

39. Associative Property: $\qquad 4 + [(-3) + 6] = $ _____

40. Associative Property: $\qquad 6(-4 \cdot 5) = $ _____

41. Distributive Property: $\qquad -2(8 + 4) = $ _____

42. Distributive Property: $\qquad 6(-4 + 8) = $ _____

43. Commutative Property: $\qquad -4 + 3 = $ _____

44. Commutative Property: $\qquad -5(2) = $ _____

45. Associative Property: $\qquad 4(-3 \cdot 7) = $ _____

46. Associative Property: $\qquad -1 + [4 + (-2)] = $ _____

In Exercises 47–56, perform the indicated operations.

47. $3 + (-7) + (-6) + 8 + (-9)$

48. $4 + (-6) + 3 + (-2) + (-12)$

49. $9 + (-5) + 7 + (-12)$

50. $-2 + (-8) + (-6) + 11$

51. $4(-5)(-7)$

52. $(-9)(-8)(-1)$

53. $(-5)(-1)(-6)(-2)$

54. $-3(-2)(5)(-4)$

55. $(-5)(-2)(-2)(-2)$

56. $8(-1)(-2)(-3)$

1.9 Powers of Integers

Because we have learned to multiply signed numbers, we can now consider products in which the same number is repeated as a factor.

The shortened notation for a product such as $3 \cdot 3 \cdot 3 \cdot 3$ is 3^4. That is, by definition, $3^4 = 3 \cdot 3 \cdot 3 \cdot 3$. In the expression 3^4, 3 is called the **base**, and 4 is called the **exponent**. The number 4 (the exponent) indicates that 3 (the base) is to be used as a *factor* four times. The entire symbol 3^4 is called an **exponential expression** and is commonly read as "three to the fourth power." See Figure 1.9.1.

Read as "three to the fourth power" $\longrightarrow 3^4 = 81$

with Exponent labeling the 4 and Base labeling the 3

FIGURE 1.9.1

A WORD OF CAUTION

$$3^4 \neq 3 \cdot 4$$
$$3^4 = \underbrace{3 \cdot 3 \cdot 3 \cdot 3}_{\text{Four factors}} = 81$$

☑

A WORD OF CAUTION When you write an exponential number, be sure your exponents look like exponents. For example, be sure that 3^4 doesn't look like 34. ☑

We usually read b^2 as "b squared" rather than as "b to the second power"; likewise, we usually read b^3 as "b cubed" rather than as "b to the third power."

Even Power   If a base has an exponent that is an even number, we say that the expression is an **even power** of the base. For example, 3^2, 5^4, and $(-2)^6$ are even powers.

Odd Power   If a base has an exponent that is an odd number, we say that the expression is an **odd power** of the base. For example, 3^5, 10^3, and $(-4)^5$ are odd powers.

Example 1   Examples of powers of positive integers:

a. $2^3 = 2 \cdot 2 \cdot 2 = 8$

b. $4^2 = 4 \cdot 4 = 16$

c. $1^4 = 1 \cdot 1 \cdot 1 \cdot 1 = 1$ ∎

The exponent in an expression always applies *only* to the symbol immediately preceding it.

Recall from Section 1.8 that an odd number of negative signs in a product gives a negative product and an even number of negative signs in a product gives a positive product. Therefore;

> An *even* power of a *negative* number is *positive*, and an *odd* power of a *negative* number is *negative* (see Example 2).

Example 2   Examples of powers of negative integers:

a. $(-3)^2 = (-3)(-3) = 9$
b. $(-1)^4 = (-1)(-1)(-1)(-1) = 1$ Notice that an *even* power of a *negative* number is *positive*

c. $(-2)^3 = (-2)(-2)(-2) = -8$
d. $(-1)^5 = (-1)(-1)(-1)(-1)(-1) = -1$ Notice that an *odd* power of a *negative* number is *negative*
e. $(-36)^3 = (-36)(-36)(-36) = -46,656$ ∎

Example 3   Rewrite each of the following with no exponents.

a. $(-6)^2 = (-6)(-6) = 36$ The exponent applies to whatever is in the parentheses (the -6)

b. $-6^2 = -(6)^2 = -(6 \cdot 6) = -36$ The exponent applies only to the 6

c. $(-1)^4 = (-1)(-1)(-1)(-1) = 1$ The exponent applies to the -1

d. $-1^4 = -(1 \cdot 1 \cdot 1 \cdot 1) = -1$ The exponent applies only to the 1   ∎

A WORD OF CAUTION   Students often think that expressions such as $(-6)^2$ and -6^2 are the same. They are *not* the same. The exponent applies only to the symbol immediately preceding it.

$(-6)^2 = (-6)(-6) = 36$ The exponent applies to whatever is inside the parentheses, because the exponent is immediately to the right of a set of parentheses

$-6^2 = -6 \cdot 6 = -36$ The exponent applies only to the 6   ☑

Example 4   Find the value of each expression.

a. $-(-3)^4$ (Read as "the negative of the fourth power of negative three.")
Solution We can think as follows: $(-3)^4 = 81$; therefore,

$$-(-3)^4 = -(81) = -81$$

or we can think

One negative here
Four more negatives here
There are an odd number of negative signs altogether; therefore, the answer is negative
$-(-3)^4 = -81$

If we write the problem in its expanded form, $-(-3)(-3)(-3)(-3)$, we *see* the five negative signs.

b. $-(-2)^3$ (Read as "the negative of the third power of negative two.")
Solution We can think as follows: $(-2)^3 = -8$; therefore,

$$-(-2)^3 = -(-8) = 8$$

or we can think

One negative here
Three more negatives here
There are an even number of negative signs altogether; therefore, the answer is positive
$-(-2)^3 = 8$

If we write the problem in its expanded form, $-(-2)(-2)(-2)$, we *see* the four negative signs.   ∎

The Exponent 1

THE EXPONENT 1

If a represents any real number, then

$$a^1 = a$$

Example 5   Examples of 1 as an exponent:

a. $5^1 = 5$

b. $(-8)^1 = -8$ ∎

Powers of Zero

POWERS OF 0

If a represents any positive real number, then

$$0^a = 0$$

Example 6   Examples of powers of zero:

a. $0^2 = 0 \cdot 0 = 0$, or, using the rule above, $0^2 = 0$.

b. $0^5 = 0 \cdot 0 \cdot 0 \cdot 0 \cdot 0 = 0$, or, using the rule above, $0^5 = 0$. ∎

You may find it helpful to memorize the following powers:

$0^2 = 0$	$7^2 = 49$	$0^3 = 0$	$0^4 = 0$
$1^2 = 1$	$8^2 = 64$	$1^3 = 1$	$1^4 = 1$
$2^2 = 4$	$9^2 = 81$	$2^3 = 8$	$2^4 = 16$
$3^2 = 9$	$10^2 = 100$	$3^3 = 27$	$3^4 = 81$
$4^2 = 16$	$11^2 = 121$	$4^3 = 64$	
$5^2 = 25$	$12^2 = 144$	$5^3 = 125$	$2^5 = 32$
$6^2 = 36$	$13^2 = 169$		$2^6 = 64$

Cases where 0 appears as an exponent, such as 5^0, are discussed in Chapter 5.

EXERCISES 1.9

Set I   Find the value of each of the following expressions.

1. 3^3 **2.** 2^4 **3.** $(-5)^2$ **4.** $(-6)^3$

5. 7^2 **6.** 3^4 **7.** 0^3 **8.** 0^4

9. $(-10)^1$ **10.** $(-10)^2$ **11.** 10^3 **12.** 10^4

13. $(-10)^5$ **14.** $(-10)^6$ **15.** 2^1 **16.** 2^5

17. $(-2)^6$ **18.** $(-2)^7$ **19.** 2^8 **20.** 25^2

21. 40^3 **22.** 0^4 **23.** $(-12)^3$ **24.** $(-15)^2$

25. $(-1)^5$ **26.** $(-1)^7$ **27.** -2^2 **28.** -3^2

29. $(-1)^{99}$ **30.** $(-1)^{98}$ **31.** $-(-1)^5$ **32.** $-(-1)^6$

33. $-(-9)^2$ **34.** $-(-5)^2$ **35.** 0^8 **36.** 0^3

37. $(12.7)^2$ **38.** $(15.4)^2$ **39.** $(0.156)^2$ **40.** $(0.087)^2$

Set II   Find the value of each of the following expressions.

1. $(-2)^3$	**2.** 6^2	**3.** $(-10)^2$	**4.** -10^2
5. 0^5	**6.** $(-1)^{35}$	**7.** $-(-5)^2$	**8.** $(-1)^{50}$
9. 10^5	**10.** 8^2	**11.** 4^3	**12.** -3^4
13. $(-3)^4$	**14.** 20^3	**15.** 24^2	**16.** $(-4)^2$
17. -4^2	**18.** $(-1)^{43}$	**19.** 0^{23}	**20.** $-(-8)^2$
21. $(-1)^{132}$	**22.** 1^{43}	**23.** $(-2)^5$	**24.** 30^2
25. 0^{42}	**26.** $(-7)^2$	**27.** $(-3)^3$	**28.** -3^3
29. $-(-3)^3$	**30.** $-(-3^3)$	**31.** $-(-1)^7$	**32.** $-(-2)^4$
33. $-(-2)^6$	**34.** $-(-4)^2$	**35.** 0^2	**36.** 1^3
37. 16^2	**38.** 0^1	**39.** $(0.894)^2$	**40.** $(0.095)^2$

1.10 Roots of Integers

1.10A Finding Square Roots of Integers by Inspection

Just as subtraction is the inverse operation of addition and division is the inverse operation of multiplication, finding *roots* is the inverse operation of raising to powers. Thus, finding the *square root* of a number is the inverse operation of *squaring* a number.

Principal Square Root   Every positive real number has both a positive and a negative square root; the *positive* square root is called the **principal square root**.

Example 1   The number 9 has two square roots: $+3$ and -3.

$+3$ is a square root of 9 because $3^2 = 9$.

-3 is a square root of 9 because $(-3)^2 = 9$.

3 is the *principal* square root of 9 because it is the positive one.   ■

The Square Root Symbol   The notation for the principal square root of p is $\sqrt{p}$, which is read "the square root of p." The entire expression $\sqrt{p}$ is called a **radical expression** or, more simply, a **radical**. The parts of a square root are shown in Figure 1.10.1.

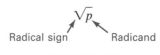

Radical sign   Radicand

FIGURE 1.10.1

When we are asked to find $\sqrt{p}$, we must find some *positive* number whose *square* is p. For example, "Find $\sqrt{9}$" means we must find a *positive* number whose *square* is 9. The answer, of course, is 3. Therefore, $\sqrt{9} = 3$.

A WORD OF CAUTION   Because $\sqrt{p}$ *always* represents the *principal square root*, when $p \geq 0$, $\sqrt{p}$ is *always* positive or zero.   ☑

In this section and in Section 1.10B, the *radicand* will always be the square of some integer. You will find the problems easier to do if you have memorized the squares of the whole numbers 0 through 13 (see page 50).

Example 2 Find the square root of 25; that is, find $\sqrt{25}$.
Solution $\sqrt{25} = 5$ Because $5^2 = 25$ ∎

Example 3 Find the square root of 16; that is, find $\sqrt{16}$.
Solution $\sqrt{16} = 4$ Because $4^2 = 16$ ∎

Example 4 Find the following square roots by inspection.

a. $\sqrt{4} = 2$ Because $2^2 = 4$

b. $\sqrt{9} = 3$ Because $3^2 = 9$

c. $\sqrt{36} = 6$ Because $6^2 = 36$

d. $\sqrt{0} = 0$ Because $0^2 = 0$

e. $\sqrt{1} = 1$ Because $1^2 = 1$ ∎

Example 5 Find $-\sqrt{16}$.
Solution This is a two-step problem. We must first find $\sqrt{16}$, and then we must find the *negative* of that number. By inspection, we find that $\sqrt{16} = 4$. Then $-(\sqrt{16}) = -(4) = -4$. ∎

NOTE   Square roots of negative numbers are *imaginary numbers*, not real numbers. If the problem in Example 5 had been $\sqrt{-16}$, the answer would have been "Not a real number," since no real number exists whose square is -16. ☑

Example 6 Find $\sqrt{-36}$ or write "Not a real number."
Solution There is no real number whose square is -36. Therefore, the answer is "Not a real number." ∎

A WORD OF CAUTION   In finding $\sqrt{16}$, students often make one of these two errors:

Incorrect method		*Correct method*
$\sqrt{16} = 4 = 2$	$4 \neq 2$	$\sqrt{16} = 4$
$\sqrt{16} = \sqrt{4} = 2$	$\sqrt{16} \neq \sqrt{4}$	

☑

A WORD OF CAUTION   The *square* of 16 is 256 (that is, $16^2 = 256$). The *square root* of 16 is 4 (that is, $\sqrt{16} = 4$). ☑

EXERCISES 1.10A

Set I   Find the following square roots by inspection, or write "Not a real number."

1. $\sqrt{16}$ **2.** $\sqrt{25}$ **3.** $-\sqrt{4}$ **4.** $-\sqrt{9}$

5. $\sqrt{81}$ **6.** $\sqrt{36}$ **7.** $\sqrt{100}$ **8.** $\sqrt{144}$

9. $-\sqrt{81}$ **10.** $-\sqrt{121}$ **11.** $\sqrt{-4}$ **12.** $\sqrt{-25}$

13. $-\sqrt{-1}$ **14.** $-\sqrt{-9}$

Set II   Find the following square roots by inspection, or write "Not a real number."

1. $-\sqrt{100}$ **2.** $-\sqrt{144}$ **3.** $\sqrt{49}$ **4.** $\sqrt{121}$

5. $\sqrt{64}$ **6.** $-\sqrt{36}$ **7.** $\sqrt{1}$ **8.** $-\sqrt{25}$

9. $\sqrt{9}$ **10.** $-\sqrt{169}$ **11.** $\sqrt{-49}$ **12.** $-\sqrt{49}$

13. $-\sqrt{-100}$ **14.** $-\sqrt{-81}$

1.10B Finding Square Roots of Positive Integers by Trial and Error

It is proved in higher mathematics that for all positive real numbers a and b, if $a > b$, then $\sqrt{a} > \sqrt{b}$, and if $a < b$, then $\sqrt{a} < \sqrt{b}$. We can use these facts in finding square roots.

Example 7 Find $\sqrt{196}$ by trial and error.

Solution We know that $\sqrt{100} = 10$, because $10^2 = 100$. Because $196 > 100$, $\sqrt{196} > 10$.

Try 12; $12^2 = 144$. Because $144 < 196$, 12 is too small.

Try 13; $13^2 = 169$. Because $169 < 196$, 13 is too small.

Try 14; $14^2 = 196$. Therefore, $\sqrt{196} = 14$. ∎

This is what we mean by "finding the square root by trial and error."

It is also proved in higher mathematics courses that if b is between a and c, when a, b, and c are positive real numbers, then $\sqrt{b}$ is between $\sqrt{a}$ and $\sqrt{c}$.

Example 8 Find $\sqrt{576}$ by trial and error. The answer is an integer.

Solution
$$\sqrt{400} = 20 \qquad \text{Because } 20^2 = 400$$
$$\sqrt{900} = 30 \qquad \text{Because } 30^2 = 900$$

Since 576 is between 400 and 900, $\sqrt{576}$ *must* be between 20 and 30. Also, since the last digit of 57 6 is 6, the last digit of the square root must be 4 or 6 (if the square root is an integer), because $4 \cdot 4 = 1\,6$ and $6 \cdot 6 = 3\,6$. Therefore, 24 and 26 are the only possible square roots. Try 24: $24^2 = 576$. Therefore, $\sqrt{576} = 24$. ∎

EXERCISES 1.10B

Set I   Find the following square roots by trial and error. All the answers are integers.

   1. $\sqrt{529}$ **2.** $\sqrt{361}$ **3.** $\sqrt{441}$ **4.** $\sqrt{625}$

   5. $\sqrt{289}$ **6.** $\sqrt{324}$ **7.** $\sqrt{729}$ **8.** $\sqrt{1,296}$

Set II   Find the following square roots by trial and error. All the answers are integers.

   1. $\sqrt{400}$ **2.** $\sqrt{484}$ **3.** $\sqrt{676}$ **4.** $\sqrt{1,024}$

   5. $\sqrt{225}$ **6.** $\sqrt{784}$ **7.** $\sqrt{1,444}$ **8.** $\sqrt{2,809}$

1.10C Finding Square Roots by Table or Calculator

Square roots of positive numbers can be found by using a calculator with a square root key $\boxed{\sqrt{x}}$ and sometimes by using tables. (See Table I, inside back cover.)

Example 9 Find $\sqrt{710,649}$ by using a calculator.

Solution On some calculators, you press the following keys in the order shown (consult your calculator manual for details if this doesn't work):

$$\boxed{7}\ \boxed{1}\ \boxed{0}\ \boxed{6}\ \boxed{4}\ \boxed{9}\ \boxed{\sqrt{x}}$$

The calculator display shows 843. Therefore, $\sqrt{710,649} = 843$. (We cannot find $\sqrt{710,649}$ by using Table I.) ∎

Irrational Numbers

We often need to *approximate* the square root of some number that is *not* the square of a rational number. Such a number is called an **irrational number**. The decimal form of an irrational number does not terminate and does not repeat. All irrational numbers are real numbers; they can all be graphed on the number line.

Example 10 Approximate $\sqrt{3}$. Use Table I, or use a calculator and round off the answer to three decimal places.

Solution We find the approximate value of $\sqrt{3}$ by referring to Table I, which gives the roots rounded off to three decimal places.

Locate 3 in the column headed *N*.
Read the value of $\sqrt{3}$ to the right of 3
in the column headed $\sqrt{N}$.
We see that $\sqrt{3} \approx 1.732$*

N	$\sqrt{N}$
1	1.000
2	1.414
3	1.732
4	2.000
5	2.236

To approximate $\sqrt{3}$ by using a calculator, press the following keys in order: $\boxed{3}$ $\boxed{\sqrt{x}}$. The display probably shows 1.7320508. (Your calculator may not show the same number of digits.) If we round off this answer to three decimal places, we have

$$\sqrt{3} \approx 1.732 \quad \blacksquare$$

Example 11 Approximate $\sqrt{94}$. Use a calculator and round off the answer to three decimal places, or use Table I.

Solution To find $\sqrt{94}$ by using a calculator, press the following keys in the order shown: $\boxed{9}$ $\boxed{4}$ $\boxed{\sqrt{x}}$. The display probably shows 9.6953597. When 9.6953597 is rounded off to three decimal places, we get 9.695. To use Table I, proceed as shown below.

Locate 94 in the column headed *N*.
Then read the value of $\sqrt{94} \approx 9.695$
in the column headed $\sqrt{N}$.

N	$\sqrt{N}$
81	9.000
82	9.055
92	9.592
93	9.644
94	9.695
95	9.747
96	9.798

There is a "paper-and-pencil" method for calculating square roots, which has become obsolete due to the widespread use of calculators. Therefore, it is not discussed in this book.

---

*The symbol "$\approx$" (read "is approximately equal to") is used to show that two numbers are *approximately* equal to each other.

EXERCISES 1.10C

Set I   In Exercises 1–8, approximate each square root by using a calculator and rounding off answers to three decimal places or by using Table I, inside the back cover.

1. $\sqrt{13}$ **2.** $\sqrt{18}$ **3.** $\sqrt{37}$ **4.** $\sqrt{50}$

5. $\sqrt{79}$ **6.** $\sqrt{60}$ **7.** $\sqrt{86}$ **8.** $\sqrt{92}$

In Exercises 9–12, find each square root by using a calculator.

9. $\sqrt{466,489}$ **10.** $\sqrt{674,041}$ **11.** $\sqrt{272,484}$ **12.** $\sqrt{89,401}$

Set II   In Exercises 1–8, approximate each square root by using a calculator and rounding off answers to three decimal places or by using Table I, inside the back cover.

1. $\sqrt{31}$ **2.** $\sqrt{69}$ **3.** $\sqrt{97}$ **4.** $\sqrt{184}$

5. $\sqrt{178}$ **6.** $\sqrt{145}$ **7.** $\sqrt{78}$ **8.** $\sqrt{125}$

In Exercises 9–12, find each square root by using a calculator.

9. $\sqrt{178,929}$ **10.** $\sqrt{373,321}$ **11.** $\sqrt{88,804}$ **12.** $\sqrt{35,344}$

1.10D   Higher Roots

Roots other than square roots are called **higher roots**. The parts of the symbol for higher roots are shown in Figure 1.10.2.

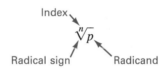

FIGURE 1.10.2

When there is no index written, the index is understood to be 2, and the radical will then be a *square root*. Some examples of higher roots are $\sqrt[3]{8}$, $\sqrt[4]{55}$, and $\sqrt[5]{-32}$.

Principal Roots   *Whenever the radical symbol is used, mathematicians agree that it is to stand for the principal root.*

When the index is an *even* number, we say the index is **even**. When the index is an *odd* number, we say the index is **odd**.

Principal higher roots are summarized as follows:

PRINCIPAL ROOTS

The symbol $\sqrt[n]{p}$ always represents the *principal n*th root of *p*.

If the *radicand* is *positive*, the principal root is positive.

If the *radicand* is *negative*:

    1. when the index is odd, the principal root is negative;

    2. when the index is even, the principal root is *not a real number*.

Some Symbols Used to Indicate Roots   The symbol $\sqrt[3]{p}$ indicates the *cubic root* of p. When the *index* of the radical is a 3, we must find a number whose *cube* is p. You will find such problems easier to do if you have memorized the cubes of the first few whole numbers (see page 50).

Example 12   Find the indicated roots.

a. $\sqrt[3]{8}$. We must find a number whose *cube* is 8. That is, we must solve $(?)^3 = 8$. If we have memorized that $2^3 = 8$, then we know that the answer is 2. If we haven't memorized that the cube of 2 is 8, we must use the "trial-and-error" method. Does $1^3 = 8$? No. Does $2^3 = 8$? Yes. Therefore, $\sqrt[3]{8} = 2$.

b. $\sqrt[3]{-8}$. We must find a number whose *cube* is -8. We know that the principal root will be negative, because the index is odd and the radicand is negative. Because $(-2)^3 = -8$, $\sqrt[3]{-8} = -2$. ∎

The symbol $\sqrt[4]{p}$ indicates the *fourth root* of p. When the *index* is a 4, we must find a number whose *fourth power* is p.

Example 13   Find $\sqrt[4]{16}$. We must find a positive number whose *fourth* power is 16. Let's use the "trial-and-error" method. Does $1^4 = 16$? No. Does $2^4 = 16$? Yes; $2^4 = 2 \cdot 2 \cdot 2 \cdot 2 = 16$. Therefore, $\sqrt[4]{16} = 2$. ∎

The symbol $\sqrt[5]{p}$ indicates the *fifth* root of p, and so forth.

Example 14   Find $\sqrt[5]{-1}$. We must find a *negative* number whose *fifth* power is -1. Does $(-1)^5 = -1$? Yes. Therefore, $\sqrt[5]{-1} = -1$. ∎

Example 15   Find $\sqrt[4]{-16}$. The radicand is negative and the index is even. Therefore, the answer is "Not a real number." (There is no *real* number whose fourth power is -16.) ∎

Example 16   Examples of roots preceded by a minus sign:

a. $-\sqrt{169} = -(13) = -13$

b. $-\sqrt[3]{8} = -(2) = -2$

c. $-\sqrt[3]{-8} = -(-2) = 2$

d. $-\sqrt{-4}$ is not a real number, because the radicand is negative and the index is even (the index is understood to be 2). ∎

The cubic root of a number that is *not* the cube of some rational number is an *irrational* number. Thus, $\sqrt[3]{35}$ and $\sqrt[3]{-17}$ are irrational numbers; their decimal forms will not terminate and will not repeat.

Recall from Section 1.2 that when the decimal form of a rational number is a repeating decimal, it is customary to place a bar above the digit or group of digits that repeats.

Example 17   Determine which of the following numbers are rational numbers, which are irrational numbers, which are real numbers, and which are *not* real numbers: $\sqrt{5}$, $\sqrt[3]{17}$, $2.\overline{52}$, 0, $\frac{3}{7}$, -4, $-\sqrt[3]{8}$, $\sqrt{-9}$, and $2.828427125\ldots$ (never terminates or repeats).

Solution The rational numbers are $2.\overline{52}$ (the digits "52" repeat), 0, $\frac{3}{7}$, -4, and $-\sqrt[3]{8}$ ($-\sqrt[3]{8} = -2$, which is rational).

The irrational numbers are $\sqrt{5}$ (there is no rational number whose square is 5), $\sqrt[3]{17}$ (there is no rational number whose cube is 17), and $2.828427125\ldots$.

The real numbers are $\sqrt{5}$, $\sqrt[3]{17}$, $2.\overline{52}$, 0, $\frac{3}{7}$, -4, $-\sqrt[3]{8}$, and $2.828427125 \ldots$ (all rational numbers and all irrational numbers are real numbers).

One number is *not* real; it is $\sqrt{-9}$ (there is no real number whose square is -9). We cannot represent $\sqrt{-9}$ on the real-number line because it is not a real number. ■

All roots of positive numbers and zero and all *odd* roots of negative numbers are real numbers and therefore can be represented by points on the number line. We show a few such points on the number line in Figure 1.10.3.

FIGURE 1.10.3

The relationships among the kinds of real numbers discussed above are shown in Figure 1.10.4.

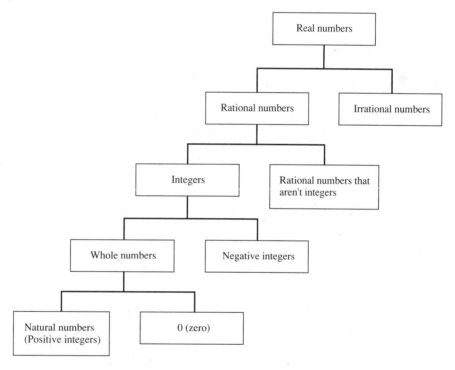

FIGURE 1.10.4

EXERCISES 1.10D

Set I In Exercises 1–20, find each of the indicated roots by trial and error, or write "Not a real number."

1. $\sqrt[3]{64}$ **2.** $\sqrt[4]{81}$ **3.** $\sqrt[3]{27}$ **4.** $\sqrt[3]{125}$

5. $-\sqrt[3]{27}$ **6.** $-\sqrt{25}$ **7.** $-\sqrt[4]{1}$ **8.** $-\sqrt[5]{-32}$

9. $\sqrt[3]{-125}$ **10.** $\sqrt[3]{-8}$ **11.** $-\sqrt[4]{-16}$ **12.** $-\sqrt{-16}$

13. $\sqrt[3]{-1,000}$ **14.** $\sqrt[3]{-64}$ **15.** $\sqrt[7]{-1}$ **16.** $\sqrt[5]{-32}$

17. $\sqrt[6]{729}$ **18.** $-\sqrt[3]{-216}$ **19.** $\sqrt{-25}$ **20.** $\sqrt{-36}$

In Exercises 21 and 22, determine which of the numbers are rational numbers, which are irrational numbers, which are real numbers, and which are *not* real numbers.

21. $\sqrt[3]{13}$, $\frac{1}{2}$, -12, $\sqrt{-15}$, $0.\overline{26}$, $0.196732468 \ldots$ (never terminates or repeats)

22. $0.67249713 \ldots$ (never terminates or repeats), $\sqrt{-27}$, 18, $\frac{11}{32}$, $0.\overline{37}$, $\sqrt[3]{12}$

Set II   In Exercises 1–20, find each of the indicated roots by trial and error, or write "Not a real number."

1. $-\sqrt[3]{-8}$	**2.** $-\sqrt[3]{-64}$	**3.** $\sqrt[5]{-1}$	**4.** $\sqrt[4]{16}$
5. $\sqrt[3]{1,000}$	**6.** $\sqrt[5]{32}$	**7.** $\sqrt[3]{216}$	**8.** $\sqrt[6]{64}$
9. $\sqrt[4]{256}$	**10.** $\sqrt[5]{243}$	**11.** $\sqrt[4]{-81}$	**12.** $-\sqrt[4]{81}$
13. $\sqrt[5]{-100,000}$	**14.** $-\sqrt[3]{1}$	**15.** $\sqrt[7]{128}$	**16.** $\sqrt[4]{625}$
17. $\sqrt[3]{343}$	**18.** $\sqrt[5]{-243}$	**19.** $\sqrt[6]{-64}$	**20.** $-\sqrt[4]{10,000}$

In Exercises 21 and 22, determine which of the numbers are rational numbers, which are irrational numbers, which are real numbers, and which are *not* real numbers.

21. -13, $-\frac{51}{22}$, $0.\overline{72}$, $\sqrt{-16}$, $\sqrt[3]{-125}$, $0.26925163 \ldots$ (never terminates or repeats)

22. $\sqrt[3]{3}$, $1.29715698 \ldots$ (never terminates or repeats), $3.\overline{13}$, $\frac{23}{52}$, 45, $\sqrt{-100}$

1.11 Grouping Symbols; Order of Operations

Earlier in this chapter, we learned how to perform the six operations of arithmetic—addition, subtraction, multiplication, division, raising to powers, and finding roots—on signed numbers. However, most of the problems dealt with only one operation. In this section, we will concentrate on evaluating expressions in which several operations are used.

Grouping Symbols   Operations indicated *within* grouping symbols should be performed before the operations outside the grouping symbols. We've already mentioned that parentheses ( ), brackets [ ], and braces { } can be used as grouping symbols. *These grouping symbols* (which must always be used in pairs) *all have the same meaning*; different grouping symbols can be used in the same expression.

A fraction bar is also a grouping symbol (see Example 1), as is the "bar" part of a radical sign (see Example 2).

NOTE   We've seen several expressions that contain "grouping symbols" within which no operation is indicated. In the expression $5(-3)$, for example, no operation is indicated *within* the grouping symbols; that is, there is no operation to perform in "-3."   ☑

Example 1   Evaluate $\dfrac{-4 + (-2)}{8 - 5}$.

Solution The fraction bar is a grouping symbol for the addition indicated *above* the bar as well as for the subtraction indicated *below* the bar.

$$\dfrac{-4 + (-2)}{8 - 5} \quad \longleftarrow \text{We must add } -4 \text{ and } -2 \text{ first}$$
$$\phantom{\dfrac{-4 + (-2)}{8 - 5}} \longleftarrow \text{We must subtract 5 from 8 next}$$

$$= \quad \dfrac{-6}{3}$$

$$= \quad -2 \quad \blacksquare$$

Example 2   Evaluate $\sqrt{16 + 9}$.

Solution The "bar" part of the radical sign is a grouping symbol.

$$\sqrt{16 + 9} \qquad \text{We must add 16 and 9 } before \text{ we take any square roots}$$

$$= \quad \sqrt{25}$$

$$= \quad 5 \quad \blacksquare$$

Order of Operations   When two or more operations are indicated in a problem (or when the same operation is indicated more than once) and when there are no grouping symbols to tell us which operation to perform first, what should we do? If the operation symbols in a problem are all addition symbols, we can do the additions in any order, and if they are all multiplication symbols, we can do the multiplications in any order. But what about other cases? We'll consider several problems.

First, let's consider the problem $12 - 6 - 2$. If we do the subtraction on the right first, we have

$$12 - 6 - 2 \overset{?}{=} 12 - (6 - 2) = 12 - 4 = 8$$

but if we do the subtraction on the left first, we have

$$12 - 6 - 2 \overset{?}{=} (12 - 6) - 2 = 6 - 2 = 4$$

Which answer is right? (The correct solution is given in Example 3.)

Next, let's consider the problem $36 \div 6 \div 2$. If we do the division on the left first, we have

$$36 \div 6 \div 2 \overset{?}{=} (36 \div 6) \div 2 = 6 \div 2 = 3$$

whereas if we do the division on the right first, we have

$$36 \div 6 \div 2 \overset{?}{=} 36 \div (6 \div 2) = 36 \div 3 = 12$$

Which answer is correct? (The problem is done correctly in Example 4.)

Finally, let's consider the problem $5 + 4 \cdot 6$. If we do the addition first, we have

$$5 + 4 \cdot 6 \overset{?}{=} (5 + 4) \cdot 6 = 9 \cdot 6 = 54$$

If we do the multiplication first, we have

$$5 + 4 \cdot 6 \overset{?}{=} 5 + (4 \cdot 6) = 5 + 24 = 29$$

Which is right? (The correct solution is given in Example 5.)

So that problems involving more than one operation will have only one correct answer, mathematicians have agreed upon the following order of operations:

ORDER OF OPERATIONS

1. If there are operations indicated inside grouping symbols, those operations within the grouping symbols should be performed first. A fraction bar and the bar of a radical sign are grouping symbols.

2. The evaluation then proceeds *in this order:*

First: Powers and roots are done.

Next: Multiplication and division are done *in order from left to right.*

Last: Addition and subtraction are done *in order from left to right.*

Example 3　Evaluate $12 - 6 - 2$.
Solution There are no grouping symbols, no powers or roots indicated, and no multiplications or divisions indicated. Therefore, we go to the last step and perform the subtractions *in order from left to right.* We have

$$12 - 6 - 2 = (12 - 6) - 2 = 6 - 2 = 4$$

Therefore, the only correct answer for $12 - 6 - 2$ is 4.　■

The expression $12 - 6 - 2$ must be evaluated by doing the subtractions from left to right because subtraction is not associative. It is possible to think of $12 - 6 - 2$ as the *sum* $12 + (-6) + (-2)$ because *any* subtraction problem can be changed to an addition problem; if the expression is considered as a sum, then the *addition* can be done in *any* order, because addition *is* commutative and associative.

Example 4　Evaluate $36 \div 6 \div 2$.
Solution There are no grouping symbols and no powers or roots indicated. We must perform the divisions *in order from left to right.* We have

$$36 \div 6 \div 2 = (36 \div 6) \div 2 = 6 \div 2 = 3$$

Therefore, the only correct answer for $36 \div 6 \div 2$ is 3.　■

The expression $36 \div 6 \div 2$ must be evaluated by doing the divisions from left to right because division is not associative. If the expression is considered as a *product*— that is, as $36 \cdot \dfrac{1}{6} \cdot \dfrac{1}{2}$—then the multiplication can be done in any order, because multiplication *is* commutative and associative.

Example 5　Evaluate $5 + 4 \cdot 6$.
Solution There are no grouping symbols and no powers or roots indicated. The multiplication must be performed *before* the addition. We then have

$$5 + 4 \cdot 6 = 5 + (4 \cdot 6) = 5 + 24 = 29$$

Therefore, the only correct answer for $5 + 4 \cdot 6$ is 29.　■

Example 6 Evaluate $8 - 6 - 4 + 7$.

Solution If we consider the problem as an addition *and* subtraction problem, the expression must be evaluated from left to right. If, however, the expression is considered as a *sum*, the addition can be done in any order.

─── Both methods are correct ───

Considered as addition and subtraction and evaluated left to right	*Changed to addition and added in any order*
$8 - 6 - 4 + 7$	$8 + (-6) + (-4) + (7)$
$= \quad 2 \quad - 4 + 7$	$= 8 + (7) + (-6) + (-4)$
$= \quad -2 \quad + 7$	$= \quad 15 \quad + \quad (-10)$
$= \quad\quad 5$	$= \quad\quad 5$ ∎

Example 7 Evaluate each of the following expressions, using the correct order of operations.

a. $7 + 3 \cdot 5$ Multiplication must be done before addition

$= 7 + \quad 15$

$= 22$

b. $4^2 + \sqrt{25} - 6$ Powers and roots must be done first

$= 16 + \quad 5 - 6$ Addition must be done next because addition and subtraction are done left to right

$= \quad 21 \quad\quad - 6$

$= 15$

c. $16 \div 2 \cdot 4$ Division must be done first because multiplication and division are done left to right

$= \quad 8 \quad \cdot 4$

$= 32$

d. $(-8) \div 2 - (-4)$ Division must be done before subtraction

$= \quad -4 \quad - (-4)$ Subtraction must be changed to addition

$= \quad -4 \quad + \quad 4$

$= 0$

e. $\sqrt[3]{-8}(-3)^2 - 2(-6)$ Roots and powers must be done first

$= -2(9) - 2(-6)$ Multiplications must be done before subtraction

$= -18 - (-12)$ Subtraction must be changed to addition

$= -18 + \quad 12$

$= -6$

f. $12\sqrt{25} + 28 \div 4$ There is an understood multiplication sign between the 12 and the $\sqrt{25}$

$= 12 \cdot 5 + 28 \div 4$ Roots were done first

$= \quad 60 \quad + \quad 7$ Multiplications and divisions were done next

$= 67$ Addition was done last

g. $\quad 5 \cdot (-4) \div 2 \cdot (3 - 8)$ The expression *within* the parentheses—$(3 - 8)$—must be evaluated first

$= 5 \cdot (-4) \div 2 \cdot (-5)$ The multiplication on the left must be done next

$= \quad\quad -20 \div 2 \cdot (-5)$ The division on the left must be done next

$= \quad\quad\quad\quad -10 \cdot (-5)$

$= 50$

h. $\quad 5 \cdot 3^2$ Powers must be done *before* multiplication

$= 5 \cdot 9$

$= 45$ ∎

A WORD OF CAUTION   Remember that an exponent applies *only* to the immediately preceding symbol. That is, $5 \cdot 3^2 = 5 \cdot 3 \cdot 3$, *not* $15 \cdot 15$.   ☑

When grouping symbols appear within other grouping symbols, evaluate the expression inside the *inner* grouping symbols first (see Examples 8 and 9).

Example 8   Evaluate each of the following expressions.

a. $\quad 10 - [3 - (2 - 7)]$ $2 - 7$ must be evaluated first: $2 - 7 = 2 + (-7) = -5$

$= 10 - [3 - (-5)]$ Subtraction must be changed to addition

$= 10 - [3 + (+5)]$

$= 10 - [8]$

$= 2$

b. $\quad 20 - 2\{5 - [3 - 5(6 - 2)]\}$ $6 - 2$ must be evaluated first

$= 20 - 2\{5 - [3 - 5(4)]\}$ Multiplication must be done before subtraction

$= 20 - 2\{5 - [3 - 20]\}$ $3 - 20$ must be evaluated next: $3 - 20 = 3 + (-20) = -17$

$= 20 - 2\{5 - [-17]\}$ Subtraction must be changed to addition

$= 20 - 2\{5 + [+17]\}$ The operation within the braces must be performed next

$= 20 - 2\{22\}$ Multiplication must be done before subtraction

$= 20 - 44$ Subtraction must be changed to addition

$= 20 + (-44)$

$= -24$ ∎

A WORD OF CAUTION   A common error in a problem like the one in Example 8b is to start the problem by subtracting 2 from 20 as follows:

$$20 - 2\{5 - [3 - 5(6 - 2)]\} = 18\{5 - [3 - 5(6 - 2)]\}$$

This is incorrect; we must multiply the expression inside the braces by 2 *before* we do the subtraction. In fact, we were *never* able to subtract 2 from 20; rather, we had to subtract $2\{22\}$, or 44, from 20.   ☑

Example 9   Evaluate $27 \div (-3)^2 - 5\left\{6 - \dfrac{8-4}{5}\right\}$.

Solution

$$27 \div (-3)^2 - 5\left\{6 - \frac{8-4}{5}\right\}$$

The innermost grouping symbol here is the fraction bar; we must subtract 4 from 8 first

$$= 27 \div (-3)^2 - 5\left\{\boxed{6} - \frac{4}{5}\right\}$$

Remember that $6 = \dfrac{6 \cdot 5}{1 \cdot 5} = \dfrac{30}{5}$

$$= 27 \div (-3)^2 - 5\left\{\frac{30}{5} - \frac{4}{5}\right\}$$

$$= 27 \div (-3)^2 - 5\left\{\frac{26}{5}\right\}$$

We must raise to powers before we divide, and $(-3)^2 = 9$

$$= 27 \div 9 \quad - \frac{\overset{1}{\cancel{5}}}{1}\left\{\frac{26}{\underset{1}{\cancel{5}}}\right\}$$

We must divide and multiply before we subtract

$$= \quad 3 \quad - \quad 26$$

Subtraction must be changed to addition

$$= \quad 3 \quad + (-26)$$

$$= -23 \quad \blacksquare$$

Example 10   Evaluate $\sqrt{16} + \sqrt{9}$.

Solution We must find the square roots before we add:

$$\sqrt{16} + \sqrt{9} = 4 + 3 = 7 \quad \blacksquare$$

NOTE   If you compare this problem with the problem from Example 2, you'll see that $\sqrt{16+9} \neq \sqrt{16} + \sqrt{9}$. In general, $\sqrt{a+b} \neq \sqrt{a} + \sqrt{b}$ and $\sqrt{a-b} \neq \sqrt{a} - \sqrt{b}$. ☑

Example 11   Evaluate $\sqrt{13^2 - 12^2}$.

Solution In the expression $\sqrt{13^2 - 12^2}$, the bar of the radical sign acts like a set of grouping symbols; therefore, we must simplify $13^2 - 12^2$ before we take the square root, and in order to do this we must raise to powers first and then subtract:

$$\sqrt{13^2 - 12^2} = \sqrt{169 - 144} = \sqrt{25} = 5 \quad \blacksquare$$

A WORD OF CAUTION   A common error is to write

$$\sqrt{13^2 - 12^2} = 13 - 12$$

We saw in Example 11 that $\sqrt{13^2 - 12^2} = 5$, and $13 - 12 = 1$, not 5. In general, $\sqrt{a^2 + b^2} \neq a + b$, and $\sqrt{a^2 - b^2} \neq a - b$. ☑

Example 12   Evaluate $(2.5)^2 \div (5.6 - 11.4)$. Use a calculator for the division, and round off the answer to three decimal places.

Solution

$$(2.5)^2 \div (5.6 - 11.4)$$

The subtraction inside the parentheses must be performed first

$$= (2.5)^2 \div (-5.8)$$

Powers must be done next

$$= 6.25 \div (-5.8)$$

The quotient will be negative

$$\approx -1.077586207$$

The calculator display

$$\approx -1.078$$

Rounded off to three decimal places   ∎

EXERCISES 1.11

Set I   In Exercises 1–46, evaluate each expression. Be sure to perform the operations in the correct order.

1. $12 - 8 - 6$

2. $15 - 9 - 4$

3. $17 - 11 + 13 - 9$

4. $12 - 8 + 14 - 6$

5. $7 + 2 \cdot 4$

6. $10 + 3 \cdot 6$

7. $9 - 3 \cdot 2$

8. $14 - 8 \cdot 3$

9. $10 \div 2 \cdot 5$

10. $20 \cdot 15 \div 5$

11. $12 \div 6 \div 2$

12. $24 \div 12 \div 6$

13. $(-12) \div 2 \cdot (-3)$

14. $(-18) \div (-3) \cdot (-6)$

15. $8 \cdot 5^2$

16. $6 \cdot 2^4$

17. $(-485)^2 \cdot 0 \cdot (-5)^2$

18. $(-589)^2 \cdot 0 \cdot (-3)^2$

19. $12 \cdot 4 + 16 \div 8$

20. $4 \cdot 3 + 15 \div 5$

21. $28 \div 4 \cdot 2(6)$

22. $48 \div 16 \cdot 2(-8)$

23. $(-2)^2 + (-4)(5) - (-3)^2$

24. $(-5)^2 + (-2)(6) - (-4)^2$

25. $2 \cdot 3 + 3^2 - 4 \cdot 2$

26. $100 \div 5^2 \cdot 6 + 8 \cdot 75$

27. $(10^2)\sqrt{16} + 5(4) - 80$

28. $(5^2)\sqrt{9} + 4(6) - 60$

29. $2 \cdot (-6) \div 3 \cdot (8 - 4)$

30. $5 \cdot (-4) \div 2 \cdot (9 - 4)$

31. $24 - [(-6) + 18]$

32. $17 - [(-9) + 15]$

33. $[12 - (-19)] - 16$

34. $[21 - (-14)] - 29$

35. $[11 - (5 + 8)] - 24$

36. $[16 - (7 + 12)] - 22$

37. $20 - [5 - (7 - 10)]$

38. $16 - [8 - (2 - 7)]$

39. $\dfrac{7 + (-12)}{8 - 3}$

40. $\dfrac{(-14) + (-2)}{9 - 5}$

41. $15 - \{4 - [2 - 3(6 - 4)]\}$

42. $17 - \{6 - [9 - 2(2 - 7)]\}$

43. $32 \div (-2)^3 - 5\left\{7 - \dfrac{6 - 2}{5}\right\}$

44. $36 \div (-3)^2 - 6\left\{4 - \dfrac{9 - 7}{3}\right\}$

45. $\sqrt{3^2 + 4^2}$

46. $\sqrt{13^2 - 5^2}$

In Exercises 47–52, use a calculator to evaluate each expression. Round off the answers to three decimal places.

47. $\sqrt{16.3^2 - 8.35^2}$

48. $\sqrt{23.9^2 + 38.6^2}$

49. $(1.5)^2 \div (-2.5) + \sqrt{35}$

50. $(-0.25)^2(-10)^3 + \sqrt{54}$

51. $18.91 - [64.3 - (8.6^2 + 14.2)]$

52. $[\sqrt{101.4} - (73.5 - 19.6^2)] \div 38.2$

Set II   In Exercises 1–46, evaluate each expression. Be sure to perform the operations in the correct order.

1. $18 - 10 - 5$

2. $13 - 9 + 16 - 8$

3. $2 + 5 \cdot 3$

4. $15 \div 3 \cdot 5$

5. $-18 \div (-\sqrt{81})$

6. $18 \div 6 \div 3$

7. $10 \cdot 15^2 - 4^3$

8. $(-10)^2 \cdot 10 + 0(-20)$

9. $(785)^3(0) + 1^5$

10. $(-5)^2 - (2)(-6) + (-2)^2$

11. $2 + 3(100) \div 25 - 10$

12. $7^2 + \sqrt{64} - 14$

13. $(-12) \div 6 - (-4)$

14. $(10^2)\sqrt{4} - 5 + 36$

15. $15 \cdot 4^2$

16. $8\sqrt{4} - 16 \div 4 \cdot 4$

17. $(863)^4 \cdot 0 \cdot (-23)^2$

18. $\sqrt{64} + \sqrt{36}$

19. $19 + 5 \cdot 10$

20. $\sqrt{64 + 36}$

21. $18 \div 9 \div 3$

22. $24 \div 12 \div 3$

23. $2 \cdot 3^2 + 4 \div 4$

24. $-5^2 - 6^2$

25. $63 - 12 - 38$

26. $35 - 5 \cdot 2$

27. $6\sqrt{9} - 18 \div 6 \cdot 3$

28. $8^2 + 12^2$

29. $4 \cdot (11 - 17) \div 3 \cdot (-2)$

30. $[18 - (-15)] - 13$

31. $33 - [(-16) + 11]$

32. $[22 - (7 + 12)] - 14$

33. $26 - [8 - (6 - 15)]$

34. $\dfrac{(-2) + (-7)}{18 + (-15)}$

35. $23 - \{6 - [5 - 2(8 - 3)]\}$

36. $28 \div (-2)^2 - 6\left\{8 - \dfrac{9 - 4}{3}\right\}$

37. $\sqrt{10^2 - 6^2}$

38. $23 - (45 - 73)$

39. $\dfrac{3 + (-15)}{13 - 7}$

40. $\dfrac{(-8) - (-12)}{2 - 8}$

41. $37 - \{23 - [8 - 9] - 1\}$

42. $13 - \{11 - [1 - 19] - 36\}$

43. $48 \div (-4)^2 - 3\left\{9 - \dfrac{8 - 4}{2}\right\}$

44. $75 \div (-5)^2 - 5\left\{6 - \dfrac{9 - 1}{4}\right\}$

45. $\sqrt{8^2 + 6^2}$

46. $\sqrt{3^2 + 4^2}$

In Exercises 47–52, use a calculator to evaluate each expression. Round off the answers to three decimal places.

47. $\sqrt{2.18^2 + 41.6^2}$

48. $\sqrt{1.34^2 + 4.1^2}$

49. $(63.1)^2 \div (-3.8) + \sqrt{89}$

50. $\sqrt{34.5^2 - 17.8^2}$

51. $(1.7)^2 \div (-3.2) + \sqrt{43}$

52. $[\sqrt{126.3} - (89.7 - 46.5^2)] \div 52.6$

1.12 Factorization of Integers

Recall that the numbers that are multiplied together to give a product are called the *factors* of that product. The tests for divisibility that follow are quite useful when we want to find the factors of a whole number.

Tests for Divisibility

We can use the following tests to determine whether a whole number is divisible by 2, 3, or 5.

Divisibility by 2 A whole number is divisible by 2 if its last digit is 0, 2, 4, 6, or 8.

Divisibility by 3 A whole number is divisible by 3 if the sum of its digits is divisible by 3.

Divisibility by 5 A whole number is divisible by 5 if its last digit is 0 or 5.

While there are tests for divisibility by other numbers, those tests are not included in this section.

Example 1   Examples of the use of the tests of divisibility:

 a. 1 2 , 30 0 , 2,03 4 , and 57 8 are divisible by 2, because the last digit of each number is a 0, 2, 4, 6, or 8.

 b. 132 is divisible by 3 because the sum of the digits, $1 + 3 + 2$, is 6, and 6 is divisible by 3.

 c. 5,162 is not divisible by 3 because $5 + 1 + 6 + 2 = 14$, and then $1 + 4 = 5$, and 5 is not divisible by 3.

 d. 25 0 and 75 5 are both divisible by 5, because the last digit of each is a 0 or a 5. ∎

Finding the Integral Factors of a Number

The "Plus or Minus" Symbol The symbol "±" is read *"plus or minus."* Thus, "±7" is read "plus or minus 7," and it represents either +7 or −7.

Factoring an Integer In Section 1.5, we learned that the numbers multiplied together to give a product are called the *factors* of that product. For example, in the statement $2(-3) = -6$, 2 and −3 can be called the *factors* or *divisors* of −6. Every positive integer greater than 1 always has at least two positive factors or divisors: 1 and the number itself. A systematic method for finding all the factors of a number is demonstrated in Examples 2 and 3. When we are asked for *all the factors* of a *positive* integer, it's usually understood that we are to write only the *positive* factors of the number (see Example 2).

Example 2   Find all the factors (or divisors) of 12.
Solution The smallest factor or divisor of 12 is 1, and the largest is 12. Let's write those numbers as follows:

We now try to divide 12 by positive integers between 1 and 12. Is 2 a factor of 12? Yes, because when we divide 12 by 2, the remainder is 0. The quotient is 6; therefore, 2 and 6 are *both* factors of 12. We insert the 2 and the 6 between 1 and 12:

We now try to divide 12 by positive integers between 2 and 6. Is 3 a factor of 12? Yes, because when we divide 12 by 3, the remainder is 0. The quotient is 4, and so 3 and 4 are both factors of 12. We insert them between 2 and 6:

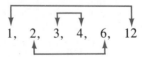

Because there are no positive integers between 3 and 4, we need search no more. The positive factors of 12 are 1, 2, 3, 4, 6, and 12. ∎

When we're asked for *all the integral factors* of a number, we must find *all* the integers, both positive and negative, that are factors of the number. We do this by first finding all the positive integers that are factors of the number and then attaching "±" to the left of each of these integers (see Example 3).

Example 3   Find all the integral factors of each of the following numbers.

a. 8
Solution 8 = 1 · 8 and 8 = 2 · 4. "Pair up" the factors:

The integral factors of 8 are ±1, ±2, ±4, and ±8.

b. 36
Solution 36 = 1 · 36, 36 = 2 · 18, 36 = 3 · 12, 36 = 4 · 9, and 36 = 6 · 6.
Because 36 is the square of an integer, when we pair up the factors, there will be a single, unpaired number left in the center:

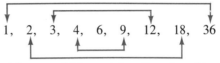

Therefore, the integeral factors of 36 are ±1, ±2, ±3, ±4, ±6, ±9, ±12, ±18, and ±36. ■

Prime and Composite Numbers

Prime Numbers   A **prime number** is a natural number greater than 1 that cannot be written as a product of two natural numbers except as the product of itself and 1. That is, a prime number has no natural-number factors other than itself and 1.

A partial list of prime numbers is 2, 3, 5, 7, 11, 13, 17, 19, 23, 29 . . . . There is no largest prime number.

Composite Numbers   A **composite number** is a natural number that does have natural-number factors other than itself and 1.

NOTE   The number 1 is neither prime nor composite.   ☑

Example 4   Examples of prime and composite numbers:

a. 9 is a composite number because 1, 3, and 9 are factors of 9, so 9 has a factor other than itself and 1.

b. 17 is a prime number because 1 and 17 are the only natural-number factors of 17.

c. 45 is a composite number; it has the factors 3, 5, 9, and 15 besides 1 and 45.   ■

Prime Factorization of Natural Numbers   The prime factorization of a prime number is simply the number itself. If we are asked to write the **prime factorization** of a composite number, we must write the number as a product of two or more numbers *that are prime numbers*. The composite number 18, for example, can be written in factored form as 1 · 18, 2 · 9, or 3 · 6. These are all "factorizations" of 18, but not *prime* factorizations, because each product contains at least one factor that is not a prime number. See Example 5 for the correct prime factorization of 18.

The prime factorization of a composite number is unique except for the order in which the factors are written.

Example 5   Find the prime factorization of 18.
Solution

$$18 = 2 \cdot 9 = \boxed{2 \cdot 3 \cdot 3} = \boxed{2 \cdot 3^2}$$
$$18 = 3 \cdot 6 = \boxed{3 \cdot 2 \cdot 3} = \boxed{2 \cdot 3^2}$$

These *are* prime factorizations because all the factors are prime numbers

Note that the two ways we factored 18 led to the *same* prime factorization ($2 \cdot 3^2$). This is so because the prime factorization is unique. ■

A systematic method for finding the prime factorization is demonstrated in Examples 6 and 7.

Example 6   Find the prime factorization of each of the following numbers.

a. 24

Solution We first try to divide 24 by the smallest prime, 2. This number *does* divide exactly into 24 and gives a quotient of 12. We try 2 once again, this time as a divisor of the quotient, 12. Two *does* divide evenly into 12 and gives a quotient of 6. We try 2 yet again, this time as a divisor of the quotient, 6. Two *does* divide evenly into 6 and gives a quotient of 3, which is itself a prime number, so the process ends. (The process ends when each factor is a prime number.)

The work can be conveniently arranged by placing the quotient *under* the number we're dividing into, as follows:

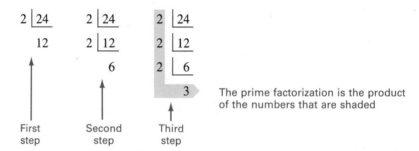

2	24
	12

First step

2	24
2	12
	6

Second step

2	24
2	12
2	6
	3

Third step

The prime factorization is the product of the numbers that are shaded

Therefore, $24 = 2 \cdot 2 \cdot 2 \cdot 3 = 2^3 \cdot 3$, and $2^3 \cdot 3$ is the prime factorization of 24.

b. 20
Solution

2	20
2	10
	5

Prime factorization of $20 = 2 \cdot 2 \cdot 5 = 2^2 \cdot 5$

c. 36
Solution

2	36
2	18
3	9
	3

Prime factorization of $36 = 2 \cdot 2 \cdot 3 \cdot 3 = 2^2 \cdot 3^2$

d. 315

Solution

$$
\begin{array}{r|l}
3 & 315 \\
3 & 105 \\
5 & 35 \\
& 7
\end{array}
$$

Prime factorization of $315 = 3 \cdot 3 \cdot 5 \cdot 7 = 3^2 \cdot 5 \cdot 7$ ∎

When we're trying to find the prime factors of a number, we do not need to try any prime that has a square greater than that number (see Example 7).

Example 7   Find the prime factorization of 131.

Solution

Primes in order of size

2	Does not divide 131
3	Does not divide 131
5	Does not divide 131
7	Does not divide 131
11	Does not divide 131
13	Does not divide 131; larger primes need not be tried because $13^2 = 169$, and 169 is greater than 131

Therefore, the prime factorization of 131 is simply 131, since 131 is a prime number. ∎

EXERCISES 1.12

Set I   In Exercises 1–16, find all the integral factors of each number.

1. 4	**2.** 9	**3.** 10	**4.** 14
5. 15	**6.** 16	**7.** 18	**8.** 20
9. 21	**10.** 22	**11.** 27	**12.** 28
13. 33	**14.** 34	**15.** 44	**16.** 45

In Exercises 17–32, state whether each number is prime or composite. To justify your answer, give the set of all positive integral factors for each number.

17. 5	**18.** 8	**19.** 13	**20.** 15
21. 12	**22.** 11	**23.** 21	**24.** 23
25. 55	**26.** 41	**27.** 49	**28.** 31
29. 51	**30.** 42	**31.** 111	**32.** 101

In Exercises 33–48, find the prime factorization of each number.

33. 14	**34.** 15	**35.** 21	**36.** 22
37. 26	**38.** 27	**39.** 29	**40.** 31
41. 32	**42.** 33	**43.** 34	**44.** 35
45. 84	**46.** 75	**47.** 144	**48.** 180

49. List any prime numbers greater than 17 and less than 37 that yield a remainder of 1 when divided by 5.

50. List any prime numbers greater than 19 and less than 41 that yield a remainder of 2 when divided by 5.

Set II   In Exercises 1–16, find all the integral factors of each number.

1. 6	**2.** 23	**3.** 24	**4.** 26
5. 30	**6.** 32	**7.** 38	**8.** 46
9. 81	**10.** 105	**11.** 49	**12.** 56
13. 63	**14.** 85	**15.** 111	**16.** 42

In Exercises 17–32, state whether each number is prime or composite. To justify your answer, give the set of all positive integral factors for each number.

17. 17	**18.** 14	**19.** 18	**20.** 19
21. 61	**22.** 63	**23.** 81	**24.** 73
25. 39	**26.** 83	**27.** 100	**28.** 29
29. 129	**30.** 105	**31.** 89	**32.** 97

In Exercises 33–48, find the prime factorization of each number.

33. 16	**34.** 18	**35.** 28	**36.** 30
37. 65	**38.** 78	**39.** 120	**40.** 112
41. 81	**42.** 49	**43.** 43	**44.** 72
45. 51	**46.** 38	**47.** 111	**48.** 88

49. List any prime numbers greater than 13 and less than 29 that yield a remainder of 5 when divided by 7.

50. List any prime numbers less than 41 and greater than 11 that yield a remainder of 1 when divided by 6.

1.13 Preparation for Solving Word Problems

We will begin the formal discussion of solving word problems *by using algebra* in Chapter 4. However, a review of a few general problem-solving techniques for solving word problems in *arithmetic* might be helpful at this time. Many of these suggestions can be applied to *any* kind of problem.

1. Read the problem carefully. Be sure you understand the problem.

2. Identify what is given and what is being asked for. Draw a picture, if possible. Ask yourself if you already have enough information to solve the problem. Do you need more facts? Do you need some special formula?

3. Ask yourself if you know which operation or operations to use. Will addition solve the problem? Subtraction? Multiplication? Division? If you're not sure, *try something*. Try addition, for example. Do not be afraid to make a mistake at this point.

4. Solve the problem, using the given numbers and facts and the operation(s) you've decided to try.

5. *Check your answer*. This is where you make sure you didn't make a mistake in step 3! Is your answer a reasonable one for the problem you are solving? If not, recheck your calculations. If you still get the same unreasonable answer, then analyze the problem again. Should you have used a different operation? Should you have used a different formula?

6. Be sure you have answered *all* the questions asked in the problem.

In this section, you will not be asked to solve any word problems. You will instead be asked whether or not enough information is given to allow you to solve the problem and what operation you would use to solve the problem.

Example 1   Determine whether enough information has been given to allow you to solve the problem. If so, what operations are needed to solve the problem? If not, what other information do you need?

a. Rachael bought some books for $4.95 each. What was the total cost?
Solution Not enough information is given. We must know *how many* books she bought before we can calculate the total cost.

b. Lil is going to cut a 12-ft length of wire into fourteen pieces of equal length. How long will each piece be?
Solution Enough information is given. We would *divide* to solve the problem.   ∎

EXERCISES 1.13   In each exercise of both sets, do *not* solve the problem! Instead, determine whether enough information has been given for the problem to be solved. If so, what operation is needed to solve the problem? If not, what other information do you need?

Set I   **1.** Ken bought a shirt for $27.99, a pair of slacks for $28.39, and a tie for $15.65. How much did he spend for these items?

2. Eric bought several pencils for 37¢ each. What was the total amount he spent?

3. The members of the math department are going to buy a gift for their student worker; they will share the cost equally. If the gift costs $35, what will each person's share be?

4. Jeffrey borrowed some money from Chad on Tuesday, and on Friday he paid back $5.50. How much does he still owe?

5. Seven members of a service club agree to contribute $200 each toward scholarships for deserving students. How much did those members contribute altogether?

6. The temperature at 7 A.M. was $-4°$ F, and at 1 P.M. it was $10°$ F. What was the rise in temperature?

Set II   **1.** Judy bought a gift for each of the court reporters in her agency. If each gift cost $24.95, how much did she spend altogether?

2. Karen bought a skirt for $29.73, a blouse for $27.75, and a tank top for $18.89. How much did she spend for these items?

3. Colin bought seven small toy cars that cost $1.29 each. How much did he spend altogether?

4. Ellen is serving equal numbers of cookies to her preschool students. If the bag of cookies contains 100 cookies, how many cookies will each child get?

5. The mileage on the odometer of Monica's car read 23,472 (miles) Tuesday morning and 23,743 (miles) that evening. How many miles was her car driven during the day?

6. The temperature at 5 A.M. was $-20°$ F. By noon it had risen $43°$ F. What was the temperature at noon?

1.14 Review: 1.8–1.13

Commutative, Associative, and Distributive Properties 1.8

Addition is commutative:

$$a + b = b + a$$

Multiplication is commutative:

$$a \cdot b = b \cdot a$$

Subtraction and division are not commutative.

Addition is associative:

$$a + (b + c) = (a + b) + c$$

Multiplication is associative:

$$a \cdot (b \cdot c) = (a \cdot b) \cdot c$$

Subtraction and division are not associative.

Multiplication is distributive over addition and subtraction:

$$a(b + c) = (ab) + (ac) \quad \text{and} \quad a(b - c) = (ab) - (ac)$$

Powers of Integers 1.9

Exponent ⟶

$$3^4 = 3 \cdot 3 \cdot 3 \cdot 3 = 81$$

Base ⟶

Roots of Integers 1.10

$$\sqrt[3]{-8} = -2 \qquad \text{Because } (-2)^3 = (-2)(-2)(-2) = -8$$

Cubic root of -8

Finding roots is an inverse operation of raising to powers.

Irrational Numbers 1.10

Numbers such as $\sqrt{7}$ (there is no rational number whose square is 7) and $\sqrt[3]{18}$ (there is no rational number whose cube is 18) are irrational numbers. The *decimal form* of any irrational number is always a nonrepeating, nonterminating decimal. All irrational numbers are *real* numbers.

Grouping Symbols
1.11

( ) Parentheses

[ ] Brackets

{ } Braces

Fraction bar⟶ $\dfrac{3-2}{5+7}$

Bar of radical sign ⟶

$\sqrt{5-1}$

Order of Operations
1.11

1. If there are any grouping symbols in an expression, that part of the expression within a set of grouping symbols is evaluated first.

2. The evaluation then proceeds in this order:

First: Powers and roots are done.

Next: Multiplication and division are done *in order from left to right.*

Last: Addition and subtraction are done *in order from left to right.*

Prime Numbers
1.12

A **prime number** is a natural number greater than 1 that cannot be written as a product of two natural numbers except as the product of itself and 1. That is, a prime number has no natural-number factors other than itself and 1.

Composite Numbers
1.12

A **composite number** is a natural number that does have natural-number factors other than itself and 1.

The number 1 is neither prime nor composite.

Prime Factorization of
Natural Numbers
1.12

The **prime factorization** of a natural number greater than 1 is the indicated product of all of its factors that are themselves prime numbers. The prime factorization is unique, except for the order in which the factors are written.

Solving Word Problems
1.13

See the suggestions in Section 1.13.

Review Exercises 1.14 Set I

1. State whether each of the following is true or false; if the statement is true, give the reason.

a. $[(-2) \cdot 3] \cdot 4 = (-2)(3 \cdot 4)$

b. $5 + (-2) = (-2) + 5$

c. $5 - (-2) = (-2) - 5$

d. $a + (b + c) = (a + b) + c$

e. $(c \cdot d) \cdot e = e \cdot (c \cdot d)$

f. $5 + (x + 7) = (x + 7) + 5$

g. $7(12 + 6) = (7 \cdot 12) + (7 \cdot 6)$

2. Find the additive inverse of -23.

3. What is the additive identity?

In Exercises 4–13, perform the indicated operations, or write "Not a real number."

4. 1^4 **5.** 0^3 **6.** $(-4)^2$ **7.** -5^2

8. $\sqrt{121}$ **9.** $-\sqrt[4]{16}$ **10.** $\sqrt{-10}$ **11.** $\sqrt[3]{-8}$

12. $-\sqrt[5]{-32}$ **13.** 16^2

14. Determine which of the following numbers are rational numbers, which are irrational numbers, which are real numbers, and which are *not* real numbers:

$$-\sqrt{5}, \quad -31, \quad 5.3892531 \ldots \text{ (never terminates or repeats)},$$

$$\frac{7}{12}, \quad 4.\overline{15}, \quad \sqrt[3]{71}$$

15. Use a calculator or Table I to find the decimal approximation of $\sqrt{31}$. (Round off the answer to three decimal places.)

In Exercises 16–24, evaluate each expression.

16. $18 - 22 - 15 + 6$ **17.** $11 - 7 \cdot 3$

18. $(11 - 7) \cdot 3$ **19.** $15 \div 5 \cdot 3$

20. $36 - [(-7) - 15]$ **21.** $6 - [8 - (3 - 4)]$

22. $(35)^2(0) + 18 \div (-6)$ **23.** $\dfrac{6 + (-14)}{3 - 7}$

24. $\sqrt{10^2 - 8^2}$

25. List any prime numbers greater than 19 and less than 31 that yield a remainder of 2 when divided by 7.

26. a. Find the prime factorization of 270.

b. List all the integral factors of 270.

27. Complete each statement by using the property indicated.

a. Distributive Property: $8(-7 + 3) = $ _____

b. Associative Property: $3(-4 \cdot 6) = $ _____

c. Commutative Property: $-3 + (-4) = $ _____

28. Determine whether enough information has been given for the following problem to be solved. If so, what operation is needed to solve the problem? If not, what other information do you need?

Michael plans to buy eight books. Some cost \$3.95 each and the rest cost \$5.95 each. What will the total cost be?

Review Exercises 1.14 Set II

NAME _____

1. State whether each of the following is true or false. If the statement is true, give the reason.

    a. $6 - (-2) = (-2) - 6$

    b. $5 \cdot (2 \cdot 3) = (5 \cdot 2) \cdot 3$

    c. $(-7) + (4 + 3) = [(-7) + (4)] + 3$

    d. $-3(4 + 6) = (-3 \cdot 4) + (-3 \cdot 6)$

    e. $(-6) \div (-3) = (-3) \div (-6)$

2. What is the multiplicative identity?

3. What is the additive inverse of $\frac{2}{3}$?

In Exercises 4–13, perform the indicated operations or write "Not a real number."

4. $\sqrt{81}$ **5.** 4^3 **6.** 1^3

7. 0^5 **8.** $\sqrt[3]{-125}$ **9.** -11^2

10. $(-2)^2$ **11.** $-\sqrt[3]{1{,}000}$ **12.** $\sqrt{-49}$

13. $-\sqrt[7]{-1}$

14. Determine which of the following numbers are rational numbers, which are irrational numbers, which are real numbers, and which are *not* real numbers:

$$\sqrt{-169}, \quad 0.\overline{60}, \quad -53, \quad -\frac{5}{17},$$

    $4.35671937 \ldots$ (never terminates or repeats), $\sqrt[3]{21}$

ANSWERS

1a. _____

   b. _____

   c. _____

   d. _____

   e. _____

2. _____

3. _____

4. _____

5. _____

6. _____

7. _____

8. _____

9. _____

10. _____

11. _____

12. _____

13. _____

14. Rational numbers _____

Irrational numbers _____

Real numbers _____

Numbers that are not real _____

15. Use a calculator or Table I to find the decimal approximation of $\sqrt{53}$. (Round off the answer to three decimal places.)

In Exercises 16–24, evaluate each expression.

16. $25 - 8 - 1$ **17.** $48 \div 8 \div 2$ **18.** $8 + 2 \cdot 6$

19. $18 \div 6 \cdot 3$ **20.** $\dfrac{16 + 8}{4 + 2}$ **21.** $5 \cdot 3^2$

22. $\sqrt{17^2 - 15^2}$ **23.** $8\sqrt{16} + 8$ **24.** $12 - [14 - (6 - 8)]$

25. List any prime numbers greater than 17 and less than 37 that yield a remainder of 4 when divided by 5.

26. a. Find the prime factorization of 240.

    b. List the integral factors of 240.

27. Complete each statement by using the property indicated.

    a. Commutative Property: $7(-3) = $ _____

    b. Associative Property: $-4 + (-2 + 6) = $ _____

    c. Distributive Property: $4(6 - 2) = $ _____

28. Determine whether enough information has been given for the following problem to be solved. If so, what operation is needed to solve the problem? If not, what other information do you need?

The temperature at 1 P.M. was 5° F, and at 11 P.M. it was $-16°$ F. What was the change in temperature?

15. _____

16. _____

17. _____

18. _____

19. _____

20. _____

21. _____

22. _____

23. _____

24. _____

25. _____

26a. _____

b. _____

27a. _____

b. _____

c. _____

28. _____

_____

Chapter 1 Diagnostic Test

The purpose of this test is to see how well you understand the operations with integers and other signed numbers. We recommend that you work this diagnostic test *before* your instructor tests you on this chapter. Allow yourself about 50 minutes to do this test.

Complete solutions for all the problems on this test, together with section references, are given in the answer section at the end of the book. We suggest that you study the sections referred to for the problems you do incorrectly.

In Problems 1–3, write "True" if the statement is always true; otherwise, write "False."

1. a. The smallest natural number is 1.

 b. "$(2 + 3) + 7 = 2 + (3 + 7)$" illustrates the commutative property of addition.

 c. $\sqrt{-25}$ is not a real number.

 d. Division is associative.

2. a. 0 is a real number.

 b. $12 \div 6 = 6 \div 12$

 c. $-4 < -3$

 d. $2(5 \cdot 6) = (2 \cdot 5) \cdot (2 \cdot 6)$

3. a. $\sqrt{16}$ is an irrational number.

 b. Multiplication is distributive over addition.

 c. "$9 \cdot 1 = 9$" illustrates the multiplicative identity property.

 d. The additive identity element is 1.

4. Determine which of the numbers $-\sqrt[3]{19}$, -63, $0.\overline{38}$, $\dfrac{2}{23}$, $0.25375913 \ldots$ (never terminates or repeats), and $\sqrt{-121}$ are rational numbers, which are irrational numbers, which are real numbers, and which are *not* real numbers.

5. a. Subtract 24 from 5.   b. Subtract 8 from -17.

In Problems 6–23, perform the indicated operation, or write either "Not defined" or "Not a real number."

6. a. $4 + (-18)$ b. $15 + (-8.62)$

7. a. $-6(0)$ b. $-\dfrac{2}{3}\left(-\dfrac{1}{4}\right)$

8. a. $-10 - 5$ b. $3 - 22$

9. a. -7^2 b. $(-7)^2$

10. a. $3 \div 0$ b. $\dfrac{15}{-21}$

11. a. $-\dfrac{5}{8} \div \left(-\dfrac{1}{2}\right)$ b. $\dfrac{0}{4.7}$

12. a. $\sqrt{49}$ b. $\sqrt[3]{64}$

13. a. $3 - (-12)$ b. $0 - 6$

14. a. $2 \cdot 5^2$ b. $\sqrt{17^2 - 15^2}$

15. $23 - 19 - 3 + 11$

16. $(-6)(-5)(-2)\left(-\dfrac{1}{3}\right)$

17. $54 \div 9 \cdot 6$

18. $8 + 9 \cdot 7$

19. $6 \cdot 4^2 - 4$

20. $\dfrac{7 - 15}{-2 + 6}$

21. $5\sqrt{36} - 6(-5)$

22. $(2^3 - 8)(8^2 - 9^2)$

23. $\{-10 - [5 + 2(4 - 7)]\} - 3$

24. a. List all the integral factors of 54.

b. Write the prime factorization of 54.

25. Determine whether enough information has been given for the following problem to be solved. If so, what operation is needed to solve the problem? If not, what other information do you need?

At 5 A.M. the temperature in Denver, Colorado, was $-20°$ F. At noon the temperature was $38°$ F. What was the rise in temperature?

2 Simplifying and Evaluating Algebraic Expressions

In this chapter, we first give some of the basic definitions of algebra. We then discuss using the distributive property in algebra, combining like terms, and removing grouping symbols. We also discuss evaluating algebraic expressions and formulas and solving word problems by using formulas.

2.1 Basic Definitions

We begin this chapter with a number of definitions used in algebra.

Variable In algebra, we use letters to represent numbers, and we usually call these letters **variables**. A variable is a letter, object, or symbol that acts as a placeholder for a number that is unknown. The variable may assume different values in a particular problem or discussion.

Constant A **constant** is an object or symbol that does *not* change its value in a particular problem or discussion. It is usually represented by a number symbol, but it can be represented by one of the first few letters of the alphabet. Thus, in the expression $2x - 7y + 3$, the constants are 2, -7, and 3. In the expression $ax + by + c$, the constants are understood to be a, b, and c.

Algebraic Expression An **algebraic expression** consists of numbers, letters, signs of operation (such as $+$, $-$, etc.), and signs of grouping. (Not all of these need be present.)

Powers of Variables In Section 1.9, we discussed bases, exponents, and powers of signed numbers.

$$\overset{\text{Exponent}}{3^4} = 3 \cdot 3 \cdot 3 \cdot 3 = 81$$

Base

The same definitions are used in expressions with variables.

$$\overset{\text{Exponent}}{x^4} = x \cdot x \cdot x \cdot x$$

Base

x^4 is read as "the fourth power of x," or as "x to the fourth power"

Example 1 Examples of algebraic expressions:

a. 8   This algebraic expression consists of a single number

b. z This algebraic expression consists of a single letter (variable)

c. $2x - 3y$

d. $5x^3 - 7x^2 + 4$ e. $\dfrac{5s^2 - 2t^2}{\sqrt{3st}}$

f. $\sqrt{b^2 - 4ac}$ g. $(x - y)^2 + (t - u)^2$ ■

Example 2 List (a) the constants and (b) the variables for

$$2x + 7y - 3z$$

Solution

a. The constants are 2, 7, and -3. (Notice that the negative sign in front of the 3 is considered to be part of the constant.)

b. The variables are x, y, and z. ■

Recall from Sections 1.5 and 1.12 that the numbers multiplied together to give a product are called the *factors* of that product.

Example 3   Identify the factors in each algebraic expression.

a. $2x$ The first factor is 2, and the second factor is x.

b. $7xyz$ The first factor is 7, the second factor is x, the third factor is y, and the fourth factor is z. ∎

Terms   A **term** of an algebraic expression consists of *factors only*. This means that the only operations indicated within a *term* are multiplication and division $\left(\text{we include division here because the division } a \div b \text{ can be interpreted as the multiplication } a \cdot \dfrac{1}{b}\right)$. Exception: An expression *within grouping symbols* is considered as a single term even though it may contain one or more terms *within* the grouping symbols (see Examples 4c, 4d, and 4e). If a term is preceded by a minus sign, that sign is *part of* the term.

Example 4   Identify the terms in each algebraic expression.

a. $2s + 5t$

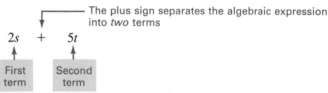

The plus sign separates the algebraic expression into *two* terms

$2s \quad + \quad 5t$

First term   Second term

The first term is $2s$; it consists of the factors 2 and s. The second term is $5t$; it consists of the factors 5 and t.

b. $3x^2y - 5xy^3 + 7xy$

The minus sign and plus sign separate the expression into *three* terms

This minus sign is *part of* the second term

$3x^2y \quad - \quad 5xy^3 \quad + \quad 7xy$

First term   Second term   Third term

The first term is $3x^2y$, the second term is $-5xy^3$, and the third term is $+7xy$, or just $7xy$.

c. $3x^2 - 9x(2y + 5z)$

The part inside the parentheses is considered as a single unit or term, and since the part inside the parentheses is being *multiplied* by $-9x$, the entire expression $-9x(2y + 5z)$ is considered to be one term.

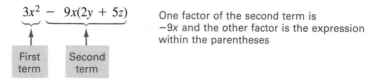

$3x^2 \quad - \quad 9x(2y + 5z)$

First term   Second term

One factor of the second term is $-9x$ and the other factor is the expression within the parentheses

Therefore, the first term is $3x^2$ and the second term is $-9x(2y + 5z)$.

81

d. $\dfrac{2 - x}{xy} + 5(2x^2 - y)$

This expression has only two terms, since the fraction bar is a grouping symbol.

$$\underbrace{\dfrac{2 - x}{xy}}_{\substack{\uparrow \\ \text{First} \\ \text{term}}} + \underbrace{5(2x^2 - y)}_{\substack{\uparrow \\ \text{Second} \\ \text{term}}} \qquad \dfrac{2 - x}{xy} \text{ can be written as } (2 - x) \cdot \dfrac{1}{x} \cdot \dfrac{1}{y}$$

Therefore, the first term is $\dfrac{2 - x}{xy}$ and the second term is $5(2x^2 - y)$.

e. $(3z + 2x - 1)$

An expression within parentheses is considered as a single term. Therefore, the algebraic expression $(3z + 2x - 1)$ has just one term. ∎

Numerical Coefficients   When a term consists of a number and one or more variables as *factors*, the factor that is a *number* is said to be the **numerical coefficient** of the term. For example, the numerical coefficient of the term $2x$ is 2, and the numerical coefficient of the term $7xyz$ is 7. When a term consists of variables only, the numerical coefficient is understood to be 1 (see Examples 5e and 5f).

Example 5   Identify the numerical coefficients.

Term	Numerical coefficient
a. $6w$	6
b. $-12xy^2$	-12
c. $\dfrac{3xy}{4} = \dfrac{3}{4}xy$	$\dfrac{3}{4}$
d. $-\dfrac{z}{5} = -\dfrac{1}{5}z$	$-\dfrac{1}{5}$
e. xy	1

↑ Even though no number is written, the numerical coefficient is understood to be 1 because $xy = 1xy$

f. $-s$	-1

↑ The numerical coefficient is understood to be -1 because $-s = -1s$ ∎

Like Terms   Terms that have equal variable parts are called **like terms**; also, terms consisting only of numbers are like terms.

Example 6   Examples of like terms:

a. $3x$, $4x$, x, $\frac{1}{2}x$, $0.7x$ are like terms. They are called "x-terms."

b. $2x^2$, $10x^2$, $\frac{3}{4}x^2$, $2.3x^2$ are like terms. They are called "x^2-terms."

c. $5xy$, $2xy$, xy, $\frac{2}{3}yx$, $5.6xy$ are like terms (because $yx = xy$). They are called "xy-terms."

d. 5, -3, $\frac{1}{4}$, 2.6 are like terms. They are called "constant terms." ∎

Unlike Terms   Terms that do not have equal variable parts are called **unlike terms**.

Example 7   Examples of unlike terms:

a. $2x$, $3y$ are unlike terms. The variables are different.

b. $5x^2$, $7x$ are unlike terms. The variable x has different exponents.

c. $4x^2y$, $10xy^2$ are unlike terms.   ■

We will see in Section 2.2B that it is possible to *combine* (that is, to write as a single term) only *like terms*.

EXERCISES 2.1

Set I   In Exercises 1–4, list (a) the different constants and (b) the different variables.

1. $2x + 4y + 2$ **2.** $7s + 3t + 7$ **3.** $7u - 8v + 2v$ **4.** $3x - 5y - 2x$

In Exercises 5 and 6, determine whether x is a factor of or a term of the expression.

5. a. $3xyz$ b. $3 + x$ c. $4x(y + 2)$ d. $5x$

6. a. $x + 7$ b. $7x$ c. $7xy$ d. $x(2 + y)$

In Exercises 7 and 8, state whether $7x$ is a factor of or a term of the expression.

7. a. $7x + 3$ b. $y + 7x$ c. $7x(3 + y)$ d. $7x + (3 + y)$

8. a. $7x(y + 6)$ b. $7xy$ c. $1 + 7x$ d. $(y + 6) + 7x$

In Exercises 9–18, (a) determine the number of terms and (b) write the second term if there is one.

9. $7xy$ **10.** $5ab$

11. $E - 5F - 3$ **12.** $R - 2T - 6$

13. $3x^2y + \dfrac{2x + y}{3xy} + 4(3x^2 - y)$ **14.** $5xy^2 + \dfrac{5x - y}{7xy} + 3(x^2 - 4y)$

15. $5u^2 - 6u(2u + v^2)$ **16.** $3E^3 - 2E(8E + F^2)$

17. $[(x + y) - (x - y)]$ **18.** $\{x - [y - (x - y)]\}$

In Exercises 19–26, write the numerical coefficient of the first term.

19. $3x + 7y$ **20.** $4R + 3T$ **21.** $x^2 - 3xy$ **22.** $x^2 + 5xy$

23. $-b + 4a$ **24.** $-y^2 - x^2$ **25.** $\dfrac{4xy}{5} + \dfrac{2a}{3}$ **26.** $\dfrac{r}{2} - \dfrac{s}{3} + \dfrac{t}{4}$

In Exercises 27–32, for each set of terms determine whether or not the terms are *all* like terms.

27. $7xy^2$, $13x^2y$, $4xy^2$, $-4x^2y$ **28.** $9x^2y$, $42x^2y$, $13xy^2$, $6xy^2$

29. $7x^5$, $12x^5$, $-3x^5$ **30.** $9y^3$, $-4y^3$, $-23y^3$

31. $8uv$, $12vu$, $-3uv$ **32.** $-6st$, $3st$, $-29ts$

Set II   In Exercises 1–4, list (a) the different constants and (b) the different variables.

1. $4x - 7y + 4$ **2.** $8u - 5v + 3u$

3. $9s + 3t + 9u$ **4.** $-8x + 2y - 5z$

In Exercises 5 and 6, determine whether x is a factor of or a term of the expression.

5. a. xy b. $x + y$ c. $x + 4y$ d. $x(y - 3)$

6. a. $5y + x$ b. $8x$ c. $x + 8$ d. $9x(y - 1)$

In Exercises 7 and 8, state whether $5x$ is a factor of or a term of the expression.

7. a. $7 + 5x$ b. $5xy$ c. $5x(5 + y)$ d. $5x + (3 + y)$

8. a. $5x - 2$ b. $5x(y - 1)$ c. $5x + y - 1$ d. $7 + 5x$

In Exercises 9–18, (a) determine the number of terms and (b) write the second term if there is one.

9. $-2x$ **10.** $8(-2x)$

11. $3 - 4y + 2z$ **12.** $3 - (4x + 2z)$

13. $5 - xy$ **14.** $5xy$

15. $3x^3 + 2(y - 3z)$ **16.** $8x^2 - 3(x + y) - z$

17. $[a + (b + c)]$ **18.** $x + y + z$

In Exercises 19–26, write the numerical coefficient of the first term.

19. $18x + 3y$ **20.** $-18(x + y)$

21. $a - b + c$ **22.** $-x - y$

23. $-x^2 + 5xy$ **24.** $-y^2 + 3(x + y)$

25. $-2a^2 - 5ab + b^3$ **26.** $\dfrac{a}{6} + \dfrac{b}{4} - \dfrac{2c}{3}$

In Exercises 27–32, for each set of terms determine whether or not the terms are *all* like terms.

27. $9w^2z$, $11wz^2$, $4w^2z$, $-5wz^2$ **28.** 15, -45, 13, -6

29. $-8y^9$, $15y^9$, $-4y^9$ **30.** $17xy$, $-12xy$, $4yx$

31. $27vw$, $-11wv$, $-7vw$ **32.** $12xy$, $4yz$, $-5zw$

2.2 Using the Distributive Property

2.2A Using the Distributive Property to Remove Grouping Symbols

We discussed the distributive property in Section 1.8; in several problems in that section, we verified that multiplication is distributive over addition and subtraction. In this section, we will use the distributive property to remove grouping symbols in a multiplication problem in which one factor contains more than one term.

Now that we know the correct order of operations, the distributive property can be restated with fewer sets of parentheses:

MULTIPLICATION IS DISTRIBUTIVE OVER ADDITION AND SUBTRACTION

$$a(b + c) = ab + ac$$

$$a(b - c) = ab - ac$$

We now know that on the right sides of the equations, the multiplications must be performed before the additions and subtractions. Therefore, we no longer need parentheses around the ab and the ac.

In arithmetic, when we're simplifying an expression such as $8(3 + 5)$, we can perform the operation inside the grouping symbols first, writing $8(3 + 5) = 8(8) = 64$, or we can apply the distributive property. To do this, we multiply each term inside the grouping symbols by the factor that is outside them and then add the products, as shown in Examples 1 and 2.

Example 1 Use the distributive property to remove the parentheses in $8(3 + 5)$; then combine the products.

Solution

$$8(3 + 5) = \textcircled{8}\,(3 + 5)$$

Each term inside the parentheses is multiplied by the factor outside the parentheses; then these products are added

First product   Second product

$$= 8(3) + 8(5) = 24 + 40 = 64$$

Notice that we obtained the same answer as when we added 3 and 5 first. ∎

In Example 2, we show on the left the use of the property $a(b - c) = ab - ac$; on the right, we interpret the subtraction problem as an addition problem and use the property $a(b + c) = ab + ac$. Both methods are acceptable.

Example 2 Use the distributive property to remove the parentheses; then combine the products.

a. $2(3 - 7)$

Solution

$$\textcircled{2}(3 - 7) = 2(3) - 2(7)$$
$$= 6 - 14$$
$$= -8$$

Alternate method

$$2(3 - 7) = \textcircled{2}(3 + [-7])$$
$$= 2(3) + (2)(-7)$$
$$= 6 + (-14)$$
$$= -8$$

Note also that $2(3 - 7) = 2(-4) = -8$.

b. $-5(2 - 7)$

Solution   *Alternate method*

$$-5(2 - 7) = -5(2) - (-5)(7) \qquad -5(2 - 7) = -5(2 + [-7])$$

$$= -10 - (-35) \qquad\qquad\qquad = -5(2) + (-5)(-7)$$

$$= -10 + 35 \qquad\qquad\qquad\quad = -10 + 35$$

$$= 25 \qquad\qquad\qquad\qquad\quad = 25$$

Note also that $-5(2 - 7) = -5(-5) = 25$. ∎

When we need to remove grouping symbols in *algebra*, we very often cannot simplify the expression inside the grouping symbols, and therefore we *must* use the distributive property to remove the grouping symbols (see Example 3).

Example 3   Use the distributive property to remove the parentheses.

a. $2(x + y)$
Solution Notice that x and y are not *like terms*; therefore, $x + y$ cannot be written as a single term.

$$2(x + y) = (2)(x) + (2)(y)$$

$$= 2x + 2y \qquad \text{2x and 2y are not like terms;}$$
$$\text{they cannot be combined}$$

b. $a(x - y)$
Solution

$$a(x - y) = (a)(x) - (a)(y)$$

$$= ax - ay \qquad \text{ax and ay are not like terms;}$$
$$\text{they cannot be combined} ∎$$

When we're using the distributive property, the factor that contains two terms can be on the right, as in $a(b + c)$, or on the left, as in $(b + c)a$. We use the commutative property of multiplication to prove that $(b + c)a = ba + ca$:

$$(b + c)a = a(b + c) \qquad \text{Commutative property of multiplication}$$

$$= ab + ac \qquad \text{Distributive property}$$

$$= ba + ca \qquad \text{Commutative property of multiplication}$$

Therefore, $(b + c)a = ba + ca$.

In Example 4, we show that in using the distributive property when the factor that contains two terms is on the left, we multiply each term inside the grouping symbols by the factor on the *right*.

Example 4   Use the distributive property to remove the parentheses.

a. $(5x + 3y)(2)$
Solution

$$(5x + 3y)(2) = (5x)(2) + (3y)(2) \quad \text{Distributive property}$$

$$= (2)(5x) + (2)(3y) \quad \text{Commutative property of multiplication}$$

$$= (2 \cdot 5)x + (2 \cdot 3)y \quad \text{Associative property of multiplication}$$

$$= 10x + 6y$$

In practice, we usually do not show the steps in which the commutative and associative properties are used. That is, we write only $(5x + 3y)(2) = (5x)(2) + (3y)(2) = 10x + 6y$, or even simply $(5x + 3y)(2) = 10x + 6y$.

b. $(2 + 5y)(-4)$
Solution

$$(2 + 5y)(-4) = (2)(-4) + (5y)(-4) \quad \text{Distributive property}$$

$$= -8 + (-4)(5y) \quad \text{Commutative property of multiplication}$$

$$= -8 + (-4 \cdot 5)y \quad \text{Associative property of multiplication}$$

$$= -8 + (-20y)$$

$$= -8 - 20y \quad \text{Can you verify that } -8 - 20y \text{ is equivalent to } -8 + (-20y)?$$

Again, you need not show the steps in which the commutative and associative properties are used. ■

A WORD OF CAUTION   In Example 4b, the parentheses around the -4 are *absolutely necessary*. The problem $(2 + 5y) - 4$ is a *subtraction* problem; 4 is to be subtracted from $2 + 5y$. The problem $(2 + 5y)(-4)$ is a *multiplication* problem; $2 + 5y$ is to be multiplied by -4.   ☑

A WORD OF CAUTION   At the end of Example 4b, it would be incorrect to write

$$-8 - 20y = -28y$$

because -8 and $-20y$ are not like terms; therefore, $-8 - 20y$ cannot be written as a single term.   ☑

Extensions of the Distributive Property

The distributive property can be extended to include any number of terms inside the parentheses. That is,

$$a(b + c - d + e + \cdots) = ab + ac - ad + ae + \cdots$$

and

$$(b + c + d + \cdots)a = ba + ca + da + \cdots$$

Example 5   Use the distributive property to remove the grouping symbols.

a. $-4(3x - 7y + 4 - 8z)$
    Solution

$$-4(3x - 7y + 4 - 8z) = (-4)(3x) - (-4)(7y) + (-4)(4) - (-4)(8z)$$
$$= -12x - (-28y) + (-16) - (-32z)$$
$$= -12x + (+28y) + (-16) + (+32z) \longleftarrow \text{Subtractions changed to additions}$$
$$= -12x + 28y - 16 + 32z$$

b. $(3x + 4y - 7)(-2)$
    Solution

$$(3x + 4y - 7)(-2) = (3x)(-2) + (4y)(-2) - (7)(-2)$$
$$= -6x + (-8y) - (-14)$$
$$= -6x + (-8y) + (+14) \qquad \text{Subtraction changed to addition}$$
$$= -6x - \quad 8y + \quad 14 \quad \blacksquare$$

A WORD OF CAUTION   It is incorrect to write (or think)

$$2(3 \cdot 4) = (2 \cdot 3)(2 \cdot 4)$$

The distributive property applies only when this symbol is a plus or minus sign

We can easily verify that $2(3 \cdot 4) \neq (2 \cdot 3)(2 \cdot 4)$.

$$2(3 \cdot 4) = 2(12) = 24$$
but
$$(2 \cdot 3)(2 \cdot 4) = (6)(8) = 48$$

Therefore, $2(3 \cdot 4) \neq (2 \cdot 3)(2 \cdot 4)$.   ☑

EXERCISES 2.2A

Set I   Use the distributive property to remove the parentheses. Do not try to combine the terms.

1. $5(a + 6)$ **2.** $4(x + 10)$ **3.** $7(x + y)$ **4.** $5(m + n)$

5. $3(m - 4)$ **6.** $3(a - 5)$ **7.** $4(x - y)$ **8.** $9(m - n)$

9. $a(6 + x)$ **10.** $b(7 + y)$ **11.** $-2(x - 3)$ **12.** $-3(x - 5)$

13. $(x - 4)(6)$ **14.** $(3 - 2x)(-5)$

15. $-3(x - 2y + 2)$ **16.** $-2(x - 3y + 4)$

Set II   Use the distributive property to remove the parentheses. Do not try to combine the terms.

1. $3(x + 4)$ **2.** $2(m - 5)$ **3.** $x(4 + y)$ **4.** $-3(x - 4)$

5. $(M + N)(3)$ **6.** $(x - y)(-4)$ **7.** $8(a - b)$ **8.** $(5 - y)(-2x)$

9. $x(y + 2)$ **10.** $(7 + a)(3b)$ **11.** $-6(x + 2y)$ **12.** $(x + 2y)(-6)$

13. $(y - 2)(4)$ **14.** $(8 - 2y)(-3)$

15. $-7(s - 4t - 5)$ **16.** $-3(4 - 5u + 7v)$

2.2B Combining Like Terms

In Section 2.2A, we used the distributive property to write a product as a sum of terms; that is, we rewrote the product $a(b + c)$ as the sum of terms $ab + ac$. (We often call this *distributing*, *multiplying out*, or *simplifying*.) We can also use the distributive property to write the sum of terms $ab + ac$ as the product $a(b + c)$. This makes it possible for us to combine *like terms* (see Example 6). We call writing $ab + ac$ as $a(b + c)$ *factoring*; we will discuss factoring in much more detail in Chapter 7.

Example 6   Combine the like terms in each expression.

a. $3x + 5x$

Solution $3x$ and $5x$ are like terms.

$$3x + 5x = (3 + 5)x \qquad \text{Distributive property (factoring)}$$
$$= 8x \qquad \text{Simplifying the expression inside the parentheses}$$

b. $7y - 3y$

Solution $7y$ and $3y$ are like terms.

$$7y - 3y = (7 - 3)y \qquad \text{Distributive property}$$
$$= 4y \qquad \text{Simplifying the expression inside the parentheses}$$

c. $z - 9z$

Solution Recall that when no numerical coefficient is written, it is understood to be 1. Therefore, the numerical coefficient of z in the first term is understood to be 1.

$$z - 9z = 1z - 9z$$
$$= (1 - 9)z$$
$$= (1 + [-9])z \qquad \text{Changing the subtraction to addition}$$
$$= -8z \quad \blacksquare$$

If you look carefully at all the problems in Example 6, you can see that you would get the same answer if you followed the rules given in the box below. For *any* method, it is understood that *only like terms can be combined* and that if any term has no written numerical coefficient the coefficient is understood to be 1.

TO COMBINE (ADD) LIKE TERMS

Add the numerical coefficients *of the like terms*; the number so obtained is the numerical coefficient of the sum. The variable part of the sum is the same as the variable part of any *one* of the like terms.

When we say "Add the numerical coefficients," it is understood that if any subtractions are indicated, those subtractions will be mentally changed to additions so that the problem can be treated *as an addition problem*.

In practice, it is neither necessary nor desirable to show all the steps that are shown in Example 6. In Example 6a you can simply write $3x + 5x = 8x$, in Example 6b you can write just $7y - 3y = 4y$, and in Example 6c you can write $z - 9z = -8z$.

When we combine like terms, we often need to change the order in which the terms appear. The commutative and associative properties of addition guarantee that when we do this the sum remains unchanged.

Example 7   Combine the like terms in each expression.

a. $3x - 7x + 10x$

Solution Think of this problem as $3x + (-7)x + 10x$.

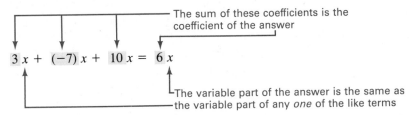

b. $9x + 5y - 3x$

Solution The only terms that can be combined are $9x$ and $-3x$ (they are the only like terms). We can "collect like terms" by rearranging the terms as shown:

$$9x + 5y - 3x$$
$$= (9x - 3x) + 5y \qquad \text{Collecting like terms}$$
$$= \quad 6x \quad + 5y \qquad \text{Combining like terms}$$

c. $12a - 7b - 9a + 4b$

Solution

$$12a - 7b - 9a + 4b$$
$$= (12a - 9a) + (-7b + 4b) \qquad \text{Collecting like terms}$$
$$= \quad 3a \quad + \quad (-3b) \qquad \text{Combining like terms}$$
$$= 3a - 3b$$

d. $7x - 2y + 9 - 11x + 3 - 4y$

Solution

$$7x - 2y + 9 - 11x + 3 - 4y$$
$$= (7x - 11x) + (-2y - 4y) + (9 + 3) \qquad \text{Collecting like terms}$$
$$= \quad -4x \quad + \quad (-6y) \quad + \quad 12 \qquad \text{Combining like terms}$$
$$= -4x - 6y + 12$$

e. $-5x^2y + 7xy^2 - 7x^2y$

Solution Be careful here! The terms $-5x^2y$ and $-7x^2y$ are the only like terms.

$$-5x^2y + 7xy^2 - 7x^2y$$
$$= 7xy^2 + (-5x^2y - 7x^2y) \qquad \text{Collecting like terms}$$
$$= 7xy^2 + \quad (-12x^2y) \qquad \text{Combining like terms}$$
$$= 7xy^2 - 12x^2y$$

f. $3x^2 - 5x + 5 - 2x^2 + 7x + 11$

Solution $3x^2 - 5x + 5 - 2x^2 + 7x + 11$
$$= (3x^2 - 2x^2) + (-5x + 7x) + (5 + 11) \qquad \text{Collecting like terms}$$
$$= \quad x^2 \quad + \quad 2x \quad + \quad 16 \qquad \text{Combining like terms} \quad \blacksquare$$

We will consider combining like terms again in Section 5.3, after we have discussed exponents in more detail.

EXERCISES 2.2B

Set I   In each exercise, combine the like terms.

1. $15x - 3x$ **2.** $12a - 9a$ **3.** $5a - 12a$

4. $10x - 24x$ **5.** $2a - 5a + 6a$ **6.** $3y - 4y + 5y$

7. $5x - 8x + x$ **8.** $3a - 5a + a$ **9.** $3x + 2y - 3x$

10. $4a - 2b + 2b$ **11.** $4y + y - 10y$ **12.** $3x - x - 5x$

13. $3mn - 5mn + 2mn$ **14.** $5cd - 8cd + 3cd$ **15.** $2xy - 5yx + xy$

16. $8mn - 7nm + 3nm$ **17.** $8x^2y - 2x^2y$ **18.** $10ab^2 - 3ab^2$

19. $a^2b - 3a^2b$ **20.** $x^2y^2 - 5x^2y^2$ **21.** $5ab + 2c - 2ba$

22. $7xy - 3z - 4yx$ **23.** $5xyz^2 - 2xyz^2 - 4xyz^2$

24. $7a^2bc - 4a^2bc - a^2bc$ **25.** $5u - 2u + 10v$

26. $8w - 4w + 5v$ **27.** $8x - 2y - 4x$

28. $9x - 8y + 2x$ **29.** $7x^2y - 2xy^2 - 4x^2y$

30. $4xy^2 - 5x^2y - 2xy^2$ **31.** $5x^2 - 3x + 7 - 2x^2 + 8x - 9$

32. $7y^2 + 4y - 6 - 9y^2 - 2y + 7$ **33.** $12.67 \text{ sec} + 9.08 \text{ sec} - 6.73 \text{ sec}$

34. $158.7 \text{ ft} + 609.5 \text{ ft} - 421.8 \text{ ft} - 263.4 \text{ ft}$

Set II   In each exercise, combine the like terms.

1. $12a - 4a$ **2.** $5x - 10x$ **3.** $18x - 25x$

4. $-4y - 12y$ **5.** $9z + 14z - 28z$ **6.** $8x - x + 3x$

7. $3x - 2x + 10x$ **8.** $7y - 5y - 2y$ **9.** $7ab + c - 2ba$

10. $8xy - 2xz - 2yz$ **11.** $6x + x - 9x$ **12.** $7x + 3x - 10x$

13. $15xy - 3xy + 6xy$ **14.** $3ab - 2bc + 4ac$ **15.** $8ab + 2ba - 4ab$

16. $6x - 6 + 4y - 3$ **17.** $23ab^2 - 17ab^2$ **18.** $16xy^2 + 5x^2y$

19. $s^3t^2 - 7s^3t^2$ **20.** $x^2y^4 + 3x^4y^2$ **21.** $x - 4xy + 3yx$

22. $5a - 4 + 3b - 3$ **23.** $7xy^2z - 2y^2xz + 5zy^2x$

24. $2x^3 - 2x^2 + 3x - 5x$ **25.** $5y^2 - 3y^3 + 2y - 4y$

26. $4x - 3y + 7 - 2x + 4 - 6y$ **27.** $3b - 5a - 9 - 2a + 4 - 5b$

28. $x - 3x^3 + 2x^2 - 5x - 4x^2 + x^3$ **29.** $y - 2y^2 - 5y^3 - y + 3y^3 - y^2$

30. $a^2b - 11ab + 12ab^2 - 3a^2b + 4ab$ **31.** $xy^2 + y - 5x^2y + 3xy^2 + x^2y$

32. $4z^2 - 6z + 5 - z^2 + 9z - 10$

33. $37.9 \text{ cm} - 13.5 \text{ cm} + 24.8 \text{ cm} - 19.3 \text{ cm}$

34. $7.2 \text{ m} - 3.65 \text{ m} + 8.002 \text{ m}$

2.2C   Simplifying an Algebraic Expression

Removing Grouping Symbols

An algebraic expression is not simplified unless all grouping symbols have been removed. The following rules, which result from applications of the distributive property, can be used to remove grouping symbols that contain more than one term.

REMOVING GROUPING SYMBOLS
THAT CONTAIN MORE THAN ONE TERM

1. If a set of grouping symbols containing more than one term is preceded by or followed by a *factor*, use the distributive property (see Examples 8d and 8f).

2. If a set of grouping symbols containing more than one term is *not* preceded by and *not* followed by a factor and is:

    a. preceded by no sign at all, drop the grouping symbols (see Examples 8a and 8e).

    b. preceded by a plus sign, drop the grouping symbols if the first term inside the grouping symbols has an *understood* plus sign (see Example 8b); drop the grouping symbols *and* the plus sign in front of them if the first term inside the grouping symbols has a *written* sign (see Example 8h).

    c. preceded by a minus sign, insert a 1 between the sign and the grouping symbol and then use the distributive property (see Examples 8c and 8g).

3. If grouping symbols occur within other grouping symbols, remove the innermost grouping symbols first (see Example 9).

Example 8 Remove the grouping symbols in each expression.

    a. $(3x - 5) + 2y$

    Solution The parentheses are neither preceded by nor followed by a factor, and they are preceded by no sign at all. (This is an addition problem.) Applying Rule 2a of the above rules, we have

$$(3x - 5) + 2y = 3x - 5 + 2y$$

    b. $3x + (4y + 7)$

    Solution The parentheses are preceded by a plus sign and are not followed by a factor; the sign of $4y$ is an *understood* plus sign. (This is an addition problem.) Applying Rule 2b, we have

$$3x + (4y + 7) = 3x + 4y + 7$$

    c. $-(8 - 6x)$

    Solution The parentheses are neither preceded by nor followed by a factor; they are preceded by a minus sign. (We are to find the negative of $8 - 6x$.) Applying Rule 2c we have

$$-(8 - 6x) = -1(8 - 6x) = (-1)(8) + (-1)(-6x) = -8 + 6x$$

              └— Inserting 1 as a *factor* does not change the value of the expression, since $1(8 - 6x) = 8 - 6x$

    d. $-2x(4 - 5z)$

    Solution The parentheses are preceded by a factor. (This is a multiplication problem.) Using Rule 1 (applying the distributive property), we have

$$-2x(4 - 5z) = (-2x)(4) + (-2x)(-5z) = -8x + 10xz$$

e. $(3x - 5) - y$

Solution The parentheses are neither preceded by nor followed by a factor. (This is a subtraction problem; the y is being subtracted.) Applying Rule 2a, we have

$$(3x - 5) - y = 3x - 5 - y$$

f. $(3x - 5)(-y)$

Solution The parentheses are followed by a factor. Using Rule 1 (applying the distributive property), we have

$$(3x - 5)(-y) = (3x)(-y) + (-5)(-y) = -3xy + 5y$$

NOTE   Notice that in Example 8e, y is being *subtracted* from $(3x - 5)$, whereas in Example 8f, $-y$ is being *multiplied* by $(3x - 5)$.   ☑

g. $2x - (8 - 6z)$

Solution This is a subtraction problem. Applying Rule 2c, we have

$$2x - (8 - 6z) = 2x - 1(8 - 6z)$$
$$= 2x + (-1)(8) + (-1)(-6z)$$
$$= 2x - 8 + 6z$$

h. $17x + (-5y + z)$

Solution The parentheses are neither preceded by nor followed by a factor; they are preceded by a plus sign, and the first term inside the parentheses is written. Applying Rule 2b, we have

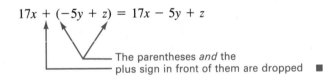

$$17x + (-5y + z) = 17x - 5y + z$$

The parentheses *and* the plus sign in front of them are dropped   ■

When grouping symbols occur within other grouping symbols, it is usually easier to remove the innermost grouping symbols first (see Rule 3 and Example 9).

Example 9   Remove the grouping symbols in each expression.

a. $x - [y + (a - b)]$

Solution

$$x - [y + (a - b)] = x - [y + a - b] \qquad \text{Applying Rule 2b}$$
$$= x - 1[y + a - b] \qquad \text{Applying Rule 2c}$$
$$= x + (-1)(y) + (-1)(a) + (-1)(-b) \quad \longleftarrow \text{Using the distributive property}$$
$$= x - y - a + b$$

b. $3 + 2[a - 5(x - 4y)$

Solution

$$3 + 2[a - 5(x - 4y)]$$
$$= 3 + 2[a - 5x + 20y] \qquad \text{Using the distributive property to remove the parentheses}$$
$$= 3 + 2a - 10x + 40y \qquad \text{Using the distributive property to remove the brackets}$$

93

c. $(3a - b) - 2\{x - [(y - 2) - z]\}$
   Solution

$$(3a - b) - 2\{x - [(y - 2) - z]\}$$

$$= (3a - b) - 2\{x - [y - 2 - z]\}$$ Using Rule 2a to remove the inner parentheses

$$= (3a - b) - 2\{x - 1[y - 2 - z]\}$$ Applying Rule 2c

$$= (3a - b) - 2\{x - y + 2 + z\}$$ Using the distributive property to remove the brackets

$$= 3a - b - 2x + 2y - 4 - 2z$$ Using Rule 2a to remove the parentheses; using the distributive property to remove the braces ∎

Simplifying Algebraic Expressions

To simplify an algebraic expression, we remove all grouping symbols and combine all like terms (see Example 10).

Example 10 Examples of simplifying algebraic expressions:

a. $x + 4(y + 3z + x)$

$$= x + 4(y) + 4(3z) + 4(x)$$ Using the distributive property

$$= x + 4y + 12z + 4x$$ Simplifying each term

Like terms

$$= x + 4x + 4y + 12z$$ Collecting like terms

$$= 5x + 4y + 12z$$ Combining like terms

b. $8x + 3x(4y + 1 - 6z)$

$$= 8x + (3x)(4y) + (3x)(1) - (3x)(6z)$$

$$= 8x + 12xy + 3x - 18xz$$

Like terms

$$= 8x + 3x + 12xy - 18xz$$ Collecting like terms

$$= 11x + 12xy - 18xz$$ Combining like terms ∎

A WORD OF CAUTION In Example 10b, it is incorrect to write

$$8x + 3x(4y + 1 - 6z) = 11x(4y + 1 - 6z)$$

Remember: Multiplication must be done *before* addition. ☑

We will discuss simplifying algebraic expressions more in Chapter 5.

EXERCISES 2.2C

Set I In Exercises 1–34, remove the grouping symbols.

1. $8 + (a - b)$ **2.** $7 + (m - n)$ **3.** $5 - (x - y)$

4. $6 - (a - b)$ **5.** $12 - 3(m - n)$ **6.** $14 - 5(x - y)$

7. $(R - S) - 8$ **8.** $(x - y) - 2$ **9.** $(R - S)(-8)$

10. $(x - y)(-2)$ **11.** $(3x - 2y)(-5z)$ **12.** $(5a - 7b)(-2c)$

13. $(3x - 2y) - 5z$ **14.** $(5a - 7b) - 2c$ **15.** $10 - 2(a - b)$

16. $12 - 5(x - 2y)$ **17.** $2(x - y) + 3$ **18.** $4(a - 2b) + 5$

19. $2a - 3(x - y)$ **20.** $5x - 2(a - b)$ **21.** $a - [x - (b - c)]$

22. $a - [x - (-b + c)]$ **23.** $4 - 2[a - 3(x - y)]$

24. $8 - 3[x - 2(3a - b)]$ **25.** $7 + 6[x - 3(a - b)]$

26. $3 + 5[y - 4(2x - a)]$ **27.** $12 - 4\{a - (b - 7c)\}$

28. $15 - 7\{x - (y + 3z)\}$ **29.** $3(a - 2x) - 2(y - 3b)$

30. $2(2x - b) - 3(y - 5a)$ **31.** $-10[-2(x - 3y) + a] - b$

32. $-5[-3(2x - y) + a] - c$ **33.** $(a - b) - \{[x - (3 - y)] - R\}$

34. $(x - y) - \{[a - (c - 5)] - b\}$

In Exercises 35–42, simplify each algebraic expression.

35. $a + 5(b + 3c + a)$ **36.** $x + 9(y + 7z + x)$

37. $9y - 2\{y - 4(y + 1)\}$ **38.** $13x - 5\{x - 4(3 + x)\}$

39. $4x - (5 - 2y + 8x)$ **40.** $8c - (2a - 1 + 3b + 12c)$

41. $6x^2 + 5 - (3x + 4 - x^2)$ **42.** $9a^2 + 7 - (12a^2 + 10 - 5a)$

Set II   In Exercises 1–34, remove the grouping symbols.

1. $5 + (x + y)$ **2.** $-3 + (a - b)$ **3.** $3 - (x - y)$

4. $-6 - (s + t)$ **5.** $12 - 2(a + b)$ **6.** $8 - 2(x - y)$

7. $(a - b) - c$ **8.** $(s + 3t)(-2u)$ **9.** $(a - b)(-c)$

10. $(x - y)(-2)$ **11.** $(8u - 5v)(-7w)$ **12.** $(x - y) - 2$

13. $(-4x - 3y) - 9z$ **14.** $(-4x - 3y)(-9z)$ **15.** $2x - 3(a - b)$

16. $5x - (y + z)$ **17.** $4(x - 2y) - 3a$ **18.** $a - (b - c)$

19. $-(x - y) + (2 - a)$ **20.** $(3 + x) - 3(a - b)$ **21.** $(a + b)(2) - 6$

22. $2 - 6(a + b)$ **23.** $x - [a + (y - b)]$ **24.** $y - [x - (a - b)]$

25. $8 + 4[y - 2(x - b)]$ **26.** $6 + 7[a - 5(3b - c)]$

27. $15 - 8\{x - (y - 3z)\}$ **28.** $14 - 9\{a - (b + 4c)\}$

29. $x - [-(y - b) + a]$ **30.** $[8 + 2(a - b) - x] - y$

31. $P - \{x - [y - (4 - z)]\}$ **32.** $P - \{x - [y - (z - 4)]\}$

33. $-\{-[-(-2 - x) + y]\} - a$ **34.** $-(-\{-[x - 3] + y\} - z)$

In Exercises 35–42, simplify each algebraic expression.

35. $x + 6(z + 8y + x)$ **36.** $a - 6(a + 9b - c)$

37. $6x - 3\{x - 5(x + 3)\}$ **38.** $11y - 9\{y - 3(4 + y)\}$

39. $6a - (8a - 3b + 4)$ **40.** $7x - (4w - 6 + 6y + 13x)$

41. $4x^2 + 2 - (6x + 8 - x^2)$ **42.** $7y^2 + 6 - (16y^2 + 15 - 4y)$

2.3 Evaluating Algebraic Expressions

In this section, we use our knowledge of the operations on signed numbers to help us find the value of algebraic expressions that contain variables when values for those variables are given.

TO EVALUATE AN EXPRESSION THAT CONTAINS VARIABLES

1. Replace each variable by its numerical value, usually enclosing the number in parentheses.

2. Carry out all arithmetic operations, using the correct order of operations (see Section 1.11).

Example 1 Find the value of $3x - 5y$ if $x = 10$ and $y = 4$.
Solution Recall from Section 1.5 that $3x$ means 3 times x, and $5y$ means 5 times y.

$$3x - 5y$$
$$= 3(10) - 5(4) \qquad \text{Replacing each variable by its numerical value}$$
$$= 30 - 20 = 10 \qquad \text{Carrying out the arithmetic operations} \quad \blacksquare$$

A WORD OF CAUTION When you replace a variable by a number, enclose the number in parentheses to avoid making the following common errors.

a. Evaluate $3x$ when $x = -2$.

Common errors	*Correct method*
$3x = 3 - 2 = 1$	$3x = 3(-2) = -6$
$3x = 3 - 2 = -6$	

Even though the final answer is correct here, *two* errors have been made!

b. Evaluate $4x^2$ when $x = -3$

Common errors	*Correct method*
$4x^2 = 4 - 3^2 = 4 - 9 = -5$	$4x^2 = 4(-3)^2 = 4 \cdot 9 = 36$
$4x^2 = 4 - 3^2 = 4 + 9 = 13$	
$4x^2 = 4 - 3^2 = 4(9) = 36$	

Even though the final answer is correct here, *two* errors have been made! ☑

Example 2 Evaluate $3x - 5y$ if $x = -4$ and $y = -6$.
Solution

$$3x - 5y$$
$$= 3(-4) - 5(-6) \qquad \text{Replacing } x \text{ by } -4 \text{ and } y \text{ by } -6$$
$$= -12 + 30$$
$$= 18 \quad \blacksquare$$

Example 3   Find the value of $\dfrac{2a - b}{10c}$ if $a = -1$, $b = 3$, and $c = -2$.

Solution

Remember, this bar is a grouping symbol

$$\frac{2a - b}{10c} = \frac{2(-1) - (3)}{10(-2)} = \frac{-2 - 3}{-20} = \frac{-5}{-20} = \frac{1}{4} = 0.25 \quad \blacksquare$$

Example 4   Evaluate $\dfrac{5hgk}{2m}$ if $h = -2$, $g = 3$, $k = -4$, and $m = 6$.

Solution

$$\frac{5hgk}{2m} = \frac{5(-2)(3)(-4)}{2(6)} = \frac{120}{12} = 10 \quad \blacksquare$$

Example 5   Find the value of $2a - [b - (3x - 4y)]$ if $a = -3$, $b = 4$, $x = -5$, and $y = 2$.

Solution

$$2a \quad - [b - \quad (3x - 4y)]$$

$$= 2(-3) - [(4) - \{3(-5) - 4(2)\}] \qquad \text{Notice that braces and brackets were used to clarify the grouping}$$

$$= 2(-3) - [4 - \{-15 - 8\}]$$

$$= 2(-3) - [4 - \{-23\}]$$

$$= 2(-3) - [4 + 23]$$

$$= \quad -6 \quad - \quad (+27)$$

$$= \quad -6 \quad + \quad (-27)$$

$$= -33 \quad \blacksquare$$

Example 6   Evaluate $b - \sqrt{b^2 - 4ac}$ when $a = 3$, $b = -7$, and $c = 2$.

Solution

$$b - \sqrt{b^2 - 4ac} \quad \longleftarrow \text{This bar is a grouping symbol for } b^2 - 4ac$$

$$= (-7) - \sqrt{(-7)^2 - 4(3)(2)}$$

$$= (-7) - \sqrt{49 - 24}$$

$$= (-7) - \sqrt{25}$$

$$= (-7) - \quad 5$$

$$= -12 \quad \blacksquare$$

A WORD OF CAUTION   In Example 6, *two* errors are made if you write

$$b - \sqrt{b^2 - 4ac} = -7 - \sqrt{-7^2 - 4(3)(2)} = -7 - \sqrt{49 - 24}$$

It is incorrect to omit the parentheses around -7 here; also, $-7^2 = -49$, not 49.   ☑

Example 7   Evaluate the following when $x = 5$.

a. $7 + 3x$.   When $x = 5$, $7 + 3x = 7 + 3(5) = 7 + 15 = 22$.

b. $10x$.   When $x = 5$, $10x = 10(5) = 50$. $\blacksquare$

NOTE   We can see from Example 7 that $7 + 3x \neq 10x$.   ☑

EXERCISES 2.3

Set I   In Exercises 1–28, evaluate each expression when $a = 3$, $b = -5$, $c = -1$, $x = 4$, and $y = -7$.

1. ab **2.** xy **3.** $a + b$ **4.** $x + y$ **5.** $3b$

6. $15b$ **7.** $9 - 6b$ **8.** $8 + 7b$ **9.** b^2 **10.** $-y^2$

11. $2a - 3b$ **12.** $3x - 2y$ **13.** $x - y - 2b$

14. $a - b - 3y$ **15.** $3b - ab + xy$ **16.** $4c + ax - by$

17. $x^2 - y^2$ **18.** $b^2 - c^2$ **19.** $4 + a(x + y)$

20. $5 - b(a + c)$ **21.** $2(a - b) - 3c$ **22.** $3(a - x) - 4b$

23. $3x^2 - 10x + 5$ **24.** $2y^2 - 7y + 9$ **25.** $a^2 - 2ab + b^2$

26. $x^2 - 2xy + y^2$ **27.** $\dfrac{3x}{y + b}$ **28.** $\dfrac{4a}{c - b}$

In Exercises 29–38, find the value of each expression when $E = -1$, $F = 3$, $G = -5$, $H = -4$, and $K = 0$. (Round off the answer in Exercise 37 to three decimal places.)

29. $\dfrac{E + F}{EF}$ **30.** $\dfrac{G + H}{GH}$ **31.** $\dfrac{(1 + G)^2 - 1}{H}$ **32.** $\dfrac{1 - (1 + E)^2}{F}$

33. $2E - [F - (3K - H)]$ **34.** $3H - [K - (4F - E)]$

35. $G - \sqrt{G^2 - 4EH}$ **36.** $H - \sqrt{H^2 - 4EK}$

 37. $\dfrac{\sqrt{2H - 5G}}{0.2F^2}$ **38.** $\dfrac{\sqrt{5HG}}{0.5E^2}$

Set II   In Exercises 1–36, evaluate each expression when $a = -2$, $b = 4$, $c = -5$, $x = 3$, and $y = -1$.

1. bc **2.** ax **3.** $b + c$ **4.** $a + x$ **5.** $8y$

6. $-3a$ **7.** $11 - 3y$ **8.** $9 - 12a$ **9.** a^2 **10.** $-a^2$

11. $4c - 5y$ **12.** $a - b - 2c$ **13.** $6x - xy + ab$

14. $c^2 - x^2$ **15.** $7 - x(a + b)$ **16.** $3(x - y) - 4c$

17. $b^2 - 4ac$ **18.** $b^2 - 2bc + c^2$ **19.** $\dfrac{5b}{x - y}$

20. $\dfrac{a - b}{ab}$ **21.** $\dfrac{(1 - x)^2 - 1}{y}$ **22.** $3a - [b - (5x - y)]$

23. $x - \sqrt{x^2 - 4ay}$ **24.** $\dfrac{b - a}{ab}$ **25.** $a^2 + 2ab + b^2$

26. $(a + b)^2$ **27.** $a^2 + b^2$ **28.** $x^2 - y^2$

29. $(x - y)^2$ **30.** $x^2 - 2xy + y^2$

31. $\dfrac{a}{b} + \dfrac{c}{x}$ **32.** $\dfrac{a + c}{b + x}$ **33.** $\dfrac{a + c}{b}$ **34.** $\dfrac{a}{b} + \dfrac{c}{b}$ **35.** $\dfrac{x}{b} - \dfrac{a}{y}$ **36.** $\dfrac{x - a}{b - y}$

In Exercises 37 and 38, evaluate each expression when $a = -19.32$, $b = 25.73$, and $c = 47.02$. Round off answers to four decimal places.

37. $\dfrac{\sqrt{3b - 4a}}{0.7c^2}$ **38.** $\dfrac{-b - \sqrt{b^2 - 4ac}}{2a}$

2.4 Using Formulas

In algebra, an **equation** is a statement that two algebraic expressions are equal. The following is an example of an equation:

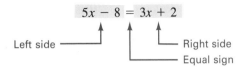

$$5x - 8 = 3x + 2$$

Left side ——————→ | | | ←—————— Right side
Equal sign

An equation is made up of three parts:

1. The equal sign ($=$)

2. The expression to the left of the equal sign, called the left side

3. The expression to the right of the equal sign, called the right side

In the equation $5x - 8 = 3x + 2$, the variable x represents an unknown number. Letters other than x may be used in equations to represent unknown numbers.

A **formula** is often an equation that expresses a mathematical or scientific fact. One reason for studying algebra is as preparation for using formulas. Students will encounter formulas in many courses they take, as well as in real-life situations. In the examples and exercises, we have listed the subject areas in which the formulas are used.

Formulas are evaluated in the same way any expression is evaluated. The equal sign is a very important part of the formula, and it must not be dropped or ignored. Write the *full* equation in every step.

Example 1   Given the formula $A = \frac{1}{2}bh$, find A when $b = 17$ and $h = 12$.   (Geometry)
Solution This is the formula for finding the area of a triangle, where A is the area, b is the base, and h is the height, or altitude.

$$A = \frac{1}{2}bh$$

$$A = \frac{1}{2}(17)(12)$$ Substitute 17 for b and 12 for h

$$A = \frac{(17)(12)}{2}$$ Notice that we write "$A = $" in every step

$$A = 102$$

Therefore, $A = 102$ when $b = 17$ and $h = 12$. ■

Example 2   Given the formula $A = P(1 + rt)$, find A when $P = 1,000$, $r = 0.08$, and $t = 1.5$.
          (Business)

Solution This is the formula for finding the amount of money in a savings account, where t is the number of years, r is the rate, P is the amount originally invested, A is the amount after t years, and the interest is *simple* interest.

$$A = P(1 + rt)$$

$$A = 1,000[1 + (0.08)(1.5)]$$ Brackets were used to clarify the grouping

$$A = 1,000[1 + 0.12]$$

$$A = 1,000[1.12]$$

$$A = 1,120$$

Therefore, $A = 1,120$ when $P = 1,000$, $r = 0.08$, and $t = 1.5$. ∎

Example 3 Given the formula $s = \frac{1}{2}gt^2$, find s when $g = 32$ and $t = 5\frac{1}{2}$. (Physics)

Solution This is the formula for the distance a freely falling object falls, where s is the distance, g is the force due to gravity, and t is the time.

$$s = \frac{1}{2}gt^2$$

$$s = \frac{1}{2}(32)\left(\frac{11}{2}\right)^2 \qquad 5\tfrac{1}{2} = \tfrac{11}{2}$$

$$s = \frac{1}{2}\left(\frac{32}{1}\right)\left(\frac{121}{4}\right)$$

$$s = 484$$

Therefore, $s = 484$ when $g = 32$ and $t = 5\frac{1}{2}$. ∎

Example 4 Given the formula $C = \frac{5}{9}(F - 32)$, find C when $F = -13$ (Science)

Solution This is the formula for converting temperature from degrees Fahrenheit to degrees Celsius.

$$C = \frac{5}{9}(F - 32)$$

$$C = \frac{5}{9}(-13 - 32)$$

$$C = \frac{5}{9}(-45)$$

$$C = -25$$

Therefore, $C = -25$ when $F = -13$. ∎

Example 5 Given the formula $T = \pi\sqrt{\dfrac{L}{g}}$, find T when $\pi \approx 3.14$, $L = 96$, and $g = 32$. (Physics)

Solution This is the formula for the time for a single swing of a pendulum, where T is the time, L is the length, g is the force due to gravity, and $\pi \approx 3.14$. (Recall that $\approx$ means "is approximately equal to.")

$$T = \pi\sqrt{\frac{L}{g}}$$

$$T \approx (3.14)\sqrt{\frac{96}{32}}$$

$$T \approx (3.14)\sqrt{3}$$

$$T \approx (3.14)(1.732) \qquad \sqrt{3} \approx 1.732 \text{ by a calculator or by Table I}$$

$$T \approx 5.44 \qquad \text{Rounded off to two decimal places}$$

Therefore, $T \approx 5.44$ when $\pi \approx 3.14$, $L = 96$, and $g = 32$. ∎

Example 6   Given the formula

$$S = \frac{a(1 - r^n)}{1 - r}$$

find S when $a = -4$, $r = \frac{1}{2}$, and $n = 3$.   (Mathematics)

Solution This is the formula for finding the sum of a geometric series, where S is the sum, a is the first term, n is the number of terms, and r is the common ratio.

$$S = \frac{a(1 - r^n)}{1 - r}$$

$$S = \frac{(-4)\left[1 - \left(\frac{1}{2}\right)^3\right]}{1 - \frac{1}{2}} \qquad \left(\frac{1}{2}\right)^3 = \left(\frac{1}{2}\right)\left(\frac{1}{2}\right)\left(\frac{1}{2}\right) = \frac{1}{8}$$

$$S = \frac{(-4)\left(1 - \frac{1}{8}\right)}{1 - \frac{1}{2}}$$

$$S = \frac{(-4)\left[\frac{7}{8}\right]}{\frac{1}{2}} = -\frac{7}{2} \div \frac{1}{2} = -\frac{7}{2}\left(\frac{2}{1}\right) = -7$$

$$S = -7$$

Therefore, $S = -7$ when $a = -4$, $r = \frac{1}{2}$, and $n = 3$.   ∎

EXERCISES 2.4

Set I   Use the formula, substituting the given values for the variables.

Geometry   The area of a triangle is given by the formula

$$A = \frac{1}{2}bh$$

where A is the area, b is the base, and h is the height.

1. Find A when $b = 15$ and $h = 14$.

2. Find A when $b = 27$ and $h = 36$.

Electricity   Ohm's law states that

$$I = \frac{E}{R} *$$

where I is the current, E is the electromotive force, and R is the resistance.

3. Find I when $E = 110$ and $R = 22$.

4. Find I when $E = 220$ and $R = 33$.

---

*When the electromotive force is measured in *volts*, the formula is often written as $I = \frac{V}{R}$.

Business   The formula for simple interest is

$$I = prt$$

where I is the interest, p is the principal, r is the rate, and t is the time.

5. Find I when $p = 600$, $r = 0.09$, $t = 4.5$.

6. Find I when $p = 700$, $r = 0.08$, $t = 2.5$.

Chemistry   The formula to change degrees Celsius to degrees Fahrenheit is

$$F = \frac{9}{5}C + 32$$

where F is degrees Fahrenheit and C is degrees Celsius.

7. Find F when $C = 25$.

8. Find F when $C = -25$.

Geometry   The area of a circle is given by the formula

$$A = \pi r^2$$

where A is the area and r is the radius.

9. Find A when $\pi \approx 3.14$ and $r = 10$.

10. Find A when $\pi \approx 3.14$ and $r = 20$.

Physics   The distance a free-falling object travels is given by the formula

$$s = \frac{1}{2}gt^2$$

where s is the distance, g is the force of gravity, and t is the time.

11. Find s when $g = 32$ and $t = 3$.

12. Find s when $g = 32$ and $t = 5$.

Business   The value of an item that is depreciated each year is given by the formula

$$V = C - Crt$$

where V is the present value, C is the original cost, r is the rate of depreciation, and t is the time.

13. Find V when $C = 500$, $r = 0.1$, $t = 2$.

14. Find V when $C = 1,000$, $r = 0.08$, $t = 5$.

Statistics   The standard deviation of a binomial distribution is given by the formula

$$\sigma^* = \sqrt{npq}$$

where σ is the standard deviation, n is the number of trials, p is the probability of success, and q is the probability of a failure.

---

*"σ" is the Greek letter *sigma*; it is used in statistics to represent a quantity called the *standard deviation*.

15. Find σ when $n = 100$, $p = 0.9$, $q = 0.1$.

16. Find σ when $n = 100$, $p = 0.8$, $q = 0.2$.

Chemistry   The formula to change degrees Fahrenheit to degrees Celsius is

$$C = \frac{5}{9}(F - 32)$$

where C is degrees Celsius and F is degrees Fahrenheit.

17. Find C when $F = -4$.

18. Find C when $F = 5$.

Nursing   The formula to determine the dosage for a child is

$$C = \frac{a}{a + 12} \cdot A$$

where C is the child's dosage, a is the age of the child, and A is the adult dosage.

19. Find C when $a = 6$ and $A = 30$.

20. Find C when $a = 4$ and $A = 48$.

Geometry   The volume of a sphere is given by the formula

$$V = \frac{4}{3}\pi r^3$$

where V is the volume and r is the radius.

21. Find V when $\pi \approx 3.14$ and $r = 3$.

22. Find V when $\pi \approx 3.14$ and $r = 6$.

Business   The formula for *compound* interest is

$$A = P(1 + i)^n$$

where A is the total amount, P is the principal, i is the interest rate per period, and n is the number of periods. Round off answers to two decimal places.

23. Find A when $P = 1,000$, $i = 0.06$, $n = 2$.

24. Find A when $P = 2,000$, $i = 0.08$, $n = 3$.

Set II   Use the formula, substituting the given values for the variables.

Formula		Problem
$A = \frac{1}{2}bh$	Geometry	**1.** Find A when $b = 24$ and $h = 19$
		2. Find A when $b = 14$ and $h = 18$
$I = \dfrac{E}{R}$	Electricity	**3.** Find I when $E = 110$ and $R = 11$
		4. Find I when $E = 220$ and $R = 20$
$I = prt$	Business	**5.** Find I when $p = 800$, $r = 0.06$, and $t = 3.5$
		6. Find I when $p = 600$, $r = 0.05$, and $t = 5.5$

$s = \dfrac{1}{2}gt^2$ Physics $\left\{\begin{array}{l}\\\\\end{array}\right.$
 7. Find s when $g = 32$ and $t = 1\frac{1}{2}$
 8. Find s when $g = 32$ and $t = 3\frac{1}{2}$

$A = P(1 + i)^n$ Business $\left\{\begin{array}{l}\\\\\end{array}\right.$
 9. Find A when $P = 500$, $i = 0.07$, and $n = 2$
 10. Find A when $P = 1{,}000$, $i = 0.06$, and $n = 3$

$A = \pi r^2$ Geometry $\left\{\begin{array}{l}\\\\\end{array}\right.$
 11. Find A when $\pi \approx 3.14$ and $r = 50$
 12. Find A when $\pi \approx 3.14$ and $r = 15$

$F = \dfrac{9}{5}C + 32$ Chemistry $\left\{\begin{array}{l}\\\\\end{array}\right.$
 13. Find F when $C = -15$
 14. Find F when $C = 35$

$A = p(1 + rt)$ Business $\left\{\begin{array}{l}\\\\\end{array}\right.$
 15. Find A when $p = 800$, $r = 0.08$, and $t = 2.5$
 16. Find A when $p = 1{,}200$, $r = 0.05$, and $t = 6.5$

$C = \dfrac{5}{9}(F - 32)$ Chemistry $\left\{\begin{array}{l}\\\\\end{array}\right.$
 17. Find C when $F = 14$
 18. Find C when $F = -22$

$V = \dfrac{4}{3}\pi r^3$ Geometry $\left\{\begin{array}{l}\\\\\end{array}\right.$
 19. Find V when $\pi \approx 3.14$ and $r = 15$
 20. Find V when $\pi \approx 3.14$ and $r = 12$

$C = \dfrac{a}{a + 12} \cdot A$ Nursing $\left\{\begin{array}{l}\\\\\end{array}\right.$
 21. Find C when $a = 8$ and $A = 20$
 22. Find C when $a = 5$ and $A = 34$

$V = C - Crt$ Business $\left\{\begin{array}{l}\\\\\end{array}\right.$
 23. Find V when $C = 2{,}000$, $r = 0.05$, and $t = 10$
 24. Find V when $C = 1{,}500$, $r = 0.06$, and $t = 8$

2.5 Solving Word Problems by Using Formulas

Although you cannot yet solve an algebraic equation and have not yet been given the suggestions for solving word problems using algebra, you should be able to solve the word problems in this section. We include here only the kinds of problems solved in Section 2.4. You are given some facts in "story form," and you are to solve the problem by using the formula given with the problem or one of the formulas given in Section 2.4. (For your convenience, these formulas are listed on the facing page of the inside back cover.)

Unfortunately, there is no good method, in most cases, to see whether the answer is a reasonable one, so in our examples we will not show a check for the solutions.

Example 1   Find the area of a triangle if the length of its base is 12 in. and its height (altitude) is 6 in.

Solution We were told in Section 2.4 that the formula for the area of a triangle is $A = \dfrac{1}{2}bh$; we can draw a picture as shown

here. We're told that the base of the triangle is 12 in., so we'll let $b = 12$ in.; we're told that the height of the triangle is 6 in., so we'll let $h = 6$ in. Substituting into the formula, we have

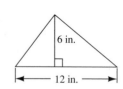

$$A = \frac{1}{2}bh$$

$$A = \frac{1}{2}(12 \text{ in.})(6 \text{ in.}) \qquad \text{Substituting 12 in. for } b \text{ and 6 in. for } h$$

$$A = 36 \text{ in.}^2, \text{ or } 36 \text{ sq. in.}$$

Therefore, the area of the triangle is 36 sq. in. ■

Example 2   Anne has just invested $1,000 at 6% simple interest. How much interest will she have earned in 3 years?

Solution The formula for finding simple interest is $I = prt$, where I is the interest, p is the principal, r is the rate, and t is the time. For this kind of problem, r *must* be expressed in decimal form; $6\% = 0.06$. Therefore, we'll let $r = 0.06$. The principal is the amount invested, $1,000; therefore, $p = \$1,000$. We're interested in finding how much interest Anne will earn in 3 years; therefore, $t = 3$.

$$I = prt$$

$$I = (\$1,000)(0.06)(3) \qquad \text{Substituting \$1,000 for } p, 0.06 \text{ for } r, \text{ and 3 for } t$$

$$I = \$180$$

Therefore, she will earn $180 interest in 3 years. ■

Example 3   If an adult should take 200 milligrams (mg) of a certain medicine, what would the correct dosage be for an 8-year-old child?

Solution The formula for determining the dosage for a child is $C = \dfrac{a}{a + 12} \cdot A$, where C is the child's dosage, a is the age of the child, and A is the adult dosage. The adult dosage is 200 mg; therefore, we'll let $A = 200$. The child is 8 years old; therefore, we'll let $a = 8$.

$$C = \frac{a}{a + 12} \cdot A$$

$$C = \left(\frac{8}{8 + 12}\right)(200) \qquad \text{Substituting 8 for } a \text{ and 200 for } A$$

$$C = \left(\frac{8}{20}\right)(200) \qquad \text{The fraction bar is a grouping symbol}$$

$$C = 80$$

Therefore, the child's dosage would be 80 mg. ■

EXERCISES 2.5

Set I   Solve the following problems by selecting and using one of the formulas from Section 2.4.

1. Water usually boils at 100° Celsius. What is this temperature in degrees Fahrenheit?

2. Tom bought a computer that cost $3,000. If, for income tax purposes, the rate of depreciation is 0.09, what will his computer be worth after 4 years?

3. Barbara invested $2,000 in an account that pays 7% simple interest. How much money will be in the account at the end of 5 years?

4. How many cubic inches of air are in a balloon if the radius of the balloon is 6 in. (Use $\pi \approx 3.14$, and round off the answer to the nearest cubic inch.)

5. If a water balloon is dropped from a tall office building, how many feet will it have fallen after 3 seconds, assuming that the force of gravity is 32 ft per second^2? (When you substitute into the formula given in Example 3 of Section 2.4 and use 32 ft per second^2 for the force due to gravity, the answer is in feet.)

6. What is the area of a triangle if its altitude is 5 ft and its base is 3 ft?

Set II Solve the following problems by selecting and using one of the formulas from Section 2.4.

1. Water usually freezes at 32° Fahrenheit. What is this temperature in degrees Celsius?

2. Suppose we are tossing a coin and we consider *heads* a "success" and *tails* a "failure." When we toss a coin, the probability of its landing "heads up" is $\frac{1}{2}$, and the probability of its landing "tails up" is $\frac{1}{2}$. What is the standard deviation if we toss the coin 100 times (that is, the number of "trials" is 100)?

3. Mr. Liao invested $5,000 at 8% simple interest 6 years ago. How much interest has he earned?

4. If a pendulum is 128 ft long, what will be the time (in seconds) for one swing of the pendulum, assuming that the force of gravity is 32 ft per second^2? (Use $\pi \approx 3.14$. When you substitute into the formula given in Example 5 of Section 2.4 and use 32 ft per second^2 for g, the answer is in seconds.)

5. Mrs. Waite bought a copy machine that cost $2,500. If, for income tax purposes, the rate of depreciation is 0.08, what will her copy machine be worth after 3 years?

6. If a hammer is accidentally dropped from a skyscraper that's under construction, how many feet will it have fallen after 4 seconds, assuming that the force of gravity is 32 ft per second^2? (When you substitute into the formula given in Example 3 of Section 2.4 and use 32 ft per second^2 for the force due to gravity, the answer is in feet.)

2.6 Review: 2.1–2.5

Definitions 2.1

Variable A **variable** is a letter, object, or symbol that acts as a placeholder for a number that is unknown. The variable may assume different values in a particular problem or discussion.

Constant A **constant** is an object or symbol that does *not* change its value in a particular problem or discussion.

Algebraic Expression An **algebraic expression** consists of numbers, letters, signs of operation, and signs of grouping.

Powers of Variables The same definition of powers given in Section 1.9 is used in expressions with variables. Thus, $x^4 = x \cdot x \cdot x \cdot x$.

Terms A **term** of an algebraic expression consists of factors only. An expression within grouping symbols is considered as a single term even though it may contain one or more terms within the grouping symbols. A minus sign is part of the term that follows it.

Numerical Coefficients When a term consists of a number and one or more variables as *factors*, the factor that is a *number* is said to be the **numerical coefficient** of the term. When a term consists of variables only, the numerical coefficient is understood to be 1.

Like Terms Terms having equal variable parts are called **like terms**. Only the numerical coefficients may differ.

Unlike Terms Terms that do not have equal variable parts are called **unlike terms**.

Using the Distributive Property 2.2

The **distributive property** can be used to remove grouping symbols in a multiplication problem in which at least one factor contains more than one term.

$$a(b + c) = ab + ac \qquad \text{and} \qquad a(b - c) = ab - ac$$

The **distributive property** can be used to combine like terms.

$$ab + ac = a(b + c)$$

To Combine Like Terms 2.2

Add the numerical coefficients of the *like terms*; the number so obtained is the numerical coefficient of the sum. The variable part of the sum is the same as the variable part of any *one* of the like terms.

To Remove Grouping Symbols 2.2

1. If a set of grouping symbols containing more than one term is preceded by or followed by a factor, use the distributive property.

2. If a set of grouping symbols containing more than one term is *not* preceded by and *not* followed by a factor and is:

   a. preceded by no sign at all, drop the grouping symbols.

   b. preceded by a plus sign, drop the grouping symbols if the first term inside the grouping symbols is an *understood* plus sign; drop the grouping symbols *and* the plus sign in front of them if the first term inside the grouping symbols has a *written* sign.

   c. preceded by a minus sign, insert a 1 between the sign and the grouping symbol and then use the distributive property.

3. When grouping symbols occur within other grouping symbols, remove the innermost grouping symbols first.

Simplifying Algebraic Expressions 2.2

To simplify an algebraic expression, remove all grouping symbols and combine all like terms.

To Evaluate an Expression or Formula 2.3 and 2.4

1. Replace each variable by its numerical value.

2. Carry out all arithmetic operations using the correct order of operations.

Solving Word Problems by Using Formulas 2.5

See the examples in Section 2.5.

Review Exercises 2.6 Set I

In Exercises 1–4, (a) determine the number of terms and (b) write the second term if there is one.

1. $a^2 + 2ab + b^2$

2. $6x - 2(x^2 + y^2)$

3. $2(m^2 - n^2)$

4. $\dfrac{x + x^2}{4} - 2(x^3 + 1)$

5. Is $2x$ a factor of or a term of the expression $3 + (2x)$?

6. Is $7y$ a factor of or a term of the expression $8(7y)$?

In Exercises 7–10, combine the like terms.

7. $18x + 23x$

8. $-42xy^2 - 12xy^2$

9. $12ab - 3a + 5b - 21ba$

10. $4 + 2x - y - 7$

In Exercises 11–16, write each expression in simplest form.

11. $8(3x - y)$

12. $-5(a + b)$

13. $(3s - 2t)(-2)$

14. $(3x - 2t) - 2$

15. $7 - 3[x - 4(2 + x)]$

16. $8x - 7\{x - (3 + 5x)\}$

In Exercises 17–22, find the value of each expression when $x = -2$, $y = 3$, and $z = -4$.

17. $3x - y + z$

18. $7 - y(x + z)$

19. $5x^2 - 3x + 10$

20. $x^2 + 2xy - y^2$

21. $x - 2[y - x(y + z)]$

22. $\dfrac{(x - y)^2 + z^2}{x - 2z}$

In Exercises 23–28, use the formula, substituting the given values for the variables.

23. $C = \dfrac{a}{a + 12} \cdot A$ Find C when $a = 8$ and $A = 35$

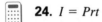 **24.** $I = Prt$ Find I when $P = 400$, $r = 0.10$, and $t = 2.75$

25. $C = \dfrac{5}{9}(F - 32)$ Find C when $F = 15\dfrac{1}{2}$

 26. $A = P(1 + i)^n$ Find A when $P = 500$, $n = 3$, and $i = 0.1$

27. $\sigma = \sqrt{npq}$ Find σ when $n = 400$, $p = 0.5$, and $q = 0.5$

28. $A = p(1 + rt)$ Find A when $p = 500$, $r = 0.12$, and $t = 10$

In Exercises 29 and 30, solve each problem by selecting and using one of the formulas from Section 2.4.

29. Find the amount of money in a savings account after three years if $2,500 was invested at 6% simple interest.

30. What is the current if the electromotive force is 220 and the resistance is 22?

Review Exercises 2.6 Set II

NAME _____

In Exercises 1–4, (a) determine the number of terms and (b) write the second term if there is one.

ANSWERS

1. $8x^3 - 2x^2 + 5x + 3$

2. $7x + (3x^4 - 1)$

3. $8a - \dfrac{2x + 1}{2} - 4$

4. $\dfrac{p + q}{2} + \dfrac{p}{4} + 3(p^2 - q)$

5. Is $8a$ a term of or a factor of the expression $8ab$?

6. Is $-3x$ a term of or a factor of the expression $1 - 3x$?

In Exercises 7–10, combine the like terms.

7. $12y - 28y$

8. $3x^2y - 8x^2y$

9. $14x^2y - 8x^2y + 3xy^2$

10. $7st - 12s + 3t - 15st + s$

In Exercises 11–16, write each expression in simplest form.

11. $-3(4a + b)$

12. $(7x + y)(-2)$

13. $2p - (5 - 6p) + 4$

14. $2 + 3(4x - 5)$

15. $5(2h - 3k) - 2(h - 2k)$

16. $8 - 3(2x^2 + 3x) - (7x + 2x^2 - 1)$

1a. _____

b. _____

2a. _____

b. _____

3a. _____

b. _____

4a. _____

b. _____

5. _____

6. _____

7. _____

8. _____

9. _____

10. _____

11. _____

12. _____

13. _____

14. _____

15. _____

16. _____

In Exercises 17–22, find the value of each expression when $x = -3$, $y = 5$, and $z = -2$.

17. $4x - y + 2z$

18. $11 - y(x - z)$

19. $3z^2 - 8z + 13$

20. $x^2 + 3xy - y^2$

21. $z - 5[z(x - y) - 2x]$

22. $\dfrac{x^2 - (y - z)^2}{2z - x}$

In Exercises 23–28, use the formula, substituting the given values for the variables.

23. $A = \pi r^2$ Find A when $r = 30$ and $\pi \approx 3.14$

24. $I = Prt$ Find I when $P = 500$, $r = 0.09$, and $t = 1.75$

25. $C = \dfrac{5}{9}(F - 32)$ Find C when $F = 59$

26. $A = P(1 + i)^n$ Find A when $P = 2{,}000$, $n = 4$, and $i = 0.15$

27. $A = \dfrac{1}{2}bh$ Find A when $b = 4\dfrac{1}{2}$ and $h = 1\dfrac{1}{3}$

28. $F = \dfrac{9}{5}C + 32$ Find F when $C = -10$

In Exercises 29 and 30, solve each problem by selecting and using one of the formulas from Section 2.4.

29. If a crowbar is accidentally dropped from the top of a building, how many feet will it have fallen after 3 seconds, assuming that the force of gravity is 32 ft per second^2? (When you substitute into the formula given in Example 3 of Section 2.4 and use 32 ft per second^2 for the force due to gravity, the answer is in feet.)

30. Find the amount of interest earned in three years if $2,500 was invested at 6% simple interest.

17. _____

18. _____

19. _____

20. _____

21. _____

22. _____

23. _____

24. _____

25. _____

26. _____

27. _____

28. _____

29. _____

30. _____

Chapter 2 Diagnostic Test

The purpose of this test is to see how well you understand the basic definitions used in algebra, simplifying algebraic expressions, evaluating algebraic expressions and formulas, and solving word problems by using formulas. We recommend that you work this diagnostic test *before* your instructor tests you on this chapter. Allow yourself about 50 minutes.

Complete solutions for all the problems on this test, together with section references, are given in the answer section at the end of the book. For the problems you do incorrectly, study the sections referred to.

1. Determine whether $5x$ is a factor of or a term of the expression.

    a. $5xy$ b. $5x + y$ c. $3 + 5x$

    d. $(3 + 2y) + 5x$ e. $(3 + 2y)(+5x)$

In Problems 2 and 3, (a) determine the number of terms; (b) write the second term; (c) write the coefficient of the first term.

2. $6x^3 + \dfrac{7x + 1}{3} - 4y$ **3.** $-y^4 + 3x - 2z + 4$

In Problems 4 and 5, remove the grouping symbols.

4. $5 - (x - y)$ **5.** $-2[-4(3c - d) + a] - b$

In Problems 6 and 7 combine like terms.

6. $4x - 3x + 5x$ **7.** $2a - 5b - 7 - 3b + 4 - 5a$

In Problems 8–11, write each expression in simplest form.

8. $5x - 3(y - x)$ **9.** $(x - 4)(-5)$

10. $(x - 4) - 5$ **11.** $3 + 5[3x - (2y - x)]$

In Problems 12–15, find the value of each expression when $a = -4$, $b = -7$, $c = -2$, $x = -6$, and $y = 5$.

12. $3c - by + cx$ **13.** $4x - [a - (3c - b)]$

14. $x^2 + 2xy + y^2$ **15.** $(x + y)^2$

In Problems 16–19, use the formula, substituting the given values for the variables.

16. $C = \dfrac{5}{9}(F - 32)$ Find C when $F = 68$

17. $A = \pi r^2$ Find A when $\pi \approx 3.14$ and $r = 3$

18. $A = P(1 + rt)$ Find A when $P = 500$, $r = 0.10$, and $t = 3.5$

19. $V = C - Crt$ Find V when $C = 800$, $r = 0.06$, and $t = 10$

20. Solve the following problem by using the formula $V = \dfrac{1}{3}\pi r^2 h$, where V is the volume of a right circular cone, r is the radius of the base, h is the altitude, and $\pi \approx 3.14$.

    Find the volume of a right circular cone if its radius is 4 ft and its altitude is 3 ft.

Cumulative Review Exercises: Chapters 1 and 2

In Exercises 1–27, perform the indicated operations. If an operation cannot be performed, give a reason.

1. $-16(-2)$

2. $-13 + (-8)$

3. $-5 - (-11)$

4. $28 \div (-7)$

5. $-15(0)(4)$

6. $(-2)^3$

7. $-27 + (10)$

8. $\sqrt{16}$

9. $\dfrac{-48}{-12}$

10. $0 \div (-6)$

11. $(25)^2$

12. $-15(3)$

13. -2^4

14. Subtract -9 from -13.

15. $\dfrac{-12}{0}$

16. $\sqrt[3]{8}$

17. $\sqrt{64}$

18. 0^4

19. $8 - 17$

20. $-20 - 9$

21. $24 \div 12 \cdot 2$

22. $6 \cdot 2^2$

23. $8 - 6 \cdot 5$

24. $64 \div 16 \div 4$

25. $17 - 9 - 2$

26. $4 + 17 \times 2$

27. $0 \div 5$

28. Using the formula $C = \dfrac{5}{9}(F - 32)$, find C when $F = 21\dfrac{1}{2}$.

In Exercises 29–38, write "True" if the statement is always true. Otherwise, write "False."

29. 1.73 is a rational number.

30. 2/3 is a real number.

31. 1.86235173 . . . (never terminates or repeats) is an irrational number.

32. 0 is an irrational number.

33. 0 is a real number.

34. $-\frac{3}{5}$ is the additive inverse of $\frac{3}{5}$.

35. 0 is the multiplicative identity.

36. $52y$ is a term of $6x + 52y$.

37. $4xy$ is a term of $4xyz$.

38. $6s$ is a factor of $6s + 3t$.

In Exercises 39 and 40, write each expression in simplest form.

39. $5x - (z - 2x)$

40. $8 - 3(x - 4 + y)$

3 Equations and Inequalities

The main reason for studying algebra is to equip oneself with the tools necessary for solving problems. Most mathematical problems are solved by the use of equations. In this chapter, we show how to solve simple equations. Methods for solving more difficult equations will be given in later chapters.

3.1 Conditional Equations, Solutions and Solution Sets, and Equivalent Equations

Recall from Section 2.4 that an equation is a statement that two quantities are equal; it is made up of three parts:

1. The equal sign ($=$)

2. The expression to the left of the equal sign, called the left side

3. The expression to the right of the equal sign, called the right side

Conditional Equations

An equation may be a *true* statement, such as $7 = 7$; a *false* statement, such as $8 = 3$; or a statement that is sometimes true and sometimes false, such as $x = -2$. An equation that is true for some values of the variable and false for other values is called a **conditional equation**. The equation $x = -2$ is a conditional equation, because it is a true statement if the value of x is -2 and a false statement otherwise.

A Solution of an Equation

A **solution** of an equation is any value of that variable that, when substituted for the variable, makes the two sides of the equation equal.

Example 1   Determine whether -2 is a solution of the equation $x = -2$.
Solution Substituting -2 for x, we have $-2 = -2$, which is a *true* statement. Therefore, -2 is a solution of the equation $x = -2$. ∎

Example 2   Determine whether 5 is a solution of the equation $x = -2$.
Solution Substituting 5 for x, we have $5 = -2$, which is a *false* statement. Therefore, 5 is *not* a solution of the equation $x = -2$. ∎

The Solution Set of an Equation

The **solution set** of an equation is the set of *all* the solutions of the equation. (Recall from Section 1.1 that we enclose the elements of a set within braces.) Thus, the solution set of the equation $x = -2$ is $\{-2\}$, because -2 is the only value of x that makes the statement $x = -2$ true.

Equivalent Equations

Equations that have the same solution set are called **equivalent equations**.

Example 3   Examples of equivalent equations:

a. The solution set of the equation $x + 2 = 0$ is $\{-2\}$, because $(-2) + 2 = 0$ is a true statement, and -2 is the only value of x that makes the statement $x + 2 = 0$ true. Since $\{-2\}$ was also the solution set for the equation $x = -2$, $x + 2 = 0$ and $x = -2$ are equivalent equations.

b. The solution set of the equation $x = 4$ is $\{4\}$, because $4 = 4$ is a true statement, and 4 is the only value of x that makes the statement $x = 4$ true. The solution set of the equation $x - 4 = 0$ is $\{4\}$, because $4 - 4 = 0$ is a true statement, and 4 is the only value of x that makes the statement $x - 4 = 0$ true. Therefore, $x = 4$ and $x - 4 = 0$, are equivalent equations. ∎

EXERCISES 3.1

Set I

1. Is -4 a solution of the equation $x + 2 = 3$?

2. Is 3 a solution of the equation $x - 5 = 2$?

3. Is 4 a solution of the equation $x + 1 = 5$?

4. Is -4 a solution of the equation $x + 4 = 0$?

5. Is -3 in the solution set for $2 + x = -1$?

6. Is -5 in the solution set for $2 + x = -3$?

7. Is 3 in the solution set for $5 + x = 1$?

8. Is 4 in the solution set for $-1 + x = 1$?

Set II

1. Is 5 a solution of the equation $x + 3 = 4$?

2. Is -3 a solution of the equation $x + 5 = 2$?

3. Is -4 a solution of the equation $x + 1 = -3$?

4. Is 4 a solution of the equation $x + 3 = 7$?

5. Is 3 in the solution set for $2 + x = 5$?

6. Is 5 in the solution set for $2 - x = -3$?

7. Is -1 in the solution set for $5 + x = 1$?

8. Is 2 in the solution set for $-1 + x = 1$?

3.2 Solving Equations by Using the Addition and Subtraction Properties of Equality

In this section and in Section 3.3, we solve equations that contain only one variable and in which the variable is not raised to any power. Such an equation is *solved* when we have found an equivalent equation that is in the form $x = a$, where a is some number. When the equation is in this form, we say that we have *isolated* the variable; that is, we have the variable by itself on one side of the equal sign and a single number (no variables) on the other side.

We can use a number of properties of equality in writing an equation equivalent to the given equation. We list three of those properties here, and we will list two others in Section 3.3.

THE ADDITION PROPERTY OF EQUALITY

If the same number is added to both sides of an equation, the new equation is equivalent to the original equation.

THE SUBTRACTION PROPERTY OF EQUALITY

If the same number is subtracted from both sides of an equation, the new equation is equivalent to the original equation.

THE SYMMETRIC PROPERTY OF EQUALITY

The equation $a = b$ is equivalent to the equation $b = a$.

The symmetric property permits us to rewrite an equation such as $4 = x$ as $x = 4$.

We call the solution we obtain by using the properties of equality the *apparent* solution because, while it *appears* to be the solution to the equation, we may have made errors in obtaining it. In this section, we solve equations of the form $x + a = b$ by reasoning as follows: Since $x + a$ means a is being *added* to x, and since the sum of a number and its *negative* is 0, we can isolate x (get x by itself) by using the addition property of equality and adding $-a$ to both sides of the equation. (Because subtraction is the inverse operation of addition, we could, equivalently, use the subtraction property of equality and isolate x by *subtracting a* from both sides of the equation.) When we are solving equations, we must *always write the complete new equation under the previous equation*.

Example 1 Solve the equation $2 + x = 10$ and graph the solution on the number line.

Solution Since $2 + x$ means 2 is being added to x, we can isolate x by adding -2 to both sides of the equation (or, equivalently, by subtracting 2 from both sides). Notice that when we set up the problem vertically, we align only like terms in each column.

$$
\begin{array}{rl}
2 + x = & 10 \\
\underline{-2 \qquad} & \underline{-2} \\
0 + x = & 8 \\
x = & 8
\end{array}
$$

- 2 and -2 are like terms; -2 is written directly under 2
- 10 and -2 are like terms; -2 is written directly under 10

Adding -2 to both sides (*or* subtracting 2 from both sides)

This step is usually not shown

$0 + x = x$

8 is the apparent solution
Do not omit the equal sign

Notice that the equation $0 + x = 8$ was written *under* the equation $2 + x = 10$, and the equation $x = 8$ was written *under* the equation $0 + x = 8$.

NOTE   If we made no errors, the solution is 8, and the solution set is $\{8\}$; "$x = 8$" is not the solution—it is an equation that is equivalent to the equation "$2 + x = 10$." ☑

Graph We must graph the number 8.

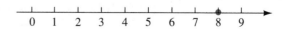

A WORD OF CAUTION   In Example 1, the addition property of equality does permit us to add 10 to both sides; however, this will *not* isolate x.

$$2 + x = 10$$
$$\underline{ 10 \qquad\quad 10}$$
$$12 + x = 20$$

└─ We do not have x by itself on the left side

The addition property of equality also permits us to add -10 to both sides of the equation in Example 1; however, this will *not* isolate x.

$$2 + x = 10$$
$$\underline{-10 \qquad\qquad -10}$$
$$-8 + x = 0$$

└─ We do not have x by itself on the left side

Furthermore, in Example 1, it is incorrect to write

$$2 + x = 10 = x = 8$$

└─ An equal sign here implies that $x = 10$, which is *not* true, and also implies that $10 = 8$, which is false ☑

Example 2   Solve the equation $8 = H - 4$ and graph the solution on the number line.
Solution Since $H - 4$ means -4 is being added to H, we can isolate H by adding 4 to both sides of the equation.

$$8 = H - 4$$
$$\underline{+\,4 \qquad\quad +\,4}$$
$$12 = H$$
$$H = 12$$

Adding $+4$ to both sides gets H by itself on the right side

The symmetric property of equality permits us to rewrite $12 = H$ as $H = 12$

The apparent solution is 12

Graph We graph the number 12.

Checking the Solution of an Equation
We cannot be sure an apparent solution is *actually* a solution until we have checked it.

TO CHECK THE SOLUTION OF AN EQUATION

1. Replace the variable in the given equation by the apparent solution.

2. Perform the indicated operations on both sides of the equal sign.

3. If the resulting number on each side of the equal sign is the same, the solution is correct.

Example 3   Solve and check $x - 5 = 3$ and graph the solution on the number line.
Solution

We isolate x by adding the negative of -5 to both sides

$$
\begin{array}{rcl}
x \;\boxed{- 5} & = & 3 \\
+\,5 & & +\,5 \\
\hline
x & = & 8
\end{array}
$$

     Adding 5 to both sides

     The apparent solution is 8

Check

$$x - 5 = 3$$

The question mark indicates that we don't know yet whether the two sides are equal

$$(8) - 5 \overset{?}{=} 3$$ x replaced by the apparent solution, 8

$$3 = 3$$ The solution is correct

Therefore, the solution *is* 8.
Graph We graph the number 8.

Example 4   Solve and check $\frac{1}{4} + x = 3\frac{1}{4}$ and graph the solution on the number line.
Solution

We isolate x by adding the negative of $\frac{1}{4}$ to both sides

$$
\begin{array}{rcl}
\dfrac{1}{4} + x & = & 3\dfrac{1}{4} \\[2mm]
-\dfrac{1}{4} & & -\dfrac{1}{4} \\
\hline
x & = & 3
\end{array}
$$

   Adding $-\frac{1}{4}$ to both sides (or subtracting $\frac{1}{4}$ from both sides)

Check

$$\frac{1}{4} + x = 3\frac{1}{4}$$

$$\frac{1}{4} + (3) \overset{?}{=} 3\frac{1}{4}$$ x replaced by the apparent solution, 3

$$3\frac{1}{4} = 3\frac{1}{4}$$

Therefore, the solution is 3.
Graph We graph the number 3.

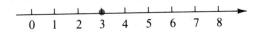

Example 5   Solve and check $9.08 = x - 5.47$ and graph the solution on the number line.
     Solution We isolate x by adding 5.47 to both sides of the equation.

$$9.08 = x - 5.47$$

$$\underline{+\ 5.47 \qquad\quad +\ 5.47}$$ Adding 5.47 to both sides

$$14.55 = x$$

or $$x = 14.55$$ Symmetric property

Check

$$9.08 = x - 5.47$$

$$9.08 \overset{?}{=} 14.55 - 5.47$$ Substituting 14.55 for x

$$9.08 = 9.08$$

Therefore, the solution is 14.55.

Graph

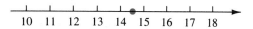

Notice that in all the equations in this section, the coefficient of the variable was always an understood 1 and some number was being added to or subtracted from the variable; we were able to isolate the variable by using the addition property of equality or the subtraction property of equality.

EXERCISES 3.2

Set I   Solve and check the following equations and graph each solution on the number line.

1. $x + 5 = 8$ **2.** $x + 4 = 9$ **3.** $x - 3 = 4$

4. $x - 7 = 2$ **5.** $3 + x = -4$ **6.** $2 + x = -5$

7. $x + 4 = 21$ **8.** $x + 15 = 24$ **9.** $x - 35 = 7$

10. $x - 42 = 9$ **11.** $9 = x + 5$ **12.** $11 = x + 8$

13. $12 = x - 11$ **14.** $14 = x - 15$ **15.** $-17 + x = 28$

16. $-14 + x = 33$ **17.** $-28 = -15 + x$ **18.** $-47 = -18 + x$

19. $x + \dfrac{1}{2} = 2\dfrac{1}{2}$ **20.** $x + \dfrac{3}{4} = 5\dfrac{3}{4}$ **21.** $5.6 + x = 2.8$

22. $3.04 + x = 2.96$ **23.** $7.84 = x - 3.98$ **24.** $4.99 = x - 2.08$

Set II   Solve and check the following equations and graph each solution on the number line.

1. $x + 7 = 12$ **2.** $x - 4 = -3$ **3.** $x - 5 = 8$

4. $3 + x = 1$ **5.** $6 + x = -9$ **6.** $-2 + x = 5$

7. $x + 11 = 25$ **8.** $-5 + x = 0$ **9.** $x - 18 = 13$

10. $5 + x = 5$ **11.** $14 = x + 6$ **12.** $3 = x - 1$

13. $17 = x - 11$ **14.** $x = 5 - 8$ **15.** $-21 + x = -41$

16. $-3 + x = 4$ **17.** $-51 = -37 + x$ **18.** $-3 + x = -4$

19. $x + 2\dfrac{1}{4} = 8\dfrac{1}{4}$ **20.** $x - 3\dfrac{1}{2} = 5\dfrac{1}{2}$ **21.** $8.4 + x = 6.2$

22. $3.5 + x = 1.07$ **23.** $5.36 = x - 4.82$ **24.** $x - 1.35 = -4.2$

3.3 Solving Equations by Using the Multiplication and Division Properties of Equality

3.3A Solving Equations of the Form $ax = b$ or $\dfrac{ax}{b} = c$

In this section, the equations still have only one term that contains the variable, but the numerical coefficient of the variable is no longer a 1. We solve such equations by using one of the following two properties of equality:

THE MULTIPLICATION PROPERTY OF EQUALITY

If both sides of an equation are multiplied by the same nonzero number,* the new equation is equivalent to the original equation.

THE DIVISION PROPERTY OF EQUALITY

If both sides of an equation are divided by the same nonzero number,† the new equation is equivalent to the original equation.

We can solve equations of the form $ax = b$ ($a \neq 0$) by reasoning as follows: Since ax means that a is being *multiplied* by x, since division is the inverse operation of multiplication, and since any number divided by itself is 1, we can isolate x by using the division property of equality and dividing both sides of the equation by a (see Example 1, Method 1). Because the product of a number and its reciprocal (its multiplicative inverse) is also 1, we can, instead, use the multiplication property of equality to isolate the variable (see Example 1, Method 2).

Example 1   Solve and check the equation $2x = 10$ and graph the solution on the number line.
Method 1 Using the division property of equality: We isolate x by dividing both sides of the equation by the coefficient of x.

We isolate x by dividing both sides by 2

$$2\,x = 10$$

$$\frac{\overset{1}{\cancel{2}}x}{\underset{1}{\cancel{2}}} = \frac{10}{2} \qquad \tfrac{2}{2} = 1, \text{ and } 1 \cdot x = x$$

$$x = 5$$

---

*If we multiply both sides of an equation by zero, we always get the equation 0 = 0, which is usually *not* equivalent to the original equation.

†We cannot divide both sides of the equation by zero, since division by zero is not permitted.

Method 2 Using the multiplication property of equality: We isolate x by multiplying both sides of the equation by the *reciprocal* of the coefficient of x $\left(\text{the reciprocal of 2 is } \tfrac{1}{2}\right).$

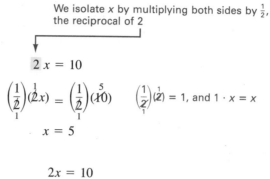

We isolate x by multiplying both sides by $\tfrac{1}{2}$, the reciprocal of 2

$$2\,x = 10$$

$$\left(\tfrac{1}{2}\right)(2x) = \left(\tfrac{1}{2}\right)(10) \qquad \left(\tfrac{1}{2}\right)(2) = 1, \text{ and } 1 \cdot x = x$$

$$x = 5$$

Check

$$2x = 10$$

$$2(5) \overset{?}{=} 10 \qquad \text{We substitute 5 for } x$$

$$10 = 10$$

Therefore, the solution is 5.
Graph

A WORD OF CAUTION   Notice the difference between the equations (a) $2x = 10$ and (b) $2 + x = 10$. In (a), 2 and x are *factors*; that is, 2 is multiplied by x. We isolate x by dividing both sides of the equation by 2 $\left(\text{or by multiplying both sides of the equation by } \tfrac{1}{2}\right).$ In (b), 2 and x are *terms*; that is, 2 is added to x. We isolate x by adding -2 to both sides of the equation (or by subtracting 2 from both sides of the equation). ☑

Example 2   Solve and check the equation $9x = -27$ and graph the solution on the number line.
Solution

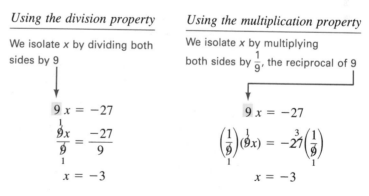

Using the division property	*Using the multiplication property*
We isolate x by dividing both sides by 9	We isolate x by multiplying both sides by $\tfrac{1}{9}$, the reciprocal of 9

$$\begin{array}{cc} 9\,x = -27 & 9\,x = -27 \\[4pt] \dfrac{9x}{9} = \dfrac{-27}{9} & \left(\tfrac{1}{9}\right)(9x) = -27\left(\tfrac{1}{9}\right) \\[4pt] x = -3 & x = -3 \end{array}$$

Check

$$9x = -27$$

$$9(-3) \overset{?}{=} -27$$

$$-27 = -27$$

Therefore, the solution is -3.
Graph

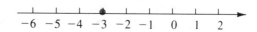

Example 3   Solve and check the equation $-12x = 8$ and graph the solution on the number line.
Solution

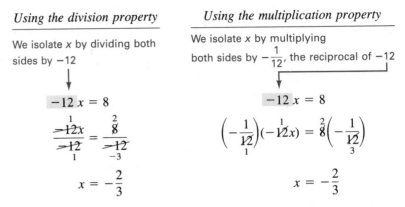

Using the division property	*Using the multiplication property*

We isolate x by dividing both sides by -12

We isolate x by multiplying both sides by $-\frac{1}{12}$, the reciprocal of -12

$$-12\,x = 8$$

$$\frac{-12x}{-12} = \frac{8}{-12}$$

$$x = -\frac{2}{3}$$

$$-12\,x = 8$$

$$\left(-\frac{1}{12}\right)(-12x) = 8\left(-\frac{1}{12}\right)$$

$$x = -\frac{2}{3}$$

Check

$$-12x = 8$$

$$-12\left(-\frac{2}{3}\right) \stackrel{?}{=} 8$$

$$8 = 8$$

Graph We graph the solution, $-\frac{2}{3}$.

We can solve equations of the form $\dfrac{ax}{b} = c$ $(a \neq 0,\ b \neq 0)$ by reasoning as follows:

Because $\dfrac{ax}{b}$ means that x is being multiplied by $\dfrac{a}{b}$ and because the product of any number and its reciprocal is 1, we can isolate the variable by using the multiplication property of equality and multiplying both sides of the equation by the *reciprocal* of $\dfrac{a}{b}$, which is $\dfrac{b}{a}$ (see Example 4, Method 1). (Because division is the inverse operation of multiplication, it is also possible to divide both sides of the equation by $\dfrac{a}{b}$; we don't demonstrate that method in this section.)

Such problems could also be solved in two separate steps. We can "clear fractions" by multiplying both sides of the equation by the denominator, b. Then we can use the methods presented earlier in this section to finish solving the equation (see Example 4, Method 2). We recommend that you become comfortable with both methods.

In Examples 4–6, we solve some fairly simple equations that contain fractions.

Example 4   Solve and check $\dfrac{2x}{3} = -4$ and graph the solution on the number line.

Method 1 The coefficient of x is $\frac{2}{3}$.

$$\frac{2x}{3} = -4$$

$$\left(\frac{3}{2}\right)\left(\frac{2x}{3}\right) = \left(\frac{3}{2}\right)(-4) \qquad \text{We multiply both sides by } \tfrac{3}{2}, \text{ the reciprocal of } \tfrac{2}{3}$$

$$x = -6$$

Method 2

$$\frac{2x}{3} = -4$$

$$\frac{1}{\cancel{3}}\left(\frac{2x}{\cancel{3}}\right) = 3\,(-4)$$ We clear fractions by multiplying both sides by 3

$$2x = -12$$ We can now multiply both sides by $\frac{1}{2}$ or divide both sides by 2

$$\frac{\overset{1}{\cancel{2}}x}{\underset{1}{\cancel{2}}} = \frac{\overset{6}{-\cancel{12}}}{\underset{1}{\cancel{2}}}$$

$$x = -6$$

Check

$$\frac{2x}{3} = -4$$

$$\frac{2(-6)}{3} \overset{?}{=} -4$$

$$-4 = -4$$

Graph We graph the solution, -6.

Example 5 Solve and check $\dfrac{x}{3} = -5$ and graph the solution on the number line.

Solution The coefficient of x is $\frac{1}{3}$.

$$\frac{x}{3} = -5$$

$$\frac{1}{\cancel{3}}\left(\frac{x}{\cancel{3}}\right) = 3\,(-5)$$ We multiply both sides by 3, the reciprocal of $\frac{1}{3}$

$$x = -15$$

Check

$$\frac{x}{3} = -5$$

$$\frac{-15}{3} \overset{?}{=} -5$$

$$-5 = -5$$

Graph We graph the solution, -15.

Example 6   Solve and check $-8 = \dfrac{x}{6}$ and graph the solution on the number line.

Solution The coefficient of x is $\frac{1}{6}$.

$$-8 = \frac{x}{6}$$

$$6\,(-8) = \overset{1}{\cancel{6}} \left(\frac{x}{\underset{1}{\cancel{6}}} \right)$$ We multiply both sides by 6, the reciprocal of $\frac{1}{6}$

$$-48 = x$$ We now use the symmetric property of equality

$$x = -48$$

Check

$$-8 = \frac{x}{6}$$

$$-8 \overset{?}{=} \frac{-48}{6}$$

$$-8 = -8$$

Graph We graph the solution, -48.

$$\begin{array}{ccccccccc} & & & & \bullet & & & & \\ -52 & -51 & -50 & -49 & -48 & -47 & -46 & -45 & -44 \end{array}$$

EXERCISES 3.3A

Set I   Solve and check the following equations. Graph each solution on the number line.

1. $2x = 8$ **2.** $3x = 15$ **3.** $21 = 7x$ **4.** $42 = 6x$

5. $11x = 33$ **6.** $12x = 48$ **7.** $\dfrac{x}{3} = 4$ **8.** $\dfrac{x}{5} = 3$

9. $\dfrac{x}{5} = -2$ **10.** $\dfrac{x}{6} = -4$ **11.** $4 = \dfrac{x}{7}$ **12.** $3 = \dfrac{x}{8}$

13. $-13 = \dfrac{x}{9}$ **14.** $-15 = \dfrac{x}{8}$ **15.** $\dfrac{x}{10} = 3.14$ **16.** $\dfrac{x}{5} = 7.8$

Set II   Solve and check the following equations. Graph each solution on the number line.

1. $5x = 35$ **2.** $8x = 32$ **3.** $24 = 6x$ **4.** $56 = 7x$

5. $12x = 36$ **6.** $8 = 24x$ **7.** $\dfrac{x}{7} = 3$ **8.** $\dfrac{x}{2} = 13$

9. $\dfrac{x}{4} = -5$ **10.** $-5 = \dfrac{x}{6}$ **11.** $6 = \dfrac{x}{3}$ **12.** $\dfrac{x}{2} = -4$

13. $-12 = \dfrac{x}{6}$ **14.** $16 = \dfrac{x}{3}$ **15.** $\dfrac{x}{8} = 12.5$ **16.** $\dfrac{x}{7} = -1$

3.3B Solving Equations of the Form $ax + b = c$

We now combine the methods we learned in Section 3.3A with the methods we learned in Section 3.2.

TO SOLVE AN EQUATION OF THE FORM $ax + b = c$

All numbers on the same side as the variable must be removed.

1. First, remove those numbers being added to or subtracted from the term containing the variable by using the addition (or subtraction) property of equality.

2. Then either

   a. multiply both sides of the equation by the reciprocal of the coefficient of x;

   or

   b. multiply both sides of the equation by the denominator (if there is one), and then divide both sides of the equation by the coefficient of the variable.

Notice that after we have used step 1, we will use *either* step 2a *or* step 2b.

Example 7   Solve and check the equation $2x + 3 = 11$ and graph the solution on the number line.
Solution The numbers 2 and 3 must be removed from the side with the x.

Step 1. Since the 3 is added to the term containing the variable, it is removed first.

$$\begin{array}{rcl} 2x + 3 &=& 11 \\ -\ 3 && -\ 3 \\ \hline 2x &=& 8 \end{array}$$ Adding -3 to both sides

Step 2b. Since 2 is the coefficient of x, it is removed by dividing both sides by 2.

$$2x = 8$$

$$\frac{\overset{1}{\cancel{2}}x}{\underset{1}{\cancel{2}}} = \frac{8}{2}$$ Dividing both sides by 2

$$x = 4$$

Check

$$2x + 3 = 11$$
$$2(4) + 3 \overset{?}{=} 11$$
$$8 + 3 \overset{?}{=} 11$$
$$11 = 11$$

Graph We graph the solution, 4

Example 8   Solve the equation $3x - 2 = 10$ and graph the solution on the number line.
Solution

Step 1. $\begin{aligned} 3x - 2 &= 10 \\ + 2 \quad &\ \ + 2 \\ \hline 3x \quad &= \quad 12 \end{aligned}$ Adding 2 to both sides

Step 2b. $\dfrac{\overset{1}{\cancel{3}}x}{\underset{1}{\cancel{3}}} = \dfrac{12}{3}$ Dividing both sides by 3

$x = 4$

You should verify that the solution, 4, checks.
Graph

Example 9   Solve and check the equation $-12 = 3x + 15$ and graph its solution on the number line.
Solution 15 and 3 must be removed from the side with the x.

Step 1. $\begin{aligned} -12 &= 3x + 15 \\ -15 \quad &\quad\ \ - 15 \\ \hline -27 &= 3x \end{aligned}$ Adding −15 to both sides

Step 2b. $\dfrac{-27}{3} = \dfrac{\overset{1}{\cancel{3}}x}{\underset{1}{\cancel{3}}}$ Dividing both sides by 3

$-9 = x$ Now we use the symmetric property of equality

$x = -9$

Check

$$-12 = 3x + 15$$
$$-12 \overset{?}{=} 3(-9) + 15$$
$$-12 \overset{?}{=} -27 + 15$$
$$-12 = -12$$

Graph We graph the solution, -9.

Example 10   Solve and check the equation $4 = \dfrac{2x}{5} - 6$ and graph the solution on the number line.

Solution In this example, three numbers (-6, 5, and 2) must be removed from the right side of the equation.

Step 1. Since the -6 is *added*, remove it first.

$$4 = \frac{2x}{5} - 6$$
$$\underline{+6 \qquad\qquad + 6}$$ Adding +6 to both sides
$$10 = \frac{2x}{5}$$

Step 2a. Since the coefficient of x is $\frac{2}{5}$, we can finish solving the equation in one step by multiplying both sides of the equation by $\frac{5}{2}$, the reciprocal of $\frac{2}{5}$.

$$\left(\frac{\overset{}{\cancel{5}}}{\cancel{2}}\right)(\cancel{10}) = \left(\frac{\cancel{5}}{\cancel{2}}\right)\left(\frac{\cancel{2}}{\cancel{5}}\right)x \qquad \text{We use the multiplication property of equality}$$

$$25 = x \qquad\qquad\qquad \text{Now we use the symmetric property of equality}$$

$$x = 25$$

Instead of using Step 2a, we could finish solving the equation $10 = \frac{2x}{5}$ in *two* steps by using Step 2b.

Step 2b. $5(10) = \overset{1}{\cancel{5}}\left(\dfrac{2x}{\underset{1}{\cancel{5}}}\right)$ Multiplying both sides by the denominator, 5

$$50 = 2x$$

$$\frac{50}{2} = \frac{\overset{1}{\cancel{2}}x}{\underset{1}{\cancel{2}}} \qquad \text{Dividing both sides by 2, the coefficient of } x$$

$$25 = x \qquad \text{Now we use the symmetric property of equality}$$

$$x = 25$$

Check

$$4 = \frac{2x}{5} - 6$$

$$4 \overset{?}{=} \frac{2(\overset{5}{\cancel{25}})}{\underset{1}{\cancel{5}}} - 6$$

$$4 \overset{?}{=} 10 - 6$$

$$4 = 4$$

Graph We graph the solution, 25.

      We feel that the two-step method is a little easier for most students. Therefore, we will show only that method for Example 11.

Example 11   Solve and check $\dfrac{3x}{4} + 2 = 11$ and graph the solution on the number line.

Solution The numbers 2, 4, and 3 must be removed from the left side of the equation.

Step 1. Since the 2 is added, remove it first.

$$\frac{3x}{4} + 2 = 11$$

$$\underline{\quad -2 \quad -2\quad} \qquad \text{Adding } -2 \text{ to both sides}$$

$$\frac{3x}{4} \quad = \quad 9$$

Step 2b. $\overset{1}{\cancel{4}}\left(\dfrac{3x}{\cancel{4}}\right) = 4(9)$ Multiplying both sides by the denominator, 4

$$3x = 36$$

$$\dfrac{\overset{1}{\cancel{3}}x}{\underset{1}{\cancel{3}}} = \dfrac{36}{3}$$ Dividing both sides by 3, the coefficient of x

$$x = 12$$

Check

$$\dfrac{3x}{4} + 2 = 11$$

$$\dfrac{3(\overset{3}{\cancel{12}})}{\underset{1}{\cancel{4}}} + 2 \overset{?}{=} 11$$

$$9 + 2 \overset{?}{=} 11$$

$$11 = 11$$

Graph We graph the solution, 12.

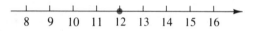

Changing Signs in an Equation   Multiplying both sides of an equation by -1 is equivalent to changing the sign of *every term* in the equation. For example, consider the equation $5 - x = 7$:

Changing the sign of every term	*Multiplying both sides by* -1
$5 - x = 7$	$5 - x = 7$
$-5 + x = -7$	$(-1)(5 - x) = (-1)7$
	$-5 + x = -7$

Because multiplying both sides of an equation by -1 doesn't change its solution (we're simply using one of the properties of equality), changing the sign of every term of an equation will not change its solution.

If we're solving an equation in which the coefficient of x is negative, we can solve the equation by methods already described (see Examples 12 and 13, Method 1). Since we're always solving for a *positive x*, however, some students find it easier to first make the coefficient of x positive by changing the sign of every term of the equation (see Examples 12 and 13, Method 2). Both methods are acceptable.

Example 12   Solve $5 - x = 7$.

Solution

Method 1 $5 - x = 7$

$\underline{-5 -5}$ Adding -5 to both sides

$-x = 2$ This equation has not yet been solved for x; the coefficient of x is -1, not 1

$(-1)(-x) = (-1)(2)$ Multiplying both sides of the equation by -1, the reciprocal of -1

$x = -2$

Method 2 $5 - x = 7$

$-5 + x = -7$	Changing the sign of *every* term
$+5 \qquad +5$	Adding 5 to both sides
$x = -2$	Same equation as for Method 1

You should verify that the solution is -2. ∎

Example 13   Solve and check $3 - 2x = 9$ and graph the solution on the number line.
Solution
Method 1

Step 1. $3 - 2x = 9$
$$\underline{-3 \qquad\qquad -3} \qquad \text{Adding } -3 \text{ to both sides}$$
$$- 2x = 6$$

Step 2b. $\dfrac{\overset{1}{(\cancel{-2})}x}{\underset{1}{(\cancel{-2})}} = \dfrac{6}{-2}$ Since the coefficient of x is -2, we divide both sides by -2

$$x = -3$$

Method 2 $3 - 2x = 9$

Step 1. $-3 + 2x = -9$ Changing the sign of every term; now the coefficient of x is positive
$$\underline{\quad 3 \qquad\qquad 3 \quad} \qquad \text{Adding 3 to both sides}$$
$$2x = -6$$

Step 2b. $\dfrac{\overset{1}{\cancel{2}x}}{\underset{1}{\cancel{2}}} = \dfrac{\overset{3}{-\cancel{6}}}{\underset{1}{\cancel{2}}}$ Dividing both sides by 2

$$x = -3$$

Check

$$3 - 2x = 9$$
$$3 - 2(-3) \overset{?}{=} 9$$
$$3 + 6 \overset{?}{=} 9$$
$$9 = 9$$

Graph We graph the solution, -3.

∎

We will look at an alternate way to solve equations containing fractions in Section 8.9.

EXERCISES 3.3B

Set I   Solve and check the following equations. Graph each solution on the number line.

1. $4x + 1 = 9$ **2.** $5x + 2 = 12$ **3.** $6x - 2 = 10$

4. $7x - 3 = 4$ **5.** $2x - 15 = 11$ **6.** $3x - 4 = 14$

7. $4x + 2 = -14$ **8.** $5x + 5 = -10$ **9.** $14 = 9x - 13$

10. $25 = 8x - 15$ **11.** $12x + 17 = 65$ **12.** $11x + 19 = 41$

13. $8x - 23 = 31$ **14.** $6x - 33 = 29$ **15.** $14 - 4x = -28$

16. $18 - 6x = -44$ **17.** $8 = 25 - 3x$ **18.** $10 = 27 - 2x$

19. $-73 = 24x + 31$ **20.** $-48 = 36x + 42$

21. $18x - 4.8 = 6$ **22.** $15x - 7.5 = 8$

23. $2.5x - 3.8 = -7.9$ **24.** $3.75x + 0.125 = -0.125$

25. $\dfrac{x}{4} + 6 = 9$ **26.** $\dfrac{x}{5} + 3 = 8$

27. $\dfrac{x}{10} - 5 = 13$ **28.** $\dfrac{x}{20} - 4 = 12$ **29.** $-14 = \dfrac{x}{6} - 7$

30. $-22 = \dfrac{x}{8} - 11$ **31.** $7 = \dfrac{2x}{5} + 3$ **32.** $9 = \dfrac{3x}{4} + 6$

33. $4 - \dfrac{7x}{5} = 11$ **34.** $3 - \dfrac{2x}{9} = 13$ **35.** $-24 + \dfrac{5x}{8} = 41$

36. $-16 + \dfrac{9x}{4} = 29$ **37.** $41 = 25 - \dfrac{4x}{5}$ **38.** $54 = 14 - \dfrac{8x}{7}$

Set II   Solve and check the following equations. Graph each solution on the number line.

1. $3x + 2 = 14$ **2.** $5x - 3 = -13$ **3.** $7x - 4 = 10$

4. $8x + 4 = -20$ **5.** $8x - 12 = 12$ **6.** $6x - 18 = -18$

7. $4x + 11 = -13$ **8.** $6x + 13 = -5$ **9.** $16 = 5x - 14$

10. $14 = 7x - 28$ **11.** $11x + 19 = 63$ **12.** $8x - 4 = 24$

13. $12x - 17 = 13$ **14.** $11x + 5 = 16$ **15.** $19 - 10x = -26$

16. $8 - 3x = -1$ **17.** $27 = 32 - 15x$ **18.** $17 = 14 - 3x$

19. $-67 = 18x + 29$ **20.** $-12 = 5x + 3$

21. $5x - 3.6 = 5$ **22.** $4x - 4.26 = 8$

23. $4.1x - 7.4 = -11.5$ **24.** $0.3x + 5.09 = -8.2$

25. $\dfrac{x}{3} + 5 = 7$ **26.** $\dfrac{x}{2} - 4 = -1$

27. $\dfrac{x}{10} - 4 = 11$ **28.** $\dfrac{x}{9} + 3 = 2$ **29.** $-16 = \dfrac{x}{7} - 9$

30. $5 = 3 + \dfrac{x}{2}$ **31.** $8 = \dfrac{3x}{5} + 4$ **32.** $\dfrac{2x}{4} - 1 = -7$

33. $6 - \dfrac{4x}{5} = 12$ **34.** $8 - \dfrac{2x}{3} = 4$ **35.** $-19 + \dfrac{7x}{4} = 23$

36. $-5 - \dfrac{3x}{4} = 4$ **37.** $38 = 14 - \dfrac{6x}{11}$ **38.** $5 = 2 - \dfrac{7x}{3}$

3.4 Solving Equations in Which Simplification of Algebraic Expressions Is Necessary

3.4A Equations in Which the Variable Appears on Both Sides

All the equations discussed in Chapter 3 so far have had the variable on only one side of the equation.

**TO SOLVE AN EQUATION IN WHICH
THE VARIABLE APPEARS ON BOTH SIDES**

1. First combine like terms (if there are any) on each side of the equation.

2. Remove the term containing the variable from one side of the equation by adding the negative of that term to both sides.

3. Solve the resulting equation by the methods given in Sections 3.2 and 3.3.

Example 1 Solve and check the equation $6x - 15 = -23 + 2x$ and graph the solution on the number line.

Solution We first remove the entire term $2x$ from the right side of the equation.

$$
\begin{array}{rl}
6x - 15 = -23 + 2x & \\
\underline{-2x \qquad\qquad\quad - 2x} & \text{Adding } -2x \text{ to both sides} \\
4x - 15 = -23 & \text{We solve this equation by the method} \\
 & \text{given in Section 3.3B} \\
\underline{+ 15 \quad +15} & \text{Adding 15 to both sides} \\
4x \quad\; = -8 & \\
\dfrac{\overset{1}{4}x}{\underset{1}{4}} = \dfrac{-8}{4} & \text{Dividing both sides by 4} \\
x = -2 &
\end{array}
$$

Check

$$
6x - 15 = -23 + 2x
$$
$$
6(-2) - 15 \overset{?}{=} -23 + 2(-2)
$$
$$
-12 - 15 \overset{?}{=} -23 - 4
$$
$$
-27 = -27
$$

The same answer is obtained whether the x-term is removed from the left side or the right side. We now solve the same problem by removing the x-term from the *left* side.

Alternate solution

$$
\begin{array}{rl}
6x - 15 = -23 + 2x & \\
\underline{-6x \qquad\qquad\quad - 6x} & \text{Adding } -6x \text{ to both sides} \\
-15 = -23 - 4x & \\
\underline{+23 \quad +23} & \text{Adding 23 to both sides} \\
8 = \quad\; - 4x &
\end{array}
$$

$$\frac{8}{-4} = \frac{\overset{1}{\cancel{-4}}x}{\cancel{-4}_1}$$ Dividing both sides by -4

$$-2 = x$$ This equation is equivalent to $x = -2$.

Graph We graph the solution, -2.

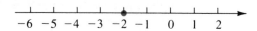

Example 2   Solve and check the equation $4x - 5 - x = 13 - 2x - 3$ and graph the solution on the number line.
Solution

$$4x - 5 - x = 13 - 2x - 3$$

$$
\begin{array}{r}
3x - 5 = 10 - 2x \\
\underline{+2x \qquad\qquad + 2x} \\
5x - 5 = 10 \\
\underline{+ 5 \quad\ \ 5} \\
5x \quad\ \ = 15
\end{array}
$$

Combining like terms on each side

Adding $2x$ to both sides so an x-term remains on only one side

Adding 5 to both sides

$$\frac{\overset{1}{\cancel{5}}x}{\underset{1}{\cancel{5}}} = \frac{15}{5}$$ Dividing both sides by 5

$$x = 3$$

Check

$$4x - 5 - x = 13 - 2x - 3$$
$$4(3) - 5 - (3) \overset{?}{=} 13 - 2(3) - 3$$
$$12 - 5 - 3 \overset{?}{=} 13 - 6 - 3$$
$$4 = 4$$

Graph We graph the solution, 3.

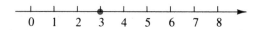

When you're checking the solution of an equation that has a solution in rounded-off decimal form, the two sides usually will not be exactly equal, but they should be close (see Example 3).

Example 3   Using a calculator, solve and check $8.23 - 4.61x = 3.65$. Round off the answer to three decimal places. Do not graph the solution.
Solution Let's begin by changing the sign of every term of the equation. This gives us

Step 1.
$$
\begin{array}{r}
-8.23 + 4.61x = -3.65 \\
\underline{+8.23 \qquad\qquad\quad +8.23} \\
4.61x = +4.58
\end{array}
$$
Adding 8.23 to both sides

Step 2b.
$$\frac{\overset{1}{\cancel{4.61}}x}{\underset{1}{\cancel{4.61}}} = \frac{4.58}{4.61}$$
Dividing both sides by 4.61

$$x \approx 0.9934924$$ Using a calculator

$$x \approx 0.993$$ Rounded off

Check

$$8.23 - 4.61x = 3.65$$

$$8.23 - 4.61(0.993) \overset{?}{=} 3.65$$

$$8.23 - 4.57773 \overset{?}{=} 3.65 \qquad \text{Using a calculator}$$

$$3.65227 \approx 3.65$$

The two sides are not exactly equal to each other, but they are close. The solution is approximately 0.993. ∎

EXERCISES 3.4A

Set I Solve and check each of the following equations and graph each solution for Exercises 1–22 on the number line. In Exercises 23 and 24, round off each answer to three decimal places.

1. $3x + 11 = 14x$ **2.** $5x + 14 = 19x$ **3.** $9x - 7 = 2x$

4. $16x - 3 = 13x$ **5.** $2x - 7 = x$ **6.** $5x - 8 = x$

7. $5x = 3x - 4$ **8.** $7x = 4x - 9$ **9.** $9 - 2x = x$

10. $8 - 5x = 3x$ **11.** $3x - 4 = 2x + 5$ **12.** $5x - 6 = 3x + 6$

13. $6x + 7 = 3 + 8x$ **14.** $4x + 28 = 7 + x$ **15.** $7x - 8 = 8 - 9x$

16. $5x - 7 = 7 - 9x$ **17.** $3x - 7 - x = 15 - 2x - 6$

18. $5x - 2 - x = 4 - 3x - 27$ **19.** $8x - 13 + 3x = 12 + 5x - 7$

20. $9x - 16 + 6x = 11 + 4x - 5$ **21.** $7 - 9x - 12 = 3x + 5 - 8x$

22. $13 - 11x - 17 = 5x + 4 - 10x$

 23. $7.84 - 1.15x = 2.45$

 24. $6.09 - 3.75x = 5.45x$

Set II Solve and check each of the following equations and graph each solution for Exercises 1–22 on the number line. In Exercises 23 and 24, round off each answer to three decimal places.

1. $4x + 15 = 19x$ **2.** $8x - 3 = 5x$ **3.** $9x - 42 = 7x$

4. $8 + x = 17x$ **5.** $3x - 10 = x$ **6.** $4x + 7 = x$

7. $6x = 2x - 8$ **8.** $9x = 3x + 18$ **9.** $12 - 5x = x$

10. $21 - 6x = x$ **11.** $5x - 7 = 4x + 6$ **12.** $7x + 2 = 8x - 2$

13. $8x + 5 = 14 + 11x$ **14.** $6x - 2 = 4x - 2$ **15.** $9x - 13 = 13 - 4x$

16. $-4x + 3 = 3x + 3$ **17.** $6x - 2 - x = 21 - 3x - 7$

18. $7x + 3 - 2x = 5 - 3x$ **19.** $4x + 14 + 2x = 12 - 3x - 8$

20. $8x - 3 - 5x = 5 - 2x - 9$ **21.** $16 - 7x - 4 = 5x + 6 - 4x$

22. $8 + 4x - 3 = 2x + 5 - 7x$

 23. $8.42 - 2.35x = 1.25x$

 24. $2.67x + 3.4 = -5.33x$

3.4B Equations Containing Grouping Symbols

When grouping symbols appear in an equation, first remove them and then solve the resulting equation by the methods discussed in the previous sections.

The complete procedure for solving an equation in one variable that has no exponents is as follows:

TO SOLVE AN EQUATION IN ONE VARIABLE

1. Remove grouping symbols.

2. Combine like terms on each side of the equation.

3. If the variable appears on both sides of the equation, remove the term that contains the variable from one side by adding the negative of that term to both sides.

4. Remove all numbers that appear on the same side as the variable.

First, remove those numbers being added to or subtracted from the term containing the variable.

Next, complete the solution by using the one-step method (multiplying both sides by the reciprocal of the coefficient of the variable)

or complete the solution by using the two-step method (clearing fractions and then dividing both sides by the coefficient of the variable).

5. Check the solution in the original equation.

Example 4 Solve and check the equation $10x - 2(3 + 4x) = 7 - (x - 2)$ and graph the solution on the number line.

Solution

$$
\begin{aligned}
10x - 2(3 + 4x) &= 7 - (x - 2) \\
10x - 6 - 8x &= 7 - x + 2 \qquad &&\text{Removing grouping symbols} \\
2x - 6 &= 9 - x \qquad &&\text{Combining like terms on each side} \\
\underline{+x} \qquad &\quad \underline{+ x} \qquad &&\text{Getting the } x\text{-term on only one side} \\
3x - 6 &= 9 \\
\underline{+ 6} \quad &\quad \underline{+6} \qquad &&\text{Adding 6 to both sides} \\
3x \quad &= 15 \\
\frac{\overset{1}{\cancel{3}}x}{\underset{1}{\cancel{3}}} &= \frac{15}{3} \qquad &&\text{Dividing both sides by 3} \\
x &= 5
\end{aligned}
$$

Check

$$
\begin{aligned}
10x - 2(3 + 4x) &= 7 - (x - 2) \\
10(5) - 2(3 + 4 \cdot 5) &\overset{?}{=} 7 - (5 - 2) \\
10(5) - 2(3 + 20) &\overset{?}{=} 7 - (3) \\
10(5) - 2(23) &\overset{?}{=} 7 - 3 \\
50 - 46 &\overset{?}{=} 4 \\
4 &= 4
\end{aligned}
$$

Graph We graph the solution, 5.

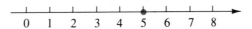

Example 5 Solve $7y - 3(2y - 5) = 6(2 + 3y) - 31$. (The check is left to the student.)
Solution

$$7y - 3(2y - 5) = 6(2 + 3y) - 31$$

$7y - 6y + 15 = 12 + 18y - 31$	Removing grouping symbols
$y + 15 = 18y - 19$	Combining like terms on each side
$\underline{-y \qquad\qquad -y}$	Getting the y-term on only one side
$15 = 17y - 19$	
$\underline{+19 \qquad\quad +19}$	Adding 19 to both sides
$34 = 17y$	

$$\frac{34}{17} = \frac{\overset{1}{\cancel{17}}y}{\underset{1}{\cancel{17}}} \qquad \text{Dividing both sides by 17}$$

$$2 = y \qquad \text{Reducing } 34/17$$

$$y = 2 \qquad \text{The symmetric property of equality}$$

A check confirms that the solution is 2. ■

When the variable appears on both sides of the equation, it is possible to get the variable on one side of the equal sign and the constant on the other side in one step rather than in two steps. This is shown in Example 6, where we add $+15x - 18$ to both sides in one step.

Example 6 Solve $5(2 - 3x) - 4 = 5x + [-(2x - 10) + 8]$. (The check is left to the student.)
Solution

$$5(2 - 3x) - 4 = 5x + [-(2x - 10) + 8]$$
$$10 - 15x - 4 = 5x + [-2x + 10 + 8]$$
$$10 - 15x - 4 = 5x + [-2x + 18] \qquad \text{Removing grouping symbols}$$
$$10 - 15x - 4 = 5x - 2x + 18$$

$-15x + 6 = 3x + 18$	Combining like terms on both sides
$\underline{+15x - 18 \quad +15x - 18}$	Adding $+15x - 18$ to both sides
$-12 = 18x$	

$$\frac{-12}{18} = \frac{\overset{1}{\cancel{18}}x}{\underset{1}{\cancel{18}}} \qquad \text{Dividing both sides by 18}$$

$$-\frac{2}{3} = x \qquad \text{Reducing } -12/18$$

$$x = -\frac{2}{3} \qquad \text{The symmetric property of equality}$$

A check confirms that the solution is $-\frac{2}{3}$. ■

EXERCISES 3.4B

Set I Solve and check the following equations and graph each solution on the number line.

1. $5x - 3(2 + 3x) = 6$

2. $7x - 2(5 + 4x) = 8$

3. $6x + 2(3 - 8x) = -14$

4. $4x + 5(4 - 5x) = -22$

5. $7x + 5 = 3(3x + 5)$

6. $8x + 6 = 2(7x + 9)$

7. $9 - 4x = 5(9 - 8x)$

8. $10 - 7x = 4(11 - 6x)$

9. $3y - 2(2y - 7) = 2(3 + y) - 4$

10. $4z - 3(5z - 14) = 5(7 + z) - 9$

11. $6(3 - 4x) + 12 = 10x - 2(5 - 3x)$

12. $7(2 - 5x) + 27 = 18x - 3(8 - 4x)$

13. $2(3x - 6) - 3(5x + 4) = 5(7x - 8)$

14. $4(7z - 9) - 7(4z + 3) = 6(9z - 10)$

15. $6(5 - 4h) = 3(4h - 2) - 7(6 + 8h)$

16. $5(3 - 2k) = 8(3k - 4) - 4(1 + 7k)$

17. $2[3 - 5(x - 4)] = 10 - 5x$

18. $3[2 - 4(x - 7)] = 26 - 8x$

19. $3[2h - 6] = 2\{2(3 - h) - 5\}$

20. $6(3h - 5) = 3\{4(1 - h) - 7\}$

21. $5(3 - 2x) - 10 = 4x + [-(2x - 5) + 15]$

22. $4(2 - 6x) - 6 = 8x + [-(3x - 11) + 20]$

23. $9 - 3(2x - 7) - 9x = 5x - 2[6x - (4 - x) - 20]$

24. $14 - 2(7 - 4x) - 4x = 8x - 3[2x - (5 - x) - 30]$

25. $-2\{5 - [6 - 3(4 - x)] - 2x\} = 13 - [-(2x - 1)]$

26. $-3\{10 - [7 - 5(4 - x) - 8]\} = 11 - [-(5x - 4)]$

In Exercises 27–30, round off answers to three decimal places.

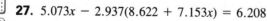

27. $5.073x - 2.937(8.622 + 7.153x) = 6.208$

28. $21.35 - 27.06x = 34.19(19.22 - 37.81x)$

29. $8.23x - 4.07(6.75x - 5.59) = 3.84(9.18 - x) - 2.67$

30. $11.28(15.93x - 24.66) - 35.42(29.05 - 41.84x) = 22.41(32.56x - 16.29)$

Set II Solve and check the following equations and graph each solution on the number line.

1. $4x - 5(3 + 2x) = 3$

2. $5x + 3(2 - x) = 8$

3. $8x + 3(4 - 5x) = -16$

4. $9x - 4(x + 3) = 3$

5. $9x + 12 = 2(4x + 5)$

6. $8x + 6 = 5(2x - 4)$

7. $10 - 6x = 4(8 - 7x)$

8. $7 - 3x = 5(7 - 2x)$

9. $2y - 3(5y - 8) = 2(5 + y) - 10$

10. $3x - 5(2x - 3) = 4(2 - x) + 7$

11. $5(6 - 3z) + 18 = -9z - 3(4 - 2z)$

12. $3(3 - 2x) + 5 = 9x - 4(3x - 1)$

13. $3(2z - 6) - 2(6z + 4) = 5(z + 8)$

14. $2(5 - 3x) + 6 = 8x - 5(4 - 2x)$

15. $7(3 - 5h) = 4(3h - 2) - 6(7 + 9h)$

16. $4(2 - x) = 5(3x - 2) - 4(2 + 5x)$

17. $3[2 - 4(k - 6)] = 12 - 6k$

18. $2[2 - 3(x - x)] = 3 - 8x$

19. $4[2x - 5] = 3\{6(7 - x) - 12\}$

20. $5(2x - 3) = 3\{6(1 - x) - 5\} + 10$

21. $5(1 - 2x) - 3 = 4x + [-(2x - 8) + 6]$

22. $20 = 18 - \{-2[3z - 2(z - 1)]\}$

23. $10 - 3(7 - 3x) - 2x = 6x - 5[3x - (4 - x) - 5]$

24. $12 = -\{-3[4z - 2(z - 2)]\}$

25. $-2\{3 - [2 - 4(5 - x) - 7]\} = 12 - [-(3x - 2)]$

26. $6(2 - 3y) - 5 = 5y + [-(2y - 7) + 14]$

In Exercises 27–30, round off answers to three decimal places.

27. $61.25 - 23.04x = 16.19(18.32 - 1.06x)$

28. $7.209x - 4.395(6.281 + 9.154x) = 8.013$

29. $21.82(39.51x - 62.46) - 24.53(50.29 - 48.14x) = 14.28(65.23x - 92.61)$

30. $5.06(18.13x - 4.021) - 6.12(3.062 - 4.31x) = 42.12(31.16x - 10.04)$

3.5 Conditional Equations, Identities, and Equations with No Solution

There are many different kinds of equations. In this section, we discuss three types: conditional equations, identities, and equations with no solution.

Conditional Equations   As mentioned in Section 3.1, a *conditional equation* is an equation whose two sides are equal only when certain numbers are substituted for the variable. All equations given so far in this chapter have been conditional equations.

Identities   If the two sides of an equation are equal when *any* permissible number is substituted for the variable, the equation is called an **identity**. Therefore, an identity has an endless number of solutions.

Example 1   Verify that 0, -5, 0.5, and 7 are solutions of the identity $2(5x - 7) = 10x - 14$.
Check for 0

$$2(5[0] - 7) \overset{?}{=} 10[0] - 14$$

$$2(0 - 7) \overset{?}{=} 0 - 14$$

$$2(-7) \overset{?}{=} -14$$

$$-14 = -14$$

Check for −5

$$2(5[-5] - 7) \overset{?}{=} 10[-5] - 14$$
$$2(-25 - 7) \overset{?}{=} -50 - 14$$
$$2(-32) \overset{?}{=} -64$$
$$-64 = -64$$

Check for 0.5

$$2(5[0.5] - 7) \overset{?}{=} 10[0.5] - 14$$
$$2(2.5 - 7) \overset{?}{=} 5 - 14$$
$$2(-4.5) \overset{?}{=} -9$$
$$-9 = -9$$

Check for 7

$$2(5[7] - 7) \overset{?}{=} 10[7] - 14$$
$$2(35 - 7) \overset{?}{=} 70 - 14$$
$$2(28) \overset{?}{=} 56$$
$$56 = 56$$

NOTE   The two sides of the equation $2(5x - 7) = 10x - 14$ are equal if *any* real number is substituted for x. ■

Equations with No Solution   If *no* number will make the two sides of an equation equal, we say that the equation is an **equation with no solution**.

Example 2   Consider the equation $x + 1 = x + 2$.

Try 0 as a solution: $0 + 1 \neq 0 + 2$.

Try 1 as a solution: $1 + 1 \neq 1 + 2$.

Try 4 as a solution: $4 + 1 \neq 4 + 2$.

Try −6 as a solution: $-6 + 1 \neq -6 + 2$.

Will *any* number work? No. *Unequal* numbers have been added to the same number, x; therefore, the sums cannot be equal. ■

Usually, we cannot determine whether an equation is a conditional equation, an identity, or an equation with no solution simply by looking at it. Instead, we try to solve the equation by using the methods shown in the preceding sections. In those sections, the equations always reduced to the form $x = a$. In this section, however, three outcomes are possible:

1. If the equation *can* be reduced to the form $x = a$, where a is some real number, the equation is a *conditional equation*.

2. If the variable drops out and the two sides of the equation reduce to the same constant so that we obtain a *true* statement (for example, $0 = 0$), the equation is an *identity*.

3. If the variable drops out and the two sides of the equation reduce to unequal constants so that we obtain a *false* statement (for example, $0 = 3$), the equation is an *equation with no solution*.

Example 3 Solve $4x - 2(3 - x) = 12$, or identify the equation as either an identity or an equation with no solution.

Solution

$$4x - 2(3 - x) = 12$$

$$4x - 6 + 2x = 12$$

$$
\begin{array}{rl}
6x - 6 = & 12 \qquad \text{Combining like terms} \\
+\ 6 = & +6 \qquad \text{Adding 6 to both sides} \\
\hline
6x \quad = & 18
\end{array}
$$

$$\frac{\overset{1}{\cancel{6}}x}{\underset{1}{\cancel{6}}} = \frac{18}{6} \qquad \text{Dividing both sides by 6}$$

$$x = 3 \qquad \text{Conditional equation (single solution)}$$

Because the equation reduced to the form $x = a$, we know that it is a conditional equation. A check confirms that the solution is 3. ∎

Example 4 Solve $2(5x - 7) = 10x - 14$, or identify the equation as either an identity or an equation with no solution.

Solution

$$2(5x - 7) = 10x - 14$$

$$
\begin{array}{rl}
10x - 14 = & 10x - 14 \\
-10x & -10x \qquad \text{Adding } -10x \text{ to both sides} \\
\hline
-14 = & -14 \qquad \text{True}
\end{array}
$$

When we tried to isolate x, all the x's dropped out. Because the two sides of the equation reduced to the same constant and we obtained a *true statement* $(-14 = -14)$, the equation is an identity. (The solution set is the set of *all* real numbers.) ∎

Example 5 Solve $3(2x - 5) = 2x + 4(x - 1)$, or identify the equation as either an identity or an equation with no solution.

Solution

$$3(2x - 5) = 2x + 4(x - 1)$$

$$6x - 15 = 2x + 4x - 4$$

$$
\begin{array}{rl}
6x - 15 = & 6x - 4 \\
-6x & -6x \qquad \text{Adding } -6x \text{ to both sides} \\
\hline
-15 = & -4 \qquad \text{False}
\end{array}
$$

When we tried to isolate x, all the x's dropped out. Because the two sides of the equation reduced to different constants and we obtained a *false statement* $(-15 = -4)$, the equation is an equation with no solution. (The solution set is the empty set, $\{\ \}$.) ∎

EXERCISES 3.5

Set I   Find the solution of each conditional equation. Identify any equation that is *not* a conditional equation as either an identity or an equation with no solution.

1. $x + 3 = 8$ **2.** $4 - x = 6$

3. $2x + 5 = 7 + 2x$ **4.** $10 - 5y = 8 - 5y$

5. $6 + 4x = 4x + 6$ **6.** $7x + 12 = 12 + 7x$

7. $5x - 2(4 - x) = 6$ **8.** $8x - 3(5 - x) = 7$

9. $6x - 3(5 + 2x) = -15$ **10.** $4x - 2(6 + 2x) = -12$

11. $4x - 2(6 + 2x) = -15$ **12.** $6x - 3(5 + 2x) = -12$

13. $7(2 - 5x) - 32 = 10x - 3(6 + 15x)$

14. $6(3 - 4x) + 10 = 8x - 3(2 - 3x)$

15. $2(2x - 5) - 3(4 - x) = 7x - 20$

16. $3(x - 4) - 5(6 - x) = 2(4x - 21)$

17. $2[3 - 4(5 - x)] = 2(3x - 11)$

18. $3[5 - 2(7 - x)] = 6(x - 7)$

In Exercises 19 and 20, if the equation is conditional, round off the solution to three decimal places.

19. $460.2x - 23.6(19.5x - 51.4) = 1{,}213.04$

20. $46.2x - 23.6[19.5x - 51.4) = 213.04$

Set II   Find the solution of each conditional equation. Identify any equation that is *not* a conditional equation as either an identity or an equation with no solution.

1. $7 - x = 11$ **2.** $6y - 8 = 3 + 6y$

3. $7x - 4 = 3 + 7x$ **4.** $3(2 - x) = 5(2x + 1)$

5. $8 - 6x = -6x + 8$ **6.** $6(2x - 1) = 3(4x + 2) - 12$

7. $4x - 3(2 - x) = 8$ **8.** $7h - 3(5 - h) = 10$

9. $9x - 3(7 + 3x) = -21$ **10.** $4(3 - 4x) - 5 = 8(1 - 2x) - 1$

11. $2(x - 4) - (3 + 2x) = 3$ **12.** $2(7k + 9) - 18 = 14k$

13. $9(3 + 4x) - 17 = 14x + 2(6 + 11x)$

14. $3(2y - 7) - 2(5 - y) = 8y - 31$

15. $8(3x - 2) - (5 + 16x) = 8x - 5$

16. $5(4x - 3) + 6 = 2(3 + 10x)$

17. $5[2 - 3(2 - x)] = 7(2x - 1)$

18. $4(5x - 9) = 3[2 - 4(6 - x)]$

In Exercises 19 and 20, if the equation is conditional, round off the solution to three decimal places.

19. $460.2x - 23.6(19.5x - 51.4) = 213.04$

20. $3.76x - 1.02(5.21x - 10.7) = 21.45$

3.6 Graphing and Solving Inequalities in One Variable

Basic Definitions

An *equation* is a statement that two expressions are *equal*. An **inequality** is a statement that two expressions are *not equal*.

An inequality has three parts:

$$3x + 4 > 2 - x$$

Left side ——————

Inequality symbol; other inequality symbols can be used here

Right side

In this text, we will discuss solving only those inequalities that have the symbols $\neq$, $>$, $<$, $\leq$, or $\geq$.

The following three inequality symbols were introduced in Section 1.1:

"Unequal to" symbol ($\neq$) $a \neq b$ is read "a is unequal to b."

"Greater than" symbol ($>$) $a > b$ is read "a is greater than b."

"Less than" symbol ($<$) $a < b$ is read "a is less than b."

Example 1   Examples of reading the inequalities $\neq$, $>$, and $<$:

a. $5 \neq x - 3$ is read "5 is unequal to x minus 3."

b. $3x - 4 > 7$ is read "$3x$ minus 4 is greater than 7."

c. $2x < 5 - x$ is read "$2x$ is less than 5 minus x." ■

"Greater Than or Equal To" Symbol ($\geq$)   The inequality $a \geq b$ is read "a is greater than *or* equal to b." This means if $\begin{cases} \text{either } a > b \\ \text{or} \quad a = b \end{cases}$ is true, then $a \geq b$ is true.

Example 2   Examples of the meaning of $\geq$:

a. $5 \geq 1$ is true because $5 > 1$ is true.

b. $1 \geq 1$ is true because $1 = 1$ is true. ■

Example 3   Examples of reading $\geq$:

a. $-9 \geq -16$ is read "negative 9 is greater than or equal to negative 16."

b. $x + 6 \geq 10$ is read "x plus 6 is greater than or equal to 10." ■

"Less Than or Equal To" Symbol ($\leq$)   The inequality $a \leq b$ is read "a is less than *or* equal to b." This means if $\begin{cases} \text{either } a < b \\ \text{or} \quad a = b \end{cases}$ is true, then $a \leq b$ is true.

Example 4   Examples of the meaning of $\leq$:

a. $2 \leq 3$ is true because $2 < 3$ is true.

b. $3 \leq 3$ is true because $3 = 3$ is true. ■

Example 5   Examples of reading ≤:

a. $-7 \le 0$ is read "negative 7 is less than or equal to 0."

b. $7 \le 5x - 2$ is read "7 is less than or equal to $5x$ minus 2." ∎

The Solution and Solution Set of an Inequality

A **conditional inequality** is an inequality that is true for some values of the variable and false for others. (Examples 1b, 1c, 3b, and 5b are examples of conditional inequalities.)

A **solution** of a conditional inequality is any number that, when substituted for the variable, makes the inequality a true statement. The **solution set** of an inequality is the set of all numbers that are solutions of the inequality. While the equations we have solved in this chapter have had just one solution (except for identities), an inequality usually has many solutions.

The Graph of the Solution Set of an Inequality in One Variable

The solution set of an inequality in one variable can be graphed on the number line.

Example 6   Graph the solution set of each of the following inequalities:

a. $x \ge 3$

Solution The solution set is the set of all real numbers greater than or equal to 3.

The solid circle and the arrow together indicate that the 3 and all numbers to the *right* of the 3 are solutions

b. $x < 1$

Solution The solution set is the set of all real numbers less than 1.

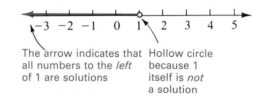

The arrow indicates that   Hollow circle
all numbers to the *left*   because 1
of 1 are solutions   itself is *not*
                             a solution

c. $x \ne -2$

Solution The solution set is the set of all real numbers except -2.

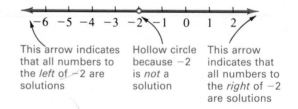

This arrow indicates   Hollow circle   This arrow
that all numbers to   because -2 indicates that
the *left* of -2 are is *not* a   all numbers to
solutions   solution   the *right* of -2
                                         are solutions   ∎

The Sense of an Inequality

The **sense** of an inequality symbol refers to whether the symbol is a *greater than* or a *less than* symbol.

$$a > b \brace c > d \quad \text{Same sense (both are >)} \qquad a < b \brace c > d \quad \text{Opposite sense (one is <, one is >)}$$

$$a \le b \brace c \ge d \quad \text{Opposite sense (one is ≤, one is ≥)} \qquad a \le b \brace c \le d \quad \text{Same sense (both are ≤)}$$

The Properties of Inequalities

Earlier in this chapter, we solved equations by adding the same number to both sides of the equation, multiplying both sides of the equation by the same number, and so on. We solve simple *inequalities* that contain one of the inequality symbols listed above by using one of the following properties of inequalities:

THE ADDITION AND SUBTRACTION PROPERTIES OF INEQUALITIES

If an inequality contains one of the symbols $>$, $<$, $\geq$, or $\leq$, the sense of the inequality is unchanged if the same number is added to or subtracted from both sides of the inequality. For example,

$$\text{if } a < b, \text{ then } a + c < b + c \text{ and } a - c < b - c$$

Senses are the same

where a, b, and c are real numbers. (See Examples 7a and 7b.)

THE MULTIPLICATION AND DIVISION PROPERTIES OF INEQUALITIES

If an inequality contains one of the symbols $>$, $<$, $\geq$, or $\leq$, the sense of the inequality is unchanged if both sides of the inequality are multiplied or divided by the same *positive* number. For example,

$$\text{if } a < b \text{ and } c > 0, \text{ then } ac < bc \text{ and } \frac{a}{c} < \frac{b}{c}$$

where a, b, and c are real numbers. (See Examples 7c and 7d.) However, the sense of the inequality is *changed* if both sides of the inequality are multiplied or divided by the same *negative* number. For example,

Senses are opposite

$$\text{if } a < b \text{ and } c < 0, \text{ then } ac > bc \text{ and } \frac{a}{c} > \frac{b}{c}$$

Senses are opposite

where a, b, and c are real numbers. (See Examples 7e and 7f.)

Example 7 Verify that the properties of inequalities are valid for the inequality $8 < 12$.

a. Add $+6$ to both sides.

$$\begin{array}{r} 8 < 12 \\ +6 \quad + 6 \\ \hline 14 \; ? \; 18 \end{array}$$

$14 < 18$

$14 < 18$ The sense of the inequality is *unchanged* when we add the same number to both sides of the inequality.

b. Subtract 3 from both sides.

$$\begin{array}{r} 8 < 12 \\ -3 \quad -3 \\ \hline 5 \; ? \; 9 \end{array}$$

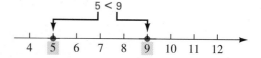

$$5 < \quad 9$$ The sense of the inequality is *unchanged* when we subtract the same number from both sides of the inequality.

c. Multiply both sides by 2 (a *positive* number).

$$\begin{array}{r} 8 < 12 \\ \times 2 \quad \times 2 \\ \hline 16 \; ? \; 24 \end{array}$$

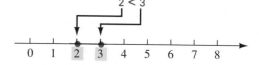

$$16 < \quad 24$$ The sense of the inequality is *unchanged* when we multiply both sides of the inequality by the same *positive* number.

d. Divide both sides by 4 (a *positive* number).

$$8 < 12$$
$$\frac{8}{4} \; ? \; \frac{12}{4}$$
$$2 \; ? \; 3$$

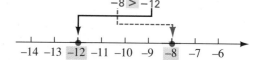

$$2 < 3$$ The sense of the inequality is *unchanged* when we divide both sides of the inequality by the same *positive* number.

e. Multiply both sides by -1 (a *negative* number).

$$\begin{array}{r} 8 < 12 \\ \times(-1) \quad \times(-1) \\ \hline -8 \; ? \; -12 \end{array}$$

$$-8 > -12$$
↑
Senses are
opposite

The sense of the inequality is *changed* when we multiply both sides of the inequality by the same *negative* number.

f. Divide both sides by -2 (a *negative* number).

$$8 < 12$$
$$\frac{8}{-2} \; ? \; \frac{12}{-2}$$
$$-4 \; ? \; -6$$

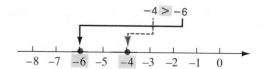

$$-4 > -6$$
↑
Senses are
opposite ∎

The sense of the inequality is *changed* when we divide both sides of the inequality by the same *negative* number.

We can see that when we multiply or divide both sides of an inequality that contains a $<$, $\leq$, $>$, or $\geq$ symbol by a *negative number*, the *sense* of the inequality changes.

Solving Inequalities

When we solve an inequality, we must find all the values of the variable that satisfy the inequality. Therefore, in the problems in this book, we want our final inequality to be in the form $x < a$, $x > a$, $x \leq a$, $x \geq a$, or $x \neq a$, where a is some number. To accomplish this, we use the following procedure:

TO SOLVE AN INEQUALITY THAT CONTAINS $<$, $\leq$, $>$, OR $\geq$

Proceed in the same way used to solve equations, with the exception that *the sense must be changed when multiplying or dividing both sides of the inequality by a negative number.*

Then, if necessary, use the facts that $a < b$ can be rewritten as $b > a$, $a \geq b$ can be rewritten as $b \leq a$, and so on, to get the inequality into one of the forms $x < a$, $x \leq a$, $x > a$, or $x \geq a$.*

TO SOLVE AN INEQUALITY THAT CONTAINS $\neq$

Proceed in the same way used to solve equations, writing $\neq$ in each step rather than $=$.

Example 8   Solve $x + 3 < 7$ and graph the solution set on the number line.
Solution

$$
\begin{array}{rl}
x + 3 < & 7 \\
\underline{-3 \quad -3} & \qquad \text{Adding } -3 \text{ to both sides} \\
x \quad < & 4
\end{array}
$$

This means that if we replace x by any number less than 4 in the given inequality, we get a true statement. For example, if we replace x by 3 (which is less than 4),

$$
\begin{array}{c}
x + 3 < 7 \\
3 + 3 \overset{?}{<} 7 \\
6 < 7 \qquad \text{True}
\end{array}
$$

Graph   We graph the solution set, the set of all real numbers less than 4.

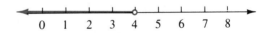

We will not show any checks for Examples 9–12.

Example 9   Solve $2x - 5 > 1$ and graph the solution set on the number line.
Solution

$$
\begin{array}{rl}
2x - 5 > & 1 \\
\underline{+5 \quad +5} & \qquad \text{Adding 5 to both sides} \\
2x \quad > & 6
\end{array}
$$

$$
\frac{2x}{2} > \frac{6}{2} \qquad \begin{array}{l} \text{Dividing both sides by 2} \\ \text{Sense is not changed} \end{array}
$$

$$
x > 3
$$

The solution set is the set of all real numbers greater than 3.

---

*Your instructor may permit you to leave answers in a form such as *a < x*.

Graph

```
  ├──┼──┼──○━━┿━━┿━━┿━━┿━━►
  0  1  2  3  4  5  6  7  8
```

Example 10 Solve $3x - 2(2x - 7) \leq 2(3 + x) - 4$ and graph the solution set on the number line.
Solution

$$3x - 2(2x - 7) \leq 2(3 + x) - 4$$

$$3x - 4x + 14 \leq 6 + 2x - 4 \qquad \text{Removing grouping symbols}$$

$$\begin{array}{rl}
-x + 14 \leq & 2 + 2x \qquad \text{Combining like terms}\\
\underline{-2x - 14} & \underline{-14 - 2x} \qquad \text{Adding } -2x - 14 \text{ to both sides}\\
-3x \quad\; \leq & -12
\end{array}$$

$$\frac{-3x}{-3} \geq \frac{-12}{-3} \qquad \begin{array}{l}\text{Dividing both sides by } -3\\ \textit{Sense is changed}\end{array}$$

$$x \geq 4$$

Alternate solution The problem can also be done as follows:

$$3x - 2(2x - 7) \leq 2(3 + x) - 4$$

$$3x - 4x + 14 \leq 6 + 2x - 4$$

$$\begin{array}{rl}
-x + 14 \leq & 2 + 2x\\
\underline{+x - \;2} & \underline{-2 + \;x} \qquad \text{Adding } +x - 2 \text{ to both sides}\\
12 \leq & 3x
\end{array}$$

$$\frac{12}{3} \leq \frac{3x}{3} \qquad \begin{array}{l}\text{Dividing both sides by } 3\\ \text{Sense is not changed}\end{array}$$

$$4 \leq x$$

$$x \geq 4 \qquad x \geq 4 \text{ has the same meaning as } 4 \leq x$$

The solution set is the set of all real numbers greater than or equal to 4.
Graph

```
  ├──┼──┼──┼──●━━┿━━┿━━┿━━►
  0  1  2  3  4  5  6  7  8
```

Example 11 Solve $4(y - 2) + 3 \geq 3(y + 3)$ and graph the solution set on the number line.
Solution

$$4(y - 2) + 3 \geq 3(y + 3)$$

$$4y - 8 + 3 \geq 3y + 9 \qquad \text{Removing parentheses}$$

$$\begin{array}{rl}
4y - 5 \geq & 3y + 9 \qquad \text{Combining like terms}\\
\underline{-3y + 5} & \underline{-3y + 5} \qquad \text{Adding } -3y + 5 \text{ to both sides}\\
y \quad\; \geq & 14
\end{array}$$

The solution set is the set of all real numbers greater than or equal to 14.
Graph

In Example 12, remember that when the inequality contains the $\neq$ sign, we proceed in exactly the same manner as for solving equations.

Example 12   Solve $4(x - 3) - 5 \neq 2x - 7$ and graph the solution set on the number line.
Solution

$$4(x - 3) - 5 \neq 2x - 7$$

$$4x - 12 - 5 \neq 2x - 7 \qquad \text{Removing parentheses}$$

$$\begin{array}{rcl} 4x - 17 & \neq & 2x - 7 \\ -2x + 17 & & -2x + 17 \\ \hline 2x & \neq & 10 \end{array}$$

Combining like terms
Adding $-2x + 17$ to both sides

$$\frac{2x}{2} \neq \frac{10}{2} \qquad \text{Dividing both sides by 2}$$

$$x \neq 5 \qquad \text{Notice that we wrote "}\neq\text{" in every step.}$$

The solution set is the set of all real numbers except 5.
Graph

EXERCISES 3.6

Set I   Solve each of the following inequalities and graph each solution set on the number line.

1. $x - 5 < 2$

2. $x - 4 < 7$

3. $5x + 4 \leq 19$

4. $3x + 5 \leq 14$

5. $6x + 7 > 3 + 8x$

6. $4x + 28 > 7 + x$

7. $2x - 9 > 3(x - 2)$

8. $3x - 11 > 5(x - 1)$

9. $6(3 - 4x) + 12 \geq 10x - 2(5 - 3x)$

10. $7(2 - 5x) + 27 \geq 18x - 3(8 - 4x)$

11. $4(6 - 2x) \neq 5x - 2$

12. $6(2x - 5) + 29 \neq 3x - 7(11 - 4x)$

13. $2[3 - 5(x - 4)] < 10 - 5x$

14. $3[2 - 4(x - 7)] < 26 - 8x$

15. $7(x - 5) - 4x > x - 8$

16. $8(x + 2) \neq 24 - 2(x - 1)$

17. $3x - 5(x + 2) \leq 4x + 8$

18. $5[6 - 2(3 - x)] - 3 < 3x + 4$

In Exercises 19 and 20, round off the decimal approximations to three decimal places.

19. $12.85x - 15.49 \geq 22.06(9.66x - 12.74)$

20. $7.12(3.65x - 8.09) + 5.76 < 5.18x - 6.92(4.27 - 3.39x)$

Set II   Solve each of the following inequalities and graph each solution set on the number line.

1. $x + 3 \geq -4$

2. $4 + x > -5$

3. $x + 4 < 7$

4. $5x + 3 < 18$

5. $5x + 7 > 13 + 11x$

6. $7x - 5 \leq 2x - 25$

7. $5x - 6 \leq 3(2 + 3x)$ **8.** $3x - 2 \geq -2(11 - x)$

9. $5(3 - 2x) + 25 \geq 4x - 6(10 - 3x)$

10. $3(x + 3) - 4 \neq 7x - 3(2 - x)$

11. $2(5 - 3x) \neq 7 - 4x$ **12.** $6(3 - 2x) - 3(x + 1) < 0$

13. $4[2 - 3(x - 5)] < 2 - 6x$ **14.** $2(3x - 4) + (x - 1) > 5$

15. $8(x - 3) - (x + 4) < 2x + 2$ **16.** $3[5 + 3(4 + x)] \leq 4(2x - 3)$

17. $5x - 3(3x - 4) \geq 2x + 7$ **18.** $5 - (3 - x) \neq 3(2 + x)$

In Exercises 19 and 20, round off the decimal approximations to three decimal places.

  19. $821.4x - 395.2 \geq 604.1(542.8x - 193.7)$

  20. $4.01x + 62.1 \leq 3.04(5.143 - 6.21x)$

3.7 Review: 3.1–3.6

Parts of an Equation
3.1

$$\boxed{5x - 8} = \boxed{3x + 2}$$

Left side ⟶   ⟵ Right side
          Equal sign

Conditional Equations
3.1

A **conditional equation** is an equation whose two sides are equal only when certain numbers (called *solutions*) are substituted for the variable.

Solution of an Equation
3.1

A **solution** of an equation is a number that, when substituted for the variable, makes the two sides of the equation equal.

Solution Set of an Equation
3.1

The **solution set** of an equation is the set of all the solutions of the equation.

Equivalent Equations
3.1

Equations that have the same solution set are **equivalent equations**.

The Properties of Equality
3.2, 3.3

Addition: If the same number is added to both sides of an equation, the new equation is equivalent to the original equation.

Subtraction: If the same number is subtracted from both sides of an equation, the new equation is equivalent to the original equation.

Division: If both sides of an equation are divided by the same nonzero number, the new equation is equivalent to the original one.

Multiplication: If both sides of an equation are multiplied by the same nonzero number, the new equation is equivalent to the original equation.

Symmetry: The equation $a = b$ is equivalent to the equation $b = a$.

To Solve an Equation
3.2, 3.3, 3.4

1. Remove grouping symbols.

2. Combine like terms on each side of the equation.

3. If the variable appears on both sides of the equation, remove the term that contains the variable from one side by adding the negative of that term to both sides.

4. Remove all numbers that appear on the same side as the variable.

   First, remove those numbers being added to or subtracted from the term containing the variable.

   Next, complete the solution by using the one-step method (multiplying both sides by the reciprocal of the coefficient of the variable)

   or complete the solution by using the two-step method (clearing fractions and then dividing both sides by the coefficient of the variable).

3.5 5. Three outcomes are possible:

   If the equation reduces to the form $x = a$, it is a *conditional equation*.

   If the variable drops out and the two sides reduce to the same constant so that we're left with a *true* statement, the equation is an *identity*.

   If the variable drops out and the two sides reduce to unequal constants so that we're left with a *false* statement, the equation has *no solution*.

6. If the equation was a conditional equation, check the solution in the *original* equation.

Inequalities
3.6

An **inequality** is a statement that two expressions are not equal.

Graphing the Solution Set of an Inequality on the Number Line
3.6

To graph the solution set of the inequality $x > c$:

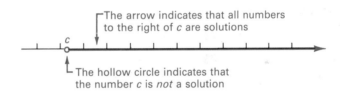

To graph the solution set of the inequality $x \le b$:

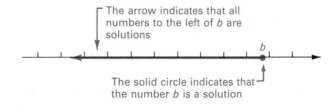

The solution sets of other types of inequalities, for example, $x < d$ and $x \ge e$, are graphed using the same procedures.

To Solve an Inequality
3.6

If it contains $<$, $\le$, $>$, or $\ge$, proceed in the same way used to solve equations, with the exception that *the sense must be changed when multiplying or dividing both sides by a negative number*.

If it contains $\ne$, proceed in exactly the same way used to solve equations, writing $\ne$ in each step rather than $=$.

Review Exercises 3.7 Set I

In Exercises 1–20, find the solution of each conditional equation and check the solution. Identify any equation that is not a conditional equation as either an *identity* or *an equation with no solution*. Graph the solution of each conditional equation on the number line.

1. $3x - 5 = 4$

2. $22 - 8x = 6$

3. $2 = 20 - 9x$

4. $5x - 3 = 5x + 4$

5. $7.5 = \dfrac{A}{10}$

6. $\dfrac{D}{9} - 12 = 8$

7. $7 - 2(M - 4) = 5$

8. $20 - 3(4 - 5x) = 8 + 15x$

9. $6R - 8 = 6(2 - 3R)$

10. $7P - 15 = 7(3 - 2P)$

11. $56T - 18 = 7(8T - 4)$

12. $65 - 77S = 11(5 - 7S)$

13. $15(4 - 5V) = 16(4 - 6V) + 10$

14. $10 - 5(2x - 3) = 5(5 - 2x)$

15. $5x - 7(4 - 2x) + 8 = 10 - 9(11 - x)$

16. $18 - 6(5x - 4) - 13x = 11(12 - 3x) - 7$

17. $2[-7y - 3(5 - 4y) + 10] = 10y - 12$

18. $5[-13z - 8(4 - 2z) + 20] = 15z - 17$

19. $4[-24 - 6(3x - 5) + 22x] = 0$

20. $3[-53 - 7(4x - 9) + 18x] = 0$

In Exercises 21–26, solve each inequality and graph the solution set on the number line.

21. $x + 7 > 2$

22. $x + 5 > 1$

23. $x - 3 \le -8$

24. $x - 6 \le -10$

25. $2(x - 4) - 5 \ge 7 + 3(2x - 1)$

26. $10 - 3(x + 2) \ge 9 - 2(4 - 3x)$

Review Exercises 3.7 Set II

In Exercises 1–20, find the solution of each conditional equation and check the solution. Identify any equation that is not a conditional equation as either an *identity* or *an equation with no solution*. Graph the solution of each of the first six conditional equations on the number line. (Use the number lines at the bottom of this page. *You* insert the exercise number and label the points on the number line.)

ANSWERS

1. $7x - 8 = 6$

2. $12 - 6x = -12$

1. _____

2. _____

3. $3 = 18 - 5x$

4. $5 - 3x = 9 - 3x$

3. _____

4. _____

5. _____

5. $\dfrac{x}{4} = 8.6$

6. $-27 = \dfrac{z}{6} - 18$

6. _____

7. _____

7. $-9 - 3(N - 8) = 30$

8. $19 - 5(3 - 2x) = 4 + 10x$

8. _____

9. _____

9. $12R - 9 = 8(3 - 4R)$

10. $42H + 12 = 6(7H - 2)$

10. _____

11. _____

11. $4(3x - 2) + 3x = 5(2 + 3x)$

12. $2(9x - 5) - 2 = 3(6x - 5)$

12. _____

13. _____

14. _____

13. $11(3 - 4V) = 8(5 - 6V) + 17$

14. $7x - (3 - x) = 2(4x + 1) - 5$

15. $4 - 5(x - 7) = 10(4 - x)$　　　**16.** $2 + 3(4 - x) = 6x - 13$

17. $25 - 2[4(x - 2) - 12x + 4] = 16x + 33$

18. $2\{3[10 - 4(3 - x) + x] - 5\} = 0$

19. $15 - 5[2(8x - 4) - 14x + 8] = 25$

20. $2\{3[2(5 - V) + 4V] - 20\} = 0$

15. _____

16. _____

17. _____

18. _____

19. _____

20. _____

21. _____

22. _____

23. _____

24. _____

25. _____

26. _____

In Exercises 21–26, solve each inequality and graph the solution set on the number line. (*You* label the points on the number line.)

21. $x + 5 \geq -6$

22. $x - 7 < 2$

23. $3x - 2 > 6 - x$

24. $7x - 2(5 + 4x) \leq 8$

25. $5(6 - 2x) + 3 \geq 2(x + 1) + 11$

26. $3[2 - (4 - x)] + 7 < 4x$

Chapter 3 Diagnostic Test

The purpose of this test is to see how well you understand solving simple equations and inequalities. We recommend that you work this diagnostic test *before* your instructor tests you on this chapter. Allow yourself about 50 minutes.

Complete solutions for all the problems on this test, together with section references, are given in the answer section in the back of the book. For the problems you do incorrectly, study the sections referred to.

In Problems 1–18, find and check the solution of each conditional equation. Identify any equation that is not a conditional equation as either an identity or an equation with no solution.

1. $x - 5 = 3$ (Graph the solution on the number line.)

2. $5y + 7 = 22$ (Show your check on this problem.)

3. $5z - 7 = 13$

4. $14 + 3x = 7 - 4x$

5. $7x - 4 = 3x + 4(x - 1)$

6. $8 = 4y - 1$

7. $3x - 4 = x + 2(x - 1)$

8. $17 - 3z = -1$

9. $\dfrac{x}{6} = 5.1$

10. $-6 = \dfrac{w}{7}$

11. $\dfrac{x}{4} - 5 = 3$

12. $6x + 1 = 17 - 2x$

13. $3z - 21 + 5z = 4 - 6z + 17$ (Graph the solution on the number line.)

14. $5k - 9(7 - 2k) = 6$ (Show your check on this problem.)

15. $10x - 2(5x - 7) = 14$

16. $2y - 4(3y - 2) = 5(6 + y) - 7$

17. $7(3z + 4) = 14 + 3(7z - 1)$

18. $3[7 - 6(x - 2)] = -3 + 2x$ (Graph the solution on the number line.)

19. Solve the inequality $4x + 5 > -3$ and graph the solution set on the number line.

20. Solve the inequality $5x - 2 \le 10 - x$ and graph the solution set on the number line.

Cumulative Review Exercises: Chapters 1–3

In Exercises 1–4, evaluate each expression or write either "Not defined" or "Not a real number."

1. $\dfrac{-7}{0}$

2. $\sqrt{8^2 + 6^2}$

3. $7\sqrt{16} - 5(-4)$

4. $25 - \{-16 - [(11 - 7) - 8]\}$

In Exercises 5 and 6, use the formula, substituting the given values for the variables.

5. $V = \dfrac{4}{3}\pi r^3$ Find V when $\pi \approx 3.14$ and $r = 9$

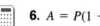

6. $A = P(1 + i)^n$ Find A when $P = 3{,}000$, $i = 0.10$, and $n = 2$

In Exercises 7–10, simplify each algebraic expression.

7. $15x - [9y - (7x - 10y)]$

8. $5(2a - 3b) - 6(4a + 7b)$

9. $8 - 3[2x - (1 - 4x)]$

10. $8x + 3\{2 + 7(4 - x)\} - 2$

In Exercises 11–17, solve each conditional equation. Identify any equation that is not a conditional equation as either an identity or an equation with no solution.

11. $2x + 1 = 2x + 7$

12. $3 = 33 - 10x$

13. $10 - 4(2 - 3x) = 2 + 12x$

14. $6 - 4(N - 3) = 2$

15. $\dfrac{C}{7} - 15 = 13$

16. $12(4W - 5) = 9(7W - 8) - 13$

17. $9 - 3(x - 2) = 3(5 - x)$

In Exercises 18–20, solve each inequality.

18. $2x - 5 < 3$

19. $3 - x \geq 4$

20. $5x - 2 \leq 8x + 4$

In Exercises 21–25, write "True" if the statement is always true. Otherwise, write "False."

21. $-\dfrac{3}{4}$ is a real number.

22. $5.\overline{23}$ is an irrational number.

23. -5 is a natural number.

24. All irrational numbers are real numbers.

25. All real numbers are rational numbers.

Critical Thinking

Each of the following problems has an error. Can you find it?

1. Evaluate $2^2 \cdot 2^3$.

$$2^2 \cdot 2^3 = 4^5 = 1,024$$

2. Evaluate $\sqrt{5^2 + 12^2}$.

$$\sqrt{5^2 + 12^2}$$
$$= \sqrt{25 + 144}$$
$$= \sqrt{25} + \sqrt{144}$$
$$= 5 + 12 = 17$$

3. Evaluate $2^3 - 3 \cdot 4 \div 2$.

$$2^3 - 3 \cdot 4 \div 2$$
$$= 8 - 3 \cdot 4 \div 2$$
$$= 5 \cdot 4 \div 2$$
$$= 20 \div 2 = 10$$

4. Simplify $x - 2[x - (4 - x) - 8]$.

$$x - 2[x - (4 - x) - 8]$$
$$= x - 2[x - 4 + x - 8]$$
$$= x - 2[2x - 12]$$
$$= x - 4x - 24$$
$$= -3x - 24$$

5. Solve $7 - 4x = -9$.

$$7 - 4x = -9$$
$$\underline{-7 \qquad\qquad -7}$$
$$4x = -16$$
$$\frac{4x}{4} = \frac{-16}{4}$$
$$x = -4$$

4 Word Problems

As we mentioned in Chapter 3, the main reason for studying algebra is to equip yourself with the tools necessary to solve mathematical problems. Most such problems are expressed in words. In this chapter, we show methods for solving some traditional word problems. The skills learned in this chapter can be applied to solving mathematical problems encountered in many fields of study as well as in real-life situations.

4.1 Problem Solving—Translating Word Expressions into Algebraic Expressions

While we can't give you a definite set of rules that will enable you to solve all word problems, in Section 4.3 we suggest a procedure that should help you get started. In this chapter, we discuss several different "types" of word problems (money problems, mixture problems, distance-rate-time problems, and so forth) in separate sections, because we feel that this technique is most helpful to beginning students. However, the general *method* of attacking word problems is the same for *all* types of word problems; it is this *method* that you should concentrate on.

In Section 1.13, we gave several suggestions for solving word problems in arithmetic. We suggest that you review those suggestions at this time, because many of them apply to the solution of word problems in algebra.

4.1A Key Word Expressions and Their Corresponding Algebraic Operations

In solving word problems, it is helpful to break them up into smaller expressions. In this section, we show how you can change these small *word* expressions into *algebraic* expressions. Below is a list of key word expressions and their corresponding algebraic operations.

+	−	×	÷	=
add	subtract	multiply	divided by	is equal to
sum	difference	times	quotient	equals
plus	minus	product		is
increased by	decreased by			
more than	diminished by			
	less than			
	subtracted from			

NOTE   Because subtraction is not commutative, care must be taken to put the numbers in a subtraction word problem in the correct order. For example, while the statements "*m* minus *n*," "the difference of *m* and *n*," and "*m* decreased by *n*" are translated as $m - n$ (see Example 1c), the expressions "*m* subtracted from *n*" and "*m* less than *n*" are translated as $n - m$ (see Example 1d).   ☑

Example 1   Change each of the following word expressions into an algebraic expression.

a. "The sum of *A* and *B*"
   Solution $A + B$

b. "The product of *l* and *w*"
   Solution lw

c. "Two decreased by *C*"
   Solution $2 - C$

d. "Two less than C"
   Solution $C - 2$

e. "Three times the square of x, plus ten"
   Solution $3x^2 + 10$

f. "Five subtracted from the quotient of S divided by T"
   Solution $\dfrac{S}{T} - 5$ ■

EXERCISES 4.1A

Set I   Change each of the following word expressions into an algebraic expression.

1. The sum of x and 10   **2.** A added to B

3. Five less than A **4.** B diminished by C

5. The product of 6 and z **6.** A multiplied by B

7. x decreased by 7   **8.** Nine increased by A

9. Four less than 3 times x **10.** Six more than twice x

11. Subtract the product of u and v from x.

12. Subtract x from the product of P and Q.

13. The product of 5 and the square of x

14. The product of 10 and the cube of x

15. The square of the sum of A and B

16. The square of the quotient of A divided by B

17. The sum of x and 7, divided by y

18. T divided by the sum of x and 9

19. The product of x and the difference, 6 less than y

20. The product of A and the sum, 3 plus B

Set II   Change each of the following word expressions into an algebraic expression.

1. Ten added to x **2.** The product of s and t

3. Three less than w **4.** x diminished by 4

5. The sum of u and v **6.** Five increased by x

7. Five decreased by y **8.** Ten plus x

9. Seven more than z **10.** Five times b

11. Twice F, subtracted from 15

12. The quotient of A divided by the sum of C and 10

13. The sum of x and the square of y

14. The quotient of the sum of A and C divided by B

15. The sum of the squares of A and B

16. The square of the sum of x and 4

17. The sum of x and y, divided by z

18. Twice the sum of x and y

19. The product of 7 and the sum of x and y

20. Three less than 5 times y

4.1B Translating Word Expressions into Algebraic Expressions

A word expression often contains unknown numbers. In algebra, we must translate such an expression into an algebraic expression, and we must represent the unknown numbers by *variables*. In this chapter, when there are two or more unknowns, we must express each unknown in terms of the same variable.

TO CHANGE A WORD EXPRESSION INTO AN ALGEBRAIC EXPRESSION

1. Identify which number or numbers are unknown.

2. Represent *one* of the unknown numbers by a variable. Express any other unknown number in terms of the same variable.

3. Change the word expression into an algebraic expression, using the variable(s) in place of the unknown number(s).

Example 2   Change the word expression "twice Albert's salary" into an algebraic expression.
Solution

Step 1. Albert's salary is the unknown number.

Step 2. Let S represent Albert's salary.

Step 3. Then $2S$ is the algebraic expression for "twice Albert's salary." ∎

Example 3   Change the word expression "the cost of five stamps" into an algebraic expession.
Solution

Step 1. The cost of one stamp is the unknown number.

Step 2. Let c represent the cost of one stamp.

Step 3. Since one stamp costs c cents, 5 stamps will cost 5 times c cents, or $5c$ cents. Therefore, $5c$ is the algebraic expression for "the cost of five stamps." ∎

Example 4   In the word expression "Mary is 10 years older than Nancy," represent both unknown numbers in terms of the same variable.
First Solution

Step 1. There are two unknown numbers in this expression: Mary's age and Nancy's age.

Step 2. Let N represent Nancy's age.

Then $N + 10$ represents Mary's age, because Mary is 10 years older than Nancy.

Second Solution

Step 1. There are two unknown numbers in this expression: Mary's age and Nancy's age.

Step 2. Let M represent Mary's age.

Then $M - 10$ represents Nancy's age, because Nancy is 10 years younger than Mary. ∎

Example 5   In the word expression "The sum of two numbers is 10," represent both unknown numbers in terms of the same variable.
Solution

Step 1. There are two unknown numbers.

Step 2. Let $x =$ one of the unknown numbers

Then $10 - x =$ the other unknown number

Notice that $x + (10 - x) = 10$; the sum of the two numbers is 10. ∎

Recall from Section 1.2 that integers that follow one another in sequence (without interruption) are called *consecutive integers*. Consecutive integers can be represented by variables. If the problem deals with two or more consecutive integers, we can let x equal the first integer, $x + 1$ equal the second one, $x + 2$ equal the third one, and so on.

Adding 2 to any even integer gives the next even integer; for example, $8 + 2 = 10$, $-16 + 2 = -14$, and so on. Therefore, if a problem deals with consecutive *even* integers, let x equal the first integer, $x + 2$ equal the second one, $x + 4$ equal the third one, and so on.

Adding 2 to any odd integer gives the next odd integer; for example, $9 + 2 = 11$, $-5 + 2 = -3$, and so forth. Therefore, if a problem deals with consecutive *odd* integers, let x equal the first integer, $x + 2$ equal the second one, $x + 4$ equal the third one, and so on.

Example 6   In the word expression "The sum of three consecutive integers," represent the sum of the three integers in terms of the same variable.
Solution

Step 1. There are three unknown integers.

Step 2. Let $x =$ the first integer
$x + 1 =$ the second integer
$x + 2 =$ the third integer

Step 3. Then the sum of the three integers is $x + (x + 1) + (x + 2)$. ∎

Example 7   In the word expression "The sum of three consecutive odd integers," represent the sum of the three integers in terms of the same variable.
Solution

Step 1. There are three unknown *odd* integers.

Step 2. Let $x =$ the first odd integer
$x + 2 =$ the second odd integer
$x + 4 =$ the third odd integer

Step 3. Then the sum of the three integers is $x + (x + 2) + (x + 4)$. ∎

Word problems sometimes deal with geometric figures. Several formulas relating to geometric figures are given below. The *perimeter* of a geometric figure is the sum of the lengths of all its sides.

Rectangle

Area: $A = lw$

Perimeter: $P = 2l + 2w$

Square

Area: $A = s^2$

Perimeter: $P = 4s$

Triangle

Area: $A = \frac{1}{2}bh$

where b is the base and h is the altitude

Perimeter: $P = a + b + c$

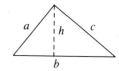

Parallelogram

Area: $A = bh$

where b is the base and h is the altitude

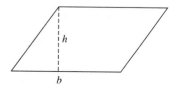

Circle

Area: $A = \pi r^2$

Circumference: $C = 2\pi r$

where $\pi \approx 3.14$ and r is the radius

Rectangular Solid

Volume: $V = lwh$

Surface area: $S = 2(lw + lh + wh)$

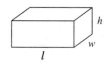

Cube

Volume: $V = s^3$

Surface area: $S = 6s^2$

Sphere

Volume: $V = \frac{4}{3}\pi r^3$

Surface area: $S = 4\pi r^2$

where $\pi \approx 3.14$ and r is the radius

Cylinder (*Right Circular Cylinder*)

Volume: $V = \pi r^2 h$

where $\pi \approx 3.14$, r is the radius,

and h is the height

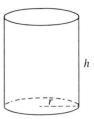

Cone (*Right Circular Cone*)

Volume: $V = \frac{1}{3}\pi r^2 h$

where $\pi \approx 3.14$, r is the radius,

and h is the height

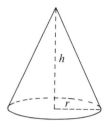

Example 8   In the word expression "The height of a cone is 2 ft more than the radius," represent the height and the radius in terms of the same variable.
First Solution

Step 1. There are two unknowns: the height and the radius.

Step 2. Let r = the radius.

Then $r + 2$ = the height, because the height is 2 ft more than the radius.

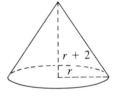

Second Solution

Step 1. There are two unknowns: the height and the radius.

Step 2. Let h = the height.

Then $h - 2$ = the radius, because the radius is 2 ft less than the height. ■

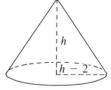

Example 9   If the length of a rectangular solid is 5 yd more than the height and the width is 2 yd less than the height, express the *volume* in terms of one variable.
Solution

Step 1. There are four unknowns: the height, the length, the width, and the volume.

Step 2. Let h = the height

$h + 5$ = the length

$h - 2$ = the width

Step 3. Then the volume is $h(h + 5)(h - 2)$.

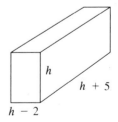

NOTE   It is also possible to let l equal the length and then express the volume in terms of l, or to let w equal the width and then express the volume in terms of w. ■

EXERCISES 4.1B

Set I   In Exercises 1–36, if there is only one unknown number, represent it by a variable and then change the word expression into an algebraic expression; if there is more than one unknown number, represent all unknowns in terms of the *same* variable and then change the word expression into an algebraic expression.

1. Fred's salary plus seventy-five dollars

2. Jaime's salary minus forty-two dollars

3. Two less than the number of children in Mr. Moore's family

4. The number of players on Jerry's team plus two more players

5. Four times Joyce's age

6. One-fourth of Rene's age

7. Twenty times the cost of a record, increased by eighty-nine cents

8. Seventeen cents less than 5 times the cost of a ballpoint pen

9. One-fifth the cost of a hamburger

10. The length of the building divided by 8

11. Five times the speed of the car plus 100 miles per hour

12. Twice the speed of a car diminished by 40 miles per hour

13. Ten less than 5 times the square of an unknown number

14. Eight more than 4 times the cube of an unknown number

15. The length of a rectangle is 12 cm more than its width.

16. The altitude of a triangle is 7 cm less than its base.

17. Take the quotient of 7 divided by an unknown number away from 50.

18. Add the quotient of an unknown number divided by 9 to 15.

19. Eleven feet more than twice the length of a rectangle

20. One inch less than the diameter of a circle

21. The combined weight of Walter and Carlos is 320 lb.

22. The combined weight of Teresa and Lucy is 224 lb.

23. 2.2 times the weight in kilograms

24. 0.62 times the distance in kilometers

25. The sum of 8 and an unknown number is divided by the square of that unknown number.

26. The sum of the square of an unknown number and 11 is divided by the unknown number.

27. The sum of 32 and nine-fifths the Celsius temperature

28. Five-ninths times the result of subtracting 32 from the Fahrenheit temperature

29. The sum of three consecutive integers

30. The sum of two consecutive integers

31. The sum of four consecutive odd integers

32. The sum of three consecutive even integers

33. Six plus the radius of a sphere

34. Five plus the radius of a circle

35. The height of a cylinder is 2 times the radius.

36. The radius of a cone is 4 times the height.

37. If the radius of a cylinder is 4 cm less than its height, express the volume of the cylinder in terms of one variable.

38. If the base of a triangle is 5 m less than the altitude, express the area of the triangle in terms of one variable.

39. If the height of a cone is 4 times the radius, express the volume of the cone in terms of one variable.

40. If the radius of a cylinder is 6 times the height, express the volume of the cylinder in terms of one variable.

Set II   In Exercises 1–36, if there is only one unknown number, represent it by a variable and then change the word expression into an algebraic expression; if there is more than one unknown number, represent all unknowns in terms of the *same* variable and then change the word expression into an algebraic expression.

1. The cost of a television set plus thirty-eight dollars

2. The sum of two numbers is 60.

3. Eight less than the length of a certain bridge

4. One-third of Carol's age

5. Eight times the width of a certain window

6. The cost of seven stamps

7. Three times the cost of a videotape increased by forty-five cents

8. Henry is five years younger than his brother Brian.

9. Two-thirds the cost of a compact disc

10. Nine less than one-fourth of Jo's height

11. Ten times the height of a building minus 32 m

12. Mrs. Lopez is twenty-one years older than her daughter Flora.

13. Five less than the product of 4 and x

14. One-half the sum of 7 and an unknown number

15. The altitude of a triangle is 8 cm less than its base.

16. Eight added to the sum of 10 and an unknown number

17. Subtract 51 from the sum of x and y.

18. The sum of two numbers is -22.

19. Three meters less than 3 times the height of the building

20. The speed of the train divided by 6

21. The combined age of Esther and Marge is 43.

22. Twice the result of subtracting an unknown number from 5

23. 8.6 times the length in kilometers

24. Pete has fifty-three dollars less than Ann.

25. The square of the result of subtracting 12 from an unknown number

26. The sum of the squares of 3 and an unknown number

27. The square of the sum of 3 and an unknown number

28. The product of 4 and the square of the length of a side of a square

29. The sum of four consecutive integers

30. The product of two consecutive integers

31. The sum of four consecutive even integers

32. The product of two consecutive odd integers

33. Fourteen plus the radius of a circle

34. Five times the radius of a sphere

35. The radius of a cylinder is 2 times the height.

36. The length of a rectangle is 3 times the width.

37. If the height of a cone is 6 cm more than its radius, express the volume of the cone in terms of one variable.

38. If the length of a rectangle is 9 yd more than its width, express the perimeter of the rectangle in terms of one variable.

39. If the altitude of a triangle is 3 times the base, express the area of the triangle in terms of one variable.

40. If the height of a cylinder is 5 ft less than the radius, express the volume of the cylinder in terms of one variable.

4.2 Translating Word Problems into Equations

In this section, we show how to translate an English sentence into an algebraic equation. (We will not yet be *solving* word problems.)

TO TRANSLATE A WORD PROBLEM INTO AN EQUATION

1. Represent one unknown number by x. Represent any *other* unknowns in terms of x (see Section 4.1).

2. Break up the word problem into small pieces.

3. Represent each piece by an algebraic expression (see Section 4.1).

4. Arrange the algebraic expressions into an equation.

In Examples 1–4, translate each sentence into an equation.

Example 1   Fifteen plus twice an unknown number is 37.
Solution

Step 1. Let x = the unknown number. ◄──── Do not omit this step

Step 2. Fifteen   plus   twice an unknown number   is   37

Step 3.   15   + $2x$ =   37

Step 4.   Equation $15 + 2x = 37$ ■

Example 2   Three times an unknown number is equal to 12 increased by the unknown number.
Solution

Step 1. Let x = the unknown number.

Step 2.	Three	times	an unknown number	is equal to	12	increased by	the unknown number
Step 3.	3		x	=	12	+	x

Step 4. $3x = 12 + x$ ■

Example 3   One-third of an unknown number is 7.
Solution

Step 1. Let x = the unknown number.

Step 2. One-third   of   an unknown number   is   7

Step 3. $\frac{1}{3}$ x =   7

Step 4. $\frac{1}{3}x = 7$ ■

Example 4   Twice the sum of 6 and an unknown number is equal to 20.
Solution

Step 1. Let x = the unknown number.

Step 2. Twice   the sum of 6 and an unknown number   is   20

Step 3.   2 · $(6 + x)$ =   20

Step 4. $2(6 + x) = 20$ ■

Example 5   Write the equation that represents these facts: A piece of wire 63 cm long is to be cut into two pieces. One piece is to be 15 cm longer than the other piece.
Solution (There are *two* unknowns.)

Step 1. Let x = the length of the shorter piece (in centimeters)

              $x + 15$ = the length of the longer piece

Step 2. Sum of lengths of two pieces   is   63

Step 3. $x + (x + 15)$ =   63

Step 4. $x + (x + 15) = 63$ ■

Example 6   Write an equation that represents these facts: The sum of three consecutive odd integers is 75.

Solution

Step 1. Let x = the first odd integer

$x + 2$ = the second odd integer

$x + 4$ = the third odd integer

Step 2. $\boxed{\text{Sum of three consecutive odd integers}}$ is $\boxed{75}$

Steps 3 and 4. $x + (x + 2) + (x + 4)$ $=$ 75 ∎

Example 7   Write an equation that represents these facts: The length of a rectangle is 5 ft more than the width, and the perimeter is 44 ft. (The formula for the perimeter of a rectangle is $P = 2l + 2w$.)

Solution 1

Step 1. Let w = the width of the rectangle

$w + 5$ = the length of the rectangle

Step 2. $\boxed{\text{The perimeter}}$ is $\boxed{44}$

Steps 3 and 4. $2(w + 5) + 2w$ $=$ 44

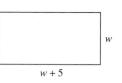

Solution 2

Step 1. Let l = the length of the rectangle

$l - 5$ = the width of the rectangle

Step 2. $\boxed{\text{The perimeter}}$ is $\boxed{44}$

Steps 3 and 4. $2l + 2(l - 5)$ $=$ 44 ∎

EXERCISES 4.2

Set I   In each exercise, write the equation that represents the given facts. Do not solve the equation. Be sure to state what your variable represents.

1. Thirteen more than twice an unknown number is 25.

2. Twenty-five more than 3 times an unknown number is 34.

3. Five times an unknown number, decreased by 8, is 22.

4. Four times an unknown number, decreased by 5, is 15.

5. Seven minus an unknown number is equal to the unknown number plus 1.

6. Six plus an unknown number is equal to 12 decreased by the unknown number.

7. One-fifth of an unknown number is 4.

8. An unknown number divided by 12 equals 6.

9. When 4 is subtracted from one-half of an unknown number, the result is 6.

10. When 5 is subtracted from one-third of an unknown number, the result is 4.

11. Twice the sum of 5 and an unknown number is equal to 26.

12. Four times the sum of 9 and an unknown number is equal to 18.

13. When the sum of an unknown number and itself is multiplied by 3, the result is 24.

14. Five times the sum of an unknown number and itself is 40.

15. A 75-m rope is to be cut into two pieces. One piece is to be 13 m longer than the other piece.

16. A 46-m wire is to be cut into two pieces. One piece is to be 8 m shorter than the other piece.

17. A 72-cm piece of wire is to be cut into two pieces. One piece is to be twice as long as the other piece.

18. A 72-cm piece of wire is to be cut into two pieces. One piece is to be 3 times as long as the other piece.

19. Ingrid buys eight more cans of peaches than cans of pears. Altogether, she buys forty-two cans of these two fruits.

20. Jane buys eleven more cans of peas than cans of corn. Altogether, she buys forty-nine cans of these two vegetables.

21. The sum of four consecutive integers is 106.

22. The sum of three consecutive integers is -72.

23. The length of a rectangle is 4 cm more than its width, and the perimeter is 36 cm.

24. The width of a rectangle is 6 ft less than its length, and the perimeter is 64 ft.

Set II   In each exercise, write the equation that represents the given facts. Do not solve the equation. Be sure to state what your variable represents.

1. Fifteen more than twice an unknown number is 27.

2. Twice the sum of an unknown number and 9 is 46.

3. Four times an unknown number, decreased by 9, is 19.

4. When 5 is subtracted from one-half of an unknown number, the result is 19.

5. Eighteen minus an unknown number is equal to 4 plus the unknown number.

6. When 7 is added to an unknown number, the result is twice that unknown number.

7. Three-eighths of an unknown number is 27.

8. When a number is decreased by 5, the difference is half of the number.

9. When 8 is subtracted from two-thirds of an unknown number, the result is 16.

10. When 3 times an unknown number is subtracted from 20, the result is the unknown number.

11. Three times the result of subtracting an unknown number from 8 is 12.

12. Four times the result of adding an unknown number to itself is 96.

13. When the sum of an unknown number and itself is multiplied by 4, the result is 56.

14. When 6 is subtracted from 5 times an unknown number, the result is the same as when 4 is added to 3 times the unknown number.

15. A 7-yd piece of fabric is to be cut into two pieces. One piece is to be 3 yd longer than the other piece.

16. Terri buys 3 times as many bottles of catsup as jars of mustard. Altogether, she buys twelve containers of these two products.

17. A 42-m rope is to be cut into two pieces. One piece is to be 6 times as long as the other piece.

18. Alice buys 5 times as many skeins (balls) of pink yarn as skeins of white yarn. Altogether, she buys eighteen skeins of these two colors.

19. Todd buys four more cans of car wax than of rubbing compound. Altogether, he buys eighteen cans of these two products.

20. A 30-m cable is to be cut into two pieces. One piece is to be 8 m longer than the other.

21. The sum of four consecutive integers is 2.

22. The sum of three consecutive even integers is 78.

23. The length of a rectangle is 19 yd more than its width, and the perimeter is 62 yd.

24. The lengths of all three sides of a triangle are equal, and the perimeter is 42 cm.

4.3 Solving Word Problems by Using Algebra

In Section 4.2, we showed how to change the words of a written problem into an equation. In this section, we show how to solve word problems, some of which lead to inequalities rather than to equations.

In the following examples of solving mathematical word problems, we use the notation "Step 1," "Step 2," and so forth for the steps you will be *writing*.

SUGGESTIONS FOR SOLVING WORD PROBLEMS

Read To solve a word problem, first read it very carefully. *Be sure you understand the problem.* Read it several times, if necessary.

Think Determine what *type* of problem it is, if possible.* Determine what is unknown. What is being asked for is often found in the last sentence of the problem, which may begin with "What is the . . ." or "Find the . . . ." Is there enough information given so that you *can* solve the problem? Do you need a special formula? What operation(s) must be used?

Sketch Draw a sketch *with labels*, if possible.

Step 1. Represent one unknown number by a variable, and declare its meaning in a sentence of the form "Let $x = \ldots$." Then reread the problem to see how you can represent any other unknown numbers in terms of the same variable.

Reread Reread the entire word problem, breaking it up into small pieces.

Step 2. Translate each English phrase into an algebraic expression and fit these expressions together into an equation or inequality.

Step 3. Using the methods described in Chapter 3, solve the equation or inequality.

Step 4. Solve for *all* the unknowns asked for in the problem.

Step 5. Check the solution(s) *in the word statement.*

Step 6. State your results clearly.

---

*As we mentioned, we discuss several general types of word problems, such as money problems, mixture problems, and so on, in this chapter.

Example 1   Seven increased by 3 times an unknown number is 13. What is the unknown number?
Solution

Step 1. Let x = the unknown number.

Reread	Seven	increased by	3 times	an unknown number	is	13
Step 2.	7	+	3 ·	x	=	13

Step 3.
$$7 + 3x = 13$$
$$\underline{-7 \qquad = -7}\qquad \text{Adding } -7 \text{ to both sides}$$
$$3x = 6$$

$$\frac{3x}{3} = \frac{6}{3}\qquad \text{Dividing both sides by 3}$$

Step 4.
$$x = 2$$

Step 5. *Check*

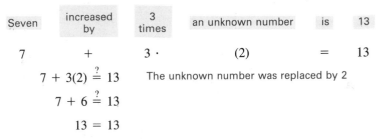

	Seven	increased by	3 times	an unknown number	is	13
	7	+	3 ·	(2)	=	13

$$7 + 3(2) \overset{?}{=} 13 \qquad \text{The unknown number was replaced by 2}$$
$$7 + 6 \overset{?}{=} 13$$
$$13 = 13$$

Step 6. Therefore, the unknown number is 2. ∎

NOTE   To check a word problem, you must check the solution in the *word statement*. Any error that may have been made in writing the equation will not be discovered if you simply substitute the solution into the equation.   ☑

Example 2   Four times an unknown number is equal to twice the sum of 5 and that unknown number. Find the unknown number.
Solution

Step 1. Let x = the unknown number.

Reread	Four times	an unknown number	is equal to	twice the sum of 5 and that unknown number
Step 2.	4 ·	x	=	$2 \cdot (5 + x)$

Step 3.
$$4x = 2(5 + x)$$

$$4x = 10 + 2x \qquad \text{Using the distributive property}$$
$$\underline{-2x \qquad\;\; - 2x}\qquad \text{Adding } -2x \text{ to both sides}$$
$$2x = 10$$

$$\frac{2x}{2} = \frac{10}{2}\qquad \text{Dividing both sides by 2}$$

Step 4. $x = 5$

Step 5. *Check* Four times 5 is 20. The sum of 5 and 5 is 10, and twice 10 is 20.

Step 6. Therefore, the unknown number is 5. ∎

Example 3   When 7 is subtracted from one-half of an unknown number, the result is 11. What is the unknown number?
Solution

Step 1. Let x = the unknown number.

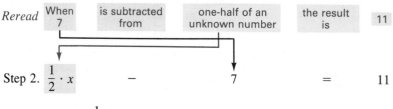

Reread

| When 7 | is subtracted from | one-half of an unknown number | the result is | 11 |

Step 2. $\frac{1}{2} \cdot x$ $-$ 7 $=$ 11

Step 3. $\frac{1}{2}x - 7 = 11$

                     $\underline{ 7 \quad 7}$ Adding 7 to both sides

                     $\frac{1}{2}x \quad = 18$

                 $2\left(\frac{1}{2}x\right) = 2(18)$ Multiplying both sides by 2

Step 4. $x = 36$

Step 5. *Check* One-half of 36 is 18. When 7 is subtracted from 18, the result is 11.

Step 6. Therefore, the unknown number is 36. ∎

It is shown in geometry that the sum of the measures of the angles of a triangle is always 180°. Example 4 uses this property.

Example 4   If the measure of one angle of a triangle is 73° and the measure of another angle is 54°, what is the measure of the third angle?
Solution

Step 1. Let x = the number of degrees of the third angle.

Reread

| The sum of all the angles | equals | 180 |

Step 2. $73 + 54 + x$ $=$ 180

Step 3. $73 + 54 + x =$ 180

                 $127 + x =$ 180 Combining like terms
                 $\underline{-127}$ $\underline{-127}$ Adding −127 to both sides

Step 4. $x =$ 53

Step 5. *Check* $73° + 54° + 53° = 180°$

Step 6. Therefore, the measure of the third angle is 53°. ∎

Example 5   The length of a rectangle is 25 in. and the perimeter is 84 in. Find the width of the rectangle.
Think The formula for the perimeter of a rectangle is $P = 2l + 2w$.

Solution

Step 1. Let w = the width of the rectangle.

Sketch:

$$2l + 2w \quad = \quad P$$

Reread The length is 25, the width is w the perimeter is 84

Step 2. $\quad 2(25) + 2w \quad = \quad$ 84

Step 3. $2(25) + 2w = 84$

$$
\begin{array}{rl}
50 + 2w = & 84 \qquad \text{Simplifying} \\
\underline{-50 \qquad\quad -50} & \qquad \text{Adding } -50 \text{ to both sides} \\
2w = & 34
\end{array}
$$

$$\frac{2w}{2} = \frac{34}{2} \qquad \text{Dividing both sides by 2}$$

Step 4. $\qquad\qquad w = 17$

Step 5. *Check* The perimeter is $2(25 \text{ in.}) + 2(17 \text{ in.}) = 50 \text{ in.} + 34 \text{ in.} = 84 \text{ in.}$

Step 6. Therefore, the width of the rectangle is 17 in. ■

Example 6 illustrates the solution of a word problem that leads to an inequality.

Example 6   In an English class, any student needs at least 730 points in order to earn an A, and the final exam is worth 200 points. If Shirley has 545 points just before the final exam, what range of scores will give her an A for the course?
Solution

Step 1. Let x = Shirley's score on the final.

Reread   Score going into final   plus   score on final   is greater than or equal to   730

Step 2.   545   + x $\geq$ 730

Step 3.
$$
\begin{array}{rl}
545 + x \geq & 730 \\
\underline{-545 \qquad\quad -545} & \qquad \text{Adding } -545 \text{ to both sides}
\end{array}
$$
Step 4. $\qquad\qquad x \geq 185$

Step 5. *Check* Shirley's score on the final exam can't be greater than 200, since the final exam is worth only 200 points. We will show the check for three different numbers:

Score of 185: $185 + 545 = 730$, and $730 \geq 730$

Score of 192: $192 + 545 = 737$, and $737 \geq 730$

Score of 200: $200 + 545 = 745$, and $745 \geq 730$

Any other numbers between 185 and 200 would also have checked.

Step 6. Therefore, the range of scores that will give Shirley an A for the course is any score greater than or equal to 185 (and less than or equal to 200, since the final exam is worth 200 points). ■

EXERCISES 4.3

Set I In each exercise, set up the problem algebraically and solve. Be sure to state what your variable represents.

In Exercises 1–10, find the unknown number and check your solution.

1. When twice an unknown number is added to 13, the sum is 25.

2. When 25 is added to 3 times an unknown number, the sum is 34.

3. Five times an unknown number, decreased by 8, is 22.

4. Four times an unknown number, decreased by 5, is 15.

5. Seven minus an unknown number is equal to the unknown number plus 1.

6. Six plus an unknown number is equal to 12 decreased by the unknown number.

7. When 4 is subtracted from one-half of an unknown number, the result is 6.

8. When 5 is subtracted from one-third of an unknown number, the result is 4.

9. Twice the sum of 5 and an unknown number is equal to 26.

10. Four times the sum of 9 and an unknown number is equal to 20.

In Exercises 11–28, solve for the unknowns and check your solutions.

11. A 36-cm piece of wire is to be cut into two pieces. One of the pieces is to be 10 cm longer than the other piece. Find the length of each piece.

12. A 10-yd piece of fabric is to be cut into two pieces. One of the pieces is to be 2 yd longer than the other piece. Find the length of each piece.

13. Rebecca buys seven more packages of dried apples than packages of dried apricots. If she buys fifteen packages of these two fruits altogether, how many of each does she buy?

14. Susan buys three more skeins of blue yarn than skeins of green yarn. If she buys thirteen skeins of these two colors altogether, how many of each color does she buy?

15. Find the area of this triangle:

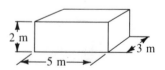

16. Find the volume and the surface area of this rectangular solid:

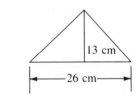

17. Find the approximate volume and surface area of this sphere (round off the answers to two decimal places):

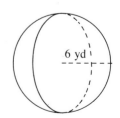

18. Find the approximate area and circumference of this circle (round off the answers to two decimal places):

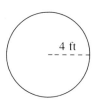

4 ft

19. The sum of three consecutive odd integers is 177. Find the integers.

20. The sum of three consecutive even integers is -144. Find the integers.

21. Four times the sum of the first and third of three consecutive integers is 140 more than the second integer. Find the integers.

22. Three times the sum of the first and third of three consecutive integers is 75 more than the second integer. Find the integers.

23. If the measure of one angle of a triangle is $62°$ and the measure of another angle is $47°$, what is the measure of the third angle?

24. If the measure of one angle of a triangle is $18°$ and the measure of another angle is $37°$, what is the measure of the third angle?

25. The length of a rectangle is twice the width, and the perimeter is 102 ft. Find the length and the width.

26. The length of a rectangle is 4 times the width, and the perimeter is 160 cm. Find the length and the width.

27. The width of a rectangle is 5 ft, and the perimeter is 44 ft. Find the length of the rectangle.

28. The length of a rectangle is 7 yd, and the perimeter is 38 yd. Find the width of the rectangle.

In Exercises 29–36, find the unknown number and check your solution. In Exercises 35 and 36, round off the answers to two decimal places.

29. When twice the sum of 4 and an unknown number is added to the unknown number, the result is the same as when 10 is added to the unknown number.

30. When 6 is subtracted from 5 times an unknown number, the result is the same as when 4 is added to 3 times the unknown number.

31. Three times the sum of 8 and twice an unknown number is equal to 4 times the sum of 3 times the unknown number and 8.

32. Five times the sum of 4 and 6 times an unknown number is equal to 4 times the sum of twice the unknown number and 10.

33. When 3 times the sum of 4 and an unknown number is subtracted from 10 times the unknown number, the result is equal to 5 times the sum of 9 and twice the unknown number.

34. When twice the sum of 5 and an unknown number is subtracted from 5 times the sum of 6 and twice the unknown number, the result is equal to zero.

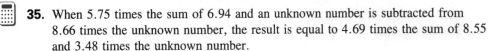

35. When 5.75 times the sum of 6.94 and an unknown number is subtracted from 8.66 times the unknown number, the result is equal to 4.69 times the sum of 8.55 and 3.48 times the unknown number.

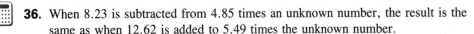

36. When 8.23 is subtracted from 4.85 times an unknown number, the result is the same as when 12.62 is added to 5.49 times the unknown number.

In Exercises 37–40, find all possible solutions for each problem.

37. The sum of an unknown number and 18 is to be at least 5. What is the range of values that the unknown number can have?

38. The sum of an unknown number and 12 is to be at least 47. What is the range of values that the unknown number can have?

39. A rope less than 80 m long is to be cut into two pieces. One piece must be 37 m long. What will the length of the other piece be?

40. A chain more than 25 m long is to be cut into two pieces. One piece must be 8 m long. What will the length of the other piece be?

Set II In each exercise, set up the problem algebraically and solve. Be sure to state what your variable represents.

In Exercises 1–10, find the unknown number and check your solution.

1. When 4 times an unknown number is added to 21, the sum is 105.

2. When 7 is added to an unknown number, the result is twice that unknown number.

3. Three times an unknown number, decreased by 12, is 6.

4. When 3 times an unknown number is subtracted from 20, the result is the unknown number.

5. Eighteen minus an unknown number is equal to the unknown number minus 16.

6. One-fifth of an unknown number is 4.

7. When 3 is subtracted from one-third of an unknown number, the result is 7.

8. An unknown number divided by 12 equals 6.

9. Five times the sum of an unknown number and itself is 40.

10. When an unknown number is subtracted from 12, the difference is one-third of the number.

In Exercises 11–28, solve for the unknowns and check your solutions.

11. A 12-yd piece of fabric is to be cut into two pieces. One of the pieces must be 3 yd longer than the other piece. Find the length of each piece.

12. Tom buys 5 times as many cassette tapes as compact discs. If he purchases twelve of these items altogether, how many of each kind does he buy?

13. Irene buys four more packages of unsalted crackers than packages of salted crackers. If she buys sixteen packages of crackers altogether, how many of each kind does she buy?

14. A 42-m piece of cord is to be cut into two pieces. If one piece must be 15 m long, what will the length of the other piece be?

15. Find the area and perimeter of this rectangle:

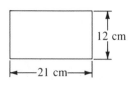

16. Find the area and perimeter of this square:

|←—14 yd—→|

17. Find the approximate volume of this cone (round off the answer to two decimal places):

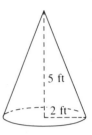

5 ft

2 ft

18. Find the approximate volume of this cylinder (round off the answer to two decimal places):

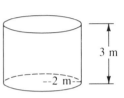

3 m

2 m

19. The sum of three consecutive even integers is −102. Find the integers.

20. The sum of three consecutive odd integers is 27. Find the integers.

21. Twice the sum of the first and third of three consecutive integers is 102 more than the second integer. Find the integers.

22. Three times the sum of the first and third of three consecutive odd integers is 55 more than the second integer. Find the integers.

23. If the measure of one angle of a triangle is 90° and the measure of another angle is 37°, what is the measure of the third angle?

24. If the measures of the three angles of a triangle are equal, find the measure of the angles.

25. The length of a rectangle is 5 times the width, and the perimeter is 204 ft. Find the length and the width.

26. The length of a rectangle is 4 m, and the area is 20 sq. m. Find the width of the rectangle.

27. The length of a rectangle is 7 cm, and the perimeter is 24 cm. Find the width of the rectangle.

28. The base of a triangle is 4 yd, and the area is 38 sq. yd. Find the altitude of the triangle.

In Exercises 29–36, find the unknown number and check your solution.

29. When 4 times the sum of 5 and an unknown number is added to the unknown number, the result is the same as when 32 is added to the unknown number.

30. Twice the result of subtracting an unknown number from 5 is 8.

31. Twice the sum of 3 and twice an unknown number is equal to 5 times the sum of the unknown number and 1.

32. When an unknown number is subtracted from 11, the result is the same as when the unknown number is added to 3.

33. If the sum of an unknown number and 12 is subtracted from 4 times the unknown number, the result is the unknown number less 4.

34. When 4 times an unknown number is subtracted from 16, the result is twice the sum of 12 and twice the unknown number.

35. When 3.48 times the sum of 9.06 and an unknown number is subtracted from 5.37 times the unknown number, the result is equal to 4.65 times the sum of 2.83 and 8.34 times the unknown number. Round off the answer to two decimal places.

36. Five times the sum of 18 and an unknown number is 2 less than the unknown number.

In Exercises 37–40, find all possible solutions for each problem.

37. The sum of an unknown number and 7 is to be at least 3. What is the range of values that the unknown number can have?

38. Five times an unknown number, plus 7, is greater than 42. What is the range of values that the unknown number can have?

39. A piece of wire less than 35 cm long is to be cut into two pieces. One piece must be 13 cm long. What is the range of values for the length of the other piece?

40. A rope less than 45 m long is to be cut into two pieces. One piece must be 4 times as long as the other piece. What is the range of values for the *shorter* piece?

4.4 Money Problems

In this section, we discuss a type of word problem commonly referred to as a "coin problem." Not all problems in this section deal with coins, but the method of solving the problems is essentially the same.

4.4A Getting Ready to Solve Money Problems

One important relationship used in solving coin or money problems is the following:

$$\begin{pmatrix} \text{The value per} \\ \text{item of one} \\ \text{kind of item} \end{pmatrix} \times \begin{pmatrix} \text{The number} \\ \text{of} \\ \text{those items} \end{pmatrix} = \begin{pmatrix} \text{The total value} \\ \text{of that kind} \\ \text{of item} \end{pmatrix}$$

Example 1   If you have eight nickels, the total number of cents is

$$5(8) = 40$$

Number of nickels

Value of one nickel ■

Example 2   If you have seven quarters, the total number of cents is

$$25(7) = 175$$

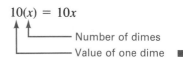

    Number of quarters
    Value of one quarter ■

Example 3   If you have x dimes, the total number of cents is

$$10(x) = 10x$$

    Number of dimes
    Value of one dime ■

Example 4   If you have y 35¢ candy bars, the total value in cents is

$$35(y) = 35y$$

    Number of candy bars
    Value of one candy bar ■

Another important relationship used in solving coin or money problems is this:

$$\left(\begin{array}{c}\text{The value}\\ \text{of the first}\\ \text{kind of item}\end{array}\right) + \left(\begin{array}{c}\text{The value}\\ \text{of the second}\\ \text{kind of item}\end{array}\right) = \left(\begin{array}{c}\text{The total value}\\ \text{of both kinds}\\ \text{of items}\end{array}\right)$$

The relationship can, of course, be extended to more than two kinds of items. We use this relationship in Examples 5–8.

Example 5   If you have two adults' movie tickets costing $5.00 each and four children's tickets costing $2.50 each, the total value in dollars is

Cost of one adult's ticket
Number of adults' tickets

$$5.00(2) + 2.50(4) = 10 + 10 = 20$$

Number of children's tickets
Cost of one child's ticket ■

Example 6   If you have x 10¢ stamps and y 12¢ stamps, the total value in cents is

Cost of one 10¢ stamp
Number of 10¢ stamps

$$10(x) + 12(y) = 10x + 12y$$

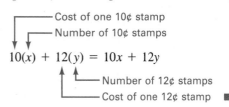

Number of 12¢ stamps
Cost of one 12¢ stamp ■

EXERCISES 4.4A

Set I   In each exercise, express the total amount either as a number or as an algebraic expression.

1. If you have seven nickels, what amount of money is this?

2. If you have eleven quarters, what amount of money is this?

3. If you have nine 50¢ pieces, what amount of money is this?

4. If you have twelve dimes, what amount of money is this?

5. If you have x quarters, what amount of money is this?

6. If you have y nickels, what amount of money is this?

7. Find the total cost of seven adults' tickets and five children's tickets if adults' are $1.50 each and children's are 75¢ each.

8. If adults' tickets are $2.50 each and children's tickets are $1.25 each, what will four adults' and nine children's tickets cost?

9. If adults' tickets are $3.50 each and children's tickets are $1.90 each, what will x adults' and y children's tickets cost?

10. Find the total cost of x adults' and y children's tickets if adults' tickets cost $2.75 each and children's tickets cost $1.50 each.

11. Find the total cost of x 25¢ stamps and y 6¢ stamps.

12. Find the total cost of x 4¢ stamps and y 2¢ stamps.

Set II   In each exercise, express the total amount either as a number or as an algebraic expression.

1. If you have nine nickels, what amount of money is this?

2. If you have eleven dimes, what amount of money is this?

3. If you have fifteen dimes, what amount of money is this?

4. If you have twenty-three quarters, what amount of money is this?

5. If you have x 50¢ coins, what amount of money is this?

6. If you have x dimes and $(x + 4)$ nickels, what amount of money is this?

7. If adults' tickets cost $4.75 each and children's tickets cost $2.75 each, what will five adults' and three children's tickets cost?

8. If adults' tickets cost $5.25 each and children's tickets cost $3.00 each, what will x adults' tickets and $4x$ children's tickets cost?

9. Find the total cost of x adults' and y children's tickets if adults' tickets cost $4.50 each and children's tickets cost $2.25 each.

10. Find the total cost of x adults' and $(x + 6)$ children's tickets if adults' tickets cost $5.00 each and children's tickets cost $2.50 each.

11. Find the total cost of x 13¢ stamps and y 15¢ stamps.

12. Find the total cost of x 22¢ stamps and $5x$ 18¢ stamps.

4.4B   Solving Money Problems

We now combine the suggestions given in Section 4.4A with the suggestions for solving all word problems given in Section 4.3.

Example 7   A piggy bank contains seventeen coins with a total value of $1.15. If the coins are all nickels and dimes, how many of each kind of coin are there?

Think This is a money problem; let's express everything in terms of cents. There are two unknown numbers and, therefore, two choices for the variable. We can let D be the number of dimes or N be the number of nickels. We'll let D be the number of dimes; if there are seventeen coins altogether, then there must be $(17 - D)$ nickels.

Solution

Step 1. Let D = the number of dimes

$17 - D$ = the number of nickels

Think The value of each dime is 10¢; therefore, the value of D dimes is $10D$ cents. The value of each nickel is 5¢; therefore, the value of $(17 - D)$ nickels is $5(17 - D)$ cents.

Reread   Value of dimes   plus   value of nickels   is   115¢

Step 2. $10D$ $+$ $5(17 - D)$ $=$ 115

Step 3. $10D + 5(17 - D) = 115$

$10D + 85 - 5D = 115$ Using the distributive property

$5D + 85 = 115$ Combining like terms
$\underline{\quad - 85 \quad - 85}$ Adding -85 to both sides
$5D \quad = \quad 30$

$\dfrac{5D}{5} = \dfrac{30}{5}$ Dividing both sides by 5

Step 4. $D = 6$ Number of dimes

$17 - D = 17 - 6 = 11$ Number of nickels

Step 5. *Check*   6 dimes: $6(10¢) = 60¢$

11 nickels: $11(5¢) = \underline{55¢}$
$115¢ = \$1.15$

Step 6. Therefore, there are six dimes and eleven nickels. ∎

We solved the problem in Example 7 by letting D be the number of dimes. We suggest that you solve the problem by letting N be the number of nickels. Your final answer should match the answer in Example 7.

Example 8   Dianne has $3.20 in nickels, dimes, and quarters. If she has seven more dimes than quarters and 3 times as many nickels as quarters, how many of each kind of coin does she have?

Think This is a money problem; let's express everything in terms of cents. There are three unknown numbers and, therefore, three choices for the variable. We can let N be the number of nickels, D be the number of dimes, or Q be the number of quarters. Because the problem expressed both the number of dimes and the number of nickels in terms of the number of quarters, it will be easiest to let Q be the number of quarters.

Solution

Step 1. Let Q = the number of quarters

$Q + 7$ = the number of dimes   Because there are seven more dimes than quarters

$3Q$ = the number of nickels   Because there are 3 times as many nickels as quarters

Think The value of each quarter is 25¢, so the value of Q quarters is $25Q$ cents. The value of $(Q + 7)$ dimes is $10(Q + 7)$ cents; the value of $(3Q)$ nickels is $5(3Q)$ cents.

Reread	Value of quarters	plus	value of dimes	plus	value of nickels	is	320¢
Step 2.	$25Q$	$+$	$10(Q + 7)$	$+$	$5(3Q)$	$=$	320

Step 3. $\qquad 25Q + 10(Q + 7) + 5(3Q) = 320$

$\qquad\qquad 25Q + 10Q + 70 + 15Q = 320 \qquad$ Simplifying

$$50Q + 70 = 320 \qquad \text{Combining like terms}$$
$$\underline{-70 \quad -70} \qquad \text{Adding } -70 \text{ to both sides}$$
$$50Q = 250$$

$$\frac{50Q}{50} = \frac{250}{50} \qquad \text{Dividing both sides by 50}$$

Step 4. $\qquad\qquad\qquad\qquad Q = 5 \qquad$ Number of quarters

$\qquad\qquad\qquad Q + 7 = 5 + 7 = 12 \qquad$ Number of dimes

$\qquad\qquad\qquad\qquad 3Q = 3(5) = 15 \qquad$ Number of nickels

Step 5. *Check* $\qquad$ 5 quarters: $5(25¢) = 125¢$

$\qquad\qquad\qquad$ 12 dimes: $12(10¢) = 120¢$

$\qquad\qquad\qquad$ 15 nickels: $15(5¢) = \underline{75¢}$

$\qquad\qquad\qquad\qquad\qquad\qquad 320¢ = \$3.20$

Step 6. Therefore, there are five quarters, twelve dimes, and fifteen nickels. ∎

EXERCISES 4.4B

Set I   Set up each problem algebraically, solve, and check. Be sure to state what your variables represent.

1. Bill has thirteen coins in his pocket that have a total value of 95¢. If these coins consist of nickels and dimes, how many of each kind are there?

2. Miko has eleven coins that have a total value of 85¢. If the coins are only nickels and dimes, how many of each kind are there?

3. Jennifer has twelve coins that have a total value of $2.20. The coins are nickels and quarters. How many of each kind of coin are there?

4. Brian has eighteen coins consisting of nickels and quarters. If the total value of the coins is $2.50, how many of each kind of coin does he have?

5. Derek has $4.00 in nickels, dimes, and quarters. If he has four more quarters than nickels and 3 times as many dimes as nickels, how many of each kind of coin does he have?

6. Staci has $5.50 in nickels, dimes, and quarters. If she has seven more dimes than nickels and twice as many quarters as dimes, how many of each kind of coin does she have?

7. Michael has $2.25 in nickels, dimes, and quarters. If he has three fewer dimes than quarters and as many nickels as the sum of the dimes and quarters, how many of each kind of coin does he have?

8. Muriel has $2.57 in dimes, nickels, and pennies. If she has five fewer pennies than nickels and as many dimes as the sum of the nickels and pennies, how many of each kind of coin does she have?

9. The total receipts for a concert were $19,800 for the 1,080 tickets sold. The promoters sold orchestra seats for $21 each, box seats for $30 each, and balcony seats for $12 each. If there were 5 times as many balcony seats as box seats sold, how many of each kind were sold?

10. The total receipts for a football game were $1,443,200. General admission tickets cost $20 each, reserved seat tickets were $32 each, and box seat tickets $40 each. If there were twice as many general admission as reserved seat tickets sold and 4 times as many reserved as box seat tickets sold, how many of each kind were sold?

11. Christy spent $3.80 for sixty stamps. She bought only 2¢, 10¢, and 12¢ stamps. If she bought twice as many 10¢ stamps as 12¢ stamps, how many of each kind did she buy?

12. Mark spent $9.80 for 100 stamps. He bought only 6¢, 8¢, and 12¢ stamps. If there were 3 times as many 6¢ stamps as 8¢ stamps, how many of each kind did he buy?

Set II Set up each problem algebraically, solve, and check. Be sure to state what your variables represent.

1. Don spent $2.36 for twenty-two stamps. If he bought only 10¢ stamps and 12¢ stamps, how many of each kind did he buy?

2. Karla has fifteen coins with a total value of $2.55. If the coins are only dimes and quarters, how many of each kind of coin does she have?

3. Jill has fifteen coins with a total value of $2.70. If the coins are only dimes and quarters, how many of each kind of coin does she have?

4. Several families went to a movie together. They spent $24.75 for eight tickets. If adults' tickets cost $4.50 each and children's tickets cost $2.25 each, how many of each kind of ticket were bought?

5. Rachelle has $4.75 in nickels, dimes, and quarters. If she has four more nickels than dimes and twice as many quarters as dimes, how many of each kind of coin does she have?

6. Jason has $4.80 in nickels, dimes, and quarters. If he has three more dimes than quarters and 3 times as many nickels as quarters, how many of each kind of coin does he have?

7. Kevin has $4.75 in nickels, dimes, and quarters. If he has two more dimes than nickels and twice as many quarters as nickels, how many of each kind of coin does he have?

8. Anne spent $2.88 for fifteen stamps. She bought only 22¢ and 15¢ stamps. How many of each kind did she buy?

9. A class received $233 for selling 200 tickets to the school play. If students' tickets cost $1 each and nonstudents' tickets cost $2 each, how many nonstudent tickets were sold?

10. Heather spent $3.02 for stamps. She bought only 22¢ and 15¢ stamps, and she bought seven more 22¢ stamps than 15¢ stamps. How many of each kind did she buy?

11. Tricia spent \$5.72 for twenty-six stamps. She bought only 22¢, 18¢, and 25¢ stamps. If she bought twice as many 22¢ stamps as 18¢ stamps and two more 25¢ stamps than 18¢ stamps, how many of each kind did she buy?

12. The total receipts for a concert were \$26,000. Some tickets cost \$20 each, some cost \$28 each, and the rest cost \$36 each. If there were eighty more \$28 tickets sold than \$36 tickets, and 10 times as many \$20 as \$36 tickets, how many of each kind were sold?

4.5 Ratio and Rate Problems

Ratio

A **ratio** is the quotient of one quantity divided by another quantity of the same kind. "The ratio of a to b" is written as $\frac{a}{b}$, and a and b are called the **terms** of the ratio. The terms of a ratio can be any kind of number, the only restriction being that the denominator cannot be zero. The ratio of a to b is also sometimes written as $a : b$, but this representation hides the fact that a ratio is a fraction or a rational number. (Notice the word *ratio* in the word *rational*.) We will emphasize the fraction meaning of ratio.

A WORD OF CAUTION   "The ratio of a to b" is *not* $\frac{b}{a}$. It is $\frac{a}{b}$. The number that is *before* the word *to* is always in the numerator. The number that is *after* the word *to* is always in the denominator.   ☑

We can give three different meanings to an expression such as $\frac{3}{4}$:

1. 3 of the 4 equal parts a unit has been divided into (fraction meaning)

2. $3 \div 4$ (division meaning)

3. The ratio of 3 to 4 (ratio meaning)

The meaning chosen depends on how the expression is used. The ratio meaning is used when we *compare* two numbers by division.

Example 1   An algebra class has thirteen men and seventeen women.

a. The ratio of men to women is $\frac{13}{17}$.

b. The ratio of women to men is $\frac{17}{13}$.   ■

The key to solving ratio problems is to use the given ratio to help represent the unknown numbers. We know from arithmetic that $\frac{3}{4} = \frac{6}{8}, \frac{12}{16}, \frac{15}{20}, \ldots$; in algebra, we can write this fact as $\frac{3}{4} = \frac{3x}{4x}$, where x represents any real number except 0. Therefore, if we're given some facts about two numbers and are also told that those numbers are in the ratio of 3 to 4, we can let the smaller number be $3x$ and the larger number be $4x$ (see Examples 2–5).

In general, when we're following the suggestions given in Section 4.3 for solving word problems, instead of representing the unknown number by x in step 1, we will substitute the following rule:

TO REPRESENT THE UNKNOWNS IN A RATIO PROBLEM

Multiply each term of the ratio by x and let the resulting products represent the unknowns.

A WORD OF CAUTION   In ratio problems, you are not finished when you have found x. You must multiply the value of x by the terms of the ratio in order to find the unknown numbers. ☑

Example 2   Two numbers are in the ratio of 3 to 5. Their sum is 32. Find the numbers. (The same problem could have been worded as follows: "Divide 32 into two parts whose ratio is $3:5$.")

Solution

Think This is a ratio problem.

$3 \ :5$ The ratio

$3x:5x$ Multiply each term of the ratio by x and
               let the resulting *products* represent the unknowns

Step 1.   Let $5x$ = one number.

               Let $3x$ = the other number.

Reread Their sum   is   32

Step 2. $3x + 5x = 32$

Step 3. $8x = 32$

               $\dfrac{8x}{8} = \dfrac{32}{8}$

               $x = 4$ *Warning:* We are not finished when we have found x; the unknown numbers are $3x$ and $5x$

Step 4. $3x = 3(4) = 12$ Smaller number

               $5x = 5(4) = 20$ Larger number

Step 5. *Check* The ratio of 12 to 20 is $\dfrac{12}{20} = \dfrac{3}{5}$, and $12 + 20 = 32$.

Step 6. Therefore, the numbers are 12 and 20. ■

Example 3   The three sides of a triangle are in the ratio $2:3:4$. The perimeter is 63 ft. Find the lengths of the three sides.

Solution

Think This is a ratio problem. The formula for the perimeter of a triangle is $P = a + b + c$.

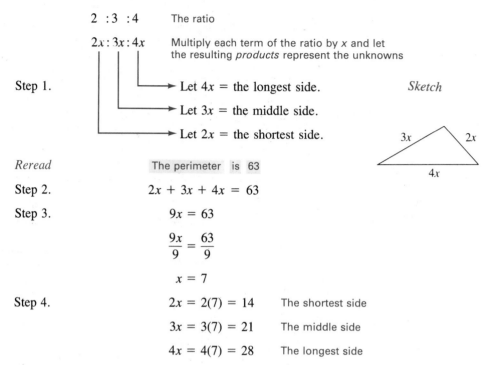

	2 : 3 : 4	The ratio
	$2x : 3x : 4x$	Multiply each term of the ratio by x and let the resulting *products* represent the unknowns

Step 1.
Let $4x$ = the longest side.
Let $3x$ = the middle side.
Let $2x$ = the shortest side.

Sketch

Reread   The perimeter is 63

Step 2. $2x + 3x + 4x = 63$

Step 3. $9x = 63$

$$\frac{9x}{9} = \frac{63}{9}$$

$$x = 7$$

Step 4. $2x = 2(7) = 14$ The shortest side

$3x = 3(7) = 21$ The middle side

$4x = 4(7) = 28$ The longest side

Step 5. *Check* Because $14 = 2(\,7\,)$, $21 = 3(\,7\,)$, and $28 = 4(\,7\,)$, 14, 21, and 28 are in the ratio $2:3:4$. The perimeter is 14 ft + 21 ft + 28 ft, or 63 ft.

Step 6. Therefore, the lengths of the three sides are 14 ft, 21 ft, and 28 ft. ∎

Example 4   Sixty-six hours of a student's week are spent in study, in class, and in work. The times spent in these activities are in the ratio $4:2:5$. How many hours are spent in each activity?
Solution

Think This is a ratio problem. Total hours per week on the three activities is 66 hr.

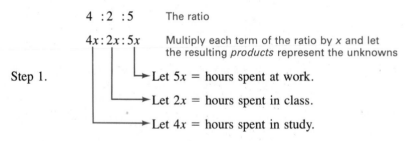

	4 : 2 : 5	The ratio
	$4x : 2x : 5x$	Multiply each term of the ratio by x and let the resulting *products* represent the unknowns

Step 1.
Let $5x$ = hours spent at work.
Let $2x$ = hours spent in class.
Let $4x$ = hours spent in study.

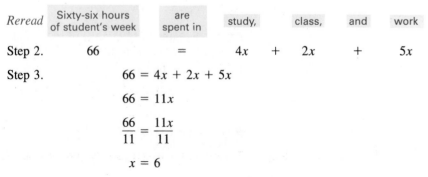

Reread	Sixty-six hours of student's week	are spent in	study,	class,	and	work
Step 2.	66	=	$4x$	+ $2x$	+	$5x$

Step 3. $66 = 4x + 2x + 5x$

$66 = 11x$

$$\frac{66}{11} = \frac{11x}{11}$$

$$x = 6$$

Step 4.

$$4x = 4(6) = 24 \qquad \text{Study hours}$$

$$2x = 2(6) = 12 \qquad \text{Class hours}$$

$$5x = 5(6) = 30 \qquad \text{Work hours}$$

Step 5. *Check* Because $24 = 4(6)$, $12 = 2(6)$, and $30 = 5(6)$, 24, 12, and 30 are in the ratio $4:2:5$. Also, 24 hr $+ 12$ hr $+ 30$ hr $= 66$ hr.

Step 6. Therefore, the student spends 24 hr per week studying, 12 hr in class, and 30 hr working. ■

Example 5   The length and width of a rectangle are in the ratio of 7 to 5. The perimeter is to be greater than 72. What is the range of values for the *length*?
Solution

Think This is a ratio problem. The formula for the perimeter of a rectangle is $P = 2l + 2w$.

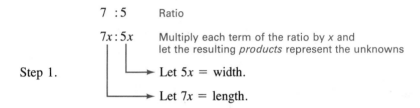

7 :5   Ratio

$7x : 5x$ Multiply each term of the ratio by x and let the resulting *products* represent the unknowns

Step 1.   └──► Let $5x =$ width.

└──────► Let $7x =$ length.

Reread	The perimeter	is greater than	72		*Sketch*
Step 2.	$2(7x) + 2(5x)$	$>$	72		
Step 3.	$14x + 10x$	$>$	72		
	$24x$	$>$	72		
	$\dfrac{24x}{24}$	$>$	$\dfrac{72}{24}$		
	x	$>$	3		
Step 4.	$7x$	$>$	$7(3)$	The length is $7x$	
	$7x$	$>$	21	The length must be greater than 21	

(Sketch shows a rectangle with side labeled $5x$ and bottom labeled $7x$.)

Step 5. *Check* We will show the check for two values substituted for the length.

Length $= 28$: If the length is 28, $7x = 28$, so $x = 4$ and the width must be $5(4)$, or 20. The perimeter, then, is

$$2(28) + 2(20) = 56 + 40 = 96$$

which is greater than 72.

Length $= 47$: If the length is 47, $7x = 47$, so $x = \frac{47}{7}$ and the width must be $5\left(\frac{47}{7}\right)$, or $\frac{235}{7}$. The perimeter is

$$2(47) + 2\left(\frac{235}{7}\right) = 94 + \frac{470}{7} = \frac{1128}{7} \approx 161.14$$

which is greater than 72.

Step 6. Therefore, the length must be greater than 21. ■

We will consider ratios again in Section 8.9.

Rate

We often find it necessary or desirable to express in fraction form the quotient of one quantity divided by a quantity of a *different* kind. These fractions are technically called *rates*, although many authors do not distinguish between a ratio and a rate. You are undoubtedly familiar with rates such as $\frac{\text{miles}}{\text{hour}}$ (miles per hour), $\frac{\text{miles}}{\text{gallon}}$ (miles per gallon), $\frac{\text{dollars}}{\text{hour}}$ (dollars per hour), and so on.

Example 6   A man earns $103 for every 8 hours he works. Express the rate of dollars to hours as a fraction.

Solution The rate is $\frac{103 \text{ dollars}}{8 \text{ hours}}$. ■

Example 7   Gloria can drive about 533 miles on 14 gallons of gasoline. Express the rate of miles to gallons as a fraction.

Solution The rate of miles to gallons is $\frac{533 \text{ miles}}{14 \text{ gallons}}$. ■

When you solve problems involving rates, it is helpful to cancel the units whenever possible (see Example 8).

Example 8   Mrs. Norton's car uses gasoline at the rate of $16 \frac{\text{miles}}{\text{gallon}}$. How many gallons of gasoline will she need in order to drive 464 miles?

Think This is a rate problem. Multiplying and "canceling units" will give

$$\left(\frac{\text{miles}}{\text{gallon}}\right)(\text{gallons}) = \text{miles}$$

Solution

Step 1. Let x = the number of gallons of gasoline.

Step 2. $\left(16 \frac{\text{miles}}{\text{gallon}}\right)(x \text{ gallons}) = 464$ miles   We cancel "gallons"

We now drop the word "miles"

Step 3. $16x$ miles = 464 miles

$$16x = 464$$

$$\frac{16x}{16} = \frac{464}{16}$$

Step 4. $x = 29$

Step 5. *Check* $\left(16 \frac{\text{miles}}{\text{gallon}}\right)(29 \text{ gallons}) = 464$ miles

Step 6. Therefore, she will need 29 gallons of gasoline. ■

We will consider rates again in Sections 4.7, 8.9, and 8.11.

EXERCISES 4.5

Set I   In Exercises 1–14, set up each problem algebraically, solve, and check. Be sure to state what your variable represents.

1. Two numbers are in the ratio of 3 to 4. Their sum is 35. Find the numbers.

2. Two numbers are in the ratio of 7 to 3. Their sum is 130. Find the numbers.

3. The length and width of a rectangle are in the ratio of 9 to 4. The perimeter is 78. Find the length and width.

4. The length and width of a rectangle are in the ratio 7:2. The perimeter is 72. Find the length and width.

5. The amounts Beverly spends for food, rent, and clothing are in the ratio 4:5:1. She spends an average of $850 a month for these three items. What can she expect to spend for each of these items in a month?

6. Frank spends $12,150 for tuition, housing, and food each school year. The amounts spent for these three items are in the ratio 4:4:1. How much does he spend for each item?

7. Divide 88 into two parts whose ratio is 6 to 5.

8. Divide 99 into two parts whose ratio is 2 to 7.

9. The three sides of a triangle are in the ratio 4:5:6. The perimeter is 90 ft. Find the lengths of the three sides.

10. The three sides of a triangle are in the ratio 5:6:7. The perimeter is 72 yd. Find the lengths of the three sides.

11. An uncle divided $27,000 among his three nephews in the ratio 2:3:4. How much did each receive?

12. A pension fund of $150,000 is invested in stocks, bonds, and real estate in the ratio 5:1:4. How much is invested in each?

 13. When mixing the concrete for his patio floor, Mr. Mora used a cement-sand-gravel ratio of $1:2\frac{1}{2}:4$. If he used 5 cubic yards of gravel, how much sand and cement did he use?

14. In one week's time, a restaurant served 865 dinners. For every five dinners served, three were seafood dinners. How many seafood dinners were served during the week?

In Exercises 15–18, set up each problem algebraically and solve.

15. The length and width of a rectangle are in the ratio of 7 to 4. The perimeter is less than 88. What is the range of values for the width?

16. The length and width of a rectangle are in the ratio of 8 to 5. The perimeter is greater than 78. What is the range of values for the length?

17. The three sides of a triangle are in the ratio 2:4:7. The perimeter is greater than 117. What is the range of values for the length of the longest side?

18. The three sides of a triangle are in the ratio 3:5:6. The perimeter is less than 112. What is the range of values for the length of the shortest side?

19. Ruby drives 360 miles in 7 hours. Express the rate of miles to hours as a fraction.

20. Justin bicycles 35 miles in 4 hours. Express the rate of miles to hours as a fraction.

21. Susan can crochet three afghans in 25 days. Express the rate of afghans to days as a fraction.

22. Cathy types 500 words in 11 minutes. Express the rate of words to minutes as a fraction.

In Exercises 23–26, Set up each problem algebraically and solve.

23. If $805 was spent for carpeting that cost $23 \frac{\text{dollars}}{\text{sq. yd}}$, how many square yards were purchased?

24. If Florence earned $432 one week and if her pay was $12 \frac{\text{dollars}}{\text{hour}}$, how many hours did she work?

25. Mr. Lee's car uses gasoline at the rate of $23 \frac{\text{miles}}{\text{gallon}}$. How much gasoline will he use in driving 368 miles?

26. If Jennifer uses wallpaper that has a rate of coverage of $56 \frac{\text{sq. ft}}{\text{roll}}$, how many rolls will she need to cover 672 sq. ft?

Set II In Exercises 1–14, set up each problem algebraically, solve, and check. Be sure to state what your variable represents.

1. Two numbers are in the ratio of 3 to 5. Their sum is 40. Find the numbers.

2. The length and width of a rectangle are in the ratio of 7 to 4. The perimeter is 88. Find the length and width.

3. The length and width of a rectangle are in the ratio of 9 to 5. The perimeter is 140. Find the length and width.

4. A wire 135 cm long was cut into two pieces. The ratio of the lengths of the pieces is 4 to 5. Find the length of each piece.

5. The amounts Diane spends for food, rent, and clothing are in the ratio $4:7:2$. She spends an average of $975 a month for these three items. What can she expect to spend for each of these items in a *year*?

6. A farmer plants 162 acres of corn and soy beans in the ratio of 5 to 4. How many acres of each does he plant?

7. Divide 126 into two parts whose ratio is 5 to 13.

8. Separate 231 into two parts whose ratio is 17 to 4.

9. The three sides of a triangle are in the ratio $4:5:7$. The perimeter is 128. Find the lengths of the three sides.

10. A man cuts a 65-in. board into two parts whose ratio is $9:4$. Find the lengths of the pieces.

11. The gold solder used in a crafts class has gold, silver, and copper in the ratio $5:3:2$. How much gold is there in 30 g of this solder?

12. A civil service office tries to employ people with no discrimination because of age. If the ratio of people over 40 to those under 40 is $3:4$, how many out of 259 people hired would have to be over 40?

13. Fast Eddie's Pro Shop sells four types of bowling balls: type A, B, C, and D. Sales of the balls totaled $3,600 for the year. If the balls sold in the ratio

$3:2:1:4$, how much of the sales, in dollars, can be attributed to each type of ball?

14. Alice is making an afghan that requires thirty-two skeins of yarn. The colors brown, beige, and rust are to be used in the ratio $5:8:3$. How many skeins of each color should she buy?

In Exercises 15–18, set up each problem algebraically and solve.

15. The length and width of a rectangle are in the ratio of 9 to 4. The perimeter is greater than 156. What is the range of values for the width?

16. The ratio of Diane's age to Michelle's age is 3 to 2. The sum of their ages is less than 90 (years). What is the range of values for Diane's age?

17. The three sides of a triangle are in the ratio $8:9:12$. The perimeter is less than 174. What is the range of values for the shortest side?

18. The amounts Carol spends for clothing, entertainment, and food are in the ratio $3:2:6$. If she can spend less than $308 a month for these three items altogether, what range of values should she allow for entertainment?

19. George drives 323 miles in 6 hours. Express the rate of miles to hours as a fraction.

20. A machine produces sixty-three parts in 2 hours. Express the rate of parts to hours as a fraction.

21. Alan paid $682 for 23 sq. yd of carpeting. Express the rate of dollars to square yards as a fraction.

22. A plane flies 361 miles in 3 hours. Express the rate of miles to hours as a fraction.

In Exercises 23–26, set up each problem algebraically and solve.

23. If Ralph earned $561 one week and if his pay was $17 \dfrac{\text{dollars}}{\text{hour}}$, how many hours did he work?

24. If Elizabeth's car uses gasoline at the rate of $29 \dfrac{\text{miles}}{\text{gallon}}$, how much gasoline will she use in driving 493 miles?

25. If Dean uses paint that has a rate of coverage of $325 \dfrac{\text{sq. ft}}{\text{gallon}}$, how many gallons will he use in painting 2,275 sq. ft?

26. If $212 was spent for fencing that cost $4 \dfrac{\text{dollars}}{\text{ft}}$, how many feet of fencing were purchased?

4.6 Percent Problems

The Meaning of Percent

Percent means "per hundred" (*centum* means 100 in Latin). For example, $5\% = \frac{5}{100}$, or $5\% = 0.05$. (If necessary, refer to Appendix B for the method for changing a number from a percent to a decimal, from a common fraction to a percent, and so forth.)

Example 1   A 5-gal paint can contains 2 gal of paint. Therefore, the paint can is $\frac{2}{5}$ full. Express this fraction as a percent.
Solution
Method 1

$$\frac{2}{5} = \frac{2 \cdot 20}{5 \cdot 20} = \frac{40}{100} = 40\%$$

↑
└── Multiplying 5 by 20 makes the denominator 100

Method 2

$$\frac{2}{5} = 0.4 = 0.40 = 40\%$$

We can say "The can is $\frac{2}{5}$ full," or "The can is 40% full," or "The can is 0.4 full." ∎

In algebra, word problems involving percents can be done by letting some variable represent the unknown number and using the fact that the word *of* in such problems indicates *multiplication*. The percent must be changed to its decimal form or to its common fraction form before the problem can be solved.

The checks will not be shown in this section, nor will we usually show "Step 1," "Step 2," and so forth.

Example 2   What number is 8% of 40?
Solution Let x = the unknown number.

What number	is	8%	of	40
x	=	(0.08)	·	(40)

↑
└── 8% changed to its decimal form

$$x = 3.2$$

Therefore, 3.2 is 8% of 40. ∎

Example 3   16 is 40% of what number?
Solution Let x = the unknown number.

16	is	40%	of	what number
16	=	(0.40)	·	x

↑
└── 40% changed to its decimal form

$$\frac{16}{0.40} = x \qquad \text{Dividing both sides by 0.40}$$

$$x = 40$$

Therefore, 16 is 40% of 40. ∎

Example 4   15 is what percent of 60?

Solution Let x = the fractional part (x must be found and then converted to its percent form).

15	is	what percent	of	60

$$15 \quad = \quad x \quad \cdot \quad 60$$

$$\frac{15}{60} \quad = \quad x \qquad\qquad \text{Dividing both sides by 60}$$

$$x \quad = \quad \frac{15}{60} = \frac{1}{4} = \frac{25}{100} = 0.25 = 25\% \qquad \text{Finding } x \text{ and converting it to a percent}$$

Therefore, 15 is 25% of 60. ∎

Example 5   What number is 175% of 80?

Solution Let x = the unknown number.

What number	is	175%	of	80

$$x \quad = \quad (1.75) \quad \cdot \quad (80)$$

175% changed to its decimal form

$$x \quad = \quad 140$$

Therefore, 140 is 175% of 80. ∎

Example 6   25 is what percent of 5?

Solution Let x = the fractional part (x must be found and then converted to its percent form).

25	is	what percent	of	5

$$25 \quad = \quad x \quad \cdot \quad 5$$

$$\frac{25}{5} \quad = \quad x \qquad\qquad \text{Dividing both sides by 5}$$

$$x \quad = \quad 5 = \frac{5}{1} = \frac{500}{100} = 500\% \qquad \text{Finding } x \text{ and converting it to a percent}$$

Therefore, 25 is 500% of 5. ∎

Example 7   25.2 is 140% of what number?

Solution Let x = the unknown number.

25.2	is	140%	of	what number

$$25.2 \quad = \quad (1.40) \quad \cdot \quad x$$

140% changed to its decimal form

$$\frac{25.2}{1.40} \quad = \quad x \qquad\qquad \text{Dividing both sides by 1.40}$$

$$x \quad = \quad 18$$

Therefore, 25.2 is 140% of 18. ∎

Example 8   In an examination, a student worked fifteen problems correctly. This was 75% of the problems. Find the total number of problems on the examination.

Think This problem could have been written "15 is 75% of what number?"

Solution Let x = the unknown number.

15	is	75%	of	what number

$$15 \ = \ (0.75) \ \cdot \ x$$

└── 75% changed to its decimal form

$$\frac{15}{0.75} = x \qquad \text{Dividing both sides by 0.75}$$

$$x \ = \ 20$$

Therefore, twenty problems were on the examination. ∎

Example 9   Mr. Delgado, a salesman, makes a 6% commission on all items he sells. One week he made $390. What were his gross sales for the week?

Think This problem could have been written "390 is 6% of what number?"

Solution Let x = the unknown number.

390	is	6%	of	what number

$$390 \ = \ (0.06) \ \cdot \ x$$

└── 6% changed to its decimal form

$$\frac{390}{0.06} = x \qquad \text{Dividing both sides by 0.06}$$

$$x \ = \ 6,500$$

Therefore, his gross sales for the week were $6,500. ∎

Markup

In a business that buys and sells merchandise, the merchandise must be sold at a price high enough to return to the merchant (1) the price paid for the goods; (2) the expenses, salaries, rents, taxes, and so on; and (3) a reasonable profit. To accomplish this, the cost of each item must be *marked up* before it is sold. We use markup based on cost. Selling price, cost, and markup are related by the following formula:

$$\text{Selling price} = \text{cost} + \text{markup}$$
$$S \qquad = \ C \ + \ M$$

Example 10   If a business pays $75 for an item, what is the selling price of the item if it is marked up 40%?

Think The cost, C, is $75. We must first find the amount of the markup (40% of the cost, or 40% of $75). Then we must use the formula $S = C + M$.

Solution Let M = the markup.

$$M = 0.40(\$75) = \$30$$

$$S = C + M$$

$$S = \$75 + \$30$$

$$S = \$105$$

Therefore, the selling price is $105. ∎

EXERCISES 4.6

Set I   Set up each of the following problems algebraically and solve, even if you could do the problem using methods learned in an arithmetic class. Be sure to state what your variables represent.

1. 15 is 30% of what number?

2. 16 is 20% of what number?

3. 115 is what percent of 250?

4. 330 is what percent of 225?

5. What is 25% of 40?

6. What is 45% of 65?

7. 15% of what number is 127.5?

8. 32% of what number is 256?

9. What percent of 8 is 17?

10. What percent of 6 is 12?

11. 63% of 48 is what number?

12. 87% of 49 is what number?

13. 750 is 125% of what number?

14. 325 is 130% of what number?

15. 23 is what percent of 16?

16. 57 is what percent of 23?

17. What is 200% of 12?

18. What is 300% of 9?

19. 15% of a number is 37.5. What is the number?

20. What is 27% of $135?

21. 42 is $66\frac{2}{3}$% of what number?

22. 36 is $16\frac{2}{3}$% of what number?

23. A team wins 80% of its games. If it wins sixty-eight games, how many games has it played?

24. Seventy percent of the 46,000 burglaries in a city were committed by persons who had previously been convicted at least three times for the same crime. If all of these criminals had been kept in jail, how many burglaries could have been prevented?

25. In a class of forty-two students, seven students received a grade of B. What percent of the class received a grade of B?

26. John's weekly gross pay is $250, but 23% of his check is withheld. How much is withheld?

27. Fifty-four out of 210 civil service applicants pass their exams. What percent of the applicants pass?

28. A 4,200-lb automobile contains 462 lb of rubber. What percent of the car's total weight is rubber?

29. A merchant pays $125 for an item. What is the selling price of this item if it is marked up 35%?

30. A suit costing a merchant $72 is marked up 20%. What is its selling price?

31. A truckload of fifteen steers and eighteen heifers was hauled to market. The average weights of the steers and heifers were 1,027 lb and 956 lb, respectively. Due to weight loss during shipment, a 3% deduction is made from the total live weight, and the seller is paid on this reduced weight. If the producer received 84¢ per pound for the steers and 78¢ per pound for the heifers, how much was his check from the buyer?

32. A timber company paid $632,000 for 1.27 million board feet of timber in the Mt. Hood National Forest. The timber had been appraised at $456,000. What percent of the appraised value was the amount that the company paid above the appraised value? (Round off the answer to the nearest tenth of a percent.)

Set II Set up each of the following problems algebraically and solve, even if you could do the problem using methods learned in an arithmetic class. Be sure to state what your variables represent.

1. 24 is 40% of what number?

2. 650 is what percent of 325?

3. 34 is 68% of what number?

4. 335 is 134% of what number?

5. What is 55% of 82?

6. 102 is what percent of 17?

7. 23% of what number is 13.11?

8. 156% of what number is 195?

9. 39 is what percent of 27? (Round off your answer to one decimal place.)

10. 86 is what percent of 344?

11. 91% of 64 is what number?

12. 63% of what number is 10.71?

13. 930 is 124% of what number?

14. What is 134% of 335?

15. 67 is what percent of 32?

16. 89 is what percent of 20?

17. What is 400% of 19?

18. What is $16\frac{2}{3}$% of 78?

19. 12% of a number is 9.36. What is the number?

20. 136% of what number is 10.88?

21. 28 is $33\frac{1}{3}$% of what number?

22. 29 is $12\frac{1}{2}$% of what number?

23. A team wins 105 games. This is 70% of the games played. How many games were played?

24. 85% of the members of the golf club voted for Brian for president. He received 238 votes. How many members are in the club?

25. Michelle is on a 1,200-calorie-a-day diet. She consumed 360 calories at breakfast. What percent of her entire daily calorie allowance was this?

26. Heather sleeps 15 hr a day (including naps). What percent of the day does she sleep?

27. Rosie's weekly salary is $450. If her deductions amount to $126, what percent of her salary is take-home pay?

28. Trisha found a television set on sale. The original price was $379, but it was on sale for 30% off. What was the sale price?

29. A camera costing a merchant $244 is marked up 30%. What is its selling price?

30. A saleswoman makes a 7% commission on all the items she sells. One week she made $504. What were her gross sales for the week?

31. The largest meteorite found in the United States was discovered in a forest near Willamette, Oregon, in 1902. This 13.5-ton meteorite is about 91% iron and 8.3% nickel. How much iron and how much nickel does this meteorite contain?

32. The largest meteorite found in the world was discovered in southwest Africa. This 60.2-ton meteorite is about 81% iron and 17.5% nickel. How much iron and how much nickel does this meteorite contain?

4.7 Distance-Rate-Time Problems

Distance-rate-time problems are used in any field involving motion. A physical law relating *distance* traveled *d*, *rate* of travel *r*, and *time* of travel *t* is

$$d = r \cdot t \quad \text{or} \quad r \cdot t = d$$

For example, you know that if you are driving your car at an average speed of 50 mph, then

you travel a distance of 100 miles in 2 hours: $50\dfrac{\text{mi}}{\text{hr}}(2 \text{ hr}) = 100 \text{ mi}$

you travel a distance of 150 miles in 3 hours: $50\dfrac{\text{mi}}{\text{hr}}(3 \text{ hr}) = 150 \text{ mi}$

and so on.

Example 1   If Mrs. Petersen drives at an average rate of 47 mph (miles per hour), how long will it take her to drive 329 miles?

Think This is a rate problem. We use the formula $r \cdot t = d$. Multiplying and "canceling units" will give $\left(\dfrac{\text{miles}}{\text{hour}}\right)(\text{hours}) = \text{miles}$.

Solution

Step 1. Let *x* = the number of hours.

Step 2. $\left(47\dfrac{\text{miles}}{\text{hour}}\right)(x \text{ hours}) = 329 \text{ miles}$ We multiply rate times time and cancel "hours"

We now drop the word "miles"

Step 3. $\qquad\qquad 47x \text{ miles} = 329 \text{ miles}$

$$47x = 329$$

$$\frac{47x}{47} = \frac{329}{47} \qquad \text{Dividing both sides by 47}$$

Step 4. $\qquad\qquad\qquad x = 7$

Step 5. *Check* $\left(47\dfrac{\text{miles}}{\text{hour}}\right)(7 \text{ hours}) = 329 \text{ miles}$

Step 6. Therefore, it will take her 7 hours to drive 329 miles. ∎

Many distance-rate-time problems involve two *different* rates. For such problems, you may find the "chart" method helpful. This method adds a few steps to the basic steps used in solving all word problems. The chart method is shown in the following box.

"CHART" METHOD FOR SOLVING DISTANCE-RATE-TIME PROBLEMS

Read Read the problem carefully. *Be sure you understand the problem.*

Think Is this a distance-rate-time problem? Are two different rates mentioned? If so, I can use a chart.

Step 1. Represent one unknown number by a variable and declare its meaning in a sentence of the form "Let $x = \ldots$." Express other unknowns in terms of x, if possible.

Chart Draw a blank chart, as follows:

	r	$\cdot$	t	$=$	d
One object					
Other object					

Fill in as many of the blanks as possible, using the information given.

Reread Fill in the remaining boxes, using x and the formula $r \cdot t = d$.

Step 2. Are the distances equal? If so, you obtain the equation by setting the two d-values from the chart equal to each other. Do the distances differ by some constant? If so, write the equation using this information. Are the *times* equal, or do you know something about the *sum* of the times? If so, write the equation using this information.

Step 3. Using the methods described in Chapter 3, solve the equation.

Step 4. Solve for *all* the unknowns asked for in the problem.

Step 5. Check the solution(s) *in the word statement.*

Step 6. State your results clearly.

Sometimes it's easier to solve a problem if we let x represent some unknown *other* than the one asked for in the problem. In Example 2, for instance, we're asked for a *distance*, but it's easier to solve the problem if we let x be a *rate of speed.* (After we've found x, we must, of course, be sure to find the distance.)

Example 2   Mr. Maxwell takes 1 hr to drive to work in the morning, but he takes $1\frac{1}{2}$ hr to return home over the same route during the evening rush hour. If his average morning speed is 10 mph faster than his average evening speed, how far is it from his home to his work?
Think This is a distance-rate-time problem (the formula is $r \cdot t = d$), and there are two different rates mentioned. Let's use a chart, letting the first row be the row for "Going to work" and the second row be the row for "Returning from work." We can let x represent the distance or one of the rates. (This problem is easier to do if we let x be the slower rate.)
Solution

Step 1. Let x = speed returning from work (the slower speed)

$x + 10$ = speed going to work (the faster speed)

Chart

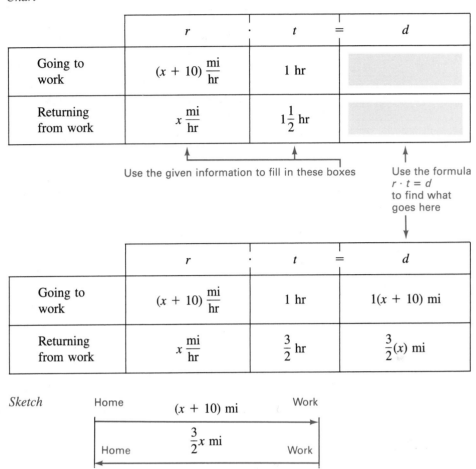

	r $\cdot$	t $=$	d
Going to work	$(x + 10) \dfrac{\text{mi}}{\text{hr}}$	1 hr	
Returning from work	$x \dfrac{\text{mi}}{\text{hr}}$	$1\dfrac{1}{2}$ hr	

Use the given information to fill in these boxes

Use the formula
$r \cdot t = d$
to find what
goes here

	r $\cdot$	t $=$	d
Going to work	$(x + 10) \dfrac{\text{mi}}{\text{hr}}$	1 hr	$1(x + 10)$ mi
Returning from work	$x \dfrac{\text{mi}}{\text{hr}}$	$\dfrac{3}{2}$ hr	$\dfrac{3}{2}(x)$ mi

Sketch

Home $(x + 10)$ mi   Work

Home $\dfrac{3}{2}x$ mi   Work

We now make use of the "unused fact" that the two distances are equal:

Distance going to work	$=$	Distance returning from work

Step 2. $1(x + 10) \;=\; \dfrac{3}{2}(x)$

Step 3. $x + 10 \;=\; \dfrac{3}{2}x$

$\underline{-x \qquad\qquad -x}$ Adding $-x$ to both sides

$10 \;=\; \dfrac{1}{2}x$

$2(10) \;=\; 2\left(\dfrac{1}{2}x\right)$ Multiplying both sides by 2

$x \;=\; 20$ Speed returning from work (in mph)

$x + 10 \;=\; 30$ Speed going to work (in mph)

Step 4. However, we were asked to find the *distance* from his home to his work.

$$\text{Distance } returning = (\text{rate})(\text{time}) = \left(20\dfrac{\text{mi}}{\text{hr}}\right)\left(\dfrac{3}{2}\text{hr}\right) = 30 \text{ mi}$$

Step 5. *Check* The distance *going* $= \left(30\dfrac{\text{mi}}{\text{hr}}\right)(1 \text{ hr}) = 30$ mi. The distances are equal.

Step 6. Therefore, it is 30 mi from Mr. Maxwell's home to his work. ■

We often encounter problems in which a boat is running in moving water. In this case, if the water is moving at w miles per hour and if the speed of the boat in still water is b miles per hour, then when the boat is going *upstream* (*against* the current, or against the movement of the water), the actual speed of the boat will be $b - w$ miles per hour; if it's going *downstream* (*with* the current), the actual speed will be $b + w$ miles per hour (see Example 3).

Similarly, if the wind is blowing at w miles per hour and an airplane is flying with a speed that would be s miles per hour in still air, then when the plane flies *with* the wind, its actual speed will be $s + w$ miles per hour, and when it's flying *against* the wind, its actual speed will be $s - w$ miles per hour.

Example 3   A boat cruises downstream for 4 hr before heading back. After traveling upstream for 5 hr, it is still 16 mi short of the starting point. If the speed of the stream is 4 mph, find the speed of the boat in still water.

Think This is a distance-rate-time problem (the formula is $r \cdot t = d$), and there are two different rates mentioned. Let's use a chart, letting the first row be the row for "Downstream" and the second row be the row for "Upstream." Let's try letting x be the speed of the boat in still water. Then, since the speed of the stream is 4 mph, the speed of the boat going downstream will be $x + 4$ mph, and the speed upstream will be $x - 4$ mph.

Solution

Step 1. Let $\quad x =$ speed of boat in still water

$\qquad x + 4 =$ speed of boat downstream

$\qquad x - 4 =$ speed of boat upstream

Chart

	r	$\cdot$	t	$=$	d	
Downstream	$(x + 4)\dfrac{\text{mi}}{\text{hr}}$		4 hr		$(x + 4)(4)$ mi	$r \cdot t = d$ $(x + 4)(4) = d$
Upstream	$(x - 4)\dfrac{\text{mi}}{\text{hr}}$		5 hr		$(x - 4)(5)$ mi	$r \cdot t = d$ $(x - 4)(5) = d$

Sketch

$\longleftarrow$ (x + 4)(4) mi $\longrightarrow$ Downstream

$\leftarrow$16 mi$\rightarrow$$\leftarrow$(x − 4)(5) mi $\longrightarrow$ Upstream

(This time the distances are *not* equal to each other. There is a 16-mi difference between them.)

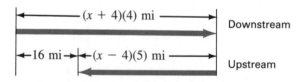

Distance traveled downstream	=	16	plus	distance traveled upstream

Step 2. $\quad (x + 4)(4) \qquad = \quad 16 \quad + \qquad (x - 4)(5)$

Step 3.

$$4x + 16 = 16 + 5x - 20$$

$$
\begin{array}{ll}
4x + 16 = 5x - 4 & \text{Combining like terms} \\
\underline{-4x + 4 \quad -4x + 4} & \text{Adding } -4x + 4 \text{ to both sides} \\
20 = x &
\end{array}
$$

Step 4.

$$x = 20 \text{ mph} \qquad \text{Speed of boat in still water}$$

Step 5. *Check*

Speed of boat downstream = (20 + 4) mph = 24 mph

Speed of boat upstream = (20 − 4) mph = 16 mph

$$\text{Distance downstream} = \left(24\frac{\text{mi}}{\text{hr}}\right)(4 \text{ hr}) = 96 \text{ mi}$$

$$\text{Distance upstream} = \left(16\frac{\text{mi}}{\text{hr}}\right)(5 \text{ hr}) = 80 \text{ mi}$$

Distance downstream = 16 mi plus distance upstream

$$96 \text{ mi} \overset{?}{=} 16 \text{ mi} + 80 \text{ mi}$$

$$96 \text{ mi} = 96 \text{ mi}$$

Step 6. Therefore, the speed of the boat in still water is 20 mph. ∎

Example 4 Teri drove from her home to San Diego to pick up her husband, Nick. She encountered heavy traffic all the way and averaged 45 mph. Nick encountered no traffic and averaged 63 mph on the trip home. If the total driving time for the round trip was 12 hr, how far is their home from San Diego?

Think This is a distance-rate-time problem (the formula is $d = r \cdot t$), and this time we *know* both rates and also know something about the *sum* of the times. Let's use a chart, letting the first row be the row for "Teri" and the second row be the row for "Nick." Let's try letting x equal Teri's time.

Solution

Step 1. Let $\quad x =$ the number of hours Teri drove

$\qquad\qquad 12 - x =$ the number of hours Nick drove

Chart

	r	$\cdot$	t	$=$	d
Teri	$45\dfrac{\text{mi}}{\text{hr}}$		x hr		$(45)(x)$ mi
Nick	$63\dfrac{\text{mi}}{\text{hr}}$		$(12 - x)$ hr		$(63)(12 - x)$ mi

Sketch

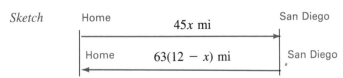

We now use the unused fact that the two distances are equal:

Step 2. $45x = 63(12 - x)$

Step 3. $45x = 756 - 63x$

$$\frac{63x \qquad\qquad 63x}{108x = 756}$$ Adding $63x$ to both sides

$$\frac{108x}{108} = \frac{756}{108}$$ Dividing both sides by 108

$x = 7$ Number of hours Teri drove

$12 - x = 5$ Number of hours Nick drove

Find the distance:

Step 4. $d = \left(45\frac{\text{mi}}{\text{hr}}\right)(7 \text{ hr}) = 315 \text{ mi}$ Distance Teri drove

Step 5. *Check* Distance Nick drove: $d = \left(63\frac{\text{mi}}{\text{hr}}\right)(5 \text{ hr}) = 315$ mi. The distances are equal, and the sum of the driving times is 12 hr.

Step 6. Therefore, their home is 315 mi from San Diego. ∎

EXERCISES 4.7

Set I   Set up each problem algebraically, solve, and check. Be sure to state what your variables represent.

1. If Robbie drives at an average rate of 51 mph, how long will it take him to drive 408 mi?

2. If Janine bicycles at an average rate of 8 mph, how long will it take her to go 72 mi?

3. Jennie drives to her son's house in 5 hr when there is heavy traffic. When there is no traffic, the same trip takes 4 hr. She can drive an average of 10 mph faster when there is no traffic than when there is heavy traffic.

a. What is her average speed when there is no traffic?

b. What is the distance from her house to her son's house?

4. Bud occasionally has to drive to a company that is some distance from his home. If there is no traffic, the trip takes 5 hr, but in heavy traffic it takes 6 hr. He can drive an average of 10 mph faster when there is no traffic than when there is heavy traffic.

a. What is his average speed when there is no traffic?

b. What is the distance from his home to the company?

5. Mr. Robinson left San Diego at 7 A.M., heading toward Sacramento. His neighbor Mr. Reid left at 8 A.M. on the same highway, also heading toward Sacramento. By driving 9 mph faster, Mr. Reid overtook Mr. Robinson at noon.

a. Find Mr. Robinson's average speed.

b. Find Mr. Reid's average speed.

c. Find the distance each traveled before they met.

6. Mr. Curtis left Fillmore at 6 A.M., heading toward San Francisco. His neighbor Mr. Castillo left Fillmore at 7 A.M., also heading toward San Francisco. By driving 9 mph faster, Mr. Castillo overtook Mr. Curtis at noon.

    a. Find Mr. Curtis's average speed.

    b. Find Mr. Castillo's average speed.

    c. Find the distance each traveled before they met.

7. A boat cruises downstream for 3 hr before heading back. It takes 4 hr going upstream for the boat to get back to its starting point. If the speed of the stream is 3 mph, find the speed of the boat in still water.

8. A boat cruised downstream for 5 hr before stopping for the night. The next day, it took 7 hr for the boat to get back to its starting point. If the speed of the stream both days was 4 mph, find the speed of the boat in still water.

9. A boat cruised downstream for 3 hr before heading back. After traveling 4 hr back upstream, the boat was still 6 mi short of the starting point. If the speed of the stream was 4 mph, find the speed of the boat in still water.

10. A boat cruised downstream for 4 hr before heading back. After traveling 5 hr back upstream, the boat was still 3 mi short of the starting point. If the speed of the stream was 3 mph, find the speed of the boat in still water.

11. Matthew drove from his home to a friend's house on a Saturday. Because traffic was very light, his average speed was 54 mph. He returned home the next day, and his average speed was 48 mph. His total driving time was 17 hr.

    a. How long did it take him to get to his friend's house?

    b. How far was it from his home to his friend's home?

12. Brian bicycled from his home to a picnic area at Lake Perris. Because much of the path was uphill, he averaged only 16 mph. On the return trip, he averaged 24 mph. His total bicycling time was 5 hr.

    a. How long did it take him to get to the picnic area?

    b. How far was it from his home to the picnic area?

Set II   Set up each problem algebraically, solve, and check. Be sure to state what your variables represent.

1. If Mrs. Gragson drives at an average rate of 49 mph, how long will it take her to drive 441 miles?

2. If Melanie drives 432 miles in 9 hours, what is her average rate of speed?

3. If driving conditions are poor, Sherma can drive to her sister's house in Arizona in 12 hr. However, if conditions are good, her average speed is 9 mph faster, and she can make the trip in 10 hr.

    a. What is her average speed when conditions are poor?

    b. What is her average speed when conditions are good?

    c. What is the distance from her house to her sister's house?

4. Tom drives to his favorite ski area in 5 hr when there is no traffic. When there is heavy traffic, his average speed is 10 mph slower than when there is no traffic, and the same trip takes him 6 hr.

    a. What is his average speed when there is heavy traffic?

    b. What is the distance from his house to the ski area?

5. Mark leaves his house at 8 A.M., heading toward Steve's house. Mark's brother, Brian, leaves their house at 9 A.M., also heading toward Steve's house. By bicycling 4 mph faster, Brian overtakes Mark at 3 P.M.

   a. What is Mark's average speed?

   b. What is Brian's average speed?

   c. Find the distance traveled before Brian overtakes Mark.

6. Lori hiked from her campsite to Green Lake. She hiked at a rate of 2 mph going to the lake and 3 mph coming back. The trip to the lake took 3 hr longer than the trip back.

   a. How long did it take her to hike to the lake?

   b. What was the distance between her campsite and the lake?

7. An airplane flies from point A to point B in 2 hr, flying with the wind. It takes the plane 3 hr to make the return trip. If the speed of the wind is 30 mph, find the speed of the plane in still air.

8. Todd and Jo live 102 mi apart. Both leave their homes by bicycle, riding toward one another. They meet 2 hr later. Todd bicycles 5 mph faster than Jo.

   a. What is Jo's average speed?

   b. What is Todd's average speed?

9. Nick drove his boat upstream for 3 hr before heading back. After cruising back downstream for 2 hr, he was only 3 mi short of his starting point. If the speed of the stream is 4 mph, find the speed of his boat in still water.

10. Danny and Cathy live 60 mi apart. Both leave their homes at 10 A.M. by bicycle, riding toward one another. They meet at 2 P.M. If Cathy's average speed is two-thirds of Danny's, how fast does each cycle?

11. Leon hiked from his camp to a lookout point at the rate of 3 mph. On the trip back to his camp, he hiked at the rate of 4 mph. He hiked a total of 7 hr.

    a. How long did it take him to hike to the lookout point?

    b. How far was it from his camp to the lookout point?

12. The Wright family sails their houseboat upstream for 4 hr. After lunch, they motor downstream for 2 hr. At that time, they are still 12 mi away from the marina where they began. If the speed of the houseboat in still water is 15 mph, what is the speed of the stream? How far upstream did the Wrights travel?

4.8 Mixture Problems

Mixture problems usually involve mixing two or more dry ingredients; because the cost of each item is important to people solving mixture problems, such problems can also be regarded as *money* problems.

Students often find a chart helpful in solving mixture problems. The general method is the same as the chart method shown in Section 4.7 for solving distance-rate-time problems; however, the chart itself is different, and we use the following relationships rather than a specific formula.

THREE IMPORTANT RELATIONSHIPS NECESSARY
TO SOLVE MIXTURE PROBLEMS

1. $\begin{pmatrix} \text{Unit cost of} \\ \text{one} \\ \text{ingredient} \end{pmatrix} \times \begin{pmatrix} \text{Amount of} \\ \text{that} \\ \text{ingredient} \end{pmatrix} = \begin{pmatrix} \text{Total cost of} \\ \text{that} \\ \text{ingredient} \end{pmatrix}$ We use this relationship *down the columns* of the chart

2. $\begin{pmatrix} \text{Amount of} \\ \text{ingredient A} \end{pmatrix} + \begin{pmatrix} \text{Amount of} \\ \text{ingredient B} \end{pmatrix} = \begin{pmatrix} \text{Total amount} \\ \text{of} \\ \text{mixture} \end{pmatrix}$ We use this relationship *in row 2* of the chart

3. $\begin{pmatrix} \text{Total cost} \\ \text{of} \\ \text{ingredient A} \end{pmatrix} + \begin{pmatrix} \text{Total cost} \\ \text{of} \\ \text{ingredient B} \end{pmatrix} = \begin{pmatrix} \text{Total cost} \\ \text{of} \\ \text{mixture} \end{pmatrix}$ We use this relationship *in the last row* of the chart.

The last row of the chart gives us the equation to use. The chart is as follows:

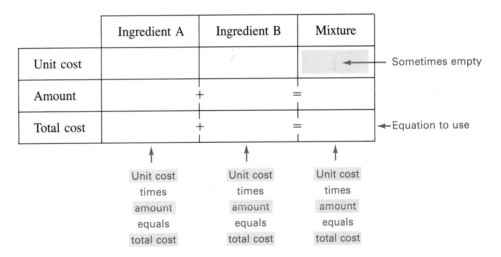

We will demonstrate the use of the chart in the examples.

Example 1   A wholesaler makes up a 50-lb mixture of two kinds of coffee. Brand A costs $3.30 per pound and Brand B costs $3.70 per pound. How many pounds of each kind of coffee must be used if the mixture is to cost $3.58 per pound?

Think This is a mixture problem. If we let x be the number of pounds of Brand A and if there are 50 lb in the mixture, then there must be $(50 - x)$ lb of Brand B. *The bottom row of the chart gives us the equation to use.*

Solution

Step 1. Let x = number of pounds of Brand A

   $50 - x$ = number of pounds of Brand B

Chart

	Brand A		Brand B		Mixture
Unit cost	3.30		3.70		3.58
Amount	x	$+$	$(50 - x)$	$=$	50
Total cost	$(3.30)x$	$+$	$3.70(50 - x)$	$=$	$3.58(50)$

Notice that

$$\boxed{\text{Amount of Brand A}} + \boxed{\text{Amount of Brand B}} = \boxed{\text{Amount of mixture}}$$

$$x \quad + \quad (50 - x) \quad = \quad 50$$

and that

$$\boxed{\text{Total cost of Brand A}} + \boxed{\text{Total cost of Brand B}} = \boxed{\text{Total cost of mixture}}$$

Step 2. $3.30x \quad + \quad 3.70(50 - x) \quad = \quad 3.58(50)$

Step 3. $3.30x + 3.70(50 - x) = 3.58(50)$

$330x + 370(50 - x) = 358(50)$	Multiplying both sides by 100
$330x + 18{,}500 - 370x = 17{,}900$	Removing parentheses
$18{,}500 - 40x = 17{,}900$	Combining like terms
$\underline{-17{,}900 + 40x \qquad -17{,}900 + 40x}$	Adding $-17{,}900 + 40x$ to both sides
$600 \quad = \quad 40x$	
$\dfrac{600}{40} = \dfrac{40x}{40}$	Dividing both sides by 40

Step 4. $x = 15$ — Number of pounds of Brand A

$50 - x = 35$ — Number of pounds of Brand B

Step 5. *Check*

$$\begin{array}{l} 15 \text{ lb @ } \$3.30 = \$\ 49.50 \\ \underline{35 \text{ lb @ } \$3.70 = \$129.50} \qquad 50 \text{ lb @ } \$3.58 = \$179.00 \\ \qquad\qquad \text{Total} = \$179.00 \leftarrow \text{Total cost} \end{array}$$

Step 6. Therefore, the wholesaler must use 15 lb of brand A and 35 lb of Brand B. ∎

Example 2 A 10-lb mixture of walnuts and almonds costs $35.20. If the walnuts cost $4.50 per pound and the almonds cost $3.10 per pound, how many pounds of each kind are in the mixture?

Think This is a mixture problem, and this time we're given the total cost of the mixture rather than the unit cost.

Solution

Step 1. Let x = number of pounds of walnuts

$10 - x$ = number of pounds of almonds

Chart

	Walnuts		Almonds		Mixture
Unit cost	4.50		3.10		////////
Amount	x	$+$	$(10 - x)$	$=$	10
Total cost	$(4.50)x$	$+$	$3.10(10 - x)$	$=$	35.20

Step 2. $4.50x + 3.10(10 - x) = 35.20$

Step 3. $\quad 45x + 31(10 - x) = 352$ Multiplying both sides by 10

$\qquad 45x + 310 - 31x = 352$ Simplifying

$$14x + 310 = 352 \qquad \text{Combining like terms}$$
$$\underline{\quad -310 \qquad -310} \qquad \text{Adding } -310 \text{ to both sides}$$
$$14x \qquad = \quad 42$$

$$\frac{14x}{14} = \frac{42}{14} \qquad \text{Dividing both sides by 14}$$

Step 4. $\qquad\qquad\qquad x = 3$ Number of pounds of walnuts

$\qquad\qquad 10 - x = 7$ Number of pounds of almonds

Step 5. *Check*

$$3 \text{ lb @ } \$4.50 = \$13.50$$
$$7 \text{ lb @ } \$3.10 = \underline{\$21.70}$$
$$\text{Total} = \$35.20 \qquad \text{Total cost of mixture}$$

Step 6. Therefore, there are 3 lb of walnuts and 7 lb of almonds. ∎

Example 3   Mrs. Reid needs to mix 20 lb of macadamia nuts that cost $8.10 per pound with pecans that cost $5.40 per pound. How many pounds of pecans should she use if the mixture is to cost $6.48 per pound?

Think This is a mixture problem, and this time we're given the amount of one ingredient but not the total amount of the mixture.

Solution

Step 1. Let x = number of pounds of pecans.

Chart

	Pecans		Macadamia nuts		Mixture
Unit cost	5.40		8.10		6.48
Amount	x	+	20	=	$x + 20$
Total cost	$(5.40)x$	+	8.10(20)	=	$6.48(x + 20)$

Step 2. $5.40x + 8.10(20) = 6.48(x + 20)$

Step 3. $\quad 540x + 810(20) = 648(x + 20)$ Multiplying both sides by 100

$\qquad 540x + 16{,}200 = 648x + 12{,}960$ Simplifying

$\underline{-540x - 12{,}960 \quad -540x - 12{,}960}$ Adding $-540x - 12{,}960$ to both sides

$\qquad\qquad 3{,}240 = 108x$

$$\frac{3{,}240}{108} = \frac{108x}{108} \qquad \text{Dividing both sides by 108}$$

Step 4. $\qquad\qquad\qquad x = 30$ Number of pounds of pecans

Step 5. *Check*

$$20 \text{ lb @ } \$8.10 = \$162 \qquad \text{Value of macadamia nuts}$$
$$\underline{30 \text{ lb @ } \$5.40 = \underline{\$162}} \qquad \text{Value of pecans}$$
$$50 \text{ lb} \leftarrow \text{Totals} \rightarrow \$324$$

$$\text{Price per pound of mixture} = \frac{324 \text{ dollars}}{50 \text{ pounds}} = \$6.48 \text{ per pound}$$

Step 6. Therefore, Mrs. Reid should use 30 lb of pecans in the mixture. ∎

NOTE   Most money problems, such as those in Section 4.4, can be done by using a chart. ☑

EXERCISES 4.8

Set I   Set up each problem algebraically, solve, and check. Be sure to state what your variables represent.

1. A grocer makes a 10-lb mixture of granola and dried apple chunks, and he wants the mixture to cost $3.00 per pound. If the granola costs $2.20 per pound and the apple chunks cost $4.20 per pound, how many pounds of each should he use?

2. Jim wants to make up a 50-lb mixture of macadamia nuts at $8.50 per pound and peanuts at $3.50 per pound. How many pounds of each should he use if the mixture is to cost $5.80 per pound?

3. Alice wants to make up a 30-lb mixture of apples that will be worth 78¢ per pound. One kind of apple costs 95¢ per pound, and the other kind costs 65¢ per pound. How many pounds of each kind should she use?

4. A grocer must make a 60-lb mixture of peanut brittle and English toffee that will be worth $5.60 per pound. If peanut brittle costs $4.20 per pound and English toffee costs $6.60 per pound, how many pounds of each should he use?

5. Margie makes a 27-lb mixture of gumdrops and caramels that is to be worth $66.00. If the gumdrops cost $2.80 per pound and the caramels cost $2.20 per pound, how many pounds of each should she use?

6. A grocer makes a 40-lb mixture of two kinds of coffee that is worth $151. If Brand C costs $4.20 per pound and Brand D costs $3.20 per pound, how many pounds of each kind does he use?

7. Dorothy needs a mixture of candy and nuts. How many pounds of candy at $3.00 per pound should she mix with 50 lb of nuts at $2.40 per pound to obtain a mixture worth $2.50 per pound?

8. How many pounds of cashews at $7.60 per pound should be mixed with 30 lb of macadamia nuts at $8.20 per pound in order to obtain a mixture worth $7.96 per pound?

Set II   Set up each problem algebraically, solve, and check. Be sure to state what your variables represent.

1. A 40-lb mixture of peanuts and cashews is to be made; it is to be worth $5.55 per pound. If peanuts cost $2.25 per pound and cashews cost $7.53 per pound, how many pounds of each should be used?

2. A grocer needs to make a 25-lb mixture of coffee worth $3.52 per pound. If Brand A costs $3.80 per pound and Brand B costs $3.30 per pound, how many pounds of each kind should he use?

3. Walt needs a mixture of walnuts and almonds to be worth $3.92 per pound. If walnuts cost $4.50 per pound and almonds cost $3.50 per pound, how many pounds of each should he use to obtain a 50-lb mixture?

4. Herb makes a 25-lb mixture of peanuts and raisins; it is worth $2.88 per pound. If peanuts cost $3.69 per pound and raisins cost $1.44 per pound, how many pounds of each did he use?

5. A 35-lb mixture of two kinds of coffee is worth $118.00. If Brand A costs $3.70 per pound and Brand B costs $3.20 per pound, how many pounds of each kind of coffee was used?

6. Sherma made a 32-lb mixture of nougats and peppermint candies that was worth $70.00. The nougats cost $2.50 per pound and the peppermints cost $2.10 per pound. How many pounds of each did she use?

7. How much Brand A coffee costing $3.95 per pound should be mixed with 25 lb of Brand B coffee costing $3.60 per pound in order to obtain a mixture worth $3.70 per pound?

8. How many pounds of Delicious apples at 95¢ per pound should be mixed with 30 lb of Spartan apples at 90¢ per pound in order to obtain a mixture worth 92¢ per pound?

4.9 Solution Problems

Another type of mixture problem involves the mixing of liquids. Such problems are often called *solution problems* or *solution mixture problems* because a mixture of two or more liquids is, under certain conditions, a *solution.**

Before we discuss solution mixture problems, we will discuss chemical solutions in general. A "30% solution of alcohol" is a mixture that is 30% pure alcohol and 70% (that is, 100% − 30%) water. Examples 1–3 demonstrate the use of the methods discussed in Section 4.6 as they apply to solution mixture problems. (The checks for these examples will not be shown, nor will we write "Step 1," "Step 2," and so on.)

Example 1   How many liters of pure alcohol are there in 2 ℓ of a 30% solution of alcohol?
Solution We must find 30% of 2.

Let x = the amount of pure alcohol.

$$x = (0.30)(2) = 0.6$$ Number of liters of pure alcohol present.   ∎

Example 2   In a 5% glycerin solution, there are 3 ml of pure glycerin present. How many milliliters of solution are there?
Solution We must answer this question: 3 is 5% of what number?

Let x = the unknown number.

$$3 = (0.05) \cdot x$$

$$\frac{3}{0.05} = \frac{0.05x}{0.05}$$ Dividing both sides by 0.05

$$x = 60$$ Number of milliliters of solution   ∎

*A solution is a homogeneous mixture of two or more substances.

Example 3   Fifty milliliters of an alcohol solution contain 6 ml of pure alcohol. What percent of the solution is alcohol?

Solution We must find what percent 6 is of 50.

Let x = the fractional part.

$$6 = x \cdot (50)$$

$$\frac{6}{50} = \frac{50x}{50} \qquad \text{Dividing both sides by 50}$$

$$x = \frac{6}{50} = 0.12 = 12\% \quad \blacksquare$$

Let us now discuss mixtures that involve the mixing of solutions. Solution problems can be solved by using a method similar to that used for solving other mixture problems. The general method is similar to the chart method shown for mixture problems in Section 4.8. We use the following relationships rather than a specific formula.

THREE RELATIONSHIPS NECESSARY TO SOLVE SOLUTION MIXTURE PROBLEMS

1. $\begin{pmatrix} \text{Strength} \\ \text{of one} \\ \text{solution} \end{pmatrix} \times \begin{pmatrix} \text{Amount} \\ \text{of that} \\ \text{solution} \end{pmatrix} = \begin{pmatrix} \text{Total amount of} \\ \text{the given substance} \\ \text{in the solution} \end{pmatrix}$ We use this relationship *down the columns of the chart*

2. $\begin{pmatrix} \text{Amount of} \\ \text{solution A} \end{pmatrix} + \begin{pmatrix} \text{Amount of} \\ \text{solution B} \end{pmatrix} = \begin{pmatrix} \text{Amount of} \\ \text{mixture} \end{pmatrix}$ We use this relationship in *row 2 of the chart*

3. $\begin{pmatrix} \text{Total amount of} \\ \text{given substance} \\ \text{in solution A} \end{pmatrix} + \begin{pmatrix} \text{Total amount of} \\ \text{given substance} \\ \text{in solution B} \end{pmatrix} = \begin{pmatrix} \text{Total amount of} \\ \text{given substance} \\ \text{in the mixture} \end{pmatrix}$ We use this relationship *in the last row of the chart*

The chart is set up as follows:

	Solution A	Solution B	Mixture
Strength (percent)				← In decimal form
Amount		+	=	
Total amount of pure substance		+	=	

↑   ↑   ↑

Strength   Strength   Strength
times   times   times
amount   amount   amount
equals   equals   equals
total amount   total amount   total amount

In solution problems, we can find the total amount of the pure substance present in one ingredient by multiplying the amount of that ingredient by its *percent*, if the percent is changed to its decimal form (see Example 1). Therefore, in each column (vertical line), the total amount of the substance is found by multiplying the percent (changed to a decimal or common fraction) by the amount. The bottom row of the chart gives the equation.

Example 4 How many liters of a 20% alcohol solution must be added to 3 ℓ of a 90% alcohol solution to make an 80% solution?

Think This is a solution mixture problem. We'll let x be the number of liters of the 20% solution.

Solution

Step 1. Let x = number of liters of 20% solution.

Chart

	20% solution	90% solution	Mixture
Strength (percent)	0.20	0.90	0.80
Amount	$x \quad +$	$3 \quad =$	$x + 3$
Total amount of pure substance	$0.20x \quad +$	$0.90(3) \quad =$	$0.80(x + 3)$

$$\underbrace{\text{Amount of alcohol in 20\% solution}}_{} + \underbrace{\text{Amount of alcohol in 90\% solution}}_{} = \underbrace{\text{Amount of alcohol in 80\% solution}}_{}$$

Step 2. $\qquad 0.20x \quad + \quad 0.90(3) \quad = \quad 0.80(x + 3)$

Step 3. $\qquad 2x + 9(3) = 8(x + 3)$ Multiplying both sides by 10

$$\begin{aligned} 2x + 27 &= \quad 8x + 24 & \text{Simplifying} \\ \underline{-2x - 24} & \quad \underline{-2x - 24} & \text{Adding } -2x - 24 \text{ to both sides} \\ 3 &= 6x \end{aligned}$$

$$\frac{3}{6} = \frac{6x}{6} \qquad \text{Dividing both sides by 6}$$

Step 4. $\qquad x = \dfrac{1}{2}$ Number of liters of 20% alcohol

Step 5. *Check*

$\left(\dfrac{1}{2}\ell\right)(0.20) = 0.10\,\ell$ Amount of pure alcohol in 20% solution

$(3\,\ell)(0.90) = \underline{2.70\,\ell}$ Amount of pure alcohol in 90% solution

$\qquad\qquad\qquad\quad 2.80\,\ell$ Total number of liters of pure alcohol

$3\,\ell + \dfrac{1}{2}\,\ell = 3.5\,\ell$ Total number of liters in mixture

$(3.5\,\ell)(0.80) = 2.80\,\ell$ Amount of pure alcohol in mixture

Step 6. Therefore, 1/2 ℓ of 20% alcohol must be added. ∎

NOTE   If we add *pure* alcohol to an alcohol solution to get a stronger solution, we are adding a *100% solution* of alcohol. If we add *water* to an alcohol solution to get a weaker solution, we are adding a *0% solution*.   ☑

Example 5   How much water should be added to a 10% solution of alcohol to obtain 20 ml of an 8% solution?

Think This is a solution mixture problem, and we're adding *water*. The percent of pure alcohol in water is 0%. There are to be 20 ml of solution in the mixture; therefore, if we let x be the number of milliliters of water, then there must be $(20 - x)$ ml of the 10% solution.

Solution

Step 1. Let x = number of milliliters of water

$20 - x$ = number of milliliters of 10% solution

Chart

	10% solution	Water	Mixture
Strength (percent)	0.10	0.00	0.08
Amount	$20 - x$ $+$	x $=$	20
Total amount of pure substance	$0.10(20 - x)$ $+$	0 $=$	$0.08(20)$

Step 2. $0.10(20 - x) + 0 = 0.08(20)$

Step 3. $10(20 - x) = 8(20)$ Multiplying both sides by 100

$$200 - 10x = 160$$ Simplifying
$$\underline{-160 + 10x \quad -160 + 10x}$$ Adding $-160 + 10x$ to both sides
$$40 = 10x$$

$$\frac{40}{10} = \frac{10x}{10}$$ Dividing both sides by 10

Step 4. $x = 4$ Number of milliliters of water

$20 - x = 16$ Number of milliliters of 10% solution

Step 5. *Check* In 16 ml of a 10% solution, there are 1.6 ml of pure alcohol.

In 20 ml of an 8% solution, there are 1.6 ml of pure alcohol.

Step 6. Therefore, we must add 4 ml of water.   ∎

EXERCISES 4.9

Set I   Set up each problem algebraically, solve, and check. Be sure to state what your variables represent.

1. How many milliliters of pure alcohol are there in 16 ml of a 20% solution of alcohol?

2. How many liters of pure sulfuric acid are there in 2 ℓ of a 20% solution?

3. There are 43 g of sulfuric acid in 500 g of solution. Find the percent of acid in the solution.

4. Five hundred grams of a solution contain 27 g of a drug. Find the percent of drug strength.

5. A 40% solution of hydrochloric acid contains 24 ml of pure acid. How many milliliters of solution are there?

6. A 30% solution of potassium chloride contains 15 ml of pure potassium chloride. How many milliliters of solution are there?

7. How many cubic centimeters (cc) of a 20% solution of sulfuric acid must be mixed with 100 cc of a 50% solution to make a 25% solution of sulfuric acid?

8. How many pints of a 2% solution of disinfectant must be mixed with 5 pt of a 12% solution to make a 4% solution of disinfectant?

9. How many milliliters of water must be added to 500 ml of a 40% solution of sodium bromide to reduce it to a 25% solution?

10. How many milliliters of water must be added to 600 ml of a 30% solution of antifreeze to reduce it to a 25% solution?

11. How many liters of pure alcohol must be added to 10 ℓ of a 20% solution of alcohol to make a 50% solution?

12. How many liters of pure alcohol must be mixed with 20 ℓ of a 30% alcohol solution in order to obtain a 44% solution?

Set II   Set up each problem algebraically, solve, and check. Be sure to state what your variables represent.

1. How many liters of pure acetic acid are there in 3 ℓ of a 15% solution of acetic acid?

2. There are 90 ml of pure sulfuric acid in a 15% sulfuric acid solution. How many milliliters of solution are there?

3. There are 66 g of hydrochloric acid in 120 g of solution. What is the percent of hydrochloric acid in the solution?

4. How many milliliters of pure antifreeze are there in 1,600 ml of a 20% solution of antifreeze?

5. A 35% glycerin solution contains 28 ℓ of pure glycerin. How many liters of the glycerin solution are there?

6. There are 48 ml of pure acetic acid in 64 ml of an acetic acid solution. What is the percent of acetic acid in the solution?

7. If 100 gal of 75% glycerin solution is made up by combining a 30% glycerin solution with a 90% glycerin solution, how much of each solution must be used?

8. How many milliliters of a 25% solution of potassium chloride must be added to 8 ml of pure potassium chloride to obtain a 35% solution?

9. How much water must be added to 60 ml of a 40% alcohol solution to obtain a 24% solution?

10. If 1,600 cc of 10% dextrose solution is made up by combining a 20% dextrose solution with a 4% dextrose solution, how much of each solution must be used?

11. How many liters of pure alcohol must be added to 15 ℓ of a 60% solution in order to obtain a 70% solution?

12. How many liters of water should be added to 20 ℓ of a 40% solution of hydrochloric acid to reduce it to a 25% solution?

4.10 Miscellaneous Word Problems

In this section, we provide you with a number of different kinds of word problems to solve. There may be some money problems, some percent problems, and so forth, but some problems are different from any you have seen so far. All the problems can (and should) be solved by using the algebraic methods that have been covered in this book.

The only way to learn to solve word problems is to attempt to solve lots of them! In this section, we want you to try your skills at general problem solving. Therefore, the odd and even problems are not "matched," nor are any examples given. You may need to refer to the formulas in Sections 2.4 or 4.1B.

EXERCISES 4.10

Set I Set up each problem algebraically, solve, and check. Be sure to state what your variables represent.

1. Jeffrey invested some money at 7% simple interest per year. After 3 years, he had earned $756 interest. How much did he invest originally?

2. Forty percent of the members of the Fifty-Plus Club are men. Last Thursday, 20% of the male members of the club and 40% of the female members attended a dinner. If there were 64 people at the members-only dinner, how many members does the club have?

3. The volume of a right circular cylinder is 100π cubic inches, and the radius is 5 in. What is the height (or altitude)?

4. A metal tank is in the shape of a rectangular box, and it contains some water. Right now it is one-third full. If we add 300 cubic inches of water, it will be three-fourths full. What is the volume of the tank?

5. Michael invested $5,000 for 1 year. He invested part of the money at 5% interest (per year) and the rest at 6% interest. If the amount of interest he earned from these investments in one year was $272, how much of the $5,000 did he invest at 5% interest?

6. Cassandra can reach Adam's house from her house in 2 hr if she averages 45 mph. If Bill lives 30 mi further from Cassandra than Adam does, what is the distance between Cassandra's house and Bill's house?

Set II Set up each problem algebraically, solve, and check. Be sure to state what your variables represent.

1. Mr. Curtis wants to enclose a rectangular area with 900 ft of fencing. If the rectangle is to be twice as long as it is wide, what must the length and width of the rectangle be?

2. One year ago, Chad invested some money at 6% simple interest. If there is now $901 in his account, how much did he invest originally?

3. A metal tank is in the shape of a right circular cylinder, and it contains some water. Right now it is one-fourth full. If we add 80 cubic inches of water, it will be one-third full. What is the volume of the tank?

4. A basket contained only blue and pink balls, and 30% of those balls were blue. Then 25% of the blue balls and 75% of the pink balls were removed. If 32 balls remained in the basket, how many balls were in the basket originally?

5. The volume of a right circular cone is 240π cubic inches. If the radius of the cone is 6 in., what is the height?

6. Margie is filling an 18-cubic-foot planter with a mixture of potting soil and mushroom compost. She wants to use twice as many cubic feet of potting soil as compost. The potting soil comes in $1\frac{1}{2}$-cubic-foot bags, and the compost comes in 2-cubic-foot bags. How many bags of potting soil should she buy? How many bags of compost?

4.11 Review: 4.1–4.10

Method for Solving Word Problems 4.1–4.10

Read To solve a word problem, first read it very carefully. *Be sure you understand the problem.* Read it several times, if necessary.

Think Determine what *type* of problem it is, if possible. Determine what is unknown. Is there enough information given so that you *can* solve the problem? Do you need a special formula? What operations must be used?

Sketch Draw a sketch with labels, if possible.

Step 1. Represent one unknown number by a variable and declare its meaning in a sentence of the form "Let $x = \ldots$." Then reread the problem to see how you can represent any other unknown numbers in terms of the same variable.

Reread Reread the entire word problem, breaking it up into small pieces.

Step 2. Translate each English phrase into an algebraic expression and fit these expressions together into an equation or inequality.

Step 3. Using the methods described in Chapter 3, solve the equation or inequality.

Step 4. Solve for *all* the unknowns asked for in the problem.

Step 5. Check the solution(s) *in the word statement*.

Step 6. State your results clearly.

(For more details on the "chart" method for distance-rate-time problems, see Section 4.7; for mixture problems, see Section 4.8; for solution mixture problems, see Section 4.9.)

Review Exercises 4.11 Set I

Set up each problem algebraically, solve, and check. Be sure to state what your variables represent.

1. Two numbers are in the ratio of 6 to 7. Their sum is 52. Find the numbers.

2. What is 245% of $450?

3. 77.5 is 31% of what number?

4. Six is what percent of 16?

5. The rent on a $380-a-month apartment was raised 5%. Find the new rent.

6. Solder used for soldering zinc is composed of lead and tin in the ratio $3:5$. How many ounces of each are present in 24 oz of solder?

7. The three sides of a triangle are in the ratio $3:4:5$. The perimeter is 108. Find the lengths of the three sides of the triangle.

8. Raul has twenty-two coins with a total value of $5.00. If these coins are dimes and 50¢ pieces, how many of each kind are there?

9. Roger bought $2.10 worth of stamps. He bought only 2¢, 10¢, and 12¢ stamps. If there were twice as many 12¢ stamps as 10¢ stamps and twice as many 2¢ stamps as 12¢ stamps, how many of each did he buy?

10. The total receipts for a football game were $492,000. General admission tickets cost $15 each, reserved seat tickets $24, and box seat tickets $30. If they sold 4,500 more reserved seats than box seats and 4 times as many general admissions as reserved seats, how many of each did they sell?

11. An affirmative action committee demands that 24% of the 1,800 entering freshmen be minority students. How many more than the actual 256 minority freshmen entering would satisfy the committee?

12. After sailing downstream for 2 hr, it takes a boat 7 hr to return to its starting point. If the speed of the boat in still water is 9 mph, what is the speed of the stream?

13. Mrs. Koontz left Downey at 5 A.M., heading toward Washington. Her neighbor Mrs. Fowler left at 6 A.M., also heading toward Washington. By driving 5 mph faster, Mrs. Fowler overtook Mrs. Koontz at 3 P.M. the same day.

    a. What was Mrs. Koontz's average speed?

    b. What was Mrs. Fowler's average speed?

    c. How far had they driven before Mrs. Fowler overtook Mrs. Koontz?

14. A dealer makes up a 30-lb mixture of nuts costing 85¢ and 95¢ a pound. How many pounds of each must be used in order for the mixture to cost 91¢ a pound?

15. How many cubic centimeters of water must be added to 500 cc of a 25% solution of potassium chloride to reduce it to a 5% solution?

Review Exercises 4.11 Set II

Set up each problem algebraically, solve, and check. Be sure to state what your variables represent.

1. A man cuts a 5-ft board into two parts whose ratio is $1:3$. Find the lengths of the pieces (in inches).

1. _____

   _____

2. What is 175% of $350?

2. _____

3. _____

3. Ninety-three is what percent of 124?

4. _____

5. _____

6. _____

4. A 26% solution of hydrochloric acid contains 62.4 ml of pure acid. How many milliliters of solution are there?

7. _____

   _____

   _____

   _____

5. There are 45 ml of sulfuric acid in 300 ml of solution. Find the percent of acid in the solution.

6. Mr. Edmonson's new contract calls for an 8% raise in salary. His present salary is $32,000. What will his new salary be?

7. A disc jockey plays folk, rock, country-western, and Latino records in the ratio $2:5:3:4$. If he played 126 records in a week, how many of each type were included?

8. Alice has sixty coins with a total value of $13.35. The coins are dimes, quarters, and 50¢ pieces. If there are 4 times as many dimes as quarters, how many coins of each kind does she have?

9. A 10-lb mixture of nuts and raisins costs $25. If raisins cost $1.90 per pound and nuts $3.40 per pound, how many pounds of each are used?

10. Randy wants to mix dried figs with dried apricots to make an 8-lb mixture costing $2.70 per pound. If dried figs cost $1.80 per pound and dried apricots cost $4.20 per pound, how many pounds of each are used?

11. Ms. Rennie invested part of $18,000 at 16% and the remainder at 20%. Her total yearly income from these investments is $3,120. How much is invested at each rate? (The interest was simple interest.)

12. Todd paddles his kayak downstream for 4 hr. After a lunch break, he paddles back upstream, but it takes him 7 hr to get back to his starting point. The speed of the stream is 3 mph.

   a. How fast does Todd's kayak move in still water?

   b. How far downstream did he travel?

13. A 100-lb mixture of two different kinds of apples costs $72.50. If one kind of apple costs 90¢ a pound and the other costs 65¢ a pound, how many pounds of each kind were used?

14. How many pounds of Spanish peanuts costing $2.50 per pound should be mixed with 35 lb of other peanuts costing $2.60 per pound in order for the mixture to be worth $2.57 per pound?

15. How many cubic centimeters of a 50% phenol solution must be added to 400 cc of a 5% solution to make it a 10% solution?

8. _____

_____

_____

9. _____

_____

10. _____

_____

11. _____

_____

12a. _____

b. _____

13. _____

_____

14. _____

15. _____

Chapter 4 Diagnostic Test

The purpose of this test is to see how well you understand solving word problems. We recommend that you work this diagnostic test *before* your instructor tests you on this chapter. Allow yourself about 50 minutes.

Complete solutions for all the problems on this test, together with section references, are given in the answer section in the back of the book. For the problems you do incorrectly, study the sections referred to.

Set up each problem algebraically, solve, and check. Be sure to state what your variables represent.

1. A solution contains 28 ml of antifreeze. If it is a 40% solution, how many milliliters of the solution are there?

2. A 20-ml solution of alcohol contains 17 ml of pure alcohol. What is the percent of alcohol in the solution?

3. When 16 is added to 3 times an unknown number, the sum is 37. Find the unknown number.

4. Cheryl has twenty-five coins in her purse with a total value of $1.65. If these coins consist of nickels and dimes, how many of each kind are there?

5. A 50-lb mixture of granola and dried apricots is to be worth $2.34 per pound. If the granola costs $2.20 per pound and the apricots cost $2.70 per pound, how many pounds of each should be used?

6. Kevin leaves his home at 7 A.M., driving toward St. George. His brother Jason leaves their home one hour later, also driving toward St. George. By driving 9 mph faster than Kevin, Jason overtakes Kevin at 1 P.M. the same day.

a. What is Kevin's average speed?

b. What is Jason's average speed?

c. How far do they drive before Jason overtakes Kevin?

7. How many pounds of nuts costing $3.50 per pound should be mixed with 30 lb of macadamia nuts costing $8.00 per pound in order to obtain a mixture worth $4.85 per pound?

8. Two hundred milliliters of a 45% alcohol solution are to be made by combining a 30% solution with an 80% solution.

a. How many milliliters of the 30% solution should be used?

b. How many milliliters of the 80% solution should be used?

9. The three sides of a triangle are in the ratio 7:9:11. The perimeter is 135 m. Find the lengths of the three sides.

10. An item costing a merchant $240 is marked up 30%. What is its selling price?

Cumulative Review Exercises: Chapters 1–4

In Exercises 1–3, evaluate each expression if possible.

1. $\dfrac{0}{0}$

2. $3\sqrt{36} - 4^2(-5)$

3. $46 - 2\{4 - [3(5 - 8) - 10]\}$

In Exercises 4 and 5, use the formula, substituting the values of the variables given with the formula.

4. $A = \dfrac{h}{2}(b + B)$ Find A when $h = 5$, $b = 7$, and $B = 11$.

5. $S = 4\pi r^2$ Find S when $\pi \approx 3.14$ and $r = 5$.

6. List any prime numbers greater than 7 and less than 31 that yield a remainder of 1 when divided by 5.

In Exercises 7 and 8, write each expression in simplest form.

7. $8 + 12\{3 - 2(x + 4)\}$

8. $3x - 2[5x - (-3x + 7)] - 4$

In Exercises 9–13, solve each equation.

9. $32 - 4x = 15$

10. $-11 = \dfrac{x}{3}$

11. $5w - 12 + 7w = 6 - 8w - 3$

12. $4(2y - 5) = 16 + 3(6y - 2)$

13. $6z - 22 = 2[8 - 4(5z - 1)]$

14. Is 0 a rational number?

15. Is 17 a real number?

16. Is addition associative?

17. What is the multiplicative identity?

In Exercises 18–20, set up each problem algebraically, solve, and check. Be sure to state what your variables represent.

18. Two numbers are in the ratio of 11 to 9. Their sum is 160. Find the numbers.

19. Leona has fifteen coins with a total value of $1.75. If these coins are all nickels and quarters, how many of each kind of coin are there?

20. Jodi left her home in Lafayette at 6 A.M., heading up the coast. Her husband, Bud, left at 7 A.M., also heading up the coast. By driving 10 mph faster, Bud overtook Jodi at noon. How fast was Bud driving?

5 Exponents

In this chapter, we discuss the properties of exponents that will be used throughout the remainder of the book. We also continue with our discussion of simplifying algebraic expressions and discuss scientific notation. We do not give proofs of the rules of exponents given in this chapter; they either require proof by mathematical induction, a technique not discussed in this book, or are true by definition.

5.1 Multiplying Exponential Numbers

The following argument does not *prove* the first rule of exponents; however, we hope it will convince you that the rule is true. Consider the product $x^3 \cdot x^2$. (Notice that x^3 and x^2 are both powers of the *same* base, x.)

$$x^3 \cdot x^2 = (xxx)(xx) = xxxxx$$

3 factors of x
+2 factors of x
5 factors of x

Also consider the problem x^{3+2}. $x^{3+2} = x^5 = xxxxx$. Because $x^3 \cdot x^2$ and x^{3+2} both equal the same number (both equal $xxxxx$), $x^3 \cdot x^2 = x^{3+2}$.

The general rule for multiplying exponential numbers *when the bases are the same* is given in Rule 5.1.

RULE 5.1   FIRST RULE OF EXPONENTS

$$x^a \cdot x^b = x^{a+b}$$

In words: To multiply powers of the same base, *add* the exponents.

Rule 5.1 can be extended to include more factors:

$$x^a x^b x^c \cdots = x^{a+b+c+\cdots}$$

NOTE   When multiplying powers of the same base, add the exponents but keep the same base. Do not multiply the bases.   ☑

Example 1   Rewrite each of the following with the base appearing only once (if possible).

a. $x^5 \cdot x^2 = x^{5+2} = x^7$

b. $x \cdot x^2 = x^{1+2} = x^3$

$x = x^1$

When no exponent is written, the exponent is understood to be 1.

c. $x^3 \cdot x^7 \cdot x^4 = x^{3+7+4} = x^{14}$

d. $10^7 \cdot 10^5 = 10^{7+5} = 10^{12} = 1{,}000{,}000{,}000{,}000$

12 zeros

e. $2 \cdot 2^3 \cdot 2^2 = 2^{1+3+2} = 2^6 = 64$

f. $a^4 \cdot a \cdot a^5 = a^{4+1+5} = a^{10}$

g. $a^x \cdot a^y = a^{x+y}$

h. $x^3 \, y^2$.

Rule 5.1 does not apply because the bases are different. Therefore, $x^3 y^2$ cannot be rewritten with the base appearing only once.

i. $x^2 + x^3$

Rule 5.1 does not apply because the operation is addition rather than multiplication. Therefore, $x^2 + x^3$ cannot be rewritten with the base appearing only once. ■

A WORD OF CAUTION   In Example 1d, notice that $10^7 \cdot 10^5 \neq 100^{12}$, and in Example 1e, notice that $2 \cdot 2^3 \cdot 2^2 \neq 8^6$; that is, we *do not multiply the bases*. ☑

EXERCISES 5.1

Set I   Rewrite each of the following with the base appearing only once (if possible).

1. $x^3 \cdot x^4$	**2.** $x^2 \cdot x^9$	**3.** $y \cdot y^3$	**4.** $z \cdot z^4$
5. $m^2 \cdot m$	**6.** $a^3 \cdot a$	**7.** $10^2 \cdot 10^3$	**8.** $10^4 \cdot 10^3$
9. $5 \cdot 5^4 \cdot 5^5$	**10.** $3 \cdot 3^2 \cdot 3^3$	**11.** $x \cdot x^3 \cdot x^4$	**12.** $y \cdot y^5 \cdot y^3$
13. $x^2 y^5$	**14.** $a^3 b^2$	**15.** $3^2 \cdot 5^3$	**16.** $2^3 \cdot 3^2$
17. $a^4 + a^2$	**18.** $x^3 + x^4$	**19.** $a^x \cdot a^w$	**20.** $x^a \cdot x^b$
21. $x^y \cdot y^x$	**22.** $a^b \cdot b^a$	**23.** $x^2 y^3 x^5$	**24.** $z^3 z^4 w^2$
25. $a^2 b^3 a^5$	**26.** $x^8 y x^4$	**27.** $s^7 + s^4$	**28.** $t^3 + t^8$
29. $7^3 \cdot 7^5$	**30.** $6^4 \cdot 6^{15}$		

Set II   Rewrite each of the following with the base appearing only once (if possible).

1. $x^2 \cdot x^5$	**2.** $y \cdot y^4$	**3.** $u^2 \cdot u$	**4.** $10^3 \cdot 10^4$
5. $3 \cdot 3^2 \cdot 3^2$	**6.** $a \cdot a^3 \cdot a^2$	**7.** $x^3 y^3$	**8.** $x^a \cdot y^b$
9. $x^2 \cdot x^y \cdot x^3$	**10.** $3^2 \cdot 4^3$	**11.** $a^4 \cdot a^2$	**12.** $x y^2 x^3$
13. $x^6 \cdot x^9$	**14.** $y^7 \cdot y$	**15.** $8^2 \cdot 3^3$	**16.** $a^4 \cdot b^8$
17. $y^2 + y^{14}$	**18.** $y^2 \cdot y^{14}$	**19.** $b^x \cdot b^y$	**20.** $b^x \cdot b$
21. $s^x \cdot t^y$	**22.** $s^2 + s^5$	**23.** $a^4 \cdot b^3 \cdot c^2$	**24.** $x + x^4$
25. $x \cdot x^7$	**26.** $x y^3 x^4$	**27.** $x^4 + x^2$	**28.** $a^5 + a$
29. $4^3 \cdot 4^{12}$	**30.** $6^8 + 6^4$		

5.2 Simplifying Products of Factors

In this section, we discuss simplifying a product of factors when there is a *single term* inside each set of grouping symbols. We use the following rules:

> ## TO SIMPLIFY A PRODUCT OF FACTORS WHEN EACH FACTOR CONTAINS ONLY ONE TERM
>
> 1. *Multiply the numerical coefficients.* Remember that the sign of the coefficient of any term is part of the numerical coefficient.
>
> 2. *Multiply the variables.* Each variable should appear only once, and the variables are usually written in alphabetical order. Use Rule 5.1 to determine the exponent on each variable.
>
> 3. The final answer is the product of the results of steps 1 and 2.

NOTE   We *cannot* use the distributive property for such problems. The distributive property can be used only when at least one set of grouping symbols contains two or more terms (see Section 5.3). ☑

In Example 1, we make use of the associative and commutative properties of multiplication and rearrange the factors so that the numerical coefficients are together, the x's are together, and so on. In practice, this step need not be shown.

Example 1   Simplify the following products.

a. $(6x)(-2x^3)$

Step 1: $(6)(-2) = -12$

$= (6)(-2)\,(x \cdot x^3)$

Step 2: Use Rule 5.1; $x \cdot x^3 = x^1 \cdot x^3 = x^{1+3} = x^4$

$= -12x^4$

b. $(-2a^2b)(5a^3b^3)$

Step 1: $(-2)(5) = -10$

$= (-2)(5)\,(a^2 \cdot a^3)(b \cdot b^3)$

Step 2: Use Rule 5.1

$= -10a^5b^4$

c. $(-5y^2)(2y)(-4y^3)$

$= (-5)(2)(-4)(y^2yy^3)$

$= 40y^6$

d. $(-4xy^2)(-3x^2y^3)(-2x^3y)$

$= (-4)(-3)(-2)(xx^2x^3)(y^2y^3y)$

$= -24x^6y^6$

e. $(3xy^2)^3$

$= (3xy^2)(3xy^2)(3xy^2)$

$= (3 \cdot 3 \cdot 3)(xxx)(y^2y^2y^2)$

$= 27x^3y^6$ ■

Recall from Section 1.8 that the product of an odd number of negative factors is negative and the product of an even number of negative factors is even. In Example 1, you can determine the sign of the answer by using these properties if you wish. Notice that in Examples 1a, 1b, and 1d we had an odd number of negative factors and the signs of the answers were negative, and that in Example 1c we had an even number of negative factors and the sign of the answer was positive.

A term containing exponents is considered *simplified* when each different base appears only once and when the exponent on that base is a single positive integer. For example, $5x^3y$ is simplified, whereas $5xyx^2$ and $5x^{1+2}y$ are not simplified.

EXERCISES 5.2

Set I Simplify the following products of factors.

1. $(-2a)(4a^2)$
2. $(-3x)(5x^3)$
3. $(-5h^2)(-6h^3)$
4. $(-6k^3)(-8k)$
5. $(-5x^3)^2$
6. $(-7y^4)^2$
7. $(-2a^3)(-4a)(3a^4)$
8. $(-6b^2)(-2b)(-4b^3)$
9. $(-9m)(m^5)(-2m^2)$
10. $(4n^4)(-7n^2)(-n)$
11. $(5x^2)(-7y)$
12. $(-3x^3)(4y)$
13. $(-6m^3n^2)(-4mn^2)$
14. $(-8h^4k)(5h^2k^3)$
15. $(2x^{10}y^2)(-3x^{12}y^7)$
16. $(-2a^2b^5)(-5a^{10}b^{10})$
17. $(3xy^2)^2$
18. $(4x^2y^3)^2$
19. $(5x^4y^5z)(-y^4z^7)$
20. $(-7E^2F^5G^8)(3F^6G^{10})$
21. $(2^3RS^2)(-2^2R^5T^4)$
22. $(-3^2xy)(3^3x^8z^5)$
23. $(-c^2d)(5d^1e^3)(-4c^5e^2)$
24. $(3m^1n^2)(-m^3r^4)(-8n^5r)$
25. $(-2x^2)(-3xy^2z)(-7yz)$
26. $(14y^2)(-5x^3z^4)(-6xyz)$
27. $(3xy)(x^2y^2)(-5x^3y^3)$
28. $(xy)(-2xz)(3yz)$
29. $(-2ab)(5bc)(ac)$
30. $(3xyz)(-2x^2y)(-yz^2)$
31. $(-5x^2y)(2yz^3)(-xz)$
32. $(2a^2b^3)(-3ab^2)(-a^2b^2)$
33. $(-5x^2y^2z)(x^5y^3z)(7xyz^5)$
34. $(-4R^2S^3T^4)(-8RS^3T^4)(-R^6S^1T^5)$
35. $(-3h^2k)(7)(-m^3k^5)(-mh^4)$
36. $(-km^2)(-6n^3)(8)(-k^2m)$

Set II Simplify the following products of factors.

1. $(-3x^2)(8x^3)$
2. $(3a^2)(-2a^4)$
3. $(-7x^3)(-12x^4)$
4. $(-5a^2b)(3ab^2)$
5. $(4x^2)^3$
6. $(3x^2)^2$
7. $(-4b^2)(-2b)(-6b^5)$
8. $(-3t^4)(5t^3)(-10t)$
9. $(-4y^2)(y^7)(-3y)$
10. $(6x^3)(-2x)(7x^4)$
11. $(15x^3)(-3y^2)$
12. $(-12a^4)(-5b)$
13. $(-4c^2d)(-13cd^3)$
14. $(-6e^5f^3)(11e^2f^3)$
15. $(2mn^2)(-4m^2n^2)(2m^3n)$
16. $(5x^2y)(-2xy^2)(-3xy)$

17. $(2m^5n)(-4mn^3)(3m^2n^6)$ **18.** $(7ab^4)(2a^2b)(-5a^2b^3)$

19. $(8a^4)(-a^4b^3c)$ **20.** $(3x^7)(-y^3)(4z^2)$

21. $(3x^3)(-3^4x^5)$ **22.** $(-5^2a^3)(5ab^2)$

23. $(5x^2y)(-3xy^2z)(-2xz^3)$ **24.** $(-8st^2)(-s^3t)(2t^4)$

25. $(-3ab^3)(-4a^2b^4)(c^5)$ **26.** $(9x^2y^3)(-2x^4y)(-5xy^3)$

27. $(8ab)(-2ab^4)(-2^4ab^3)$ **28.** $(-3u^3v^4)(-2^2u^2)(-3^2v^5)$

29. $(-7xy^2)(-3x^2z)(-2xz^3)$ **30.** $(8ac)(-9b^2c^3)(abc)$

31. $(3s^4)(-5st^3)(-s^2t^4)$ **32.** $(-2xy^2z)^2$

33. $(2H^2K^4M)(-2^2H^4K^5N)(2^3M^6K^7N)$ **34.** $(-10x^3y^2z)(-10^2xk^{10}z^1)(-10ky)$

35. $(-2e^1f^2)(-3e^3h)(-6fh)(-5e^2h)$ **36.** $(-4x^2y)(-7y^2z^2)(-5xy^4z)(-2x^3z^5)$

5.3 Using the Distributive Property to Simplify Algebraic Expressions

When we first discussed using the distributive property to remove grouping symbols in Section 2.2, the examples and exercises were carefully selected so that Rule 5.1 was never needed. In this section, we continue our discussion of simplifying algebraic expressions by combining what we have learned in Sections 5.1 and 5.2 with our knowledge of the distributive property.

In Section 5.2, we learned how to find products of factors when each factor consists of only one term. If one or more factors contains *more* than one term, we must use the distributive property. When we remove grouping symbols by using the distributive property, we must be sure to multiply *each term* inside the grouping symbols by the other factor.

In the "Alternate method" shown in Examples 1c, 1d, and 1e, we treat subtraction problems as addition problems.

Example 1   Use the distributive property to remove the parentheses. (In all parts of this example, it is not necessary to show the intermediate steps that we show; you can write just the final answer.)

Solutions

a. $x(x^2 + y) = (x)(x^2) + (x)(y)$

$\qquad\qquad = x^3 + xy$

b. $3x(4x + x^3y) = (3x)(4x) + (3x)(x^3y)$

$\qquad\qquad\quad = 12x^2 + 3x^4y$

c. $(-5a)(4a^3 - 2a^2b + b^2) = (-5a)(4a^3) - (-5a)(2a^2b) + (-5a)(b^2)$

$\qquad\qquad\qquad\qquad\quad = -20a^4 + 10a^3b - 5ab^2$

Alternate method

$-5a(4a^3 - 2a^2b + b^2) = (-5a)(4a^3 + [-2a^2b] + b^2)$

$\qquad\qquad\qquad\qquad = (-5a)(4a^3) + (-5a)(-2a^2b) + (-5a)(b^2)$

$\qquad\qquad\qquad\qquad = -20a^4 + 10a^3b - 5ab^2$

d. $(2x - 5)(-4x)) = (2x)(-4x) - (5)(-4x)$

$\qquad = -8x^2 + 20x$

Alternate method

$(2x - 5)(-4x) = (2x + [-5])(-4x))$

$\qquad = (2x)(-4x) + (-5)(-4x)$

$\qquad = -8x^2 + 20x$

e. $(-2x^2 + xy - 5y^2)(-3xy)) = (-2x^2)(-3xy) + (xy)(-3xy) - (5y^2)(-3xy)$

$\qquad = 6x^3y - 3x^2y^2 + 15xy^3$

Alternate method

$(-2x^2 + xy - 5y^2)(-3xy) = (-2x^2 + xy + [-5y^2])(-3xy))$

$\qquad = (-2x^2)(-3xy) + (xy)(-3xy) + (-5y^2)(-3xy)$

$\qquad = 6x^3y - 3x^2y^2 + 15xy^3$ ■

Simplifying Algebraic Expressions

To simplify an algebraic expression, we remove all grouping symbols, simplify each term (each variable can appear only once in a term), and combine all like terms. It is important to write the *complete* expression in every step and, as you are learning, to make only one type of change per step (see Example 2).

Example 2   Simplify each of the following algebraic expressions.

a. $5x^2(3x^2 - 2x + 1) - 3x(7x^2 + x - 6)$

$= 5x^2(3x^2) + 5x^2(-2x) + 5x^2(1) - 3x(7x^2) - 3x(x) - 3x(-6)$ Using the distributive property

$= 15x^4 - 10x^3 + 5x^2 - 21x^3 - 3x^2 + 18x$ Simplifying each term

$= 15x^4 - 10x^3 - 21x^3 + 5x^2 - 3x^2 + 18x$ Collecting like terms

$= 15x^4 - 31x^3 + 2x^2 + 18x$ Combining like terms

b. $5x - 3[6 - 4(3x - y)]$ Do *not* subtract 4 from 6; multiplication must be done before subtraction

$= 5x - 3[6 - 4(3x) - 4(-y)]$ Using the distributive property to remove the innermost grouping symbols

$= 5x - 3[6 - 12x + 4y]$ Simplifying each term inside the brackets

$= 5x - 3(6) - 3(-12x) - 3(4y)$ Using the distributive property to remove the brackets

$= 5x - 18 + 36x - 12y$ Simplifying each term

$= 5x + 36x - 18 - 12y$ Collecting like terms

$= 41x - 18 - 12y$ Combining like terms

c. $8 - 3(4x^2 - 2x[3x + 7])$ Do *not* subtract 3 from 8

  $= 8 - 3(4x^2 - 2x[3x] - 2x[7])$ Using the distributive property to remove the innermost grouping symbols

  $= 8 - 3(\,4x^2 - 6x^2\, -14x)$ Simplifying each term inside the parentheses

  ———————————————— Like terms

  $= 8 - 3(-2x^2 - 14x)$ Combining like terms within the grouping symbols

  $= 8 - 3[-2x^2] - 3[-14x]$ Using the distributive property to remove the parentheses

  $= 8 + 6x^2 + 42x$ Removing the brackets

d. $7b + 2b(3a + 1 - 5b)$

  $= 7b + 2b(3a) + 2b(1) + 2b(-5b)$ Using the distributive property

  $= 7b + 6ab + 2b - 10b^2$ Simplifying each term

  ———————————————— Like terms

  $= 7b + 2b + 6ab - 10b^2$ Collecting like terms

  $= 9b + 6ab - 10b^2$ Combining like terms ∎

A WORD OF CAUTION   In Example 4d, it is incorrect to say

$$7b + 2b(3a + 1 - 5b) = 9b(3a + 1 - 5b)$$

Remember: Multiplication must be done *before* addition.   ☑

NOTE   You need not show all the steps that we showed in the examples unless your instructor requires you to do so.   ☑

EXERCISES 5.3

Set I   Simplify the following algebraic expressions.

1. $-3(2x^2 - 4x + 5)$ **2.** $-5(3x^2 - 2x - 7)$

3. $4x(3x^2 - 6)$ **4.** $3x(5x^2 - 10)$

5. $-2x(5x^2 + 3x - 4)$ **6.** $-4x(2x^2 - 5x + 3)$

7. $(y^2 - 4y + 3)(7)$ **8.** $(-9 + z - 2z^2)(-7)$

9. $(2x^2 - 3x + 5)(4x)$ **10.** $(3w^2 + 2w - 8)(5w)$

11. $x(xy - 3)$ **12.** $a(ab - 4)$

13. $3a(ab - 2a^2)$ **14.** $4x(3x - 2y^2)$

15. $(-2x + 4x^2y)(-3y)$ **16.** $(-3a + 2a^2z)(-2z)$

17. $-2xy(x^2y - y^2x - y - 5)$ **18.** $-3ab(8 - a^2 - b^2 + ab)$

19. $(3x^3 - 2x^2y + y^3)(-2xy)$ **20.** $(4z^3 - z^2y - y^3)(-2yz)$

21. $(2xy^2z - 7x^2z^2)(-5xz^3)$ **22.** $(3a^2bc^2 - 4ab^3c)(-3ac^2)$

23. $(5x^2y^3z - 2xz^3 + y^4)(-4xz^2)$ **24.** $(-2xy^2z^2)(6x^2y - 3yz - 4xz^2)$

25. $6 - 3(4 - 3z - 2z^3)$ **26.** $9 - 2(3x - 4x^2 + 3x)$

27. $5x - 2x(3x^2 + 7x - 3)$ **28.** $8y - 3y(4y - 9 + 2y)$

29. $(3x^2 - 4x + 2) - 5x$ **30.** $(14y^2 + 2y - 4) + 3y$

31. $(3x^2 - 4x + 2)(-5x)$ **32.** $(14y^2 + 2y - 4)(+3y)$

33. $(3x^3)(2x \cdot y)$ **34.** $(5x^2)(6x \cdot z^2)$

35. $4x^2(2x^2 - 3x + 7) - 4x(x^3 + 2x - 5)$

36. $3y(4y + 2y^2 - 5y^3) - 3y(y^2 + y - 1)$

37. $8x^2(7x^2 - 3x + 1) - 5x(6x^2 - 8x + 9)$

38. $7a^3(6a^2 + 9a - 1) - 5a^2(9a^2 - 8a + 3)$

39. $3x - 5[8 - 2(4x - y)]$ **40.** $8x - 2[6 - 3(x - 4y)]$

41. $3 - 2(7x^2 - 3x[8x + 2])$ **42.** $7 - 4(3y^2 - 2y[y - 4])$

43. $8x - 2x[6 - 3(2x - 1)]$ **44.** $7y - 2y(8 - 2[3y + 4])$

45. $7x + 4x(3y - 5 + 8x)$ **46.** $9y + 8y(4z + 7 - 9y)$

Set II   Simplify the following algebraic expressions.

1. $-5(3x^2 + 4x - 7)$ **2.** $-2(5x^4 - 3x^3 + 4x)$

3. $6x(5x^2 - 4)$ **4.** $8x^2(7x^3 + 2x^2 - 3)$

5. $-3x(4x^2 - 6x + 8)$ **6.** $(9 - 5x)(-2)$

7. $(3z^2 - 5z + 4)(-6)$ **8.** $(2u - 5u^2 + 11)(6)$

9. $4x(3xy^2 - 2x^2y)$ **10.** $(6s^2t - 5st)(-3t)$

11. $a(2ab - b)$ **12.** $-3x^2(-5x^2 + 2x - 1)$

13. $5x(-2xy + y)$ **14.** $-2xy^2(3xy^2 + 5xy - y)$

15. $(-3x + 2y)(-3x)$ **16.** $(5x + 7y - z)(-3y)$

17. $-4xy(3x + 5xy - y^2 + 2y)$ **18.** $-x^2(2xy - 3xy^2 + x^2y + y)$

19. $(4x^3 - x^2 + 1)(-4x^2)$ **20.** $(-3x^2 + x - 1)(4x)$

21. $(x^2 - 4xy + y^2)(-2x)$ **22.** $3x^2y(x^3 - 3x^2y + 3xy^2 - y^3)$

23. $(-5a^2bc)(2b^2c - 5ac^3 - 3a^2b^3)$ **24.** $(2a^3bc^2 - 3ac + b^3)(-5ab^2)$

25. $9 - 2(3 - 5z - 7z^3)$ **26.** $x(xy - 6)$

27. $7x - 4x(6x^2 + 8x - 2)$ **28.** $4a(ab - b^2)$

29. $(6x^2 - 3x + 5) - 7x$ **30.** $(12y^2 + 4y - 3) + 6y$

31. $(6x^2 - 3x + 5)(-7x)$ **32.** $(12y^2 + 4y - 3)(+6y)$

33. $(7y^3)(3x \cdot y)$ **34.** $6 - 2(4y^2 - 2y[y - 5])$

35. $6x^2(7x^2 - 2x + 5) - 3x(x^3 + 2x - 8)$

36. $4x^3(x^3 - 2x^2 + 3x - 1)$

37. $4x^2(8x^2 - 7x + 9) - 8x(5x^2 - 9x + 7)$

38. $8x^4(5x^3 + 8x - 1) - 8x^2(7x^5 - 8x^3 + 2x^2)$

39. $7x - 4[7 - 5(3x - y)]$ **40.** $-8(2a + 3b - 1)$

41. $5 - 3(8x^2 - 5x[3x + 6])$ **42.** $(8x^2)(5z \cdot x^2)$

43. $7x - 3x[9 - 4(5x - 3)]$ **44.** $6y - 3y(9 - 5[6y + 7])$

45. $8c + 6c(7a + 4 - 9b)$ **46.** $15x - 3x(6y - 5x - 7)$

5.4 More Rules of Exponents

5.4A Raising Exponential Numbers and Products to a Power

Raising an Exponential Number to a Power

Consider the expressions $(x^4)^2$ and $x^{4 \cdot 2}$.

$$(x^4)^2 = (x^4)(x^4) = x^{4+4} = x^8$$

$$x^{4 \cdot 2} = x^8$$

Since the expressions $(x^4)^2$ and $x^{4 \cdot 2}$ both equal x^8, they must equal each other. Therefore, $(x^4)^2 = x^{4 \cdot 2}$.

The general rule for raising an exponential number to a power is given in Rule 5.2.

RULE 5.2   SECOND RULE OF EXPONENTS

$$(x^a)^b = x^{ab}$$

In words: To raise an exponential number to a power, *multiply* the exponents.

Example 1   Remove the parentheses in each of the following expressions.

a. $(x^5)^4 = x^{5 \cdot 4} = x^{20}$

b. $(x)^4 = (x^1)^4 = x^{1 \cdot 4} = x^4$

c. $(y^4)^3 = y^{4 \cdot 3} = y^{12}$

d. $(10^6)^2 = 10^{6 \cdot 2} = 10^{12} = 1,\underbrace{000,000,000,000}_{12 \text{ zeros}}$

e. $(2^a)^b = 2^{a \cdot b} = 2^{ab}$ ∎

Raising a Product of Factors to a Power

Next, consider the expression $(xy)^3$.

$$(xy)^3 = (xy)(xy)(xy) = (xxx)(yyy) = x^3 y^3$$

The general rule for raising a product of factors to a power is given in Rule 5.3.

RULE 5.3   THIRD RULE OF EXPONENTS

$$(xy)^a = x^a y^a$$

In words: To raise a product of factors to a power, raise *each factor* to that power.

Example 2   Remove the parentheses in each of the following expressions.

   a. $(xy)^5 = x^5y^5$

   b. $(2x)^3 = 2^3x^3$

Be sure to raise *numbers* in the parentheses to the power also. Remember that

$$(2x)^3 = (2x)(2x)(2x) = (2 \cdot 2 \cdot 2)(xxx) = 2^3x^3 \quad \blacksquare$$

A WORD OF CAUTION $2x^3 \neq (2x)^3$. In the expression $2x^3$, the exponent applies only to the x. That is, $2x^3 = 2xxx$. In the expression $(2x)^3$, the exponent applies to everything inside the parentheses. ☑

Example 3   Evaluate each expression.

   ┌────── This exponent applies to everything in the parentheses: (-2)

   a. $5(-2)^2 = 5(4) = 20$

   ┌────── This exponent applies only to the 5

   b. $-2 \cdot 5^2 = -2 \cdot 25 = -50$

   ┌────── This exponent applies to everything in parentheses: $(-2 \cdot 5)$

   c. $(-2 \cdot 5)^2 = (-10)^2 = 100 \quad \blacksquare$

EXERCISES 5.4A

Set I   In Exercises 1–14, remove the parentheses.

   1. $(y^2)^5$ **2.** $(N^3)^4$ **3.** $(x^8)^2$ **4.** $(z^4)^7$ **5.** $(2^4)^2$

   6. $(3^4)^2$ **7.** $(xy)^5$ **8.** $(ab)^4$ **9.** $(2c)^6$ **10.** $(3x)^4$

   11. $(x^4)^7$ **12.** $(v^3)^8$ **13.** $(10^2)^3$ **14.** $(10^7)^2$

In Exercises 15–20, evaluate each expression.

   15. $(-2 \cdot 3)^2$ **16.** $(-3 \cdot 4)^2$ **17.** $2(-3)^3$

   18. $3(-2)^3$ **19.** $-4 \cdot 5^2$ **20.** $-3 \cdot 4^2$

Set II   In Exercises 1–14, remove the parentheses.

   1. $(a^3)^4$ **2.** $(2x)^4$ **3.** $(b^9)^4$ **4.** $(x^3)^2$ **5.** $(2^2)^4$

   6. $(2x)^5$ **7.** $(abc)^4$ **8.** $(2xy)^5$ **9.** $(3x)^4$ **10.** $(st)^8$

   11. $(u^7)^4$ **12.** $(5x)^2$ **13.** $(10^3)^3$ **14.** $(x^6)^4$

In Exercises 15–20, evaluate each expression.

   15. $(-3 \cdot 5)^2$ **16.** $(-5 \cdot 2)^3$ **17.** $5(-2)^5$

   18. $-4 \cdot 3^2$ **19.** $-6 \cdot 2^2$ **20.** $7(-2)^2$

5.4B   Exponential Numbers and Division

Dividing Exponential Numbers When the Bases Are the Same

Consider the division problem $x^5 \div x^3$, or $\dfrac{x^5}{x^3}$ (assuming $x \neq 0$).

$$\frac{x^5}{x^3} = \frac{xxxxx}{xxx} = \frac{xxx \cdot xx}{xxx \cdot 1} = \frac{xxx}{xxx} \cdot \frac{xx}{1} = 1 \cdot \frac{xx}{1} = xx = x^2$$

└── The value of this fraction
is 1 (for $x \neq 0$)

Consider x^{5-3}. $x^{5-3} = x^2$. Since $\frac{x^5}{x^3}$ and x^{5-3} both equal the same number (x^2), they must equal each other. Therefore, if $x \neq 0$, $\frac{x^5}{x^3} = x^{5-3}$.

The general rule for dividing one exponential number by another *when the bases are the same* is given in Rule 5.4.

RULE 5.4   FOURTH RULE OF EXPONENTS

If $x \neq 0$, $\frac{x^a}{x^b} = x^{a-b}$.

In words: To divide one exponential number by another *when the bases are the same*, subtract the exponent of the divisor (the exponent in the denominator) from the exponent of the dividend (the exponent in the numerator).

In Example 4, assume $x \neq 0$, $r \neq 0$, $y \neq 0$, $a \neq 0$, and $b \neq 0$.

Example 4   Rewrite each expression with the base appearing only once (if possible).

a. $\dfrac{x^6}{x^2} = x^{6-2} = x^4$ b. $\dfrac{r^{12}}{r^5} = r^{12-5} = r^7$

c. $\dfrac{y^3}{y} = \dfrac{y^3}{y^1} = y^{3-1} = y^2$

d. $\dfrac{10^7}{10^3} = 10^{7-3} = 10^4 = \underbrace{10{,}000}_{\text{4 zeros}}$

e. $\dfrac{x^5}{y^2}$

Rule 5.4 does not apply when the bases are different. Therefore, $\dfrac{x^5}{y^2}$ cannot be rewritten with the base appearing only once.

f. $x^4 \ominus x^2$

Rule 5.4 does not apply, because the operation is subtraction rather than division. Therefore, $x^4 - x^2$ cannot be rewritten with the base appearing only once.

g. $\dfrac{8x^3}{2x} = \dfrac{\overset{4}{\cancel{8}}}{\underset{1}{\cancel{2}}} \cdot \dfrac{x^3}{x} = 4 \cdot x^{3-1} = 4x^2$

h. $\dfrac{6a^3b^4}{9ab^2} = \dfrac{\overset{2}{\cancel{6}}}{\underset{3}{\cancel{9}}} \cdot \dfrac{a^3}{a} \cdot \dfrac{b^4}{b^2} = \dfrac{2}{3} \cdot a^{3-1} \cdot b^{4-2} = \dfrac{2}{3}a^2b^2$ or $\dfrac{2a^2b^2}{3}$

i. $\dfrac{2^a}{2^b} = 2^{a-b}$

j. $\dfrac{x^5 - y^3}{x^2}$

The expression is usually left unchanged. It could be changed as follows:

$$\frac{x^5 - y^3}{x^2} = \frac{x^5}{x^2} - \frac{y^3}{x^2} = x^3 - \frac{y^3}{x^2}$$

Expressions of this form are discussed in Sections 6.5 and 8.2 ■

A WORD OF CAUTION We show examples of a common mistake students make in division. (Assume $x \neq 0$.)

Correct method	Incorrect method

a. $\dfrac{2 + 6}{2} = \dfrac{8}{2} = 4$ $\qquad$ $\dfrac{\cancel{2}^{\,1} + 6}{\cancel{2}_{\,1}} = \dfrac{1 + 6}{1} = 7$

b. $\dfrac{x^2 + y}{x^2} = \dfrac{x^2}{x^2} + \dfrac{y}{x^2}$ $\qquad$ $\dfrac{\cancel{x^2}^{\,1} + y}{\cancel{x^2}_{\,1}} = 1 + y$

$\qquad\qquad = 1 + \dfrac{y}{x^2}$

This expression is usually left in the form $\dfrac{x^2 + y}{x^2}$. ☑

NOTE To see why the restriction $x \neq 0$ is included in Rule 5.4, consider the example $\dfrac{0^5}{0^2}$. $\dfrac{0^5}{0^2} = \dfrac{0 \cdot 0 \cdot 0 \cdot 0 \cdot 0}{0 \cdot 0} = \dfrac{0}{0}$, which cannot be determined. For this reason, x cannot be zero in Rule 5.4. ☑

In this book, unless otherwise noted, we will assume that none of the variables has a value that makes a denominator zero.

In Example 4, we were careful to have the exponent in the numerator be larger than the exponent in the denominator, so that the quotient always had a positive exponent. When the exponent in the denominator is larger than the exponent in the numerator, the quotient has a negative exponent. Negative exponents are discussed in Section 5.5

Example 5 Evaluate each of the following expressions. (Compare parts a and b and note the differences when *negative* values are raised to an even or an odd power.)

———— This exponent applies only to the 5

a. $\dfrac{-5^2}{(-5)^2} = \dfrac{-5 \cdot 5}{(-5)(-5)} = \dfrac{-25}{25} = -1$

———— This exponent applies to everything in the parentheses

b. $\dfrac{-2^3}{(-2)^3} = \dfrac{-2 \cdot 2 \cdot 2}{(-2)(-2)(-2)} = \dfrac{-8}{-8} = 1$ ■

Raising Fractions to Powers

Consider the problem $\left(\dfrac{x}{y}\right)^3$.

$$\left(\frac{x}{y}\right)^3 = \left(\frac{x}{y}\right)\left(\frac{x}{y}\right)\left(\frac{x}{y}\right) = \frac{xxx}{yyy} = \frac{x^3}{y^3}.$$

The general rule for raising a fraction to a power is given in Rule 5.5.

RULE 5.5   FIFTH RULE OF EXPONENTS

If $y \neq 0$, $\left(\dfrac{x}{y}\right)^n = \dfrac{x^n}{y^n}$.

In words: To raise a fraction to a power, raise the numerator to the power and also raise the denominator to the power.

Example 6   Remove the parentheses.

a. $\left(\dfrac{a}{b}\right)^5 = \dfrac{a^5}{b^5}$

b. $\left(\dfrac{3}{c}\right)^3 = \dfrac{3^3}{c^3} = \dfrac{27}{c^3}$ Be sure to raise *numbers* in parentheses to the power also ∎

The General Rule of Exponents

Rules 5.2, 5.3, 5.4, and 5.5 can be combined into the following general rule:

RULE 5.6   SIXTH (GENERAL) RULE OF EXPONENTS

$$\left(\dfrac{x^a y^b}{z^c}\right)^n = \dfrac{x^{an} y^{bn}}{z^{cn}}$$

None of the variables can have a value that makes the denominator zero.

In applying Rule 5.6, notice the following:

1. x, y, and z are *factors* of the expression within the parentheses. That is, they are *not* separated by plus or minus signs.

2. The exponent of each factor within the parentheses is multiplied by the exponent outside the parentheses.

Example 7   Simplify each of the following expressions.

a. $(x^2 y^3)^2 = x^{2 \cdot 2} y^{3 \cdot 2} = x^4 y^6$ Each exponent inside the parentheses must be multiplied by 2

The exponents on the 5 and the x are understood to be 1

Those 1s must be multiplied by 3

b. $(5xy^2 z^5)^3 = (5^1 x^1 y^2 z^5)^3 = 5^{1 \cdot 3} x^{1 \cdot 3} y^{2 \cdot 3} z^{5 \cdot 3} = 5^3 x^3 y^6 z^{15} = 125 x^3 y^6 z^{15}$

c. $(x^2 + y^3)^4$

Rule 5.6 *cannot* be used here because the plus sign means that x^2 and y^3 are terms, not factors, of the expression being raised to the fourth power. We cannot simplify $(x^2 + y^3)^4$ at this time. ∎

Example 8   In the following examples, assume that none of the factors appearing in any denominator is zero.

a. $\left(\dfrac{x^2 y^5}{z^3}\right)^4 = \dfrac{x^{2\cdot4} y^{5\cdot4}}{z^{3\cdot4}} = \dfrac{x^8 y^{20}}{z^{12}}$

⌐――― Simplify the expression within parentheses first whenever possible

b. $\left(\dfrac{x^5 y^6}{x^3 y^3}\right)^4 = (x^{5-3} y^{6-3})^4 = (x^2 y^3)^4 = x^{2\cdot4} y^{3\cdot4} = x^8 y^{12}$

c. $\left(\dfrac{3 b^3}{2 c^4}\right)^2 = \left(\dfrac{3^1 b^3}{2^1 c^4}\right)^2 = \dfrac{3^{1\cdot2} b^{3\cdot2}}{2^{1\cdot2} c^{4\cdot2}} = \dfrac{3^2 b^6}{2^2 c^8} = \dfrac{9 b^6}{4 c^8}$ ∎

See Section 5.7 for a summary of all the rules of exponents.

EXERCISES 5.4B

Set I   Assume that none of the factors appearing in any denominator is zero. In Exercises 1–24, rewrite each expression with the base appearing only once, if possible.

1. $\dfrac{x^7}{x^2}$　　　　2. $\dfrac{y^8}{y^6}$　　　　3. $x^4 - x^2$　　　　4. $s^8 - s^3$

5. $\dfrac{a^5}{a}$　　　　6. $\dfrac{b^7}{b}$　　　　7. $\dfrac{10^{11}}{10}$　　　　8. $\dfrac{5^6}{5}$

9. $\dfrac{x^8}{y^4}$　　　　10. $\dfrac{a^4}{b^3}$　　　　11. $\dfrac{6 x^2}{2x}$　　　　12. $\dfrac{9 y^3}{3y}$

13. $\dfrac{a^3}{b^2}$　　　　14. $\dfrac{x^5}{y^3}$　　　　15. $\dfrac{10 x^4}{5 x^3}$　　　　16. $\dfrac{15 y^5}{9 y^2}$

17. $\dfrac{12 h^4 k^3}{8 h^2 k}$　　　18. $\dfrac{16 a^5 b^3}{12 a b^2}$　　　19. $\dfrac{x^{5a}}{x^{3a}}$　　　20. $\dfrac{M^{6x}}{M^{2x}}$

21. $\dfrac{a^4 - b^3}{a^2}$　　　22. $\dfrac{x^6 + y^4}{y^2}$　　　23. $\dfrac{-5 x^8}{10 x^2}$　　　24. $\dfrac{-4 y^9}{12 y^4}$

In Exercises 25–42, remove the parentheses.

25. $\left(\dfrac{s}{t}\right)^7$　　　26. $\left(\dfrac{x}{y}\right)^9$　　　27. $\left(\dfrac{2}{x}\right)^4$　　　28. $\left(\dfrac{3}{z}\right)^2$

29. $\left(\dfrac{x}{2}\right)^6$　　　30. $\left(\dfrac{c}{5}\right)^3$　　　31. $(a^2 b^3)^2$　　　32. $(x^4 y^5)^3$

33. $(2 z^3)^2$　　　34. $(3 w^2)^3$　　　35. $\left(\dfrac{x y^4}{z^2}\right)^2$　　　36. $\left(\dfrac{a^3 b}{c^2}\right)^3$

37. $\left(\dfrac{5 y^3}{2 x^2}\right)^4$　　38. $\left(\dfrac{6 b^4}{7 c^2}\right)^3$　　39. $\left(\dfrac{x^3 y^7}{x y^2}\right)^3$　　40. $\left(\dfrac{a^6 b^8}{a^4 b^3}\right)^3$

41. $\left(\dfrac{-2 x^4 y^6}{x y^2}\right)^3$　　42. $\left(\dfrac{-2 s^5 t^4}{s^3 t}\right)^3$

In Exercises 43 and 44, evaluate each expression.

43. $\dfrac{(-4)^2}{-4^2}$　　　　　　　　　　44. $\dfrac{-9^2}{(-9)^2}$

Set II   Assume that none of the factors appearing in any denominator is zero. In Exercises 1–24, rewrite each expression with the base appearing only once, if possible.

1. $\dfrac{x^9}{x^3}$

2. $\dfrac{a^4}{b^2}$

3. $s^6 - s^4$

4. $\dfrac{s^6}{s^4}$

5. $\dfrac{y^7}{y}$

6. $\dfrac{3^5}{3}$

7. $\dfrac{8^6}{8^2}$

8. $\dfrac{15^9}{15^4}$

9. $\dfrac{a^9}{b^4}$

10. $\dfrac{a^6}{a^4}$

11. $\dfrac{12b^6}{3b^3}$

12. $\dfrac{12x^3}{10x}$

13. $\dfrac{m^4}{n^2}$

14. $\dfrac{18z^6}{10z^4}$

15. $\dfrac{25a^3b^4}{15ab^3}$

16. $\dfrac{18a^8}{2a^4}$

17. $\dfrac{9x^4y^8}{6xy^4}$

18. $6x^3y^2 - 4x^2y$

19. $\dfrac{H^{4n}}{H^{2n}}$

20. $z^{4a} - z^{2a}$

21. $\dfrac{x^3 - y^2}{x^2}$

22. $\dfrac{a^4 - b^5}{b^3}$

23. $\dfrac{-3s^7}{12s^4}$

24. $\dfrac{-21t^6}{-3t^2}$

In Exercises 25–42, remove the parentheses.

25. $\left(\dfrac{u}{v}\right)^4$

26. $\left(\dfrac{8}{c}\right)^2$

27. $\left(\dfrac{3}{a}\right)^3$

28. $\left(\dfrac{x}{2}\right)^5$

29. $\left(\dfrac{t}{5}\right)^2$

30. $\left(\dfrac{a}{c}\right)^7$

31. $(h^3k^4)^2$

32. $(2x^4y^2)^3$

33. $(5a^3)^2$

34. $(8xy^3)^2$

35. $\left(\dfrac{x^3y}{z^2}\right)^3$

36. $\left(\dfrac{s^4t^2}{u^3}\right)^3$

37. $\left(\dfrac{3a^5}{2x^2}\right)^4$

38. $\left(\dfrac{5}{2t^5}\right)^2$

39. $\left(\dfrac{x^4y^7}{x^3y^3}\right)^5$

40. $\left(\dfrac{6a^9b^3}{2ab}\right)^3$

41. $\left(\dfrac{-3x^6y^8}{xy^3}\right)^3$

42. $\left(\dfrac{-2m^8n^9}{m^3n}\right)^4$

In Exercises 43 and 44, evaluate each expression.

43. $\dfrac{-2^4}{(-2)^4}$

44. $\dfrac{-3^3}{(-3)^3}$

5.5 Zero and Negative Exponents

The Zero Exponent

RULE 5.7   SEVENTH RULE OF EXPONENTS
If $x \neq 0$, $x^0 = 1$.

In words: Any nonzero number raised to the zero power equals 1.

To see why we define x^0 to be 1, let's consider dividing one exponential number by another when the bases are the same *and* when the exponents are equal. In particular, let's consider the problem $x^4 \div x^4$, or $\dfrac{x^4}{x^4}$.

$$\frac{x^4}{x^4} = \frac{xxxx}{xxxx} = 1 \qquad \text{A number divided by itself is 1}$$

If we apply Rule 5.4 to the problem $\frac{x^4}{x^4}$, we have $\frac{x^4}{x^4} = x^{4-4} = x^0$. Since x^0 and 1 both equal $\frac{x^4}{x^4}$, we would *like* x^0 to equal 1. Therefore, if $x \neq 0$, we *define* x^0 to be 1 (see Rule 5.7).

Example 1 Examples of zero as an exponent:

    a. $a^0 = 1$ Provided $a \neq 0$

    b. $H^0 = 1$ Provided $H \neq 0$

    c. $10^0 = 1$

    d. $5^0 = 1$

    e. $6x^0 = 6 \cdot 1 = 6$ Provided $x \neq 0$; the 0 exponent applies only to x ■

NOTE To see why we included the statement "$x \neq 0$" in Rule 5.7, let's apply Rule 5.4 to the problem $\frac{0}{0}$:

$$\frac{0}{0} = \frac{0^1}{0^1} = 0^{1-1} = 0^0$$

But $\frac{0}{0}$ is undefined. Therefore, 0^0 must also be undefined.   ☑

Negative Exponents

RULE 5.8a   EIGHTH RULE OF EXPONENTS

$$\text{If } x \neq 0,\ x^{-n} = \frac{1}{x^n}.$$

To show why we define x^{-n} to be $\frac{1}{x^n}$, let's consider dividing one exponential number by another when the bases are the same *and* when the exponent of the divisor is *greater than* the exponent of the dividend. In particular, let's consider the problem $x^3 \div x^5$, or $\frac{x^3}{x^5}$.

$$\frac{x^3}{x^5} = \frac{xxx}{xxxxx} = \frac{xxx \cdot 1}{xxx \cdot xx} = \frac{xxx}{xxx} \cdot \frac{1}{xx} = 1 \cdot \frac{1}{xx} = \frac{1}{x^2}$$

The value of this fraction is 1

However, if we use Rule 5.4, we have

$$\frac{x^3}{x^5} = x^{3-5} = x^{-2}$$

Since x^{-2} and $\dfrac{1}{x^2}$ both equal $\dfrac{x^3}{x^5}$, we would like the following statement to be true:

$$x^{-2} = \frac{1}{x^2}$$

Also, according to Rule 5.1, $x^2 \cdot x^{-2} = x^{2+(-2)}$, and $x^{2+(-2)} = x^0 = 1$ (Rule 5.7). Then, since the product of x^2 and x^{-2} is 1, x^{-2} must be the multiplicative inverse of x^2. Therefore, we *define* x^{-n} to equal $\dfrac{1}{x^n}$ when $x \neq 0$.

A WORD OF CAUTION x^{-n} is not necessarily a negative number.   ☑

When we use Rule 5.8a, the exponent n can be either positive or negative.

Example 2   Remove the negative exponents, using Rule 5.8a.

a. $x^{-5} = \dfrac{1}{x^5}$

b. $a^{-3} = \dfrac{1}{a^3}$

c. $10^{-4} = \dfrac{1}{10^4}$ ■

The next two rules follow from Rule 5.8a.

RULE 5.8b

If $x \neq 0$, $\dfrac{1}{x^{-n}} = x^n$.

RULE 5.8c

If $x \neq 0$ and $y \neq 0$, $\left(\dfrac{x}{y}\right)^{-n} = \left(\dfrac{y}{x}\right)^n$.

Rules 5.8b and 5.8c can both be used to remove negative exponents (see Examples 3 and 4), and Rules 5.8a, 5.8b, and 5.8c combined lead to the following rule for removing negative exponents when only *factors* are involved:

A *factor* can be moved either from the numerator to the denominator or from the denominator to the numerator simply by so moving it and changing the sign of its exponent. This does not change the sign of the *expression*. If a factor has no exponent, the exponent is understood to equal 1.

As we mentioned earlier, there is a summary of all the rules of exponents in Section 5.7.

Example 3 Remove the negative exponents, using Rule 5.8b.

a. $\dfrac{1}{x^{-4}} = x^4$

b. $\dfrac{1}{3w^{-2}} = \dfrac{1}{3} \cdot \dfrac{1}{w^{-2}} = \dfrac{1}{3} \cdot w^2 = \dfrac{w^2}{3}$ ∎

Example 4 Remove the negative exponent and the parentheses from $\left(\dfrac{a}{b}\right)^{-4}$.

$$\left(\dfrac{a}{b}\right)^{-4} = \left(\dfrac{b}{a}\right)^{4} = \dfrac{b^4}{a^4} \qquad \text{Using Rules 5.8c and 5.5} \quad ∎$$

Rules 5.8a, 5.8b, and 5.8c can also be used to insert negative exponents (see Example 5).

Example 5 Rewrite $\dfrac{1}{y^3}$ without fractions.

$$\dfrac{1}{y^3} = y^{-3} \qquad \text{Using Rule 5.8a} \quad ∎$$

NOTE If a single number or variable appears in the numerator or the denominator of a fraction, that number can still be consider a factor of the numerator or denominator. For example, $\dfrac{5}{x} = \dfrac{1 \cdot 5}{1 \cdot x}$. Therefore, we can say that 5 is a factor of the numerator and x is a factor of the denominator of $\dfrac{5}{x}$. ☑

Example 6 Remove the negative exponents in each of the following expressions.

a^{-3} is a factor of $a^{-3}b^4$

a. $a^{-3}b^4 = \dfrac{a^{-3}}{1} \cdot \dfrac{b^4}{1} = \dfrac{1}{a^3} \cdot \dfrac{b^4}{1} = \dfrac{b^4}{a^3}$

The factor a^{-3} was moved from the numerator to the denominator by changing the sign of its exponent

y^{-4} and z^{-2} are factors of $y^{-4}w^5z^{-2}$

b. $y^{-4}w^5z^{-2} = \dfrac{y^{-4}}{1} \cdot \dfrac{w^5}{1} \cdot \dfrac{z^{-2}}{1} = \dfrac{1}{y^4} \cdot \dfrac{w^5}{1} \cdot \dfrac{1}{z^2} = \dfrac{w^5}{y^4z^2}$

These shaded steps need not be written; we use them in these examples to show you why this method works

x^{-2} is a factor of x^{-2}

c. $\dfrac{x^{-2}}{y} = \dfrac{x^{-2}}{1} \cdot \dfrac{1}{y} = \dfrac{1}{x^2} \cdot \dfrac{1}{y} = \dfrac{1}{x^2y}$

d. $\dfrac{a^{-2}b^4}{c^5d^{-3}} = \dfrac{a^{-2}}{1} \cdot \dfrac{b^4}{1} \cdot \dfrac{1}{c^5} \cdot \dfrac{1}{d^{-3}} = \dfrac{1}{a^2} \cdot \dfrac{b^4}{1} \cdot \dfrac{1}{c^5} \cdot \dfrac{d^3}{1} = \dfrac{b^4d^3}{a^2c^5}$

e. $\dfrac{e^2f}{g^{-1}} = \dfrac{e^2}{1} \cdot \dfrac{f^1}{1} \cdot \dfrac{1}{g^{-1}} = \dfrac{e^2}{1} \cdot \dfrac{f^1}{1} \cdot \dfrac{g^1}{1} = e^2fg$ ∎

Example 7   Rewrite each expression without fractions, using negative exponents when necessary.

a. $\dfrac{h^5}{k^{-4}} = \dfrac{h^5}{1} \cdot \dfrac{1}{k^{-4}} = \dfrac{h^5}{1} \cdot \dfrac{k^4}{1} = h^5 k^4$

b. $\dfrac{a^3}{b^2} = \dfrac{a^3}{1} \cdot \dfrac{1}{b^2} = \dfrac{a^3}{1} \cdot \dfrac{b^{-2}}{1} = a^3 b^{-2}$

c. $\dfrac{z^3}{x} = \dfrac{z^3}{1} \cdot \dfrac{1}{x} = \dfrac{z^3}{1} \cdot \dfrac{x^{-1}}{1} = z^3 x^{-1}$

————   d. $\dfrac{a}{bc^2} = \dfrac{a^1}{1} \cdot \dfrac{1}{b^1} \cdot \dfrac{1}{c^2} = \dfrac{a^1}{1} \cdot \dfrac{b^{-1}}{1} \cdot \dfrac{c^{-2}}{1} = ab^{-1}c^{-2}$ ∎

A WORD OF CAUTION   An expression that is a *term* of, rather than a factor of, a numerator cannot be moved from the numerator to the denominator of a fraction simply by moving it and changing the sign of its exponents. For example, $\dfrac{a^{-2} + b^5}{c^4} \neq \dfrac{b^5}{a^2 c^4}$.

The plus sign indicates that a^{-2} is a *term* rather than a factor of the numerator

$$\dfrac{a^{-2} + b^5}{c^4} = \dfrac{\dfrac{1}{a^2} + b^5}{c^4}$$

Expressions of this kind will be simplified in Section 8.7   ☑

Using the Rules of Exponents with Positive, Zero, and Negative Exponents

All the rules of exponents also apply to expressions that have zero and negative exponents.

Example 8   Apply the rules of exponents to simplify each of the following expressions.

a. $a^4 \cdot a^{-3} = a^{4+(-3)} = a^1 = a$ Rule 5.1

b. $x^{-5} \cdot x^2 = x^{-5+2} = x^{-3} = \dfrac{1}{x^3}$ Rules 5.1 and 5.8a

c. $(y^{-2})^{-1} = y^{(-2)(-1)} = y^2$ Rule 5.2

d. $(x^2)^{-4} = x^{2(-4)} = x^{-8} = \dfrac{1}{x^8}$ Rules 5.2 and 5.8a

e. $\dfrac{y^{-2}}{y^{-6}} = y^{(-2)-(-6)} = y^{-2+6} = y^4$ Rule 5.4

f. $\dfrac{z^{-4}}{z^{-2}} = z^{(-4)-(-2)} = z^{-4+2} = z^{-2} = \dfrac{1}{z^2}$ Rules 5.4 and 5.8a

g. $h^3 h^0 h^{-2} = h^{3+0+(-2)} = h^1 = h$ Rule 5.1

h. $(7x)^{-2} = 7^{-2} x^{-2} = \dfrac{1}{7^2 x^2} = \dfrac{1}{49x^2}$ Rules 5.3 and 5.8a

i. $\left(\dfrac{5}{d}\right)^{-3} = \left(\dfrac{d}{5}\right)^3 = \dfrac{d^3}{5^3} = \dfrac{d^3}{125}$ Rules 5.8c and 5.5

————   j. $\left(\dfrac{a}{2bc}\right)^0 = 1$ Rule 5.7   ∎

An expression such as $\dfrac{x^3}{x^5}$ can be simplified in either of two ways:

$$\frac{x^3}{x^5} = x^{3-5} = x^{-2} = \frac{1}{x^2}$$

or

$$\frac{x^3}{x^5} = \frac{1}{x^5 \cdot x^{-3}} = \frac{1}{x^{5-3}} = \frac{1}{x^2}$$

We moved the factor x^3 to the denominator and changed the sign of its exponent

Example 9 Simplify the following expressions, using the rules of exponents; write the results using only positive exponents.

a. $\dfrac{12x^{-2}}{4x^{-3}} = \dfrac{\cancel{12}^{3}}{\cancel{4}_{1}} \cdot \dfrac{x^{-2}}{x^{-3}} = \dfrac{3}{1} \cdot \dfrac{x^{-2-(-3)}}{1} = 3 \cdot x^{-2+(+3)} = 3x^1 = 3x$

b. $\dfrac{5a^4b^{-3}}{10a^{-2}b^{-4}} = \dfrac{\cancel{5}^{1}}{\cancel{10}_{2}} \cdot \dfrac{a^4}{a^{-2}} \cdot \dfrac{b^{-3}}{b^{-4}} = \dfrac{1}{2} \cdot a^{4-(-2)}b^{-3-(-4)} = \dfrac{1}{2}a^{4+(+2)}b^{-3+(+4)}$

$\qquad = \dfrac{1}{2}a^6b$ or $\dfrac{a^6b}{2}$

c. $\dfrac{9xy^{-3}}{15x^3y^{-4}} = \dfrac{\cancel{9}^{3}}{\cancel{15}_{5}} \cdot \dfrac{x}{x^3} \cdot \dfrac{y^{-3}}{y^{-4}} = \dfrac{3}{5} \cdot \dfrac{x^{1-3}}{1} \cdot \dfrac{y^{-3-(-4)}}{1} = \dfrac{3}{5}x^{1+(-3)}y^{-3+(+4)} = \dfrac{3}{5}x^{-2}y^1$

$\qquad = \dfrac{3}{5} \cdot \dfrac{1}{x^2} \cdot y = \dfrac{3y}{5x^2}$ ∎

Evaluating Expressions with Numerical Bases
Example 10 demonstrates the evaluation of expressions with numerical bases.

Example 10 Evaluate each expression.

After applying the rules of exponents, we often evaluate the power of a number

a. $10^3 \cdot 10^2 = 10^5 = 10 \cdot 10 \cdot 10 \cdot 10 \cdot 10 = \overbrace{100,000}^{5\ zeros}$

b. $10^{-2} = \dfrac{1}{10^2} = \dfrac{1}{10 \cdot 10} = \dfrac{1}{\underbrace{100}_{2\ zeros}}$

c. $(2^3)^{-1} = 2^{-3} = \dfrac{1}{2^3} = \dfrac{1}{2 \cdot 2 \cdot 2} = \dfrac{1}{8}$

d. $\dfrac{5^0}{5^2} = \dfrac{1}{5 \cdot 5} = \dfrac{1}{25}$

e. $(-2)^{-5} = \dfrac{1}{(-2)^5} = \dfrac{1}{-32}$ or $-\dfrac{1}{32}$

f. $(-3)^{-4} = \dfrac{1}{(-3)^4} = \dfrac{1}{81}$ ∎

A WORD OF CAUTION A common mistake students make is shown by the following examples.

Correct Method	*Incorrect Method*
a. $2^3 \cdot 2^2 = 2^{3+2}$	$2^3 \cdot 2^2 = (2 \cdot 2)^{3+2} = 4^5 = 1{,}024$
$= 2^5 = 32$	
b. $10^2 \cdot 10 = 10^{2+1}$	$10^2 \cdot 10 = (10 \cdot 10)^{2+1} = 100^3$
$= 10^3 = 1{,}000$	$= 1{,}000{,}000$

In words: When multiplying powers of the same base, add the exponents; do *not* multiply the bases as well. ☑

Simplified Form of Terms with Exponents

A term with exponents is considered simplified when each different base appears only once and the exponent on each base is a single positive integer.

Example 11 Simplify each expression.

a. $x^2 \cdot x^7 = x^9$

b. $\dfrac{x^5y^2}{x^3y} = x^2y$

c. $(x^2)^3 = x^6$

d. $\dfrac{x^5y^2}{x^4y} = \dfrac{xy^2}{y}$

This expression is not considered to be completely simplified because the base y appears twice. Continuing the simplification:

$$\frac{xy^2}{y} = xy$$

The expression is now completely simplified. ■

Example 12 Rule 5.6 also applies to expressions with zero and negative exponents. In the following examples, assume that none of the factors appearing in any denominator is zero.

a. $(x^3y^{-1})^5 = x^{3\cdot5}y^{(-1)5} = x^{15}y^{-5} = \dfrac{x^{15}}{y^5}$

The rules of exponents apply to numerical bases as well as to variable bases

b. $\left(\dfrac{2a^{-3}b^2}{c^5}\right)^3 = \dfrac{2^{1\cdot3}a^{(-3)3}b^{2\cdot3}}{c^{5\cdot3}} = \dfrac{2^3a^{-9}b^6}{c^{15}} = \dfrac{8b^6}{a^9c^{15}}$

c. $\left(\dfrac{3^2c^{-4}}{d^3}\right)^{-1} = \dfrac{3^{2(-1)}c^{(-4)(-1)}}{d^{3(-1)}} = \dfrac{3^{-2}c^4}{d^{-3}} = \dfrac{c^4d^3}{3^2} = \dfrac{c^4d^3}{9}$

d. $\left(\dfrac{3^{-7}x^{10}}{y^{-4}}\right)^0 = 1$

e. $\left(\dfrac{x^5y^4}{x^3y^7}\right)^2 = (x^{5-3}y^{4-7})^2 = (x^2y^{-3})^2 = x^4y^{-6} = \dfrac{x^4}{y^6}$

Simplify the expression within the parentheses first whenever possible

f. $(5^0h^{-2})^{-3} = (1h^{-2})^{-3} = (h^{-2})^{-3} = h^{(-2)(-3)} = h^6$

g. $\left(\dfrac{10^{-2} \cdot 10^5}{10^4}\right)^3 = \left(\dfrac{10^3}{10^4}\right)^3 = (10^{-1})^3 = 10^{(-1)(3)} = 10^{-3} = \dfrac{1}{10^3}$

h. $(-3x)^{-2} = \dfrac{1}{(-3x)^2} = \dfrac{1}{(-3)^2 x^2} = \dfrac{1}{9x^2}$ The exponent applies to $(-3x)$

i. $-3x^{-2} = -3(x^{-2}) = -3\left(\dfrac{1}{x^2}\right) = \dfrac{-3}{x^2}$ The exponent applies to the x only ∎

EXERCISES 5.5

Set I   Assume that none of the factors appearing in any denominator is zero. In Exercises 1–28, simplify each expression. Write answers using only positive exponents.

1. x^{-4} **2.** y^{-7} **3.** $\dfrac{1}{a^{-4}}$ **4.** $\dfrac{1}{b^{-5}}$

5. $r^{-4}st^{-2}$ **6.** $r^{-5}s^{-3}t$ **7.** $(xy)^{-2}$ **8.** $(ab)^{-4}$

9. $\dfrac{h^2}{k^{-4}}$ **10.** $\dfrac{m^3}{n^{-2}}$ **11.** $\dfrac{x^{-4}}{y}$ **12.** $\dfrac{a^{-5}}{b}$

13. $ab^{-2}c^0$ **14.** $x^{-3}y^0z$ **15.** $x^{-3} \cdot x^4$ **16.** $y^6 \cdot y^{-2}$

17. $10^3 \cdot 10^{-2}$ **18.** $2^{-3} \cdot 2^2$ **19.** $(x^2)^{-4}$ **20.** $(z^3)^{-2}$

21. $(a^{-2})^3$ **22.** $(b^{-5})^2$ **23.** $\dfrac{y^{-2}}{y^5}$ **24.** $\dfrac{z^{-2}}{z^2}$

25. $\dfrac{10^2}{10^{-5}}$ **26.** $\dfrac{2^3}{2^{-2}}$ **27.** $\left(\dfrac{x}{y}\right)^{-3}$ **28.** $\left(\dfrac{s}{t}\right)^{-2}$

In Exercises 29–34, write each expression without fractions, using negative exponents if necessary.

29. $\dfrac{1}{x^2}$ **30.** $\dfrac{1}{y^3}$ **31.** $\dfrac{h}{k}$

32. $\dfrac{m}{n}$ **33.** $\dfrac{x^2}{yz^5}$ **34.** $\dfrac{a^3}{b^2c}$

In Exercises 35–50, evaluate each expression.

35. $10^4 \cdot 10^{-2}$ **36.** $3^{-2} \cdot 3^3$ **37.** 10^{-4} **38.** 2^{-3}

39. $5^0 \cdot 7^2$ **40.** $4^3 \cdot 2^0$ **41.** $\dfrac{10^0}{10^2}$ **42.** $\dfrac{5^2}{5^0}$

43. $\dfrac{10^{-3} \cdot 10^2}{10^5}$ **44.** $\dfrac{2^3 \cdot 2^{-4}}{2^2}$ **45.** $(10^2)^{-1}$ **46.** $(2^{-3})^2$

47. $(-5)^{-3}$ **48.** $(-4)^{-3}$ **49.** $(-12)^{-2}$ **50.** $(-13)^{-2}$

In Exercises 51–94, simplify each expression. Write answers using only positive exponents.

51. $\dfrac{a^3b^0}{c^{-2}}$ **52.** $\dfrac{d^0e^2}{f^{-3}}$ **53.** $\dfrac{p^4r^{-1}}{t^{-2}}$ **54.** $\dfrac{u^5v^{-2}}{w^{-3}}$

55. $\dfrac{8x^{-3}}{12x}$ **56.** $\dfrac{15y^{-2}}{10y}$ **57.** $\dfrac{20h^{-2}}{35h^{-4}}$ **58.** $\dfrac{35k^{-1}}{28k^{-4}}$

59. $\dfrac{15m^0n^{-2}}{5m^{-3}n^4}$ **60.** $\dfrac{14x^0y^{-3}}{12x^{-2}y^{-4}}$ **61.** $x^{3m} \cdot x^{-m}$ **62.** $y^{-2n} \cdot y^{5n}$

63. $(x^{3b})^{-2}$ **64.** $(y^{2a})^{-3}$ **65.** $\dfrac{x^{2a}}{x^{-5a}}$ **66.** $\dfrac{a^{3x}}{a^{-5x}}$

67. $(m^{-2}n)^4$ **68.** $(p^{-3}r)^5$ **69.** $(x^{-2}y^3)^{-4}$ **70.** $(w^{-3}z^4)^{-2}$

71. $\left(\dfrac{M^{-2}}{N^3}\right)^4$ **72.** $\left(\dfrac{R^5}{S^{-4}}\right)^3$ **73.** $\left(\dfrac{a^2b^{-4}}{b^{-5}}\right)^2$ **74.** $\left(\dfrac{x^{-2}y^2}{x^{-3}}\right)^3$

75. $\left(\dfrac{mn^{-1}}{m^3}\right)^{-2}$ **76.** $\left(\dfrac{ab^{-2}}{a^2}\right)^{-3}$ **77.** $\left(\dfrac{x^4}{x^{-1}y^{-2}}\right)^{-1}$ **78.** $\left(\dfrac{x^3}{x^{-2}y^{-4}}\right)^{-1}$

79. $(10^0k^{-4})^{-2}$ **80.** $(6^0z^{-5})^{-2}$ **81.** $\left(\dfrac{r^7s^8}{r^9s^6}\right)^0$ **82.** $\left(\dfrac{t^5u^6}{t^8u^7}\right)^0$

83. $(-3x)^{-2}$ **84.** $(-5y)^{-2}$ **85.** $-3x^{-2}$ **86.** $-5y^{-2}$

87. $(-2z)^{-3}$ **88.** $(-2a)^{-5}$ **89.** $-2z^{-3}$ **90.** $-2a^{-5}$

91. $\left(\dfrac{3m^{-1}}{9n^3}\right)^4$ **92.** $\left(\dfrac{5x^{-1}}{20y^2}\right)^2$ **93.** $\left(\dfrac{6x^{-3}}{9y^{-2}}\right)^{-3}$ **94.** $\left(\dfrac{9a^{-2}}{12b^{-3}}\right)^{-3}$

Set II Assume that none of the factors appearing in any denominator is zero. In Exercises 1–28, simplify each expression. Write answers using only positive exponents.

1. a^{-3} **2.** $\dfrac{1}{x^{-2}}$ **3.** $\dfrac{1}{s^{-3}}$ **4.** x^0y^{-4}

5. $r^{-2}st^{-4}$ **6.** $8^0x^3y^{-3}$ **7.** $h^{-3}k^5$ **8.** $3x^{-1}$

9. $\dfrac{x}{y^{-3}}$ **10.** $(x^{-2})^2$ **11.** $\dfrac{a^{-2}}{b}$ **12.** $(y^4)^{-2}$

13. mn^0p^{-4} **14.** $(y^{-4})^0$ **15.** $y^{-7}\cdot y^9$ **16.** $(3^0)^{-4}$

17. $10^{-4}\cdot 10^3$ **18.** $x^{-8}\cdot x^4$ **19.** $(n^4)^{-1}$ **20.** $x^{-8}\cdot y^4$

21. $(z^{-3})^2$ **22.** $\dfrac{x^0}{y^{-4}}$ **23.** $\dfrac{y^{-1}}{y^4}$ **24.** $\dfrac{x^{-2}}{y^{-3}}$

25. $\dfrac{10^3}{10^{-2}}$ **26.** $\dfrac{5^0x^{-4}}{x^2}$ **27.** $p^2p^0p^{-3}$ **28.** $\left(\dfrac{s}{t}\right)^{-6}$

In Exercises 29–34, write each expression without fractions, using negative exponents if necessary.

29. $\dfrac{1}{c^4}$ **30.** $\dfrac{8}{x^{-3}}$ **31.** $\dfrac{x}{y}$ **32.** $\dfrac{5}{ax^2}$ **33.** $\dfrac{h^4}{kt^3}$ **34.** $\dfrac{x^{-1}}{y^2}$

In Exercises 35–50, evaluate each expression.

35. $2^{-3}\cdot 2^5$ **36.** $5^4\cdot 5^{-6}$ **37.** 8^{-2} **38.** $6^0\cdot 3^{-4}$

39. $3^0\cdot 5^2$ **40.** $(8\cdot 2^4)^0$ **41.** $\dfrac{10^0}{10^{-2}}$ **42.** $\left(\dfrac{1}{2}\right)^{-3}$

43. $\dfrac{10^{-2}\cdot 10^3}{10^4}$ **44.** $\dfrac{5^0\cdot 5^3}{5^{-2}}$ **45.** $(2^{-4})^2$ **46.** $(3^3)^{-1}$

47. $(-10)^{-3}$ **48.** $(-3)^{-4}$ **49.** $(-11)^{-2}$ **50.** $(-2)^{-3}$

In Exercises 51–94, simplify each expression. Write answers using only positive exponents.

51. $\dfrac{x^0y^2}{z^{-5}}$ **52.** $\dfrac{3^4x^{-4}}{3x^2}$ **53.** $\dfrac{u^{-1}v^2}{w^{-3}}$ **54.** $\dfrac{2x^{-3}}{2^3x}$

55. $\dfrac{16h^{-2}}{10h}$ **56.** $\left(\dfrac{x}{2}\right)^{-4}$ **57.** $\dfrac{24m^{-1}}{18m^{-3}}$ **58.** $(4x)^{-2}$

59. $\dfrac{18w^0z^{-4}}{16w^{-2}z^2}$ **60.** $\dfrac{2^{-3}}{x}$ **61.** $x^{2n} \cdot x^{-n}$ **62.** $\left(\dfrac{2}{x}\right)^{-3}$

63. $(k^{-2c})^2$ **64.** $(2y)^{-4}$ **65.** $\dfrac{y^{4n}}{y^{-3n}}$ **66.** $2y^{-4}$

67. $(m^{-3}n)^3$ **68.** $3 \cdot 5^2$ **69.** $(w^2z^{-4})^{-3}$ **70.** $(2x^3)^{-4}$

71. $\left(\dfrac{R^4}{S^{-2}}\right)^3$ **72.** $\left(\dfrac{x^4}{2y^{-3}}\right)^{-3}$ **73.** $\left(\dfrac{a^{-3}b^2}{a^{-4}}\right)^3$ **74.** $\left(\dfrac{2x^2y^{-3}}{x^{-3}y}\right)^{-4}$

75. $\left(\dfrac{m^{-1}n^3}{m}\right)^{-2}$ **76.** $\left(\dfrac{a^3b^{-2}}{b^3c^{-1}}\right)^{-3}$ **77.** $\left(\dfrac{x^2}{x^{-3}y^{-2}}\right)^{-1}$ **78.** $\left(\dfrac{ab^{-3}}{2b^{-2}}\right)^{-3}$

79. $(2^0h^2)^{-3}$ **80.** $(5^3x^{-2}y^{-4})^0$ **81.** $\left(\dfrac{5t^{-1}u^2}{t^4u^{-3}}\right)^0$ **82.** $\left(\dfrac{6^0a^2b^{-3}}{2^{-1}ab}\right)^{-2}$

83. $(-4a)^{-2}$ **84.** $(-3y)^{-3}$ **85.** $-4a^{-2}$ **86.** $-3y^{-3}$

87. $(-2b)^{-5}$ **88.** $(-a)^{-4}$ **89.** $-2b^{-5}$ **90.** $-a^{-4}$

91. $\left(\dfrac{2p^{-1}}{8q^3}\right)^2$ **92.** $\left(\dfrac{4a^{-2}}{20b^3}\right)^3$ **93.** $\left(\dfrac{3x^{-4}}{12y^{-3}}\right)^{-4}$ **94.** $\left(\dfrac{12c^{-3}}{8d^{-4}}\right)^{-3}$

5.6   Scientific Notation

Now that we have discussed zero and negative exponents, we can introduce *scientific notation*, a notation that is used in many sciences. Scientific notation gives us a shorter way of writing very large numbers, such as Avogadro's number ($\approx 602{,}000{,}000{,}000{,}000{,}000{,}000{,}000$, used in chemistry), and very small numbers, such as Boltzmann's constant ($0.000\,000\,000\,000\,000\,138$, used in physics) (see Example 6). We can also use scientific notation to simplify the arithmetic in a problem such as $\dfrac{30{,}000{,}000 \times 0.0005}{0.0000006 \times 80{,}000}$ (see Example 7).

Scientific calculators usually express answers that are very large or very small in scientific notation (see Examples 8 and 9). Also, you should be aware that most calculators will give the *wrong answer* if very large or very small numbers are entered *unless* the numbers are entered in scientific notation (see Example 10).

Scientific Notation   A positive number written in **scientific notation** is written in the form $a \times 10^n$, where a is a number greater than or equal to 1 but less than 10 and n is an integer. For example, 8.021×10^4 is correctly written in scientific notation, because 8.021 is a number between 1 and 10 and 10^4 is a power of 10. Since a is to be greater than or equal to 1 but less than 10, it must have *exactly one digit to the left of its decimal point*.

When we change a number to scientific notation, we are using the facts that multiplying a number by 1 doesn't change its value and that $10^n \times 10^{-n} = 1$; therefore, when we change a number to scientific notation, we do not change its value. Consider the following examples.

Example 1   Change 0.0712 to scientific notation.

Solution When 0.0712 is written in scientific notation, the decimal point must be between the 7 and the 1; that is, the decimal point must be moved two places to the right.

Because the decimal point is to be moved two places to the right, we'll multiply 0.0712 by $10^{\,2} \times 10^{-2}$.

$$0.0712 = 0.0712 \times \overbrace{10^2 \times 10^{-2}}^{\text{This product is 1}} = \underbrace{(0.0712 \times 10^2)}_{} \times 10^{-2} = \underbrace{7.12 \times 10^{-2}}_{\text{This number is in scientific notation}} \quad \blacksquare$$

Example 2   Change 27,400 to scientific notation.

Solution When 27,400 is written in scientific notation, the decimal point must be between the 2 and the 7; that is, the decimal point must be moved four places to the left. Because the decimal point is to be moved four places to the left, we'll multiply 27,400 by $10^{-\,4} \times 10^4$.

$$27,400 = 27,400 \times \overbrace{10^{-4} \times 10^4}^{\text{This product is 1}} = \underbrace{(27,400 \times 10^{-4})}_{} \times 10^4 = \underbrace{2.74 \times 10^4}_{\text{This number is in scientific notation}} \quad \blacksquare$$

In practice, we usually do not show any of the intermediate steps shown in Examples 1 and 2; instead, we use the following method.

Writing a Positive Number in Scientific Notation

Step 1. Finding a

a. Replace the decimal point with a caret ($_\wedge$). (If there is no decimal point, put the caret just to the right of the last digit of the number.)

b. Place a decimal point in the number so that there is exactly one nonzero digit to the left of the decimal point. (The number just written is a.)

Step 2. Finding the correct power of 10.

a. The number of digits separating the caret and the decimal point in step 1 gives the numerical part of the exponent of 10. If the decimal point and the caret coincide (lie one on top of the other), the exponent is zero.

b. The sign of the exponent of 10 is *positive* if the caret is to the right of the decimal point and *negative* if the caret is to the left of the decimal point.

The number in scientific notation is the product of the two numbers found in steps 1 and 2. (We drop the caret in the final answer.)

These rules imply that if the number to be converted to scientific notation is greater than or equal to 10, the exponent of 10 will be positive; if the number is less than 1, the exponent of 10 will be negative. If the number is between 1 and 10, the exponent of 10 will be zero.

Example 3 Write the following decimal numbers in scientific notation.

Decimal notation	Finding a	Scientific notation
a. 0.00753	$0_\wedge 007.53 \times 10^?$	7.53×10^{-3}

In the middle column, we replaced the decimal point with a caret and placed a decimal point just after the 7. The caret is *three* digits to the *left* of the decimal point; therefore, the exponent that will replace the question mark is -3, as shown in the third column.

b. 86,100,000	$8.6100000_\wedge \times 10^?$	8.61×10^7
c. 2.49	$2_\wedge 49 \times 10^?$	2.49×10^0
d. 6,410	$6.410_\wedge \times 10^?$	6.41×10^3
e. 0.0003015	$0_\wedge 0003.015 \times 10^?$	3.015×10^{-4}

a, the number between 1 and 10 ⟶ ⌐ ⌐ ⟵ Power of 10

Notice that in parts a and e the caret is to the left of the decimal point and the exponent on the 10 is negative and that in parts b and d the caret is to the right of the decimal point and the exponent on the 10 is positive. In part c, the decimal point and the caret coincide, and the exponent on the 10 is zero. ∎

To change from scientific notation to decimal notation, simply multiply by the power of 10 (see Example 4).

Example 4 Convert each of the following to decimal notation.

a. $8.601 \times 10^4 = 8.601 \times 10,000 = 86,010$

b. $4.23 \times 10^{-3} = 4.23 \times \dfrac{1}{10^3} = 4.23 \times \dfrac{1}{1,000} = 0.00423$ ∎

It is sometimes necessary to convert a number such as 672.3×10^2 or 0.0791×10^{-3} to scientific notation or to decimal notation, as shown in Example 5.

Example 5 Convert each number to scientific notation and then to decimal notation.

a. 672.3×10^2
Solution

$$672.3 = 6.723 \times 10^2$$

Therefore,

$$672.3 \times 10^2 = (6.723 \times 10^2) \times 10^2$$
$$= 6.723 \times (10^2 \times 10^2)$$
$$= 6.723 \times 10^4 \quad \text{Scientific notation}$$
$$= 67,230 \quad \text{Decimal notation}$$

b. 0.0791×10^{-3}
Solution

$$0.0791 = 7.91 \times 10^{-2}$$

Therefore,

$$0.0791 \times 10^{-3} = (7.91 \times 10^{-2}) \times 10^{-3}$$

$$= 7.91 \times (10^{-2} \times 10^{-3})$$

$$= 7.91 \times 10^{-5} \qquad \text{Scientific notation}$$

$$= 0.0000791 \qquad \text{Decimal notation} \quad \blacksquare$$

Example 6   Express (a) Avogadro's number and (b) Boltzmann's constant in scientific notation.
Solution

a. Avogadro's number is approximately 602,000,000,000,000,000,000,000.

The caret is to the right of the decimal point;
23 digits between caret and decimal point

$$602{,}000{,}000{,}000{,}000{,}000{,}000{,}000 = 6.020\,000\,000\,000\,000\,000\,000\,00_{\wedge} \times 10^{23}$$

$$= 6.02 \times 10^{23}$$

b. Boltzmann's constant is $0.000\,000\,000\,000\,000\,138$.

The caret is to the left of the decimal point;
16 digits between caret and decimal point

$$0.000\,000\,000\,000\,000\,138 = 0_{\wedge}000\,000\,000\,000\,000\,1.38 \times 10^{-16}$$

$$= 1.38 \times 10^{-16} \quad \blacksquare$$

Example 7   Use scientific notation in solving this problem: $\dfrac{30{,}000{,}000 \times 0.0005}{0.0000006 \times 80{,}000}$.

Solution

$$\frac{30{,}000{,}000 \times 0.0005}{0.0000006 \times 80{,}000} = \frac{(3 \times 10^{7}) \times (5 \times 10^{-4})}{(6 \times 10^{-7}) \times (8 \times 10^{4})} \qquad \begin{array}{l}\text{Writing each factor in}\\ \text{scientific notation}\end{array}$$

$$= \frac{(\overset{1}{\cancel{3}} \times 5) \times (10^{7} \times 10^{-4})}{(\underset{2}{\cancel{6}} \times 8) \times (10^{-7} \times 10^{4})} \qquad \text{Collecting the powers of 10}$$

$$= \frac{5}{16} \times \frac{10^{3}}{10^{-3}} \qquad \begin{array}{l}\text{Simplifying what's in the}\\ \text{parentheses } and \text{ writing the}\\ \text{problem as a product of two}\\ \text{fractions}\end{array}$$

$$= 0.3125 \times (10^{3} \times 10^{3}) \qquad \begin{array}{l}\text{Writing 5/16 in decimal form and}\\ \text{writing } 1/10^{-3} \text{ as } 10^{3}\end{array}$$

$$= (3.125 \times 10^{-1}) \times 10^{6} \qquad \begin{array}{l}\text{Changing 0.3125 to scientific}\\ \text{notation}\end{array}$$

$$= 3.125 \times 10^{5} \qquad \text{The answer in scientific notation}$$

$$= 312{,}500 \qquad \text{The answer in decimal notation} \quad \blacksquare$$

As we mentioned, when a scientific calculator is used and answers are very large or very small, the calculator display will probably be in scientific notation. On the calculator, however, numbers in scientific notation are displayed in a different (and possibly misleading) way. The calculator display 5.06 ^{03} does *not* mean 5.06 to the third power. It means 5.06×10^{3}. (The calculator displays 5.06 03 and 5.06 E 3 also mean 5.06×10^{3}.)

Example 8   Find 80,000,000 × 300,000, using a scientific calculator.
Solution The display probably shows 2.4^{13} , 2.4 13 , or 2.4 E 13 . These displays all mean 2.4×10^{13}. ■

Example 9   Find 0.0000008 ÷ 400, using a scientific calculator.
Solution The display probably shows $2.^{-09}$, 2. −09 , or 2. E −09 . These displays all mean 2.0×10^{-9}. ■

    Very large and very small numbers must be entered into a calculator in scientific notation. The keystrokes used to indicate operations vary, of course, with the brand of calculator. We will assume, in Example 10, that the key for indicating that you're entering a number in scientific notation is marked EE , but it could be marked EXP . You must consult your calculator manual for details about *your* calculator.

Example 10   Given that the formula for simple interest is $I = prt$, use a calculator to find the (simple) interest for one year on $2,600,000,000,000 if the interest rate is $8\frac{3}{4}\%$.
Solution P is 2,600,000,000,000; $t = 1$; and r is 0.0875 (0.0875 is $8\frac{3}{4}\%$ in decimal form). If we change P to scientific notation, we have

$$2,600,000,000,000 = 2.600,000,000,000_{\wedge} \times 10^{12} = 2.6 \times 10^{12}$$

$$I = (2.6 \times 10^{12})(0.0875)(1)$$

Keystrokes	*Display*
2 . 6 EE 1 2 × . 0 8 7 5 =	2.275 11

This calculator display means 2.275×10^{11}, or 227,500,000,000. Therefore, the interest is $227,500,000,000. ■

A WORD OF CAUTION   In Example 10, watch your calculator display carefully if you start entering the problem as follows:

~~2 6 0 0 0 0 0 0 0 0 0 0~~

You can see that the calculator *ignores* the last few zeros, and you will get an *incorrect* answer. ☑

    The even-numbered word problems at the end of Exercises 5.6 are not "matched" to the odd-numbered problems.

EXERCISES 5.6

Set I   In Exercises 1–6, write each number in decimal notation.

1. 8.06×10^3 **2.** 3.14×10^4 **3.** 1.32×10^{-3}

4. 8.2×10^{-4} **5.** 5.26×10^0 **6.** 9.11×10^0

In Exercises 7–18, write each number in scientific notation.

7. 35,300   **8.** 825,000   **9.** 0.00312   **10.** 0.000145

11. 8.97   **12.** 2.497   **13.** 0.815   **14.** 0.274

15. 0.0002   **16.** 0.006   **17.** 45   **18.** 12

In Exercises 19 and 20, use scientific notation to solve the problem.

19. $\dfrac{5,000,000 \times 0.000003}{0.0006 \times 20,000}$ **20.** $\dfrac{0.0000004 \times 700,000,000}{32,000,000 \times 0.00005}$

In Exercises 21–24, perform the indicated operations with a scientific calculator, and express each answer correctly in scientific notation.

21. $860,000 \times 630,000$

22. $0.0000009 \div 3,000$

23. $\sqrt{0.00000081}$

24. $\sqrt{0.00000225}$

In Exercises 25–28, use a scientific calculator.

25. By definition (in chemistry and physics), 1 mole of any substance contains approximately 6.02×10^{23} molecules. How many molecules will 700 moles of hydrogen contain? (Express your answer in scientific notation.)

26. The indebtedness of one of the developing nations is $120,000,000,000. If the interest rate on this debt is 7.6% simple interest per year, what is the interest on this debt for one year?

27. If the spacecraft Voyager traveled 4,400,000,000 miles in 12 years, what was its average speed in miles per hour? (Assume that each year has 365 days, and round off the answer to the nearest mile per hour.)

28. The speed of light is about 186,000 miles per second. How many miles does light travel in one day? (Express your answer in scientific notation and in decimal notation.)

Set II   In Exercises 1–6, write each number in decimal notation.

1. 5.23×10^4

2. 6.34×10^2

3. 7.12×10^{-4}

4. 2.3×10^{-3}

5. 7.32×10^0

6. 3.71×10^0

In Exercises 7–18, write each number in scientific notation.

7. $87,600$

8. $25,000,000$

9. 0.00631

10. 0.00614

11. 3.69

12. 3.9

13. 0.153

14. 0.456

15. 0.0052

16. 0.00003

17. 28

18. 2.9

In Exercises 19 and 20, use scientific notation to solve the problem.

19. $\dfrac{80,000,000 \times 0.00006}{0.00016 \times 300,000}$

20. $\dfrac{0.0000034 \times 600,000,000}{17,000,000 \times 0.00003}$

In Exercises 21–24, perform the indicated operations with a scientific calculator, and express each answer correctly in scientific notation.

21. $60,000 \times 730,000,000$

22. $0.00000012 \div 4,000$

23. $\sqrt{0.00000036}$

24. $\sqrt{0.00000289}$

In Exercises 25–28, use a scientific calculator.

25. The indebtedness of one of the developing nations is $52,000,000,000. If the interest rate on this debt is 7.6% simple interest per year, what is the interest on this debt for two years?

26. The planet Neptune is about 2,700,000,000 miles from the earth. What is this distance in kilometers? (One mile $\approx$ 1.61 km.)

27. A unit used in measuring the length of light waves is the Ångström. One micron is a millionth of a meter, and one Ångström is one ten-thousandth of a micron. One Ångström is what part of a meter? (Express your answer in scientific notation.)

28. The speed of light is about 186,000 miles per second. How long will it take light to reach us from the planet Neptune? (Assume that Neptune is about 2,700,000,000 miles from the earth, and round off your answer to the nearest hour.)

5.7 Review: 5.1–5.6

Rules of Exponents

5.1 Rule 5.1 $x^a \cdot x^b = x^{a+b}$

5.4 Rule 5.2 $(x^a)^b = x^{ab}$

5.4 Rule 5.3 $(xy)^a = x^a y^a$

5.4 Rule 5.4 $\dfrac{x^a}{x^b} = x^{a-b}$, if $x \neq 0$

5.4 Rule 5.5 $\left(\dfrac{x}{y}\right)^n = \dfrac{x^n}{y^n}$, if $y \neq 0$

5.4 Rule 5.6 $\left(\dfrac{x^a y^b}{z^c}\right)^n = \dfrac{x^{an} y^{bn}}{z^{cn}}$, if $z \neq 0$

5.5 Rule 5.7 $x^0 = 1$, if $x \neq 0$

5.5 Rule 5.8a $x^{-n} = \dfrac{1}{x^n}$, if $x \neq 0$

5.5 Rule 5.8b $\dfrac{1}{x^{-n}} = x^n$, if $x \neq 0$

5.5 Rule 5.8c $\left(\dfrac{x}{y}\right)^{-n} = \left(\dfrac{y}{x}\right)^n$, if $y \neq 0$ and $x \neq 0$

To Simplify a Product of Factors When Each Factor Has Only One Term
5.2

1. *Multiply the numerical coefficients.* The sign of the coefficient of any term is *part of* the numerical coefficient.

2. *Multiply the variables.* Each variable should appear only once, and the variables are usually written in alphabetical order. Use Rule 5.1 to determine the exponent on each variable.

3. The final answer is the product of the results of steps 1 and 2.

Simplified Form of Expressions with Exponents
5.2

A term with exponents is considered *simplified* when each different base appears only once and when the exponent on that base is a single positive integer.

To Simplify a Product of Factors When One Factor Has More Than One Term
5.3

Use the distributive property:

$$a(b + c) = ab + ac$$

To Simplify an Algebraic Expression
5.1, 5.2, 5.3, 5.4, and 5.5

Remove all grouping symbols, simplify each term, and combine all like terms.

Scientific Notation
5.6

A number is correctly expressed in **scientific notation** if it is in the form $a \times 10^n$, where a is greater than or equal to 1 but less than 10 and n is any integer.

Review Exercises 5.7 Set I

Assume that none of the factors appearing in any denominator is zero. In Exercises 1–6, write each product in simplest form.

1. $m^2 m^3$

2. yy^7

3. $2 \cdot 2^2$

4. $x^{2y} x^{5y}$

5. $10 \cdot 10^y$

6. $a^5 \cdot a^{-3}$

In Exercises 7–34, write each expression in simplest form.

7. $5x(x^2 + 7)$

8. $(5x^2 y + 3x - 1)(-2x)$

9. $(-5ef^2)(-f^7 g^3)(-10e^4 g^2)(-2ef)$

10. $(3x)^3$

11. $(x^2 y^3)^4$

12. $(n^{-5})^{-3}$

13. $(p^{-3})^5$

14. $(2c)^{-4}$

15. $(k^{-7})^0$

16. $(s^4 t^{-1})^{-3}$

17. $(-2x^4)^2$

18. $(-5a^{-4})^3$

19. $(-10)^{-3}$

20. $\dfrac{r^7}{r^5}$

21. $\dfrac{x^{-4}}{x^5}$

22. $\left(\dfrac{c}{d}\right)^4$

23. $\dfrac{m^0}{m^{-3}}$

24. $x^4 + x^2$

25. $\dfrac{n^{-6}}{n^0}$

26. $\left(\dfrac{x^2 y^3}{2z^4}\right)^5$

27. $\left(\dfrac{a^{-4}}{b^3 c^0}\right)^{-5}$

28. $x^{-5} \cdot x^{-3}$

29. $\left(\dfrac{r^{-6}}{s^5 t^{-3}}\right)^4$

30. $\left(\dfrac{6x^{-5} y^8}{3x^2 y^{-4}}\right)^0$

31. $(5a^3 b^{-4})^{-2}$

32. $(9x^8 y^3)(-7x^4 y^5)$

33. $2x(3x^2 - x) - (3x^2 - 4)$

34. $2m^2 n(3mn^2 - 2n) - 5mn^2(2m - 3m^2 n)$

In Exercises 35–37, write each expression without fractions, using negative exponents if necessary.

35. $\dfrac{a^3}{b^2}$

36. $\dfrac{m^2}{n^{-3}}$

37. $\dfrac{u^{-4} v^3}{10^2 w^{-5}}$

In Exercises 38–42, evaluate each expression.

38. 4^{-2}

39. $(10^{-2})^2$

40. $\dfrac{2^0}{2^{-3}}$

41. $4^0 \cdot 3^2$

42. $\dfrac{(-8)^2}{-8^2}$

43. Express 45,300 in scientific notation.

44. Express 0.03156 in scientific notation.

Review Exercises 5.7 Set II

Assume that none of the factors appearing in any denominator is zero. In Exercises 1–6, write each product in simplest form.

1. $x^4y^2x^6$

2. $3 \cdot 3^7$

3. a^2b^7

4. m^xm^y

5. $x^{-7} \cdot x^4$

6. $4^2 \cdot 4^4$

In Exercises 7–34, write each expression in simplest form.

7. $3x(2x - 1)$

8. $9a^2b(ab + 1)$

9. $(3x^4 - x^3 - x^2)(-2x)$

10. $7 + 4(3x^4 - 5 - x^2)$

11. $(-x^2)(3xy^4)(5y^3)$

12. $(-9w^2z^3)(7z^4)(-wx^2z^4)$

13. $(6e^3)^0$

14. $(2xy^2)^5$

15. $(m^{-2})^4$

16. $(m^{-1}n^2)^{-3}$

17. $c^{-5} \cdot d^0$

18. $(-2t^{-3})^2$

19. $(3a^{-2}b)^{-4}$

20. $m^{3x} \cdot m^{-x}$

21. $\dfrac{y^6}{y^2}$

1. _____

2. _____

3. _____

4. _____

5. _____

6. _____

7. _____

8. _____

9. _____

10. _____

11. _____

12. _____

13. _____

14. _____

15. _____

16. _____

17. _____

18. _____

19. _____

20. _____

21. _____

22. $\left(\dfrac{y}{2}\right)^5$

23. $\dfrac{x^{-2}}{x^2}$

24. $x^5 - x^2$

25. $\left(\dfrac{x}{2}\right)^{-4}$

26. $x^3 y^5$

27. $\left(\dfrac{r^{-3}}{s^2 t^{-2}}\right)^2$

28. $\left(\dfrac{3p}{m^2 n^{-1}}\right)^{-3}$

29. $\left(\dfrac{5a^2 b}{a^{-3} b^4}\right)^{-2}$

30. $\left(\dfrac{18x^{-3} y}{15x^2 y^{-2}}\right)^0$

31. $\left(\dfrac{x^8 y}{x^5}\right)^2$

32. $(2c^{-4} d^2)^{-2}$

33. $-5x^{-3}$

34. $10k(4k^2 - 3k - 5) - 2k(5k^2 - 4k + 10)$

In Exercises 35–37, write each expression without fractions, using negative exponents if necessary.

35. $\dfrac{z}{y^{-2}}$

36. $\dfrac{h^2}{k^2}$

37. $\dfrac{u^{-3} v^2}{5^2 w}$

In Exercises 38–42, evaluate each expression.

38. 5^{-2}

39. $(2^{-3})^2$

40. $\dfrac{10^{-2}}{10^0}$

41. $\dfrac{(-11)^2}{-11^2}$

42. $2^4 \cdot 3^0$

43. Express 0.00297 in scientific notation.

44. Express 120,000,000 in scientific notation.

22. _____

23. _____

24. _____

25. _____

26. _____

27. _____

28. _____

29. _____

30. _____

31. _____

32. _____

33. _____

34. _____

35. _____

36. _____

37. _____

38. _____

39. _____

40. _____

41. _____

42. _____

43. _____

44. _____

Chapter 5 Diagnostic Test

The purpose of this test is to see how well you understand working with exponents. We recommend that you work this diagnostic test *before* your instructor tests you on this chapter. Allow yourself about 50 minutes.

Complete solutions for all the problems on this test, together with section references, are given in the answer section in the back of the book. For the problems you do incorrectly, study the sections referred to.

In Problems 1 and 2, find each product.

1. $(-3xy)(5x^3y)(-2xy^4)$ **2.** $2xy^2(x^2 - 3y - 4)$

In Problems 3–17, simplify each expression. Assume that none of the factors appearing in any denominator is zero.

3. $x^3 \cdot x^4$ **4.** $(x^2)^3$ **5.** $\dfrac{x^5}{x^2}$ **6.** x^{-4}

7. x^2y^{-3} **8.** $\dfrac{a^{-3}}{b}$ **9.** $\dfrac{x^{5a}}{x^{3a}}$ **10.** $(4^{3x})^0$

11. $(x^2y^4)^3$ **12.** $(a^{-3}b)^2$ **13.** $\left(\dfrac{p^3}{q^2}\right)^2$ **14.** $\left(\dfrac{x}{y^2}\right)^{-3}$

15. $\left(\dfrac{4x^{-2}}{2x^{-3}}\right)^{-1}$

16. $3h(2k^2 - 5h) - h(2h - 3k^2)$

17. $x(x^2 + 2x + 4) - 2(x^2 + 2x + 4)$

18. Write the expression $\dfrac{a^3}{b}$ without fractions, using negative exponents if necessary.

In Problems 19–24, evaluate each expression.

19. $2^3 \cdot 2^2$ **20.** $10^{-4} \cdot 10^2$ **21.** 5^{-2} **22.** $(2^{-3})^2$

23. $\dfrac{10^{-3}}{10^{-4}}$ **24.** $(5^0)^2$

25. Write each number in scientific notation.

a. 1.326 b. 0.527

Cumulative Review Exercises: Chapters 1–5

In Exercises 1–6, write "true" if the statement is always true; otherwise, write "false."

1. The additive identity is 1.

2. The additive inverse of $\frac{8}{5}$ is $-\frac{8}{5}$.

3. Subtraction is commutative.

4. Division is associative.

5. $3{,}870{,}000 = 3.87 \times 10^{-6}$ in scientific notation.

6. $3x$ is a factor of the expression $3x + 7$.

In Exercises 7–11, evaluate each expression.

7. $\dfrac{14 - 23}{-8 + 11}$

8. $\sqrt{13^2 - 5^2}$

9. $6(-3) - 4\sqrt{36}$

10. $\{-12 - [7 + (3 - 9)]\} - 15$

11. $\dfrac{0}{-11}$

In Exercises 12 and 13, use the formula, substituting the values of the variables given with the formula.

12. $C = \dfrac{5}{9}(F - 32)$ \quad Find C when $F = -13$.

13. $A = P(1 + rt)$ \quad Find A when $P = 1{,}200$, $r = 0.15$, $t = 4$.

14. Solve and check the equation $3x - 7(2 - x) = 8x - 22$.

In Exercises 15–18, simplify each given expression.

15. $(-5p)(4p^3)$

16. $(-4h^1j^5)(-j^3k^4)(-5hk)(-10j^2h^7)$

17. $7 - 4(3xy - 3)$

18. $3xy^2(8x - 2xy) - 5xy(10xy - 3xy^2)$

In Exercises 19 and 20, set up each problem algebraically, solve, and check.

19. Several families went to a movie together. They spent $16.25 for nine tickets. If adults' tickets cost $2.50 and children's tickets cost $1.25, how many of each kind of ticket was bought?

20. When twice the sum of 11 and an unknown number is subtracted from 6 times the sum of 8 and twice the unknown number, the result is 6. What is the unknown number?

6 Polynomials

In this chapter, we look in detail at a particular type of algebraic expression called a *polynomial*. Polynomials have the same importance in algebra that whole numbers have in arithmetic. Just as much of the work in arithmetic involves operations with whole numbers, much of the work in algebra involves operations with polynomials.

Because a polynomial is a special kind of algebraic expression, we have already discussed some of the work with polynomials. In this chapter, we review and extend these concepts.

6.1 Basic Definitions

Because of its importance, we repeat here the definition of a *term* of an algebraic expression. A *term* of an algebraic expression consists only of factors; each plus and minus sign is part of the term that follows it. *Exception:* An expression within grouping symbols is considered as a single term, even though it may contain one or more terms inside the grouping symbols.

Polynomials

A **polynomial in one variable** is an algebraic expression that has only terms of the form ax^n, where a stands for any real number, n stands for any whole number, and x stands for any variable. For example, $8x^4 + 3x^2 - 3x$ is a polynomial in x, because each of its terms is of the form ax^n.

A polynomial with only one term is called a **monomial**, a polynomial with two unlike terms is called a **binomial**, and a polynomial with three unlike terms is called a **trinomial**. We will use the general term **polynomial** for polynomials with four or more terms.

Example 1   The following algebraic expressions are all polynomials in one variable.

  a. $-2x$ This polynomial is a monomial in x

  b. $7y + 3y^3$ This polynomial is a binomial in y

  c. 8 This polynomial is a monomial; it is of the form $8x^0$, because $8x^0 = 8 \cdot 1 = 8$

  d. $z^7 - 2z^2 + 3$ This polynomial is a trinomial in z

  e. $x^3 + 3x^2 + 3x + 1$ This is a polynomial in x

  f. $\dfrac{1}{x^{-5}}$ This is a polynomial because, in simplified form, it becomes x^5 ∎

If an algebraic expression in simplified form contains terms with negative (or fractional*) exponents *on the variables,* or if it contains simplified terms with variables in a denominator or under a radical sign, then the algebraic expression is *not* a polynomial.

Example 2   The following algebraic expressions are *not* polynomials.

  a. $3x^{-5}$ This is *not* a polynomial because it has a negative exponent on a variable and, in simplified form, it would have a variable in the denominator

  b. $\dfrac{1}{x + 2}$ This is *not* a polynomial because it has a variable in the denominator

  c. $\sqrt{3 + x}$ This is *not* a polynomial because the variable is under a radical sign ∎

An algebraic expression with two variables is a **polynomial in two variables** if (1) it contains no negative or fractional exponents on the variables, (2) no variables are in denominators, and (3) no variables are under radical signs.

---

*Fractional exponents will not be discussed in this book.

Example 3   The following algebraic expressions are polynomials in two variables.

a. $xy^3 \sqrt{7}$ This polynomial is a monomial; note that *constants* can be under radical signs

b. $-4xy + \frac{1}{2}x^2y$ This polynomial is a binomial; note that *constants* can be in denominators

c. $(x + y)^2 - 3(x + y) + 4$ This is a polynomial in *x* and *y*

d. $x^3y^2 \; - 2x \; + 3y^2 \; - 1$ This is a polynomial in *x* and *y*

$-1 = -1x^0y^0$
$3y^2 = 3x^0y^2$
$-2x = -2x^1y^0$

e. $7uv^4 - 5u^2v + 2u$ Polynomials can be in any variables; this is a polynomial in *u* and *v* ∎

Degree of a Term of a Polynomial   If a polynomial has only one variable, then the **degree of any term** of that polynomial is the exponent on the variable in that term. If a polynomial has more than one variable, the degree of any term of that polynomial is the *sum* of the exponents on the variables in that term.

Example 4   Find the degree of each term.

a. $5x^3$ 3rd degree

b. $6x^2y$ 3rd degree because $6x^2y = 6x^2y^1$

c. 14 0 degree because $14 = 14x^0$

d. $-2u^3vw^2$ 6th degree because $-2u^3vw^2 = -2u^3v^1w^2$

$$3 + 1 + 2 = 6$$ ∎

Degree of a Polynomial   The **degree of a polynomial** is defined to be the degree of its highest-degree term. Therefore, to find the degree of a polynomial, first find the degree of each of its terms. The *largest* of these numbers will be the degree of the polynomial.*

Example 5   Find the degree of each polynomial.

a. $9x^3 - 7x + 5$

— 0-degree term
— 1st-degree term
— 3rd-degree term ◄— Highest-degree term

Therefore, $9x^3 - 7x + 5$ is a third-degree polynomial.

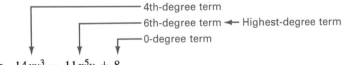

— 4th-degree term
— 6th-degree term ◄— Highest-degree term
— 0-degree term

b. $14xy^3 - 11x^5y + 8$

Therefore, $14xy^3 - 11x^5y + 8$ is a sixth-degree polynomial.

c. $6a^2bc^3 + 12ab^6c^2$

This is a ninth-degree polynomial, because the term with the highest degree $(12ab^6c^2)$ is of degree 9. ∎

*Mathematicians define the zero polynomial, 0, as having no degree.

✳**Leading Coefficient** The **leading coefficient** of a polynomial is defined to be the numerical coefficient of its highest-degree term.

Example 6 Name the leading coefficient for each polynomial from Example 5.

a. The leading coefficient of $9x^3 - 7x + 5$ is 9.

b. The leading coefficient of $14xy^3 - 11x^5y + 8$ is -11.

c. The leading coefficient of $6a^2bc^3 + 12ab^6c^2$ is 12. ∎

Descending and Ascending Powers If the exponents on one variable get smaller as we read from left to right, the polynomial is arranged in **descending powers** of that variable. If the exponents on one variable get larger as we read from left to right, the polynomial is arranged in **ascending powers** of that variable. For example,

$$8x^3 - 3x^2 + 5x^1 + 7$$

Exponents get smaller from left to right

$7 = 7x^0$

Therefore, $8x^3 - 3x^2 + 5x + 7$ is arranged in descending powers of x.

Example 7 Arrange $5 - 2x^2 + 4x$ in descending powers of x.
Solution $-2x^2 + 4x + 5$ ∎

A polynomial with more than one variable can be arranged in descending or ascending powers of any one of its variables.

Example 8 Arrange $3x^3y - 5xy + 2x^2y^2 - 10$ (a) in descending powers of x, then (b) in descending powers of y.
Solution

a. $3x^3y + 2x^2y^2 - 5xy - 10$ Arranged in descending powers of x

b. $2x^2y^2 + 3x^3y - 5xy - 10$ Arranged in descending powers of y

Since y is to the same power in both terms, the higher-degree *term* is written first ∎

EXERCISES 6.1

Set I In Exercises 1–16, if the expression is a polynomial, (a) find the degree of the first term and (b) find the degree of the polynomial. If it is *not* a polynomial, write "Not a polynomial."

1. $3x + 2x^2$

2. $4y + 5y^3$

3. $\dfrac{1}{x} - 3x^2 + 3$

4. $8x^4 - \dfrac{3}{2x} - 2x$

5. $3xy^2 + x^3y^3 - 3x^2y - y^3$

6. $6mn^2 + 8m^3 - 12m^2n - n^3$

7. $xy + 3x^4 + 5$

8. $xy + 5x^3 + 2$

9. $x^{-2} + 5x^{-1} + 4$

10. $y^{-3} + y^{-2} + 6$

11. $\sqrt{x - 3} + x^3$

12. $y^2 + \sqrt{2 + x}$

13. $x^2\sqrt{5} - xy + 3$

14. $x^3\sqrt{6} - xy + 2$

15. $\dfrac{3}{2x^2 - 3x + 1}$

16. $\dfrac{8}{3x^3 - 2x - 5}$

In Exercises 17–20, write each polynomial in descending powers of the indicated variable and find the leading coefficient.

17. $7x^3 - 4x - 5 + 8x^5$ Powers of x

18. $10 - 3y^5 + 4y^2 - 2y^3$ Powers of y

19. $8xy^2 + xy^3 - 4x^2y$ Powers of y

20. $3x^3y + x^4y^2 - 3xy^3$ Powers of x

Set II   In Exercises 1–16, if the expression is a polynomial, (a) find the degree of the first term and (b) find the degree of the polynomial. If it is *not* a polynomial, write "Not a polynomial."

1. $8x + 5x^3$

2. $4y^3 + \dfrac{5}{4x}$

3. $7x^3 + \dfrac{9}{2x} - 3x^2$

4. $5x^{-3} - 2x + 4$

5. $x^3y - 8x^4y^3 + xy - y^5$

6. $8xyz^2 + 3x^2y - z^3$

7. $xy + 8x^3 + 5$

8. $7uv + 8$

9. $3x^{-4} + 2x^{-2} + 6$

10. $\dfrac{1}{2x^2 - 5x}$

11. $x^2 - \sqrt{2x - 5}$

12. $x^2\sqrt{7} - 3 + 4x$

13. $5 - xy + x^2\sqrt{2}$

14. $\dfrac{5}{2x^4 + 3x^2 - 3}$

15. $\dfrac{2}{6x^2 - 5x + 8}$

16. $\sqrt{9x + 5} - x^2$

In Exercises 17–20, write each polynomial in descending powers of the indicated variable and find the leading coefficient.

17. $17a - 15a^3 + a^{10} - 4a^5$ Powers of a

18. $3x^2y + 8x^3 + y^3 - xy^5$ Powers of y

19. $5st - 9rs^2t - rt^2 - 3rs$ Powers of r

20. $9x^2y^4 - 8x^5y^2 + 3x^4 - 3$ Powers of x

6.2 Addition and Subtraction of Polynomials

Addition of Polynomials

Polynomials can be added horizontally by removing the grouping symbols and combining like terms, as was done in Section 2.2C. (In most *addition* problems, this means that you can simply "drop the grouping symbols.")

It is often helpful to underline like terms with the same kind of marking before you combine them (see Example 1).

Example 1   Add the polynomials; write the answers in descending powers of x.

a. $(3x^2 + 5x - 4) + (2x + 5) + (x^3 - 4x^2 + x)$

$= 3x^2 + 5x - 4 + 2x + 5 + x^3 - 4x^2 + x$ Using rules 2a and 2b from Section 2.2C

$= x^3 - x^2 + 8x + 1$ Combining like terms

b. $(5x^3y^2 - 3x^2y^2 + 4xy^3) + (4x^2y^2 - 2xy^2) + (-7x^3y^2 + 6xy^2 - 3xy^3)$

$= \underline{5x^3y^2} - \underline{\underline{3x^2y^2}} + 4xy^3 + \underline{\underline{4x^2y^2}} - 2xy^2 - \underline{7x^3y^2} + 6xy^2 - 3xy^3$ Using rules 2a and 2b from Section 2.2C

$= -2x^3y^2 + x^2y^2 + xy^3 + 4xy^2$

c. $\underline{4x^2y} + \underline{8xyx} + \underline{6yx^2} - 3xy^2$ The underlined terms *are* like terms, since $xyx = x^2y$ and $yx^2 = x^2y$

$= \underline{4x^2y} + \underline{8x^2y} + \underline{6x^2y} - 3xy^2$

$= 18x^2y - 3xy^2$ ∎

In vertical addition, which is sometimes desirable, it is important to have all like terms lined up vertically.

TO ADD POLYNOMIALS VERTICALLY

1. Arrange the polynomials under one another *so that like terms are in the same vertical column.*

2. Find the sum of the terms in each vertical column.

Example 2   Add $(3x^2 + 2x - 1)$, $(2x + 5)$, and $(4x^3 + 7x^2 - 6)$ vertically.
Solution

$$
\begin{array}{r}
3x^2 + 2x - 1 \\
2x + 5 \\
4x^3 + 7x^2 - 6 \\
\hline
4x^3 + 10x^2 + 4x - 2
\end{array}
$$
∎

Example 3   Add $(8x^2y - 3xy^2 + xy - 2)$, $(4xy^2 - 7xy)$, and $(-5x^2y + 9)$ vertically.
Solution

$$
\begin{array}{r}
8x^2y - 3xy^2 + xy - 2 \\
4xy^2 - 7xy \\
-5x^2y + 9 \\
\hline
3x^2y + xy^2 - 6xy + 7
\end{array}
$$
∎

Subtraction of Polynomials

We subtract polynomials the same way we subtract signed numbers: Change the signs in the polynomial being subtracted and then add the resulting polynomials.

TO SUBTRACT POLYNOMIALS

If the problem is to be set up *vertically*, write the polynomial being subtracted *under* the polynomial it is being subtracted from with like terms in the same vertical column.

For horizontal *or* vertical subtraction:

1. Change the subtraction symbol to an addition symbol *and* change the sign of each term in the polynomial *being subtracted.*

2. Find the sum of the resulting polynomials.

Example 4 Subtract the following polynomials as indicated.

This is the polynomial being subtracted

a. $(-4x^3 + 8x^2 - 2x - 3) - (4x^3 - 7x^2 + 6x + 5)$

Solution

$$(-4x^3 + 8x^2 - 2x - 3) - (4x^3 - 7x^2 + 6x + 5)$$

Changing the subtraction symbol to an addition symbol *and*

$$= (-4x^3 + 8x^2 - 2x - 3) + (-4x^3 + 7x^2 - 6x - 5)$$

changing the sign of every term in the polynomial being subtracted

$$= -4x^3 + 8x^2 - 2x - 3 - 4x^3 + 7x^2 - 6x - 5 \qquad \text{Removing the grouping symbols}$$

$$= -8x^3 + 15x^2 - 8x - 8 \qquad \text{Combining like terms}$$

b. Subtract $(-4x^2y + 10xy^2 + 9xy - 7)$ from $(11x^2y - 8xy^2 + 7xy + 2)$.

This is the polynomial being subtracted

Solution

$$(11x^2y - 8xy^2 + 7xy + 2) - (-4x^2y + 10xy^2 + 9xy - 7)$$

Changing subtraction to addition

$$= (11x^2y - 8xy^2 + 7xy + 2) + (+4x^2y - 10xy^2 - 9xy + 7)$$

$$= 11x^2y - 8xy^2 + 7xy + 2 + 4x^2y - 10xy^2 - 9xy + 7 \qquad \text{Removing the grouping symbols}$$

$$= 15x^2y - 18xy^2 - 2xy + 9 \qquad \text{Combining like terms}$$

c. Subtract $(2x^2 - 5x + 3)$ from the sum of $(8x^2 - 6x - 1)$ and $(4x^2 + 7x - 9)$.

Solution

$$[(8x^2 - 6x - 1) + (4x^2 + 7x - 9)] - (2x^2 - 5x + 3)$$

$$= [8x^2 - 6x - 1 + 4x^2 + 7x - 9] + (-2x^2 + 5x - 3) \qquad \begin{array}{l}\text{Changing}\\ \text{subtraction}\\ \text{to addition}\end{array}$$

$$= 12x^2 + x - 10 - 2x^2 + 5x - 3 \qquad \begin{array}{l}\text{Combining like terms inside the brackets}\\ \text{and removing all grouping symbols}\end{array}$$

$$= 10x^2 + 6x - 13 \qquad \text{Combining like terms}$$

d. $(x^2 + 5) - [(x^2 - 3) + (2x^2 - 1)]$

Solution

$$(x^2 + 5) - [(x^2 - 3) + (2x^2 - 1)]$$

$$= (x^2 + 5) - [x^2 - 3 + 2x^2 - 1] \qquad \text{Removing innermost parentheses}$$

$$= (x^2 + 5) - [3x^2 - 4] \qquad \text{Combining like terms inside brackets}$$

$$= (x^2 + 5) + [-3x^2 + 4] \qquad \begin{array}{l}\text{Changing the subtraction problem to an}\\ \text{addition problem}\end{array}$$

$$= x^2 + 5 - 3x^2 + 4 \qquad \text{Removing grouping symbols}$$

$$= -2x^2 + 9 \qquad \text{Combining like terms} \quad \blacksquare$$

In Examples 5 and 6, we show vertical subtraction. Because subtraction in long division problems is always done vertically, it is essential that you know how to subtract vertically.

Example 5   Subtract the lower polynomial from the upper one: $5x^2 + 3x - 6$
$$3x^2 - 5x + 2$$

Solution

$$
\begin{array}{c}
5x^2 + 3x - 6 \\
-(3x^2 - 5x + 2)
\end{array}
\rightarrow
\begin{array}{c}
5x^2 + 3x - 6 \\
+(-3x^2 + 5x - 2) \\
\hline
2x^2 + 8x - 8
\end{array}
$$

Change the sign of *each* term in the polynomial being subtracted, then *add* the resulting terms ∎

Example 6   Subtract $(3x^2 - 2x - 7)$ from $(x^3 - 2x - 5)$ vertically.

Solution

$$
\begin{array}{c}
x^3 \qquad - 2x - 5 \\
-(3x^2 - 2x - 7)
\end{array}
\rightarrow
\begin{array}{c}
x^3 \qquad\quad - 2x - 5 \\
+(- 3x^2 + 2x + 7) \\
\hline
x^3 \quad - 3x^2 \qquad + 2
\end{array}
$$

Change signs and *add* ∎

NOTE   Your instructor may require you to change the signs *mentally* and may not allow you to show the sign changes. ☑

EXERCISES 6.2

Set I   Perform the indicated operations.

1. $(2m^2 - m + 4) + (3m^2 + m - 5)$

2. $(5n^2 + 8n - 7) + (6n^2 - 6n + 10)$

3. $(2x^3 - 4) + (4x^2 + 8x) + (-9x + 7)$

4. $(5 + 8z^2) + (4 - 7z) + (z^2 + 7z)$

5. $(3x^2 + 4x - 10) - (5x^2 - 3x + 7)$

6. $(2a^2 - 3a + 9) - (3a^2 + 4a - 5)$

7. Subtract $(-5b^2 + 4b + 8)$ from $(8b^2 + 2b - 14)$.

8. Subtract $(-8c^2 - 9c + 6)$ from $(11c^2 - 4c + 7)$.

9. $(6a - 5a^2 + 6) + (4a^2 + 6 - 3a)$

10. $(2b + 7b^2 - 5) + (4b^2 - 2b + 8)$

11. Subtract $(5a + 3a^2 - 4)$ from $(4a^2 + 6 - 3a)$.

12. Subtract $(2b + b^2 - 7)$ from $(8 + 3b^2 - 7b)$.

13. Add: $17a^3 \qquad + 4a - 9$
$$8a^2 - 6a + 9$$

14. Add: $- b^3 + 5b^2 - 8$
$$-20b^4 + 2b^3 \qquad\quad + 7$$

15. Add: $14x^2y^3 - 11xy^2 + 8xy$
$$-9x^2y^3 + \;\; 6xy^2 - 3xy$$
$$7x^2y^3 - \;\; 4xy^2 - 5xy$$

16. Add: $12a^2b - \;\; 8ab^2 + \;\; 6ab$
$$-7a^2b + 11ab^2 - \;\; 3ab$$
$$4a^2b - \quad ab^2 - 13ab$$

In Exercises 17–20, subtract the lower polynomial from the upper one.

17. Subtract:
$$15x^3 - 4x^2 \quad\quad + 12$$
$$\underline{\quad 8x^3 \quad\quad + 9x - 5}$$

18. Subtract:
$$-14y^2 + 6y - 24$$
$$\underline{7y^3 + 14y^2 - 13y \quad\quad}$$

19. Subtract:
$$10a^2b - 6ab + 5ab^2$$
$$\underline{3a^2b + 6ab - 7ab^2}$$

20. Subtract:
$$14m^3n^2 - 9m^2n^2 - 6mn$$
$$\underline{-8m^3n^2 - 5m^2n^2 + 3mn}$$

21. $(7m^8 - 4m^4) + (4m^4 + m^5) + (8m^8 - m^5)$

22. $(8h - 4h^6) + (5h^7 + 3h^6) + (9h - 5h^7)$

23. $(6r^3t + 14r^2t - 11) + (19 - 8r^2t + r^3t) + (8 - 6r^2t)$

24. $(13m^2n^2 + 4mn + 23) + (17 + 4mn - 9m^2n^2) + (-29 - 8mn)$

25. $(7x^2y^2 - 3x^2y + xy + 7) - (3x^2y^2 - 5xy + 4 + 7x^2y)$

26. $(4x^2y^2 + x^2y - 5xy - 4) - (9 - 5x^2y^2 - xy + 3x^2y)$

27. $(x^2 + 4) - [(x^2 - 5) - (3x^2 + 1)]$

28. $(3x^2 - 2) - [(4 - x^2) - (2x^2 - 1)]$

29. Subtract $(2x^2 - 4x + 3)$ from the sum of $(5x^2 - 2x + 1)$ and $(-4x^2 + 6x - 8)$.

30. Subtract $(6y^2 + 3y - 4)$ from the sum of $(-2y^2 + y - 9)$ and $(8y^2 - 2y + 5)$.

31. Subtract the sum of $(x^3y + 3xy^2 - 4)$ and $(2x^3y - xy^2 + 5)$ from the sum of $(5 + xy^2 + x^3y)$ and $(-6 - 3xy^2 + 4x^3y)$.

32. Subtract the sum of $(2m^2n - 4mn^2 + 6)$ and $(-3m^2n + 5mn^2 - 4)$ from the sum of $(5 + m^2n - mn^2)$ and $(3 + 4m^2n + 2mn^2)$.

33. $(7.239x^2 - 4.028x + 6.205) + (-2.846x^2 + 8.096x + 5.307)$

34. $29.62x^2 + 35.78x - 19.80) + (7.908x^2 - 29.63x - 32.84)$

Set II   Perform the indicated operations.

1. $(2x^2 - 3x + 1) + (4x^2 + 5x - 3)$

2. $(5z + 7) - (7z^2 - 8)$

3. $(8x^3 - 4x^2) + (x^2 - 3x) + (8 - 2x^2)$

4. $(3x^2 + 5x) + (2x^3 - 6x^2) - (3 - 5x)$

5. $(9x^2 - 2x + 3) - (12x^2 - 8x - 9)$

6. $(3 - x^3 + 5x) - (-8x - x^2 + 8x)$

7. Subtract $(8y^3 - 3y)$ from $(y^2 - 3y + 12)$.

8. Subtract $(-3x^3 + 2x^2 - 3x + 2)$ from $(1 - x - x^2 - x^3)$.

9. $(9c + 2c^2 - 8) + (3 - 17c - 8c^2)$

10. $(8x^3 - x + 2) - (x^2 - x + 2)$

11. Subtract $(9 - z + z^2)$ from $(-3z^2 - z + 9)$.

12. Subtract $(2x - 7x^2 + 3x^3)$ from $(-x + 9x^2 - 1 + x^3)$.

13. Add: $\begin{aligned} 8x^3 \qquad + 3x - 7 \\ 5x^2 - 5x + 7 \end{aligned}$ **14.** Add: $\begin{aligned} 4x^3 + 7x^2 - 5x + 4 \\ 2x^3 - 5x^2 + 5x - 6 \end{aligned}$

15. Add: $\begin{aligned} 3y^4 - 2y^3 + 4y + 10 \\ -5y^4 + 2y^3 + 4y - 6 \\ 7y^4 \qquad\qquad - 6y - 8 \end{aligned}$ **16.** Add: $\begin{aligned} 18x^2y - 3xy^2 + 4xy \\ -6x^2y + 8xy^2 - 9xy \\ -4x^2y - 9xy^2 + xy \end{aligned}$

In Exercises 17–20, subtract the lower polynomial from the upper one.

17. Subtract: $\begin{aligned} 10x^3 - 5x^2 + 6x - 1 \\ -2x^3 + 3x^2 + 9x - 5 \end{aligned}$

18. Subtract: $\begin{aligned} 5z^3 \qquad\quad - 7z + 8 \\ 8z^3 - 10z^2 + 7z \end{aligned}$

19. Subtract: $\begin{aligned} 7x^2y^2 - 8xy^2 + xy - 6 \\ 2x^2y^2 - 6xy^2 - 5xy + 9 \end{aligned}$

20. Subtract: $\begin{aligned} -3x^2 + 3x \qquad\quad \\ 5x^2 - 2x + 6 \end{aligned}$

21. $(8x^6 - 3x^4) + (x - 7x^4) + (3x^4 - x)$

22. $(5x^2 - 3x + 1) - (5x^2 - 3x + 1)$

23. $(7x^2y + 4xy^2 - 5) + (8xy^2 - 7x^2y + xy) + (-2 - xy^2)$

24. $(5x^2y - 4x + 3y^2) + (-yx^2 + 6x - y^2)$

25. $(9x^4y - x^2y^2 - 5 + 8xy^3) - (3x^2y^2 - 4xy^3 + 7 - 6x^3)$

26. $(3y^2z - 4x^2y^2 + 5) - (5x^2y^2 - 4y^2z)$

27. $(3x^2 + 5) - [(x^2 - 3) - (2x^2 + 8)]$

28. $(x^2 - 3xy + 4) - [x^2 - 3xy - (6 + x^2)]$

29. Subtract $(3x^2 - 7x - 1)$ from the sum of $(x^2 - x + 2)$ and $(x^2 - 3x - 6)$.

30. Subtract $(2y^2 + 3y - 9)$ from the sum of $(y - y^2 + 2)$ and $(5y - y^2 + 7)$.

31. Subtract the sum of $(xy^2 - 2x^2y - 2)$ and $(3x^2y - 4xy^2 - 6)$ from the sum of $(3 - x^2y)$ and $(2xy^2 - 5x^2y - 8)$.

32. Subtract $(-3m^2n^2 + 2mn - 7)$ from the sum of $(6m^2n^2 - 8mn + 9)$ and $(-10m^2n^2 + 18mn - 11)$.

33. $(5.416x - 34.54x^2 + 7.806) + (51.75x^2 - 1.644x - 9.444)$

34. $(5.886x^2 - 3.009x + 7.966) - [4.961x^2 - 54.51x - (7.864 - 1.394x^2)]$

6.3 Simplifying and Multiplying Polynomials

6.3A Simplifying Polynomials

We simplify a polynomial the same way we simplify an algebraic expression: We remove all grouping symbols, simplify each term, and combine all like terms. In Section 6.2, we discussed removing grouping symbols when addition and subtraction were indicated. In this section, we discuss removing grouping symbols when multiplication is indicated.

When we multiply a monomial by a monomial, we use the methods learned in Section 5.2 (see Example 1).

Example 1   Multiply $(3x^3y)(2xy)(-5xy^2)$.
Solution We have a *monomial* inside each set of parentheses.

$$(3x^3y)(2xy)(-5xy^2) = (3)(2)(-5)(x^3 \cdot x \cdot x)(y \cdot y \cdot y^2) = -30x^5y^4 \quad \blacksquare$$

When we multiply a polynomial with more than one term by a monomial, we use the distributive property (see Examples 2 and 3).

Example 2   Multiply $(3x^3y)(2xy - 5xy^2)$.
Solution We have a *binomial* inside the second set of parentheses.

$$(3x^3y)(2xy - 5xy^2) = 6x^4y^2 - 15x^4y^3 \quad \blacksquare$$

Example 3   Simplify the following polynomials.

a. $a - 2(a + b)$

$= a - 2a - 2b$ Using the distributive property, we multiply each term inside the parentheses by -2

$= -a - 2b$ Combining like terms

It often helps to underline like terms when combining them.

b. $3(5x - 2y) - 4(3x - 6y)$

$= 15x - 6y - 12x + 24y$ Using the distributive property, we multiply each term inside the first set of parentheses by 3 and each term inside the second set of parentheses by -4

$= 15x - 12x - 6y + 24y$ Collecting like terms

$= \quad 3x \quad + \quad 18y$ Combining like terms

c. $x(x^2 + xy + y^2) - y(x^2 + xy + y^2)$

$= x^3 + x^2y + xy^2 - x^2y - xy^2 - y^3$ Using the distributive property (Watch the signs!)

$= x^3 + x^2y - x^2y + xy^2 - xy^2 - y^3$ Collecting like terms

$= x^3 + \quad 0 \quad + \quad 0 \quad - y^3$ Combining like terms

$= x^3 - y^3$

d. $2x^2y(4xy - 3xy^2) - 5xy(3x^2y^2 - 2x^2y)$ Use the distributive property (Watch the signs!)

$= 8x^3y^2 - 6x^3y^3 - 15x^3y^3 + 10x^3y^2$

$= 8x^3y^2 + 10x^3y^2 - 6x^3y^3 - 15x^3y^3$ Collecting like terms

$= \quad 18x^3y^2 \quad - \quad 21x^3y^3$ Combining like terms

Sometimes it is helpful to combine like terms within a pair of grouping symbols *before* removing that pair of grouping symbols, as is done in Examples 3e and 3f.

e. $-8[-5(3x - 2) + 13] - 11x$

$= -8[-15x + 10 + 13] - 11x$ Removing innermost grouping symbol first

$= -8[-15x + 23] \quad - 11x$ Combining like terms inside the brackets

$= +120x \quad - 184 \quad - 11x$ Using the distributive property

$= \quad 109x \quad - 184$ Combining like terms

f. $(8x + 10y) - 2\{[4x - 5(8 - y)] - 15\}$

$= (8x + 10y) - 2\{[4x - 40 + 5y] - 15\}$ Removing the inner grouping symbols

$= (8x + 10y) - 2\{4x - \underline{40} + 5y - \underline{15}\}$

$= (8x + 10y) - 2\{4x \qquad + 5y - 55\}$ Combining like terms inside the braces

$= \underline{8x} + 10y - \underline{8x} \qquad -10y + 110$ Removing the grouping symbols

$= 110$ Combining like terms ∎

EXERCISES 6.3A

Set I   Simplify.

1. $x - 3(x + y)$ **2.** $y - 2(y + z)$

3. $2a - 4(a - b)$ **4.** $3c - 2(c - d)$

5. $u(u^2 + 2u + 4) - 2(u^2 + 2u + 4)$ **6.** $x(x^2 - 3x + 9) + 3(x^2 - 3x + 9)$

7. $x^2(x^2 + y^2) - y^2(x^2 + y^2)$ **8.** $w^2(w^2 - 4) + 4(w^2 - 4)$

9. $2x(3x^2 - 5x + 1) - 4x(2x^2 - 3x - 5)$

10. $3x(4x^2 - 2x - 3) - 2x(3x^2 - x + 1)$

11. $-3(a - 2b) + 2(a - 3b)$ **12.** $-2(m - 3n) + 4(m - 2n)$

13. $-5(2x - 3y) - 10(x + 5y)$ **14.** $-4(3s - 7t) - 8(3s - 4t)$

15. $3xy(2x)(-5y)$ **16.** $2ac(5b)(-3c)$

17. $3xy(2x - 5y)$ **18.** $2ac(5b - 3c)$

19. $(3xy + 2x)(-5y)$ **20.** $(2ac + 5b)(-3c)$

21. $(3xy + 2x) - 5y$ **22.** $(2ac + 5b) - 3c$

23. $x^2y(3xy^2 - y) - 2xy^2(4x - x^2y)$ **24.** $ab^2(2a - ab) - 3ab(2ab - ab^2)$

25. $2h(3h^2 - k) - k(h - 3k^3)$ **26.** $4x(2y^2 - 3x) - x(2x - 3y^2)$

27. $3f(2f^2 - 4g) - g(2f - g^2)$ **28.** $4a(b^2 - 2b) - b(ab - a)$

29. $2x - [3a + (4x - 5a)]$ **30.** $2y - [5c + (3y - 4c)]$

31. $5x + [-(2x - 10) + 7]$ **32.** $4x + [-(3x - 5) + 4]$

33. $25 - 2[3g - 5(2g - 7)]$ **34.** $40 - 3[2h - 8(3h - 10)]$

35. $-2\{-3[-5(-4 - 3z) - 2z] + 30z\}$

36. $-3\{-2[-3(-5 - 2z) - 3z] - 40z\}$

37. $(4x)(3x)^2$ **38.** $(5y)^2(2y)$ **39.** $(9x - 2y)(-5x)$

40. $(3c - 7d)(-9d)$ **41.** $(2xy^2)^2(3y)$ **42.** $(3a^2b)^2(5b)$

43. $(9x - 2y) - 5x$ **44.** $(3c - 7d) - 9d$

Set II   Simplify.

1. $a - 4(a + b)$ **2.** $2x - 3(x - y)$

3. $5x - 2(4x - 3y)$ **4.** $2(a - 3b) - 3(4a - b)$

5. $2h(4h^2 - k) - 3k(h - 2k^2)$ **6.** $x(x^2 + 2x + 4) - 2(x^2 + 2x + 4)$

7. $a^2(a^2 - b^2) + b^2(a^2 - b^2)$ **8.** $2x^2y(3xy^2 - 2y) - 3xy^2(2x^2y - 5x)$

9. $3xy^2(2x - xy + 4) + x(3xy^3 - 12y^2)$

10. $5r^2s(3r - 5 - 2s) + 5r^2s(5 + 2s)$

11. $-5(x - 3y + z) - 2(3y - x - z)$ **12.** $(a + b + c) - 4$

13. $-8(3x - 2y) + (3x - 4y)(-3)$ **14.** $-8(3x - 2y) + (3x - 4y) - 3$

15. $8xy(3x)(-2y)$ **16.** $4x(5x)^2$

17. $8xy(3x - 2y)$ **18.** $4 + 3(2x - 7)$

19. $(8xy + 3x)(-2y)$ **20.** $8 + 3x(9x - 2)$

21. $(8xy + 3x) - 2y$ **22.** $(5st - 3s) - 9s$

23. $a^2b(3ab - b^2 + a^2b) - (5a^3b^2 + a^2b^3)$

24. $3 + 2(4x^4 - 3x^3 + 2x^2) - (4x^3 - 2x^2)$

25. $5x(4x^2 - 3x + 1) - 3(2x^2 + x - 1)$

26. $4x^0(3x^3y - 4xy^2 + 3y^4) + 2(xy^2 - 12x^3y)$

27. $5x(2x^2 + x - 1) + (3x^2 - 4)$ **28.** $(3x - 1)^0 - (5x^0 + 5y - 2)$

29. $8x - [3x - (5x - 2)]$ **30.** $9 - 3[5x - (8x - 2\{x + 3\})]$

31. $6x + [-2(5 - x) + 3]$ **32.** $7^0 - \{3x[2 - (x^2 - 1) + 5]\}$

33. $15 + 3[2 - (5 + x^2)]$ **34.** $21 - 5[3 + 2(x - 5) - x]$

35. $-3\{-2[-1(3 - x) + 5x]\}$ **36.** $-5\{3[8 + 2(x - 6) - 5x] + 1\}$

37. $(9t)(3t)^2$ **38.** $(7a)^2(2a)$ **39.** $(8x - 3y)(-2x)$

40. $(c - 8d)(-6d)$ **41.** $(6xy^2)^2(2y)$ **42.** $(2a^2b)^3(1b)$

43. $(6a - b) - 4b$ **44.** $(7 - x) - 9x$

6.3B   Products of Two Binomials

Since we often need to find the product of two binomials, it is helpful to be able to find their product by inspection (that is, without writing anything down except the answer). First, however, we show the step-by-step procedure for multiplying two binomials; it is necessary to use the distributive property *more than once*.

Example 4   Multiply $(3x + 2y)(4x + 5y)$.

Solution We first treat $3x + 2y$ as if it were a single number.

Step 1. $(3x + 2y)(4x + 5y) = (3x + 2y)(4x) + (3x + 2y)(5y)$

Step 2.   We now use the distributive property again on $(3x + 2y)(4x)$ and on $(3x + 2y)(5y)$:

This step need not be shown

$$= (3x)(4x) + (2y)(4x) + (3x)(5y) + (2y)(5y)$$

Step 3. $= 12x^2 + 8xy + 15xy + 10y^2$

Step 4. We combine like terms: $= 12x^2 + 23xy + 10y^2$ ∎

Let us agree on some terminology so we can more easily discuss finding products of two binomials by inspection.

In step 3 of finding the product $(3x + 2y)(4x + 5y)$, the two *middle terms* are $8xy$ and $15xy$. Since $8xy$ is the product of the two "inside" terms, we call it the *inner product*, and since $15xy$ is the product of the two "outside" terms, we call it the *outer product*.

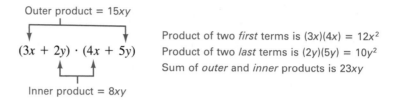

Outer product = $15xy$

$(3x + 2y) \cdot (4x + 5y)$

Inner product = $8xy$

Product of two *first* terms is $(3x)(4x) = 12x^2$
Product of two *last* terms is $(2y)(5y) = 10y^2$
Sum of *outer* and *inner* products is $23xy$

When a multiplication problem is in the form $(ax + by)(cx + dy)$, as is Example 4, we can quickly find the product by using the following rules:

TO MULTIPLY $(ax + by)(cx + dy)$

1. The *first term* of the product is the product of the first terms of the binomials.

2. The *middle term* of the product is the sum of the inner and outer products.

3. The *last term* of the product is the product of the last terms of the binomials.

NOTE   Plus (or minus) signs between the *terms* of the product are essential.   ☑

When we use this method of multiplying binomials, we find the product of the two *F*irst terms, the *O*uter product, the *I*nner product, and the product of the two *L*ast terms. For this reason, this procedure is often called the *FOIL* method.

Example 5   Multiply $(x + 2)(x - 5)$.
Solution
Method 1

$$(x + 2)(x - 5)$$
$$= (x + 2)(x) + (x + 2)(-5) \qquad \text{Using the distributive property}$$
$$= (x)(x) + (2)(x) + (x)(-5) + (2)(-5) \qquad \text{Using the distributive property again}$$
$$= x^2 + \underline{2x} - \underline{5x} - 10$$
$$= x^2 - 3x - 10 \qquad \text{Combining like terms}$$

Method 2 Because the problem is in the form $(ax + by)(cx + dy)$, we can find the product by using the FOIL method.

First	Outer	Inner	Last
$(x + 2)(x - 5)$	$(x + 2)(x - 5)$	$(x + 2)(x - 5)$	$(x + 2)(x - 5)$
x^2	$-5x$	$2x$	-10
x^2	$-3x$		-10 ∎

When Method 2 (the FOIL method) is used, we can add the inner and outer products mentally. No intermediate steps need to be shown. (In the next few examples, we will continue to show steps that don't have to be shown.)

Example 6   Multiply $(x - 3)(x - 4)$.
Solution
Method 1

$$(x - 3)(x - 4) = (x - 3)(x) + (x - 3)(-4)$$
$$= x^2 - 3x + (-4x) + 12$$
$$= x^2 - 7x + 12$$

Method 2

First	Outer	Inner	Last
$(x - 3)(x - 4)$	$(x - 3)(x - 4)$	$(x - 3)(x - 4)$	$(x - 3)(x - 4)$
x^2	$-4x$	$-3x$	$+12$
x^2	$-7x$		$+12$

Therefore, $(x - 3)(x - 4) = x^2 - 7x + 12$. ■

Example 7   Multiply $(3x + 2)(4x - 5)$.
Solution
Method 1

$$(3x + 2)(4x - 5) = (3x + 2)(4x) + (3x + 2)(-5)$$
$$= 12x^2 + 8x - 15x - 10$$
$$= 12x^2 - 7x - 10$$

Method 2

$(3x + 2)(4x - 5)$	$(3x + 2)(4x - 5)$	$(3x + 2)(4x - 5)$
	$8x$	
	$-15x$	
$12x^2$	$- 7x$	-10

Therefore, $(3x + 2)(4x - 5) = 12x^2 - 7x - 10$. ■

In the remaining examples, we show Method 2 only.

Example 8   Multiply $(5x - 4y)(6x + 7y)$.
Solution

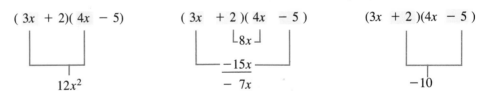

$(5x - 4y)(6x + 7y)$	$(5x - 4y)(6x + 7y)$	$(5x - 4y)(6x + 7y)$
	$-24xy$	
	$+35xy$	
$30x^2$	$+11xy$	$-28y^2$

Therefore, $(5x - 4y)(6x + 7y) = 30x^2 + 11xy - 28y^2$. ■

Example 9   Simplify $(3x - 8y)^2$.

Solution Because raising to a power is repeated multiplication, we have

$$(3x - 8y)^2 = (3x - 8y)(3x - 8y)$$

$$-24xy$$

$$-24xy$$

$$9x^2 - 48xy + 64y^2$$

Therefore, $(3x - 8y)^2 = 9x^2 - 48xy + 64y^2$. ∎

A WORD OF CAUTION   A common error is to write

$$(3x - 8y)^2 = 9x^2 - 64y^2 \qquad \text{or} \qquad (3x - 8y)^2 = 9x^2 + 64y^2$$

Example 9 shows that neither of these answers is correct. In Section 6.4B, we discuss a short method for finding the correct answer for $(3x - 8y)^2$.   ☑

We can use the FOIL method to find the product of two binomials even when the problem is *not* in the form $(ax + by)(cx + dy)$. In such problems, however, the inner and outer products are not necessarily like terms (see Examples 11 and 12).

Example 10   Multiply $(2x^2 - 5x)(3x + 7)$.
Solution

The inner and outer products are like terms and must be combined

$$(2x^2 - 5x)(3x + 7) = 6x^3 + \overbrace{14x^2 - 15x^2} - 35x$$

$$= 6x^3 - x^2 - 35x \quad ∎$$

Example 11   Multiply $(3x + 7)(4y + 2)$.
Solution

The inner and outer products are not like terms; they cannot be combined

$$(3x + 7)(4y + 2) = 12xy + \overbrace{6x + 28y} + 14 \quad ∎$$

Example 12   Multiply $(2x^2 + 7)(4x + 2)$.
Solution

The inner and outer products are not like terms; they cannot be combined

$$(2x^2 + 7)(4x + 2) = 8x^3 + \overbrace{4x^2 + 28x} + 14 \quad ∎$$

EXERCISES 6.3B

Set I   Simplify. Be careful! Some problems are products of monomials, some are products of a monomial and a polynomial, some are products of two binomials, and some are polynomials raised to a power.

1. $(7x^3y)(-2y^2)(4x)$ **2.** $(-4ab^2)(3a^4)(2b^3)$ **3.** $(x + 3)(x - 2)$

4. $(a - 4)(a + 3)$ **5.** $2x(3x^2 + 7)$ **6.** $3y(4y^3 - 2)$

7. $(y + 8)(y - 9)$ **8.** $(z - 3)(z + 10)$ **9.** $(5x^2 - 2y)(3x + y)$

10. $(2s + 3t)(s^2 - 4t)$ **11.** $(5a^2 - b^2)(2a)$ **12.** $(7c^2 - d^2)(3c)$

13. $(x + 1)(x + 4)$ **14.** $(x + 3)(x + 1)$ **15.** $(a + 5)(a + 2)$

16. $(a + 7)(a + 1)$ **17.** $(m - 4)(m + 2)$ **18.** $(n - 3)(n + 7)$

19. $(y + 9)(y - 12)$ **20.** $(z - 5)(z - 11)$ **21.** $(x + y)(-ab)$

22. $(x - y)(ab)$ **23.** $(x + y)(a - b)$ **24.** $(x - y)(a + b)$

25. $(4x)^2$ **26.** $(3x)^2$ **27.** $(4 + x)^2$

28. $(3 + x)^2$ **29.** $(b - 4)^2$ **30.** $(b - 6)^2$

31. $(3x + 1)(x + 2)$ **32.** $(2x + 3)(x + 2)$ **33.** $(2x + 4)(4x - 3)$

34. $(3x + 5)(2x - 1)$ **35.** $(4x - 6)(5x - 2)$ **36.** $(5x - 4)(3x - 4)$

37. $(2x + 5)^2$ **38.** $(3x + 4)^2$ **39.** $(3x^2 - 2)(2x + 1)$

40. $(8a^2 + 1)(3a - 2)$ **41.** $(4x - y)(2x + 7y)$ **42.** $(3x - 2y)(4x + 5y)$

43. $(7x - 10y)(7x - 10y)$ **44.** $(4u - 9v)(4u - 9v)$ **45.** $(3a + 2b)^2$

46. $(2x - 6y)^2$ **47.** $(4c - 3d)(4c + 3d)$ **48.** $(5e + 2f)(5e - 2f)$

Set II   Simplify. Be careful! Some problems are products of monomials, some are products of a monomial and a polynomial, some are products of two binomials, and some are polynomials raised to a power.

1. $(8x)(2y^2)(-3x)$ **2.** $(8x)(2y^2 - 3x)$ **3.** $(a + 2)(a - 5)$

4. $(a + 2)(a)(-5)$ **5.** $3m(4m^3 - 5)$ **6.** $3m(4m^3)(-5)$

7. $(x - 4)(x + 7)$ **8.** $(x^2 - 4)(x + 7)$ **9.** $(3m^2 + 2n)(m - n)$

10. $(3m + 2n)(m - n)$ **11.** $(2x^2 - y^2)(12x)$ **12.** $(2x - y)^2$

13. $(x + 2)(x + 3)$ **14.** $(y - 1)(y - 1)$ **15.** $(m + 3)(m + 5)$

16. $(m - 3)(m + 5)$ **17.** $(h - 6)(h + 3)$ **18.** $(h + 6)(h - 3)$

19. $(w + 7)(w - 8)$ **20.** $(w - 7)(w + 8)$ **21.** $(a + b)(-cd)$

22. $(a + b)(a - b)$ **23.** $(a + b)(c - d)$ **24.** $(a + b)^2$

25. $(7x)^2$ **26.** $(2 + y)^2$ **27.** $(7 + x)^2$

28. $(2y)^2$ **29.** $(y - 5)^2$ **30.** $(-5y)^2$

31. $(5x + 1)(x + 3)$ **32.** $(2a + 5b)(a + b)$ **33.** $(3c + 2)(c + 1)$

34. $(4x^2 + 1)(x + 2)$ **35.** $(y - 5)(y - 5)$ **36.** $(4x - 7)^2$

37. $(a + 4)^2$ **38.** $(5x - 3)^2$ **39.** $(5x^2 - 3)(2x + 3)$

40. $(x - 1)(x^2 + 2)$ **41.** $(11x + 10y)(3x - 4y)$ **42.** $(10x - 7y)(8x + 9y)$

43. $(2x - 3y)(2x + 3y)$ **44.** $(2x - 3y)^2$ **45.** $(5m + 2n)^2$

46. $(2y + 5z)^2$ **47.** $(2y + 5z)(2y - 5z)$ **48.** $(3x + 4)(2x - 5)$

6.3C   Multiplying a Polynomial by a Polynomial

When we're finding the product of two polynomials and both polynomials have two or more terms, it is necessary to use the distributive property more than once.

Example 13 Multiply $(3x^2 - 4x + 6)(5x - 2)$.

Solution 1 We multiply each term of the right factor by $(3x^2 - 4x + 6)$:

$$(3x^2 - 4x + 6)(5x - 2) = (3x^2 - 4x + 6)(5x) + (3x^2 - 4x + 6)(-2)$$
$$= 15x^3 - 20x^2 + 30x - 6x^2 + 8x - 12$$
$$= 15x^3 - 26x^2 + 38x - 12$$

Solution 2 We multiply each term of the left factor by $(5x - 2)$:

$$(3x^2 - 4x + 6)(5x - 2) = 3x^2(5x - 2) - 4x(5x - 2) + 6(5x - 2)$$
$$= 15x^3 - 6x^2 - 20x^2 + 8x + 30x - 12$$
$$= 15x^3 - 26x^2 + 38x - 12 \quad \blacksquare$$

The multiplication can, however, be conveniently arranged vertically if we're careful to line up like terms in the same vertical column as we perform the multiplications.

It helps to compare this procedure with the one used in arithmetic for multiplying whole numbers.

$$
\begin{array}{r}
56 \\
23 \\
\hline
168 \\
112 \\
\hline
1288
\end{array}
$$

168   Product of 56 and 3

112   Product of 56 and 2

1288   Notice that the second line in this arithmetic example is positioned so that the tens digits are lined up and the hundreds digits are lined up

Example 14 Multiply $(3x^2 - 4x + 6)(5x - 2)$, setting up the problem vertically.

Solution We usually place the polynomial that has more terms above the other polynomial.

Multiplying from right to left

$$
\begin{array}{r}
3x^2 - 4x + 6 \\
5x - 2 \\
\hline
-6x^2 + 8x - 12 \\
15x^3 - 20x^2 + 30x \\
\hline
15x^3 - 26x^2 + 38x - 12
\end{array}
$$

This is $(3x^2 - 4x + 6)(-2)$

This is $(3x^2 - 4x + 6)(5x)$

Notice that we have like terms in the same vertical column $\blacksquare$

Multiplying from left to right

$$
\begin{array}{r}
3x^2 - 4x + 6 \\
5x - 2 \\
\hline
15x^3 - 20x^2 + 30x \\
-6x^2 + 8x - 12 \\
\hline
15x^3 - 26x^2 + 38x - 12
\end{array}
$$

The vertical method can be used to multiply two binomials (see Example 15).

Example 15 Multiply $(2x^2 - 5x)(3x + 7)$, setting up the problem vertically. (This is the same problem as in Example 10.)

Solution

Multiplying from right to left

$$
\begin{array}{r}
2x^2 - 5x \\
3x + 7 \\
\hline
14x^2 - 35x \\
6x^3 - 15x^2 \\
\hline
6x^3 - x^2 - 35x
\end{array}
$$

This is $(2x^2 - 5x)(7)$

This is $(2x^2 - 5x)(3x)$

Multiplying from left to right

$$
\begin{array}{r}
2x^2 - 5x \\
3x + 7 \\
\hline
6x^3 - 15x^2 \\
14x^2 - 35x \\
\hline
6x^3 - x^2 - 35x
\end{array}
$$
$\blacksquare$

Example 16 Multiply $(2m + 3m^2 - 5)(2 + m^2 - 3m)$.

Solution The multiplication is simplified by first arranging the polynomials in descending powers of m.

Multiplying from right to left

$$
\begin{array}{r}
3m^2 + 2m - 5 \\
m^2 - 3m + 2 \\
\hline
6m^2 + 4m - 10 \\
-9m^3 - 6m^2 + 15m \\
3m^4 + 2m^3 - 5m^2 \\
\hline
3m^4 - 7m^3 - 5m^2 + 19m - 10
\end{array}
$$

Multiplying from left to right

$$
\begin{array}{l}
3m^2 + 2m - 5 \\
m^2 - 3m + 2 \\
\hline
3m^4 + 2m^3 - 5m^2 \\
- 9m^3 - 6m^2 + 15m \\
6m^2 + 4m - 10 \\
\hline
3m^4 - 7m^3 - 5m^2 + 19m - 10 \quad \blacksquare
\end{array}
$$

Example 17 Multiply $(a^2 + 3a + 9)(a - 3)$.

Solution

Multiplying from right to left

$$
\begin{array}{r}
a^2 + 3a + 9 \\
a - 3 \\
\hline
-3a^2 - 9a - 27 \\
a^3 + 3a^2 + 9a \\
\hline
a^3 - 27
\end{array}
$$

Multiplying from left to right

$$
\begin{array}{l}
a^2 + 3a + 9 \\
a - 3 \\
\hline
a^3 + 3a^2 + 9a \\
- 3a^2 - 9a - 27 \\
\hline
a^3 - 27 \quad \blacksquare
\end{array}
$$

Missing Terms Consider the polynomial $ax^3 + bx + c$. Because there is no x^2 term written, we say that we have a *missing term*; that is, the coefficient of x^2 is an understood zero. When we multiply and divide polynomials, it is usually desirable to write in any missing terms with a coefficient of zero. That is, we would write $x^3 + x - 1$ as $x^3 + 0x^2 + x - 1$ (see Example 18).

Example 18 Multiply $(x^3 - 1 + x)(2x^2 + 2 - x)$.

Solution We write both polynomials in descending powers of x.

Multiplying from right to left

Multiplying from left to right

Note that $0x^2$ was written in to save a place for the x^2 terms that arise in the multiplication

Multiplying from right to left

$$
\begin{array}{r}
x^3 + 0x^2 + x - 1 \\
2x^2 - x + 2 \\
\hline
2x^3 + 0x^2 + 2x - 2 \\
-x^4 + 0x^3 - x^2 + x \\
2x^5 + 0x^4 + 2x^3 - 2x^2 \\
\hline
2x^5 - x^4 + 4x^3 - 3x^2 + 3x - 2
\end{array}
$$

Multiplying from left to right

$$
\begin{array}{l}
x^3 + 0x^2 + x - 1 \\
2x^2 - x + 2 \\
\hline
2x^5 + 0x^4 + 2x^3 - 2x^2 \\
- x^4 + 0x^3 - x^2 + x \\
2x^3 + 0x^2 + 2x - 2 \\
\hline
2x^5 - x^4 + 4x^3 - 3x^2 + 3x - 2 \quad \blacksquare
\end{array}
$$

Powers of Polynomials

We can raise any polynomial to any power by using repeated multiplication, just as we found $(3x - 8y)^2$ in Example 9. (In Section 6.4B, we will show an alternate method for squaring a binomial.)

Example 19 Simplify $(a - b)^3$.

Solution To simplify $(a - b)^3$, we must remove the grouping symbols; therefore, we must multiply $a - b$ by itself three times.

$$(a - b)^3 = \underbrace{(a - b)(a - b)(a - b)}$$

First find   Then multiply that
$(a - b)^2$ product by $(a - b)$

$$\begin{array}{r} a - b \\ a - b \\ \hline -\ ab + b^2 \\ a^2 - \ ab \\ \hline a^2 - 2ab + b^2 \end{array} \qquad \begin{array}{r} a^2 - 2ab + b^2 \\ a \ - \ b \\ \hline -\ a^2b + 2ab^2 - b^3 \\ a^3 - 2a^2b + \ ab^2 \\ \hline a^3 - 3a^2b + 3ab^2 - b^3 \end{array} \ \blacksquare$$

A WORD OF CAUTION Students often make this error:

$$(x + 3)^2 = x^2 + 9$$

However,

$$(x + 3)^2 \neq x^2 + 3^2$$

x and 3 are *terms*

For example, suppose x is 4. Then $(x + 3)^2 = (4 + 3)^2 = 7^2 = 49$, but $x^2 + 3^2 = 4^2 + 3^2 = 16 + 9 = 25$, and $49 \neq 25$.

Correct Method

$$\begin{aligned} (x + 3)^2 &= (x + 3)(x + 3) \\ &= (x + 3)x + (x + 3)(3) && \text{Distributive property} \\ &= x^2 + 3x + 3x + 9 && \text{Distributive property} \\ &= x^2 + \boxed{6x} + 9 && \text{Combining like terms} \end{aligned}$$

This is the term that students sometimes leave out

Again, if x is 4, $x^2 + 6x + 9 = 4^2 + 6(4) + 9 = 16 + 24 + 9 = 49$.
    It *is* true that

$$(3x)^2 = 3^2x^2 = 9x^2 \qquad \text{By the rules for exponents}$$

3 and x are *factors*

EXERCISES 6.3C

Set I   Simplify.

1. $(x - 3)(2x^2 + x - 1)$ **2.** $(x - 2)(3x^2 + x - 1)$

3. $(-3x^2y + xy^2 - 4y^3)(-2xy)$ **4.** $(-4xy^2 - x^2y + 3x^3)(-3xy)$

5. $(x^2 + x + 1)(x^2 + x + 1)$ **6.** $(x^2 - x - 1)(x^2 - x - 1)$

7. $(4z)(z^2 - 4z + 16)$ **8.** $(-5a)(a^2 + 5a + 25)$

9. $(z + 4)(z^2 - 4z + 16)$ **10.** $(a - 5)(a^2 + 5a + 25)$

11. $(4 - 3z^3 + z^2 - 5z)(4 - z)$ **12.** $(3 + 2v^2 - v^3 + 4v)(2 - v)$

13. $(2x^2 - x + 4)(x^2 - 5x - 3)$ **14.** $(6a^2 + 2a - 3)(2a^2 - 3a + 5)$

15. $(7 - 2y + 3y^2)(8y + 4y^2 - 3)$ **16.** $(x - 4 + 6x^2)(9 - 3x + x^2)$

17. $(5x - 2)^2$

18. $(2x - 5)^2$

19. $(x + y)^2(x - y)^2$

20. $(x - 2)^2(x + 2)^2$

21. $(x + 2)^3$

22. $(x + 3)^3$

23. $(x^2 + 2x - 3)^2$

24. $(y^2 - 4y - 5)^2$

25. $(x + 3)^4$

26. $(x + 2)^4$

Set II   Simplify.

1. $(x - 1)(5x^2 + x - 1)$

2. $(x + 1)(2x^2 + x + 1)$

3. $(3xy^2 - 5x^2y + 4)(-2x^2y)$

4. $(a^3 - 3a^2b + 3ab^2 - b^3)(-5a^2b)$

5. $(x^2 + 2x + 1)(x^2 + 2x + 1)$

6. $(x^2 - 2x + 1)(x^2 - 2x - 1)$

7. $(3x)(x^2 - 3x + 9)$

8. $(-2x)(4 + 2x + x^2)$

9. $(3 + x)(x^2 - 3x + 9)$

10. $(2 - x)(4 + 2x + x^2)$

11. $(2 - 2x^3 + x^2 - 3x)(3 - x)$

12. $(x + 3x^3 + 4)(5 - x)$

13. $(7 + 2x^3 + 3x)(4 - x)$

14. $(4 + a^4 + 3a^2 - 2a)(a + 3)$

15. $(5y^2 - 2y - 6)(2y^2 - 4y + 3)$

16. $(2 + 3x^2 - 4x)(6x - 7 + 2x^2)$

17. $(3x + 4)^2$

18. $(x^2 - 2x + 1)(x^2 - 2x + 1)$

19. $(x - 1)^2(x + 1)^2$

20. $(y + 2)^2(y - 2)^2$

21. $(x + 1)^3$

22. $(2w - 3)^3$

23. $(x^2 + 4x + 4)^2$

24. $(z^2 - 3z - 4)^2$

25. $(x + 1)^4$

26. $(x - 1)^4$

6.4 Special Products

Some products are especially important and have special formulas. You *must* learn these formulas and how to use them so you will be able to do the factoring problems that are so important in all of higher mathematics.

6.4A The Product of the Sum and Difference of Two Terms

Because $a + b$ is the *sum of two terms* and $a - b$ is the *difference of two terms*, we call the product $(a + b)(a - b)$ the *product of the sum and difference of two terms*.

RULE 6.1

$$(a + b)(a - b) = a^2 - b^2$$

Proof:

$$(a + b)(a - b) = (a + b)a + (a + b)(-b) \qquad \text{Distributive property}$$

$$= a^2 + ba - ab - b^2$$

$$= a^2 - b^2$$

Rule 6.1 can be stated in words as follows:

> The product of the sum and difference of two terms is equal to the square of the first term minus the square of the second term.

When we use Rule 6.1 to find a product, we say that we are finding the product by inspection.

Example 1   Find the products by inspection.

   a. $(x + 2)(x - 2) = (x)^2 - (2)^2 = x^2 - 4$

   b. $(2x + 3y)(2x - 3y) = (2x)^2 - (3y)^2 = 4x^2 - 9y^2$

   c. $(10x^2 - 7y^3)(10x^2 + 7y^3) = (10x^2)^2 - (7y^3)^2 = 100x^4 - 49y^6$

   d. $(5a^3b^2 + 6cd^4)(5a^3b^2 - 6cd^4) = (5a^3b^2)^2 - (6cd^4)^2 = 25a^6b^4 - 36c^2d^8$ ■

EXERCISES 6.4A

Set I   Find the products by inspection.

   1. $(x + 3)(x - 3)$ **2.** $(z + 4)(z - 4)$

   3. $(w - 6)(w + 6)$ **4.** $(y - 5)(y + 5)$

   5. $(5a + 4)(5a - 4)$ **6.** $(6a - 5)(6a + 5)$

   7. $(2u + 5v)(2u - 5v)$ **8.** $(3m - 7n)(3m + 7n)$

   9. $(4b - 9c)(4b + 9c)$ **10.** $(7a - 8b)(7a + 8b)$

   11. $(2x^2 - 9)(2x^2 + 9)$ **12.** $(10y^2 - 3)(10y^2 + 3)$

   13. $(1 + 8z^3)(1 - 8z^3)$ **14.** $(9v^4 - 1)(9v^4 + 1)$

   15. $(5xy + z)(5xy - z)$ **16.** $(10ab + c)(10ab - c)$

   17. $(7mn + 2rs)(7mn - 2rs)$ **18.** $(8hk + 5ef)(8hk - 5ef)$

Set II   Find the products by inspection.

   1. $(h - 7)(h + 7)$ **2.** $(u + 9)(u - 9)$

   3. $(3m + 5)(3m - 5)$ **4.** $(2x - 3y)(2x + 3y)$

   5. $(6a - 7b)(6a + 7b)$ **6.** $(3x + 7)(3x - 7)$

   7. $(8x - 2y)(8x + 2y)$ **8.** $(5x - 12y)(5x + 12y)$

   9. $(7a - 9b)(7a + 9b)$ **10.** $(13c + d)(13c - d)$

   11. $(4h^2 - 5)(4h^2 + 5)$ **12.** $(8x^2 + 1)(8x^2 - 1)$

   13. $(1 + 9k^3)(1 - 9k^3)$ **14.** $(3x^2y - 8xy^2)(3x^2y + 8xy^2)$

   15. $(4w + 3xy)(4w - 3xy)$ **16.** $(9ab - 2c)(9ab + 2c)$

   17. $(5uv - 8ef)(5uv + 8ef)$ **18.** $(9st + 4z^2)(9st - 4z^2)$

6.4B The Square of a Binomial

Rules 6.2 and 6.3 help us quickly square a binomial.

RULE 6.2   THE SQUARE OF A BINOMIAL SUM

$$(a + b)^2 = a^2 + 2ab + b^2$$

RULE 6.3   THE SQUARE OF A BINOMIAL DIFFERENCE

$$(a - b)^2 = a^2 - 2ab + b^2$$

Proof of Rule 6.2:

$$(a + b)^2 = (a + b)(a + b) = a^2 + 2ab + b^2$$

$$a^2 + 2ab + b^2$$

Proof of Rule 6.3:

$$(a - b)^2 = (a - b)(a - b) = a^2 - 2ab + b^2$$

$$a^2 - 2ab + b^2$$

Rules 6.2 and 6.3 can be stated in words as follows:

TO SQUARE A BINOMIAL

1. The *first term* of the result is the square of the first term of the binomial.

2. The *middle term* of the result is twice the product of the two terms of the binomial.

3. The *last term* of the result is the square of the last term of the binomial.

Although the square of a binomial *can* be found by using the methods learned in Section 6.3, you are strongly urged to use Rule 6.2 or 6.3 in solving such problems.

Example 2   Simplify each of the following.

a. $(m + n)^2 = (m)^2 + 2(m)(n) + (n)^2 = m^2 + 2mn + n^2$

b. $(a - 3)^2 = (a)^2 - 2(a)(3) + (3)^2 = a^2 - 6a + 9$

c. $(2x - 5)^2 = (2x)^2 - 2(2x)(5) + (5)^2 = 4x^2 - 20x + 25$ ■

A WORD OF CAUTION   Using the rules for exponents, students remember that $(ab)^2 = a^2b^2$

Here, a and b are *factors*

They try to apply this rule of exponents to the expression $(a + b)^2$. However, $(a + b)^2$ cannot be found simply by squaring a and b.

Here, a and b are *terms*

Correct method	*Incorrect method*
$(a + b)^2 = (a + b)(a + b)$	$(a + b)^2 = a^2 + b^2$
$= a^2 + 2ab + b^2$	

When squaring a binomial, do not forget this middle term ☑

EXERCISES 6.4B

Set I   Simplify.

1. $(x - 1)^2$ **2.** $(x - 5)^2$ **3.** $(x + 3)^2$ **4.** $(x + 4)^2$

5. $(4x - 1)^2$ **6.** $(7x - 1)^2$ **7.** $(12x + 1)^2$ **8.** $(11x + 1)^2$

9. $(2s + 4t)^2$ **10.** $(3u + 7v)^2$ **11.** $(5x - 3y)^2$ **12.** $(4x - 7y)^2$

13. $(3x + 2z)^2$ **14.** $(2x + 7s)^2$

Set II   Simplify.

1. $(x + 1)^2$ **2.** $(x - 12)^2$ **3.** $(x + 15)^2$ **4.** $(x - 9)^2$

5. $(9x - 1)^2$ **6.** $(8x + 1)^2$ **7.** $(15y + 1)^2$ **8.** $(10z - 2)^2$

9. $(6x + 3y)^2$ **10.** $(5x + 12u)^2$ **11.** $(10x - 3t)^2$ **12.** $(12x + 5y)^2$

13. $(11x + 2y)^2$ **14.** $(6x - 5a)^2$

6.5 Division of Polynomials

6.5A Division of a Polynomial by a Monomial

The rule for dividing a polynomial by a monomial is based on a property of fractions you should recall from arithmetic, $\dfrac{a}{b} = a \cdot \dfrac{1}{b}$, and on the distributive property. Consider this example:

$$\frac{4x^3 - 6x^2}{2x} = \frac{1}{2x} \cdot \frac{4x^3 - 6x^2}{1} = \frac{1}{2x}(4x^3 - 6x^2)$$

$$= \left(\frac{1}{2x}\right)(4x^3) + \left(\frac{1}{2x}\right)(-6x^2) \quad \text{By the distributive property}$$

$$= \frac{4x^3}{2x} + \frac{-6x^2}{2x}$$

$$= 2x^2 - 3x$$

The rule for dividing a polynomial by a monomial is as follows:

TO DIVIDE A POLYNOMIAL BY A MONOMIAL

Divide *each* term of the polynomial by the monomial, simplify each term, and then add the resulting quotients.

Example 1 Divide each of the following.

a. $\dfrac{6 + 8}{2} = \dfrac{6}{2} + \dfrac{8}{2} = 3 + 4 = 7$

b. $\dfrac{4x + 2}{2} = \dfrac{4x}{2} + \dfrac{2}{2} = 2x + 1$

c. $\dfrac{9x^3 - 6x^2 + 12x}{3x} = \dfrac{9x^3}{3x} + \dfrac{-6x^2}{3x} + \dfrac{12x}{3x} = 3x^2 - 2x + 4$

d. $\dfrac{4x^4 - 8x^3 + 16x}{-4x} = \dfrac{4x^4}{-4x} + \dfrac{-8x^3}{-4x} + \dfrac{16x}{-4x} = -x^3 + 2x^2 - 4$

e. $\dfrac{15x^4y^2z + 20xy^3z - 10xyz^2}{5xyz} = \dfrac{15x^4y^2z}{5xyz} + \dfrac{20xy^3z}{5xyz} + \dfrac{-10xyz^2}{5xyz}$

$= 3x^3y + 4y^2 - 2z$

f. $\dfrac{4a^2bc^2 - 6ab^2c^2 + 6abc}{-6abc} = \dfrac{4a^2bc^2}{-6abc} + \dfrac{-6ab^2c^2}{-6abc} + \dfrac{6abc}{-6abc}$

$= -\dfrac{2}{3}ac + bc - 1$ ∎

EXERCISES 6.5A

Set I Perform the indicated divisions.

1. $\dfrac{3x + 6}{3}$ 2. $\dfrac{10x + 15}{5}$ 3. $\dfrac{4 + 8x}{4}$ 4. $\dfrac{5 - 10x}{5}$

5. $\dfrac{6x - 8y}{2}$ 6. $\dfrac{5x - 10y}{5}$ 7. $\dfrac{2x^2 + 3x}{x}$ 8. $\dfrac{4y^2 - 3y}{y}$

9. $\dfrac{15x^3 - 5x^2}{5x^2}$ 10. $\dfrac{12y^4 - 6y^2}{6y^2}$ 11. $\dfrac{3a^2b - ab}{ab}$ 12. $\dfrac{5mn^2 - mn}{mn}$

13. $\dfrac{8x^7 + 4x^5 - 12x^3}{4x^2}$ 14. $\dfrac{6y^6 + 18y^4 - 12y^3}{6y^2}$

15. $\dfrac{5x^5 - 4x^3 + 10x^2}{-5x^2}$ 16. $\dfrac{7y^4 - 5y^3 + 14y^2}{-7y^2}$

17. $\dfrac{-15x^2y^2z^2 - 30xyz}{-5xyz}$ 18. $\dfrac{-24a^2b^2c^2 - 16abc}{-8abc}$

19. $\dfrac{13x^4y^2 - 26x^2y^3 + 39x^2y^2}{13x^2y^2}$ 20. $\dfrac{21m^2n^5 - 35m^3n^2 - 14m^2n^2}{7m^2n^2}$

Set II   Perform the indicated divisions.

1. $\dfrac{9x + 12}{3}$ 2. $\dfrac{15y + 20}{5}$ 3. $\dfrac{10 + 20x}{10}$ 4. $\dfrac{8 - 10y}{8}$

5. $\dfrac{5x - 15y}{5}$ 6. $\dfrac{4x - 6}{2}$ 7. $\dfrac{3x^2 - 6x}{3x}$ 8. $\dfrac{8x^3 - 10x}{4x}$

9. $\dfrac{8x^3 - 4x^2}{4x^2}$ 10. $\dfrac{15x^3 - 30x}{15x}$ 11. $\dfrac{6ab^2 - ab}{ab}$ 12. $\dfrac{12xy^2 - 6x^2y}{9xy}$

13. $\dfrac{42w^8 + 14w^4 - 28w^3}{7w^2}$ 14. $\dfrac{15x^7 + 20x^4 - 35x^3}{5x^3}$

15. $\dfrac{8x^4 - 4x^2 + 12x^3}{-8x^2}$ 16. $\dfrac{9y^2 - 3y^3 + 18y^5}{-9y^2}$

17. $\dfrac{-16a^3b^2c^3 - 32abc}{-16abc}$ 18. $\dfrac{-18s^2t^3u^2 - 9stu^2}{-9stu}$

19. $\dfrac{12a^5b^3 - 24a^3b^3 + 48a^2b^2}{12a^2b^2}$ 20. $\dfrac{15x^5y^2 - 25x^2y^3 - 10x^2y^2}{15x^2y^2}$

6.5B Division of a Polynomial by a Polynomial

The method used to divide one polynomial (the dividend) by a polynomial (the divisor) with two or more terms is similar to the method used to divide one whole number by another (using long division) in arithmetic. (An example of long division in arithmetic is given in Appendix B.) The long-division procedure for polynomials will be demonstrated step by step; it can be summarized as follows [Studying hint: As you read the steps in the summary, follow them in Example 2]:

TO DIVIDE ONE POLYNOMIAL BY ANOTHER

1. Arrange the divisor and the dividend in descending powers of one variable. In the *dividend*, leave spaces for any *missing terms*.

2. Find the first term of the quotient by dividing the first term of the dividend by the first term of the divisor.

3. Multiply the *entire* divisor by the first term of the quotient. Place the product under the dividend, lining up like terms.

4. Subtract the product found in step 3 from the dividend, bringing down at least one term. This difference is the remainder. If the degree of the remainder is not less than the degree of the divisor, continue with steps 5–8.

5. Find the next term of the quotient by dividing the first term of the remainder by the first term of the divisor.

6. Multiply the entire divisor by the term found in step 5.

7. Subtract the product found in step 6 from the polynomial above it, bringing down at least one more term.

8. Repeat steps 5 through 7 until the remainder is 0 *or* until the degree of the remainder is less than the degree of the divisor.

9. Check your answer. (Divisor × quotient + remainder = dividend)

When the remainder is not zero, it can be preceded by an R (for Remainder) and written to the right of the quotient. An alternate method is to write the remainder in fraction form; in this case, the numerator of the fraction is the remainder, the denominator of the fraction is the divisor, and the fraction must be *added to* the quotient.

Example 2 Divide $(5x^2 + 6x^3 - 5 - 4x)$ by $(3 + 2x)$.
Solution

Step 1. First arrange the divisor and the dividend in descending powers of x.

$$2x + 3\overline{)6x^3 + 5x^2 - 4x - 5}$$

First term in quotient is $\dfrac{6x^3}{2x} = 3x^2$

$$\begin{array}{r} 3x^2 \\ \hline \end{array}$$
Step 2. $2x + 3)\ \overline{6x^3 + 5x^2 - 4x - 5}$

Multiply

Step 3. $2x + 3\overline{)6x^3 + 5x^2 - 4x - 5}$
$\qquad\qquad\ \ \underline{6x^3 + 9x^2}\qquad$ This is $3x^2(2x + 3)$

Step 4. $2x + 3\overline{)6x^3 + 5x^2 - 4x - 5}$
$\qquad\qquad\ominus\ \underline{6x^3 \ominus 9x^2}$ ← Subtract (change signs and add) and bring down a term
Remainder → $- 4x^2 - 4x$
↑ Degree is 2

The division must be continued, because the degree of the remainder is not less than the degree of the divisor.

Second term in quotient is $\dfrac{-4x^2}{2x} = -2x$

Step 5. $2x + 3\overline{)6x^3 + 5x^2 - 4x - 5}$
$\qquad\qquad\ominus\ \underline{6x^3 \ominus 9x^2}$
$\qquad\qquad\quad \underline{- 4x^2 - 4x}$

Multiply

Step 6. $2x + 3\overline{)6x^3 + 5x^2 - 4x - 5}$
$\qquad\qquad\ominus\ \underline{6x^3 \ominus 9x^2}$
$\qquad\qquad\quad -4x^2 - 4x$
$\qquad\qquad\quad \underline{-4x^2 - 6x}$ ← This is $(-2x)(2x + 3)$

$$\text{Step 7.} \quad 2x + 3\overline{\smash{\big)}6x^3 + 5x^2 - 4x - 5}$$

$$3x^2 - 2x$$

Step 7. 2x + 3)6x³ + 5x² − 4x − 5

$$\underset{\ominus}{}6x^3 \underset{\ominus}{\mp} 9x^2$$

$$-4x^2 - 4x$$

$$\underset{\oplus}{}-4x^2 \underset{\oplus}{-} 6x \longleftarrow \text{Subtract and bring down a term}$$

Remainder ⟶ $2x - 5$

⌐ Degree is 1

Step 8. The degree of the remainder is still not less than the degree of the divisor; therefore, we must repeat steps 5, 6, and 7.

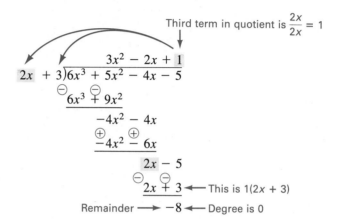

Third term in quotient is $\dfrac{2x}{2x} = 1$

$$3x^2 - 2x + 1$$

$$2x + 3\overline{\smash{\big)}6x^3 + 5x^2 - 4x - 5}$$

$$6x^3 \mp 9x^2$$

$$-4x^2 - 4x$$

$$-4x^2 - 6x$$

$$2x - 5$$

$$2x + 3 \longleftarrow \text{This is } 1(2x + 3)$$

Remainder ⟶ -8 ⟵ Degree is 0

The division is now finished, because the degree of the remainder is less than the degree of the divisor.

Step 9. *Check*

$$(2x + 3)(3x^2 - 2x + 1) + (-8) \overset{?}{=} 6x^3 + 5x^2 - 4x - 5$$

$$(6x^3 + 5x^2 - 4x + 3) + (-8) \overset{?}{=} 6x^3 + 5x^2 - 4x - 5$$

$$6x^3 + 5x^2 - 4x - 5 = 6x^3 + 5x^2 - 4x - 5$$

Answer $3x^2 - 2x + 1 \text{ R} - 8 \text{ } or \text{ } 3x^2 - 2x + 1 + \dfrac{-8}{2x + 3}$

$or \text{ } 3x^2 - 2x + 1 - \dfrac{8}{2x + 3}$ ∎

Example 3 Divide $(-3x + x^2 - 10)$ by $(2 + x)$.
Solution

Step 1. First arrange the divisor and the dividend in descending powers of x.

$$x + 2\overline{\smash{\big)}x^2 - 3x - 10}$$

First term in quotient is $\dfrac{x^2}{x} = x$

$$x$$

Step 2. $x + 2\overline{\smash{\big)}x^2 - 3x - 10}$

Step 3.
$$x + 2 \overline{)x^2 - 3x - 10} \qquad \text{Multiply}$$
$$\underline{x^2 + 2x} \qquad \text{This is } x(x + 2)$$

Step 4.
$$x + 2 \overline{)x^2 - 3x - 10}$$
$$\ominus \quad \ominus$$
$$\underline{x^2 + 2x} \qquad \text{Change signs and add}$$
$$-5x - 10 \qquad \text{Bring down next term}$$

Second term in quotient is $\dfrac{-5x}{x} = -5$

Step 5.
$$x + 2 \overline{)x^2 - 3x - 10}$$
$$\ominus \quad \ominus$$
$$\underline{x^2 + 2x}$$
$$-5x - 10$$

Multiply

Step 6.
$$x + 2 \overline{)x^2 - 3x - 10}$$
$$\ominus \quad \ominus$$
$$\underline{x^2 + 2x}$$
$$-5x - 10$$
$$\underline{-5x - 10} \qquad \text{This is } -5(x + 2)$$

Step 7.
$$x + 2 \overline{)x^2 - 3x - 10}$$
$$\ominus \quad \ominus$$
$$\underline{x^2 + 2x}$$
$$-5x - 10$$
$$\oplus \quad \oplus$$
$$\underline{-5x - 10} \qquad \text{Change signs and add}$$
$$0$$

Step 9. *Check*

$$(x + 2)(x - 5) + 0 \overset{?}{=} x^2 - 3x - 10$$
$$x^2 - 3x - 10 = x^2 - 3x - 10 \quad \blacksquare$$

NOTE   Remember that when the final remainder is zero, we can say that the divisor is a *factor* of the dividend and that the quotient is a *factor* of the dividend. In Example 3, $(x + 2)$ and $(x - 5)$ are *factors* of $x^2 - 3x - 10$.   ☑

Example 4   Solve $(6x^2 + x - 10) \div (2x + 3)$.
Solution

$$2x + 3 \overline{)6x^2 + x - 10} \qquad 3x - 4 \text{ R } 2 \text{ or } 3x - 4 + \dfrac{2}{2x + 3}$$
$$\ominus \quad \ominus$$
$$\underline{6x^2 + 9x}$$
$$-8x - 10$$
$$\oplus \quad \oplus$$
$$\underline{-8x - 12}$$
$$2 \qquad \text{Remainder}$$

Check

$$(2x + 3)(3x - 4) + 2 \overset{?}{=} 6x^2 + x - 10$$

$$(6x^2 + x - 12) + 2 \overset{?}{=} 6x^2 + x - 10$$

$$6x^2 + x - 10 = 6x^2 + x - 10 \quad \blacksquare$$

Example 5   Solve $(27x - 19x^2 + 6x^3 + 10) \div (5 - 3x)$.

Solution

$$
\begin{array}{r}
-2x^2 + 3x - 4 \quad \text{R } 30 \quad or \quad -2x^2 + 3x - 4 + \dfrac{30}{-3x + 5} \\
-3x + 5 \overline{)6x^3 - 19x^2 + 27x + 10} \\
\ominus \;6x^3 \overset{\oplus}{-} 10x^2 \\
\hline
-9x^2 + 27x \\
\overset{\oplus}{-} 9x^2 \overset{\ominus}{+} 15x \\
\hline
12x + 10 \\
\ominus \;12x \overset{\oplus}{-} 20 \\
\hline
30
\end{array}
$$

Check

$$(-3x + 5)(-2x^2 + 3x - 4) + 30 \overset{?}{=} 6x^3 - 19x^2 + 27x + 10$$

$$(6x^3 - 19x^2 + 27x - 20) + 30 \overset{?}{=} 6x^3 - 19x^2 + 27x + 10$$

$$6x^3 - 19x^2 + 27x + 10 = 6x^3 - 19x^2 + 27x + 10 \quad \blacksquare$$

In Example 6, there are two variables. Because the coefficient of a^3 is positive and the coefficient of b^3 is negative, we will arrange the divisor and dividend in descending powers of a before beginning the division. We must continue the division until the degree *in a* of the remainder is less than the degree *in a* of the divisor.

Example 6   Solve $(17ab^2 + 12a^3 - 10b^3 - 11a^2b) \div (3a - 2b)$.

Solution

$$
\begin{array}{r}
4a^2 - ab + 5b^2 \\
3a - 2b \overline{)12a^3 - 11a^2b + 17ab^2 - 10b^3} \\
\ominus \;12a^3 \overset{\oplus}{-} 8a^2b \\
\hline
-3a^2b + 17ab^2 \\
\overset{\oplus}{-} 3a^2b \overset{\ominus}{+} 2ab^2 \\
\hline
15ab^2 - 10b^3 \\
\ominus \;15ab^2 \overset{\oplus}{-} 10b^3 \\
\hline
0
\end{array}
$$

The checking is left to the student. $\blacksquare$

In Example 7, the dividend has no x^2-term and no x-term. Therefore, we write $0x^2$ and $0x$ as placeholders when we prepare for the division.

Example 7   Solve $(x^3 - 1) \div (x - 1)$.
Solution

$$
\begin{array}{r}
x^2 + x + 1 \\
x - 1{\overline{\smash{\big)}\,x^3 + 0x^2 + 0x - 1}} \\
\end{array}
$$

It is helpful to leave space for missing terms by using zeros in this way

$$
\ominus\ x^3 \oplus\ x^2 \\
\underline{} \\
x^2 + 0x \\
\ominus\ x^2 \oplus\ x \\
\underline{} \\
x - 1 \\
\ominus\ x \oplus\ 1 \\
\underline{} \\
0
$$

The checking is left to the student. ∎

In Example 8, the divisor has more than two terms; we still use the procedure outlined in the summary for long division.

Example 8   Solve $(2x^4 + x^3 - 8x^2 - 5x - 2) \div (x^2 - x - 2)$.
Solution

$$
\begin{array}{r}
2x^2 + 3x - 1 \quad R -4 \\
x^2 - x - 2{\overline{\smash{\big)}\,2x^4 + x^3 - 8x^2 - 5x - 2}} \\
\end{array}
$$

$$
\ominus\ 2x^4 \oplus\ 2x^3 \oplus\ 4x^2 \\
\underline{} \\
3x^3 - 4x^2 - 5x \\
\ominus\ 3x^3 \oplus\ 3x^2 \oplus\ 6x \\
\underline{} \\
-x^2 + x - 2 \\
\oplus\ x^2 \ominus\ x \ominus\ 2 \\
\underline{} \\
-4
$$

The checking is left to the student. ∎

EXERCISES 6.5B

Set I   Perform the indicated divisions.

1. $(x^2 + 5x + 6) \div (x + 2)$ 2. $(x^2 + 5x + 6) \div (x + 3)$

3. $(x^2 - x - 12) \div (x - 4)$ 4. $(x^2 - x - 12) \div (x + 3)$

5. $(6x^2 + 5x - 6) \div (3x - 2)$ 6. $(20x^2 + 13x - 15) \div (5x - 3)$

7. $(15v^2 + 19v + 10) \div (5v - 7$ 8. $(15v^2 + 19v - 4) \div (3v + 8)$

9. $(x + 6x^2 - 15) \div (2x - 3)$ 10. $(10 - 26x + 12x^2) \div (3x - 5)$

11. $(8x - 4x^3 + 10) \div (2 - x)$ 12. $(12x - 15 - x^3) \div (3 - x)$

13. $(6a^2 + 5ab + b^2) \div (2a + 3b)$ 14. $(6a^2 + 5ab - b^2) \div (3a - 2b)$

15. $(a^3 - 8) \div (a - 2)$ 16. $(c^3 - 27) \div (c - 3)$

17. $(x^3 - 8x - 15) \div (x - 4)$ 18. $(2x^3 + x^2 + 9) \div (x + 2)$

19. $(x^4 + 2x^3 - x^2 - 2x + 4) \div (x^2 + x - 1)$

20. $(x^4 - 2x^3 + 3x^2 - 2x + 7) \div (x^2 - x + 1)$

21. $(x^4 + 3x^3 + 6x^2 + 5x - 5) \div (x^2 + 2x + 3)$

22. $(x^4 + 3x^3 + 6x^2 + 5x + 5) \div (x^2 + x + 1)$

Set II   Perform the indicated divisions.

1. $(x^2 + 9x + 14) \div (x + 2)$ **2.** $(x^2 - 11x + 24) \div (x - 3)$

3. $(x^2 + x - 12) \div (x - 3)$ **4.** $(x^2 + x - 12) \div (x + 4)$

5. $(6m^2 - m - 30) \div (2m - 5)$ **6.** $(2x^2 - 14x - 16) \div (x - 8)$

7. $(6m^2 - m + 30) \div (3m + 7)$ **8.** $(6x^2 + 13x - 8) \div (2x + 5)$

9. $(13x + 3x^2 - 10) \div (3x - 2)$ **10.** $(7x + x^2 - 8) \div (x + 8)$

11. $(7x - 3x^3 + 5) \div (2 - x)$ **12.** $(10x^2 - 3x^3 + 100) \div (5 - x)$

13. $(8a^2 - 2ab - b^2) \div (2a - b)$ **14.** $(8x - 4x^3 - 3) \div (2 - x)$

15. $(x^3 + 1) \div (x + 1)$ **16.** $(x^3 + 8) \div (x + 2)$

17. $(x^3 - 5x + 1) \div (x - 3)$ **18.** $(x^3 + x^2 + 3) \div (x + 2)$

19. $(x^4 - x^3 - 5x^2 + 3x + 6) \div (x^2 - 2x - 1)$

20. $(x^4 + 2x^3 + 5x^2 + 4x + 4) \div (x^2 + x + 2)$

21. $(x^4 + 2x^3 + 5x^2 + 4x + 1) \div (x^2 + x + 2)$

22. $(x^4 + 2x^3 + x^2 - 4) \div (x^2 + x - 2)$

6.6 Word Expressions and Polynomials

In this section, you are asked to translate a word expression into a polynomial and then to express that polynomial in simplest form.

Example 1   Write all answers in simplest form.

a. Write the polynomial that represents the area of a rectangle if the length of the rectangle is 7 more than the width, and the width is x.
Solution If the width is x, the length is $x + 7$. Therefore, the area is $x(x + 7)$, or, in simplest form, $x^2 + 7x$.

b. Write the polynomial that represents the product of three consecutive odd integers, if the smallest integer is x.
Solution If the smallest integer is x, the second odd integer must be $x + 2$, and the third must be $x + 4$. Therefore, the product is $x(x + 2)(x + 4)$, and

$$x(x + 2)(x + 4) = x(x^2 + 6x + 8) = x^3 + 6x^2 + 8x \quad \blacksquare$$

EXERCISES 6.6

Set I   In Exercises 1–4, translate each word expression into a polynomial and then express that polynomial in simplest form.

$x(x+2) = x^2 + 2x$

1. The product of two consecutive even integers, where the smaller integer is x

2. The sum of three consecutive odd integers, where the smallest integer is x

$V = L\,wh \qquad V = 4x(x)(x+3) = 4x^2(x+3) = 4x^3 + 12x^2$

3. The volume of a rectangular box, if the width is x, the length is 4 times the width, and the height is 3 more than the width

4. The volume of a rectangular box, if the height is x, the width is 4 more than the height, and the length is 4 less than 3 times the height

5. A photograph that is 2 in. longer than it is wide is to be surrounded by a mat. The width of the mat is to be 2 in. on the two sides and 3 in. at the top and bottom. Write a polynomial that represents the area of the *mat*, if the width of the photograph is x. $A = bh$

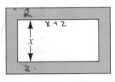

$A = (x+4)(x+8) - [x(x+2)]$
$= x^2 + 8x + 4x + 32 - x^2 - 2x$
$A = 10x - 32$

6. A photograph that is 3 in. wider than it is high is to be surrounded by a mat. The width of the mat is to be 2 in. on all four sides. Write a polynomial that represents the area of the *mat*, if the height of the photograph is x.

7. A piece of cardboard 10 in. square is to have square pieces cut out of each corner, as shown below, so that it can be made into a box, as shown. If the size of each square that will be cut out is x in. by x in., write a polynomial that represents the volume of the box.

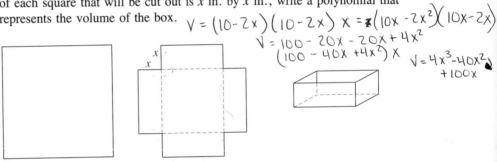

$V = (10-2x)(10-2x)x = x(10x - 2x^2)(10x - 2x)$
$V = 100 - 20x - 20x + 4x^2$
$(100 - 40x + 4x^2)x \quad V = 4x^3 - 40x^2 + 100x$

8. A piece of cardboard 12 in. wide and 14 in. long is to have square pieces cut out of each corner so that it can be made into a box, as shown for Exercise 7. If the size of each square that will be cut out is x in. by x in., write a polynomial that represents the volume of the box.

Set II   In Exercises 1–4, translate each word expression into a polynomial and then express that polynomial in simplest form.

1. The sum of four consecutive even integers, where the smallest integer is x

2. The product of three consecutive integers, where the smallest integer is x

3. The volume of a rectangular box, if the height is x, the length is 3 more than twice the height, and the width is 5 less than the height

4. The volume of a rectangular box, if the width is x, the length is 2 less than 3 times the width, and the height is 3 more than the width

5. A picture that is 4 in. longer than it is wide is to be surrounded by a mat. The width of the mat is to be 3 in. on the two sides and 4 in. at the top and bottom. Write a polynomial that represents the area of the *mat*, if the width of the picture is x.

6. A rectangular garden that is 3 times as long as it is wide is to be surrounded by a brick path. The path is to be 3 ft wide. Write a polynomial that represents the area of the *walk*, if the width of the garden is x.

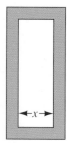

7. A piece of cardboard 18 in. wide and 24 in. long is to have square pieces cut out of each corner so that it can be made into a box, as shown for Exercise 7 of Set I. If the size of each square that will be cut out is x in. by x in., write a polynomial that represents the volume of the box.

8. The radius of a right circular cone is twice the height. Write a polynomial that represents the volume of the cone, if the height of the cone is x.

6.7 Review: 6.1–6.6

Polynomials
6.1

A **polynomial in x** is an algebraic expression that has only terms of the form ax^n, where a is any real number and n is a whole number. A polynomial cannot have negative exponents on the variables and cannot have variables in a denominator or under a radical sign.

A **monomial** is a polynomial with one term.

A **binomial** is a polynomial with two unlike terms.

A **trinomial** is a polynomial with three unlike terms.

Degree of a Polynomial
6.1

The **degree of a term** of a polynomial is the sum of the exponents of its variables. The **degree of a polynomial** is the same as that of its highest-degree term.

Operations on Polynomials
6.2

To add polynomials, add like terms.

To subtract polynomials, change the subtraction symbol to an addition symbol *and* change the sign of each term in the polynomial being subtracted; then combine any like terms.

6.3A *To multiply a polynomial by a monomial*, use the distributive property.

6.3B *To multiply two binomials*, the FOIL method may be used.

6.3C *To multiply a polynomial by a polynomial*, multiply the first polynomial by each term of the second polynomial, then add the results.

Special products:

6.4A $(a + b)(a - b) = a^2 - b^2$ Sum and difference of two terms

6.4B $(a + b)^2 = a^2 + 2ab + b^2$
$(a - b)^2 = a^2 - 2ab + b^2$ Square of a binomial

6.5A *To divide a polynomial by a monomial*, divide each term in the polynomial by the monomial, then add the results.

6.5B *To divide a polynomial by a polynomial*, see Section 6.5B.

6.6 Word expressions often translate into polynomials.

Review Exercises 6.7 Set I

In Exercises 1–4, if the expression is a polynomial, (a) find the degree of the polynomial and (b) find the degree of the first term. If the expression is not a polynomial, write "Not a polynomial."

1. $5xy^2 + 3x$ **2.** $5u^2 + \dfrac{3}{u + v}$ **3.** $\sqrt{16 + x^2}$ **4.** $4u - 2u^2v^2$

In Exercises 5 and 6, (a) write each polynomial in descending powers of the indicated variable and (b) find the leading coefficient.

5. $3x^4 - 6 + 7x^2 + x$ Powers of x

6. $x^3 + 3x^2y + y^4 + 3xy^2$ Powers of y

In Exercises 7–28, perform the indicated operations and simplify the results.

7. $(5x^2y + 3xy^2 - 4y^3) + (2xy^2 + 4y^3 + 3x^2y)$

8. $(9xy^2)(x^2y)(-2xy)$ **9.** $(9xy^2)(x^2y - 2xy)$

10. $(8a + 5a^2b - 3 + 6ab^2) - (5 + 4a - 3ab^2 + 2b^2)$

11. $\dfrac{8x + 2}{2}$ **12.** $(5x^2y - 3y)(-4xy^2)$

13. $(x - 5)(x + 7)$ **14.** $(x - 2)(x^2 + 2x + 4)$

15. $(6x^2 - 9x + 10) \div (2x - 3)$ **16.** $(x - 8)^2$

17. Subtract $(4x^3 - x + 4x^2 - 1)$ from $(x^2 + 3 - 5x)$.

18. $(2a^4 - a^3 + a^2 + a - 3) \div (2a^2 - a + 3)$

19. $(2m^2 - 5) - [(7 - m^2) - (4m^2 - 3)]$

20. Subtract $(5x^2 - 9x + 6)$ from the sum of $(2x - 8 - 7x^2)$ and $(12 - 2x^2 + 11x)$.

21. $3xy^2(4x^2y^2 - 5xy - 10)$ **22.** $(x - 2)^3$

23. $(3a - 5)(3a + 5)$ **24.** $(3 + x + y)^2$

25. $(4 - 9y^2) \div (3y + 2)$ **26.** $\dfrac{-15a^2b^3 + 4ab^2 - 10ab}{-5ab}$

27. $(3xy + z)(-z)$ **28.** $(3xy + z) - z$

In Exercises 29 and 30, subtract the lower polynomial from the upper one.

29. $7a^2 - 3ab + 5 - b^2$ **30.** $\quad 3x^3 - 5x^2 \quad\ + 2$
$\ \underline{9a^2 + 2ab - 8 + b^2}$ $\ \underline{-3x^3 + 2x^2 - x + 6}$

In Exercises 31 and 32, translate each word expression into a polynomial and then express that polynomial in simplest form.

31. The product of three consecutive even integers, where the smallest integer is x

32. The volume of a right circular cylinder, where the radius is 3 times the height

Review Exercises 6.7 Set II

NAME _____

In Exercises 1–4, if the expression is a polynomial, (a) find the degree of the polynomial and (b) find the degree of the first term. If the expression is not a polynomial, write "Not a polynomial."

1. $\sqrt{x^2 + 9}$

2. $8x^3y^2 - 5x^4$

3. $2 + 3x^4y$

4. $\dfrac{1}{x^2 + 2}$

In Exercises 5 and 6, (a) write each polynomial in descending powers of the indicated variable and (b) find the leading coefficient.

5. $5 - 3y^2 + y$ Powers of y

6. $5 - a^2b^3 + a^4b^2 + ab$ Powers of a

In Exercises 7–28, perform the indicated operations and simplify the results.

7. $(13x - 6x^3 + 14 - 15x^2) + (-17 - 23x^2 + 4x^3 + 11x)$

8. $(6x^3y - 4x^2y^2 + xy + 5) - (-4x^2y^2 + 3xy + 7 + 6x^3y)$

9. $5x^2y(3xy^3 + 4x - 2z)$

10. $(z + 3)(z^2 - 3z + 9)$

11. $(2xy^3 - 3x^2y - 4) - 5x^2y$

12. $(2xy^3 - 3x^2y - 4)(-5x^2y)$

ANSWERS

1a. _____

b. _____

2a. _____

b. _____

3a. _____

b. _____

4a. _____

b. _____

5a. _____

b. _____

6a. _____

b. _____

7. _____

8. _____

9. _____

10. _____

11. _____

12. _____

13. $(c - 2)(5c + 2)$

14. $(20a^2 - 7a + 5) \div (4a - 3)$

15. $\dfrac{5mn^2 - 10m^2n^3}{5mn^2}$

16. $(2 - x + y)^2$

17. Subtract $(-2x + 7x^3 - x^2 + 4)$ from $(x^3 + 5x - 2 + 4x^2)$.

18. Subtract $(3x^2 - 4x + 8)$ from the sum of $(3x + 5x^2 - 2)$ and $(7 - 3x - 4x^2)$.

19. $(3x^4 - 2x^3 + 2x^2 + 2x - 5) \div (3x^2 - 2x + 5)$

20. $(6b + 5)(6b - 5)$

21. $(6b - 5)^2$

22. $\dfrac{-12x^2y^2 + 4xy^3 - 3xy^2}{-3xy^2}$

13. _____

14. _____

15. _____

16. _____

17. _____

18. _____

19. _____

20. _____

21. _____

22. _____

23. $(3x^2 - 4) - [(2 - x^2) - (5x^2 + 1)]$

24. $(4x^2 - 1) \div (2x + 1)$

25. $(7xy^2)(3x)(-2y)$

26. $(7xy^2)(3x - 2y)$

27. $(15x^2 - 29xy - 14y^2) \div (3x - 7y)$

23. _____

24. _____

25. _____

26. _____

27. _____

28. _____

29. _____

30. _____

31. _____

32. _____

28. $(8x + y)(3x + 2)$

In Exercises 29 and 30, subtract the lower polynomial from the upper one.

29. $2x^3 - 4x^2 + x - 3$
    $\underline{5x^3 + 2x^2 + x - 5}$

30. $4xy^2 + 2x^2y - 3x + 4$
    $\underline{-2xy^2 - 2x^2y - 5x + 9}$

In Exercises 31 and 32, translate each word expression into a polynomial and then express that polynomial in simplest form.

31. The sum of three consecutive odd integers, where the smallest integer is x

32. The area of a parallelogram, where the base is twice the height

Chapter 6 Diagnostic Test

The purpose of this test is to see how well you understand operations with polynomials. We recommend that you work this diagnostic test *before* your instructor tests you on this chapter. Allow yourself about 50 minutes.

Complete solutions for all the problems on this test, together with section references, are given in the answer section in the back of the book. For the problems you do incorrectly, study the sections referred to.

1. In the polynomial $x^2 - 4xy^2 + 5$, find:

   a. the degree of the first term   b. the degree of the polynomial

   c. the numerical coefficient of the second term   d. the leading coefficient

   e. the degree of the third term

2. Add the polynomials.

   a.
$$\begin{array}{r} -7x^3 + 4x^2 \qquad\ + 3 \\ 3x^3 \qquad\quad\ + 6x - 5 \\ 7x^2 - 4x + 8 \\ \hline \end{array}$$

   b. $(6xy^2 - 5xy) + (17xy - 7x^2y) + (3xy^2 - y^3)$

3. a. Subtract $(8 - 2x + 5x^2)$ from $(-3x^2 - 6x + 9)$.

   b. Subtract the lower polynomial from the upper one.

$$\begin{array}{r} -6a^3 + 5a^2 \qquad\ + 4 \\ 4a^3 \qquad\quad + 6a - 7 \\ \hline \end{array}$$

In Problems 4–7, simplify each expression.

4. a. $(8x - 3) - (10x - 5) + (9 - 5x)$ **5.** a. $(9x - 7)(8x + 9)$

   b. $-6xy(3x^2 - 5xy^2 + 8y)$ b. $(3x - 8)^2$

6. $(x^2 + 3x - 5)(2x^2 - x - 4)$ **7.** $(4abc^2)(3b)(-2a^2c)$

In Problems 8 and 9, perform the divisions.

8. a. $\dfrac{8x^4 - 4x^3 + 12x^2}{4x^2}$ **9.** $(12 - 6x^2 + x^3) \div (x - 4)$

   b. $(15x^2 + x + 1) \div (3x - 1)$

10. Translate the following word expression into a polynomial (*you* state what the variable represents), and then express that polynomial in simplest form.

A piece of cardboard that is 5 in. longer than it is wide is to have square pieces cut out of each corner, as shown below, so that it can be made into a box, as shown. If the size of each square that will be cut out is 2 in. by 2 in., write a polynomial that represents the volume of the box.

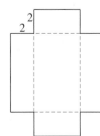
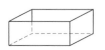

Cumulative Review Exercises: Chapters 1–6

1. Using the formula $F = \frac{9}{5}C + 32$, find the value of F when $C = -20$.

In Exercises 2 and 3, evaluate each expression.

2. $16 \div 2 \cdot 4 - 9\sqrt{25}$

3. $5[11 - 2(6 - 9)] - 4(13 - 5)$

4. Simplify: $2x(3x^2 + 6x + 8) - 4(3x^2 + 5x - 6)$

5. Evaluate the expression $-5^2 \cdot 4 - 15 \div 3\sqrt{25}$.

6. Solve the equation $5(x - 6) + 2(3 - 4x) = 6 - (2x + 8)$.

7. Find the prime factorization of 294.

In Exercises 8–11, perform the indicated operations.

8. $(3x^2 - 2x + 3)(2x - 5)$

9. $(9x - 2)(9x + 2)$

10. $(5x - 3)^2$

11. $(5x^3 - 2x^2 + 4x - 1) \div (x - 1)$

In Exercises 12–16, simplify each expression and write your answers using only positive exponents.

12. $\dfrac{10^2 \cdot 10^0}{10^{-3}}$

13. $\dfrac{x^{3c}}{x^c}$

14. $(2a^2b^{-1})^3$

15. $\left(\dfrac{6y^{-1}}{3y^3}\right)^2$

16. $\left(\dfrac{12x^3}{4x^5}\right)^{-2}$

In Exercises 17–20, set up each problem algebraically and solve. Be sure to state what your variables represent.

17. Susan worked twenty-two problems correctly on a math test that had twenty-five problems. Find her percent score.

18. A business pays \$125 for an item. What is the selling price of the item if it is then marked up 40%?

19. A 10-lb mixture of walnuts and almonds is to be worth \$3.52 per pound. If walnuts cost \$4.50 per pound and almonds cost \$3.10 per pound, how many pounds of each should be used?

20. The three sides of a triangle are in the ratio $3:5:7$. The perimeter is 75. Find the lengths of the three sides.

Critical Thinking

Each of the following problems has an error. Can you find it?

1. Simplify $2^3 \cdot 3^2$.

   $2^3 \cdot 3^2 = 6^5$

2. Simplify -5^{-2}.

   $-5^{-2} = \frac{1}{25}$

3. Simplify $(2x^2y^4)^3$.

   $(2x^2y^4)^3 = 2x^6y^{12}$

4. Simplify $(x + 4)^2$

   $(x + 4)^2 = x^2 + 16$

5. Simplify $2x(3x^2 - 4x) + 5x^2(x - 2)$.

   $2x(3x^2 - 4x) + 5x^2(x - 2)$

   $= 6x^3 - 8x^2 + 5x^3 - 10x^2$

   $= 11x^6 - 18x^4$

6. Divide, using long division:
   $(x^3 - 7x + 12) \div (x - 3)$

$$
\begin{array}{r}
x^2 - 4 \\
x - 3\overline{)x^3 - 7x + 12} \\
\ominus\ x^3 \oplus\ 3x \\
\hline
-4x + 12 \\
\oplus\ 4x \ominus\ 12 \\
\hline
0
\end{array}
$$

7. The sum of 4 times a number and 12 is 6 less than twice the number. Find the number.
   Let $x = $ the number.

   $4x + 12 = 6 - 2x$

   $6x + 12 = 6$

   $6x = -6$

   $x = -1$

7 Factoring

This chapter deals with factoring polynomials, solving equations that can be solved by factoring, and solving word problems that lead to such equations.

7.1 Greatest Common Factor (GCF)

Factoring a sum of terms means rewriting it, if possible, *as a single term* that is a *product of prime factors*. It is essential that the techniques of factoring be mastered, because factoring is used a great deal in the remainder of this course and in higher-level mathematics courses.

7.1A Factoring Expressions with a Common Monomial Factor

Finding the Greatest Common Factor (GCF)

In Section 1.12, we discussed factoring an integer. The **greatest common factor (GCF)** of two integers is the largest integer that is a factor of *both* integers.

Example 1   Find the GCF of 12 and 16.
Solution

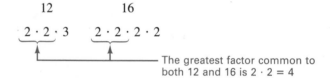

Therefore, the GCF of 12 and 16 is 4. ∎

The *greatest common factor (GCF) of a polynomial* is the largest term that is a factor of all the terms in the polynomial. We find the GCF as follows:

TO FIND THE GREATEST COMMON FACTOR (GCF)

1. Write each numerical coefficient in prime factored form. Repeated factors must be expressed in exponential form.

2. Write down each different base, numerical or variable, that is common to all terms.

3. Raise each of the bases in step 2 to the *lowest* power to which it occurs in any of the terms.

4. The *greatest common factor* (GCF) is the *product* of all the factors found in step 3. It may be positive or negative.

Example 2   Find the GCF for $15x^3 + 9x$.
Solution

$$15x^3 + 9x = 3^1 \cdot 5^1 \cdot x^3 + 3^2 \cdot x^1 \quad \longleftarrow \text{ Each numerical coefficient is in prime factored form}$$

The bases common to both terms are 3 and x. The lowest power on the 3 is a 1, and the lowest power on the x is a 1. Therefore, the GCF is $3x$ or $-3x$. ∎

Factoring Polynomials That Have a Common Factor

In Section 2.2, we used the distributive property to rewrite a product of factors as a sum of terms; that is, $a(b + c) = ab + ac$. In this section, we use the distributive property to rewrite a sum of terms as a product of factors (that is, *as a single term*) whenever possible. It is not always possible to do this, since a sum of terms is not always factorable. When we use the distributive property to get

$$ab + ac = a(b + c)$$

we say we are factoring out the greatest common factor. Notice that the right side of the equation *has only one term*.

We now consider how to factor a polynomial that contains a factor common to all the terms of the polynomial.

TO FACTOR OUT THE GCF

1. Combine like terms, if there are any.

2. Find the GCF for all the terms. It will often, but not always, be a monomial.

3. Find the *polynomial factor** by dividing each term of the polynomial being factored by the GCF. The polynomial factor will always have as many terms as the expression in step 1. It should have only *integer* coefficients.

4. Rewrite the expression as the product of the factors found in steps 2 and 3.

5. Check the result by using the distributive property to remove the parentheses; you should get back the polynomial from step 1.

Example 3   Factor $15x^3 + 9x$.

Solution We found in Example 2 that the GCF is $3x$. To find the polynomial factor, divide $15x^3$ by $3x$ and then divide $9x$ by $3x$.

$$15x^3 + 9x$$
$$= 3x(\quad + \quad)$$ The polynomial factor has as many terms as the original expression
$$= 3x(5x^2 + 3)$$

This term is $\dfrac{9x}{3x} = 3$

This term is $\dfrac{15x^3}{3x} = 5x^2$

Therefore, the factors of $15x^3 + 9x$ are $3x$ and $(5x^2 + 3)$.

$$15x^3 + 9x = \;3x\;(\;5x^2 + 3\;)$$

GCF ⟶   ⟵ Polynomial factor

The answer, $3x(5x^2 + 3)$, has *only one term*, and the expression within the parentheses has no common factor left.

Check $3x(5x^2 + 3) = (3x)(5x^2) + (3x)(3) = 15x^3 + 9x$ ∎

---

*We will call this factor the *polynomial factor* because it will be a polynomial and will always have more than one term.

A WORD OF CAUTION In Example 3, students often make one of these two errors:

$$15x^3 + 9x = 3x \cdot 5x^2 + 3$$

$$15x^3 + 9x = (3x) \cdot 5x^2 + 3$$

Neither answer is correct. There *must* be parentheses around $5x^2 + 3$. Note that

$$3x \cdot 5x^2 + 3 = 15x^3 + 3 \neq 15x^3 + 9x$$

$$(3x) \cdot 5x^2 + 3 = 15x^3 + 3 \neq 15x^3 + 9x \qquad \boxed{\checkmark}$$

Example 4   Factor $6x + 4$.
Solution

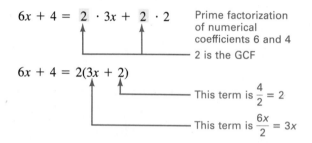

Therefore, the factors of $6x + 4$ are 2 and $(3x + 2)$.

$$6x + 4 = 2 (3x + 2)$$

GCF ⎯⎯⎯⎯⎯⎯   ⎯⎯⎯⎯ Polynomial factor

The answer, $2(3x + 2)$, has *only one term*, and the expression within the parentheses has no common factor left.
Check $2(3x + 2) = 6x + 4$ ∎

Prime (or Irreducible) Polynomials

It is important to realize that most polynomials cannot be factored. A polynomial is said to be **prime** (or **irreducible**) *over the integers* if it *cannot* be expressed as a product of polynomials of lower degree such that all the constants in the new polynomials are *integers*. In this book, when we say to factor a polynomial, we mean to factor *over the integers* (that is, with all the constants being integers).

Example 5   Factor $5x + 2y$.
Solution Although 5 and x are factors of the *first term* of $5x + 2y$, neither one is a factor of the *second term*. In fact, $5x$ and $2y$ have no common integral factors. While it is true that

$$5x + 2y = 5\left(x + \frac{2}{5}y\right)$$

and that

$$5x + 2y = 2\left(\frac{5}{2}x + y\right)$$

the parentheses in both cases contain constants that are *not integers*. Therefore, $5x + 2y$ is not factorable over the integers. It is a prime polynomial.   ∎

The Uniqueness of Factorization

We mentioned in Section 1.12 that a composite number can be expressed in prime factored form in one and only one way, except for the order in which the factors are written. A similar property holds for the factorization of polynomials.

Factorization of polynomials over the integers is *unique*; that is, a polynomial can be completely factored over the integers in one and only one way (except for the order in which the factors are written and the sign of the product).

Example 6 Factor $12x^3 - 16x^5$.
Solution

Coefficients in prime factored form

$$12x^3 - 16x^5 = 2^2 \cdot 3 \, x^3 - 2^4 \, x^5$$

The bases common to both terms are 2 and x. The lowest power that occurs on the 2 is a 2, and the lowest power that occurs on the x is a 3. Therefore, the GCF is $2^2 x^3$ or $4x^3$.

$$\text{The polynomial factor} = \frac{12x^3}{4x^3} - \frac{16x^5}{4x^3} = 3 - 4x^2$$

Therefore,

GCF ────┐   ┌──── Polynomial factor

$$12x^3 - 16x^5 = 4x^3(3 - 4x^2)$$

The answer, $4x^3(3 - 4x^2)$, has *only one term*, and $(3 - 4x^2)$ has no common factor left.
Check

$$4x^3(3 - 4x^2) = (4x^3)(3) - (4x^3)(4x^2) = 12x^3 - 16x^5$$

The factoring could also be done using $-4x^3$ as the GCF:

$$12x^3 - 16x^5 = -4x^3(-3 + 4x^2) = -4x^3(4x^2 - 3)$$

Check

$$-4x^3(-3 + 4x^2) = (-4x^3)(-3) + (-4x^3)(4x^2) = 12x^3 - 16x^5$$

These answers are all acceptable:

$4x^3(3 - 4x^2)$, $(3 - 4x^2)(4x^3)$, $-4x^3(-3 + 4x^2)$, $-4x^3(4x^2 - 3)$, $(-3 + 4x^2)(-4x^3)$, $(4x^2 - 3)(-4x^3)$ ∎

A WORD OF CAUTION   In Example 6, it is incorrect to write

$$12x^3 - 16x^5 = (-3 + 4x^2) - 4x^3$$

or

$$12x^3 - 16x^5 = (4x^2 - 3) - 4x^3$$

Both equal $-3 + 4x^2 - 4x^3$, *not* $12x^3 - 16x^5$.

If an expression has been factored, it has only one term. However, even if an expression has only one term, it still may not be factored completely. Any expression inside parentheses should always be examined carefully to see whether it can be factored further.

Example 7 Factor $2x^4 + 4x^3 - 8x^2$.
Solution

$$2x^4 + 4x^3 - 8x^2 = 2^1x^4 + 2^2x^3 - 2^3x^2 \longleftarrow \text{Coefficients in prime factored form}$$

2^1 is the lowest power of 2 that occurs in any term. x^2 is the lowest power of x that occurs in any term. Therefore, the GCF $= 2^1x^2 = 2x^2$.

$$\text{The polynomial factor} = \frac{2x^4}{2x^2} + \frac{4x^3}{2x^2} - \frac{8x^2}{2x^2} = x^2 + 2x - 4$$

Therefore,

$$\text{GCF} \qquad \text{Polynomial factor}$$
$$2x^4 + 4x^3 - 8x^2 = 2x^2(x^2 + 2x - 4)$$

Check $2x^2(x^2 + 2x - 4) = 2x^4 + 4x^3 - 8x^2$ ∎

A WORD OF CAUTION Suppose we had factored the trinomial in Example 7 as follows:

$$2x^4 + 4x^3 - 8x^2 = 2(x^4 + 2x^3 - 4x^2)$$

This *is* in factored form, since it has only one term, but it is not factored *completely*, because $x^4 + 2x^3 - 4x^2$ can still be factored. Similarly, if the trinomial is factored as

$$2x^4 + 4x^3 - 8x^2 = x^2(2x^2 + 4x - 8)$$

the factoring is not complete, since $2x^2 + 4x - 8$ is not a prime polynomial. ☑

Example 8 Factor $6a^3b^3 - 8a^2b^2 + 10a^3b$.
Solution

$$6a^3b^3 - 8a^2b^2 + 10a^3b = 2 \cdot 3a^3b^3 - 2^3a^2b^2 + 2 \cdot 5a^3b^1$$

$$\text{GCF} = 2^1 \cdot a^2 \cdot b^1 = 2a^2b$$

$$\text{The polynomial factor} = \frac{6a^3b^3}{2a^2b} - \frac{8a^2b^2}{2a^2b} + \frac{10a^3b}{2a^2b} = 3ab^2 - 4b + 5a$$

Therefore, $6a^3b^3 - 8a^2b^2 + 10a^3b = 2a^2b(3ab^2 - 4b + 5a)$.
Check $2a^2b(3ab^2 - 4b + 5a) = 6a^3b^3 - 8a^2b^2 + 10a^3b$ ∎

Example 9 Factor $3xy - 6y^2 - 3y$.
Solution

$$3xy - 6y^2 - 3y = 3xy - 2 \cdot 3y^2 - 3y$$

$$\text{GCF} = 3 \cdot y = 3y$$

$$\text{The polynomial factor} = \frac{3xy}{3y} - \frac{6y^2}{3y} - \frac{3y}{3y} = x - 2y - 1$$

Therefore, $3xy - 6y^2 - 3y = 3y(x - 2y - 1)$.
Check $3y(x - 2y - 1) = 3xy - 6y^2 - 3y$ ∎

Example 10　Factor $-15x^2y^3 - 20xy^4 + 25x^3y^2$.
Solution

$$-15x^2y^3 - 20xy^4 + 25x^3y^2 = -3 \cdot 5x^2y^3 - 2^2 \cdot 5xy^4 + 5^2x^3y^2$$

$$\text{GCF} = 5xy^2 \text{ or } -5xy^2$$

First solution, using $5xy^2$ as the GCF:

$$\text{The polynomial factor} = \frac{-15x^2y^3}{5xy^2} + \frac{-20xy^4}{5xy^2} + \frac{25x^3y^2}{5xy^2} = -3xy - 4y^2 + 5x^2$$

Therefore,

$$-15x^2y^3 - 20xy^4 + 25x^3y^2 = 5xy^2(-3xy - 4y^2 + 5x^2)$$

Second solution, using $-5xy^2$ as the GCF:

$$\text{The polynomial factor} = \frac{-15x^2y^3}{-5xy^2} + \frac{-20xy^4}{-5xy^2} + \frac{25x^3y^2}{-5xy^2} = 3xy + 4y^2 - 5x^2$$

Therefore, $-15x^2y^3 - 20xy^4 + 25x^3y^2 = -5xy^2(3xy + 4y^2 - 5x^2)$. Answers from both the first solution and the second solution are correct.
Checks

$$5xy^2(-3xy - 4y^2 + 5x^2) = -15x^2y^3 - 20xy^4 + 25x^3y^2$$
$$-5xy^2(3xy + 4y^2 - 5x^2) = -15x^2y^3 - 20xy^4 + 25x^3y^2 \quad \blacksquare$$

EXERCISES 7.1A

Set I　Factor each expression completely, or write "Not factorable."

1. $12x + 8$　　2. $6x + 9$　　3. $5a - 8$
4. $7b - 2$　　5. $2x + 8$　　6. $3x + 9$
7. $5a - 10$　　8. $7b - 14$　　9. $6y - 3$
10. $15z - 5$　　11. $9x^2 + 3x$　　12. $8y^2 - 4y$
13. $10a^3 - 25a^2$　　14. $27b^2 - 18b^4$　　15. $21w^2 - 20z^2$
16. $15x^3 - 16y^3$　　17. $2a^2b + 4ab^2$　　18. $3mn^2 + 6m^2n^2$
19. $12c^3d^2 - 18c^2d^3$　　20. $15ab^3 - 45a^2b^4$　　21. $4x^3 - 12x - 24x^2$
22. $18y - 6y^2 - 30y^3$　　23. $4x^4 - 7x^2 + 1$　　24. $3x^2 + 2x - 2$
25. $8x^3 - 6x^2 + 2x$　　26. $9y^4 + 6y^3 - 3y^2$
27. $24a^4 + 8a^2 - 40$　　28. $45b^3 - 15b^4 - 30$
29. $-14x^8y^9 + 42x^5y^4 - 28xy^3$　　30. $-21u^7v^8 - 63uv^5 + 35u^2v^5$
31. $15h^2k - 8hk^2 + 9st$　　32. $10uv^3 + 5u^2v - 4wz$
33. $-44a^{14}b^7 - 33a^{10}b^5 + 22a^{11}b^4$　　34. $-26e^8f^6 + 13e^{10}f^8 - 39e^{12}f^5$
35. $18u^{10}v^5 + 24 - 14u^{10}v^6$　　36. $30a^3b^4 - 15 + 45a^8b^7$
37. $18x^3y^4 - 12y^2z^3 - 48x^4y^3$　　38. $32m^5n^7 - 24m^8p^9 - 40m^3n^6$

Set II Factor each expression completely, or write "Not factorable."

1. $6h + 9$ **2.** $12x - 5$ **3.** $2x - 7$ **4.** $8k - 12$

5. $6h + 18$ **6.** $8x + 9$ **7.** $8k - 16$ **8.** $6 - 12x$

9. $8k - 4$ **10.** $3x - 22$ **11.** $4a^2 + 2a$ **12.** $10x + 25$

13. $8x^3 - 12x^2$ **14.** $4y - 12y^2$ **15.** $9w^2 + 16z^2$ **16.** $14a^2 - 21a^3$

17. $5xy^2 + 10x^2y$ **18.** $x^2 + 4$ **19.** $15a^2b^3 - 12ab^2$

20. $12x^2y + 9xy^2$ **21.** $16z - 8z^3 - 12z^2$ **22.** $12x^3 - 5y + 3x$

23. $8x^4 - 3x^2 + 5$ **24.** $12x^3 - 28x^2 - 8x$ **25.** $12x^4 - 6x^3 + 3x^2$

26. $4y^2 - 8y^4 + 16y^5$ **27.** $42x^5 - 6x^4 + 12x^3$ **28.** $-7a^2 + 14a^3 - 35a^4$

29. $-12ab^2 + 9a^2b - 36ab$ **30.** $20z^5 - 30z^3 - 10z^2$

31. $12h^2k - 18hk^2 - 35mp$ **32.** $30e^3f + 18 - 12ef^2$

33. $10m^2n - 21mn^3 - 13mn$ **34.** $-16u^3v^2 + 24uv^3 - 40v^4w^2$

35. $25x^2y^3 + 10 - 15x^3y^2$ **36.** $-5xy^2 + 7x - 3y$

37. $15a^3y^4 - 18ay^3 + 12a^2y$ **38.** $x^4 + 16$

7.1B Factoring Expressions with a Common Binomial Factor

Sometimes an expression has a GCF that is not a monomial. Such an expression can still be factored, using the rules given in Section 7.1A. This type of factoring is used in *factoring by grouping* (see Section 7.5)* and also in higher-level mathematics courses.

Example 11 Factor $a(x + y) + b(x + y)$.
Solution This expression has two terms, and therefore it is *not* in factored form. The common factor, $x + y$, is not a monomial; it is a binomial.

Common factor is $(x + y)$

$$a(\,x + y\,) + b(\,x + y\,) = (x + y)(\,a + b\,)$$

This term is $\dfrac{b(x + y)}{x + y} = b$

This term is $\dfrac{a(x + y)}{x + y} = a$

Therefore, $a(x + y) + b(x + y) = (x + y)(a + b)$.
Check $(x + y)(a + b) = (x + y)a + (x + y)b = a(x + y) + b(x + y)$

The answer, $(x + y)(a + b)$, has only one term, so it is in factored form. ∎

Example 12 Factor $b(a - 1) + (a - 1)$.
Solution The expression has two terms. The common binomial factor is $a - 1$.

Common factor is $(a - 1)$

$$b(\,a - 1\,) + (\,a - 1\,) = (a - 1)(\,b + 1\,) \longleftarrow \text{One term}$$

This term is $\dfrac{1(a - 1)}{a - 1}$

This term is $\dfrac{b(a - 1)}{a - 1}$

---
*Note to the Instructor: Section 7.5 *can* be covered immediately after Section 7.1B.

Therefore, $b(a - 1) + (a - 1) = (a - 1)(b + 1)$.
Check $(a - 1)(b + 1) = (a - 1)b + (a - 1)(1) = b(a - 1) + (a - 1)$ ∎

It is possible, of course, to have a common monomial factor *and* a common binomial factor in the same expression (see Example 13).

Example 13   Factor $6s^3t^2(x - y) + 8s^2t^3(x - y)$.
Solution

$$6s^3t^2(x - y) + 8s^2t^3(x - y) = 2 \cdot 3s^3t^2(x - y) + 2^3s^2t^3(x - y)$$

The expression has two terms. The greatest common *monomial* factor is $2s^2t^2$, and the greatest common *binomial* factor is $(x - y)$. The GCF, then, is the product of both of these factors, or $2s^2t^2(x - y)$.

$$2 \cdot 3s^3t^2(x - y) + 2^3s^2t^3(x - y) = \overbrace{2s^2t^2(x - y)}^{\text{GCF}}(\ 3s\ +\ 4t\ ) \longleftarrow \text{One Term}$$

This term is $\dfrac{2^3s^2t^3(x - y)}{2s^2t^2(x - y)}$

This term is $\dfrac{2 \cdot 3s^3t^2(x - y)}{2s^2t^2(x - y)}$

Therefore, $6s^3t^2(x - y) + 8s^2t^3(x - y) = 2s^2t^2(x - y)(3s + 4t)$.
Check

$$2s^2t^2(x - y)(3s + 4t) = [2s^2t^2(x - y)](3s) + [2s^2t^2(x - y)](4t)$$
$$= [2s^2t^2(3s)(x - y)] + [2s^2t^2(4t)(x - y)]$$
$$= 6s^3t^2(x - y) + 8s^2t^3(x - y)$$ ∎

EXERCISES 7.1B

Set I   Factor each expression completely, or write "Not factorable."

1. $c(s + t) + b(s + t)$ 2. $a(b + c) + d(b + c)$

3. $x(a - b) + 5(a - b)$ 4. $y(s - t) + 7(s - t)$

5. $x(u + v) - 3(u + v)$ 6. $s(t + u) - 2(t + u)$

7. $8(x - y) - a(x - y)$ 8. $7(a - b) - c(a - b)$

9. $4(s - t) + u(s - t) - v(s - t)$ 10. $a(x - y) + 5(x - y) - b(x - y)$

11. $3x^2y^3(a + b) + 9xy^2(a + b)$ 12. $10ab^3(s + t^2) + 15a^2b(s + t^2)$

13. $4u^3v^5(x - y) + 6u^4v^4(x - y)$ 14. $10x^2y^5(3a - 2b) + 15x^5y^3(3a - 2b)$

Set II   Factor each expression completely, or write "Not factorable."

1. $x(f + g) + s(f + g)$ 2. $y(u - v) - z(u - v)$

3. $u(x - y) + 4(x - y)$ 4. $d(a - b) - 3(a - b)$

5. $a(b + c) - 9(b + c)$ 6. $x(y + z) - w(y + z)$

7. $9(a - b) - c(a - b)$ 8. $y(x - z) - 9(x - z)$

9. $8(x - y) + z(x - y) - w(x - y)$ 10. $f(a + b) - g(a + b) + (a + b)$

11. $8x^3y^2(c - d) + 4x^2y^3(c - d)$ **12.** $14a^3b^3(x - 2) + 7ab^2(x - 2)$

13. $12x^4y^3(a + b) + 8x^3y^2(a + b)$ **14.** $9a^2b^4(u - v) - 12a^3b^7(u - v)$

7.2 Factoring the Difference of Two Squares

Any polynomial that can be expressed in the form $a^2 - b^2$ is called a **difference of two squares**.

7.2A Principal Square Root of a Term

We discussed principal square roots of integers in Section 1.10. In this section, we discuss finding the principal square root of an algebraic term with *even* exponents (exponents exactly divisible by 2) on the variables. We assume that all variables represent positive integers.

Example 1   Examples of finding the principal square root of a term by inspection:

a. $\sqrt{16} = 4$, because $(4)^2 = 16$.

b. $\sqrt{25x^2} = 5x$, because $(5x)^2 = 25x^2$.

c. $\sqrt{100a^4b^6} = 10a^2b^3$, because $(10a^2b^3)^2 = 100a^4b^6$. ∎

The results of Example 1 lead to the following rule:

> **TO FIND THE PRINCIPAL SQUARE ROOT OF A TERM**
> 1. The square root of the numerical coefficient is found by inspection.
> 2. The square root of each variable factor is found by dividing its exponent by 2.

Example 2   Use the information given in the preceding box to find the principal square root for each of the following terms.

a. $\sqrt{36e^8f^4} = 6e^{8/2}f^{4/2} = 6e^4f^2$

b. $\sqrt{9x^{10}y^6} = 3x^{10/2}y^{6/2} = 3x^5y^3$ ∎

In Chapter 11, we will discuss simplifying square roots when the exponents on the variables are not exactly divisible by 2.

EXERCISES 7.2A

Set I   Find the principal square root of each term.

1. $\sqrt{64}$ **2.** $\sqrt{81}$ **3.** $\sqrt{4x^2}$ **4.** $\sqrt{9y^2}$ **5.** $\sqrt{100a^8}$

6. $\sqrt{49b^6}$ **7.** $\sqrt{m^4n^2}$ **8.** $\sqrt{u^{10}v^6}$ **9.** $\sqrt{x^{10}y^4}$ **10.** $\sqrt{x^{12}y^8}$

11. $\sqrt{25a^4b^2}$ **12.** $\sqrt{100b^4c^2}$ **13.** $\sqrt{36e^8f^2}$ **14.** $\sqrt{81h^{12}k^{14}}$

15. $\sqrt{100a^{10}y^2}$ **16.** $\sqrt{121a^{24}b^4}$ **17.** $\sqrt{9a^4b^2c^6}$ **18.** $\sqrt{144x^8y^2z^6}$

Set II   Find the principal square root of each term.

1. $\sqrt{49}$ **2.** $\sqrt{144}$ **3.** $\sqrt{16z^2}$ **4.** $\sqrt{36a^8}$

5. $\sqrt{64a^6}$ **6.** $\sqrt{169x^4y^8}$ **7.** $\sqrt{x^2y^6}$ **8.** $\sqrt{400x^4}$

9. $\sqrt{h^8k^4}$ **10.** $\sqrt{4x^6y^{12}}$ **11.** $\sqrt{81m^6n^8}$ **12.** $\sqrt{25a^8b^8c^2}$

13. $\sqrt{144u^{12}v^8}$ **14.** $\sqrt{x^{14}y^{16}}$ **15.** $\sqrt{121c^{10}d^6}$ **16.** $\sqrt{4a^2b^4c^6}$

17. $\sqrt{25r^6s^8t^4}$ **18.** $\sqrt{121x^{20}y^{30}}$

7.2B Factoring the Difference of Two Squares

We now consider factoring a difference of two squares, that is, factoring $a^2 - b^2$. Factoring $a^2 - b^2$ depends on the product $(a + b)(a - b)$.

Finding the product
$$(a + b)(a - b) = a^2 - b^2$$
Finding factors

Because $(a + b)(a - b) = a^2 - b^2$, $a^2 - b^2$ *factors into* $(a + b)(a - b)$.

TO FACTOR THE DIFFERENCE OF TWO SQUARES

1. Make the following blank outline for the factors:

$(\ \blacksquare\ +\ \blacksquare\ )(\ \blacksquare\ -\ \blacksquare\ )$

One factor has +, the other has −

2. Put the principal square root of the *first* term here.

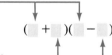

Put the principal square root of the *second* term here.

The product found in step 2 is the factored form of the difference of two squares. In symbols,

$$a^2 - b^2 = (a + b)(a - b)$$

Example 3   Factor $x^2 - 4$.
Solution

Difference of
$$x^2 - 4 = x^2 - 2^2$$
two squares

Step 1. $x^2 - 4 = (\ \blacksquare\ +\ \blacksquare\ )(\ \blacksquare\ -\ \blacksquare\ )$

Step 2. $x^2 - 4 = (x + 2)(x - 2)$

Therefore, $x^2 - 4 = (x + 2)(x - 2)$.
Check $(x + 2)(x - 2) = x^2 - 4$ ∎

Example 4   Factor $25y^4 - 9z^2$.
Solution $25y^4 - 9z^2 = (5y^2)^2 - (3z)^2$, which is in the form $a^2 - b^2$.

Step 1. $25y^4 - 9z^2 = (\ + \ )(\ - \ )$

Step 2. $25y^4 - 9z^2 = (5y^2 + 3z)(5y^2 - 3z)$

Therefore, $25y^4 - 9z^2 = (5y^2 + 3z)(5y^2 - 3z)$.
Check $(5y^2 + 3z)(5y^2 - 3z) = 25y^4 - 9z^2$ ∎

Example 5   Factor $49a^6b^2 - 81c^4d^8$.
Solution

Step 1. $49a^6b^2 - 81c^4d^8 = (\ + \ )(\ - \ )$

Step 2. $49a^6b^2 - 81c^4d^8 = (7a^3b + 9c^2d^4)(7a^3b - 9c^2d^4)$

Therefore, $49a^6b^2 - 81c^4d^8 = (7a^3b + 9c^2d^4)(7a^3b - 9c^2d^4)$.
Check $(7a^3b + 9c^2d^4)(7a^3b - 9c^2d^4) = 49a^6b^2 - 81c^4d^8$ ∎

A WORD OF CAUTION   Students often think that

$$a^2 + b^2 = (a + b)(a + b)$$

However, this is not so, because $(a + b)(a + b) = a^2 + 2ab + b^2$, not $a^2 + b^2$. Students also often think that

$$a^2 + b^2 = (a + b)(a - b)$$

However, this is not so, because $(a + b)(a - b) = a^2 - b^2$. Students often *then* think that

$$a^2 + b^2 = (a - b)(a - b)$$

But this cannot be so, because $(a - b)(a - b) = a^2 - 2ab + b^2$. In fact, a *sum* of two squares (that is, a polynomial that can be expressed in the form $a^2 + b^2$) is *not factorable* over the integers. (Exception: If the exponents on the variables are even numbers greater than 2, the polynomial may be factorable; however, we will not discuss the methods of such factoring in this book.) ☑

Example 6 Factor $x^2 + 4$.

Solution There is no common monomial factor. Then, because $x^2 + 4$ is a sum of two squares, we conclude that it is not factorable. ∎

You should begin every factoring problem by looking for and factoring out any factor common to every term. Then try to factor the polynomial in the parentheses (see Example 7).

Example 7 Factor $16x^4 - 4x^2$.

Solution There *is* a common monomial factor: $4x^2$. Therefore,

$$16x^4 - 4x^2 = 4x^2\underbrace{(4x^2 - 1)}\;\longleftarrow \text{This has been factored, but not completely}$$

Difference of two squares

$$= 4x^2(2x + 1)(2x - 1)$$

Check $4x^2(2x + 1)(2x - 1) = 4x^2(4x^2 - 1) = 16x^4 - 4x^2$ ∎

EXERCISES 7.2B

Set I Factor each polynomial completely, or write "Not factorable."

1. $m^2 - n^2$
2. $u^2 - v^2$
3. $x^2 - 9$
4. $x^2 - 25$
5. $a^2 - 1$
6. $1 - b^2$
7. $4c^2 - 1$
8. $16d^2 - 1$
9. $16x^2 - 9y^2$
10. $25a^2 - 4b^2$
11. $9h^2 - 10k^2$
12. $16e^2 - 15f^2$
13. $4x^4 - 2x$
14. $9a^4 - 3a$
15. $16x^2 + 1$
16. $25y^2 + 1$
17. $49u^4 - 36v^4$
18. $81m^6 - 100n^4$
19. $x^6 - a^4$
20. $b^2 - y^6$
21. $2x^2 - 18$
22. $3x^2 - 12$
23. $2x^2 + 9$
24. $5y^2 + 2$
25. $a^2b^2 - c^2d^2$
26. $m^2n^2 - r^2s^2$
27. $49 - 25w^2z^2$
28. $36 - 25u^2v^2$
29. $4h^4k^4 - 1$
30. $9x^4y^4 - 1$
31. $81a^4b^6 - 16m^2n^8$
32. $49c^8d^4 - 100e^6f^2$
33. $49x^4y^2 - 7x^2$
34. $25x^2y^4 - 5y^2$
35. $5x^3 - 45xy^2$
36. $11r^3 - 44rs^2$

Set II Factor each polynomial completely, or write "Not factorable."

1. $h^2 - k^2$
2. $36 - m^2$
3. $x^2 - 1$
4. $1 - 25a^2$
5. $9w^2 - 49z^2$
6. $16x^2 + 64x$
7. $25x^2 - 1$
8. $9x^2 - 5$
9. $100x^2 - y^2$
10. $2y^3 - 2y$
11. $16x^2 - 11y^2$
12. $a^2 + 16b^2$
13. $9x^4 + 3x$
14. $3x^2 - 48x$
15. $64x^2 + 1$
16. $4x^2 - 5$
17. $49m^4 - 64n^2$
18. $1 - 81y^2$
19. $e^6 - 4$
20. $f^2 + 4$
21. $7a^2 - 63$
22. $3x^3 - 48x$
23. $5y^2 + 16$
24. $3x^4 - 12x^2$
25. $r^2s^2 - t^2u^2$
26. $r^2s^2 + t^2u^2$
27. $81 - 16u^2v^2$
28. $81 + 9u^4v^2$
29. $25x^4y^4 - 1$
30. $9 - 16a^4b^4$
31. $144w^4x^6 - 121y^8z^2$
32. $20y^4 - 5y^2$
33. $36a^4b^2 - 6a^2$
34. $16s^2t^4 - s^2$
35. $6y^3 - 54x^2y$
36. $12c^3 - 48cd^2$

7.3 Factoring Trinomials

7.3A Factoring a Trinomial with a Leading Coefficient of 1

While many trinomials are not factorable, many others will factor into the product of two binomials. Recall from Section 6.1 that the leading coefficient of a polynomial is the numerical coefficient of its highest-degree term. The easiest type of trinomial to factor is one with a leading coefficient of 1.

We will first be concerned with what the *signs* of a trinomial tell us about the (possible) factors of that trinomial. Let us consider four similar products:

1. $(x + 2)(x + 5)$

2. $(x - 2)(x - 5)$

3. $(x + 2)(x - 5)$

4. $(x - 2)(x + 5)$

In product 1,

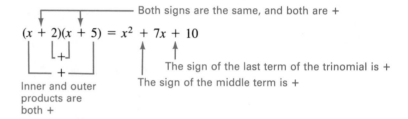

In product 2,

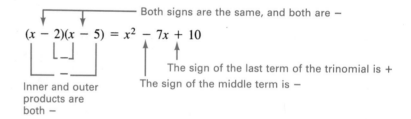

In product 3,

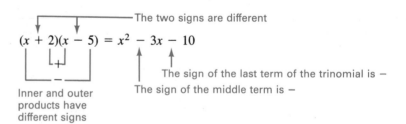

In product 4,

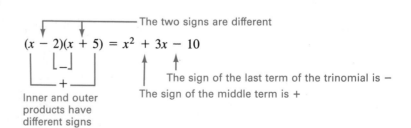

A careful examination of the four products above shows that the Rule of Signs for Factoring Trinomials given below is reasonable.

RULE OF SIGNS FOR FACTORING TRINOMIALS

Arrange the trinomial in descending powers of one variable.

1. If the sign of the last term of the trinomial is $+$, the signs of the two last terms in the binomials will be the same; then if the sign of the middle term of the trinomial is
   a. $+$, the signs of the two last terms in the binomials will both be $+$.
   b. $-$, the signs of the two last terms in the binomials will both be $-$.

2. If the sign of the last term of the trinomial is $-$, the signs of the two last terms in the binomials will be different (one $+$, one $-$), and it will be more convenient to put the signs in last.

Let's discuss the product $(x + 2)(x + 5)$ a little more.

$$(x + 2)(x + 5) = x^2 + 7x + 10$$

Therefore, $x^2 + 7x + 10$ factors into $(x + 2)(x + 5)$.

$$x^2 = \sqrt{x^2} \cdot \sqrt{x^2} = x \cdot x$$

When we try to factor a trinomial, we must *assume* that it will factor into a product of two binomials (we may decide later that it does *not*) and first make a blank outline: ( )( ).

When the leading coefficient is 1, the first term in each binomial factor is the square root of the first term of the trinomial.

Example 1   Examples of partially filled-in outlines for factoring:

a. $x^2 + 6x + 8 = (x \quad)(x \quad)$

b. $z^2 - z - 6 = (z \quad)(z \quad)$ Partially filled-in outlines

c. $m^2 + 7m + 12 = (m \quad)(m \quad)$ ∎

We continue our discussion of the product $(x + 2)(x + 5)$.

$$(x + 2)(x + 5) = x^2 + 7x + 10$$

Therefore, $x^2 + 7x + 10$ factors into $(x + 2)(x + 5)$.

$$10 = 2 \cdot 5$$

The last terms of the binomial factors must be factors of the last term of the trinomial.

$$(x + 2)(x + 5)$$
$$+ 2x$$
$$+ 5x$$
$$x^2 + 7x + 10$$

When we multiply the binomials, notice that the sum of the inner and outer products must equal the middle term of the trinomial.

The method of factoring a trinomial with a leading coefficient of 1 is summarized as follows:

TO FACTOR A TRINOMIAL WITH A LEADING COEFFICIENT OF 1

(It is assumed that like terms have been combined.)

1. Factor out the GCF.

2. Arrange the trinomial in descending powers of the variable.

3. Make a blank outline. If the Rule of Signs indicates that the signs in the binomials will be the same, put them in the outline.

4. The first term inside each set of parentheses will be the square root of the first term of the trinomial. If the third term of the trinomial has a variable, the square root of that variable must be a factor of the second term in each binomial.

5. To find the last term of each binomial:
   a. List (mentally, at least) *all* pairs of integral factors of the coefficient of the last term of the trinomial.
   b. *Select the particular pair of factors that has a sum equal to the coefficient of the middle term of the trinomial.* If no such pair of factors exists, the trinomial is not factorable.

6. Check your result by multiplying the binomials together. You should get back the trinomial from step 2.

Example 2   Factor $x^2 + 5x + 6$.
Solution

Steps 1 and 2. There is no common factor. The trinomial is in descending powers.

Step 3. The last sign in the trinomial is $+$; the Rule of Signs for factoring a trinomial tells us that the signs in the binomials will be the same. The middle term of the trinomial is $+$; therefore, both signs will be $+$.

$$(\ \ +\ \ )(\ \ +\ \ ) \longleftarrow \text{The outline}$$

Step 4. The first term in each binomial is $\sqrt{x^2} = x$:

$$(x +\ \ )(x +\ \ ) \longleftarrow \text{Partially filled-in outline}$$

Step 5. The product of the last terms of the binomials must be 6. Since $6 = 1 \cdot 6$ and $6 = 2 \cdot 3$, we could have either

$$(x + 1)(x + 6) \quad or \quad (x + 2)(x + 3)$$

The sum of the inner and outer products must equal the middle term of the trinomial, $5x$.

$$(x + 1) \quad (x + 6) \qquad\qquad (x + 2) \quad (x + 3)$$

$$+ 1x$$
$$+ 6x$$
$$+ 7x \longleftarrow \text{Incorrect middle term} \qquad + 5x \longleftarrow \text{Correct middle term}$$

$$+ 2x$$
$$+ 3x$$

Therefore,

$$x^2 + 5x + 6 = (x + 2)(x + 3) \quad \text{or} \quad (x + 3)(x + 2)$$

The order of the factors is unimportant. Notice that $(x + 2)(x + 3)$ has only *one term*; it *is* in factored form.

———— Step 6. *Check* $(x + 2)(x + 3) = x^2 + 5x + 6$ ∎

A WORD OF CAUTION   In Example 2, students often make one of these three errors:

$$x^2 + 5x + 6 = (x + 2) \cdot x + 3$$

$$x^2 + 5x + 6 = x + 2 \cdot (x + 3)$$

$$x^2 + 5x + 6 = x + 2 \cdot x + 3$$

However, *both sets of parentheses are essential*.

$$(x + 2) \cdot x + 3 = x^2 + 2x + 3 \neq x^2 + 5x + 6$$

$$x + 2 \cdot (x + 3) = x + 2x + 6 = 3x + 6 \neq x^2 + 5x + 6$$

$$x + 2 \cdot x + 3 = x + 2x + 3 = 3x + 3 \neq x^2 + 5x + 6 \quad \boxed{\checkmark}$$

Example 3   Factor $m^2 - 9m + 8$.
Solution

Steps 1 and 2. There is no common factor. The trinomial is in descending powers.

Step 3. $m^2 - 9m + 8$

         + here tells us that the signs in the binomials will be the same

         − here tells us that both signs will be −

$$(\ -\ )(\ -\ ) \longleftarrow \text{The outline}$$

Step 4. The first terms of the binomial factors must be m.

$$(m - \ )(m - \ ) \longleftarrow \text{Partially filled-in outline}$$

Step 5. The pairs of negative factors of 8 are $(-1)(-8)$ and $(-2)(-4)$. We must select the pair whose sum is -9. This is the pair -1 and -8. Therefore,

$$m^2 - 9m + 8 = (m - 1)(m - 8) = (m - 8)(m - 1)$$

———— Step 6. *Check* $(m - 1)(m - 8) = m^2 - 9m + 8$ ∎

A WORD OF CAUTION   While $(m - 2)(m - 4)$ gives us the correct *first* term and the correct *third* term for the problem in Example 3, the *middle term* is incorrect:

$$(m - 2)(m - 4) = m^2 - 6m + 8 \neq m^2 - 9m + 8 \quad \boxed{\checkmark}$$

Example 4   Factor $4a - 12 + a^2$.
Solution

Step 1. There is no common factor.

Step 2. $a^2 + 4a - 12$ Arranged in descending powers of a

— here tells us that the signs in the binomials will be different; we will put them in last

Step 3. (   )(   )◄—— The outline

Step 4. The first terms of the binomial factors must be a.

(a)(a)◄—— Partially filled-in outline

Step 5. The pairs of factors of -12 are $(1)(-12)$, $(-1)(12)$, $(2)(-6)$, $(-2)(6)$, $(3)(-4)$, and $(-3)(4)$. We must select the pair whose sum is $+4$; this is the pair -2 and 6. Therefore,

$$a^2 + 4a - 12 = (a - 2)(a + 6) = (a + 6)(a - 2)$$

Step 6. *Check* $(a - 2)(a + 6) = a^2 + 4a - 12 = 4a - 12 + a^2$ ■

Example 5   Factor $a^2 - 4a - 12$.
Solution

$$a^2 - 4a - 12$$

— here tells us that the signs in the binomials will be different

See Example 4 for pairs of factors of -12. We must select the pair whose sum is -4; this is the pair 2 and -6. Therefore,

$$a^2 - 4a - 12 = (a + 2)(a - 6) = (a - 6)(a + 2)$$

Check $(a + 2)(a - 6) = a^2 - 4a - 12$ ■

Example 6   Factor $3x^2 - 24x - 60$.
Solution This polynomial does *not* have a leading coefficient of 1; however, there is a common factor. We factor out 3 (the GCF) first.

$$3x^2 - 24x - 60 = 3(x^2 - 8x - 20)$$

The expression $3(x^2 - 8x - 20)$ *has* been factored (it has only one term); however, the polynomial factor is still factorable, and it *does* have a leading coefficient of 1:

$$x^2 - 8x - 20$$

— here tells us that the signs in the binomials will be different

The pairs of factors of -20 are $(1)(-20)$, $(-1)(20)$, $(2)(-10)$, $(-2)(10)$, $(4)(-5)$, and $(-4)(5)$. The pair whose sum is -8 is the pair 2 and -10. Therefore,

$$x^2 - 8x - 20 = (x + 2)(x - 10)$$

and
$$3x^2 - 24x - 60 = 3(x + 2)(x - 10)$$

Check $3(x + 2)(x - 10) = 3(x^2 - 8x - 20) = 3x^2 - 24x - 60$ ■

Example 7   Factor $x^4 + 9x^3 + 20x^2$.
Solution There is a common factor. We factor out x^2 (the GCF) first.

$$x^4 + 9x^3 + 20x^2 = x^2(x^2 + 9x + 20)$$

The expression $x^2(x^2 + 9x + 20)$ has been factored; however, the polynomial factor is still factorable:

$$x^2 + 9x + 20$$

+ here tells us that the signs in the binomials will be the same

+ here tells us that both signs will be +

The pairs of positive factors of 20 are $(1)(20)$, $(2)(10)$, and $(4)(5)$. The pair whose sum is 9 is the pair 4 and 5. Therefore,

$$x^4 + 9x^3 + 20x^2 = x^2(x^2 + 9x + 20) = x^2(x + 4)(x + 5)$$

Check $x^2(x + 4)(x + 5) = x^2(x^2 + 9x + 20) = x^4 + 9x^3 + 20x^2$ ■

Example 8   Factor $x^2 - 11xy + 24y^2$.
Solution

$$x^2 - 11xy + 24y^2$$

+ here tells us that both signs will be the same

− here tells us that they will both be −

Because the third term contains the variable y^2, we must be sure to have $\sqrt{y^2} = y$ as a *factor* of the second term in each binomial.

$$\sqrt{y^2} = y$$
$$(x - \boxed{}\, y)(x - \boxed{}\, y) \longleftarrow \text{Partially filled-in outline}$$

The pairs of negative factors of 24 are $(-1)(-24)$, $(-2)(-12)$, $(-3)(-8)$, and $(-4)(-6)$. The pair whose sum is -11 is the pair -3 and -8. Therefore,

$$x^2 - 11xy + 24y^2 = (x - 3y)(x - 8y)$$

$(x - 3y)(x - 8y) = x^2 - 11xy + 24y^2$ ■

Example 9   Factor $x^2 - 5x + 3$.
Solution

$$x^2 - 5x + 3$$

+ here tells us that both signs will be the same

− here tells us that they will both be −

$(x - \ \ )(x - \ \ ) \longleftarrow \text{Partially filled-in outline}$

We must find two numbers whose product is 3 and whose sum is -5. There are no such numbers. Therefore, $x^2 - 5x + 3$ is *not factorable*. ■

EXERCISES 7.3A

Set I   Factor each polynomial completely, or write "Not factorable."

1. $x^2 + 6x + 8$ **2.** $x^2 + 9x + 8$ **3.** $x^2 + 5x + 4$

4. $x^2 + 4x + 4$ **5.** $k^2 + 7k + 6$ **6.** $k^2 + 5k + 6$

7. $7u + u^2 + 10$ **8.** $11u + u^2 + 10$ **9.** $y^2 - 2y + 8$

10. $y^2 - 7y + 8$ **11.** $b^2 - 9b + 14$ **12.** $b^2 - 15b + 14$

13. $z^2 - 9z + 20$ **14.** $z^2 - 12z + 20$ **15.** $18 + x^2 - 11x$

16. $18 + x^2 - 9x$ **17.** $x^2 + 9x - 10$ **18.** $y^2 - 3y - 10$

19. $z^2 - z - 6$ **20.** $m^2 + 5m - 6$ **21.** $5x^2 + 10x$

22. $8y^3 + 4y^2$ **23.** $x^2 + 4x - 5$ **24.** $y^2 + 6y - 7$

25. $x^2 + 100$ **26.** $z^2 + 1$ **27.** $z^5 + 9z^4 - 10z^3$

28. $x^4 + 7x^3 - 8x^2$ **29.** $t^2 + 11t - 30$ **30.** $m^2 - 17m - 30$

31. $u^4 + 12u^2 - 64$ **32.** $v^4 - 30v^2 - 64$ **33.** $16 + v^2 - 8v$

34. $16 + v^2 - 10v$ **35.** $b^2 - 11bd - 60d^2$ **36.** $c^2 + 17cx - 60x^2$

37. $r^2 - 13rs - 48s^2$ **38.** $s^2 + 22st - 48t^2$ **39.** $x^4 + 2x^3 - 35x^2$

40. $x^4 + 2x^3 - 48x^2$ **41.** $14x^2 - 15x + x^3$ **42.** $8x^3 - 9x^2 + x^4$

43. $3x^2 + 6x - 24$ **44.** $5x^2 + 15x - 50$ **45.** $12 + 4x^2 - 16x$

46. $24 + 2x^2 - 14x$ **47.** $x^4 + 6x^3 + x^2$ **48.** $y^4 + 5y^3 + y^2$

49. $x^4 - 9$ **50.** $a^4 - 25$

Set II   Factor each polynomial completely, or write "Not factorable."

1. $m^2 + 7m + 12$ **2.** $x^2 + 9x + 20$ **3.** $h^2 + 11h + 18$

4. $x^2 + 13x + 12$ **5.** $a^2 + 10a + 16$ **6.** $x^2 + x + 1$

7. $10k + k^2 + 24$ **8.** $21y + y^2 + 20$ **9.** $x^2 - x + 12$

10. $x^2 - 7x + 12$ **11.** $n^2 - 9n + 18$ **12.** $x^2 - 5x + 4$

13. $z^2 - 8z + 16$ **14.** $y^2 - 2y + 9$ **15.** $15 + x^2 - 8x$

16. $13n + n^2 - 14$ **17.** $w^2 + 2w - 24$ **18.** $3z - 18 + z^2$

19. $r^2 - 9r - 20$ **20.** $y^2 - 8y - 20$ **21.** $3x^2 + 12x$

22. $c^2 + 121$ **23.** $x^2 - 4x - 45$ **24.** $a^4 + a^3 + a^2$

25. $a^2 + 25$ **26.** $5a^2 + 25$ **27.** $z^5 - 5z^4 - 6z^3$

28. $x^4 + 5x^3 + 5x^2$ **29.** $n^2 - 10n - 24$ **30.** $x^2 + 4x - 45$

31. $v^4 - 18v^2 + 45$ **32.** $u^4 - 4u^3 - 21u^2$ **33.** $36 + y^2 - 15y$

34. $u^4 - 21 - 4u^2$ **35.** $w^2 - 8wz - 48z^2$ **36.** $x^2 - 15xy + 54y^2$

37. $p^2 - 9pt - 52t^2$ **38.** $a^2 + 10ab + 16b^2$ **39.** $s^4 + 3s^3 - 28s^2$

40. $s^2 - 3st + 28t^2$ **41.** $y^3 - 30y^2 + y^4$ **42.** $y^2 - 11yz + 30z^2$

43. $4x^2 - 8x - 32$ **44.** $6x^2 + 6 + 6x$ **45.** $-6x + 2x^2 - 20$

46. $-5x + 5x^3$ **47.** $t^4 + 5t^3 + t^2$ **48.** $x^2 - 5x - 24$

49. $x^2 - 121$ **50.** $y^2 + 121$

7.3B Factoring Any Trinomial

The method we now show for factoring any trinomial is often called the *trial method*. In deciding what the variable parts of each term of each binomial should be, we use the method used in Section 7.3A.

The following method can be used for factoring any trinomial:

TO FACTOR ANY TRINOMIAL

(It is assumed that like terms have been combined.)

1. Factor out the GCF.

2. Arrange the trinomial in descending powers of the variable. (If the leading coefficient of the trinomial is negative, either arrange the trinomial in ascending powers or factor out -1 before proceeding.)

3. a. If the leading coefficient of the trinomial is 1, factor by using the techniques given in Section 7.3A.
   b. If the leading coefficient of the trinomial is unequal to 1, proceed with steps 4–7.

4. Make a blank outline and fill in the *variable parts* of each binomial. If the Rule of Signs indicates that the signs in the binomials will both be positive or both be negative, put those signs in the outline.

5. List (mentally, at least) all pairs of factors of the coefficient of the first term of the trinomial and of the last term of the trinomial.

6. By trial and error, select the pairs of factors (if they exist) from step 6 that make the sum of the inner and outer products of the binomials equal to the middle term of the trinomial. If no such pairs exist, the trinomial is *not factorable*.

7. Check your result by multiplying the binomials together. You should get back the trinomial from step 2.

Example 10 Factor $2x^2 + 7x + 5$.

Solution There are no like terms and no common factor; the trinomial is already in descending powers of x. The signs for both binomial factors are $+$.

$$(\quad \oplus \quad)(\quad \oplus \quad)$$

Each first term of the binomial factors must contain an x in order to give the x^2 in the first term of the trinomial $2x^2$.

$$(\quad x + \quad)(\quad x + \quad)$$

The only factors of 2 are 1 and 2. We put these numbers in the outline.

$$(\quad 1x + \quad)(\quad 2x + \quad)$$

We have two ways to try the factors of 5:

$$(1x + 5) \quad (2x + 1) \qquad or \qquad (1x + 1) \quad (2x + 5)$$

$$+ 10x \qquad\qquad\qquad + 2x$$
$$+ 1x \qquad\qquad\qquad\quad + 5x$$
$$+ 11x \longleftarrow \text{Incorrect} \qquad\quad + 7x \longleftarrow \text{Correct}$$
$$\text{middle} \qquad\qquad\qquad\qquad \text{middle}$$
$$\text{term} \qquad\qquad\qquad\qquad\quad \text{term}$$

The sum of the inner and outer products must equal the middle term of the trinomial. We found the correct pair by trial. Therefore, $2x^2 + 7x + 5$ factors into $(x + 1)(2x + 5)$.

Check $(x + 1)(2x + 5) = 2x^2 + 7x + 5$ ■

Example 11

Factor $13x + 5x^2 + 6$

Solution There are no like terms and no common factor.

$$13x + 5x^2 + 6 = 5x^2 + 13x + 6 \longleftarrow \text{Arranged in descending powers of } x$$
$$\sqrt{x^2}$$
$$(\quad x + \quad)(\quad x + \quad) \longleftarrow \text{Partially filled-in outline}$$
$$\longleftarrow \text{Signs are both } +$$

Next, list all pairs of factors of the leading coefficient and of the last term of the trinomial.

Factors of the *leading coefficient*	*Factors of the* *last term*
$5 = 1 \cdot 5$	$6 = 1 \cdot 6$
	$= 2 \cdot 3$

There are two pairs of factors to try for 6: (1)(6) or (2)(3). Therefore, we must try the following combinations:

$(5x + 1)(x + 6)$

$(5x + 6)(x + 1)$

$(5x + 2)(x + 3)$

$(5x + 3)(x + 2)$

In all these combinations, the product of the two first terms is $5x^2$ and the product of the two last terms is 6. However, only in the last combination is the sum of the inner and outer products equal to $13x$. Therefore,

$$13x + 5x^2 + 6 = 5x^2 + 13x + 6 = (5x + 3)(x + 2)$$

Check $(5x + 3)(x + 2) = 5x^2 + 13x + 6 = 13x + 5x^2 + 6$ ■

When the leading coefficient or the coefficient of the third term is a prime number, it is probably easiest to insert the factors of that number first (see Example 12).

Example 12

Factor $4z^2 - 3z - 7$.

Solution The polynomial is in descending powers of z; there are no like terms and no common factors. The signs will be different. Because 7 is a prime number, we will fill in the factors of 7 first.

$$(\Box z \bullet 7)(\Box z \bullet 1) \longleftarrow \text{Partially filled-in outline}$$

Now we must insert the factors of 4 (either 1 and 4 or 2 and 2) and the signs so that the sum of the inner and outer products is $-3z$. We must consider the following combinations:

$(2z + 7)(2z - 1)$ $\qquad$ $(2z - 7)(2z + 1)$

$(1z + 7)(4z - 1)$ $\qquad$ $(1z - 7)(4z + 1)$

$(4z + 7)(1z - 1)$ $\qquad$ $(4z - 7)(1z + 1)$ $\longleftarrow$ This is the only combination in which the sum of the inner and outer products is $-3z$

Therefore, $4z^2 - 3z - 7 = (4z - 7)(z + 1)$.
Check $(4z - 7)(z + 1) = 4z^2 - 3z - 7$ $\quad\blacksquare$

Example 13 $\quad$ Factor $5x^2 - x + 2$.
Solution These are the only combinations we need to consider:

$(5x - 1)(x - 2)$

$(5x - 2)(x - 1)$

In both of these combinations, the product of the two first terms is $5x^2$ and the product of the last two terms is $+2$. However, in neither one is the sum of the inner and outer products $-x$. Therefore, $5x^2 - x + 2$ is *not factorable*. $\quad\blacksquare$

Example 14 $\quad$ Factor $19xy - 15y^2 - 6x^2$.
Solution Because the coefficients of x^2 and y^2 are both negative, we factor out -1 before proceeding.

$$19xy - 15y^2 - 6x^2 = -6x^2 + 19xy - 15y^2 = -1(6x^2 - 19xy + 15y^2)$$

We now factor $6x^2 - 19xy + 15y^2$.

$$(\Box x - \Box y)(\Box x - \Box y) \longleftarrow \text{Partially filled-in outline}$$

These are the combinations we must consider:

$(6x - 15y)(1x - 1y)$ $\qquad$ $(6x - 1y)(1x - 15y)$

$(6x - 3y)(1x - 5y)$ $\qquad$ $(6x - 5y)(1x - 3y)$

$(2x - 15y)(3x - 1y)$ $\qquad$ $(2x - 1y)(3x - 15y)$

$(2x - 5y)(3x - 3y)$ $\qquad$ $(2x - 3y)(3x - 5y)$

The last combination, $(2x - 3y)(3x - 5y)$, is the only one in which the sum of the inner and outer products is $-19xy$. Therefore,

$$19xy - 15y^2 - 6x^2 = -1(6x^2 - 19xy + 15y^2) = -(2x - 3y)(3x - 5y)$$

Other acceptable answers include $(3y - 2x)(3x - 5y)$ and $(2x - 3y)(5y - 3x)$.
Check $-(2x - 3y)(3x - 5y) = -(6x^2 - 19xy + 15y^2) = 19xy - 15y^2 - 6x^2$ $\quad\blacksquare$

Example 15   Factor $10u^2 - 12u + 8u^3$.

Solution We first arrange the trinomial in descending powers of u and factor out the GCF.

$$10u^2 - 12u + 8u^3 = 8u^3 + 10u^2 - 12u = 2u(4u^2 + 5u - 6)$$

We now try to factor the polynomial in the parentheses.

$$(\ \blacksquare u \ \bullet \ \blacksquare \ )(\ \blacksquare u \ \bullet \ \blacksquare \ ) \longleftarrow \text{Partially filled-in outline}$$

The combinations to consider:

$(2u + 6)(2u - 1)$ $(2u - 6)(2u + 1)$

$(2u + 2)(2u - 3)$ $(2u - 2)(2u + 3)$

$(4u + 1)(1u - 6)$ $(4u - 1)(1u + 6)$

$(4u + 6)(1u - 1)$ $(4u - 6)(1u + 1)$

$(4u + 2)(1u - 3)$ $(4u - 2)(1u + 3)$

$(4u + 3)(1u - 2)$ $(4u - 3)(1u + 2) \longleftarrow$ This is the only combination in which the sum of the inner and outer products is $+5u$

Therefore,

$$4u^2 + 5u - 6 = (u + 2)(4u - 3)$$

$$10u^2 - 12u + 8u^3 = 2u(u + 2)(4u - 3)$$

Check

$$2u(u + 2)(4u - 3) = 2u(4u^2 + 5u - 6)$$
$$= 8u^3 + 10u^2 - 12u$$
$$= 10u^2 - 12u + 8u^3 \quad \blacksquare$$

NOTE   It is not necessary to write down all possible combinations as was done in Examples 11–15.   ☑

HINT   If the original trinomial does not have a common factor, then neither *binomial factor* can have a common factor. This fact can help you eliminate some combinations from consideration (see Example 16).

Example 16   Factor $12a^2 + 7ab - 10b^2$.

Solution The outline is as follows:

$$(\ \blacksquare a \ \bullet \ \blacksquare b)(\ \blacksquare a \ \bullet \ \blacksquare b)$$

If we first try 3 and 4 as the factors of 12, and 2 and 5 as the factors of 10, we have these combinations to consider:

★(3a + 5b)(4a − 2b)   ★(3a − 5b)(4a + 2b)

(3a + 2b)(4a − 5b)   (3a − 2b)(4a + 5b) ⟵ In this combination, the sum of the inner and outer products is +7ab

Therefore, $12a^2 + 7ab - 10b^2 = (3a - 2b)(4a + 5b)$
Check $(3a - 2b)(4a + 5b) = 12a^2 + 7ab - 10b^2$

The starred combinations both have a common factor of 2 in one of the binomials. Since the original trinomial had no common factor, these combinations need not be given any consideration. ∎

Sometimes, the first term is a constant and the last term contains the variable. In this case, we proceed in almost the same way.

Example 17   Factor $8 + 2x - x^2$.
Solution If we arranged this trinomial in descending powers of x, the first term would be *negative*. The expression is easier to factor if we leave it in ascending powers of x. The outline is as follows:

$$(\blacksquare \; \bullet \; x)(\blacksquare \; \bullet \; x)$$

If we first try 2 and 4 as the factors of 8, we have this combination:

$$(2 \; \bullet \; x)(4 \; \bullet \; x)$$

If we put the signs as

$$(2 - x)(4 + x)$$

we do *not* get the correct middle term. Let's try $(2 + x)(4 - x)$. The sum of the inner and outer products *is* $+2x$. Therefore,

$$8 + 2x - x^2 = (2 + x)(4 - x)$$

Check $(2 + x)(4 - x) = 8 + 2x - x^2$ ∎

When the first and third terms of the trinomial are perfect squares, the trinomial *may* be the square of a binomial (see Section 6.4B).

Example 18   Factor $9x^2 - 12xy + 4y^2$.
Solution Because $9x^2 = (3x)^2$ and $4y^2 = (2y)^2$, it is possible that $9x^2 - 12xy + 4y^2$ is the square of a binomial. We'll try

$$(3x - 2y)^2$$

— because middle term of the trinomial was −

$$(3x - 2y)^2 = (3x)^2 - 2(3x)(2y) + (2y)^2 = 9x^2 - 12xy + 4y^2$$

Therefore, $9x^2 - 12xy + 4y^2 = (3x - 2y)^2$ or $(3x - 2y)(3x - 2y)$.

($9x^2 - 12xy - 4y^2$ could also have been factored by using the trial method.)
Check $(3x - 2y)^2 = 9x^2 - 12xy + 4y^2$ ∎

EXERCISES 7.3B

Set I   Factor each expression completely, or write "Not factorable."

1. $3x^2 + 7x + 2$ **2.** $3x^2 + 5x + 2$ **3.** $5x^2 + 7x + 2$

4. $5x^2 + 11x + 2$ **5.** $7x + 4x^2 + 3$ **6.** $13x + 4x^2 + 3$

7. $5x^2 + 20x + 4$ **8.** $5x^2 + 11x + 4$ **9.** $5a^2 - 16a + 3$

10. $5m^2 - 8m + 3$ **11.** $3b^2 - 22b + 7$ **12.** $3u^2 - 10u + 7$

13. $5z^2 - 36z + 7$ **14.** $5z^2 - 12z + 7$ **15.** $14n - 5 + 3n^2$

16. $2k - 7 + 5k^2$ **17.** $9x^2 - 49$ **18.** $16y^2 - 1$

19. $7x^2 + 23xy + 6y^2$ **20.** $7a^2 + 43ab + 6b^2$ **21.** $7h^2 - 11hk + 4k^2$

22. $7h^2 - 16hk + 4k^2$ **23.** $3t^2 + 19tz - 6z^2$ **24.** $3w^2 - 11wx - 6x^2$

25. $49x^2 - 42x + 9$ **26.** $25x^2 - 20x + 4$ **27.** $18u^3 + 39u^2 - 15u$

28. $12y^3 + 14y^2 - 10y$ **29.** $4x^2 + 4x + 4$ **30.** $6x^2 + 6x + 18$

31. $7x^2 - 49$ **32.** $5x^2 - 25$ **33.** $6 - 17v + 5v^2$

34. $6 - 11v + 5v^2$ **35.** $20x + 3x^2 + 12$ **36.** $13x + 3x^2 + 12$

37. $45x^2 - 120x + 80$ **38.** $64x^2 - 96x + 36$ **39.** $8x^2 - 2x - 15$

40. $8x^2 - 19x - 15$ **41.** $6y^2 - 19y + 10$ **42.** $6y^2 - 23y + 10$

43. $9a^2 + 24a - 20$ **44.** $9b^2 + 3b - 20$ **45.** $6e^4 - 7e^2 - 20$

46. $10f^4 - 29f^2 - 21$ **47.** $x^2 + 144$ **48.** $y^2 + 9$

Set II   Factor each expression completely, or write "Not factorable."

1. $3y^2 + 16y + 5$ **2.** $2z^2 + 9z + 7$ **3.** $4a^2 + 9a + 5$

4. $3b^2 + b + 4$ **5.** $13x + 6x^2 + 2$ **6.** $2 + 6x^2 + 7x$

7. $2x^2 + 6x + 3$ **8.** $6x^2 + 15x + 9$ **9.** $7a^2 - 22a + 3$

10. $7a^2 - 20a - 3$ **11.** $2x^2 - 15x + 7$ **12.** $7 + 2x^2 + 9x$

13. $3t^2 - 16t + 5$ **14.** $3t^2 - 16t - 5$ **15.** $13x - 7 + 2x^2$

16. $9x + 2x^2 - 7$ **17.** $25x^2 - 16$ **18.** $16x^2 + 25$

19. $5x^2 + 13x + 6$ **20.** $10a^3 + 14a^2 - 12a$ **21.** $5x^2 - 17x + 6$

22. $5x^2 - 13x - 6$ **23.** $5x^2 + 30x + 6$ **24.** $5x^2 + 11x + 6$

25. $25s^2 - 30s + 9$ **26.** $9x^2 - 37x + 4$ **27.** $14m^3 - 20m^2 + 6m$

28. $3e^2 - 20e - 7$ **29.** $5s^2 - 5s + 5$ **30.** $5h^2 - 8h + 3$

31. $6x^2 - 36$ **32.** $36x^2 - 1$ **33.** $12 - 17u - 5u^2$

34. $8ab + 5b^2 + 3a^2$ **35.** $15x + 2x^2 + 18$ **36.** $7x^2 + 2x - 5$

37. $70k^2 - 24k + 2$ **38.** $6s^2 - 17st - 5t^2$ **39.** $12x^2 + 7x - 10$

40. $8x^2 - 6x - 9$ **41.** $5x^2 - 16x + 12$ **42.** $6t^2 - 5t - 21$

43. $4m^2 - 16m + 7$ **44.** $14x^2 - 4x - 10$ **45.** $8v^4 - 14v^2 - 15$

46. $8x^3 - 8x^2 + 8x$ **47.** $z^2 + 4$ **48.** $2x^2 - 72$

7.4 Review: 7.1–7.3

Greatest Common Factor
7.1

The **greatest common factor (GCF)** of two integers is the largest integer that is a factor of both integers.

Methods of Factoring
7.1
7.2

1. Greatest common factor (GCF)
2. Difference of two squares:

$$a^2 - b^2 = (a + b)(a - b)$$

7.3A
7.3B

3. Trinomial $\begin{cases} \text{Leading coefficient equal to 1} \\ \text{Leading coefficient unequal to 1} \end{cases}$

Review Exercises 7.4 Set I

Factor each expression completely, or write "Not factorable."

1. $8x - 4$

2. $5 + 7a$

3. $m^2 - 4$

4. $25 + n^2$

5. $x^2 + 10x + 21$

6. $y^2 - 10y + 15$

7. $2u^2 + 4u$

8. $3b - 6b^2 + 12b^3$

9. $z^2 - 7z - 18$

10. $7x - 30 - x^2$

11. $4x^2 - 25x + 6$

12. $4x^2 + 20x + 25$

13. $9k^2 - 144$

14. $100 + 81p^2$

15. $8 - 2a^2$

16. $10c^2 - 44c + 16$

17. $15u^2v - 3uv$

18. $5ab^2 - 10a^2b - 5ab$

19. $10x^2 - xy - 24y^2$

20. $8x^5y^2 - 12x^2y^4 - 16x^2y^2$

Review Exercises 7.4 Set II

Factor each expression completely, or write "Not factorable."

1. $6 - 30u$

2. $36 - m^2$

3. $8x + 5y$

4. $x^2 + 9x + 18$

5. $12v^2 - 8v$

6. $z^2 + 2z - 35$

7. $4e^2 - 27e + 18$

8. $x^2 - 8x + 8$

9. $64 - 49p^2$

10. $48 - 3a^2$

11. $16x^2 - 8x + 1$

12. $3x^2 + 6xy + 9y^2$

13. $3a^2b - 15ab^2 - 3ab$

14. $24hk^2 - 6hk$

15. $40x^2 + 10xy - 50y^2$

16. $21u^2 + 13uv - 20v^2$

1. _____

2. _____

3. _____

4. _____

5. _____

6. _____

7. _____

8. _____

9. _____

10. _____

11. _____

12. _____

13. _____

14. _____

15. _____

16. _____

17. $8m^3n^2 - 14mn^4 - 6mn^2$

18. $4x^2 + 8x + 8$

19. $28x^2 - 13xy - 6y^2$

20. $21ab^2 - 7ab$

17. _____

18. _____

19. _____

20. _____

7.5 Factoring by Grouping

If a polynomial has four terms, we can sometimes factor it by first separating its terms into two groups and factoring each group separately. Since we will still have more than one term at this point, the expression will not yet be factored. However, if we then see that each of the groups has a common factor, we will be able to factor the polynomial.

The rules that follow assume that any factor common to all *four* terms has already been factored out.

**TO FACTOR AN EXPRESSION OF FOUR TERMS
BY GROUPING TWO AND TWO**

1. Arrange the four terms into two groups of two terms each so that each group of two terms is factorable.

2. Factor each *group*. You will now have two terms. *The expression will not yet be factored.*

3. Factor the two-term expression resulting from step 2 if the two terms now have a common binomial factor.

4. If the two terms resulting from step 2 do *not* have a GCF, try a different arrangement of the original four terms.

Example 1   Factor $ax + ay + bx + by$.
Solution

GCF = a ⟶   ⟵ GCF = b

Step 1. $ax + ay + bx + by$

This polynomial is not in factored form because it has two terms

Step 2. $= a(x + y) + b(x + y)$ $(x + y)$ is the GCF of these two terms

Step 3. $= (x + y)(a + b)$

This term is $\dfrac{b(x + y)}{(x + y)} = b$

This term is $\dfrac{a(x + y)}{(x + y)} = a$

Therefore, $ax + ay + bx + by = (x + y)(a + b)$.
Check $(x + y)(a + b) = (x + y)a + (x + y)b = ax + ay + bx + by$ ■

A WORD OF CAUTION   In Example 1, students often make this error:

No sign here

$$ax + ay + bx + by = a(x + y) \quad b(x + y)$$
$$= (x + y)(a + b)$$

Even though the final answer is correct, *two* errors have been made, since

$$a(x + y)b(x + y) \neq ax + ay + bx + by$$

and $$(x + y)(a + b) \neq a(x + y)b(x + y)$$

It is often possible to group terms differently and still be able to factor the expression. *The same factors are obtained no matter what grouping is used*, because factorization over the integers is unique.

Example 2   Factor $ab - b + a - 1$.
Solution

	One grouping	*A different grouping*
Step 1.	$ab - b \ + \ a - 1$	$ab + a \ - b - 1$
Step 2.	$= b\ (a - 1) \ + 1\ (a - 1)$	$= a\ (b + 1) \ - 1\ (b + 1)$
Step 3.	$= (a - 1)(b + 1)$	$= (b + 1)(a - 1)$

Same factors

Therefore, $ab - b + a - 1 = (a - 1)(b + 1) = (b + 1)(a - 1)$.
Check $(a - 1)(b + 1) = (a - 1)b + (a - 1)(1) = ab - b + a - 1$ ∎

Example 3   Factor $2x^2 + 3x - 6xy - 9y$.
Solution 1

GCF = x GCF = $3y$ or $-3y$

Step 1. $2x^2 + 3x \ + \ (-6xy - 9y)$

Using $-3y$ as the GCF

Step 2. $= x\ (2x + 3) \ - 3y\ (2x + 3)$ ◂— $(2x + 3)$ is the common binomial factor

Step 3. $= (2x + 3)(x - 3y)$

(If we had used $3y$ as the GCF for the second group of two terms, we would have had

$$2x^2 + 3x - 6xy - 9y = x(2x + 3) + 3y(-2x - 3)$$

and we would *not* have had a common binomial factor, since $2x + 3 \neq -2x - 3$.)
Solution 2 We can also *rearrange* the terms and factor as follows:

$$2x^2 + 3x - 6xy - 9y = 2x^2 - 6xy + 3x - 9y$$

GCF = $2x$ GCF = 3

Step 1. $2x^2 - 6xy + 3x - 9y$

Step 2. $= 2x\ (x - 3y) + 3\ (x - 3y)$ $(x - 3y)$ is the common binomial factor

Step 3. $= (x - 3y)(2x + 3)$

Therefore, $2x^2 - 6xy + 3x - 9y = (x - 3y)(2x + 3)$.
Check

$$(x - 3y)(2x + 3) = (x - 3y)(2x) + (x - 3y)(3)$$
$$= 2x^2 - 6xy + 3x - 9y \quad ∎$$

In Example 4, one group of two terms does not have a common factor. Instead, it is a difference of two squares.

Example 4   Factor $a^2 - b^2 + 3a - 3b$.
Solution

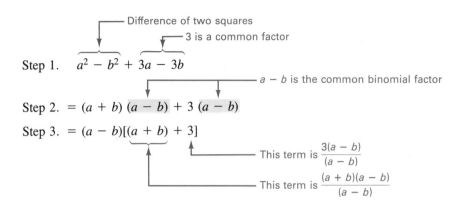

Step 1. $a^2 - b^2 + 3a - 3b$

— Difference of two squares
— 3 is a common factor

Step 2. $= (a + b)(a - b) + 3(a - b)$

— $a - b$ is the common binomial factor

Step 3. $= (a - b)[(a + b) + 3]$

This term is $\dfrac{3(a - b)}{(a - b)}$

This term is $\dfrac{(a + b)(a - b)}{(a - b)}$

Therefore, $a^2 - b^2 + 3a - 3b = (a - b)(a + b + 3)$.
Check

$$(a - b)(a + b + 3) = (a - b)a + (a - b)b + (a - b)(3)$$
$$= a^2 - ab + ab - b^2 + 3a - 3b$$
$$= a^2 - b^2 + 3a - 3b \quad \blacksquare$$

A WORD OF CAUTION   An expression is not factored until it has been written as a single term that is a product of factors. To illustrate this, let us consider Example 1 again.

$$ax + ay + bx + by$$

$$= \underbrace{a(x + y)}_{\substack{\text{First} \\ \text{term}}} + \underbrace{b(x + y)}_{\substack{\text{Second} \\ \text{term}}}$$ This expression is not in factored form because it has two terms

$$= \underbrace{(x + y)(a + b)}_{\text{Single term}}$$ Factored form of $ax + ay + bx + by$ ☑

Expressions with more than four terms may also be factored by grouping; however, in this book we consider factoring only expressions of four terms by grouping.

EXERCISES 7.5

Set I   Factor each polynomial completely, or write "Not factorable."

1. $am + bm + an + bn$

2. $cu + cv + du + dv$

3. $st + 4t + 3su + 12u$

4. $5cz + 5z + 7ct + 7t$

5. $3xr - 6yr + 4x - 8y$

6. $4ms - 6mt + 10ns - 15nt$

7. $mx - nx - my + ny$

8. $ah - ak - bh + bk$

9. $xy + x - y - 1$

10. $ad - d + a - 1$

11. $3a^2 - 6ab + 2a - 4b$

12. $2h^2 - 6hk + 5h - 15k$

13. $6e^2 - 2ef - 9e + 3f$

14. $8m^2 - 4mn - 6m + 3n$

15. $h^2 - k^2 + 2h + 2k$

16. $x^2 - y^2 + 4x + 4y$.

17. $x^3 + 3x^2 - 4x + 12$

18. $a^3 + 5a^2 - 2a + 10$

19. $a^3 - 2a^2 - 4a + 8$ **20.** $x^3 - 3x^2 - 9x + 27$

21. $10xy - 15y + 8x - 12$ **22.** $35 - 42m - 18mn + 15n$

23. $a^2 - 4 + ab - 2b$ **24.** $x^2 - 25 - xy + 5y$

Set II   Factor each polynomial completely, or write "Not factorable."

1. $cz + dz + cw + dw$ **2.** $sx + sy + tx + ty$

3. $xz + 5z + 2sx + 10s$ **4.** $4cz - 10c + 6bz - 15b$

5. $4ab - 12ac + 3b - 9c$ **6.** $5yz - 3xz + 15xy - 9x^2$

7. $hw - kw - hz + kz$ **8.** $bx - by - cx + cy$

9. $ef + f - e - 1$ **10.** $ef + f - e + 1$

11. $2s^2 - 6st + 5s - 15t$ **12.** $6cd + 4c - 15de - 10e$

13. $6a^2 - 3ab - 14a + 7b$ **14.** $10x^2 - 5xy - 4x + 2y$

15. $x^2 - y^2 + 5x - 5y$ **16.** $s^2 - t^2 + 2s + 2t$

17. $y^3 - 2y^2 - 4y - 8$ **18.** $a^3 - 3a^2 - 5a + 15$

19. $u^3 - 3u^2 - 9u + 27$ **20.** $v^3 + 2v^2 - 25v - 50$

21. $12xy - 8y + 15x - 10$ **22.** $x^3 - 2x^2 + 9x - 18$

23. $x^2 - 9 + xy + 3y$ **24.** $y^2 - 16 - xy - 4x$

7.6 The Master Product Method for Factoring Trinomials (Optional)

The Master Product method, which makes use of factoring by grouping, is a method for factoring trinomials that eliminates some of the guesswork. It can also be used to determine whether a trinomial is factorable or not. The rules that follow assume that any common factors have been factored out.

TO FACTOR A TRINOMIAL BY THE MASTER PRODUCT METHOD

Arrange the trinomial in descending powers of one variable:

$$ax^2 + bxy + cy^2$$

1. Find the Master Product (MP) by multiplying the first and last coefficients of the trinomial being factored (MP = $a \cdot c$).

2. Write the pairs of factors of the Master Product (MP).

3. Choose the pair of factors whose sum is the coefficient of the middle term (b).

4. Rewrite the given trinomial, replacing the middle term by the sum of two terms whose coefficients are the pair of factors found in step 3.

5. Factor the expression in step 4 by grouping.

6. Check your factoring by multiplying the binomial factors to see if their product is the given trinomial.

Example 1   Factor $7x + 2x^2 + 5$ by the Master Product method.
Solution

$$7x + 2x^2 + 5 = 2x^2 + 7x + 5 \longleftarrow \text{Arranged in descending powers of } x$$

Step 1. Master Product $= (2)(+5) = 10$; the middle coefficient is $+7$.

Step 2. $\qquad\qquad\qquad\qquad 10 = (-1)(-10) = (1)(10)$

$$= (-2)(-5) = \boxed{(2)(5)}$$

Step 3. The pair whose sum is $+7$ is $(2)(5)$.

Step 4. $\qquad 2x^2 + 7x + 5 = \underline{2x^2 + 2x} + \underline{5x + 5}$ $\qquad$ Replacing $7x$ with $2x + 5x$

Step 5. $\qquad\qquad\qquad\quad = 2x(x + 1) + 5(x + 1)$ $\qquad$ Factoring by grouping

$$= (x + 1)(2x + 5)$$

Therefore, $2x^2 + 7x + 5 = (x + 1)(2x + 5)$.

Step 6. *Check* $(x + 1)(2x + 5) = 2x^2 + 7x + 5 = 7x + 2x^2 + 5$ ∎

Example 2   Factor $4z^2 - 3z - 7$ by the Master Product method.
Solution

Step 1. Master Product $= (4)(-7) = -28$; middle coefficient is -3.

Step 2. $-28 = (-1)(28) = (1)(-28)$

$$= (-2)(14) = (2)(-14)$$

Step 3. $\qquad = (-4)(7) = \boxed{(4)(-7)} \longleftarrow \text{This is the pair whose sum is } -3$

Step 4. $4z^2 - 3z - 7 = \underline{4z^2 + 4z} - \underline{7z - 7}$ $\qquad$ Replacing $-3z$ with $4z - 7z$

Step 5. $\qquad\qquad\quad = 4z(z + 1) - 7(z + 1)$ $\qquad$ Factoring by grouping

$$= (z + 1)(4z - 7)$$

Therefore, $4z^2 - 3z - 7 = (z + 1)(4z - 7)$.

Step 6. *Check* $(z + 1)(4z - 7) = 4z^2 - 3z - 7$ ∎

Example 3   Factor $12a^2 + 5a - 10$.
Solution

Step 1. Master Product $= (12)(-10) = -120$; middle coefficient is $+5$.

Step 2. $\qquad\qquad\qquad -120 = (-1)(120) = (1)(-120)$

$$= (-2)(60) \ = (2)(-60)$$

$$= (-3)(40) \ = (3)(-40)$$

$$= (-4)(30) \ = (4)(-30)$$

$$= (-5)(24) \ = (5)(-24)$$

$$= (-6)(20) \ = (6)(-20)$$

$$= (-8)(15) \ = (8)(-15)$$

$$= (-10)(12) = (10)(-12)$$

None of the sums of these pairs is $+5$. Therefore, the trinomial is *not factorable*. ∎

NOTE   The Master Product method can also be used with trinomials whose leading coefficient is 1; however, we think the method presented in Section 7.3A is shorter and simpler for trinomials of that type.   ☑

EXERCISES 7.6   Do Exercises 1–30 of Exercises 7.3B, factoring by the Master Product method.

7.7 How to Select the Method of Factoring

The following procedure can be used to select the correct method for factoring a polynomial:

First, check for a common factor, no matter how many terms the expression has. If there is a common factor, factor it out. (Section 7.1)

If the expression to be factored has two terms:

1. Is it a difference of two squares? (Section 7.2B)

2. Is it a sum of two squares? (If so, it is *not* factorable.)

If the expression to be factored has three terms:

1. Is the leading coefficient 1? (Section 7.3A)

2. Is the leading coefficient unequal to 1? (Section 7.3B or 7.6)

3. Is the trinomial the square of a binomial? (Section 7.3B)

If the expression to be factored has four terms, can it be factored by grouping? (Section 7.5)

Check to see if any factor can be factored again. When the expression is completely factored, the same factors are obtained no matter what method is used.

Check the result by multiplying the factors together.

Example 1   Factor $27x^2 - 12y^2$.
Solution

$$27x^2 - 12y^2 \qquad \text{3 is the GCF}$$

$$= 3(\;9x^2 - 4y^2\;) \longleftarrow (9x^2 - 4y^2) = (3x + 2y)(3x - 2y)$$

$$= 3(3x + 2y)(3x - 2y)$$

Check $3(3x + 2y)(3x - 2y) = 3(9x^2 - 4y^2) = 27x^2 - 12y^2$ ■

Example 2   Factor $3x^3 - 27x + 5x^2 - 45$.
Solution This expression has four terms.

$$\underbrace{3x^3 - 27x}_{} + \underbrace{5x^2 - 45}_{} \qquad \text{Factor by grouping}$$

$$= 3x(x^2 - 9) + 5(x^2 - 9)$$

This factor can be factored again

$$= (\boxed{x^2 - 9})(3x + 5) \qquad (x^2 - 9) = (x + 3)(x - 3)$$

$$= (x + 3)(x - 3)(3x + 5)$$

Check

$$(x + 3)(x - 3)(3x + 5) = (x^2 - 9)(3x + 5)$$
$$= (x^2 - 9)(3x) + (x^2 - 9)(5)$$
$$= 3x^3 - 27x + 5x^2 - 45 \quad \blacksquare$$

Example 3   Factor $5x^2 + 10x - 40$.
Solution

$$5x^2 + 10x - 40 \qquad \text{5 is the GCF}$$

This factor can be factored again

$$= 5(\boxed{x^2 + 2x - 8}) \longleftarrow (x^2 + 2x - 8) = (x - 2)(x + 4)$$

$$= 5(x - 2)(x + 4)$$

Check $5(x - 2)(x + 4) = 5(x^2 + 2x - 8) = 5x^2 + 10x - 40 \quad \blacksquare$

Example 4   Factor $a^4 - b^4$.
Solution This expression has two terms and is a difference of two squares.

$$a^4 - b^4$$

This factor can be factored again

$$= (a^2 + b^2)(\boxed{a^2 - b^2}) \longleftarrow (a^2 - b^2) = (a + b)(a - b)$$

$$= (a^2 + b^2)(a + b)(a - b)$$

Check $(a^2 + b^2)(a + b)(a - b) = (a^2 + b^2)(a^2 - b^2) = a^4 - b^4 \quad \blacksquare$

EXERCISES 7.7

Set I   Factor each polynomial completely, or write "Not factorable."

1. $2x^2 - 8y^2$ **2.** $3x^2 - 27y^2$ **3.** $5a^4 - 20b^2$

4. $6m^2 - 54n^4$ **5.** $x^4 - y^4$ **6.** $a^4 - 16$

7. $4v^2 + 14v - 8$ **8.** $6v^2 - 27v - 15$ **9.** $8z^2 - 12z - 8$

10. $18z^2 - 21z - 9$ **11.** $4x^2 - 100$ **12.** $9x^2 - 36$

13. $12x^2 + 10x - 8$ **14.** $45x^2 - 6x - 24$ **15.** $ab^2 - 2ab + a$

16. $au^2 - 2au + a$ **17.** $x^4 - 81$ **18.** $16y^8 - z^4$

19. $16x^2 + 16$ **20.** $25b^2 + 100$

21. $2u^3 + 2u^2v - 12uv^2$

22. $3m^3 - 3m^2n - 36mn^2$

23. $8h^3 - 20h^2k + 12hk^2$

24. $15h^2k - 35hk^2 + 10k^3$

25. $a^5b^2 - 4a^3b^4$

26. $x^2y^4 - 100x^4y^2$

27. $2ax^2 - 8a^3y^2$

28. $3b^2x^4 - 12b^2y^2$

29. $12 + 4x - 3x^2 - x^3$

30. $45 - 9z - 5z^2 + z^3$

31. $6my - 4nz + 15mz - 5zn$

32. $10xy + 5mn - 6xy - mn$

33. $x^4 - 8x^2 + 16$

34. $y^4 - 18y^2 + 81$

35. $x^8 - 1$

36. $a^8 - b^8$

37. $x^2 - a^2 - 4x + 4a$

38. $m^2 - 25 + mn - 5n$

39. $6ac - 6bd + 6bc - 6ad$

40. $10cy - 6cz + 5dy - 3dz$

Set II Factor each polynomial completely, or write "Not factorable."

1. $3a^2 - 75b^2$

2. $7c^2 - 63b^2$

3. $4h^4 - 36b^2$

4. $9x^4 - 36y^2$

5. $x^4 - 16y^4$

6. $m^4 - 1$

7. $10x^2 + 25x - 15$

8. $6x^2 + 15x - 9$

9. $10y^2 + 14y - 12$

10. $6x^2 + 6x + 12$

11. $16x^2 - 36$

12. $25a^2 - 100$

13. $30w^2 + 27w - 21$

14. $8x^2 + 22x - 6$

15. $h^2k - 4hk + 4k$

16. $x^2y + 4xy + 4$

17. $81c^4 - 16$

18. $81c^2 + 16$

19. $5x^2 + 20$

20. $4m^3n^3 - mn^5$

21. $5wz^2 + 5w^2z - 10w^3$

22. $2t^2r^4 - 18t^4$

23. $12x^2y - 42xy^2 + 36y^3$

24. $xy + 3y - 4x - 12$

25. $6x^3y^2 - 12xy^4$

26. $5x^4y + 20x^2y^3$

27. $3bx^2 - 12b^3y^2$

28. $9x^2 + 36$

29. $45 + 9b - 5b^2 - b^3$

30. $12 + 4x - 3x^2 - x^3$

31. $3xy + 2xz - 8xw + 3xz$

32. $8wx + 5xy - 4yz - 11yz$

33. $x^4 - 2x^2 + 1$

34. $x^3 + 3x^2 - 25x - 75$

35. $y^8 - 1$

36. $x^2 + 1$

37. $a^2 - 4b^2 + 2a + 4b$

38. $10ac + 10ad - 5bc - 5bd$

39. $6ef + 3gf - 12eh - 9gh$

40. $x^2 - 9y^2 + 2x - 6y$

7.8 Solving Equations by Factoring

Factoring has many applications. In this section, we use factoring to solve equations.

A **polynomial equation** is an equation that has a polynomial on both sides of the equal sign; the polynomial on one side of the equal sign can be the zero polynomial, 0. The *degree* of the equation equals the degree of the highest-degree *term* in the equation.

Polynomial equations with a first-degree term as the highest-degree term are called **first-degree** or **linear equations**. (All the equations that were solved in Chapter 4 were first-degree equations.) Polynomial equations with a second-degree term as the highest-degree term are called **second-degree** or **quadratic equations**.

Example 1   Examples of polynomial equations:

    a. $5x - 3 = 0$ Linear (or first-degree) equation in one variable

    b. $2x^2 - 4x + 7 = 0$ Quadratic (or second-degree) equation in one variable ■

We are now ready to solve quadratic and higher-degree equations. One method of solving such equations is based on Rule 7.1, which is stated without proof.

RULE 7.1

If the product of two factors is zero, then one or both of the factors must be zero.

$$\text{If } a \cdot b = 0, \text{ then } \begin{cases} a = 0 \\ \text{or } b = 0 \\ \text{or both } a \text{ and } b = 0 \end{cases}$$

Rule 7.1 can be extended to include more than two factors; that is, if a product of factors is zero, at least one of the factors must be zero. We use Rule 7.1 in solving higher-degree equations. The method is summarized below.

TO SOLVE AN EQUATION BY FACTORING

1. Write all nonzero terms on one side of the equation by adding the same expression to both sides. *Only zero must remain on the other side.* Then arrange the polynomial in descending powers.

2. Factor the polynomial completely.

3. Set each factor equal to zero.*

4. Solve each resulting first-degree equation.

5. Check apparent solutions in the original equation.

In Examples 2–4, we already have zero on one side of the equal sign, and the polynomial has already been factored. Therefore, we proceed with step 3.

Example 2   Solve $(x - 1)(x - 2) = 0$.
Solution Since $(x - 1)(x - 2) = 0$,

Step 3. then $(x - 1) = 0$ *or* $(x - 2) = 0$.

Step 4. If $x - 1 = \quad 0$ | If $x - 2 = \quad 0$
                $\underline{+ 1 \quad + 1}$ | $\underline{+ 2 \quad + 2}$
    then $x \quad = \quad 1$ | then $x \qquad = \quad 2$

---

*If any of the factors are not first-degree polynomials, we cannot solve the equation at this time.

Step 5. *Check for x = 1* | *Check for x = 2*

$$(x - 1)(x - 2) = 0 \qquad\qquad (x - 1)(x - 2) = 0$$

$$(1 - 1)(1 - 2) \stackrel{?}{=} 0 \qquad (2 - 1)(2 - 2) \stackrel{?}{=} 0$$

$$(0)(-1) \stackrel{?}{=} 0 \qquad\qquad (1)(0) \stackrel{?}{=} 0$$

$$0 = 0 \qquad\qquad\qquad 0 = 0$$

Therefore, 1 and 2 are solutions for the equation $(x - 1)(x - 2) = 0$. ∎

Example 3 Solve $3(x + 2)(x - 1) = 0$.
Solution $3(x + 2)(x - 1) = 0$

Step 3. $3 \neq 0$

	$x + 2 =$	0		$x - 1 =$	0
Step 4.		-2	-2		$+1$	$+1$
	x	$= -2$		x	$= 1$

Step 5. We leave the checking of the solutions -2 and 1 to the student. ∎

Example 4 Solve $2x(x - 3)(x + 4) = 0$.
Solution $2x(x - 3)(x + 4) = 0$

Step 3. $2x = 0$ | $x - 3 = 0$ | $x + 4 = 0$

Step 4. $\dfrac{2x}{2} = \dfrac{0}{2}$ | $+3 \quad +3$ | $-4 \quad -4$

$x = 0$ | $x = 3$ | $x = -4$

Step 5. We leave the checking of the solutions 0, 3, and -4 to the student. ∎

Example 5 Solve $x^2 - 9x = 0$.
Solution We have 0 on one side of the equal sign. We proceed with step 2.

$$x^2 - 9x = 0$$

Step 2. $\qquad\qquad x(x - 9) = 0 \qquad$ Factoring the left side

Step 3. $\qquad\qquad x = 0 \quad\Big|\quad x - 9 = 9$
Step 4. $\qquad\qquad\qquad\qquad\qquad\quad 9 \quad 9$
$\qquad\qquad\qquad\qquad\qquad\qquad x \quad = 9$

Step 5. *Check for x = 0* | *Check for x = 9*

$$x^2 - 9x = 0 \qquad\qquad x^2 - 9x = 0$$

$$0^2 - 9(0) \stackrel{?}{=} 0 \qquad 9^2 - 9(9) \stackrel{?}{=} 0$$

$$0 - 0 \stackrel{?}{=} 0 \qquad\qquad 81 - 81 \stackrel{?}{=} 0$$

$$0 = 0 \qquad\qquad\qquad 0 = 0$$

The solutions are 0 and 9. ∎

Example 6 Solve $6x^2 = 5 - 7x$.
Solution

$$6x^2 \qquad\qquad = \quad 5 - 7x$$

Step 1. $\underline{\qquad +7x - 5 \qquad -5 + 7x} \qquad$ Adding $7x - 5$ to both sides

$$6x^2 + 7x - 5 = \quad 0$$

Step 2. $(2x - 1)(3x + 5) = 0$ — Factoring the left side

Step 3. $2x - 1 = 0$ $\qquad$ $3x + 5 = 0$

Step 4. $\dfrac{+ 1 +1}{2x = 1}$ $\qquad$ $\dfrac{- 5 -5}{3x = -5}$

$x = \dfrac{1}{2}$ $\qquad\qquad$ $x = -\dfrac{5}{3}$

Step 5. $\quad$ *Check for $x = \dfrac{1}{2}$* $\qquad\qquad$ *Check for $x = -\dfrac{5}{3}$*

$$6x^2 = 5 - 7x \qquad\qquad 6x^2 = 5 - 7x$$

$$6\left(\frac{1}{2}\right)^2 \overset{?}{=} 5 - 7\left(\frac{1}{2}\right) \qquad 6\left(-\frac{5}{3}\right)^2 \overset{?}{=} 5 - 7\left(-\frac{5}{3}\right)$$

$$6\left(\frac{1}{4}\right) \overset{?}{=} 5 - \frac{7}{2} \qquad\qquad 6\left(\frac{25}{9}\right) \overset{?}{=} 5 + \frac{35}{3}$$

$$\frac{6}{4} \overset{?}{=} \frac{10 - 7}{2} \qquad\qquad \frac{2 \cdot 25}{3} \overset{?}{=} \frac{15 + 35}{3}$$

$$\frac{3}{2} = \frac{3}{2} \qquad\qquad\qquad \frac{50}{3} = \frac{50}{3}$$

The solutions are $\frac{1}{2}$ and $-\frac{5}{3}$. ∎

A WORD OF CAUTION The product must equal zero, or no conclusions can be drawn about the factors.

Suppose $(x - 1)(x - 3) = $ 8 .

└── No conclusion can be drawn because the product ≠ 0

Students sometimes think that

$$\text{if} \qquad (x - 1)(x - 3) = 8$$
$$\text{then } x - 1 = 8 \quad \Big| \quad x - 3 = 8$$
$$x = 9 \quad \Big| \quad x = 11$$

This "solution" is incorrect, because

if $\qquad x = 9$ $\qquad\qquad$ *or* $\quad$ if $\qquad x = 11$

then $(x - 1)(x - 3)$ $\qquad\qquad$ then $(x - 1)(x - 3)$

$(9 - 1)(9 - 3) =$ $\qquad\qquad\qquad$ $(11 - 1)(11 - 3) =$

$8 \cdot 6 = 48 \neq 8$ $\qquad$ $10 \cdot 8 = 80 \neq 8$

The correct solution is

$$(x - 1)(x - 3) = 8$$
$$x^2 - 4x + 3 = 8$$

Step 1. $\qquad\qquad \dfrac{- 8 -8}{x^2 - 4x - 5 = 0}$ $\qquad$ Adding -8 to both sides

Step 2. $\qquad\qquad (x - 5)(x + 1) = 0$ $\qquad$ Factoring the left side

Step 3. $\qquad x - 5 = 0 \qquad\qquad x + 1 = 0$
Step 4. $\qquad\qquad\quad \underline{+ 5 \quad +5} \qquad\qquad \underline{-1 \quad -1}$
$\qquad\qquad\qquad\qquad\quad x \quad = \quad 5 \qquad\qquad x \qquad = -1$

Step 5. A check will verify that 5 and -1 are the solutions. ☑

Example 7   Solve $(x - 5)(x + 4) = -14$.
Solution We first remove the parentheses.

$$(x - 5)(x + 4) = -14$$

$$x^2 - x - 20 = -14 \qquad \text{Removing parentheses}$$
Step 1. $\qquad\qquad \underline{+ 14 \qquad +14} \qquad \text{Adding 14 to both sides}$
$$x^2 - x - 6 = 0$$

Step 2. $\qquad (x + 2)(x - 3) = 0 \qquad \text{Factoring the left side}$

Step 3. $\qquad\qquad\qquad x + 2 = 0 \qquad\quad x - 3 = 0$
Step 4. $\qquad\qquad\qquad \underline{-2 \quad -2} \qquad \underline{+3 \quad +3}$
$\qquad\qquad\qquad\qquad\quad x \quad = -2 \qquad\quad x \quad = 3$

Step 5. $\qquad$ *Check for $x = -2$* $\qquad$ *Check for $x = 3$*

$(x - 5)(x + 4) = -14 \qquad (x - 5)(x + 4) = -14$

$(-2 - 5)(-2 + 4) \overset{?}{=} -14 \qquad (3 - 5)(3 + 4) \overset{?}{=} -14$

$(-7)(2) \overset{?}{=} -14 \qquad (-2)(7) \overset{?}{=} -14$

$-14 = -14 \qquad\qquad -14 = -14$

The solutions are -2 and 3. ∎

Example 8   Solve $3x^3 = 4x - x^2$.
Solution

$$3x^3 \qquad\qquad = \quad 4x - x^2$$
Step 1. $\qquad\quad \underline{+ x^2 - 4x \qquad -4x + x^2} \qquad \text{Adding } x^2 - 4x \text{ to both sides}$
$$3x^3 + x^2 - 4x = 0$$

Step 2. $\qquad x(3x^2 + x - 4) = 0 \qquad \text{Factoring out the GCF}$
$\qquad\quad x(x - 1)(3x + 4) = 0 \qquad \text{Factoring the polynomial factor}$

Step 3. $\qquad\qquad\qquad x = 0 \quad\Big|\quad x - 1 = 0 \quad\Big|\quad 3x + 4 = 0$
Step 4. $\qquad\qquad\qquad\qquad\qquad\qquad \underline{+1 \quad +1} \qquad \underline{-4 \quad -4}$
$\qquad\qquad\qquad\qquad\qquad\qquad\qquad x \quad = 1 \qquad 3x \quad = -4$
$$x = -\frac{4}{3}$$

Step 5. *Check for $x = 0$* $\qquad$ *Check for $x = 1$* $\qquad$ *Check for $x = -\dfrac{4}{3}$*

$3x^3 = 4x - x^2 \qquad 3x^3 = 4x - x^2 \qquad 3x^3 = 4x - x^2$

$3(0)^3 \overset{?}{=} 4(0) - 0^2 \qquad 3(1)^3 \overset{?}{=} 4(1) - 1^2 \qquad 3\left(-\frac{4}{3}\right)^3 \overset{?}{=} 4\left(-\frac{4}{3}\right) - \left(-\frac{4}{3}\right)^2$

$3(0) \overset{?}{=} 0 - 0 \qquad 3(1) \overset{?}{=} 4 - 1 \qquad 3\left(-\frac{64}{27}\right) \overset{?}{=} -\frac{16}{3} - \frac{16}{9}$

$0 = 0 \qquad\qquad 3 = 3 \qquad\qquad -\frac{64}{9} \overset{?}{=} -\frac{48}{9} - \frac{16}{9}$

$-\frac{64}{9} = -\frac{64}{9}$

The solutions are 0, 1, and $-\frac{4}{3}$. ∎

A WORD OF CAUTION   Pay close attention to whether you're asked to *factor* a polynomial or to *solve* a polynomial equation! If you're asked to *factor* $x^2 - 9x$, the correct answer is "$x(x - 9)$," *not* "$x = 0$ or $x = 9$." (You can't attach "$= 0$" to $x^2 - 9x$.) On the other hand, if you're asked to *solve the equation* $x^2 - 9x = 0$, the correct answer is "The solutions are 0 and 9," *not* "$x(x - 9)$"; "$x(x - 9)$" is not the solution for an equation.   ☑

EXERCISES 7.8

Set I   Solve and check each of the following equations.

1. $(x - 5)(x + 4) = 0$ **2.** $(x + 7)(x - 2) = 0$ **3.** $3x(x - 4) = 0$

4. $5x(x + 6) = 0$ **5.** $(x + 10)(2x - 3) = 0$ **6.** $(x - 8)(3x + 2) = 0$

7. $x^2 + 9x + 8 = 0$ **8.** $x^2 + 6x + 8 = 0$ **9.** $x^2 - x - 12 = 0$

10. $x^2 + x - 12 = 0$ **11.** $x^2 = 64$ **12.** $x^2 = 144$

13. $6x^2 - 10x = 0$ **14.** $6y^2 - 21y = 0$ **15.** $24w = 4w^2$

16. $20m = 5m^2$ **17.** $5a^2 = 16a - 3$ **18.** $3z^2 = 22z - 7$

19. $3u^2 = 2u + 5$ **20.** $5k^2 = 34k + 7$ **21.** $(x - 2)(x - 3) = 2$

22. $(x - 3)(x - 5) = 3$ **23.** $x(x - 4) = 12$ **24.** $x(x - 2) = 15$

25. $4x(2x - 1)(3x + 7) = 0$ **26.** $5x(4x - 3)(7x - 6) = 0$

27. $2x^3 + x^2 = 3x$ **28.** $4x^3 = 10x - 18x^2$

29. $2a^3 - 10a^2 = 0$ **30.** $4b^3 - 24b^2 = 0$

Set II   Solve and check each of the following equations.

1. $(x + 3)(x - 5) = 0$ **2.** $(y - 8)(y + 9) = 0$

3. $2y(y - 7) = 0$ **4.** $7x(x + 4) = 0$

5. $(z - 6)(3z + 2) = 0$ **6.** $(5x - 10)(4x + 5) = 0$

7. $a^2 + 8a + 12 = 0$ **8.** $x^2 + 2x = 48$

9. $m^2 + 3m - 18 = 0$ **10.** $2x^2 + 11x + 15 = 0$

11. $w^2 - 24 = 5w$ **12.** $x^2 = 4x + 5$

13. $5h^2 - 20h = 0$ **14.** $k^3 = 5k^2$

15. $12t = 6t^2$ **16.** $x = 10 - 2x^2$

17. $3n^2 = 7n + 6$ **18.** $(x - 4)(x + 2) = -9$

19. $13x + 3 = -4x^2$ **20.** $y(3y - 2)(y + 5) = 0$

21. $(y - 3)(y - 6) = -2$ **22.** $10x^2 + 11x = 6$

23. $u(u - 9) = -14$

24. $(x - 2)(x + 4) = 7$

25. $2x(3x - 2)(5x + 9) = 0$

26. $8y(2y^2 + 5y - 3) = 0$

27. $21x^2 + 60x = 18x^3$

28. $12y^2 = 4y^3 + 5y$

29. $3a^2 + 18a = 0$

30. $(x - 5)(x + 6) = -10$

7.9 Word Problems Solved by Factoring

For your convenience, we repeat here the main suggestions for solving word problems. In this section, most of the equations you need to solve (see step 3) will be second-degree (quadratic) equations.

METHOD FOR SOLVING WORD PROBLEMS

Read To solve a word problem, first read it very carefully.

Think Determine what *type* of problem it is, if possible. Determine what is unknown. Do you need a special formula?

Sketch Draw a sketch *with labels*, if possible.

Step 1. Represent one unknown number by a variable, and declare its meaning in a sentence of the form "Let $x = \ldots$." Then express any other unknown numbers in terms of the same variable.

Reread Reread the entire word problem, breaking it up into small pieces.

Step 2. Translate each English phrase into an algebraic expression and fit these expressions together into an equation or inequality.

Step 3. Solve the equation or inequality.

Step 4. Solve for *all* the unknowns asked for in the problem.

Step 5. Check the solution(s) *in the word statement*.

Step 6. State your results clearly.

Example 1   The difference of two numbers is 3. Their product is 10. What are the two numbers?

Solution

Step 1. Let $\left.\begin{array}{l} x = \text{the smaller number} \\ x + 3 = \text{the larger number} \end{array}\right\}$ Since their difference is 3

Reread   Their product is 10

Step 2. $x(x + 3) = 10$ Remember, it is incorrect to say that $x = 10$ or $x + 3 = 10$

Step 3. $x^2 + 3x = 10$

$\underline{ - 10 \quad -10}$ Adding -10 to both sides

$x^2 + 3x - 10 = 0$

$(x - 2)(x + 5) = 0$ Factoring the polynomial

$$\begin{array}{rcl} x - 2 & = & 0 \\ +2 & & +2 \end{array} \quad \bigg| \quad \begin{array}{rcl} x + 5 & = & 0 \\ -5 & & -5 \end{array} \qquad \text{Setting each factor equal to zero}$$

Step 4.
$$\begin{array}{rcl} x & = & 2 \\[4pt] x + 3 & = & 5 \end{array} \quad \bigg| \quad \begin{array}{rcl} x & = & -5 \\[4pt] x + 3 & = & -2 \end{array} \qquad \begin{array}{l} \text{Smaller number} \\[10pt] \text{Larger number} \end{array}$$

Step 5. *Check for 2 and 5* The difference is $5 - 2 = 3$. The product is $5 \cdot 2 = 10$.
Check for -5 *and* -2 The difference is $-2 - (-5) = -2 + 5 = 3$. The product is $(-2)(-5) = 10$.

Step 6. Therefore the numbers 2 and 5 are a solution, and the numbers -5 and -2 are another solution. ∎

Example 2 Find two consecutive integers whose product is 19 more than their sum.

Solution

Step 1. Let $x = $ the smaller integer

$x + 1 = $ the larger integer

Then $x(x + 1)$ represents their product, and $x + (x + 1)$ represents their sum.

Reread　　　　Their product　is　19　more than　their sum

Step 2.
$$x(x + 1) \quad = \quad 19 \quad + \quad x + (x + 1)$$

Step 3.
$$x^2 + x = 19 + 2x + 1$$

$$\begin{array}{rcl} x^2 + x & = & 2x + 20 \\ -2x - 20 & & -2x - 20 \qquad \text{Adding } -2x - 20 \text{ to both sides} \\ \hline x^2 - x - 20 & = & 0 \end{array}$$

$$(x + 4)(x - 5) = 0$$

Step 4.
$$\begin{array}{rcl} x + 4 & = & 0 \\ -4 & & -4 \\ \hline x & = & -4 \\[4pt] x + 1 & = & -3 \end{array} \quad \bigg| \quad \begin{array}{rcl} x - 5 & = & 0 \\ 5 & & 5 \\ \hline x & = & 5 \\[4pt] x + 1 & = & 6 \end{array}$$

Step 5. *Check for* -4 *and* -3 Their sum is -7. Their product is 12. 12 is 19 more than -7.
Check for 5 and 6 Their sum is 11. Their product is 30. 30 is 19 more than 11.

Step 6. There are two answers: The numbers are -4 and -3, or the numbers are 5 and 6. ∎

Example 3 Find three consecutive odd integers such that the product of the first two is 21 more than 6 times the third.

Solution

Step 1. Let $x = $ the first odd integer

$x + 2 = $ the second odd integer

$x + 4 = $ the third odd integer

Reread　　The product of the first two numbers　is　21　more than　6 times the third

Step 2.
$$x(x + 2) \quad = \quad 21 \quad + \quad 6(x + 4)$$

Step 3.
$$x^2 + 2x = 21 + 6x + 24$$

$$x^2 + 2x = 45 + 6x$$

$$\underline{-6x - 45 \qquad -45 - 6x} \qquad \text{Adding } -6x - 45 \text{ to both sides}$$

$$x^2 - 4x - 45 = 0$$

$$(x + 5)(x - 9) = 0$$

Step 4.

$x + 5 = 0$

$\underline{\quad -5 \quad -5}$

$x \quad = -5$ The first integer

$x + 2 = -3$ The second integer

$x + 4 = -1$ The third integer

$x - 9 = 0$

$\underline{\qquad\quad 9 \quad 9}$

$x \quad = 9$ The first integer

$x + 2 = 11$ The second integer

$x + 4 = 13$ The third integer

Step 5.

Check for $x = -5$

$$(-5)(-3) \overset{?}{=} 21 + 6(-1)$$

$$15 \overset{?}{=} 21 - 6$$

$$15 = 15$$

Check for $x = 9$

$$(9)(11) \overset{?}{=} 21 + 6(13)$$

$$99 \overset{?}{=} 21 + 78$$

$$99 = 99$$

Step 6. There are two answers: The integers are -5, -3, and -1, or the integers are 9, 11, and 13. ■

In solving word problems about geometric figures, make a drawing of the figure and write the given information on it. Lengths of sides of geometric figures cannot be negative; therefore, any value of the variable that would make a length negative must be rejected.

Example 4   The base of a triangle is 4 cm more than its altitude. The area is 30 sq. cm. Find the altitude and the base. (30 sq. cm = 30 cm²)

Solution

Step 1. Let h = altitude (in centimeters)   *Sketch*

$h + 4$ = base

Area of triangle $= \dfrac{1}{2}$(base)(altitude)

$$= \frac{1}{2}(h + 4)h$$

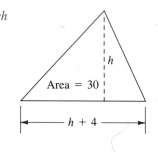

Reread   Area of triangle   is   30   sq. cm

Step 2. $\dfrac{1}{2}(h + 4)h = 30$ "Clear fractions" by multiplying both sides of the equation by 2

Step 3. $(h + 4)h = 60$

$$h^2 + 4h = 60$$

$$\underline{\qquad\quad -60 \qquad -60} \qquad \text{Adding } -60 \text{ to both sides}$$

$$h^2 + 4h - 60 = 0$$

$$(h - 6)(h + 10) = 0$$

Step 4.

$h - 6 = 0$

$\underline{\qquad 6 \quad 6}$

Altitude $h \quad = 6$

Base $h + 4 = 10$

$h + 10 = 0$

$\underline{\qquad -10 \quad -10}$

$h \quad = -10 \longleftarrow$ -10 must be rejected since a length cannot be negative

Step 5. *Check* The base is 4 cm more than the altitude. The area is $\frac{1}{2}$(6 cm)(10 cm) = 30 cm^2.

Step 6. Therefore, the triangle has an altitude of 6 cm and a base of 10 cm. ■

Example 5 One square has a side 3 ft longer than the side of a second square. If the area of the larger square is 4 times as great as the area of the smaller square, find the length of the side of each square.
Solution

Step 1. Let x = length of side of smaller square
$x + 3$ = length of side of larger square

Sketch

Area = x^2

x

Area = $(x + 3)^2$

$x + 3$

Reread   The area of the larger square   is   4 times as great as   the area of the smaller square

Step 2. $(x + 3)^2 =$ $4 \cdot$ x^2

Step 3. $(x + 3)^2 =$ $4x^2$

$x^2 + 6x + 9 =$ $4x^2$

$\underline{-4x^2}$ $\underline{-4x^2}$

$-3x^2 + 6x + 9 =$ 0

$-3(x^2 - 2x - 3) =$ 0

$-3(x - 3)(x + 1) =$ 0

Step 4. $-3 \neq 0 \mid x - 3 = 0$ $\qquad x + 1 = \quad 0$

$\qquad\qquad\qquad \underline{3 \quad 3}$ $\qquad \underline{-1 \quad -1}$

Small square $\qquad x \quad = 3$ $\qquad x \qquad = -1$ ◄—— -1 cannot be a solution of the word statement

Large square $\qquad x + 3 = 6$

Step 5. *Check* Area of smaller square is (3 ft)^2 = 9 ft^2. Area of larger square is (6 ft)^2 = 36 ft^2. The area of the larger square (36 sq. ft) is 4 times as great as the area of the smaller square (9 sq. ft).

Step 6. Therefore, the smaller square has a side of 3 ft and the larger square a side of 6 ft. ■

Example 6 The width of a rectangle is 5 cm less than its length. Its area is 10 more (numerically*) than its perimeter. What are the dimensions of the rectangle?
Solution

Step 1.

Sketch

$l - 5$

l

Let l = length

$l - 5$ = width

Area = (length)(width) = $l(l - 5)$

Perimeter = $2l + 2(l - 5)$

———
*We say that the area is "numerically" 10 more than the perimeter because the area is measured in *square centimeters* whereas the perimeter is measured in *centimeters*.

Reread   Its area   is   10 more than   its perimeter

Step 2. $l(l - 5) =$ $10 +$ $2l + 2(l - 5)$

Step 3. $l(l - 5) = 10 + 2l + 2(l - 5)$

$l^2 - 5l = 10 + 2l + 2l - 10$

$$\begin{array}{rcl} l^2 - 5l &=& 4l \\ -4l & & -4l \\ \hline l^2 - 9l &=& 0 \end{array}$$

$l(l - 9) = 0$

Step 4. $l = 0$ $l - 9 = 0$
                                           Reject; the $\underline{\quad 9 \quad 9}$
                                           length of a $l \quad = 9$ Length
                                           rectangle
                                           can't be zero $l - 5 = 4$ Width

Step 5. *Check* The width is 5 cm less than the length. The area is (4 cm)(9 cm) = 36 cm². The perimeter is 2(9 cm) + 2(4 cm) = 18 cm + 8 cm = 26 cm. 36 is 10 more than 26.

Step 6. Therefore, the rectangle has a length of 9 cm and a width of 4 cm. ∎

Example 7 The length of a rectangular solid is 2 cm more than its width. The height is 3 cm, and the volume is 72 cc (cubic centimeters). Find the width and the length.
Solution

Step 1. Let $w =$ width   *Sketch*

$w + 2 =$ length

Volume = (length)(width)(height) = $(w + 2)(w)(3)$

Reread   The volume   is   72 $w + 2$

Step 2. $(w + 2)(w)(3) =$ 72

$$\begin{array}{rcl} 3w^2 + 6w &=& 72 \\ -72 & & -72 \\ \hline 3w^2 + 6w - 72 &=& 0 \end{array}$$

Step 3.

$3(w^2 + 2w - 24) = 0$

$3(w - 4)(w + 6) = 0$

Step 4. $3 \neq 0$ $w - 4 = 0$ $w + 6 = 0$
                            $\underline{\quad 4 \quad 4}$ $\underline{\quad -6 \quad -6}$
                  Width $w \quad = 4$ $w \quad = -6$ ← Reject, since a length
                                                              can't be negative
                  Length $w + 2 = 6$

Step 5. *Check* The volume is (6 cm)(4 cm)(3 cm) = 72 cm³.

Step 6. Therefore, the width is 4 cm and the height is 6 cm. ∎

LENGTH

EXERCISES 7.9

Set I   Set up each problem algebraically and solve. Be sure to state what your variables represent.

1. The difference of two numbers is 5. Their product is 14. Find the numbers.

2. The difference of two numbers is 6. Their product is 27. Find the numbers.

3. The sum of two numbers is 12. Their product is 35. Find the numbers.

4. The sum of two numbers is −4. Their product is −12. Find the numbers.

5. The base of a triangle is 3 in. longer than the altitude. The area is 20 sq. in. Find the altitude and the base.

6. The base of a triangle is 3 cm longer than the altitude. The area is 90 sq. cm. Find the altitude and the base.

7. Find three consecutive integers such that the product of the first two plus the product of the last two is 8.

8. Find three consecutive integers such that the product of the first two plus the product of the first and third is 14.

9. Find three consecutive even integers such that twice the product of the first two is 16 more than the product of the last two.

10. Find three consecutive odd integers such that twice the product of the last two is 91 more than the product of the first two.

11. The length of a rectangle is 5 ft more than its width. Its area is 84 sq. ft. What are its dimensions?

12. The width of a rectangle is 3 ft less than its length. Its area is 28 sq. ft. What are its dimensions?

13. One square has a side 3 cm shorter than the side of a second square. The area of the larger square is 4 times as great as the area of the smaller square. Find the length of the side of each square.

14. One square has a side 4 ft longer than the side of a second square. The area of the larger square is 9 times as great as the area of the smaller square. Find the length of the side of each square.

15. The width of a rectangle is 4 yd less than its length. Its area is 17 more (numerically) than its perimeter. What are the dimensions of the rectangle?

16. The area of a square is twice its perimeter (numerically). What is the length of its side?

17. The base of a triangle is 3 in. more than its altitude. Its area is 35 sq. in. Find the base and the altitude.

18. The base of a triangle is 5 m more than its altitude. Its area is 18 sq. m. Find the base and the altitude.

19. The sum of the base and the altitude of a triangle is 19 in. The area is 42 sq. in. Find the base and the altitude.

20. The sum of the base and the altitude of a triangle is 15 cm. The area of the triangle is 27 sq. cm. Find the base and the altitude.

21. The length of a rectangular solid is 4 cm more than its height. Its width is 5 cm, and its volume is 225 cc. Find its length and its height.

22. The length of a rectangular solid is 2 in. more than its width. Its height is 6 in. and its volume is 378 cu. in. Find its length and its width.

Set II   Set up each problem algebraically and solve. Be sure to state what your variables represent.

1. The difference of two numbers is 12. Their product is 28. Find the numbers.

2. The difference of two numbers is 19. Their product is −84. Find the numbers.

3. The sum of two numbers is 10. Their product is −24. Find the numbers.

4. Find two consecutive integers whose product is 1 less than their sum.

5. The base of a triangle is 4 m longer than the altitude. The area is 48 sq. m. Find the altitude and the base.

6. The base of a triangle is 2 cm shorter than the altitude. The area is 40 sq. cm. Find the altitude and the base.

7. Find three consecutive integers such that the product of the first two minus the third is 7.

8. Find three consecutive integers such that the product of the first and third minus the second is 41.

9. Find three consecutive odd integers such that twice the product of the first two is 7 more than the product of the last two.

10. A 3-in.-wide mat surrounds a picture. The area of the *picture itself* is 176 sq. in. If the length of the outside of the mat is twice its width, what are the dimensions of the outside of the mat? What are the dimensions of the picture?

11. The length of a rectangle is 8 m more than its width. Its area is 48 sq. m. What are its dimensions?

12. The width of a rectangular box equals the length of the side of a certain cube. The length of the box is 3 cm more than its width, and the height of the box is 1 cm less than its width. The volume of the box is 9 cc more than the volume of the cube. Find the dimensions of the cube and of the box.

13. One square has a side 2 km shorter than the side of a second square. The area of the larger square is 9 times as great as the area of the smaller square. Find the length of the side of each square.

14. Bruce's vegetable garden is now square. If he forms a rectangle by increasing the length of one side by 3 ft and the length of the adjacent side by 6 ft, the area of the rectangle will be 3 times as great as the area of the square. What is the size of the vegetable garden now?

15. The width of a rectangle is 2 in. less than its length. Its area is 4 more (numerically) than its perimeter. What are the dimensions of the rectangle?

16. A pieced quilt 5 ft by 6 ft is to be surrounded by a border of uniform width. How wide should the border be if the area of the border is to be 4 sq. ft less than the area of the pieced quilt?

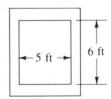

17. The base of a triangle is 5 cm more than its altitude. Its area is 7 sq. cm. What is the length of the altitude?

18. The difference of the base and altitude of a triangle is 4 ft. (The altitude is greater than the base.) The area of the triangle is 16 sq. ft. Find the base and the altitude.

19. The sum of the base and the altitude of a triangle is 17 m. The area of the triangle is 36 sq. m. Find the base and the altitude.

20. The length of a rectangular solid is 8 cm. Its width is 2 cm more than its height, and its volume is 280 cc. Find its height and its width.

21. The length of a rectangular solid is 5 cm more than its height. Its width is 3 cm, and its volume is 18 cc. Find its length and its height.

22. The length of a rectangular solid is 8 in. more than its width. Its height is 2 in. and its volume is 96 cu. in. Find its length and its width.

7.10 Review: 7.5–7.9

Factoring by Grouping
7.5

If a polynomial has four terms, it can sometimes be factored by grouping.

Master Product Method
7.6

The Master Product method can be used to factor a trinomial with a leading coefficient unequal to 1.

Factoring Completely
7.7

1. Look for a greatest common factor first.

2. Look for a difference of two squares.

3. Look for a factorable trinomial.

4. Look for four terms that can be factored by grouping.

5. Check to see if any factor already obtained can be factored again.

To Solve an Equation
by Factoring
7.8

1. Write all nonzero terms on one side of the equation by adding the same expression to both sides. Only zero must remain on the other side. Then arrange the polynomial in descending powers.

2. Factor the polynomial.

3. Set each factor equal to zero, and solve for the variable.

4. Check apparent solutions in the *original* equation.

CAUTION Pay close attention to whether you're asked to *factor* a polynomial or to *solve* a polynomial equation.

Word Problems Solved
by Factoring
7.9

Many word problems lead to second-degree (quadratic) equations that can be solved by factoring. (See the suggestions and examples in Section 7.9.)

Review Exercises 7.10 Set I

In Exercises 1–14, factor each expression completely, or write "Not factorable."

1. $5 + 15a$ **2.** $50 - 2n^2$ **3.** $ab + 2b - a - 2$

4. $y^3 + 10y^2 + 16y$ **5.** $y^2 + 5y + 5$ **6.** $3b - 6b^2$

7. $5c^2 - 22c + 8$ **8.** $mn - 5n - m + 5$ **9.** $5x^2 - 35x - 150$

10. $x^2 + 81$ **11.** $x^3 + 5x^2 + 3x + 15$ **12.** $x^3 + 3x^2 - 9x - 27$

13. $x^2 - y^2 + x - y$ **14.** $15a^2 + 15ab - 30b^2$

In Exercises 15–24, solve each equation.

15. $(x - 5)(x + 3) = 0$ **16.** $x^2 - 5x - 24 = 0$ **17.** $m^2 = 18 + 3m$

18. $x^2 - 36 = 0$ **19.** $3z^2 = 12z$ **20.** $12u^2 = 47u - 45$

21. $3x^2 + 13x = 10$ **22.** $10x^2 + 13x = 3$

23. $2u(u + 6)(u - 2) = 0$ **24.** $39y^2 = 18y^3 + 18y$

In Exercises 25–30, set up each problem algebraically and solve. Be sure to state what your variables represent.

25. The difference of two numbers is 3. Their product is 28. Find the numbers.

26. One side of a square is 6 ft longer than the side of a second square. The area of the larger square is 16 times as great as the area of the smaller square. Find the length of the side of each square.

27. The width of a rectangle is 3 m less than its length. Its area is 40 sq. m. Find the length and width of the rectangle.

28. The length of a rectangle is 6 yd more than its width. Its area is 12 more (numerically) than its perimeter. Find the dimensions of the rectangle.

29. Find two consecutive integers whose product is 1 more than their sum.

30. Find three consecutive odd integers such that 4 times the product of the first two is 3 less than the product of the last two.

Review Exercises 7.10 Set II

NAME _____

In Exercises 1–14, factor each expression completely, or write "Not factorable."

1. $8x + 10$

2. $6x^2 - 54$

3. $8x^2 + 32$

4. $4x^2y - 8xy^2 + 4xy$

5. $st - 3t - s + 3$

6. $4x^2 + 5x - 6$

7. $9 + x^2$

8. $w^2 - z^2 - w - z$

9. $x^3 + 2x^2 - 16x - 32$

10. $25x - 15 + 10x^2$

11. $a^2 - 3a + 3$

12. $6y^2 + 12y + 6$

13. $a^2 - b^2 - a + b$

14. $5x^2 + 10xy + 15$

In Exercises 15–24, solve each equation.

15. $(x + 8)(x - 6) = 0$

16. $x^2 - x - 20 = 0$

17. $z^2 = 4z + 21$

18. $m^2 - 81 = 0$

ANSWERS

1. _____

2. _____

3. _____

4. _____

5. _____

6. _____

7. _____

8. _____

9. _____

10. _____

11. _____

12. _____

13. _____

14. _____

15. _____

16. _____

17. _____

18. _____

19. $5t^2 = 40t$

20. $12m^2 = 10 - 7m$

19. _____

20. _____

21. $5h(h - 11)(h + 3) = 0$

22. $30x^3 = 87x^2 + 63x$

21. _____

22. _____

23. $14x^2 = 26x + 4$

24. $32x^2 + 96x + 72 = 0$

23. _____

24. _____

In Exercises 25–30, set up each problem algebraically and solve. Be sure to state what your variables represent.

25. The difference of two numbers is 6. Their product is 72. Find the numbers.

25. _____

_____

26. One side of a square is 8 cm longer than the side of a second square. The area of the larger square is 9 times as great as the area of the smaller square. Find the length of the side of each square.

26. _____

_____

27. The length of a rectangle is 7 yd more than its width. Its area is 4 more (numerically) than its perimeter. Find the dimensions of the rectangle.

27. _____

_____

28. Find two consecutive integers whose product is 10 less than 4 times their sum.

28. _____

_____

29. Find three consecutive even integers such that twice the product of the first two is 10 more than the last integer.

29. _____

_____

30. The width of a rectangular solid is 3 cm more than its height. Its length is 18 cm and its volume is 324 cc. Find its height and its width.

30. _____

_____

Chapter 7 Diagnostic Test

The purpose of this test is to see how well you understand factoring. We recommend that you work this diagnostic test *before* your instructor tests you on this chapter. Allow yourself about 50 minutes.

Complete solutions for all the problems on this test, together with section references, are given in the answer section in the back of this book. For the problems you do incorrectly, study the sections referred to.

In Problems 1–12, factor each polynomial completely, or write "Not factorable."

1. $8x + 12$

2. $5x^3 - 35x^2$

3. $25x^2 - 121y^2$

4. $8x^2 + 7$

5. $5a^2 - 180$

6. $z^2 + 9z + 8$

7. $m^2 + 5m - 6$

8. $11x^2 - 18x + 7$

9. $x^2 + 7x - 6$

10. $4y^2 + 19y - 5$

11. $5n - mn - 5 + m$

12. $6h^2k - 8hk^2 + 2k^3$

In Problems 13–18, solve each equation.

13. $3x^2 - 12x = 0$

14. $x^2 + 20 = 12x$

15. $3x^3 = x^2 + 10x$

16. $(x - 7)(x + 6) = -22$

17. $3x^2 = 75$

18. $2x^2 - 5x = 3$

In Problems 19 and 20, set up each problem algebraically and solve. Be sure to state what your variables represent.

19. Find three consecutive odd integers such that twice the product of the first two minus the product of the first and third is 49.

20. The length of a rectangle is 3 ft more than its width. Its area is 28 sq. ft. Find its width and its length.

Cumulative Review Exercises: Chapters 1–7

1. Evaluate $24 \div 2\sqrt{16} - 3^2 \cdot 5$.

2. Solve $5(2b - 4) = 8(3b + 5) - 4(9 + 6b)$.

3. Use the formula, substituting the values of the variables given with the formula.

$$C = \frac{a}{a + 12} \cdot A \qquad \text{Find } C \text{ if } a = 8 \text{ and } A = 35.$$

In Exercises 4 and 5, simplify each expression and write your answers using only positive exponents.

4. $(3y^{-3}z^2)^{-2}$

5. $\left(\dfrac{15x^2}{10x^3}\right)^3$

6. Write the following in scientific notation.

a. 57,300,000

b. 0.00351

In Exercises 7–10, perform the indicated operations and simplify.

7. $(2x^2 + 5x - 3) - (-4x^2 + 8x + 10) + (6x^2 + 3x - 8)$

8. $(y - 4)(3y^2 - 2y + 5)$

9. $\dfrac{12a^2 - 3a}{3a}$

10. $(8x^2 - 2x - 17) \div (2x - 3)$

In Exercises 11–16, factor each polynomial completely, or write "Not factorable."

11. $36x^2 - 18x$

12. $1 - 36t^2$

13. $x^2 - 8x + 15$

14. $7x^2 - x - 5$

15. $5k^2 - 34k - 7$

16. $3n^2 - 2n - 5$

In Exercises 17 and 18, solve each equation.

17. $x(5x + 9) = 2$

18. $5(3x + 2) = 2(1 - x) - 3$

In Exercises 19 and 20, set up each problem algebraically and solve.

19. How many pounds of cashews at $7.50 per pound should be mixed with 20 lb of peanuts at $3.50 per pound if the mixture is to be worth $5.90 per pound?

20. The length and width of a rectangle are in the ratio of 3 to 2. The area is 150 sq. m. Find the width and the length.

8 Rational Expressions

In this chapter, we define rational expressions (algebraic fractions) and discuss both how to perform necessary operations with them and how to solve equations and word problems that involve them. A knowledge of the different methods of factoring discussed in Chapter 7 is *essential* to your work with rational expressions.

8.1 Rational Expressions

A **rational expression** (also called an **algebraic fraction** or sometimes just a **fraction**) is an algebraic expression of the form $\dfrac{P}{Q}$, where P and Q are polynomials. We call P and Q the **terms** of the rational expression. We call P the **numerator** and Q the **denominator** of the rational expression.

The terms of the rational expression
$$\frac{P}{Q}$$
— Numerator
← Fraction bar
— Denominator (cannot be zero)

Excluded Values

Because a denominator may not be zero, any value of the variable or variables that makes the denominator, Q, equal to zero *must be excluded* (see Sections 1.1 and 1.6). We find excluded values as follows:

TO FIND EXCLUDED VALUES

If there are variables in the denominator, set the denominator equal to zero and solve the resulting equation; the solutions of that equation are the values of the variable that must be excluded.

If there are no variables in the denominator, no values of the variable need be excluded.

Example 1 Examples of finding excluded values for rational expressions:

    a. $\dfrac{x}{5}$ There are no variables in the denominator.
                Therefore, no values of the variable need be excluded.

    b. $\dfrac{4x + 1}{x + 2}$ We set $x + 2$ equal to zero and solve:

$$x + 2 = 0$$
$$x = -2$$

Therefore, -2 is an excluded value.

c. $\dfrac{x^2 + 1}{x^2 + 3x - 4}$ We set $x^2 + 3x - 4$ equal to zero and solve:

$$x^2 + 3x - 4 = 0$$
$$(x + 4)(x - 1) = 0$$

$$x + 4 = 0 \qquad\qquad x - 1 = 0$$
$$x = -4 \qquad\qquad\quad x = 1$$

Therefore, -4 and 1 are excluded values.

d. $\dfrac{7}{x}$ We set x equal to zero and solve: $x = 0$. Therefore, 0 is an excluded value.

e. $\dfrac{1}{2}$ There are no variables in the denominator. Therefore, no values of the variable need be excluded. Arithmetic fractions are also rational expressions; in this case, 1 and 2 are each polynomials of degree 0. ∎

NOTE   After this section, whenever a fraction is written, it is understood that the values of the variables that make the denominator zero are excluded. ☑

A WORD OF CAUTION   If you are accustomed to writing fractions with a *slanted bar* (/), you are *strongly urged* to break the habit! If you insist on using the slanted bar, be sure to put parentheses around any numerator or denominator that contains more than one term. For example, note that

$$(x - 5)/(x + 3) = \frac{x - 5}{x + 3}$$

but

$$x - 5/x + 3 = x - \frac{5}{x} + 3$$

Therefore, $(x - 5)/(x + 3) \neq x - 5/x + 3$. ☑

Equivalent Rational Expressions and the Fundamental Property of Rational Expressions

Equivalent rational expressions (equivalent fractions) are rational expressions (fractions) that have the same value. The **fundamental property of rational expressions**, which follows, allows us to reduce fractions and also to "build up" fractions in order to add and subtract them.

THE FUNDAMENTAL PROPERTY OF RATIONAL EXPRESSIONS

If P, Q, and C are polynomials, and if $Q \neq 0$ and $C \neq 0$, then

$$\frac{P \cdot C}{Q \cdot C} = \frac{P}{Q}$$

The fundamental property of rational expressions permits us to perform the following operations:

1. Multiply both numerator and denominator by the same nonzero number; that is,
$$\frac{P}{Q} = \frac{P \cdot C}{Q \cdot C}, \; C \neq 0.$$

2. Divide both numerator and denominator by the same nonzero number; that is,
$$\frac{P}{Q} = \frac{P \div C}{Q \div C}, \; C \neq 0.$$

A WORD OF CAUTION   We do *not* get a fraction equivalent to the one we started with if we *add* the same number to or *subtract* the same number from both the numerator and the denominator. For example, while $\frac{2}{3} = \frac{2 \cdot 4}{3 \cdot 4}$, $\frac{2}{3} \neq \frac{2 + 4}{3 + 4}$, and while $\frac{6}{9} = \frac{6 \div 3}{9 \div 3}$, $\frac{6}{9} \neq \frac{6 - 3}{9 - 3}$. ☑

Example 2   Determine whether the pairs of fractions are equivalent.

a. $\dfrac{1}{2}, \dfrac{1 + x}{2 + x}$ No; we can't get the second fraction from the first by multiplying or dividing 1 and 2 both by the same number.

b. $\dfrac{3}{x}, \dfrac{6}{2x}$ Yes; if we multiply both 3 and x by 2, we get 6 and $2x$.

c. $\dfrac{8x}{12y}, \dfrac{2x}{3y}$ Yes; if we divide both $8x$ and $12y$ by 4, we get $2x$ and $3y$.

d. $\dfrac{3 + x}{4 + x}, \dfrac{3}{4}$ No; we can't get the second fraction from the first by multiplying or dividing both $(3 + x)$ and $(4 + x)$ by the same number. ■

The Three Signs of a Fraction

Every fraction has three signs associated with it, even if those signs are not visible: the sign of the fraction, the sign of the numerator, and the sign of the denominator. Consider the fraction $\frac{8}{4}$:

Sign of fraction ⟶ $+\dfrac{+8}{+4}$ — Sign of numerator / Sign of denominator

Let's compare three other fractions with $+\dfrac{+8}{+4}$:

$$+\frac{-8}{-4} = +\left(\frac{-8}{-4}\right) = +(+2) = 2$$

Sign of numerator and sign of denominator are different from $+\dfrac{+8}{+4}$

$$-\frac{-8}{+4} = -\left(\frac{-8}{+4}\right) = -(-2) = 2$$

Sign of fraction and sign of numerator are different from $+\dfrac{+8}{+4}$

$$- \frac{+8}{-4} = -\left(\frac{+8}{-4}\right) = -(-2) = 2$$

Sign of fraction and sign of denominator are different from $+\frac{+8}{+4}$

Because each of these fractions equals the same number (2), the fractions must all equal each other. Therefore,

$$+\frac{+8}{+4} = +\frac{-8}{-4} = -\frac{-8}{+4} = -\frac{+8}{-4} = 2$$

It can also be shown that

$$-\frac{+8}{+4} = +\frac{-8}{+4} = +\frac{+8}{-4} = -\frac{-8}{-4} = -2$$

RULE OF SIGNS FOR RATIONAL EXPRESSIONS

If any *two* of the three signs of a rational expression are changed, the value of the rational expression is unchanged. Rational expressions obtained in this way are *equivalent rational expressions*.

This rule of signs is helpful when we are reducing some rational expressions or performing operations (adding, multiplying, and so forth) on some rational expressions.

Example 3   Find the missing term in each expression.

a. $-\frac{-3x}{2y} = \frac{?}{2y}$

Solution Because the signs of the *denominators* are the same in both fractions (they are understood to be +) and the signs of the *fractions* are different, the signs of the *numerators* must be different.

Therefore, $-\frac{-3x}{2y} = \frac{3x}{2y}$. The missing term is $3x$.

b. $\frac{x}{-5} = \frac{-x}{?}$

Solution Because the signs of the *fractions* are the same in both fractions (they are understood to be +) and the signs of the *numerators* are different, the signs of the *denominators* must be different.

Therefore, $\frac{x}{-5} = \frac{-x}{5}$. The missing term is 5. ■

Recall from Section 2.2 that

$$-(x - y) = -1(x - y) = -x + y = y - x$$

Therefore, $y - x$ can always be substituted for $-(x - y)$, and $-(x - y)$ can always be substituted for $y - x$. We use these facts in Example 4.

Example 4   Find the missing term.

a. $-\dfrac{1}{2 - x} = \dfrac{1}{?}$

Solution The signs of the numerators are the same (both are understood to be +); the signs of the fractions are different. The signs of the denominators must be different. Therefore, the new denominator must be $-(2 - x)$. If we then substitute $x - 2$ for $-(2 - x)$, we have

This step need not be shown

$$-\frac{1}{2 - x} = \frac{1}{-(2 - x)} = \frac{1}{x - 2}$$

Therefore, the missing term is $x - 2$.

b. $\dfrac{a - 1}{-2} = \dfrac{?}{2}$

Solution The signs of the fractions are both understood to be +; the signs of the denominators are different. The signs of the numerators must be different. Therefore,

This step need not be shown

$$\frac{a - 1}{-2} = \frac{-(a - 1)}{2} = \frac{1 - a}{2}$$

Therefore, the missing term is $1 - a$.

c. $\dfrac{x - y}{(u + v)(a - b)} = \dfrac{?}{(u + v)(b - a)}$

Solution The signs of the fractions are both understood to be +. The signs of the denominators are different because $b - a$ is the negative of $a - b$. The signs of the numerators must be different. Therefore,

This step need not be shown

$$\frac{x - y}{(u + v)(a - b)} = \frac{-(x - y)}{(u + v)(b - a)} = \frac{y - x}{(u + v)(b - a)}$$

Therefore, the missing term is $y - x$.   ∎

EXERCISES 8.1

Set I   In Exercises 1–8, determine what value (or values) of the variable must be excluded.

1. $\dfrac{3x + 4}{x - 2}$

2. $\dfrac{5 - 4x}{x + 3}$

3. $\dfrac{x}{10}$

4. $\dfrac{y}{20}$

5. $\dfrac{x - 4}{3x^2 - 6x}$

6. $\dfrac{3x + 2}{4x^2 - 12x}$

7. $\dfrac{3 + x}{x^2 - x - 2}$

8. $\dfrac{x - 5}{x^2 + x - 12}$

In Exercises 9–14, determine whether the pairs of rational expressions are equivalent.

9. $\dfrac{x}{2y}, \dfrac{5x}{10y}$

10. $\dfrac{a}{7b}, \dfrac{2a}{14b}$

11. $\dfrac{x}{2y}, \dfrac{x + 5}{2y + 5}$

12. $\dfrac{3c}{5d}, \dfrac{3c + 8}{5d + 8}$

13. $\dfrac{6(x + 1)}{12(3x - 2)}, \dfrac{x + 1}{2(3x - 2)}$

14. $\dfrac{9(x + 5)}{27(2x - 7)}, \dfrac{x + 5}{3(2x - 7)}$

In Exercises 15–26, find the missing term.

15. $-\dfrac{5}{6} = \dfrac{?}{6}$ **16.** $-\dfrac{8}{9} = \dfrac{?}{9}$ **17.** $\dfrac{5}{-y} = \dfrac{-5}{?}$

18. $\dfrac{2}{-x} = \dfrac{?}{x}$ **19.** $\dfrac{2-x}{-9} = \dfrac{?}{9}$ **20.** $\dfrac{5-y}{-2} = \dfrac{?}{2}$

21. $\dfrac{6-y}{5} = \dfrac{y-6}{?}$ **22.** $\dfrac{8-x}{7} = \dfrac{x-8}{?}$

23. $-\dfrac{4}{x-5} = \dfrac{4}{?}$ **24.** $-\dfrac{3}{x-4} = \dfrac{3}{?}$

25. $\dfrac{a-b}{(c+d)(5-x)} = \dfrac{?}{(c+d)(x-5)}$ **26.** $\dfrac{x-y}{(u+v)(c-2)} = \dfrac{?}{(u+v)(2-c)}$

Set II In Exercises 1–8, determine what value (or values) of the variable must be excluded.

1. $\dfrac{6x+4}{x-7}$ **2.** $\dfrac{x+8}{x}$ **3.** $\dfrac{x}{7}$ **4.** $\dfrac{7}{x}$

5. $\dfrac{x-7}{5x^2-10x}$ **6.** $\dfrac{5x-3}{x^2-2x-24}$ **7.** $\dfrac{8+x}{x^2-6x-27}$ **8.** $\dfrac{x+2}{x^2+6x+9}$

In Exercises 9–14, determine whether the pairs of rational expressions are equivalent.

9. $\dfrac{a}{5b}, \dfrac{10a}{50b}$ **10.** $\dfrac{x}{3y}, \dfrac{x+4}{3y+4}$ **11.** $\dfrac{x}{4y}, \dfrac{x+2}{4y+2}$ **12.** $\dfrac{7x}{3y}, \dfrac{7x(3+z)}{3y(3+z)}$

13. $\dfrac{5(2x-7)}{15(2x-5)}, \dfrac{2x-7}{3(2x-5)}$ **14.** $\dfrac{4(x+3)}{16(2x-5)}, \dfrac{4x+3}{32x-5}$

In Exercises 15–26, find the missing term.

15. $-\dfrac{4}{3} = \dfrac{?}{3}$ **16.** $-\dfrac{4}{3} = \dfrac{?}{-3}$

17. $\dfrac{8}{-b} = \dfrac{-8}{?}$ **18.** $\dfrac{9}{-a} = \dfrac{?}{a}$

19. $\dfrac{1-y}{-5} = \dfrac{?}{5}$ **20.** $\dfrac{5-y}{-2} = \dfrac{?}{2}$

21. $\dfrac{8-a}{3} = \dfrac{a-8}{?}$ **22.** $\dfrac{5-x}{a-b} = \dfrac{x-5}{?}$

23. $-\dfrac{7}{y-2} = \dfrac{7}{?}$ **24.** $-\dfrac{a-2}{x-3} = \dfrac{2-a}{?}$

25. $\dfrac{u-v}{(a+3)(7-x)} = \dfrac{?}{(a+3)(x-7)}$ **26.** $\dfrac{3-a}{(x+1)(y-3)} = \dfrac{?}{(3-y)(1+x)}$

8.2 Reducing Rational Expressions to Lowest Terms

A rational expression is in *lowest terms* if the greatest common factor (GCF) of its numerator and denominator is 1. In this section and in the remainder of the book, it is understood that all rational expressions are to be reduced to lowest terms unless otherwise indicated.

If the numerator and denominator of a rational expression do not have a common factor other than 1, we cannot reduce the expression.

If the numerator and denominator of a rational expression do have a common factor, we can use the fundamental property of rational expressions and divide both the numerator and denominator by any common factors; the new rational expression will, of course, be equivalent to the original one (see Section 8.1). We can reduce the rational expression to lowest terms in one step by dividing both numerator and denominator by their GCF.

In Arithmetic Because 12 and 18 have a common factor, we can reduce the fraction $\frac{12}{18}$ as follows:

$$\frac{12}{18} = \frac{2 \cdot 2 \cdot 3}{2 \cdot 3 \cdot 3} = \frac{2}{3}$$

2 and 3 are both factors of the numerator

Numerator and denominator were both divided by $2 \cdot 3$

2 and 3 are both factors of the denominator

In practice, the work is often done in one of the following ways:

In two steps:

Dividing both numerator and denominator by 2

$$\frac{12}{18} = \frac{6}{9} = \frac{2}{3}$$

Dividing both numerator and denominator by 3

This work is often shown as follows:

$$\frac{12}{18} = \frac{2}{3}$$

In one step: We determine that the GCF of 12 and 18 is 6. Then

$$\frac{12}{18} = \frac{2}{3} \quad \text{Dividing both numerator and denominator by 6}$$

In Algebra As in arithmetic, a rational expression can be reduced only when the numerator and denominator have a common *factor*. One way to do the reduction is as follows:

TO REDUCE A RATIONAL EXPRESSION TO LOWEST TERMS

1. Factor the numerator and denominator completely and find their greatest common factor (GCF).

2. Divide both numerator and denominator by their GCF.

We use this method in Example 1.

We can also reduce a rational expression by dividing both numerator and denominator by any factor common to both and then checking for any *other* common factors (see Example 2).

Example 1   Examples of reducing to lowest terms by finding the GCF of the numerator and denominator:

a. $\dfrac{4x^2y}{2xy}$ The GCF of $4x^2y$ and $2xy$ is $2xy$.

Writing $4x^2y$ as $2x(2xy)$

Using the fundamental property of rational expressions

$$\frac{4x^2y}{2xy} = \frac{2x\,(2xy)}{1\,(2xy)} = \frac{2x}{1} = 2x$$

b. $\dfrac{15ab^2c^3}{6a^4bc^2}$ The GCF of $15ab^2c^3$ and $6a^4bc^2$ is $3abc^2$.

Writing $15ab^2c^3$ as $5bc(3abc^2)$

$$\frac{15ab^2c^3}{6a^4bc^2} = \frac{5bc\,(3abc^2)}{2a^3\,(3abc^2)} = \frac{5bc}{2a^3}$$ Using the fundamental property of rational expressions

Writing $6a^4bc^2$ as $2a^3(3abc^2)$ ∎

Example 2   Examples of reducing rational expressions by dividing both numerator and denominator by *any* common factors:

Dividing both numerator and denominator by 2

Dividing both numerator and denominator by y

a. $\dfrac{4x^2y}{2xy} = \dfrac{\overset{2}{4}\overset{x}{x^2}\overset{1}{y}}{\underset{1\,1\,1}{2xy}} = \dfrac{2x}{1} = 2x$

Dividing both numerator and denominator by x

Dividing both numerator and denominator by 3

Dividing both numerator and denominator by b

b. $\dfrac{15ab^2c^3}{6a^4bc^2} = \dfrac{\overset{5\ 1\ b\ c}{15ab^2c^3}}{\underset{2\,a^3\,1\,1}{6a^4bc^2}} = \dfrac{5bc}{2a^3}$

Dividing both numerator and denominator by c^2

Dividing both numerator and denominator by a ∎

We will show a third way of reducing rational expressions like those in Examples 1 and 2 in Section 8.3.

Example 3   Examples of reducing rational expressions:

a. $\dfrac{x-3}{x^2-9} = \dfrac{\overset{1}{(x-3)}}{(x+3)\underset{1}{(x-3)}} = \dfrac{1}{x+3}$ Dividing both numerator and denominator by $(x-3)$, the GCF

b. $\dfrac{x^2-4x-5}{x^2+5x+4} = \dfrac{\overset{1}{(x+1)}(x-5)}{\underset{1}{(x+1)}(x+4)} = \dfrac{x-5}{x+4}$ Dividing both numerator and denominator by $(x+1)$, the GCF

c. $\dfrac{3x^2-5xy-2y^2}{6x^3y+2x^2y^2} = \dfrac{(x-2y)\overset{1}{(3x+y)}}{2x^2y\underset{1}{(3x+y)}} = \dfrac{x-2y}{2x^2y}$ Dividing both numerator and denominator by $(3x+y)$, the GCF

d. $\dfrac{x - y}{y - x} = \dfrac{x - y}{-(y - x)} = -\dfrac{\overset{1}{\cancel{(x - y)}}}{\underset{1}{\cancel{(x - y)}}} = -1$ Dividing both numerator and denominator by $(x - y)$, the GCF

└─ Changing the sign of the fraction and the sign of the denominator

NOTE   The *result* of Example 3d is quite important. It permits you to replace any fraction of the *form* $\dfrac{x - y}{y - x}$ with -1.   ☑

e. $\dfrac{2b^2 + ab - 3a^2}{4a^2 - 9ab + 5b^2} = \dfrac{(b - a)(2b + 3a)}{(a - b)(4a - 5b)}$

┌─── $(-1)(b - a) = (a - b)$ ───┐

$= \dfrac{(-1)(b - a)\,(2b + 3a)}{(-1)(a - b)(4a - 5b)} = \dfrac{\overset{1}{\cancel{(a - b)}}\,(2b + 3a)}{(-1)\underset{1}{\cancel{(a - b)}}(4a - 5b)}$ Dividing both numerator and denominator by $(a - b)$, the GCF

└─ Changing the signs of both numerator and denominator is equivalent to multiplying each by -1

$= \dfrac{(2b + 3a)}{(-1)(4a - 5b)}$

$= \dfrac{2b + 3a}{5b - 4a}$ Removing the parentheses

or $\dfrac{(b - a)(2b + 3a)}{(a - b)(4a - 5b)} = \dfrac{\overset{1}{\cancel{(b - a)}}(2b + 3a)}{-\underset{1}{\cancel{(b - a)}}(4a - 5b)}$ Dividing both numerator and denominator by $(b - a)$, the GCF

└─ Substituting $-(b - a)$ for $a - b$

$= \dfrac{2b + 3a}{-(4a - 5b)}$

$= \dfrac{2b + 3a}{5b - 4a}$ Removing the parentheses

or, using the results of Example 3d,

$\dfrac{(b - a)(2b + 3a)}{(a - b)(4a - 5b)} = \boxed{\dfrac{b - a}{a - b}} \cdot \dfrac{2b + 3a}{4a - 5b}$ $\dfrac{b - a}{a - b}$ is of the form $\dfrac{x - y}{y - x}$

$= -1\left(\dfrac{2b + 3a}{4a - 5b}\right)$ Replacing $\dfrac{b - a}{a - b}$ with -1

$= \dfrac{2b + 3a}{-(4a - 5b)}$ Changing the sign of the fraction and the sign of the denominator

$= \dfrac{2b + 3a}{5b - 4a}$

f. $\dfrac{z}{2z} = \dfrac{\overset{1}{\cancel{z}}}{2\underset{1}{\cancel{z}}} = \dfrac{1}{2}$

NOTE   A factor of 1 will always remain in the numerator and denominator after they have been divided by factors common to both.   ☑

A WORD OF CAUTION   A common error made in reducing fractions is to forget that the expression the numerator and denominator are divided by must be a factor of *both* (see Examples 3g and 3h).

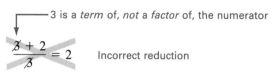

The above reduction is incorrect because

$$\frac{3 + 2}{3} = \frac{5}{3} \neq 2$$

☑

g. $\dfrac{x + 3}{x + 6}$

This rational expression cannot be reduced, since neither x nor 3 is a *factor* of the numerator or the denominator.

h.   ——x is not a factor of the numerator

$$\frac{x + y}{x}$$

This rational expression cannot be reduced.   ■

EXERCISES 8.2

Set I   Reduce each rational expression to lowest terms.

1. $\dfrac{9}{12}$

2. $\dfrac{8}{14}$

3. $\dfrac{6ab^2}{3ab}$

4. $\dfrac{10m^2n}{5mn}$

5. $\dfrac{4x^2y}{2xy}$

6. $\dfrac{12x^3y}{4xy}$

7. $\dfrac{5x - 10}{x - 2}$

8. $\dfrac{3x + 12}{x + 4}$

9. $\dfrac{7x - 21}{15x^2 - 45x}$

10. $\dfrac{12y - 18}{4y^3 - 6y^2}$

11. $\dfrac{6x^2y}{30x^3y - 18xy^2}$

12. $\dfrac{9ab^2}{18a^3b^2 - 36ab^3}$

13. $\dfrac{8x^2y + 12xy^2}{12x^3y + 18x^2y^2}$

14. $\dfrac{18s^3t + 30s^2t^3}{24s^2t^2 + 40st^4}$

15. $-\dfrac{5x - 6}{6 - 5x}$

16. $\dfrac{4 - 3z}{3z - 4}$

17. $\dfrac{5x^2 + 30x}{10x^2 - 40x}$

18. $\dfrac{4x^3 - 4x^2}{12x^2 - 12x}$

19. $\dfrac{2 + 4}{4}$

20. $\dfrac{3 + 9}{3}$

21. $\dfrac{5 + x}{5}$

22. $\dfrac{x + 8}{8}$

23. $\dfrac{x^2 - 1}{x + 1}$

24. $\dfrac{x^2 - 4}{x - 2}$

25. $\dfrac{6x^2 - x - 2}{10x^2 + 3x - 1}$

26. $\dfrac{8x^2 - 10x - 3}{12x^2 + 11x + 2}$

27. $\dfrac{x^2 - y^2}{(x + y)^2}$

28. $\dfrac{a^2 - 9b^2}{(a - 3b)^2}$

29. $\dfrac{2y^2 + xy - 6x^2}{3x^2 + xy - 2y^2}$

30. $\dfrac{10y^2 + 11xy - 6x^2}{4x^2 - 4xy - 15y^2}$

31. $\dfrac{8x^2 - 2y^2}{2ax - ay + 2bx - by}$

32. $\dfrac{3x^2 - 12y^2}{ax + 2by + 2ay + bx}$

33. $\dfrac{(-1)(z - 8)}{8 - z}$

34. $\dfrac{x - 12}{(-1)(12 - x)}$

35. $\dfrac{(-1)(a - 2b)(b - a)}{(2a + b)(a - b)}$

36. $\dfrac{8(n - 2m)}{(-1)(3n + m)(2m - n)}$

37. $\dfrac{9 - 16x^2}{16x^2 - 24x + 9}$

38. $\dfrac{25 - 9x^2}{9x^2 - 30x + 25}$

39. $\dfrac{10 + x - 3x^2}{2x^2 + x - 10}$

40. $\dfrac{12 - 19x + 5x^2}{3x^2 - 5x - 12}$

41. $\dfrac{18 - 3x - 3x^2}{6x^2 + 6x - 36}$

42. $\dfrac{16 + 4x - 2x^2}{8x^2 - 16x - 64}$

Set II Reduce each rational expression to lowest terms.

1. $\dfrac{12}{16}$

2. $\dfrac{10}{2x}$

3. $\dfrac{8mn^3}{4n^2}$

4. $\dfrac{15x^3y}{10xy^2}$

5. $\dfrac{9x^2y}{3xy}$

6. $\dfrac{2ab}{8a^2b^3}$

7. $\dfrac{8x - 12}{2x - 3}$

8. $\dfrac{8x + 5}{5 + 8x}$

9. $\dfrac{9y - 27}{12y^2 - 36y}$

10. $\dfrac{12x}{4x^3 - 6x^2}$

11. $\dfrac{5s^2t}{30s^4t^2 - 20s^2t}$

12. $\dfrac{9xy^2 - 18x^2y}{18x^3y^2 - 36xy^3}$

13. $\dfrac{27a^2b + 18ab^3}{36a^3b + 24a^2b^3}$

14. $\dfrac{10x^2 + 5xy}{12xy - 6y^2}$

15. $-\dfrac{7x - 1}{1 - 7x}$

16. $-\dfrac{3x + 4}{4 + 3x}$

17. $\dfrac{6x^2 + 9x}{12x^2 + 9x}$

18. $\dfrac{x^2 - 9}{3x + 9}$

19. $\dfrac{5 + 10}{10}$

20. $\dfrac{5 + 10x}{10}$

21. $\dfrac{7 + x}{7}$

22. $\dfrac{21 + 3x}{3}$

23. $\dfrac{x^2 - 16}{x + 4}$

24. $\dfrac{x - 1}{x^2 - 1}$

25. $\dfrac{4x^2 - 9x + 2}{4x^2 + 7x - 2}$

26. $\dfrac{x^2 - 9}{x^2 + 5x + 6}$

27. $\dfrac{x^2 - 25}{(x + 5)^2}$

28. $\dfrac{x^2 - 16}{x^2 - x - 12}$

29. $\dfrac{3y^2 + 5xy - 2x^2}{3x^2 - 8xy - 3y^2}$

30. $\dfrac{x^2 + x - 20}{x^2 + 2x - 15}$

31. $\dfrac{8x^2 - 18y^2}{2ax - 3ay + 2bx - 3by}$

32. $\dfrac{x^2 - 11x + 30}{x^2 - 9x + 20}$

33. $\dfrac{(-1)(a - 5)}{5 - a}$

34. $\dfrac{12a^3b + 6a^2b^2}{18a^2b^2 + 9ab^3}$

35. $\dfrac{(-1)(x - y)(3y - x)}{(3y + x)(y - x)}$

36. $\dfrac{15m^2n^2 - 15mn^3}{10m^2 - 10mn}$

37. $\dfrac{4 - 25x^2}{(5x - 2)^2}$

38. $\dfrac{16x^2 - y^2}{y^2 - 8xy + 16x^2}$

39. $\dfrac{8 + 10y - 12y^2}{2y^2 - 3y - 20}$

40. $\dfrac{60a^2 - 110a + 30}{15a + 5a^2 - 10a^3}$

41. $\dfrac{36 - 3x - 3x^2}{9x^2 + 9x - 108}$

42. $\dfrac{6x^3 - 54x}{108x^2 - 9x^3 - 9x^4}$

8.3 Multiplying and Dividing Rational Expressions

Multiplying Rational Expressions

Multiplication of rational expressions is defined as follows:

$$\frac{P}{Q} \cdot \frac{R}{S} = \frac{P \cdot R}{Q \cdot S}$$

where P, Q, R, and S are polynomials and $Q \neq 0$ and $S \neq 0$.

In practice, however, we can often reduce the resulting fraction. Therefore, we give the following suggestions for multiplying fractions:

TO MULTIPLY RATIONAL EXPRESSIONS

1. Factor any numerators or denominators that have more than one term.

2. Divide the numerators and denominators by any factor common to both. (The common factors can be, but do not have to be, in the same fraction.)

3. The answer is the product of the factors remaining in the numerator divided by the product of the factors remaining in the denominator. A factor of 1 will always remain in both numerator and denominator.

We can write the product as a single fraction either *before* we divide by the common factors (see Examples 1 and 2a–c) or *after* we divide by the common factors (see Examples 2d–e).

Example 1   Multiply $\dfrac{4}{9} \cdot \dfrac{3}{8}$.

Solution

Writing 4 as $2 \cdot 2$

$$\frac{4}{9} \cdot \frac{3}{8} = \frac{2 \cdot 2 \cdot 3}{3 \cdot 3 \cdot 2 \cdot 2 \cdot 2} = \frac{1(2 \cdot 2 \cdot 3)}{2 \cdot 3(2 \cdot 2 \cdot 3)} = \frac{1}{6}$$

Dividing both numerator and denominator by $2 \cdot 2 \cdot 3$

Writing 9 as $3 \cdot 3$ and 8 as $2 \cdot 2 \cdot 2$ ∎

Example 2   Examples of multiplying rational expressions:

a. $\dfrac{1}{m^2} \cdot \dfrac{m}{5} = \dfrac{1 \cdot \overset{1}{\cancel{m}}}{\underset{m}{\cancel{m^2}} \cdot 5} = \dfrac{1}{5m}$ Dividing both numerator and denominator by m

b. $\dfrac{2y^3}{3x^2} \cdot \dfrac{12x}{5y^2} = \dfrac{2\overset{y}{\cancel{y^3}} \cdot \overset{4}{\cancel{12x}}^{1}}{\underset{1}{\cancel{3x^2}} \cdot \underset{x}{\cancel{5y^2}}_{1}} = \dfrac{8y}{5x}$ Dividing both numerator and denominator by 3, x, and y^2

Dividing both numerator and denominator by 2, x, and $(x - 3)$

c. $\dfrac{x}{2x - 6} \cdot \dfrac{4x - 12}{x^2} = \dfrac{x}{2(x - 3)} \cdot \dfrac{4(x - 3)}{x^2} = \dfrac{\overset{1}{\cancel{x}} \cdot \overset{2}{\cancel{4}}\overset{1}{\cancel{(x - 3)}}}{\underset{1}{\cancel{2}}\underset{1}{\cancel{(x - 3)}} \cdot \underset{x}{\cancel{x^2}}} = \dfrac{2}{x}$

d. $\dfrac{x + 2}{6x^2} \cdot \dfrac{8x}{x^2 - x - 6} = \dfrac{\overset{1}{\cancel{(x + 2)}}}{\underset{3x}{\cancel{6x^2}}} \cdot \dfrac{\overset{4}{\cancel{8x}}}{\underset{1}{\cancel{(x + 2)}}(x - 3)} = \dfrac{4}{3x(x - 3)}$ Dividing both numerator and denominator by 2, x, and $(x + 2)$

e. $\dfrac{10xy^3}{x^2 - y^2} \cdot \dfrac{2x^2 + xy - y^2}{15x^2y} = \dfrac{\overset{2}{\cancel{10}}x\overset{y^2}{\cancel{y^3}}}{\underset{1}{\cancel{(x + y)}}(x - y)} \cdot \dfrac{\overset{1}{\cancel{(x + y)}}(2x - y)}{\underset{3}{\cancel{15}}\underset{x}{\cancel{x^2}}y}$

$= \dfrac{2y^2(2x - y)}{3x(x - y)}$ Dividing both numerator and denominator by 5, x, y, and $(x + y)$ ∎

Now that we have discussed multiplication of rational expressions, we can give an alternate method for reducing rational expressions based on the definition of multiplication. We can reduce the rational expression $\dfrac{P \cdot R}{Q \cdot S}$ as follows (assuming $Q \neq 0$ and $S \neq 0$):

Using the definition of multiplication

$$\dfrac{P \cdot R}{Q \cdot S} = \dfrac{P}{Q} \cdot \dfrac{R}{S}$$

In other words, we can now interpret a single rational expression as a multiplication and division problem and can *subtract exponents* when we're reducing a rational expression (see Example 3).

Example 3   Reduce the rational expressions from Example 1 of Section 8.2.

a. $\dfrac{4x^2y}{2xy} = \dfrac{\overset{2}{\cancel{4}}}{\underset{1}{\cancel{2}}} \cdot \dfrac{x^2}{x^1} \cdot \dfrac{y^1}{y^1} = 2 \cdot x^{2-1} \cdot y^{1-1} = 2x^1y^0 = 2x(1) = 2x$

These steps need not be shown

b. $\dfrac{15ab^2c^3}{6a^4bc^2} = \dfrac{\overset{5}{\cancel{15}}}{\underset{2}{\cancel{6}}} \cdot \dfrac{a^1}{a^4} \cdot \dfrac{b^2}{b^1} \cdot \dfrac{c^3}{c^2} = \dfrac{5}{2} \cdot a^{1-4} \cdot b^{2-1} \cdot c^{3-2} = \dfrac{5}{2}a^{-3}b^1c^1 = \dfrac{5bc}{2a^3}$

These steps need not be shown ∎

Dividing Rational Expressions

Recall from Section 1.6 that in a division problem the number we're dividing *by* is called the *divisor*, and the number we're dividing *into* is called the *dividend*; also recall from arithmetic that to divide one rational number by another we multiply the dividend by the multiplicative inverse (the reciprocal) of the divisor.

In Arithmetic

Multiplicative inverse of the divisor

$$\frac{3}{5} \div \frac{4}{7} = \frac{3}{5} \cdot \frac{7}{4} = \frac{3 \cdot 7}{5 \cdot 4} = \frac{21}{20}$$

Dividend ———↑   ↑——— Divisor

In Algebra   The rule for dividing rational expressions is as follows:

TO DIVIDE RATIONAL EXPRESSIONS

Multiply the dividend by the multiplicative inverse (the reciprocal) of the divisor.

$$\frac{P}{Q} \div \frac{S}{T} = \frac{P}{Q} \cdot \frac{T}{S}$$

where P, Q, S, and T are polynomials and $Q \neq 0$, $S \neq 0$, and $T \neq 0$.

Proof In algebra *and* in arithmetic, a division problem is equivalent to a rational expression $\left(\text{that is, } P \div Q = \dfrac{P}{Q} \right)$.

Writing the division problem in fraction form ——┐

$\dfrac{T}{S}$ is the multiplicative inverse of the divisor

$$\frac{P}{Q} \div \frac{S}{T} = \frac{\frac{P}{Q}}{\frac{S}{T}} = \frac{\frac{P}{Q}}{\frac{S}{T}} \cdot \frac{\frac{T}{S}}{\frac{T}{S}} = \frac{\frac{P}{Q} \cdot \frac{T}{S}}{\frac{S}{T} \cdot \frac{T}{S}} = \frac{\frac{P}{Q} \cdot \frac{T}{S}}{1} = \frac{P}{Q} \cdot \frac{T}{S}$$

$$\frac{S}{T} \cdot \frac{T}{S} = 1$$

The value of $\dfrac{\frac{T}{S}}{\frac{T}{S}}$ is 1, and multiplying a number by 1 does not change its value

Therefore, $\dfrac{P}{Q} \div \dfrac{S}{T} = \dfrac{P}{Q} \cdot \dfrac{T}{S}$.

Example 4   Examples of dividing rational expressions:

Changing the division problem to a multiplication problem

a. $\dfrac{4}{3x} \div \dfrac{12}{x^3} = \dfrac{4}{3x} \cdot \dfrac{x^3}{12} = \dfrac{4x^3}{3 \cdot 12x} = \dfrac{4}{3 \cdot 12} \cdot \dfrac{x^3}{x} = \dfrac{1}{9}x^{3-1} = \dfrac{1}{9}x^2$, or $\dfrac{x^2}{9}$

These steps need not be shown

The problem can also be done as follows:

$$\frac{4}{3x} \div \frac{12}{x^3} = \frac{\overset{1}{\cancel{4}}}{3\cancel{x}} \cdot \frac{\overset{x^2}{\cancel{x^3}}}{\cancel{12}} = \frac{x^2}{9}$$

b. $\dfrac{4r^3}{9s^2} \div \dfrac{8r^2s^4}{15rs} = \dfrac{4r^3}{9s^2} \cdot \dfrac{15rs}{8r^2s^4} = \boxed{\dfrac{4 \cdot 15r^4s}{9 \cdot 8r^2s^6} = \dfrac{\overset{1}{\cancel{4}} \cdot \overset{5}{\cancel{15}}}{\underset{2}{\cancel{8}} \cdot \underset{3}{\cancel{9}}} \cdot \dfrac{r^4}{r^2} \cdot \dfrac{s^1}{s^6} = \dfrac{5}{6}r^{4-2}s^{1-6}}$

$$= \frac{5}{6}r^2s^{-5} = \frac{5r^2}{6s^5}$$

└─ These steps need not be shown

The problem can also be done as follows:

$$\frac{4r^3}{9s^2} \div \frac{8r^2s^4}{15rs} = \frac{\overset{1}{\cancel{4}}\overset{r}{\cancel{r^3}}}{\underset{3}{\cancel{9}}\underset{s}{s^2}} \cdot \frac{\overset{5}{\cancel{15}}\overset{1}{r\cancel{s}}}{\underset{2}{\cancel{8}}\underset{1}{\cancel{r^2}}s^4} = \frac{5r^2}{6s^5}$$

c. $\dfrac{y^2 - x^2}{4xy - 2y^2} \div \dfrac{2x - 2y}{2x^2 + xy - y^2}$

$$= \frac{(y + x)(y - x)}{2y(\cancel{2x - y})} \cdot \frac{(x + y)\overset{1}{(\cancel{2x - y})}}{2(x - y)} = \frac{(x + y)^2(y - x)}{4y(x - y)}$$

$$= -\frac{(x + y)^2 \overset{1}{(\cancel{x - y})}}{4y\underset{1}{(\cancel{x - y})}} = -\frac{(x + y)^2}{4y}$$

└─ Changing the signs of the fraction and of the numerator

d. $\dfrac{3y^3 - 3y^2}{16y^5 + 8y^4} \div \dfrac{3y^2 + 6y - 9}{4y + 12} = \dfrac{3y^2(y - 1)}{8y^4(2y + 1)} \cdot \dfrac{4(y + 3)}{3(y^2 + 2y - 3)}$

$$= \frac{\overset{1}{\cancel{3}}\overset{1}{\cancel{y^2}}\overset{1}{(\cancel{y - 1})}}{\underset{2y^2}{\cancel{8}\cancel{y^4}}(2y + 1)} \cdot \frac{\overset{1}{\cancel{4}}\overset{1}{(\cancel{y + 3})}}{\underset{1}{\cancel{3}}\underset{1}{(\cancel{y + 3})}\underset{1}{(\cancel{y - 1})}}$$

$$= \frac{1}{2y^2(2y + 1)} \quad \blacksquare$$

A WORD OF CAUTION $2y^2(2y + 1)$ is not an acceptable answer for Example 4d. The 1 in the numerator cannot be omitted. ☑

EXERCISES 8.3

Set I Perform the indicated operations.

1. $\dfrac{5}{6} \div \dfrac{5}{3}$ 2. $\dfrac{3}{8} \div \dfrac{21}{12}$ 3. $\dfrac{4a^3}{5b^2} \cdot \dfrac{10b}{8a^2}$

4. $\dfrac{6d^2}{8c} \cdot \dfrac{12c^2}{9d^3}$ 5. $\dfrac{3x^2}{16} \div \dfrac{x}{8}$ 6. $\dfrac{4y^3}{7} \div \dfrac{8y^2}{21}$

7. $\dfrac{3x^4y^2z}{18xy} \cdot \dfrac{15z}{x^3yz^2}$ 8. $\dfrac{21a^2b^5}{4b^2c^2} \cdot \dfrac{6c^3}{7a^2b^3c}$ 9. $\dfrac{x}{x + 2} \cdot \dfrac{5x + 10}{x^3}$

10. $\dfrac{b}{b + 5} \cdot \dfrac{3b + 15}{b^4}$ 11. $\dfrac{y - 2}{y} \cdot \dfrac{6}{3y - 6}$ 12. $\dfrac{a - 1}{a} \cdot \dfrac{8}{4a - 4}$

13. $\dfrac{b^3}{a + 3} \div \dfrac{4b^2}{2a + 6}$ 14. $\dfrac{m^4}{m + 2} \div \dfrac{6m^3}{3m + 6}$

15. $\dfrac{5s - 15}{30s} \div \dfrac{s - 3}{45s^2}$

16. $\dfrac{3n - 6}{15n} \div \dfrac{n - 2}{20n^2}$

17. $\dfrac{a + 4}{a - 4} \div \dfrac{a^2 + 8a + 16}{a^2 - 16}$

18. $\dfrac{x + 8}{x - 8} \div \dfrac{x^2 + 16x + 64}{x^2 - 64}$

19. $\dfrac{5}{z + 4} \cdot \dfrac{z^2 - 16}{(z - 4)^2}$

20. $\dfrac{7}{x + 3} \cdot \dfrac{x^2 - 9}{(x - 3)^2}$

21. $\dfrac{3a - 3b}{4c + 4d} \cdot \dfrac{2c + 2d}{b - a}$

22. $\dfrac{7x - 7y}{8a + 8b} \cdot \dfrac{4a + 4b}{y - x}$

23. $\dfrac{4x - 8}{4} \cdot \dfrac{x + 2}{x^2 - 4}$

24. $\dfrac{5x - 20}{5} \cdot \dfrac{x + 4}{x^2 - 16}$

25. $\dfrac{4a + 4b}{ab^2} \div \dfrac{3a + 3b}{a^2b}$

26. $\dfrac{5x - 5y}{x^2y} \div \dfrac{4x - 4y}{xy^2}$

27. $\dfrac{a^2 - 9b^2}{a^2 - 6ab + 9b^2} \div \dfrac{a + 3b}{a - 3b}$

28. $\dfrac{x^2 - y^2}{x^2 - 2xy + y^2} \div \dfrac{x + y}{x - y}$

29. $\dfrac{x - y}{9x + 9y} \div \dfrac{x^2 - y^2}{3x^2 + 6xy + 3y^2}$

30. $\dfrac{x - 5y}{8x + 40y} \div \dfrac{x^2 - 25y^2}{4x^2 + 40xy + 100y^2}$

31. $\dfrac{a^2 - 9b^2}{6a^2 - 36ab + 54b^2} \div \dfrac{a + 3b}{2a - 6b}$

32. $\dfrac{x^2 - y^2}{8x^2 - 16xy + 8y^2} \div \dfrac{x + y}{4x - 4y}$

33. $\dfrac{2x^3y + 2x^2y^2}{6x} \div \dfrac{x^2y^2 - xy^3}{y - x}$

34. $\dfrac{5st^4 + 5s^2t^3}{10st^3} \div \dfrac{s^2t^2 - s^3t}{s - t}$

Set II   Perform the indicated operations.

1. $\dfrac{2}{3} \div \dfrac{1}{2}$

2. $\dfrac{5}{8} \div \dfrac{5}{14}$

3. $\dfrac{5x^3}{4y^2} \cdot \dfrac{8y}{10x^2}$

4. $\dfrac{8y^2}{12x} \cdot \dfrac{6x^2}{4y^3}$

5. $\dfrac{5a^2}{15} \div \dfrac{a}{3}$

6. $\dfrac{5x^3}{6} \div \dfrac{8x^2}{20}$

7. $\dfrac{6a^3b^2c^3}{24a^2b} \cdot \dfrac{18b}{a^2bc^2}$

8. $\dfrac{18x^2y^3}{10x^3y^2} \cdot \dfrac{5z^3}{9x^2y^2z}$

9. $\dfrac{s}{s + 3} \cdot \dfrac{4s + 12}{s^4}$

10. $\dfrac{x + 8}{x + 4} \cdot \dfrac{4x + 16}{x^3}$

11. $\dfrac{x + 7}{x} \cdot \dfrac{2}{3x + 21}$

12. $\dfrac{y - 3}{12y + 12} \cdot \dfrac{9}{4y - 12}$

13. $\dfrac{x^3}{y + 5} \div \dfrac{4x^2}{3y + 15}$

14. $\dfrac{x^5}{x - 3} \div \dfrac{2x^7}{2x - 6}$

15. $\dfrac{4a - 16}{12a} \div \dfrac{a - 4}{48a^3}$

16. $\dfrac{12b - 6}{12b} \div \dfrac{2b - 1}{24b^2}$

17. $\dfrac{x + 1}{x - 1} \div \dfrac{x^2 + 2x + 1}{x^2 - 1}$

18. $\dfrac{x^2 + x - 2}{x - 1} \div \dfrac{x^2 + 5x + 6}{x^2}$

19. $\dfrac{9}{z + 3} \cdot \dfrac{z^2 - 9}{(z - 3)^2}$

20. $\dfrac{z^3}{z + 4} \div \dfrac{z - 1}{z^2 + 3z - 4}$

21. $\dfrac{5x - 5y}{3x + 3y} \cdot \dfrac{4x + 4y}{y - x}$

22. $\dfrac{3 - y}{y + 1} \cdot \dfrac{4 + 5y + y^2}{y^2 + y - 12}$

23. $\dfrac{2x - 10}{8} \cdot \dfrac{x + 5}{x^2 - 25}$

24. $\dfrac{x^2 + 10x + 25}{x^2 - 25} \div \dfrac{5 - x}{x + 5}$

25. $\dfrac{5x + 5y}{xy^3} \div \dfrac{8x + 8y}{x^3y}$

26. $\dfrac{12f + 16}{15f} \div \dfrac{6f^3 + 8f^2}{20f^4}$

27. $\dfrac{s^2 - 4t^2}{s^2 - 4st + 4t^2} \div \dfrac{s + 2t}{s - 2t}$

28. $\dfrac{x^3 - 5x}{9x} \div \dfrac{4x^3 - 20x}{12x^2}$

29. $\dfrac{a - 4b}{a + 4b} \div \dfrac{a^2 - 16b^2}{2a^2 + 16ab + 32b^2}$

30. $\dfrac{2b^2c - 2bc^2}{b + c} \div \dfrac{4bc^2 - 4b^2c}{4b + 4c}$

31. $\dfrac{x^2 - 25y^2}{3x^2 - 30xy + 75y^2} \div \dfrac{x + 5y}{4x - 20y}$

32. $\dfrac{2x - 1}{y} \div \dfrac{2x^2 + x - 1}{y^2}$

33. $\dfrac{8x^2y^2 + 16xy^3}{4y} \div \dfrac{2x^3y - 2x^2y^2}{y - x}$

34. $\dfrac{a + 2}{a + 1} \cdot \dfrac{a^2 + a}{a^2 - 4}$

8.4 Adding and Subtracting Like Rational Expressions

Like Rational Expressions   **Like rational expressions** (fractions) are rational expressions that have the *same* denominator.

Example 1   Examples of like rational expressions:

a. $\dfrac{2}{3}, \dfrac{5}{3}, \dfrac{1}{3}$ are like fractions.

⤒ ⤒ ⤒ —— Same denominator

b. $\dfrac{3}{x + 2}, \dfrac{7}{x + 2}, \dfrac{-5}{x + 2}$ are like rational expressions.

⤒ ⤒ ⤒ —— Same denominator ■

Unlike Rational Expressions   **Unlike rational expressions** (fractions) are rational expressions that have *different* denominators.

Example 2   Examples of unlike rational expressions:

a. $\dfrac{1}{3}, \dfrac{1}{7}, \dfrac{3}{8}$ are unlike fractions.

⤒ ⤒ ⤒ —— Different denominators

b. $\dfrac{5}{3x}, \dfrac{5}{3x^2}, \dfrac{5}{3 + x}$ are unlike rational expressions.

⤒ ⤒ ⤒ —— Different denominators ■

Adding Like Rational Expressions

The distributive property permits us to add like rational expressions. For example,

┌── We factored $\dfrac{1}{9}$ from both terms

$$\dfrac{1}{9} + \dfrac{7}{9} = 1\left(\dfrac{1}{9}\right) + 7\left(\dfrac{1}{9}\right) = \dfrac{1}{9}(1 + 7) = \dfrac{1 + 7}{9} = \dfrac{8}{9}$$

Similarly,

We factored $\frac{1}{3}$ from all the terms

$$\frac{2}{3} + \frac{5}{3} + \frac{1}{3} = 2\left(\frac{1}{3}\right) + 5\left(\frac{1}{3}\right) + 1\left(\frac{1}{3}\right) = \frac{1}{3}(2 + 5 + 1) = \frac{2 + 5 + 1}{3} = \frac{8}{3}$$

The second and third steps need not be shown; we can simply use the following rule:

TO ADD LIKE RATIONAL EXPRESSIONS

1. Write the sum of the numerators over the denominator of the like rational expressions.

$$\frac{P}{Q} + \frac{R}{Q} = \frac{P + R}{Q} \qquad (Q \neq 0)$$

2. Reduce the resulting rational expression to lowest terms.

Subtracting Like Rational Expressions

Any subtraction of rational expressions can always be changed into an addition problem, or the following definition can be used:

$$\frac{P}{Q} - \frac{R}{Q} = \frac{P - R}{Q} \qquad (Q \neq 0)$$

Example 3   Examples of adding and subtracting like arithmetic fractions:

a. $\dfrac{11}{23} + \dfrac{5}{23} = \dfrac{11 + 5}{23} = \dfrac{16}{23}$ The answer is already in lowest terms

b. $\dfrac{7}{9} - \dfrac{1}{9} = \dfrac{7 - 1}{9} = \dfrac{\overset{2}{\cancel{6}}}{\underset{3}{\cancel{9}}} = \dfrac{2}{3}$

c. $\dfrac{3}{12} + \dfrac{1}{12} + \dfrac{4}{12} + \dfrac{7}{12} = \dfrac{3 + 1 + 4 + 7}{12} = \dfrac{\overset{5}{\cancel{15}}}{\underset{4}{\cancel{12}}} = \dfrac{5}{4}$ ■

Example 4   Examples of adding and subtracting like rational expressions:

a. $\dfrac{2}{x} + \dfrac{5}{x} = \dfrac{2 + 5}{x} = \dfrac{7}{x}$

There's an understood plus sign in front of the 4

We change the subtraction problem to an addition problem by changing the sign in front of the fraction *and* the sign of the numerator

b. $\dfrac{7}{a - 2} - \dfrac{4}{a - 2} = \dfrac{7}{a - 2} + \dfrac{-4}{a - 2} = \dfrac{7 + (-4)}{a - 2} = \dfrac{3}{a - 2}$

c. $\dfrac{3}{4a} - \dfrac{5}{4a} = \dfrac{3 - 5}{4a} = \dfrac{-2}{4a} = -\dfrac{\overset{1}{\cancel{2}}}{\underset{2}{\cancel{4}a}} = -\dfrac{1}{2a}$

d. $\dfrac{4x}{2x-y} - \dfrac{2y}{2x-y} = \dfrac{4x-2y}{2x-y} = \dfrac{2(2x-y)}{(2x-y)} = \dfrac{2(\overset{1}{\cancel{2x-y}})}{\underset{1}{\cancel{(2x-y)}}} = 2$

e. $\dfrac{15}{d-5} + \dfrac{-3d}{d-5} = \dfrac{15-3d}{d-5} = \dfrac{3(5-d)}{d-5} = -\dfrac{3\overset{1}{\cancel{(d-5)}}}{\underset{1}{\cancel{d-5}}} = -3$

Changing the sign of the fraction and the sign of the numerator

∎

In a subtraction problem, if the numerator of the fraction being subtracted contains more than one term, you *must* put parentheses around that numerator when you rewrite the problem as a single fraction (see Example 5).

Example 5   Subtract $\dfrac{5}{2x+3} - \dfrac{x+1}{2x+3}$.

Solution

$$\dfrac{5}{2x+3} - \dfrac{x+1}{2x+3} = \dfrac{5-(x+1)}{2x+3} = \dfrac{5-x-1}{2x+3} = \dfrac{4-x}{2x+3} \quad ∎$$

A WORD OF CAUTION   It is *incorrect* to do Example 5 this way:

$$\dfrac{5}{2x+3} - \dfrac{x+1}{2x+3} = \dfrac{5-x+1}{2x+3} = \dfrac{6-x}{2x+3}$$ ☑

When the denominators are not identical but are the *negatives* of each other, we can make the fractions like fractions by changing the signs of the numerator and denominator of one of the fractions (see Example 6).

Example 6   Add $\dfrac{9}{x-2} + \dfrac{5}{2-x}$.

Solution $x-2$ and $2-x$ are the negatives of each other.

Changing the sign of the numerator and the sign of the denominator

$$\dfrac{9}{x-2} + \dfrac{5}{2-x} = \dfrac{9}{x-2} + \dfrac{-5}{-(2-x)} = \dfrac{9}{x-2} + \dfrac{-5}{x-2} = \dfrac{9-5}{x-2} = \dfrac{4}{x-2} \quad ∎$$

In some problems, it is easier to change the sign of the denominator and the sign of the fraction (see Example 7).

Example 7   Subtract $\dfrac{8}{y-5} - \dfrac{3}{5-y}$.

Solution

Changing the sign of the fraction and the sign of the denominator

$$\dfrac{8}{y-5} - \dfrac{3}{5-y} = \dfrac{8}{y-5} + \dfrac{3}{-(5-y)} = \dfrac{8}{y-5} + \dfrac{3}{y-5} = \dfrac{11}{y-5} \quad ∎$$

A WORD OF CAUTION Students often confuse *addition of fractions* with *solving equations*, and they multiply both fractions by the same number. This is incorrect.

Correct method	*Incorrect method*

$$\frac{1}{3x} + \frac{4}{3x} = \frac{5}{3x} \qquad \frac{1}{3x} + \frac{4}{3x} = (3x)\left(\frac{1}{3x}\right) + (3x)\left(\frac{4}{3x}\right) = 1 + 4 = 5 \qquad \boxed{\checkmark}$$

EXERCISES 8.4

Set I   Perform the indicated operations.

1. $\dfrac{7}{a} + \dfrac{2}{a}$ **2.** $\dfrac{5}{b} + \dfrac{2}{b}$

3. $\dfrac{6}{x - y} - \dfrac{2}{x - y}$ **4.** $\dfrac{7}{m + n} - \dfrac{1}{m + n}$

5. $\dfrac{2}{3a} + \dfrac{4}{3a}$ **6.** $\dfrac{8}{5z} + \dfrac{2}{5z}$

7. $\dfrac{2y}{y + 1} + \dfrac{2}{y + 1}$ **8.** $\dfrac{10x}{2x + 3} + \dfrac{15}{2x + 3}$

9. $\dfrac{3}{x + 3} + \dfrac{x}{x + 3}$ **10.** $\dfrac{5}{y + 5} + \dfrac{y}{y + 5}$

11. $\dfrac{3x}{x - 4} - \dfrac{12}{x - 4}$ **12.** $\dfrac{7x}{x - 2} - \dfrac{14}{x - 2}$

13. $\dfrac{x - 3}{y - 2} - \dfrac{x + 5}{y - 2}$ **14.** $\dfrac{z - 4}{a - b} - \dfrac{z + 3}{a - b}$

15. $\dfrac{a + 2}{2a + 1} - \dfrac{1 - a}{2a + 1}$ **16.** $\dfrac{6x - 1}{3x - 2} - \dfrac{3x + 1}{3x - 2}$

17. $\dfrac{4x - 1}{2x + 3} + \dfrac{5x - 3}{2x + 3}$ **18.** $\dfrac{8x + 5}{3x - 2} + \dfrac{4x + 3}{3x - 2}$

19. $\dfrac{9 + 7x}{7x - 9} - \dfrac{3x - 2}{7x - 9}$ **20.** $\dfrac{4 + 5x}{5x - 4} - \dfrac{6x - 5}{5x - 4}$

21. $\dfrac{-x}{x - 2} - \dfrac{2}{2 - x}$ **22.** $\dfrac{-b}{2a - b} - \dfrac{2a}{b - 2a}$

23. $\dfrac{-15w}{1 - 5w} - \dfrac{3}{5w - 1}$ **24.** $\dfrac{-35}{6w - 7} - \dfrac{30w}{7 - 6w}$

25. $\dfrac{7z}{8z - 4} + \dfrac{6 - 5z}{4 - 8z}$ **26.** $\dfrac{5x}{9x - 3} + \dfrac{4 - 7x}{3 - 9x}$

27. $\dfrac{31 - 8x}{12 - 8x} - \dfrac{5 - 16x}{8x - 12}$ **28.** $\dfrac{13 - 30w}{15 - 10w} - \dfrac{10w + 17}{10w - 15}$

Set II   Perform the indicated operations.

1. $\dfrac{8}{y} + \dfrac{7}{y}$ **2.** $\dfrac{6}{x} - \dfrac{2}{x}$

3. $\dfrac{8}{x - y} - \dfrac{5}{x - y}$ **4.** $\dfrac{x + 1}{a + b} + \dfrac{x - 1}{a + b}$

5. $\dfrac{7}{5x} + \dfrac{8}{5x}$

6. $\dfrac{15}{7x} - \dfrac{1}{7x}$

7. $\dfrac{4x}{x + 3} + \dfrac{12}{x + 3}$

8. $\dfrac{2x + 1}{x + 4} - \dfrac{x - 3}{x + 4}$

9. $\dfrac{3}{2a + 3} + \dfrac{2a}{2a + 3}$

10. $\dfrac{9x}{3x - y} - \dfrac{3y}{3x - y}$

11. $\dfrac{5a}{a - 2b} - \dfrac{10b}{a - 2b}$

12. $\dfrac{8}{z - 4} - \dfrac{2z}{z - 4}$

13. $\dfrac{a - 5}{x - 4} - \dfrac{a + 3}{x - 4}$

14. $\dfrac{4x + y}{2x + y} + \dfrac{2x + 2y}{2x + y}$

15. $\dfrac{7x - 1}{3x - 1} - \dfrac{x + 1}{3x - 1}$

16. $\dfrac{b}{b - 2a} + \dfrac{2a}{2a - b}$

17. $\dfrac{12x - 5}{9x + 2} + \dfrac{7x - 4}{9x + 2}$

18. $\dfrac{11x + 3}{6x - 1} + \dfrac{11x + 3}{6x - 1}$

19. $\dfrac{8 + 3x}{3x - 8} - \dfrac{4x - 5}{3x - 8}$

20. $\dfrac{18 + 7x}{3x - 7} - \dfrac{5x - 6}{3x - 7}$

21. $\dfrac{-s}{s - 5} - \dfrac{5}{5 - s}$

22. $\dfrac{5}{2x + 3} - \dfrac{x + 4}{2x + 3}$

23. $\dfrac{-3m}{5 - m} - \dfrac{15}{m - 5}$

24. $\dfrac{2x}{3x - 8} + \dfrac{5x - 8}{8 - 3x}$

25. $\dfrac{8x}{12x - 15} + \dfrac{25 - 12x}{15 - 12x}$

26. $\dfrac{5x - 2}{4x - 5} - \dfrac{x + 3}{4x - 5}$

27. $\dfrac{11x - 9y}{8x - 6y} - \dfrac{13x - 9y}{6y - 8x}$

28. $\dfrac{18y - 21}{10y - 16} - \dfrac{3y + 3}{10y - 16}$

8.5 Least Common Multiple (LCM) and Least Common Denominator (LCD)

Least Common Multiple

The **least common multiple** (**LCM**) of two or more polynomials is the *smallest* polynomial that is exactly divisible by each of the given polynomials.

TO FIND THE LEAST COMMON MULTIPLE OF POLYNOMIALS

1. Factor each polynomial completely. Repeated factors must be expressed as powers.

2. Write down each different base that appears in any of the factorizations.

3. Raise each base to the *highest power* to which it occurs in *any* of the factorizations.

4. The LCM is the product of all the factors found in step 3.

In Arithmetic   Consider the following problem: Find the LCM of 12 and 15.

Step 1. Find the prime factorization of each number.

$$
\begin{array}{c|c}
2 & 12 \\
2 & 6 \\
& 3
\end{array}
\qquad
\begin{array}{c|c}
3 & 15 \\
& 5
\end{array}
$$

$$12 = 2^2 \cdot 3 \qquad 15 = 3 \cdot 5$$

Step 2. Write down each different base that appears in the prime factorizations.

$$2, 3, 5$$

Step 3. Raise each base to the *highest* power to which it occurs in any of the factorizations.

$$2^2, 3, 5$$

Step 4. LCM $= 2^2 \cdot 3 \cdot 5 = 60$

In Algebra

Example 1   Find the LCM for $15b^3$ and $20b^2$.
Solution

Step 1. $3 \cdot 5 \cdot b^3,\ 2^2 \cdot 5 \cdot b^2$ Expressions in factored form

Step 2. $2, 3, 5, b$ All the different bases

Step 3. $2^2, 3^1, 5^1, b^3$ Highest power of each base

Step 4. LCM $= 2^2 \cdot 3^1 \cdot 5^1 \cdot b^3 = 60b^3$ ∎

Least Common Denominator
The **least common denominator** (**LCD**) of two or more rational expressions is the *least common multiple* of their denominators.

Example 2   Find the LCD for the terms in $\dfrac{3}{2} + \dfrac{4}{y}$.
Solution

Step 1. The denominators are already factored.

Step 2. 2 and y are the different bases.

Step 3. $2^1, y^1$ (highest power of each base)

Step 4. LCD $= 2^1 \cdot y^1 = 2y$ ∎

Example 3   Find the LCD for the terms in $\dfrac{2}{x} + \dfrac{5}{x^2}$.
Solution

Step 1. The denominators are already factored.

Step 2. x (x is the only base.)

Step 3. x^2 (x^2 is the highest power in any denominator.)

Step 4. LCD $= x^2$ ∎

Example 4 Find the LCD for the terms in $\dfrac{7}{18x^2y} + \dfrac{5}{8xy^4}$.

Solution

Step 1. $2 \cdot 3^2 \cdot x^2 \cdot y,\ 2^3 \cdot x \cdot y^4$ Denominators in factored form

Step 2. $2,\ 3,\ x,\ y$ All the different bases

Step 3. $2^3,\ 3^2,\ x^2,\ y^4$ Highest powers

Step 4. LCD $= 2^3 \cdot 3^2 \cdot x^2 \cdot y^4 = 72x^2y^4$ ■

Example 5 Find the LCD for the terms in $\dfrac{2}{x} + \dfrac{x}{x+2}$.

Solution

Step 1. The denominators are already factored.

Step 2. $x,\ (x+2)$

Step 3. $x^1,\ (x+2)^1$

Step 4. LCD $= x(x+2)$ ■

Example 6 Find the LCD for the terms in $\dfrac{16}{a^2b} + \dfrac{a-2}{2a(a-b)} - \dfrac{b+1}{4b^3(a-b)}$.

Solution

Step 1. $a^2b,\ 2a(a-b),\ 2^2b^3(a-b)$

Step 2. $2,\ a,\ b,\ (a-b)$ All the different bases

Step 3. $2^2,\ a^2,\ b^3,\ (a-b)^1$ Highest powers

Step 4. LCD $= 2^2a^2b^3(a-b)^1 = 4a^2b^3(a-b)$ ■

Example 7 Find the LCD for the terms in $\dfrac{8}{3x-3} - \dfrac{5}{x^2+2x+1}$.

Solution

Step 1. $3x - 3 = 3(x-1);\ x^2 + 2x + 1 = (x+1)^2$

Step 2. $3,\ (x-1),\ (x+1)$

Step 3. $3^1,\ (x-1)^1,\ (x+1)^2$

Step 4. LCD $= 3(x-1)(x+1)^2$ ■

Example 8 Find the LCD for the terms in $\dfrac{2x-3}{x^2+10x+25} - \dfrac{14}{4x^2+20x} + \dfrac{4x-3}{x^2+2x-15}$.

Solution

Step 1. $x^2 + 10x + 25 = (x+5)^2$

$\qquad\quad 4x^2 + 20x = 4x(x+5) = 2^2x(x+5)$

$\qquad\quad x^2 + 2x - 15 = (x+5)(x-3)$

Step 2. $2,\ x,\ (x+5),\ (x-3)$

Step 3. $2^2,\ x,\ (x+5)^2,\ (x-3)$

Step 4. LCD $= 4x(x+5)^2(x-3)$ ■

EXERCISES 8.5

Set I   In Exercises 1–8, find the LCM of the given pairs of polynomials.

1. 12, 4   **2.** 3, 9   **3.** 3, x

4. $2y, y$ **5.** $5x, 2x$ **6.** $3z, 4z$

7. $15x^2, 12x^3y$ **8.** $18a^2bc, 16ab^2$

In Exercises 9–22, find the LCD. Do *not* add the rational expressions.

9. $\dfrac{7}{12u^3v^2} - \dfrac{11}{18uv^3}$ **10.** $\dfrac{13}{50x^3y^4} - \dfrac{17}{20x^2y^5}$

11. $\dfrac{4}{a} + \dfrac{a}{a+3}$ **12.** $\dfrac{5}{b} + \dfrac{b}{b-5}$

13. $\dfrac{x}{2x+4} - \dfrac{5}{4x}$ **14.** $\dfrac{4}{3x} + \dfrac{2x}{3x+6}$

15. $\dfrac{3}{4z^2} + \dfrac{2z}{z^2+2z+1} - \dfrac{4z}{z+1}$

16. $\dfrac{2x}{x^2-2x+1} - \dfrac{5}{x-1} + \dfrac{11}{12x^3}$

17. $\dfrac{x-4}{x^2+3x+2} + \dfrac{3x+1}{x^2+2x+1}$

18. $\dfrac{2x+3}{x^2-x-12} + \dfrac{x-4}{x^2+6x+9}$

19. $\dfrac{x^2+1}{12x^3+24x^2} - \dfrac{4x+3}{x^2-4x+4} + \dfrac{1}{x^2-4}$

20. $\dfrac{2y+5}{y^2+6y+9} - \dfrac{7y}{y^2-9} - \dfrac{11}{8y^2-24y}$

21. $\dfrac{3x+1}{6x^2+x-2} + \dfrac{x^2+1}{9x^3+12x^2+4x} + \dfrac{5x^2-1}{4x^2-4x+1}$

22. $\dfrac{5x+1}{10x^2+13x-3} + \dfrac{3x^2-1}{4x^3+12x^2+9x} + \dfrac{x-4}{25x^2-10x+1}$

Set II   In Exercises 1–8, find the LCM of the given pairs of polynomials.

1. 21, 24   **2.** 2, 5   **3.** 2, y

4. $4y^2, 6y$ **5.** $2x, 3x$ **6.** $9x^2y, 30y^2z$

7. $8y^4, 12x^2y$ **8.** $5x^3, 10x^5$

In Exercises 9–22, find the LCD. Do *not* add the rational expressions.

9. $\dfrac{5}{9x^2y} - \dfrac{7}{6xy^2}$ **10.** $\dfrac{5}{2+x^2} + \dfrac{9}{2x^2}$

11. $\dfrac{2}{x} + \dfrac{x}{x+5}$ **12.** $\dfrac{x+1}{4} - \dfrac{x+3}{12}$

13. $\dfrac{a}{3a+6} - \dfrac{5}{3a}$ **14.** $\dfrac{9y}{x^2-y^2} + \dfrac{6x}{x^2+2xy+y^2}$

15. $\dfrac{3}{2x} + \dfrac{x}{x^2 + 4x + 4} - \dfrac{1}{x + 2}$

16. $\dfrac{11}{x^2 + 16x + 64} - \dfrac{7}{2x^2 - 128} + \dfrac{1}{4x - 32}$

17. $\dfrac{x - 5}{x^2 + 4x + 3} + \dfrac{5x - 3}{x^2 + 7x + 12}$

18. $\dfrac{5}{8z^3} - \dfrac{8z}{z^2 - 4} + \dfrac{5z}{9z + 18}$

19. $\dfrac{x^2 + 3}{9x^3 + 45x^2} - \dfrac{5x + 2}{x^2 - 10x + 25} + \dfrac{1}{x^2 - 25}$

20. $\dfrac{a^2 + 1}{3a^2 + 5a - 2} - \dfrac{1}{3a^2 + 6a} + \dfrac{5a}{18a^4 - 6a^3}$

21. $\dfrac{5x + 1}{4x^2 - 11x - 3} + \dfrac{x + 3}{16x^3 + 8x^2 + x} + \dfrac{3x^2}{x^2 - 6x + 9}$

22. $\dfrac{6x^2}{6x^2 + 5x - 1} + \dfrac{3}{30x^3 - 5x^2} + \dfrac{5 + x}{10x^4 + 10x^3}$

8.6 Adding and Subtracting Unlike Rational Expressions

Because the definition of addition of fractions is $\dfrac{P}{Q} + \dfrac{R}{Q} = \dfrac{P + R}{Q}$, we can add and subtract rational expressions only when they are *like rational expressions*. The method of converting unlike rational expressions to like ones (often called "building fractions") is based on the definition of multiplication and on the identity property of multiplication. Because $\dfrac{2}{2} = 1, \dfrac{x}{x} = 1, \dfrac{x + 2}{x + 2} = 1$, and so on, multiplying one rational expression by another whose numerator and denominator are equal produces a rational expression equivalent to the original expression. For example, in arithmetic,

Multiplying by 1

$$\frac{5}{6} = \frac{5}{6} \cdot \boxed{\frac{2}{2}} = \frac{10}{12}$$

Therefore, $\dfrac{10}{12}$ is equivalent to $\dfrac{5}{6}$. Likewise, in algebra,

Multiplying by 1

$$\frac{x}{x + 2} = \frac{x}{x + 2} \cdot \boxed{\frac{x}{x}} = \frac{x^2}{x(x + 2)}$$

Therefore, $\dfrac{x^2}{x(x + 2)}$ is equivalent to $\dfrac{x}{x + 2}$.

Multiplying by 1

$$\frac{2}{x} = \frac{2}{x} \cdot \boxed{\frac{x + 2}{x + 2}} = \frac{2x + 4}{x(x + 2)}$$

Therefore, $\dfrac{2x + 4}{x(x + 2)}$ is equivalent to $\dfrac{2}{x}$.

Of course, multiplying one rational expression by another whose numerator and denominator are equal is equivalent to multiplying both the numerator and the denominator of a rational expression by the same number. That is,

These steps are equivalent

$$\frac{2}{x} = \boxed{\frac{2}{x} \cdot \frac{x + 2}{x + 2} = \frac{2(x + 2)}{x(x + 2)}} = \frac{2x + 4}{x(x + 2)}$$

Consequently, the rules for adding and subtracting unlike rational expressions are as follows:

TO ADD OR SUBTRACT UNLIKE RATIONAL EXPRESSIONS

1. Find the LCD.

2. Convert each rational expression to an equivalent rational expression that has the LCD as its denominator by performing the following operations on each rational expression:
   a. Divide the LCD by the denominator.
   b. Multiply the numerator and denominator by the quotient from step 2a.

3. Add or subtract the resulting *like* rational expressions according to the method given in Section 8.4.

4. Reduce the sum or difference to lowest terms.

In Arithmetic

Example 1   Find $\dfrac{1}{2} + \dfrac{2}{3} + \dfrac{3}{4}$.

Solution LCD $= 12$

Since $\dfrac{1}{2} = \dfrac{1 \cdot 6}{2 \cdot 6} = \dfrac{6}{12}$, $\dfrac{2}{3} = \dfrac{2 \cdot 4}{3 \cdot 4} = \dfrac{8}{12}$, and $\dfrac{3}{4} = \dfrac{3 \cdot 3}{4 \cdot 3} = \dfrac{9}{12}$, then

$$\frac{1}{2} + \frac{2}{3} + \frac{3}{4} = \frac{6}{12} + \frac{8}{12} + \frac{9}{12} = \frac{6 + 8 + 9}{12} = \frac{23}{12} \quad \blacksquare$$

In Algebra

Example 2   Add $\dfrac{2}{x} + \dfrac{5}{x^2}$.

Solution

Step 1. LCD $= x^2$ (Section 8.5, Example 3)

Step 2. $\dfrac{2}{x} = \dfrac{2 \cdot \boxed{x}}{x \cdot \boxed{x}} = \dfrac{2x}{x^2}$

We multiply numerator and denominator by x in order to obtain an equivalent rational expression whose denominator is the LCD x^2

Step 3. $\dfrac{2}{x} + \dfrac{5}{x^2} = \dfrac{2x}{x^2} + \dfrac{5}{x^2} = \dfrac{2x + 5}{x^2}$

Step 4. The rational expression cannot be reduced. ∎

A WORD OF CAUTION Students frequently reduce the rational expressions (fractions) just after they have converted them to equivalent fractions with the LCD as the denominator (see step 2). The addition then cannot be done, because the fractions no longer have the same denominator. This gives us back what we started with:

$$\frac{2}{x} + \frac{5}{x^2} = \frac{\overset{1}{2\cancel{x}}}{\underset{x}{\cancel{x^2}}} + \frac{5}{x^2} = \underbrace{\frac{2}{x} + \frac{5}{x^2}}$$

We cannot add these rational expressions ⟶

☑

Example 3 Add $\dfrac{7}{18x^2y} + \dfrac{5}{8xy^4}$.
Solution

Step 1. LCD $= 72x^2y^4$ (Section 8.5, Example 4)

$$4y^3 = \frac{72x^2y^4}{18x^2y} = \frac{\text{LCD}}{\text{denominator of fraction}}$$

Step 2. $\dfrac{7}{18x^2y} = \dfrac{7\,(4y^3)}{18x^2y\,(4y^3)} = \dfrac{28y^3}{72x^2y^4}$

$\dfrac{5}{8xy^4} = \dfrac{5\,(9x)}{8xy^4\,(9x)} = \dfrac{45x}{72x^2y^4}$

$$9x = \frac{72x^2y^4}{8xy^4} = \frac{\text{LCD}}{\text{denominator of fraction}}$$

Step 3. $\dfrac{7}{18x^2y} + \dfrac{5}{8xy^4} = \dfrac{28y^3}{72x^2y^4} + \dfrac{45x}{72x^2y^4} = \dfrac{28y^3 + 45x}{72x^2y^4}$

Step 4. The rational expression cannot be reduced. ∎

In applying step 2, if the multiplication involves any expression containing more than one term, you *must* put parentheses around that expression (see Example 4).

Example 4 Add $\dfrac{2}{x} + \dfrac{x}{x + 2}$.
Solution

Step 1. LCD $= x(x + 2)$ (Section 8.5, Example 5)

$$x + 2 = \frac{x(x + 2)}{x} = \frac{\text{LCD}}{\text{denominator of fraction}}$$

Step 2. $\dfrac{2}{x} = \dfrac{2\,(x + 2)}{x\,(x + 2)}$

Note parentheses

$\dfrac{x}{x + 2} = \dfrac{x \cdot x}{(x + 2)\,x} = \dfrac{x^2}{x(x + 2)}$

$$x = \frac{x(x + 2)}{x + 2} = \frac{\text{LCD}}{\text{denominator of fraction}}$$

Step 3. $\dfrac{2}{x} + \dfrac{x}{x + 2} = \dfrac{2(x + 2)}{x(x + 2)} + \dfrac{x^2}{x(x + 2)} = \dfrac{2x + 4 + x^2}{x(x + 2)}$

Step 4. $\dfrac{2x + 4 + x^2}{x(x + 2)} = \dfrac{x^2 + 2x + 4}{x(x + 2)}$

The rational expression cannot be reduced because $x^2 + 2x + 4$ does not factor. ■

Example 5   Subtract $3 - \dfrac{2a}{a + 2}$.

Solution

Step 1. LCD $= a + 2$

Step 2. $3 = \dfrac{3\,(a + 2)}{1\,(a + 2)} = \dfrac{3a + 6}{a + 2}$

$\dfrac{2a}{a + 2} = \dfrac{2a}{a + 2}$

Step 3. $3 - \dfrac{2a}{a + 2} = \dfrac{3a + 6}{a + 2} - \dfrac{2a}{a + 2} = \dfrac{3a + 6 - 2a}{a + 2} = \dfrac{a + 6}{a + 2}$

Step 4. The rational expression cannot be reduced. ■

A WORD OF CAUTION   It is incorrect to divide both the numerator of one fraction and the denominator of a *different* fraction by a common factor when adding or subtracting.

$$\dfrac{x + 1}{x - 2} + \dfrac{5}{x + 1} = \dfrac{\overset{1}{\cancel{x + 1}}}{x - 2} + \dfrac{5}{\underset{1}{\cancel{x + 1}}} = \dfrac{1}{x - 2} + \dfrac{5}{1}$$ ☑

Example 6   Subtract $\dfrac{z - 1}{z + 3} - \dfrac{z + 3}{z - 1}$.

Solution We cannot divide the first numerator and the second denominator by $z - 1$ or the first denominator and the second numerator by $z + 3$ because this is not a multiplication problem.

Step 1. LCD $= (z + 3)(z - 1)$

Step 2. $\dfrac{z - 1}{z + 3} = \dfrac{(z - 1)\,(z - 1)}{(z + 3)\,(z - 1)}$ ⟵

$\dfrac{z + 3}{z - 1} = \dfrac{(z + 3)\,(z + 3)}{(z - 1)\,(z + 3)}$ ⟵   Note the parentheses

Step 3. $\dfrac{z - 1}{z + 3} - \dfrac{z + 3}{z - 1} = \dfrac{(z - 1)\,(z - 1)}{(z + 3)\,(z - 1)} - \dfrac{(z + 3)\,(z + 3)}{(z - 1)\,(z + 3)}$

$= \dfrac{z^2 - 2z + 1}{(z + 3)(z - 1)} - \dfrac{z^2 + 6z + 9}{(z + 3)(z - 1)}$

These parentheses are essential

$= \dfrac{(z^2 - 2z + 1) - (z^2 + 6z + 9)}{(z + 3)(z - 1)}$

$= \dfrac{z^2 - 2z + 1 - z^2 - 6z - 9}{(z + 3)(z - 1)}$

$= \dfrac{-8z - 8}{(z + 3)(z - 1)} = \dfrac{-8(z + 1)}{(z + 3)(z - 1)}$

Step 4. The rational expression cannot be reduced. (Note: Other forms of the answer are also acceptable.) ∎

Example 7 Subtract $\dfrac{x-1}{2x^2+11x+12} - \dfrac{x-2}{2x^2+x-3}$.

Solution

Step 1. In order to find the LCD, we must factor each denominator.

$$2x^2 + 11x + 12 = (2x+3)(x+4)$$
$$2x^2 + x - 3 = (2x+3)(x-1)$$

The LCD is $(2x+3)(x+4)(x-1)$.

Step 2. $\dfrac{x-1}{2x^2+11x+12} = \dfrac{x-1}{(2x+3)(x+4)} = \dfrac{(x-1)\,(x-1)}{(2x+3)(x+4)\,(x-1)}$

$\dfrac{x-2}{2x^2+x-3} = \dfrac{x-2}{(2x+3)(x-1)} = \dfrac{(x-2)\,(x+4)}{(2x+3)(x-1)\,(x+4)}$

Step 3. $\dfrac{x-1}{2x^2+11x+12} - \dfrac{x-2}{2x^2+x-3}$

$= \dfrac{(x-1)(x-1)}{(2x+3)(x+4)(x-1)} - \dfrac{(x-2)(x+4)}{(2x+3)(x-1)(x+4)}$

$= \dfrac{x^2-2x+1}{(2x+3)(x+4)(x-1)} - \dfrac{x^2+2x-8}{(2x+3)(x+4)(x-1)}$

$= \dfrac{x^2-2x+1-(x^2+2x-8)}{(2x+3)(x+4)(x-1)}$ The parentheses are essential

$= \dfrac{x^2-2x+1-x^2-2x+8}{(2x+3)(x+4)(x-1)}$

$= \dfrac{-4x+9}{(2x+3)(x+4)(x-1)}$

Step 4. The rational expression cannot be reduced. ∎

A WORD OF CAUTION It is incorrect to multiply all the fractions by the LCD. Example 7 cannot be done as follows:

$(2x+3)(x+4)(x-1) \cdot \dfrac{x-1}{(2x+3)(x+4)} - (2x+3)(x+4)(x-1) \cdot \dfrac{x-2}{(2x+3)(x-1)}$

$= (x-1)(x-1) - (x+4)(x-2) = x^2 - 2x + 1 - (x^2+2x-8)$

$= -4x + 9$ ☑

EXERCISES 8.6

Set I Perform the indicated operations.

1. $\dfrac{3}{a^2} + \dfrac{2}{a^3}$

2. $\dfrac{5}{u} + \dfrac{4}{u^3}$

3. $\dfrac{1}{2} + \dfrac{3}{x} - \dfrac{5}{x^2}$

4. $\dfrac{2}{3} - \dfrac{1}{y} + \dfrac{4}{y^2}$

5. $\dfrac{2}{xy} - \dfrac{3}{y}$

6. $\dfrac{5}{ab} - \dfrac{4}{a}$

7. $5 + \dfrac{2}{x}$

8. $3 + \dfrac{4}{y}$

9. $\dfrac{5}{4y^2} + \dfrac{9}{6y}$

10. $\dfrac{10}{5x} + \dfrac{3}{4x^2}$

11. $3x - \dfrac{3}{x}$

12. $4y - \dfrac{5}{y}$

13. $\dfrac{3}{a} + \dfrac{a}{a+3}$

14. $\dfrac{5}{b} + \dfrac{b}{b-5}$

15. $\dfrac{x}{2x+4} + \dfrac{-5}{4x}$

16. $\dfrac{4}{3x} + \dfrac{-2x}{3x+6}$

17. $x + \dfrac{2}{x} - \dfrac{3}{x-2}$

18. $m + \dfrac{3}{m} - \dfrac{2}{m-4}$

19. $\dfrac{a+b}{b} + \dfrac{b}{a-b}$

20. $\dfrac{y+x}{x} + \dfrac{x}{y-x}$

21. $\dfrac{3}{2e-2} - \dfrac{2}{3e-3}$

22. $\dfrac{2}{3f+6} - \dfrac{1}{5f+10}$

23. $\dfrac{3}{m-2} - \dfrac{5}{2-m}$

24. $\dfrac{7}{n-5} - \dfrac{2}{5-n}$

25. $\dfrac{x+1}{x-1} - \dfrac{x-1}{x+1}$

26. $\dfrac{x+4}{x-4} - \dfrac{x-4}{x+4}$

27. $\dfrac{y}{x^2-xy} + \dfrac{x}{y^2-xy}$

28. $\dfrac{b}{ab-a^2} + \dfrac{a}{ab-b^2}$

29. $\dfrac{2x}{x-3} - \dfrac{2x}{x+3} + \dfrac{36}{x^2-9}$

30. $\dfrac{x}{x+4} - \dfrac{x}{x-4} - \dfrac{32}{x^2-16}$

31. $\dfrac{x}{x^2+4x+4} + \dfrac{1}{x+2}$

32. $\dfrac{2x}{x^2-2x+1} - \dfrac{5}{x-1}$

33. $\dfrac{2x+3}{x^2+4x-5} - \dfrac{x+5}{2x^2+x-3}$

34. $\dfrac{3x+2}{x^2+6x-7} - \dfrac{x+7}{3x^2-x-2}$

35. $\dfrac{x^2}{x^2+8x+16} - \dfrac{x+1}{x^2-16}$

36. $\dfrac{x^2}{x^2+6x+9} - \dfrac{x+2}{x^2-9}$

37. $\dfrac{x+1}{2x^3+3x^2-5x} - \dfrac{x-1}{2x^3+7x^2+5x}$

38. $\dfrac{x+1}{3x^3+x^2-4x} - \dfrac{x-1}{3x^3+7x^2+4x}$

Set II   Perform the indicated operations.

1. $\dfrac{3}{x} + \dfrac{4}{x^2}$

2. $\dfrac{5}{2x^3y} + \dfrac{3}{4xy^2}$

3. $\dfrac{1}{3} - \dfrac{1}{a} + \dfrac{2}{a^2}$

4. $\dfrac{1}{6} - \dfrac{1}{k} + \dfrac{3}{k^2}$

5. $\dfrac{5}{xy} - \dfrac{3}{x}$

6. $\dfrac{3}{b} - \dfrac{7}{ab}$

7. $2 + \dfrac{4}{z}$

8. $m - \dfrac{3}{m} + \dfrac{2}{m+4}$

9. $\dfrac{5}{6xy^2} + \dfrac{7}{8x^2y}$

10. $\dfrac{z+1}{z+2} - \dfrac{z-1}{z-2}$

11. $4m - \dfrac{5}{m}$

12. $\dfrac{2}{x+5} + \dfrac{3}{x-5}$

13. $\dfrac{2}{x} + \dfrac{x}{x+4}$

14. $\dfrac{2}{x+5} - \dfrac{3}{x-5}$

15. $\dfrac{x}{3x+6} + \dfrac{-3}{2x}$

16. $\dfrac{x+5}{x-2} - \dfrac{x-2}{x+5}$

17. $a + \dfrac{3}{a} - \dfrac{2}{a - 2}$

18. $\dfrac{2}{a + 3} - \dfrac{4}{a - 1}$

19. $\dfrac{a - b}{b} + \dfrac{b}{a + b}$

20. $\dfrac{x + 2}{x - 3} - \dfrac{x + 3}{x - 2}$

21. $\dfrac{7}{4x - 4} - \dfrac{4}{7x - 7}$

22. $\dfrac{2x + 3}{x + 8} + \dfrac{x + 8}{2x + 3}$

23. $\dfrac{5}{y - 7} - \dfrac{8}{7 - y}$

24. $\dfrac{x - y}{x} - \dfrac{x}{x + y}$

25. $\dfrac{x - 1}{x + 2} - \dfrac{x - 2}{x + 1}$

26. $\dfrac{5}{x - 3} - \dfrac{4}{3 - x}$

27. $\dfrac{c}{d^2 - cd} + \dfrac{d}{c^2 - cd}$

28. $\dfrac{x - 5}{x + 5} - \dfrac{x + 5}{x - 5}$

29. $\dfrac{x}{x - 1} - \dfrac{x}{x + 1} + \dfrac{2}{x^2 - 1}$

30. $\dfrac{x + 2}{x^2 + x - 2} + \dfrac{3}{x^2 - 1}$

31. $\dfrac{x}{x^2 + 10x + 25} + \dfrac{1}{x + 5}$

32. $\dfrac{b}{ab - a^2} - \dfrac{a}{b^2 - ab}$

33. $\dfrac{4x + 3}{x^2 + 3x - 4} - \dfrac{x + 4}{4x^2 - x - 3}$

34. $\dfrac{a^2}{4a^2 + 12a + 9} - \dfrac{a + 3}{4a^2 - 9}$

35. $\dfrac{x^2}{x^2 + 12x + 36} - \dfrac{x + 3}{x^2 - 36}$

36. $\dfrac{a^2}{9a^2 - 25} - \dfrac{a + 1}{9a^2 - 30a + 25}$

37. $\dfrac{x + 1}{3x^3 + x^2 - 4x} - \dfrac{x - 1}{3x^3 + 7x^2 + 4x}$

38. $\dfrac{5}{3x^4 - 27x^2} + \dfrac{2}{x^3 + 6x^2 + 9x} - \dfrac{3}{6x - 2x^2}$

8.7 Complex Fractions

If the numerator and/or denominator of a fraction is itself a rational expression, we call the fraction a **complex fraction**. The following are examples of complex fractions:

$$\dfrac{\dfrac{2}{x}}{3}, \quad \dfrac{a}{\dfrac{1}{c}}, \quad \dfrac{\dfrac{3}{z}}{\dfrac{5}{z}}, \quad \dfrac{\dfrac{3}{x} - \dfrac{2}{y}}{\dfrac{5}{x} + \dfrac{3}{y}}$$

The parts of a complex fraction are as follows:

$$\left.\dfrac{\dfrac{1}{x} + \dfrac{3}{y}}{\dfrac{5}{x} - \dfrac{2}{y}}\right\}$$

$\left.\rule{0pt}{12pt}\right\}$ Primary numerator

←Main fraction bar

$\left.\rule{0pt}{12pt}\right\}$ Primary denominator

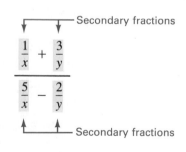

TO SIMPLIFY COMPLEX FRACTIONS

Method 1: Multiply both the numerator and the denominator of the complex fraction by the LCD of the secondary fractions; then simplify the results.

Method 2: First simplify the numerator and denominator of the complex fraction; then divide the simplified numerator by the simplified denominator.

Note that in some of the following examples the solution by method 1 is easier than that by method 2. In others, the opposite is true.

Example 1　Simplify the following arithmetic complex fraction:

$$\frac{\dfrac{1}{2}+\dfrac{3}{4}}{\dfrac{5}{6}-\dfrac{2}{3}}$$

Method 1 The LCD of the secondary denominators 2, 4, 6, and 3 is 12.

$$\frac{12}{12}\cdot\frac{\dfrac{1}{2}+\dfrac{3}{4}}{\dfrac{5}{6}-\dfrac{2}{3}}=\frac{12\left(\dfrac{1}{2}+\dfrac{3}{4}\right)}{12\left(\dfrac{5}{6}-\dfrac{2}{3}\right)}=\frac{\dfrac{12}{1}\left(\dfrac{1}{2}\right)+\dfrac{12}{1}\left(\dfrac{3}{4}\right)}{\dfrac{12}{1}\left(\dfrac{5}{6}\right)-\dfrac{12}{1}\left(\dfrac{2}{3}\right)}=\frac{6+9}{10-8}=\frac{15}{2}=7\frac{1}{2}$$

└── Multiplying by 1

Method 2

$$\frac{\dfrac{1}{2}+\dfrac{3}{4}}{\dfrac{5}{6}-\dfrac{2}{3}}=\left(\dfrac{1}{2}+\dfrac{3}{4}\right)\div\left(\dfrac{5}{6}-\dfrac{2}{3}\right)=\left(\dfrac{2}{4}+\dfrac{3}{4}\right)\div\left(\dfrac{5}{6}-\dfrac{4}{6}\right)$$

$$=\frac{5}{4}\div\frac{1}{6}=\frac{5}{\overset{}{4}_{2}}\cdot\frac{\overset{3}{6}}{1}=\frac{15}{2}=7\frac{1}{2}\quad\blacksquare$$

A WORD OF CAUTION　In Example 1, students sometimes make this error:

$$\left(\frac{1}{2}+\frac{3}{4}\right)\div\left(\frac{5}{6}-\frac{2}{3}\right)=\left(\frac{1}{2}+\frac{3}{4}\right)\times\left(\frac{6}{5}-\frac{3}{2}\right)$$

This is incorrect, since the multiplicative inverse of $\left(\dfrac{5}{6}-\dfrac{2}{3}\right)$ is not $\left(\dfrac{6}{5}-\dfrac{3}{2}\right)$. (You should verify that $\left(\dfrac{5}{6}-\dfrac{2}{3}\right)\left(\dfrac{6}{5}-\dfrac{3}{2}\right)\neq 1$.) $\quad\checkmark$

Example 2   Simplify $\dfrac{\dfrac{4b^2}{9a^2}}{\dfrac{8b}{3a^3}}$. ← Main fraction bar

Method 1 The LCD of the secondary denominators $9a^2$ and $3a^3$ is $9a^3$.

$$\dfrac{9a^3}{9a^3}\left(\dfrac{\dfrac{4b^2}{9a^2}}{\dfrac{8b}{3a^3}}\right) = \dfrac{\dfrac{9a^3}{1}\left(\dfrac{4b^2}{9a^2}\right)}{\dfrac{9a^3}{1}\left(\dfrac{8b}{3a^3}\right)} = \dfrac{4ab^2}{24b} = \dfrac{ab}{6}$$

└─ The value of this fraction is 1

Method 2

$$\dfrac{\dfrac{4b^2}{9a^2}}{\dfrac{8b}{3a^3}} = \dfrac{4b^2}{9a^2} \div \dfrac{8b}{3a^3} = \dfrac{\overset{1}{4b^2}}{\underset{3}{9a^2}} \cdot \dfrac{\overset{1}{3a^3}}{\underset{2}{8b}} = \dfrac{ab}{6}$$ ∎

Example 3   Simplify $\dfrac{\dfrac{2}{x} - \dfrac{3}{x^2}}{5 + \dfrac{1}{x}}$.

Method 1 The LCD of the secondary denominators x and x^2 is x^2.

$$\dfrac{x^2}{x^2}\left(\dfrac{\dfrac{2}{x} - \dfrac{3}{x^2}}{5 + \dfrac{1}{x}}\right) = \dfrac{\dfrac{x^2}{1}\left(\dfrac{2}{x}\right) - \dfrac{x^2}{1}\left(\dfrac{3}{x^2}\right)}{\dfrac{x^2}{1}\left(\dfrac{5}{1}\right) + \dfrac{x^2}{1}\left(\dfrac{1}{x}\right)} = \dfrac{2x - 3}{5x^2 + x}$$

Method 2

$$\dfrac{\dfrac{2}{x} - \dfrac{3}{x^2}}{5 + \dfrac{1}{x}} = \left(\dfrac{2}{x} - \dfrac{3}{x^2}\right) \div \left(5 + \dfrac{1}{x}\right)$$

$$= \left(\dfrac{2x}{x^2} - \dfrac{3}{x^2}\right) \div \left(\dfrac{5x}{x} + \dfrac{1}{x}\right)$$

$$= \dfrac{2x - 3}{x^2} \div \dfrac{5x + 1}{x}$$

$$= \dfrac{2x - 3}{\underset{x}{x^2}} \cdot \dfrac{\overset{1}{x}}{5x + 1} = \dfrac{2x - 3}{5x^2 + x}$$ ∎

Example 4 Simplify $\dfrac{1 + \dfrac{2}{a}}{1 - \dfrac{4}{a^2}}$.

Method 1 The LCD of the secondary denominators a and a^2 is a^2.

$$\frac{a^2}{a^2}\left(\frac{1 + \dfrac{2}{a}}{1 - \dfrac{4}{a^2}}\right) = \frac{\dfrac{a^2}{1}\left(\dfrac{1}{1}\right) + \dfrac{a^2}{1}\left(\dfrac{2}{a}\right)}{\dfrac{a^2}{1}\left(\dfrac{1}{1}\right) - \dfrac{a^2}{1}\left(\dfrac{4}{a^2}\right)} = \frac{a^2 + 2a}{a^2 - 4} = \frac{a(a + 2)}{(a + 2)(a - 2)} = \frac{a}{a - 2}$$

Method 2

$$\frac{1 + \dfrac{2}{a}}{1 - \dfrac{4}{a^2}} = \left(1 + \frac{2}{a}\right) \div \left(1 - \frac{4}{a^2}\right)$$

$$= \left(\frac{a}{a} + \frac{2}{a}\right) \div \left(\frac{a^2}{a^2} - \frac{4}{a^2}\right)$$

$$= \frac{a + 2}{a} \div \frac{a^2 - 4}{a^2}$$

$$= \frac{a + 2}{a} \cdot \frac{a^2}{a^2 - 4}$$

$$= \frac{a + 2}{a} \cdot \frac{a^2}{(a + 2)(a - 2)}$$

$$= \frac{a}{a - 2} \quad \blacksquare$$

Example 5 Simplify $\dfrac{\dfrac{2}{x + 3} + \dfrac{1}{x}}{\dfrac{3}{x + 3} - \dfrac{2}{x}}$.

Method 1 The LCD of the secondary denominators is $x(x + 3)$.

$$\frac{x(x + 3)}{x(x + 3)}\left(\frac{\dfrac{2}{x + 3} + \dfrac{1}{x}}{\dfrac{3}{x + 3} - \dfrac{2}{x}}\right) = \frac{\dfrac{x(x + 3)}{1}\left(\dfrac{2}{x + 3}\right) + \dfrac{x(x + 3)}{1}\left(\dfrac{1}{x}\right)}{\dfrac{x(x + 3)}{1}\left(\dfrac{3}{x + 3}\right) - \dfrac{x(x + 3)}{1}\left(\dfrac{2}{x}\right)}$$

$$= \frac{2x + (x + 3)}{3x - (x + 3)(2)}$$

$$= \frac{2x + x + 3}{3x - 2x - 6}$$

$$= \frac{3x + 3}{x - 6}$$

Method 2

$$\frac{\dfrac{2}{x+3}+\dfrac{1}{x}}{\dfrac{3}{x+3}-\dfrac{2}{x}} = \left(\frac{2}{x+3}+\frac{1}{x}\right) \div \left(\frac{3}{x+3}-\frac{2}{x}\right)$$

$$= \left(\frac{2\,x}{x\,(x+3)}+\frac{1\,(x+3)}{x\,(x+3)}\right) \div \left(\frac{3\,x}{x\,(x+3)}-\frac{2\,(x+3)}{x\,(x+3)}\right)$$

$$= \left(\frac{2x+(x+3)}{x(x+3)}\right) \div \left(\frac{3x-2(x+3)}{x(x+3)}\right)$$

$$= \left(\frac{2x+x+3}{x(x+3)}\right) \div \left(\frac{3x-2x-6}{x(x+3)}\right)$$

$$= \left(\frac{3x+3}{x(x+3)}\right) \div \left(\frac{x-6}{x(x+3)}\right)$$

$$= \left(\frac{3x+3}{\cancel{x(x+3)}_{1}}\right)\left(\frac{\overset{1}{\cancel{x(x+3)}}}{x-6}\right)$$

$$= \frac{3x+3}{x-6} \quad \blacksquare$$

EXERCISES 8.7

Set I   Simplify each complex fraction.

1. $\dfrac{\dfrac{3}{4}}{\dfrac{5}{6}}$

2. $\dfrac{\dfrac{3}{5}}{\dfrac{3}{4}}$

3. $\dfrac{\dfrac{2}{3}}{\dfrac{4}{9}}$

4. $\dfrac{\dfrac{5}{6}}{\dfrac{5}{9}}$

5. $\dfrac{\dfrac{3}{4}-\dfrac{1}{2}}{\dfrac{5}{8}+\dfrac{1}{4}}$

6. $\dfrac{\dfrac{5}{6}-\dfrac{1}{3}}{\dfrac{2}{9}+\dfrac{1}{6}}$

7. $\dfrac{\dfrac{3}{5}+2}{2-\dfrac{3}{4}}$

8. $\dfrac{\dfrac{3}{16}+5}{6-\dfrac{7}{8}}$

9. $\dfrac{\dfrac{5x^3}{3y^4}}{\dfrac{10x}{9y}}$

10. $\dfrac{\dfrac{8a^4}{5b}}{\dfrac{4a^3}{15b^2}}$

11. $\dfrac{\dfrac{18cd^2}{5a^3b}}{\dfrac{12cd^2}{15ab^2}}$

12. $\dfrac{\dfrac{8x^2y}{7z^3}}{\dfrac{12xy^2}{21z^5}}$

13. $\dfrac{\dfrac{x+3}{5}}{\dfrac{2x+6}{10}}$

14. $\dfrac{\dfrac{a-4}{3}}{\dfrac{2a-8}{9}}$

15. $\dfrac{\dfrac{x+2}{2x}}{\dfrac{x+1}{4x^2}}$

16. $\dfrac{\dfrac{x-3}{3x^2}}{\dfrac{x-9}{9x}}$

17. $\dfrac{\dfrac{a}{b}+1}{\dfrac{a}{b}-1}$

18. $\dfrac{2+\dfrac{x}{y}}{2-\dfrac{x}{y}}$

19. $\dfrac{\dfrac{1}{x}+x}{\dfrac{1}{x}-x}$

20. $\dfrac{a-\dfrac{4}{a}}{a+\dfrac{4}{a}}$

21. $\dfrac{\dfrac{c}{d}+2}{\dfrac{c^2}{d^2}-4}$

22. $\dfrac{\dfrac{x^2}{y^2}-1}{\dfrac{x}{y}-1}$

23. $\dfrac{x+\dfrac{x}{y}}{1+\dfrac{1}{y}}$

24. $\dfrac{1-\dfrac{1}{b}}{3-\dfrac{3}{b}}$

25. $\dfrac{\dfrac{1}{x^2} - \dfrac{1}{y^2}}{\dfrac{1}{x} + \dfrac{1}{y}}$

26. $\dfrac{\dfrac{1}{a^2} - \dfrac{1}{4}}{\dfrac{1}{a} - \dfrac{1}{2}}$

27. $\dfrac{\dfrac{2}{x} - \dfrac{4}{x^2}}{\dfrac{1}{x} - \dfrac{2}{x^2}}$

28. $\dfrac{\dfrac{2}{y^2} + \dfrac{1}{y}}{\dfrac{8}{y^2} + \dfrac{4}{y}}$

29. $\dfrac{\dfrac{x}{x+1} + \dfrac{4}{3x}}{\dfrac{x}{x+1} - \dfrac{3}{x}}$

30. $\dfrac{\dfrac{4x}{4x+1} + \dfrac{1}{2x}}{\dfrac{2}{4x+1} + \dfrac{2}{x}}$

31. $\dfrac{\dfrac{1}{4x} + \dfrac{x}{x-6}}{\dfrac{x-1}{x} - \dfrac{3}{x-6}}$

32. $\dfrac{\dfrac{x+1}{y} + \dfrac{1}{x-1}}{\dfrac{1}{x-1} + \dfrac{1}{2y}}$

Set II Simplify each complex fraction.

1. $\dfrac{\dfrac{5}{6}}{\dfrac{2}{9}}$

2. $\dfrac{\dfrac{3}{5}}{\dfrac{9}{10}}$

3. $\dfrac{\dfrac{5}{6}}{\dfrac{10}{18}}$

4. $\dfrac{\dfrac{8}{9}}{\dfrac{8}{27}}$

5. $\dfrac{\dfrac{7}{9} - \dfrac{1}{3}}{\dfrac{5}{2} - \dfrac{1}{4}}$

6. $\dfrac{\dfrac{5}{6} - \dfrac{1}{3}}{\dfrac{3}{2} - \dfrac{1}{4}}$

7. $\dfrac{\dfrac{3}{4} + 2}{3 - \dfrac{1}{2}}$

8. $\dfrac{\dfrac{1}{8} + 3}{4 - \dfrac{1}{2}}$

9. $\dfrac{\dfrac{3a^3}{5b^2}}{\dfrac{6a^2}{10b^3}}$

10. $\dfrac{\dfrac{4x^4}{5y^3}}{\dfrac{8x^2}{10y^4}}$

11. $\dfrac{\dfrac{16bc^2}{5a^2d}}{\dfrac{4ac^4}{10b^2d^3}}$

12. $\dfrac{\dfrac{9x^2y^3}{11wz^2}}{\dfrac{18xy^4}{22w^3z}}$

13. $\dfrac{\dfrac{x-2}{4}}{\dfrac{3x-6}{12}}$

14. $\dfrac{\dfrac{a}{b} - 2}{2 + \dfrac{a}{b}}$

15. $\dfrac{\dfrac{x+3}{3x}}{\dfrac{x+1}{6x^2}}$

16. $\dfrac{\dfrac{1}{x+2}}{\dfrac{4}{3x+6}}$

17. $\dfrac{\dfrac{4}{x} + 1}{\dfrac{4}{x} - 1}$

18. $\dfrac{2}{1 + \dfrac{1}{x}}$

19. $\dfrac{\dfrac{3}{a} - a}{\dfrac{3}{a} + a}$

20. $\dfrac{1 - \dfrac{1}{a^2}}{\dfrac{1}{a} - \dfrac{1}{a^2}}$

21. $\dfrac{\dfrac{x^2}{y^2} - 4}{\dfrac{x}{y} + 2}$

22. $\dfrac{\dfrac{1}{a} + \dfrac{1}{b}}{\dfrac{1}{ab}}$

23. $\dfrac{2 + \dfrac{2}{b}}{a + \dfrac{a}{2}}$

24. $\dfrac{\dfrac{1}{x^2} - 9}{\dfrac{1}{x} + 3}$

25. $\dfrac{\dfrac{1}{x^2} - \dfrac{4}{y^2}}{\dfrac{1}{x} + \dfrac{2}{y}}$

26. $\dfrac{\dfrac{x+1}{6x^2}}{\dfrac{x+2}{2x}}$

27. $\dfrac{\dfrac{1}{y^2} - \dfrac{1}{9}}{\dfrac{1}{y} + \dfrac{1}{3}}$

28. $\dfrac{\dfrac{a^2}{9} - \dfrac{1}{b^2}}{\dfrac{a}{3} + \dfrac{1}{b}}$

29. $\dfrac{\dfrac{x}{x+3} + \dfrac{3}{2x}}{\dfrac{x}{x+3} - \dfrac{2}{x}}$

30. $\dfrac{\dfrac{1}{6x} + \dfrac{x}{x-3}}{\dfrac{x-2}{x} - \dfrac{2}{x-3}}$

31. $\dfrac{\dfrac{3x}{3x+1}+\dfrac{1}{3x}}{\dfrac{3}{3x+1}+\dfrac{3}{x}}$ **32.** $\dfrac{\dfrac{x+4}{y}+\dfrac{1}{x-4}}{\dfrac{4}{x-4}+\dfrac{1}{4y}}$

8.8 Review: 8.1–8.7

**Rational Expressions
8.1**

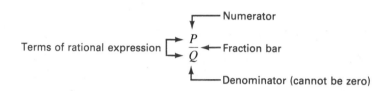

where P and Q are polynomials and $Q \neq 0$.

**Excluded Values
8.1**

Any value of the variable(s) that would make any denominator be zero must be *excluded*.

**Equivalent Rational
Expressions
8.1**

Equivalent rational expressions are rational expressions that have the same value. The fundamental property of rational expressions states that $\dfrac{P \cdot C}{Q \cdot C}$ and $\dfrac{P}{Q}$ are equivalent fractions, where P, Q, and C are polynomials and where $Q \neq 0$ and $C \neq 0$.

**The Three Signs
of a Fraction
8.1**

Every fraction has three signs associated with it: the sign of the entire fraction, the sign of the numerator, and the sign of the denominator. If any two of the three signs of a fraction are changed, the value of the fraction is unchanged.

**To Reduce a Rational
Expression to
Lowest Terms
8.2**

1. Factor the numerator and denominator completely and find their GCF.

2. Divide both numerator and denominator by their GCF.

(We can, instead, divide both the numerator and the denominator by any factor common to both and then check for other common factors.)

**To Multiply Rational
Expressions
8.3**

1. Factor any numerators or denominators that have more than one term.

2. Divide the numerators and denominators by any factor common to both.

3. The answer is the product of factors remaining in the numerator divided by the product of factors remaining in the denominator.

**To Divide Rational
Expressions
8.3**

Multiply the dividend by the multiplicative inverse (the reciprocal) of the divisor.

$$\underset{\text{Dividend}}{\underbrace{\dfrac{P}{Q}}} \div \underset{\text{Divisor}}{\underbrace{\dfrac{S}{T}}} = \dfrac{P}{Q} \cdot \dfrac{T}{S}$$

**To Add or Subtract Like
Rational Expressions
8.4**

1. Write the sum or difference of the numerators over the denominator of the like rational expressions.

$$\dfrac{P}{Q} + \dfrac{R}{Q} = \dfrac{P+R}{Q} \quad \text{and} \quad \dfrac{P}{Q} - \dfrac{R}{Q} = \dfrac{P-R}{Q}$$

2. Reduce the resulting rational expression to lowest terms.

To Find the LCM of Two or More Polynomials
8.5

1. Factor each polynomial completely. Express repeated factors as powers.

2. Write down each different base that appears in any of the factorizations.

3. Raise each base to the highest power to which it occurs in any of the factorizations.

4. The **least common multiple** (**LCM**) is the product of all the factors found in step 3.

To Find the LCD
8.5

The **least common denominator** (**LCD**) of two or more rational expressions is the LCM of their denominators.

To Add or Subtract Unlike Rational Expressions
8.6

1. Find the LCD.

2. Convert each rational expression to an equivalent rational expression having the LCD as denominator.

3. Add or subtract the resulting like rational expressions.

4. Reduce the resulting rational expression to lowest terms.

Complex Fractions
8.7

A **complex fraction** is a fraction that contains a rational expression in its numerator and/ or its denominator.

To Simplify Complex Fractions
8.7

Method 1: Multiply both the numerator and the denominator of the complex fraction by the LCD of the secondary fractions, then simplify the results.

Method 2: First simplify the numerator and denominator of the complex fraction, then divide the simplified numerator by the simplified denominator.

Review Exercises 8.8 Set I

In Exercises 1–4, determine what value (or values) of the variable must be excluded.

1. $\dfrac{2x - 1}{x + 4}$

2. $8y - \dfrac{4}{5y}$

3. $\dfrac{x - 2}{x^2 - 9}$

4. $\dfrac{x + 4}{x^3 + x^2 - 2x}$

In Exercises 5 and 6, determine whether the pairs of rational expressions are equivalent.

5. $\dfrac{x + 1}{2x + 2}, \dfrac{1}{2}$

6. $\dfrac{4x + 5}{3x + 5}, \dfrac{4x}{3x}$

In Exercises 7–9, find the missing term.

7. $\dfrac{x - 6}{7} = -\dfrac{?}{-7}$

8. $\dfrac{2}{a - b} = -\dfrac{?}{b - a}$

9. $\dfrac{3 - x}{4} = \dfrac{x - 3}{?}$

In Exercises 10–15, reduce each rational expression to lowest terms.

10. $\dfrac{6ab^4}{3a^3b^3}$

11. $\dfrac{2 + 4m}{2}$

12. $\dfrac{3 - 6n}{3}$

13. $\dfrac{a^2 - 4}{a + 2}$

14. $\dfrac{x - 5}{x^2 - 3x - 10}$

15. $\dfrac{a - b}{ax + ay - bx - by}$

In Exercises 16–28, perform the indicated operations.

16. $\dfrac{8}{k} + \dfrac{2}{k}$

17. $5 - \dfrac{3}{2x}$

18. $\dfrac{4m}{2m-3} - \dfrac{2m+3}{2m-3}$

19. $\dfrac{-5a^2}{3b} \div \dfrac{10a}{9b^2}$

20. $\dfrac{21x^3}{4y^2} \div \dfrac{-7x}{8y^4}$

21. $\dfrac{3x+6}{6} \cdot \dfrac{2x^2}{4x+8}$

22. $\dfrac{y-3}{2} - \dfrac{y+4}{3}$

23. $\dfrac{a-2}{a-1} + \dfrac{a+1}{a+2}$

24. $\dfrac{k+3}{k-2} - \dfrac{k+2}{k-3}$

25. $\dfrac{2x^2-6x}{x+2} \div \dfrac{x}{4x+8}$

26. $\dfrac{4z^2}{z-5} \div \dfrac{z}{2z-10}$

27. $\dfrac{3x+4}{2x^2+7x+6} - \dfrac{x+2}{3x^2+10x+8}$

28. $\dfrac{5x-20}{3x^3+24x^2+48x} - \dfrac{x+4}{5x^2-80}$

In Exercises 29–32, simplify the complex fractions.

29. $\dfrac{\dfrac{5k^2}{3m^2}}{\dfrac{10k}{9m}}$

30. $\dfrac{3-\dfrac{a}{b}}{2+\dfrac{a}{b}}$

31. $\dfrac{\dfrac{a^2}{b^2}-1}{\dfrac{a}{b}-1}$

32. $\dfrac{1-\dfrac{16}{x^2}}{1-\dfrac{4}{x}}$

Review Exercises 8.8 Set II

NAME _____

In Exercises 1–4, determine what value (or values) of the variable must be excluded.

ANSWERS

1. $\dfrac{x + 2}{x - 1}$

2. $3x + \dfrac{5}{2x}$

1. _____

2. _____

3. _____

3. $\dfrac{x + 2}{x^2 + 2x - 3}$

4. $\dfrac{x - 2}{2x^2 + x - 3}$

4. _____

5. _____

In Exercises 5 and 6, determine whether the pairs of rational expressions are equivalent.

6. _____

5. $\dfrac{5x}{3y}, \dfrac{5x + 1}{3y + 1}$

6. $\dfrac{3x(x - 1)}{2x(x - 1)}, \dfrac{3x}{2x}$

7. _____

8. _____

9. _____

In Exercises 7–9, find the missing term.

7. $\dfrac{6}{x - 3} = \dfrac{-6}{?}$

8. $\dfrac{4 - x}{7} = -\dfrac{?}{7}$

10. _____

11. _____

12. _____

9. $\dfrac{x - 3}{(2 + x)(5 - 2x)} = \dfrac{?}{(x + 2)(2x - 5)}$

13. _____

14. _____

In Exercises 10–15, reduce each rational expression to lowest terms.

15. _____

10. $\dfrac{6ab^3}{2ab}$

11. $\dfrac{x + 2}{x^2 - 2x - 8}$

12. $\dfrac{9 - y^2}{3 - y}$

13. $\dfrac{x + 3}{x^2 - x - 12}$

14. $\dfrac{4z^3 + 4z^2 - 24z}{2z^2 + 4z - 6}$

15. $\dfrac{6k^3 - 12k^2 - 18k}{3k^2 + 3k - 36}$

In Exercises 16–28, perform the indicated operations.

16. $\dfrac{3}{x} + \dfrac{5}{x}$

17. $4 - \dfrac{3}{2x}$

18. $\dfrac{2a}{3a+1} - \dfrac{3a-1}{3a+1}$

19. $\dfrac{x+1}{2} - \dfrac{x-3}{5}$

20. $\dfrac{y-2}{y+1} - \dfrac{y-1}{y+2}$

21. $\dfrac{15x^3}{4y^2} \div \dfrac{5x^2}{8y}$

22. $\dfrac{4x-4}{2} \cdot \dfrac{6x^2}{3x-3}$

23. $\dfrac{z^2+3z+2}{z^2+2z+1} \div \dfrac{z^2+2z-3}{z^2-1}$

24. $\dfrac{5x-5}{10} \cdot \dfrac{4x^3}{2x-2}$

25. $\dfrac{x+4}{5} - \dfrac{x-2}{3}$

26. $\dfrac{x^2+2x-3}{2x^2-x-1} \cdot \dfrac{6x+3}{3x^2+11x+6}$

27. $\dfrac{15x+3}{50x^2-20x+2} - \dfrac{4}{75x^2-3}$

28. $\dfrac{5}{6x^2+5x-1} - \dfrac{x+1}{36x^2-12x+1}$

In Exercises 29–32, simplify the complex fractions.

29. $\dfrac{\dfrac{10a^2b}{12a^4b^3}}{\dfrac{5ab^2}{16a^2b^3}}$

30. $\dfrac{2+\dfrac{a}{b}}{\dfrac{a}{b}-2}$

31. $\dfrac{\dfrac{1}{x^2}-\dfrac{9}{y^2}}{\dfrac{1}{x}-\dfrac{3}{y}}$

32. $\dfrac{\dfrac{x}{y}+2}{\dfrac{x}{y}-2}$

16. _____

17. _____

18. _____

19. _____

20. _____

21. _____

22. _____

23. _____

24. _____

25. _____

26. _____

27. _____

28. _____

29. _____

30. _____

31. _____

32. _____

Sections 8.1–8.8 Diagnostic Test

The purpose of this test is to see how well you understand operations with fractions. If you will be tested on Sections 8.1–8.8, we recommend that you work this diagnostic test *before* your instructor tests you on this material. Allow yourself about 50 minutes to do this test.

Complete solutions for all the problems on this test, together with section references, are given in the answer section in the back of the book. We suggest that you study the sections referred to for the problems you do incorrectly.

In Problems 1 and 2, determine what value(s) of the variable, if any, must be excluded.

1. $\dfrac{7x}{x - 5}$ **2.** $\dfrac{3x - 1}{x^2 + 9x}$

In Problems 3 and 4, find the missing term.

3. $\dfrac{5}{-3x} = \dfrac{-5}{?}$ **4.** $\dfrac{17}{x - 2} = \dfrac{?}{2 - x}$

In Problems 5 and 6, reduce each rational expression to lowest terms.

5. $\dfrac{5x + 1}{25x^2 + 10x + 1}$ **6.** $\dfrac{x^2 - 4y^2}{2x^2 + 3xy - 2y^2}$

7. Find the least common multiple (LCM) of $6x^3y$ and $15x^2y^2$.

In Problems 8–18, perform the indicated operations. Be sure to reduce your answers to lowest terms.

8. $\dfrac{x}{3y^2} \cdot \dfrac{6xy}{5y^3}$ **9.** $\dfrac{x}{x^2 + 6x + 8} \cdot \dfrac{3x + 12}{12x^2 + 24x}$

10. $\dfrac{4}{x - 5} \div \dfrac{8}{x^2 - 5x}$ **11.** $\dfrac{7y}{7y + y^2} \div \dfrac{1 - y^2}{7 + 8y + y^2}$

12. $\dfrac{8}{3y^2} - \dfrac{2}{3y^2}$ **13.** $\dfrac{8x}{x + 7} + \dfrac{3}{x + 7}$

14. $\dfrac{8x}{4x - 1} - \dfrac{2}{4x - 1}$ **15.** $\dfrac{3x}{x - 1} + \dfrac{4}{x + 3}$

16. $\dfrac{x - 3}{x + 5} + \dfrac{x + 5}{x - 3}$ **17.** $\dfrac{x + 14}{x^2 + 7x + 12} - \dfrac{x + 5}{x^2 + 5x + 6}$

18. $\dfrac{x}{x - 4} - \dfrac{x}{x + 4} - \dfrac{32}{x^2 - 16}$

In Problems 19 and 20, simplify the complex fractions.

19. $\dfrac{\dfrac{4x^3}{9y^4}}{\dfrac{2x}{27y^2}}$ **20.** $\dfrac{\dfrac{6}{x} - \dfrac{4}{x^2}}{5 + \dfrac{2}{x}}$

8.9 Solving Rational Equations

A **rational equation** is an equation that contains one or more rational expressions. The techniques for solving rational equations differ greatly from the techniques for performing operations (addition, subtraction, and so on) on rational expressions. In Section 3.3, we "cleared fractions" (that is, we removed the denominators) in equations that contained only *one* denominator by multiplying both sides of the equation by that denominator. If an equation has *more* than one denominator, we can clear fractions by multiplying both sides of the equation by the least common denominator (LCD).

Extraneous Roots   If any of the denominators have variables in them, the LCD will also have variables in it. If we multiply both sides of an equation by an expression containing a variable, we may get solutions that are not solutions of the original equation. We call such solutions **extraneous roots**. For this reason, when the LCD has variables in it, we must be sure to find all the *excluded values* of the variable before we attempt to solve the equation. (Recall from Section 8.1 that any value of the variable that makes a denominator zero is an excluded value.) Then any *apparent* solution to the equation that is an excluded value must be rejected; it is an *extraneous root*. (Any value of the variable that, when checked, gives a false statement—such as $3 = 0$—is also an extraneous root. We will see roots of this kind in Chapter 11.)

8.9A Solving Rational Equations That Simplify to First-Degree Equations

In this section, after we remove denominators and grouping symbols from the rational equation, the equation that remains will be a *first-degree equation* (a polynomial equation whose highest-degree term is first-degree).

We solve such equations by using the addition, subtraction, multiplication, and division properties of equality in order to isolate the variable, as outlined below.

TO SOLVE AN EQUATION WITH RATIONAL EXPRESSIONS THAT SIMPLIFIES TO A FIRST-DEGREE EQUATION

1. Find the LCD and find all excluded values.

2. Remove denominators by multiplying *both sides of the equation* (that is, by multiplying *every term*) by the LCD.

3. a. Remove all grouping symbols.
   b. Collect and combine like terms on each side of the equal sign.

4. a. Get all the terms that contain the variable on one side of the equation and all other terms on the other side.
   b. Divide both sides of the equation by the coefficient of the variable, or multiply both sides of the equation by the reciprocal of the coefficient of the variable.

5. Reject any apparent solutions that are excluded values.

6. Check any other apparent solutions in the original equation.

Example 1 Solve $\dfrac{x}{2} + \dfrac{x}{3} = 5$.

Solution

Step 1. The LCD is 6. There are no excluded values.

Step 2. $6\left(\dfrac{x}{2} + \dfrac{x}{3}\right) = 6\,(5)$ Use the distributive property on the left side

$6\left(\dfrac{x}{2}\right) + 6\left(\dfrac{x}{3}\right) = 6\,(5)$ This results in *each term* of the equation being multiplied by the LCD, 6

$\dfrac{\overset{3}{\cancel{6}}}{1}\left(\dfrac{x}{\cancel{2}}\right) + \dfrac{\overset{2}{\cancel{6}}}{1}\left(\dfrac{x}{\cancel{3}}\right) = \dfrac{6}{1}\left(\dfrac{5}{1}\right)$

Step 3a. $3x + 2x = 30$

Step 3b. $5x = 30$

Step 4b. $\dfrac{5x}{5} = \dfrac{30}{5}$ or $\left(\dfrac{1}{5}\right)(5x) = \left(\dfrac{1}{5}\right)(30)$

$x = 6$ $x = 6$

Step 5. Does not apply.

Step 6. *Check*

$$\dfrac{x}{2} + \dfrac{x}{3} = 5$$

$$\dfrac{6}{2} + \dfrac{6}{3} \overset{?}{=} 5$$

$$3 + 2 \overset{?}{=} 5$$

$$5 = 5$$

The solution is 6. ■

Example 2 Solve $\dfrac{x-4}{2} - \dfrac{x}{5} = \dfrac{1}{10}$.

Solution

Step 1. LCD = 10. There are no excluded values.

Step 2. $10\left(\dfrac{x-4}{2} - \dfrac{x}{5}\right) = 10\left(\dfrac{1}{10}\right)$ Multiplying both sides by the LCD

$\dfrac{\overset{5}{\cancel{10}}}{1}\left(\dfrac{x-4}{\cancel{2}}\right) - \dfrac{\overset{2}{\cancel{10}}}{1}\left(\dfrac{x}{\cancel{5}}\right) = \dfrac{\overset{1}{\cancel{10}}}{1}\left(\dfrac{1}{\cancel{10}}\right)$

$5(x - 4) - 2x = 1$

Step 3a. $5x - 20 - 2x = 1$

Step 3b. $3x - 20 = 1$

Step 4a. $\dfrac{+\,20 \qquad +20}{3x \qquad = \qquad 21}$

Step 4b.
$$\frac{3x}{3} = \frac{21}{3} \quad \text{or} \quad \left(\frac{1}{3}\right)(3x) = \left(\frac{1}{3}\right)(21)$$
$$x = 7 \qquad\qquad\qquad x = 7$$

Step 5. Does not apply.

Step 6. *Check*

$$\frac{x-4}{2} - \frac{x}{5} = \frac{1}{10}$$

$$\frac{7-4}{2} - \frac{7}{5} \overset{?}{=} \frac{1}{10}$$

$$\frac{3}{2} - \frac{7}{5} \overset{?}{=} \frac{1}{10}$$

$$\frac{15}{10} - \frac{14}{10} \overset{?}{=} \frac{1}{10}$$

$$\frac{1}{10} = \frac{1}{10}$$

The solution is 7. ∎

NOTE   There could not have been any extraneous roots in Examples 1 and 2 because the LCD did not contain a variable.   ☑

Example 3   Solve $\dfrac{x}{x-3} = \dfrac{3}{x-3} + 4$.

Solution

Step 1. The LCD is $x - 3$. 3 is an excluded value because it makes the denominator $x - 3$ zero.

Step 2.
$$(x-3)\left(\frac{x}{x-3}\right) = (x-3)\left(\frac{3}{x-3} + 4\right)$$

$$\frac{\overset{1}{\cancel{(x-3)}}}{1} \cdot \frac{x}{\underset{1}{\cancel{(x-3)}}} = \frac{\overset{1}{\cancel{(x-3)}}}{1} \cdot \frac{3}{\underset{1}{\cancel{(x-3)}}} + \frac{(x-3)}{1} \cdot \frac{4}{1}$$

$$x \qquad\qquad = \qquad 3 \qquad + \quad 4(x-3)$$

Step 3a. $\qquad\qquad x = 3 + 4x - 12$

Step 3b. $\qquad\qquad x = 4x - 9$

Step 4a.
$$\begin{array}{r} -4x \quad -4x \\ \hline -3x = \qquad -9 \end{array}$$

Step 4b.
$$\frac{-3x}{-3} = \frac{-9}{-3} \quad \text{or} \quad \left(-\frac{1}{3}\right)(-3x) = \left(-\frac{1}{3}\right)(-9)$$
$$x = 3 \qquad\qquad\qquad x = 3$$

Step 5. Since 3 is an excluded value, it must be rejected. Because 3 was the only apparent solution, the given equation has *no solution*.

NOTE   If we had failed to look for excluded values, the check would show that there is no solution.

$$\frac{x}{x-3} = \frac{3}{x-3} + 4$$

$$\frac{3}{3-3} \overset{?}{=} \frac{3}{3-3} + 4$$

$$\boxed{\frac{3}{0}} \overset{?}{=} \boxed{\frac{3}{0}} + 4$$

Not defined ■

EXERCISES 8.9A

Set I   Solve the equations.

1. $\dfrac{x}{3} + \dfrac{x}{4} = 7$ **2.** $\dfrac{x}{5} + \dfrac{x}{3} = 8$ **3.** $\dfrac{10}{c} = 2$ **4.** $\dfrac{6}{z} = 4$

5. $\dfrac{a}{2} - \dfrac{a}{5} = 6$ **6.** $\dfrac{b}{3} - \dfrac{b}{7} = 12$ **7.** $\dfrac{9}{2x} = 3$ **8.** $\dfrac{14}{3x} = 7$

9. $\dfrac{M-2}{5} + \dfrac{M}{3} = \dfrac{1}{5}$ **10.** $\dfrac{y+2}{4} + \dfrac{y}{5} = \dfrac{1}{4}$ **11.** $\dfrac{7}{x+4} = \dfrac{3}{x}$

12. $\dfrac{5}{x+6} = \dfrac{2}{x}$ **13.** $\dfrac{3x}{x-2} = 5$ **14.** $\dfrac{4}{x+3} = \dfrac{2}{x}$

15. $\dfrac{3x-1}{x-2} = 3 + \dfrac{2x+1}{x-2}$ **16.** $\dfrac{5x-2}{x-4} = 7 + \dfrac{4x+2}{x-4}$

17. $\dfrac{x}{x^2+1} = \dfrac{2}{1+2x}$ **18.** $\dfrac{3x}{3x^2+2} = \dfrac{1}{x+1}$

19. $\dfrac{2x-1}{3} + \dfrac{3x}{4} = \dfrac{5}{6}$ **20.** $\dfrac{3z-2}{4} + \dfrac{3z}{8} = \dfrac{3}{4}$

21. $\dfrac{2(m-3)}{5} - \dfrac{3(m+2)}{2} = \dfrac{7}{10}$ **22.** $\dfrac{5(x-4)}{6} - \dfrac{2(x+4)}{9} = \dfrac{5}{18}$

23. $\dfrac{x}{x-2} = \dfrac{2}{x-2} + 5$ **24.** $\dfrac{x}{x+5} = 4 - \dfrac{5}{x+5}$

Set II   Solve the equations.

1. $\dfrac{x}{5} + \dfrac{x}{2} = 7$ **2.** $\dfrac{x}{6} + \dfrac{x-6}{3} = 2$ **3.** $\dfrac{8}{z} = 4$

4. $\dfrac{15}{x-1} = 3$ **5.** $\dfrac{a}{4} - \dfrac{a}{3} = 1$ **6.** $\dfrac{x}{3} - \dfrac{x}{5} = 2$

7. $\dfrac{12}{5x} = 2$ **8.** $\dfrac{3}{x} + \dfrac{1}{2} = \dfrac{6}{x}$ **9.** $\dfrac{y-1}{2} + \dfrac{y}{5} = \dfrac{3}{10}$

10. $\dfrac{24}{x} = \dfrac{5}{2} + \dfrac{4}{x}$ **11.** $\dfrac{6}{y-2} = \dfrac{3}{y}$ **12.** $\dfrac{5}{x-3} = \dfrac{2}{x}$

13. $\dfrac{6x}{x-2} = 8$ **14.** $\dfrac{3}{x+7} + \dfrac{17}{2x} = \dfrac{1}{x}$

15. $\dfrac{7x - 2}{x - 1} = 5 + \dfrac{6x - 1}{x - 1}$

16. $\dfrac{3}{x + 1} - \dfrac{2}{x} = \dfrac{5}{2x}$

17. $\dfrac{x}{3x^2 + 5} = \dfrac{1}{1 + 3x}$

18. $\dfrac{4x - 1}{x - 6} = 2 + \dfrac{3x + 5}{x - 6}$

19. $\dfrac{2z - 4}{3} + \dfrac{3z}{2} = \dfrac{5}{6}$

20. $\dfrac{5}{1 + x} - \dfrac{5}{x} = \dfrac{1}{x}$

21. $\dfrac{2(m - 1)}{5} - \dfrac{3(m + 1)}{2} = \dfrac{3}{10}$

22. $\dfrac{5x}{5x^2 + 2} = \dfrac{1}{x + 2}$

23. $\dfrac{y}{y - 3} + 10 = \dfrac{3}{y - 3}$

24. $\dfrac{x}{x + 3} = 2 + \dfrac{3}{x + 3}$

8.9B Solving Proportions

A **proportion** is a statement that two ratios or two rates are equal. That is, a proportion is an equation of the form $\dfrac{P}{Q} = \dfrac{S}{T}$, where P, Q, S, and T are polynomials and $Q \neq 0$ and $T \neq 0$. (P, Q, S, and T are called the *terms* of the proportion.) When we multiply both sides of the equation $\dfrac{P}{Q} = \dfrac{S}{T}$ by QT (the LCD), we obtain the new equation $PT = QS$. (Notice that $\dfrac{P}{Q} = \dfrac{S}{T}$ is equivalent to $QT\left(\dfrac{P}{Q}\right) = QT\left(\dfrac{S}{T}\right)$, which is equivalent to $PT = QS$.) When we rewrite $\dfrac{P}{Q} = \dfrac{S}{T}$ as $PT = QS$, we often say we are "cross-multiplying."

CROSS-MULTIPLICATION PROPERTY

If $\dfrac{P}{Q} = \dfrac{S}{T}$, then $PT = QS$.

We can cross-multiply as the first step in solving a proportion; however, we must be sure to check for excluded values before we cross-multiply.

Example 4   Solve the proportion $\dfrac{4}{7} = \dfrac{3}{z}$ for z.

Solution We can use the multiplication property of equality, or we can cross-multiply. Zero is an *excluded value*, since $z = 0$ would make a denominator zero.

Method 1 The LCD is $7z$.

$$\frac{4}{7} = \frac{3}{z}$$

$$(\overset{1}{\cancel{7}}z)\left(\frac{4}{\cancel{7}}\right) = (7\cancel{z})\left(\frac{3}{\cancel{z}}\right) \xleftarrow{\text{Multiplying both sides by } 7z}$$

Method 2

$$\frac{4}{7} = \frac{3}{z}$$

$$4z = 3(7) \xleftarrow{\text{Cross-multiplying}}$$

Then, using either method,

$$4z = 21$$

$$\frac{4z}{4} = \frac{21}{4} \quad \text{Dividing both sides by 4}$$
$$\left(\text{or multiplying both sides by } \frac{1}{4}\right)$$

$$z = \frac{21}{4}$$

Checking will show that $\frac{21}{4}$ is the solution. ∎

Example 5 Solve $\frac{3}{a} = 4$.

Solution This equation *can* be treated like a proportion (see method 2). Zero is an excluded value, since $a = 0$ makes a denominator zero.

Method 1 The LCD is a. If we multiply both sides of $\frac{3}{a} = 4$ by a, we have

$$\overset{1}{a}\left(\frac{3}{\underset{1}{a}}\right) = a(4)$$

$$3 = 4a$$

Method 2 If we rewrite 4 as $\frac{4}{1}$, we have

$\frac{3}{a} = \frac{4}{1}$, which is a proportion.

$$3 \cdot 1 = 4 \cdot a \quad \text{Cross-multiplying}$$

These equations are equivalent

Then, using either method, and dividing both sides by 4 or multiplying both sides by $\frac{1}{4}$, we have

$$a = \frac{3}{4}$$

Check

$$\frac{3}{a} = 4$$

$$\frac{3}{\frac{3}{4}} \overset{?}{=} 4 \quad \text{Remember: } \frac{3}{\frac{3}{4}} = 3 \div \frac{3}{4} = 3 \cdot \frac{4}{3} = 4$$

$$4 = 4$$

The solution is $\frac{3}{4}$. ∎

Example 6   Solve $\dfrac{9x}{x - 3} = 6$.

Solution This equation *can* be treated like a proportion (see method 2). 3 is an excluded value, since $x = 3$ makes a denominator zero.

Method 1 The LCD is $x - 3$.

$$\frac{9x}{x - 3} = 6$$

$$(x - 3)\left(\frac{9x}{x - 3}\right) = (x - 3)(6)$$

$$9x = 6(x - 3)$$

└─ These equations are equivalent

Method 2

$$\frac{9x}{x - 3} = \frac{6}{1}$$

$$9x \cdot 1 = 6(x - 3)$$

Writing 6 as $\dfrac{6}{1}$ makes the equation a proportion

Cross-multiplying

Then, using either method,

$$\begin{array}{rl} 9x = & 6x - 18 \\ \underline{-6x} & \underline{-6x} \\ 3x = & -18 \end{array}$$

Removing the parentheses

Adding $-6x$ to both sides

$$x = -6$$

Dividing both sides by 3

or multiplying both sides by $\dfrac{1}{3}$

Check

$$\frac{9x}{x - 3} = 6$$

$$\frac{9(-6)}{-6 - 3} \overset{?}{=} 6$$

$$\frac{-54}{-9} \overset{?}{=} 6$$

$$6 = 6$$

The solution is -6. ∎

Example 7   Solve $\dfrac{x + 1}{x - 2} = \dfrac{7}{4}$.

Solution 2 is an excluded value, since $x = 2$ will make a denominator zero.

$$\frac{x + 1}{x - 2} = \frac{7}{4}$$

This is a proportion

$$4(x + 1) = 7(x - 2)$$

Cross-multiplying *or* multiplying both sides by the LCD

$$\begin{array}{rl} 4x + 4 = & 7x - 14 \\ \underline{-7x - 4} & \underline{-7x - 4} \\ -3x = & -18 \end{array}$$

Removing parentheses

Adding $-7x - 4$ to both sides

$$x = 6$$

Dividing both sides by -3 or multiplying both sides by $-\dfrac{1}{3}$

Check

$$\frac{x + 1}{x - 2} = \frac{7}{4}$$

$$\frac{6 + 1}{6 - 2} \stackrel{?}{=} \frac{7}{4}$$

$$\frac{7}{4} = \frac{7}{4}$$

The solution is 6. ∎

The terms of a proportion can be any kind of number, except that no denominator can be zero. In some problems, it is especially easy to *finish* solving the equation by multiplying both sides of the equation by the multiplicative inverse of the coefficient of the variable (see "Alternate method" in Examples 8 and 9). (The checking will not be shown in Examples 8–10.)

Example 8 Solve for P: $\dfrac{P}{3} = \dfrac{\frac{5}{6}}{5}$

Solution No values of the variable need be excluded.

$$\frac{P}{3} = \frac{\frac{5}{6}}{5}$$

$$5 \cdot P = \frac{\overset{1}{\cancel{3}}}{1} \cdot \frac{5}{\underset{2}{\cancel{6}}} \qquad \text{Cross-multiplying}$$

$$5 \cdot P = \frac{5}{2}$$

$$\frac{\overset{1}{\cancel{5}} \cdot P}{\underset{1}{\cancel{5}}} = \frac{\frac{5}{2}}{5} \quad \longleftarrow \text{Dividing both sides by 5}$$

$$P = \frac{5}{2} \div 5 = \frac{\overset{1}{\cancel{5}}}{2} \cdot \frac{1}{\underset{1}{\cancel{5}}} = \frac{1}{2}$$

Alternate method for finishing the problem

$$\left(\frac{1}{\cancel{5}}\right)_{\!1}^{\!1} (\cancel{5}P) = \left(\frac{1}{\cancel{5}}\right)_{\!1} \left(\frac{\overset{1}{\cancel{5}}}{2}\right)$$

Multiplying both sides by the multiplicative inverse of 5

$$P = \frac{1}{2}$$

Checking will verify that the solution is $\dfrac{1}{2}$. ∎

Example 9 Solve for x: $\dfrac{3\frac{1}{2}}{5\frac{1}{4}} = \dfrac{x}{4}$

Solution

$$\frac{3\frac{1}{2}}{5\frac{1}{4}} = \frac{x}{4}$$

$$\left(3\frac{1}{2}\right)4 = \left(5\frac{1}{4}\right)x \qquad \text{Cross-multiplying}$$

$$\frac{7}{\underset{1}{\cancel{2}}} \cdot \frac{\overset{2}{\cancel{4}}}{1} = \frac{21}{4} \cdot x$$

Alternate method for finishing the problem

$$\frac{14}{\frac{21}{4}} = \frac{\overset{1}{\cancel{\frac{21}{4}}} \cdot x}{\cancel{\frac{21}{4}}_1} \quad \longleftarrow \text{Dividing both sides by } \frac{21}{4}$$

$$\left(\frac{\overset{1}{\cancel{4}}}{\cancel{21}_1}\right)\left(\frac{\cancel{21}^1}{\cancel{4}_1}x\right) = \left(\frac{4}{\cancel{21}_3}\right)(\overset{2}{\cancel{14}}) \qquad \begin{array}{l}\text{Multiplying both} \\ \text{sides by the} \\ \text{multiplicative} \\ \text{inverse of } 21/4 \end{array}$$

$$x = 14 \div \frac{21}{4}$$

$$x = \frac{\overset{2}{\cancel{14}}}{1} \cdot \frac{4}{\cancel{21}_3} = \frac{8}{3} = 2\frac{2}{3} \qquad\qquad x = \frac{8}{3} \text{ or } 2\frac{2}{3}$$

Checking will verify that the solution is $2\frac{2}{3}$. ∎

Example 10   Solve for B: $\dfrac{0.24}{2.7} = \dfrac{4}{B}$

Solution

$$\frac{0.24}{2.7} = \frac{4}{B}$$

$$\frac{\overset{4}{\cancel{24}}}{\underset{45}{\cancel{270}}} = \frac{4}{B} \qquad \begin{array}{l}\text{We first multiply numerator and denominator by 100} \\ \text{to eliminate both decimal points; then we reduce}\end{array}$$

$$\frac{4}{45} = \frac{4}{B}$$

$$4 \cdot B = 45 \cdot 4 \qquad \text{Cross-multiplying}$$

$$\frac{\overset{1}{\cancel{4}} \cdot B}{\underset{1}{\cancel{4}}} = \frac{45 \cdot \overset{1}{\cancel{4}}}{\underset{1}{\cancel{4}}} \qquad \text{Dividing both sides by 4 or multiplying both sides by } \frac{1}{4}$$

$$B = 45$$

Checking will verify that the solution is 45. ∎

A WORD OF CAUTION   Because students can solve a proportion by cross-multiplication, they sometimes use cross-multiplication incorrectly in a *sum* or *product* of fractions. It is incorrect to cross-multiply when adding or multiplying fractions.

Incorrect application of cross-multiplication	*Correct application of cross-multiplication*	*Incorrect application of cross-multiplication*
This is a *sum*	This is an *equation*	This is a *product*
$\dfrac{3}{4} + \dfrac{2}{3} = 9 + 8$	If $\dfrac{16}{6} = \dfrac{8}{3}$,	$\dfrac{16}{6} \cdot \dfrac{8}{3} = \dfrac{16 \cdot 3}{6 \cdot 8}$
Correct sum:	then $16 \cdot 3 = 6 \cdot 8$	*Correct product:*
$\dfrac{3}{4} + \dfrac{2}{3} = \dfrac{9}{12} + \dfrac{8}{12} = \dfrac{17}{12}$	$48 = 48$	$\dfrac{16}{6} \cdot \dfrac{8}{3} = \dfrac{\overset{8}{\cancel{16}} \cdot 8}{\underset{3}{\cancel{6}} \cdot 3} = \dfrac{64}{9}$

EXERCISES 8.9B

Set I   Solve for the variable.

1. $\dfrac{x}{14} = \dfrac{-3}{7}$ **2.** $\dfrac{x}{12} = \dfrac{-5}{6}$ **3.** $\dfrac{x}{4} = \dfrac{2}{3}$ **4.** $\dfrac{x}{5} = \dfrac{6}{4}$

5. $\dfrac{8}{x} = \dfrac{4}{5}$ **6.** $\dfrac{10}{x} = \dfrac{15}{4}$ **7.** $\dfrac{4}{7} = \dfrac{x}{21}$ **8.** $\dfrac{15}{12} = \dfrac{x}{9}$

9. $\dfrac{100}{x} = \dfrac{40}{30}$ **10.** $\dfrac{144}{36} = \dfrac{96}{x}$ **11.** $\dfrac{x+1}{x-1} = \dfrac{3}{2}$

12. $\dfrac{x+1}{5} = \dfrac{x-1}{3}$ **13.** $\dfrac{2x+7}{9} = \dfrac{2x+3}{5}$ **14.** $\dfrac{2x+7}{3x+10} = \dfrac{3}{4}$

15. $\dfrac{5x-10}{10} = \dfrac{3x-5}{7}$ **16.** $\dfrac{8x-2}{3x+4} = \dfrac{3}{2}$ **17.** $\dfrac{\frac{3}{4}}{6} = \dfrac{P}{16}$

18. $\dfrac{\frac{2}{5}}{4} = \dfrac{P}{25}$ **19.** $\dfrac{A}{9} = \dfrac{3\frac{1}{3}}{5}$ **20.** $\dfrac{A}{8} = \dfrac{2\frac{1}{4}}{18}$

21. $\dfrac{7.7}{B} = \dfrac{3.5}{5}$ **22.** $\dfrac{6.8}{B} = \dfrac{17}{57.4}$ **23.** $\dfrac{P}{100} = \dfrac{\frac{3}{2}}{50}$

24. $\dfrac{P}{100} = \dfrac{\frac{7}{5}}{35}$ **25.** $\dfrac{12\frac{1}{2}}{100} = \dfrac{A}{48}$ **26.** $\dfrac{16\frac{2}{3}}{100} = \dfrac{9}{B}$

Set II   Solve for the variable.

1. $\dfrac{x}{15} = \dfrac{-2}{5}$ **2.** $\dfrac{x}{-18} = \dfrac{7}{-6}$ **3.** $\dfrac{x}{15} = \dfrac{6}{5}$

4. $\dfrac{x+2}{5} = \dfrac{5-x}{8}$ **5.** $\dfrac{81}{x} = \dfrac{9}{5}$ **6.** $\dfrac{4}{13} = \dfrac{16}{x}$

7. $\dfrac{18}{28} = \dfrac{x}{14}$ **8.** $\dfrac{x}{100} = \dfrac{75}{125}$ **9.** $\dfrac{26}{x} = \dfrac{39}{14}$

10. $\dfrac{x}{18} = \dfrac{24}{30}$ **11.** $\dfrac{x+5}{x-5} = \dfrac{27}{10}$ **12.** $\dfrac{x+8}{x-8} = 5$

13. $\dfrac{4x+3}{2} = \dfrac{4x+1}{3}$ **14.** $\dfrac{3x-7}{x-5} = 2$ **15.** $\dfrac{2x+5}{3} = \dfrac{3x-1}{2}$

16. $\dfrac{15}{22} = \dfrac{x}{33}$ **17.** $\dfrac{\frac{3}{4}}{6} = \dfrac{P}{16}$ **18.** $\dfrac{\frac{5}{6}}{\frac{1}{6}} = \dfrac{z}{2}$ **19.** $\dfrac{A}{16} = \dfrac{2\frac{1}{2}}{10}$

20. $\dfrac{1.2}{2} = \dfrac{x}{2.4}$ **21.** $\dfrac{P}{100} = \dfrac{\frac{3}{4}}{15}$ **22.** $\dfrac{8}{\frac{1}{2}} = \dfrac{x}{4}$ **23.** $\dfrac{P}{100} = \dfrac{12\frac{1}{2}}{50}$

24. $\dfrac{9}{y} = \dfrac{3\frac{1}{3}}{\frac{1}{6}}$ **25.** $\dfrac{6\frac{1}{4}}{100} = \dfrac{1}{B}$ **26.** $\dfrac{3.1}{x} = \dfrac{10}{2.5}$

8.9C Solving Rational Equations That Simplify to Second-Degree Equations

In this section, the equation that remains after we remove denominators and grouping symbols from the rational equation will usually be a *second-degree equation* (a polynomial equation whose highest-degree term is second-degree), and it can be solved by factoring, as discussed in Section 7.8. The suggestions in the following box are based on using the addition, subtraction, multiplication, and division properties of equality.

TO SOLVE AN EQUATION WITH RATIONAL EXPRESSIONS THAT SIMPLIFIES TO A SECOND-DEGREE EQUATION

1. Find the LCD and find all excluded values.

2. Remove denominators by multiplying *both sides of the equation* (that is, by multiplying *every term*) by the LCD.

3. a. Remove all grouping symbols.
  b. Collect and combine like terms on each side of the equal sign.

4. If there are second degree terms, solve by these quadratic methods:
  a. Get all nonzero terms on one side by adding the same expression to both sides. Only zero must remain on the other side. Then arrange the terms in descending powers.
  b. Factor the polynomial.*
  c. Set each factor equal to zero, and solve each resulting equation for the variable.

5. Reject any apparent solutions that are excluded values.

6. Check any other apparent solutions in the original equation.

NOTE If the equation from step 3 is a first-degree equation, solve it by using the methods discussed in Section 8.9A. ☑

Example 11 Solve $\dfrac{2}{x} + \dfrac{3}{x^2} = 1$

Solution

Step 1. LCD $= x^2$. 0 is an excluded value.

Step 2.
$$x^2\left(\frac{2}{x} + \frac{3}{x^2}\right) = x^2\,(1)$$ Multiplying both sides by x^2

$$\frac{x^2}{1}\left(\frac{2}{x}\right) + \frac{x^2}{1}\left(\frac{3}{x^2}\right) = \frac{x^2}{1}\left(\frac{1}{1}\right)$$

$\ulcorner$ Second-degree term

Step 3.
$$2x + 3 = x^2$$
Step 4a.
$$\underline{-2x - 3 \qquad\qquad -2x - 3}$$
$$0 = x^2 - 2x - 3$$

---

*If the polynomial cannot be factored, we cannot solve the equation at this time.

Step 4b. $\qquad\qquad\qquad\qquad$ $0 = (x - 3)(x + 1)$

Step 4c.

$$
\begin{array}{c|c}
x - 3 = 0 & x + 1 = 0 \\
 \underline{ 3 \quad 3} & \underline{- 1 \quad -1} \\
x = 3 & x = -1
\end{array}
$$

Step 5. Does not apply.

Step 6.

Check for x = 3	*Check for x = −1*
$\dfrac{2}{x} + \dfrac{3}{x^2} = 1$	$\dfrac{2}{x} + \dfrac{3}{x^2} = 1$
$\dfrac{2}{3} + \dfrac{3}{3^2} \overset{?}{=} 1$	$\dfrac{2}{-1} + \dfrac{3}{(-1)^2} \overset{?}{=} 1$
$\dfrac{2}{3} + \dfrac{1}{3} \overset{?}{=} 1$	$-2 + 3 \overset{?}{=} 1$
$1 = 1$	$1 = 1$

The solutions are 3 and −1. ■

Example 12 $\quad$ Solve $\dfrac{8}{x} = \dfrac{3}{x + 1} + 3$.

Solution

Step 1. LCD $= x(x + 1)$. 0 and −1 are excluded values.

Step 2. $\quad x(x + 1)\left(\dfrac{8}{x}\right) = x(x + 1)\left(\dfrac{3}{x + 1} + 3\right)$

$$\dfrac{\overset{1}{\cancel{x}(x + 1)}}{1} \cdot \dfrac{8}{\underset{1}{\cancel{x}}} = \dfrac{\overset{1}{\cancel{x(x + 1)}}}{1} \cdot \dfrac{3}{\underset{1}{\cancel{(x + 1)}}} + \dfrac{x(x + 1)}{1} \cdot \dfrac{3}{1}$$

$$8(x + 1) \qquad = \qquad 3x \qquad + \quad 3x(x + 1)$$

$\qquad\qquad\qquad\qquad\qquad\qquad\qquad\qquad$ ┌─Second-degree term

Step 3a. $\qquad\qquad 8x + 8 = 3x + \boxed{3x^2} + 3x$

Step 3b. $\qquad\qquad 8x + 8 = 3x^2 + 6x$

Step 4a. $\qquad\qquad \underline{-8x - 8 \qquad\quad - 8x - 8}$

$\qquad\qquad\qquad\quad 0 = 3x^2 - 2x - 8$

Step 4b. $\qquad\qquad\quad 0 = (3x + 4)(x - 2)$

Step 4c.

$$
\begin{array}{c|c}
3x + 4 = 0 & x - 2 = 0 \\
3x = -4 & x = 2 \\
x = -\dfrac{4}{3} &
\end{array}
$$

Step 5. Does not apply.

Step 6. Checking will verify that $-\dfrac{4}{3}$ and 2 are the solutions. ■

NOTE $\quad$ When we multiply both sides of an equation by the LCD, the denominators are removed completely. When we add or subtract fractions, the denominators can *not* be removed. $\qquad\qquad\qquad\qquad\qquad\qquad\qquad\qquad\qquad\qquad$ ☑

A WORD OF CAUTION   A common mistake students make is to confuse an *equation* such as $\frac{2}{x} + \frac{3}{x^2} = 1$ with an *addition problem* such as $\frac{2}{x} + \frac{3}{x^2}$. You can solve *only* equations; if there is no equal sign, you can't solve for x.

The equation	*The addition problem*
Both sides are multiplied by the LCD to remove fractions.	Each fraction is changed into an equivalent fraction with the LCD for a denominator

The equation

$$\frac{2}{x} + \frac{3}{x^2} = 1 \qquad \text{LCD} = x^2$$

$$\frac{x^2}{1} \cdot \frac{2}{x} + \frac{x^2}{1} \cdot \frac{3}{x^2} = \frac{x^2}{1} \cdot \frac{1}{1}$$

$$2x \quad + \quad 3 \quad = \quad x^2$$

This equation is then solved by factoring (see Example 11). Here the result is two numbers (-1 and 3) that make both sides of the given equation equal.

The addition problem

$$\frac{2}{x} + \frac{3}{x^2} \qquad \text{LCD} = x^2$$

This is 1

$$= \frac{2}{x} \cdot \frac{x}{x} + \frac{3}{x^2}$$

$$= \frac{2x}{x^2} + \frac{3}{x^2} = \frac{2x + 3}{x^2}$$

Here the result is a fraction that represents the sum of the given fractions.

The usual mistake made is to multiply both terms of *the sum* by the LCD.

$$\frac{x^2}{1} \cdot \frac{2}{x} + \frac{x^2}{1} \cdot \frac{3}{x^2} = 2x + 3$$

$$\neq \frac{2}{x} + \frac{3}{x^2}$$

The sum has been multiplied by x^2 and therefore no longer has its original value. ☑

EXERCISES 8.9C

Set I   Solve the equations.

1. $z + \frac{1}{z} = \frac{17}{z}$

2. $y + \frac{3}{y} = \frac{12}{y}$

3. $\frac{2}{x} - \frac{2}{x^2} = \frac{1}{2}$

4. $\frac{3}{x} - \frac{4}{x^2} = \frac{1}{2}$

5. $\frac{x}{x + 1} = \frac{4x}{3x + 2}$

6. $\frac{x}{3x - 4} = \frac{3x}{2x + 2}$

7. $\frac{5}{x} - 1 = \frac{x + 11}{x}$

8. $\frac{7}{x} - 1 = \frac{x + 15}{x}$

9. $\frac{1}{x - 1} + \frac{2}{x + 1} = \frac{5}{3}$

10. $\frac{2}{3x + 1} + \frac{1}{x - 1} = \frac{7}{10}$

11. $\frac{3}{2x + 5} + \frac{x}{4} = \frac{3}{4}$

12. $\frac{5}{2x - 1} - \frac{x}{6} = \frac{4}{3}$

13. $\frac{4}{x + 1} - \frac{3}{x} = \frac{1}{15}$

14. $\frac{1}{x + 1} - \frac{3}{x} = \frac{1}{2}$

15. $\dfrac{6}{x^2 - 9} = \dfrac{1}{x - 3} - \dfrac{1}{5}$

16. $\dfrac{6 - x}{x^2 - 4} = \dfrac{x}{x + 2} + 2$

Set II   Solve the equations.

1. $x + \dfrac{1}{x} = \dfrac{10}{x}$

2. $y + \dfrac{2}{y} = \dfrac{18}{y}$

3. $\dfrac{5}{x} - \dfrac{1}{x^2} = \dfrac{9}{4}$

4. $\dfrac{1}{2} - \dfrac{1}{2x} = \dfrac{6}{x^2}$

5. $\dfrac{2x}{3x + 1} = \dfrac{4x}{5x + 1}$

6. $\dfrac{x}{8} + \dfrac{x}{2} = \dfrac{10}{x}$

7. $\dfrac{8}{x} - 1 = \dfrac{x + 18}{x}$

8. $\dfrac{5}{x - 3} + \dfrac{1}{6} = \dfrac{7}{x - 2}$

9. $\dfrac{4}{x + 1} = \dfrac{3}{x} + \dfrac{1}{15}$

10. $\dfrac{6}{x + 2} - 1 = \dfrac{x - 4}{x + 2}$

11. $\dfrac{1}{x - 2} - \dfrac{4}{x + 2} = \dfrac{1}{5}$

12. $\dfrac{1}{x - 5} + \dfrac{3}{x + 2} = \dfrac{5}{6}$

13. $\dfrac{4}{3x - 1} - 1 = \dfrac{2}{x}$

14. $\dfrac{3}{1 - 2x} = \dfrac{2x + 1}{x - 2}$

15. $\dfrac{x + 4}{x^2 - 16} + \dfrac{7}{x - 4} = -1$

16. $\dfrac{x + 25}{x^2 - 25} + \dfrac{12x}{x - 5} = 1$

8.10 Literal Equations

Literal equations are equations that contain more than one variable.

Example 1   Examples of literal equations:

a. $3x + 4y = 12$
   This is an equation in two variables. We might be asked to solve it for x or for y.

b. $\dfrac{4ab}{d} = 15$
   This is an equation in three variables. We might be asked to solve it for a, for b, or for d.

c. $A = P(1 + rt)$
   This is an equation in four variables. We might be asked to solve it for P, for r, or for t. (It has already been solved for A.) ∎

Generally, when we solve a literal equation for one of its variables, the solution will contain the other variables as well as constants. We must isolate the variable we are solving for; that is, that variable must appear only once all by itself on one side of the equal sign. All other variables and all constants must be on the other side. The suggestions in the following box are based on the addition, subtraction, multiplication, and division properties of equality.

TO SOLVE A LITERAL EQUATION

1. Remove rational expressions (if there are any) by multiplying both sides of the equation by the LCD.

2. Remove grouping symbols (if there are any).

3. Collect and combine like terms. Move all terms containing the variable you are solving for to one side of the equation and all other terms to the other side.

4. Factor out the variable you are solving for (if it appears in more than one term).

5. Divide both sides of the equation by the coefficient of the variable you are solving for (or multiply both sides of the equation by the multiplicative inverse of that variable).

In Example 2a, we will compare solving a literal equation with solving a "regular" equation, so that you can see how similar the methods are.

Example 2 Solve the given equation for the indicated variable.

a. Solve $3x + 4y = 12$ for x, and solve $3x + 32 = 12$ for x.
Solutions

$$3x + 4y = 12 \qquad\qquad\qquad\qquad\qquad 3x + 32 = 12$$
$$\underline{\quad\;\; -4y \qquad -4y} \quad \text{Getting the } x\text{-term} \qquad \underline{\quad\; -32 \quad -32}$$
$$3x \qquad\quad = 12 - 4y \quad \begin{array}{l}\text{by itself on one side}\\ \text{of the equal sign}\end{array} \qquad 3x \qquad = 12 - 32$$

$$\frac{3x}{3} = \frac{12 - 4y}{3} \qquad \begin{array}{l}\text{Dividing both sides}\\ \text{by 3 (or multiplying}\\ \text{both sides by } \frac{1}{3})\end{array} \qquad \frac{3x}{3} = \frac{12 - 32}{3}$$

$$x = \frac{12 - 4y}{3} \qquad \begin{array}{l}\text{The equations have}\\ \text{been solved for } x\end{array} \qquad x = -\frac{20}{3}$$

Notice that in both cases we have *isolated x*.

b. Solve $3x + 4y = 12$ for y.
Solution

$$3x + 4y = 12$$
$$\underline{-3x \qquad\qquad\;\; -3x} \qquad \begin{array}{l}\text{Getting the } y\text{-term by itself on one side}\\ \text{of the equal sign by adding } -3x \text{ to both sides}\end{array}$$
$$4y = 12 - 3x$$

$$\frac{4y}{4} = \frac{12 - 3x}{4} \qquad \begin{array}{l}\text{Dividing both sides by 4}\\ \text{(or multiplying both sides by } \frac{1}{4})\end{array}$$

$$y = \frac{12 - 3x}{4} \qquad \text{The equation has been solved for } y$$

c. Solve $\dfrac{4ab}{d} = 15$ for a.

Solution The LCD is d.

$$\frac{4ab}{d} = 15$$

$$\overset{1}{\cancel{d}}\left(\frac{4ab}{\underset{1}{\cancel{d}}}\right) = (15)\,d \qquad \text{Multiplying both sides by } d$$

$$4ab = 15d \qquad \text{Simplifying}$$

$$\frac{4ab}{4b} = \frac{15d}{4b} \qquad \text{Dividing both sides by } 4b$$

$$a = \frac{15d}{4b} \qquad \text{The equation has been solved for } a$$

d. Solve $A = P(1 + rt)$ for t.

Solution

$$A = P(1 + rt)$$

$$A = P + Prt \qquad \text{Removing ( ) by using the distributive property}$$

$$\frac{-P \quad\; -P}{A - P = \qquad Prt} \qquad \begin{array}{l}\text{Collecting terms with the variable we are solving for } (t)\\ \text{on one side and all other terms on the other side}\end{array}$$

$$\frac{A - P}{Pr} = \frac{Prt}{Pr} \qquad \begin{array}{l}\text{The coefficient of } t \text{ was } Pr; \text{ therefore, we divide}\\ \text{both sides of the equation by } Pr\end{array}$$

$$t = \frac{A - P}{Pr} \qquad \text{The equation has been solved for } t$$

e. Solve $I = \dfrac{nE}{R + nr}$ for n.

Solution The LCD is $R + nr$.

$$I = \frac{nE}{R + nr}$$

$$(R + nr)\,(I) = \overset{1}{\cancel{(R + nr)}}\left(\frac{nE}{\underset{1}{\cancel{R + nr}}}\right) \qquad \text{Multiplying both sides by the LCD}$$

$$\begin{array}{l}\text{We reduce the rational expression on the right side}\\ \text{On the left side we use the distributive property}\end{array}$$

$$IR + Inr = nE$$

$$\frac{-\,Inr \qquad\quad -\,Inr}{IR \qquad = nE - Inr} \qquad \begin{array}{l}\text{We get all terms with } n \text{ on one side}\\ \text{by adding } -Inr \text{ to both sides}\end{array}$$

$$IR = n\,(E - Ir) \qquad \begin{array}{l}\text{Because } n \text{ appears in more than one term,}\\ \text{we factor out } n\end{array}$$

$$\frac{IR}{E - Ir} = \frac{n(E - Ir)}{E - Ir} \qquad \begin{array}{l}\text{Because the coefficient of } n \text{ is } (E - Ir),\\ \text{we divide both sides by } (E - Ir)\end{array}$$

$$n = \frac{IR}{E - Ir} \qquad \text{The equation has been solved for } n \quad \blacksquare$$

You may want to use this method of solving a literal equation for one of its variables in Section 9.3, and you will definitely need to use it in Sections 9.5 and 10.4.

EXERCISES 8.10

Set I   Solve each equation for the variable indicated after each equation.

1. $2x + y = 4$; x | **2.** $x + 3y = 6$; y

3. $y - z = -8$; z | **4.** $m - n = -5$; n

5. $2x - y = -4$; y | **6.** $3y - z = -5$; z

7. $2x - 3y = 6$; x | **8.** $3x - 2y = 6$; x

9. $2(x - 3y) = x + 4$; x | **10.** $3x - 14 = 2(y - 2x)$; x

11. $PV = k$; V | **12.** $IR = E$; R

13. $I = prt$; p | **14.** $V = \ell wh$; ℓ

15. $p = 2\ell + 2w$; ℓ | **16.** $P = 2\ell + 2w$; w

17. $y = mx + b$; x | **18.** $V = k + gt$; t

19. $S = \dfrac{a}{1 - r}$; r | **20.** $I = \dfrac{E}{R + r}$; R

21. $C = \dfrac{5}{9}(F - 32)$; F | **22.** $A = \dfrac{h}{2}(B + b)$; B

23. $L = a + (n - 1)d$; n | **24.** $A = 2\pi rh + 2\pi r^2$; h

25. $z = \dfrac{Rr}{R + r}$; R | **26.** $c = \dfrac{ab}{a + b}$; b

27. $\dfrac{1}{F} = \dfrac{1}{u} + \dfrac{1}{v}$; u | **28.** $\dfrac{1}{c} = \dfrac{1}{a} + \dfrac{1}{b}$; a

Set II   Solve each equation for the variable indicated after each equation.

1. $x + 2y = 5$; x | **2.** $3x - 5y = 8$; y

3. $x - y = -4$; y | **4.** $8x + 3y = 1$; x

5. $2x - y = -4$; x | **6.** $y - 3x = -4$; x

7. $3x - 4y = 12$; y | **8.** $5y - 2x = 10$; x

9. $3(x + 2y) = x + 2$; x | **10.** $4(3x + y) = y - 3$; y

11. $d = rt$; t | **12.** $\dfrac{3xy}{z} = 10$; z

13. $I = prt$; r | **14.** $\dfrac{x}{a} + \dfrac{y}{b} = 1$; y

15. $x = 3y + 4z$; y | **16.** $A = \dfrac{1}{2}bh$; h

17. $y = mx + b$; m | **18.** $A = P(1 + rt)$; r

19. $T = \dfrac{s}{2 - x}$; x | **20.** $\dfrac{mn}{m + n} = 1$; m

21. $A = \dfrac{h}{2}(B + b)$; h | **22.** $\dfrac{ab}{c} = a + b$; b

23. $N = x + (y + 2)z$; y | **24.** $5xy - 3 = 2(3x + y)$; y

25. $I = \dfrac{E}{R + r}; r$

26. $S = \dfrac{n}{2}(A + L); L$

27. $\dfrac{1}{F} = \dfrac{1}{u} + \dfrac{1}{v}; v$

28. $a = \dfrac{b}{2 + c}; c$

8.11 Word Problems That Involve Rational Expressions

In this section, we discuss word problems that involve rational expressions. All the kinds of word problems discussed in previous chapters can lead to equations that contain rational expressions; the general method of solving such problems is the one given in Section 7.9, except that when we solve the equation (part of step 3 of those suggestions) we will have to "clear fractions."

Example 1 The denominator of a fraction exceeds the numerator by 8. If 2 is added to the numerator and 4 is subtracted from the denominator, the value of the resulting fraction is $\frac{5}{6}$. What is the original fraction?

Solution In this example, we must let x equal the numerator *or* the denominator of the fraction—*not* the original fraction.

Step 1. Let x = the numerator of the original fraction

$x + 8$ = the denominator of the original fraction

$\dfrac{x + 2}{(x + 8) - 4}$ = the new fraction

Reread   The value of the resulting fraction   is $\dfrac{5}{6}$

Step 2. $\dfrac{x + 2}{(x + 8) - 4} = \dfrac{5}{6}$

Step 3. $\dfrac{x + 2}{x + 4} = \dfrac{5}{6}$ This is a proportion

$6(x + 2) = 5(x + 4)$ Cross-multiplying *or* multiplying both sides by the LCD

$6x + 12 = 5x + 20$

$\underline{-5x - 12 \quad -5x - 12}$

Step 4. $x = 8$ The numerator

$x + 8 = 16$ The denominator

The original fraction is apparently $\frac{8}{16}$.

Step 5. *Check* The denominator exceeds the numerator by 8.

$$\frac{8 + 2}{16 - 4} = \frac{10}{12} = \frac{5}{6}$$

Step 6. Therefore, the original fraction is $\frac{8}{16}$. ∎

Example 2 The sum of a number and its reciprocal is $\frac{25}{12}$. Find the number and its reciprocal.
Solution

Step 1.   Let x = the number

$$\frac{1}{x} = \text{its reciprocal}$$

Reread   The sum of a number and its reciprocal   is $\frac{25}{12}$

Step 2. $x + \dfrac{1}{x} = \dfrac{25}{12}$

Step 3. $(12x)\left(x + \dfrac{1}{x}\right) = (\overset{1}{\cancel{12x}})\left(\dfrac{25}{\underset{1}{\cancel{12}}}\right)$ Multiplying both sides by 12x, the LCD

$$12x^2 + 12 = 25x$$ Second-degree equation

$$\underline{\quad -25x \qquad -25x}$$ Getting all nonzero terms on one side

$$12x^2 - 25x + 12 = 0$$

$$(3x - 4)(4x - 3) = 0$$ Factoring the left side

$3x - 4 = 0$ | $4x - 3 = 0$ Setting each factor equal to zero and solving each resulting equation

$3x = 4$ | $4x = 3$

Step 4. $x = \dfrac{4}{3}$ | $x = \dfrac{3}{4}$

$\dfrac{1}{x} = \dfrac{3}{4}$ | $\dfrac{1}{x} = \dfrac{4}{3}$

Step 5. *Check for* $x = \dfrac{4}{3}$ *and* $\dfrac{1}{x} = \dfrac{3}{4}$

$$\frac{4}{3} + \frac{3}{4} = \frac{16}{12} + \frac{9}{12} = \frac{25}{12}$$

Check for $x = \dfrac{3}{4}$ *and* $\dfrac{1}{x} = \dfrac{4}{3}$

$$\frac{3}{4} + \frac{4}{3} = \frac{9}{12} + \frac{16}{12} = \frac{25}{12}$$

Step 6. Therefore, the answers are (1) the number is $\frac{4}{3}$ and its reciprocal is $\frac{3}{4}$ and (2) the number is $\frac{3}{4}$ and its reciprocal is $\frac{4}{3}$. ∎

Solving Word Problems by Using Proportions
Some word problems are easily solved by using proportions. The method for doing so is described in the following box.

SOLVING WORD PROBLEMS THAT LEAD TO PROPORTIONS

1. Represent the unknown quantity by x, and use the given conditions to form two ratios or two rates.

2. Form a proportion by setting the two ratios or rates equal to each other, being sure to put the *units* next to the numbers when you write the proportion. Be sure the units occupy corresponding positions in the two ratios (or rates) of the proportion.

Correct arrangements	*Incorrect arrangements*

$$\frac{\text{miles}}{\text{hours}} = \frac{\text{miles}}{\text{hours}} \qquad \qquad \frac{\text{miles}}{\text{hours}} = \frac{\text{hours}}{\text{miles}}$$

$$\frac{\text{hours}}{\text{miles}} = \frac{\text{hours}}{\text{miles}} \qquad \qquad \frac{\text{hours}}{\text{miles}} = \frac{\text{miles}}{\text{hours}}$$

$$\frac{\text{miles}}{\text{miles}} = \frac{\text{hours}}{\text{hours}}$$

— Miles and hours in both numerators must correspond to the first condition

— Miles and hours in both denominators must correspond to the second condition

3. After the numbers have been correctly entered into the proportion by using the units as a guide, drop the units.

4. Solve the equation for x, using the method described in Section 8.9B.

We will not show the checks or "Step 1," "Step 2," and so forth in Examples 3–6.

Example 3 The scale on an architectural drawing is stated as "1 inch equals 8 feet." What are the dimensions of a room that measures $3\frac{1}{2}$ by 4 in. on the drawing?

Solution There are really two variables. We must find the *width* of the room (it corresponds to $3\frac{1}{2}$ in.) and the *length* of the room (it corresponds to 4 in.).

Let $x =$ the width of the room (in feet).

$$\frac{1 \text{ in.}}{8 \text{ ft}} = \frac{3\frac{1}{2} \text{ in.}}{x \text{ ft}} \qquad \text{Setting the two ratios equal to each other}$$

$$\frac{1}{8} = \frac{3\frac{1}{2}}{x} \qquad \text{Dropping the units}$$

$$x = 8\left(3\frac{1}{2}\right) = 8\left(\frac{7}{2}\right) = 28 \qquad \text{Cross-multiplying}$$

Let $y =$ the length of the room (in feet).

$$\frac{1 \text{ in.}}{8 \text{ ft}} = \frac{4 \text{ in.}}{y \text{ ft}}$$

$$\frac{1}{8} = \frac{4}{y} \qquad \text{Dropping the units}$$

$$y = 8(4) = 32 \qquad \text{Cross-multiplying}$$

Therefore, the room is 28 ft by 32 ft. ∎

Example 4   Jim knows he can drive 406 mi on 14 gal of gasoline. At this rate, how far can he expect to drive on 35 gal?

Solution Let x = the number of miles on 35 gal.

$$\frac{x \text{ mi}}{35 \text{ gal}} = \frac{406 \text{ mi}}{14 \text{ gal}} \qquad \text{Setting the two rates equal to each other}$$

$$\frac{x}{35} = \frac{406}{14} \qquad \text{Dropping the units}$$

$$14x = (406)(35) \qquad \text{Cross-multiplying}$$

$$x = \frac{(406)(35)}{14} = 1{,}015$$

———

Therefore, Jim can expect to drive 1,015 mi on 35 gal of gasoline. ∎

Work Problems

Work problems are similar to rate-time-distance problems. One basic relationship used to solve work problems is the following.

Rate × Time = Amount of work

In symbols: $r \cdot t = w$

If we know the rate and the time, we can find the amount of work done. If we know the amount of work done in a certain length of time, we can find the rate. If we know the rate and the amount of work done, we can find the time.

Example 5   If Machine A produces brackets at the rate of thirty-five brackets per hour for 7 hr, how many brackets has it produced?

Solution We use the formula $w = rt$.

———

$$w = 35\frac{\text{brackets}}{\text{hr}} \times 7 \text{ hr} = 245 \text{ brackets} \quad ∎$$

Example 6   Machine B produces sprocket wheels at the rate of 175 sprocket wheels per hour. How long does it take the machine to produce 805 sprocket wheels?

Solution Because $r \cdot t = w$, $t = \dfrac{w}{r}$. Therefore,

$$t = \frac{805 \text{ sprocket wheels}}{175\dfrac{\text{spr. wh.}}{\text{hr}}} = 4.6 \text{ hr, or } 4 \text{ hr } 36 \text{ min}$$

———   ∎

The other basic relationship used to solve work problems is the following:

$$\left(\begin{array}{c}\text{Amount A}\\ \text{does}\\ \text{in time } x\end{array}\right) + \left(\begin{array}{c}\text{Amount B}\\ \text{does}\\ \text{in time } x\end{array}\right) = \left(\begin{array}{c}\text{Amount done}\\ \text{together}\\ \text{in time } x\end{array}\right)$$

Example 7   George can build a fence in 6 days. Brian can do the same job in 4 days. (a) What is George's rate? (b) What is Brian's rate? (c) How long would it take them to build the same fence if they worked together?

Solution

a. Because rate × time = work ($r \cdot t = w$), $r = \dfrac{w}{t}$. Therefore, George's rate is

$$\frac{1 \text{ fence}}{6 \text{ days}} = \frac{1}{6}\frac{\text{fence}}{\text{day}}.$$

b. Brian's rate is $\dfrac{1 \text{ fence}}{4 \text{ days}} = \dfrac{1}{4}\dfrac{\text{fence}}{\text{day}}.$

c. Step 1. Let x = the number of days to build the fence together.

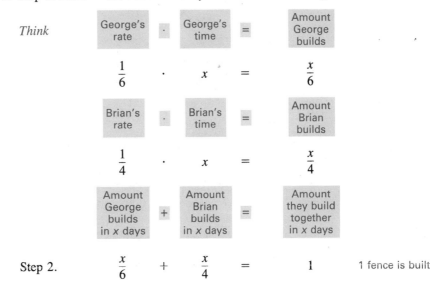

Step 2. $\dfrac{x}{6} + \dfrac{x}{4} = 1$ 1 fence is built

Step 3. LCD = 12

$$12\left(\frac{x}{6}\right) + 12\left(\frac{x}{4}\right) = 12(1) \qquad \text{Multiplying both sides by 12, the LCD}$$

$$2x + 3x = 12$$

$$5x = 12$$

Step 4. $x = \dfrac{12}{5} = 2\dfrac{2}{5}$

Step 5. *Check*

George's work: $\dfrac{1}{6}\dfrac{\text{fence}}{\text{day}} \cdot \dfrac{12}{5}\text{days} = \dfrac{2}{5}\text{ fence}$

Brian's work: $\dfrac{1}{4}\dfrac{\text{fence}}{\text{day}} \cdot \dfrac{12}{5}\text{days} = \dfrac{3}{5}\text{ fence}$

Together: $\dfrac{2}{5}\text{ fence} + \dfrac{3}{5}\text{ fence} = 1\text{ fence}$

Step 6. Therefore, it would take George and Brian $2\dfrac{2}{5}$ days to build the fence if they worked together.   ∎

Example 8   In a film-processing lab, machine A can process 5,400 ft of film in 60 min. (a) What is the rate of machine A? (b) How long does it take machine B to process 4,400 ft of film if the two machines working together can process 14,240 ft in 80 min?
Solution

a. The rate of machine A is $\dfrac{5,400 \text{ ft}}{60 \text{ min}} = 90\dfrac{\text{ft}}{\text{min}}$.

b. Step 1. Let x = the number of minutes for machine B to process 4,400 ft of film.

$$\frac{4,400 \text{ ft}}{x \text{ min}} = \text{rate of machine B}$$

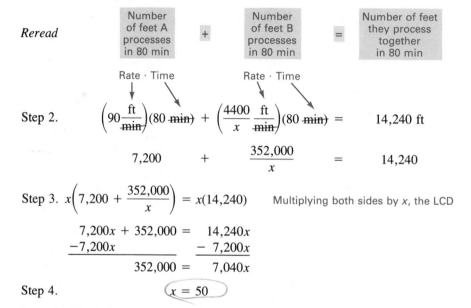

Step 2. $\left(90\dfrac{\text{ft}}{\text{min}}\right)(80 \text{ min}) + \left(\dfrac{4400}{x}\dfrac{\text{ft}}{\text{min}}\right)(80 \text{ min}) = \quad$ 14,240 ft

$$7,200 \quad + \quad \frac{352,000}{x} \quad = \quad 14,240$$

Step 3. $x\left(7,200 + \dfrac{352,000}{x}\right) = x(14,240)$ Multiplying both sides by x, the LCD

$$\begin{array}{rl} 7,200x + 352,000 = & 14,240x \\ -7,200x \qquad\qquad & -\ 7,200x \\ \hline 352,000 = & 7,040x \end{array}$$

Step 4. $x = 50$

Step 5. *Check* If machine B processes 4,400 ft of film in 50 min, its rate is $\dfrac{4,400}{50}\dfrac{\text{ft}}{\text{min}} = 88\dfrac{\text{ft}}{\text{min}}$. The amount of film it processes in 80 min is $\left(88\dfrac{\text{ft}}{\text{min}}\right)(80 \text{ min}) = 7,040 \text{ ft}$.

$$7,040 \text{ ft} + 7,200 \text{ ft} = 14,240 \text{ ft}$$

Machine B ⎯⎯⏋   ⎿⎯⎯ Machine A

Step 6. Therefore, it takes machine B 50 min to process 4,400 ft of film. ■

EXERCISES 8.11

Set I   Set up each problem algebraically, solve, and check. Be sure to state what your variables represent. (Note: These word problems do not necessarily lead to equations involving fractions.)

1. The denominator of a fraction exceeds the numerator by 6. If 4 is added to the numerator and subtracted from the denominator, the resulting fraction equals $\frac{11}{10}$. What is the original fraction?

2. The denominator of a fraction exceeds the numerator by 4. If 6 is subtracted from the numerator and added to the denominator, the resulting fraction equals $\frac{5}{13}$. What is the original fraction?

3. The denominator of a fraction is twice the numerator. If 5 is added to the numerator and subtracted from the denominator, the value of the resulting fraction is $\frac{4}{5}$. What is the original fraction?

4. The denominator of a fraction is 3 times the numerator. If 5 is added to the numerator and subtracted from the denominator, the value of the resulting fraction is $\frac{1}{2}$. What is the original fraction?

5. The sum of a number and its reciprocal is $\frac{29}{10}$. Find the number.

6. The sum of a number and its reciprocal is $\frac{130}{33}$. Find the number.

For Exercises 7 and 8, use this statement: The scale in an architectural drawing is 1 inch equals 8 feet.

7. Find the dimensions of a room that measures $2\frac{1}{2}$ by 3 in. on the drawing.

8. Find the dimensions of a room that measures $3\frac{1}{4}$ by $4\frac{1}{4}$ in. on the drawing.

9. The ratio of a woman's weight on earth compared to her weight on the moon is $6:1$. How much would a 150-lb woman weigh on the moon?

10. The ratio of a man's weight on Mars compared to his weight on earth is $2:5$. How much would a 196-lb man weigh on Mars?

11. The ratio of the weight of lead to the weight of an equal volume of aluminum is $21:5$. If an aluminum bar weighs 150 lb, what would a lead bar of the same size weigh?

12. The ratio of the weight of platinum to the weight of an equal volume of copper is $12:5$. If a platinum bar weighs 18 lb, what would a copper bar of the same size weigh?

13. On the first $6\frac{1}{2}$ hr of their shift, a fire crew built twenty-six chains of fire line. How much fire line can they build in the remainder of a 10-hr day if they work at the same pace?

14. The IRS informed a local businesswoman that for every $5,000 worth of sales she made, her tax would be $75. If the woman's sales totaled $125,000 for the year, how much was her tax?

15. A meat packer paid a cattle producer $3,420 for nine steers. Assuming all of the animals are of the same approximate weight, how much will the producer receive for sixteen steers?

16. A hog producer fed 1,320 hogs until they reached a marketing weight of 100 kg each. He received a check for their sale amounting to $59,400. How much did the producer receive per hog? What was the price paid for the hogs per kg of live weight?

17. Tricia can do a job in 8 hr. Mike can do the same job in 10 hr. How long will it take them to do the same job if they work together?

18. Joyce can paint a house in 5 days. Fred can paint the same house in 7 days. How long will it take them to paint the same house if they work together?

19. The sum of the numerator and denominator of a fraction is 42. If 2 is added to the numerator and 6 is added to the denominator, the resulting fraction is $\frac{2}{3}$. What is the original fraction?

20. The sum of the numerator and denominator of a fraction is 40. If 3 is added to the numerator and 2 is added to the denominator, the resulting fraction is $\frac{4}{5}$. What is the original fraction?

21. Ruth can proofread 220 pages of a deposition in 4 hr. How long does it take Jerry to proofread 210 pages if, when they work together, they can proofread 291 pages in 3 hr?

22. Barbara can type 95 pages of a manuscript in 3 hr. How long does it take Mike to type 110 pages if he and Barbara working together can type 322 pages in 6 hr?

23. Jim and Sandra live 63 mi apart. Both leave their homes at 7 A.M. by bicycle, riding toward one another. They meet at 10 A.M. If Jim's average speed is three-fourths of Sandra's, how fast does each cycle?

24. Rebecca and Jill live 75 mi apart. Both leave their homes at 8 A.M. by bicycle, riding toward one another. They meet at 10:30 A.M. If Jill's average speed is seven-eighths of Rebecca's, how fast does each cycle?

25. The speed of the current of a river is 5 mph. If a boat can travel 999 mi with the current in the same time it could travel 729 mi against the current, what is the speed of the boat in still water?

26. The speed of the current in a river is 4 mph. If a boat can travel 288 mi with the current in the same time it could travel 224 mi against the current, what is the speed of the boat in still water?

27. Two numbers differ by 12. One-sixth the larger exceeds one-fifth the smaller by 2. Find the numbers.

28. Two numbers differ by 8. One-seventh the larger exceeds one-eighth the smaller by 2. Find the numbers.

Set II Set up each problem algebraically, solve, and check. Be sure to state what your variables represent. (Note: These word problems do not necessarily lead to equations involving fractions.)

1. The denominator of a fraction exceeds the numerator by 20. If 7 is added to the numerator and subtracted from the denominator, the resulting fraction equals $\frac{6}{7}$. Find the fraction.

2. Suppose the wind speed is 30 mph. If an airplane can fly 240 mi against the wind in the same time that it can fly 420 mi with the wind, what is the speed of the plane in still air?

3. The denominator of a fraction is twice the numerator. If 2 is added to the numerator and subtracted from the denominator, the value of the resulting fraction is $\frac{5}{7}$. What is the original fraction?

4. When a number is subtracted from its reciprocal, the difference is $\frac{21}{10}$. What is the number?

5. The sum of a number and its reciprocal is $\frac{34}{15}$. Find the number.

6. Mrs. Summers drove her motorboat upstream a certain distance while pulling her son Brian on a water ski. She returned to the starting point pulling her other son Derek. The round trip took 25 min of skiing time. On both legs of the trip, the speedometer read 30 mph. If the speed of the current is 6 mph, how far upstream did she travel?

For Exercises 7 and 8, use this statement: The scale in an architectural drawing is 1 inch equals 8 feet.

7. Find the dimensions of a room that measures $2\frac{1}{4}$ by $3\frac{3}{4}$ in. on the drawing.

8. Find the dimensions of a room that measures $3\frac{1}{8}$ by $4\frac{3}{8}$ in. on the drawing.

9. An apartment house manager spent 22 hr painting three apartments. How much time can she expect to spend painting the remaining fifteen apartments?

10. Ralph drove 420 mi in $\frac{3}{4}$ of a day. About how far can he drive in $2\frac{1}{2}$ days?

11. The ratio of the width of a rectangle to its length is $4:9$. If its area is 900 sq. m, find the width and the length.

12. A crew of ten men takes a week to overhaul 25 trucks in a fleet of 100 trucks. How many men would it take to complete the fleet overhaul in one week?

13. A car burns $2\frac{1}{2}$ qt of oil on a 1,800-mi trip. How many quarts of oil can the owner expect to use on a 12,000-mi trip?

14. Fifteen defective axles were found in 100,000 cars of a particular model. How many defective axles would you expect to find in the 2 million cars made of that same model?

15. Mr. Sanders has 300 Leghorn hens and needs 22.86 cm of roosting space per hen. How many meters of roosting space does he need?

16. The Forest Service must determine how many acres will be needed to make an addition of twenty-two campsites. If the existing 24-acre campground accommodates fifty-five campsites, how many additional acres will be needed?

17. Karla can do a job in 9 hr. Mark can do the same job in 11 hr. How long will it take them to do the same job if they work together?

18. Ben bought 120 stamps consisting of 20¢, 15¢, and 3¢ stamps at a total cost of $12.80. He bought twice as many 20¢ stamps as 15¢ stamps and 20 more 3¢ stamps than 20¢ stamps. How many of each kind did he buy?

19. The sum of the numerator and denominator of a fraction is 68. If 1 is added to the numerator and 12 is added to the denominator, the resulting fraction is $\frac{2}{7}$. What is the original fraction?

20. The sum of two numbers is 32. One-half the smaller exceeds one-third the larger by 1. Find the numbers.

21. David can type 65 pages of a manuscript in 2 hr. How long does it take Susan to type 450 pages if, when they work together, they can type 210 pages in 3 hr?

22. The sum of the first two of three consecutive odd integers added to the sum of the last two is 60. Find the integers.

23. Kim and Jerry live 80 mi apart. Both leave their homes at 6 A.M. by bicycle, riding toward one another. They meet at 10 A.M. If Kim's average speed is two-thirds of Jerry's, how fast does each cycle?

24. It takes Jim 3 times as long as Sherry to paint a certain house. Working together, Jim and Sherry could paint the same house in 3 days.

  a. How long would it take Sherry working alone to paint the house?

  b. How long would it take Jim working alone to paint the house?

25. The speed of the current in a river is 5 mph. If a boat can travel 198 mi with the current in the same time it could travel 138 mi against the current, what is the speed of the boat in still water?

26. Eleanore paddles a kayak downstream for 3 hr. After having lunch, she paddles upstream for 5 hr. At that time she is still 6 mi short of getting back to her starting point. If the speed of the stream is 2 mph, how fast does Eleanore's kayak move in still water? How far downstream did she travel?

27. Two numbers differ by 3. One-fifth the larger exceeds one-sixth the smaller by 1. Find the numbers.

28. Machine A takes 3 times as long as machine B to do a certain job. Both machines running together can do this same job in 4 hr. How long does it take each machine working alone to do the job?

8.12 Review: 8.9–8.11

To Solve a Rational Equation 8.9

1. Find the LCD and find all excluded values.

2. Remove denominators by multiplying *both sides of the equation* (that is, by multiplying *every term*) by the LCD.

3. a. Remove all grouping symbols.

   b. Collect and combine like terms on each side of the equal sign.

First-degree equations	*Second-degree equations*
4a. Get all terms containing the variable on one side of the equation and all other terms on the other side. b. Divide both sides of the equation by the coefficient of the variable, or multiply both sides of the equation by the reciprocal of that coefficient.	4a. Get *all* nonzero terms on one side of the equation by adding the same expression to both sides. *Only zero must remain on the other side.* Then arrange the terms in descending powers. b. Factor the polynomial. c. Set each factor equal to zero and solve for the unknown.

5. Reject any apparent solutions that are excluded values.

6. Check any other apparent solutions in the original equation.

To Solve a Proportion 8.9

A **proportion** is a statement that two ratios or two rates are equal.

To solve a proportion, we can use the rules above, or we can "cross-multiply." That is, we can use this rule: If $\frac{P}{Q} = \frac{S}{T}$, then $PT = QS$.

Literal Equations 8.10

Literal equations are equations that have more than one variable.

To solve a literal equation, proceed in the same way used to solve an equation with a single variable. The solution will be expressed in terms of the other variable(s) given in the literal equation, as well as in numbers.

To Solve Word Problems Using Proportions 8.11

1. Represent the unknown quantity by a variable.

2. Write a proportion; be sure the same units occupy corresponding positions in the two ratios (or rates) of the proportion. Put the units next to the numbers when writing the proportion.

3. Drop the units.

4. Solve the equation for the variable, using the method described in Section 8.9B.

Review Exercises 8.12 Set I

In Exercises 1–7, solve for the variable.

1. $\dfrac{6}{m} = 5$

2. $x - \dfrac{3x}{5} = 2$

3. $\dfrac{z}{5} - \dfrac{z}{8} = 3$

4. $\dfrac{3}{2} = \dfrac{3x + 4}{5x - 1}$

5. $\dfrac{4}{2z} + \dfrac{2}{z} = 1$

6. $\dfrac{4}{x^2} - \dfrac{3}{x} = \dfrac{5}{2}$

7. $\dfrac{7}{2x - 1} + \dfrac{1}{18} = \dfrac{x}{6}$

In Exercises 8–11, solve for the variable indicated after the equation.

8. $2x - 7y = 14$; y

9. $\dfrac{2m}{n} = P$; n

10. $V = \dfrac{1}{3}Bh$; B

11. $\dfrac{F - 32}{C} = \dfrac{9}{5}$; C

In Exercises 12–17, set up each problem algebraically, solve, and check. Be sure to state what your variables represent.

12. The numerator of a fraction exceeds the denominator by 4. If 3 is added to the denominator and subtracted from the numerator, the value of the resulting fraction is $\frac{7}{8}$. Find the original fraction.

13. Mr. Maxwell takes 30 min to drive to work in the morning, but he takes 45 min to return home over the same route during the evening rush hour. If his average morning speed is 10 mph faster than his average evening speed, how far is it from his home to his work?

14. The sum of a number and its reciprocal is $\frac{13}{6}$. Find the number.

15. Machine A can do a job in 6 hr. How long does it take machine B to do the same job if, when the two machines work together, they get the job done in 4 hr?

16. Justin drove 324 mi in $\frac{3}{5}$ of a day. At that same rate, about how far can he drive in a day and a half?

17. Alan weaves $2\frac{1}{2}$ yd of fabric in $\frac{5}{8}$ of a day. At that same rate, how many yards of fabric can he weave in half a day?

Review Exercises 8.12 Set II

NAME _____

In Exercises 1–7, solve for the variable.

1. $\dfrac{3}{x} = 4$

2. $\dfrac{x}{3} - \dfrac{x}{2} = 2$

3. $\dfrac{x+2}{5} + \dfrac{2x}{3} = 3$

4. $\dfrac{3x}{7} = \dfrac{x-1}{5}$

5. $\dfrac{x+2}{-2} = \dfrac{3}{x-3}$

6. $\dfrac{17}{6x} + \dfrac{5}{2x^2} = \dfrac{2}{3}$

7. $\dfrac{3}{x} - \dfrac{8}{x^2} = \dfrac{1}{4}$

In Exercises 8–11, solve for the variable indicated after the equation.

8. $\dfrac{P}{V} = C;\ V$

9. $V = lwh;\ h$

10. $V^2 = 2gS;\ S$

11. $5(x - 2y) = 14 + 3(2x - y);\ y$

In Exercises 12–17, set up each problem algebraically, solve, and check. Be sure to state what your variables represent.

12. When a number is subtracted from its reciprocal, the difference is $\frac{40}{21}$. Find the number.

ANSWERS

1. _____

2. _____

3. _____

4. _____

5. _____

6. _____

7. _____

8. _____

9. _____

10. _____

11. _____

12. _____

13. When the speed of the wind is 25 mph, a certain airplane can fly only 480 mi against the wind in the same time it can fly 780 mi with the wind. Find the speed of the plane in still air.

14. The denominator of a fraction exceeds the numerator by 7. If 10 is added to the numerator and 15 is subtracted from the denominator, the value of the resulting fraction is $\frac{11}{5}$. Find the original fraction.

15. Rebecca can paint a house in 5 days. How long would it take Karla alone to paint the same house if the two women, working together, can paint the house in 3 days?

16. Jill drove 220 mi in $\frac{2}{5}$ of a day. At that same rate, about how far can she drive in half a day?

17. Susan crochets $3\frac{1}{2}$ scarfs in $2\frac{1}{3}$ days. At that same rate, how many scarfs can she crochet in 6 days?

13. _____

14. _____

15. _____

16. _____

17. _____

Chapter 8 Diagnostic Test

The purpose of this test is to see how well you understand operations with rational expressions and solving rational equations. We recommend that you work this diagnostic test *before* your instructor tests you on this chapter. Allow yourself about 50 minutes.

Complete solutions for all the problems on this test, together with section references, are given in the answer section in the back of this book. For the problems you do incorrectly, study the sections referred to.

1. What value(s) of the variable must be excluded, if any, in each of the following expressions?

   a. $\dfrac{3x}{x + 4}$

   b. $\dfrac{5x - 4}{x^2 + 4x}$

2. Find the missing term in each of the following expressions.

   a. $-\dfrac{-8}{3} = \dfrac{8}{?}$

   b. $\dfrac{3}{x - 5} = \dfrac{?}{5 - x}$

In Problems 3 and 4, reduce each rational expression to lowest terms.

3. $\dfrac{x^2 - 9}{x^2 - 6x + 9}$

4. $\dfrac{4x^2 - 23xy + 15y^2}{20x^2y - 15xy^2}$

In Problems 5–10, perform the indicated operations. Be sure to reduce rational expressions to lowest terms.

5. $\dfrac{a}{a^2 + 5a + 6} \cdot \dfrac{4a + 8}{6a^3 + 18a^2}$

6. $\dfrac{6x}{x - 2} - \dfrac{3}{x - 2}$

7. $\dfrac{2x}{x^2 - 9} \div \dfrac{4x^2}{x^2 - 6x + 9}$

8. $\dfrac{b}{b - 1} - \dfrac{b + 1}{b}$

9. $\dfrac{x + 3}{x^2 - 10x + 25} + \dfrac{x}{2x^2 - 7x - 15}$

10. $\dfrac{x}{x + 5} - \dfrac{x}{x - 5} - \dfrac{50}{x^2 - 25}$

In Problems 11 and 12, simplify each complex fraction.

11. $\dfrac{\dfrac{6x^4}{11y^2}}{\dfrac{9x}{22y^4}}$

12. $\dfrac{\dfrac{6}{x} + \dfrac{15}{x^2}}{2 + \dfrac{5}{x}}$

In Problems 13–16, solve each equation.

13. $\dfrac{x + 7}{3} = \dfrac{2x - 1}{4}$

14. $x - \dfrac{4}{x} = 3$

15. $\dfrac{3}{4x + 2} + \dfrac{x + 1}{6} = 1$

16. $\dfrac{3x - 5}{x} = \dfrac{x + 1}{3}$

17. Solve for y: $5x + 7y = 18$

In Problems 18–20, set up each problem algebraically, solve, and check. Be sure to state what your variables represent.

18. The denominator of a fraction is 6 more than the numerator. If 1 is subtracted from the numerator and 7 is added to the denominator, the value of the resulting fraction is $\frac{1}{3}$. What is the original fraction?

19. Mr. Ames drove 21 mi in $\frac{2}{5}$ hr. At that same rate, how far can he drive in $2\frac{2}{3}$ hr?

20. David can type 128 pages of a manuscript in 4 hr. How long does it take Kenneth to type 111 pages if both men working together can type 345 pages in 5 hr?

Cumulative Review Exercises: Chapters 1–8

1. Evaluate $10 - (3\sqrt{4} - 5^2)$.

2. Use the following formula, substituting the values of the variables given with the formula.

$$V = \frac{25}{8}\left(\frac{H}{D} - \frac{A}{R}\right) \quad \text{Find } V \text{ if } H = 10, D = 18, A = 1, \text{ and } R = 5.$$

3. Simplify. Write your answer using only positive exponents.

$$\left(\frac{24y^{-3}}{8y^{-1}}\right)^{-2}$$

In Exercises 4–9, perform the indicated operations and simplify.

4. $\dfrac{6x^2 - 2x}{2x}$

5. $(15z^2 + 11z + 4) \div (3z - 2)$

6. $\dfrac{5x - 5}{10} \cdot \dfrac{4x^3}{2x - 2}$

7. $\dfrac{2x}{x + 1} - \dfrac{2x - 1}{x + 1}$

8. $5 + \dfrac{3}{2x^3}$

9. $\dfrac{2x - 1}{x^2 + 9x + 20} - \dfrac{x + 5}{2x^2 + 7x - 4}$

In Exercises 10 and 11, solve each equation.

10. $\dfrac{14}{3x} + \dfrac{42}{x} = 1$

11. $\dfrac{3}{x} - \dfrac{8}{x^2} = \dfrac{1}{4}$

In Exercises 12–15, factor each expression.

12. $x^2 - x - 42$

13. $3w^2 - 48$

14. $20a^2 - 7ab - 3b^2$

15. $3x^2 - 6x - 2xy + 4y$

In Exercises 16–20, set up each problem algebraically, solve, and check. Be sure to state what your variables represent.

16. The sum of two consecutive integers is 33. What are the integers?

17. The sum of two numbers is 5. Their product is -24. What are the numbers?

18. A dealer makes up a 15-lb mixture of oranges. One kind costs 78¢ per pound, and the other costs 99¢ per pound. How many pounds of each kind must be used in order for the mixture to cost 85¢ per pound?

19. Manny has twenty coins with a total value of $1.65. If all the coins are nickels or dimes, how many of each does he have?

20. Margaret has $2.60 in nickels, dimes, and quarters. If she has 1 more dime than quarters and 3 times as many nickels as quarters, how many of each kind of coin does she have?

9 Graphing

Many algebraic relationships are easier to understand if a picture called a graph is drawn. In this chapter, we graph ordered pairs and solution sets of equations and inequalities in two variables. We also discuss writing the equation of a straight line.

9.1 The Rectangular Coordinate System

In Chapter 3, we graphed the solution of an equation in one variable on the real number line. Solutions of equations and inequalities in two variables usually do not lie on a single number line. Instead, they lie in a *plane*, which is a flat surface. (In this book, we work in the plane only and not in three-dimensional space.)

The **rectangular coordinate system** in the plane consists of a horizontal real number line called the **horizontal axis**, or **x-axis**, and a vertical real number line called the **vertical axis** or **y-axis**. These lines intersect at right angles at a point called the **origin**, and they separate the plane into four regions called **quadrants**, numbered as shown in Figure 9.1.1.

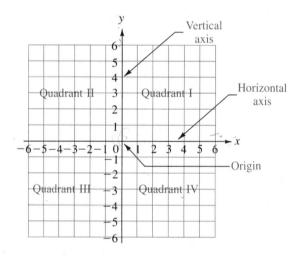

FIGURE 9.1.1   Rectangular Coordinate System

Ordered Pairs   While points on the number line can be represented by a single real number, points in the plane must be represented by an **ordered pair** of real numbers. Two ordered pairs are equal to each other if they contain the same elements *in the same order*. We enclose the elements of an ordered pair within parentheses, never braces or brackets. Thus, $(a, b) \neq (b, a)$.

The first number of the ordered pair is called the **x-coordinate** (or horizontal coordinate, first coordinate, or abscissa). The second number of the ordered pair is called the **y-coordinate** (or vertical coordinate, second coordinate, or ordinate). Consider the ordered pair (3, 2). We call 3 and 2 the *coordinates* of the point (3, 2). The first number, 3, is called the x-coordinate of the point (3, 2). The second number, 2, is called the y-coordinate of the point (3, 2).

Graph of a Point   There is exactly one point in the plane corresponding to each ordered pair of real numbers, and there is exactly one ordered pair of real numbers corresponding to each point in the plane. The *origin* is the point that corresponds to (0, 0). See Figures 9.1.2 and 9.1.3 for instructions on graphing the point (3, 2).

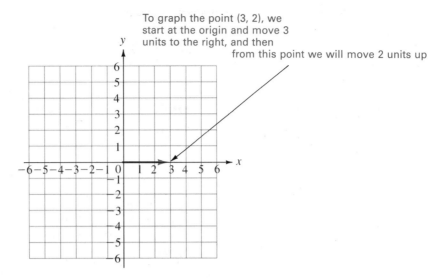

FIGURE 9.1.2

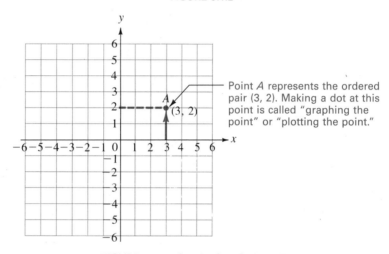

FIGURE 9.1.3   Graph of an Ordered Pair

The *x*-coordinate tells us how far the point is from the *vertical* axis (the *y*-axis). A positive *x*-coordinate indicates that the point is to the right of the *y*-axis. (See Figure 9.1.4.) A negative *x*-coordinate indicates that the point is to the left of the *y*-axis.

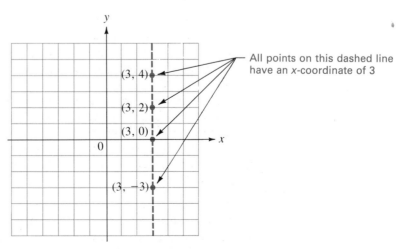

FIGURE 9.1.4

The y-coordinate tells us how far the point is from the *horizontal* axis (the x-axis). A positive y-coordinate indicates that the point is above the x-axis. (See Figure 9.1.5.) A negative y-coordinate indicates that the point is below the x-axis.

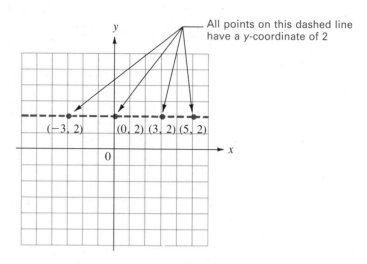

FIGURE 9.1.5

NOTE   When the order is changed in an ordered pair, we get a different point. For example, (1, 4) and (4, 1) are two different points (see Figure 9.1.6).

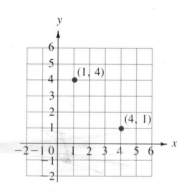

FIGURE 9.1.6

Example 1   Examples of graphing points:

a. (3, 5)   Start at the origin and move *right* 3 units; then move *up* 5 units (point *A* in Figure 9.1.7).

b. (−5, 2)   Start at the origin and move *left* 5 units; then move *up* 2 units (point *B* in Figure 9.1.7).

c. (−5, −4)   Start at the origin and move *left* 5 units; then move *down* 4 units (point *C* in Figure 9.1.7).

d. (0, −3)   Start at the origin, but because the first number is zero, do not move either right or left. Just move *down* 3 units (point *D* in Figure 9.1.7).

e. (4, −6)   Start at the origin and move *right* 4 units; then move *down* 6 units (point *E* in Figure 9.1.7).

The statement "plot the points" means the same as "graph the points."

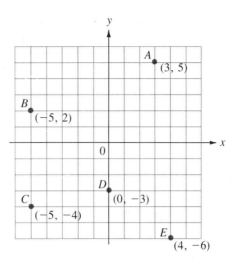

FIGURE 9.1.7 ∎

Example 2   Name the ordered pairs corresponding to points B, C, D, and E in Figure 9.1.8. Also name the quadrant in which each point lies.

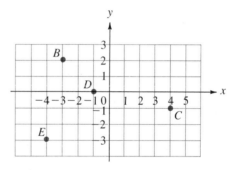

FIGURE 9.1.8

Solution The ordered pair corresponding to B is $(-3, 2)$, because B lies 3 units to the *left* of the y-axis and 2 units *above* the x-axis. B lies in quadrant II.

The ordered pair corresponding to C, which lies in quadrant IV, is $(4, -1)$, because C lies 4 units to the *right* of the y-axis and 1 unit *below* the x-axis.

The ordered pair corresponding to D is $(-1, 0)$, because D lies 1 unit to the *left* of the y-axis and *on* the x-axis. D does not lie in any quadrant; it lies on the x-axis.

The ordered pair corresponding to E is $(-4, -3)$, because E lies 4 units to the *left* of the y-axis and 3 units *below* the x-axis. E lies in quadrant III. ∎

    The points at which the sides of a triangle meet are called the *vertices* of the triangle (see Example 3).

Example 3   Draw the triangle with vertices at the following points:

$$A(-1, 2) \quad B(3, -2) \quad C(-3, -4)$$

Solution

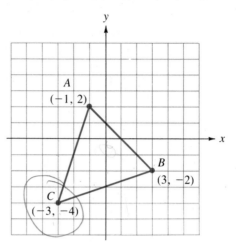

NOTE   The scales on the two axes do not have to be equal. If we needed to graph the point (3, 200), for example, it would be better if the scales were *not* equal. ☑

EXERCISES 9.1

Set I   **1.** Graph each of the following points.

a. (3, 1)   b. (−4, −2)   c. (0, 3)

d. (5, −4)   e. (4, 0)   f. (−2, 4)

2. Graph each of the following points.

a. (2, 4)   b. (2, −4)   c. (3, 0)

d. (−3, −2)   e. (0, 0)   f. (0, −4)

In Exercises 3 and 4, use Figure 9.1.9. Name the ordered pair corresponding to each point and name the quadrant in which the point lies.

3. a. *R*   b. *N*   c. *U*   d. *S*

4. a. *M*   b. *P*   c. *Q*   d. *T*

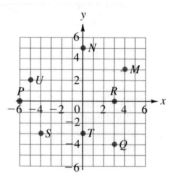

FIGURE 9.1.9

In Exercises 5–8, use Figure 9.1.10.

5. Write the *x*-coordinate of each of these points.

a. *A*   b. *C*   c. *E*   d. *F*

6. Write the *y*-coordinate of each of these points.

    a. *B*   b. *D*   c. *E*   d. *F*

7. a. What is the abscissa of point *F*?

    b. What is the ordinate of point *C*?

8. a. What is the ordinate of point *B*?

    b. What is the abscissa of point *D*?

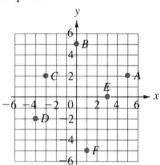

FIGURE 9.1.10

9. What is the *y*-coordinate of the origin?

10. What is the *x*-coordinate of the origin?

11. Draw the triangle with vertices at the following points:

$$A(0, 0) \quad B(3, 2) \quad C(-4, 5)$$

12. Draw the triangle with vertices at the following points:

$$A(-2, -3) \quad B(-2, 4) \quad C(3, 5)$$

Set II   **1.** Graph each of the following points.

    a. $(-2, 4)$ b. $(2, -5)$ c. $(-1, -3)$

    d. $(3, 6)$ e. $(-5, -1)$ f. $(5, -2)$

2. Graph each of the following points.

    a. $(-3, 5)$ b. $(0, 4)$ c. $(-2, 0)$

    d. $(2, -4)$ e. $(-4, -3)$ f. $(0, -2)$

In Exercises 3 and 4, use Figure 9.1.11. Name the ordered pair corresponding to each point and name the quadrant in which the point lies.

3. a. *C*   b. *D*   c. *E*   d. *F*

4. a. *K*   b. *L*   c. *W*   d. *U*

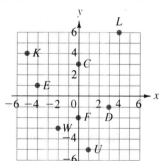

FIGURE 9.1.11

In Exercises 5–8, use Figure 9.1.12.

5. Write the x-coordinate of each of these points.

   a. S   b. T   c. U   d. V

6. Write the y-coordinate of each of these points.

   a. S   b. T   c. U   d. V

7. a. What is the abscissa of point U?

   b. What is the ordinate of point S?

8. a. What is the ordinate of point V?

   b. What is the abscissa of point T?

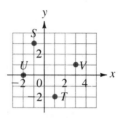

FIGURE 9.1.12

9. What name is given to the point (0, 0)?

10. Does every point on the y-axis have the same x-coordinate? If so, what is it?

11. Draw the triangle with vertices at the following points:

$$A(-6, -4) \quad B(2, -5) \quad C(5, 3)$$

12. Does every point on the x-axis have the same x-coordinate? If so, what is it?

9.2 Linear Equations in Two Variables

If an equation can be put into the form

$$Ax + By + C = 0$$

where A, B, and C are any real numbers and A and B are not both zero, it is called a **linear equation in two variables**. (Note the word *line* in the word *line*ar.)

Example 1   Examples of linear equations in two variables:

   a. $7x = 5y + 3$ The equation can be rewritten $7x - 5y - 3 = 0$.

   b. $2y = 4$ The equation can be rewritten $0x + 2y - 4 = 0$.

   c. $x = -5$ The equation can be rewritten $1x + 0y + 5 = 0$. ■

The Solutions of Equations in Two Variables

An ordered pair (a, b) is a *solution* of an equation in x and y if we get a true statement when we substitute a for x and b for y in that equation. There are always many ordered pairs that are solutions for such an equation and many others that are not solutions.

Example 2 Determine whether each ordered pair is a solution of the equation $2x - 3y = 6$.

  a. $(2, -3)$

  Solution Substituting 2 for x and -3 for y, we have

$$2x - 3y = 6$$
$$2(2) - 3(-3) \stackrel{?}{=} 6$$
$$4 + 9 \stackrel{?}{=} 6$$
$$13 = 6 \quad \text{False}$$

  $(2, -3)$ is not a solution.

  b. $(3, 0)$

  Solution Substituting 3 for x and 0 for y, we have

$$2x - 3y = 6$$
$$2(3) - 3(0) \stackrel{?}{=} 6$$
$$6 - 0 \stackrel{?}{=} 6$$
$$6 = 6 \quad \text{True}$$

  $(3, 0)$ is a solution.

  c. $(0, 0)$

  Solution Substituting 0 for x and 0 for y, we have

$$2x - 3y = 6$$
$$2(0) - 3(0) \stackrel{?}{=} 6$$
$$0 + 0 \stackrel{?}{=} 6$$
$$0 = 6 \quad \text{False}$$

  $(0, 0)$ is not a solution.

  d. $(0, -2)$

  Solution Substituting 0 for x and -2 for y, we have

$$2x - 3y = 6$$
$$2(0) - 3(-2) \stackrel{?}{=} 6$$
$$0 + 6 \stackrel{?}{=} 6$$
$$6 = 6 \quad \text{True}$$

  $(0, -2)$ is a solution. ■

Finding Solutions for Equations in Two Variables

We can find a solution for an equation in two variables when we are given a value for one of the variables (see Example 3).

Example 3   Find solutions for the equation $2x - 3y = 6$ given that (a) $x = 0$, (b) $y = 0$, (c) $x = 2$.
Solution

a. If we substitute 0 for x, we have

$$2x - 3y = 6$$
$$2(0) - 3y = 6 \qquad \text{We can now solve for } y$$
$$-3y = 6$$
$$y = -2 \qquad \text{We often say "-2 is the y-value that corresponds to $x = 0$"}$$

Therefore, $(0, -2)$ is a solution for $2x - 3y = 6$.

b. If we substitute 0 for y, we have

$$2x - 3y = 6$$
$$2x - 3(0) = 6 \qquad \text{We can now solve for } x$$
$$2x = 6$$
$$x = 3 \qquad \text{3 is the x-value that corresponds to $y = 0$}$$

Therefore, $(3, 0)$ is a solution for $2x - 3y = 6$.

c. If we substitute 2 for x, we have

$$2x - 3y = 6$$
$$2(2) - 3y = 6 \qquad \text{We can now solve for } y$$
$$\begin{array}{r} 4 - 3y = 6 \\ \underline{-4 \qquad\quad -4} \\ -3y = 2 \end{array} \qquad \text{Adding -4 to both sides}$$
$$y = -\frac{2}{3}$$

Therefore, $\left(2, -\frac{2}{3}\right)$ is a solution for $2x - 3y = 6$. ∎

If we aren't given a value for one of the variables, we can usually choose any value we wish for either variable (see Example 4).

Example 4   Find three ordered pairs that are solutions of $y = x + 1$.
Solution We can let x have any value whatever and solve the equation for y. We can also let y have any value whatever and solve the equation for x.

If $x = 5$, $y = 5 + 1 = 6$. Therefore, $(5, 6)$ is a solution.

If $x = 402$, $y = 402 + 1 = 403$. Therefore, $(402, 403)$ is a solution.

If $y = -23$, we have

$$\begin{array}{r} -23 = x + 1 \\ \underline{-1 \qquad\quad -1} \\ -24 = x \end{array}$$
$$x = -24$$

Therefore, $(-24, -23)$ is a solution.

────── NOTE   Infinitely many other answers are possible.   ∎

Example 5   Complete the following ordered pairs so they will be solutions of the equation $x + 3 = 0$; (⬚ , 5), (⬚ , −2), (⬚ , 0).
Solution The equation $x + 3 = 0$ is equivalent to the equation $x + 0y + 3 = 0$. When we substitute *any* value for y in this equation, we have

$$x + 3 = 0$$

────── Therefore, $x = -3$ for *all* values of y. The ordered pairs are $(-3, 5)$, $(-3, -2)$, and $(-3, 0)$.   ∎

In Example 5, we *cannot* let x have any value whatever, since x must always equal -3.

EXERCISES 9.2

Set I   In Exercises 1–4, determine whether the given ordered pairs are solutions of the given equation.

1. $3x + 2y = 6$

a. $(0, 0)$ b. $(0, 3)$ c. $(3, 0)$ d. $(2, 0)$

2. $5x + 2y = 10$

a. $(0, 0)$ b. $(5, 0)$ c. $(0, 5)$ d. $(2, 0)$

3. $3x + 4y = 0$

a. $(0, 0)$ b. $(0, 3)$ c. $(4, -3)$ d. $(-4, 3)$

4. $x - 2y = 0$

a. $(0, 0)$ b. $(0, 2)$ c. $(2, 0)$ d. $(-4, -2)$

In Exercises 5–10, complete the ordered pairs so they will be solutions of the given equation.

5. $4x - 3y = 12$

a. $(0, ⬚)$ b. $(⬚, 0)$ c. $(3, ⬚)$ d. $(⬚, -4)$

6. $3x + y = 3$

a. $(0, ⬚)$ b. $(⬚, 0)$ c. $(3, ⬚)$ d. $(⬚, -3)$

7. $2x - 5y = 0$

a. $(0, ⬚)$ b. $(⬚, 0)$ c. $(2, ⬚)$ d. $(⬚, 2)$

8. $3x + y = 0$

a. $(0, ⬚)$ b. $(⬚, 0)$ c. $(3, ⬚)$ d. $(⬚, -3)$

9. $x - 5 = 0$

a. $(⬚, 0)$ b. $(⬚, 5)$ c. $(⬚, 2)$ d. $(⬚, -2)$

10. $y + 3 = 0$

a. $(0, ⬚)$ b. $(3, ⬚)$ c. $(-3, ⬚)$ d. $(5, ⬚)$

Set II   In Exercises 1–4, determine whether the given ordered pairs are solutions of the given equation.

1. $2x + 3y = 6$

    a. $(0, 0)$ b. $(0, 3)$ c. $(3, 0)$ d. $(2, 0)$

2. $2x - 5y = 10$

    a. $(0, 0)$ b. $(5, 0)$ c. $(0, 5)$ d. $(2, 0)$

3. $2x + 5y = 0$

    a. $(0, 0)$ b. $(0, 5)$ c. $(5, -2)$ d. $(-5, 2)$

4. $2x - y = 0$

    a. $(0, 0)$ b. $(0, 2)$ c. $(2, 0)$ d. $(-4, -2)$

In Exercises 5–10, complete the ordered pairs so they will be solutions of the given equation.

5. $5x - 2y = 10$

    a. $(0, \ )$ b. $(\ , 0)$ c. $(2, \ )$ d. $(\ , -5)$

6. $y = -2$

    a. $(0, \ )$ b. $(-2, \ )$ c. $(3, \ )$ d. $(-10, \ )$

7. $3x - y = 0$

    a. $(0, \ )$ b. $(\ , 0)$ c. $(3, \ )$ d. $(\ , 3)$

8. $x = -4$

    a. $(\ , 0)$ b. $(\ , 2)$ c. $(\ , -4)$ d. $(\ , -30)$

9. $y + 2 = 0$

    a. $(0, \ )$ b. $(5, \ )$ c. $(-2, \ )$ d. $(3, \ )$

10. $y = 0$

    a. $(0, \ )$ b. $(3, \ )$ c. $(-3, \ )$ d. $(5, \ )$

9.3 Graphing Straight Lines

We will now combine what we learned in Section 9.1 (how to graph an ordered pair, that is, a point) with what we learned in Section 9.2 (finding ordered pairs that are solutions of an equation). If an ordered pair (a point) *satisfies an equation*, we say that the point *lies on the graph of the equation*.

NOTE   We commonly refer to "the graph of an equation" when we really mean the graph of the *solution set* of that equation. Likewise, we refer to "the graph of an inequality" when we mean the graph of the *solution set* of that inequality. ☑

Example 1   Graph four ordered pairs (points) that lie on the graph of $y = x + 1$.

Solution By choosing four values for x and finding the corresponding values for y, we obtain a set of ordered pairs (points) that satisfy the equation. We will let x be 0, 2, 5, and -3 (see the table of values). (We could have let x have *any* other real value, of course, for example, 7 or $-\frac{1}{2}$ or $\sqrt{11}$.)

Table of values

Equation: $y = x + 1$

x	y
0	1
2	3
5	6
−3	−2

When $x = 0$, $y = 0 + 1 = 1$

When $x = 2$, $y = 2 + 1 = 3$

When $x = 5$, $y = 5 + 1 = 6$

When $x = -3$, $y = -3 + 1 = -2$

Each of these ordered pairs, called "a pair of corresponding values," represents a point of the graph

In Figure 9.3.1, we plot the four points from the table of values.

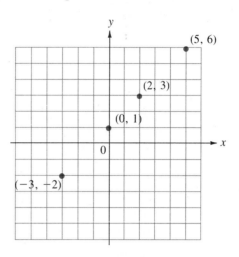

FIGURE 9.3.1

Notice that the points appear to lie in a straight line. The graph in Figure 9.3.1 is *not* yet the graph of $y = x + 1$. ∎

There are *infinitely* many points whose coordinates satisfy an equation in one or two variables, and we usually cannot graph *all* of these points—our graph would have to be infinitely large! However, we can use arrowheads to indicate that the graph extends beyond the picture we have drawn. The coordinates of every point on the graph we draw must satisfy the equation, and, within the limits of the size of our graph, we must include in our graph *all* the points whose coordinates satisfy the equation. Thus, in Figure 9.3.1, we have not graphed $y = x + 1$, because the points $(1, 2)$, $(-5, -4)$, and $(0.2, 1.2)$, for example, also satisfy the equation (verify this!), and those points were not included in the graph.

The following statement must be memorized; it will not be proved.

THE GRAPH OF A LINEAR (FIRST-DEGREE) EQUATION

The graph of any linear (first-degree) equation in one or two variables is the *straight line* that contains, within the limits of the size of the graph, all the ordered pairs, and only those ordered pairs, that satisfy the given equation.

Therefore, when we graph a first-degree equation in one or two variables, after we have graphed a few points that lie on the line (see the NOTE below), *we must connect those points with a straight line*, and *we must put arrowheads at each end of the line.* Only in this manner can we indicate that we're including *all* the ordered pairs that satisfy the equation.

NOTE   A straight line can be drawn if we know any two points on that line. Although only two points are *necessary*, it is advisable to graph a third point as a *checkpoint*. If the equation is a first-degree equation and if the three points do *not* lie in a straight line, some mistake has been made either in calculating the coordinates of the points or in graphing the points.   ☑

Example 2   Graph $y = x + 1$.

Solution Because the equation is a first-degree equation, we know that its graph is a straight line. In Example 1, we graphed four points (one more than we really need) that lie on the line. We must now connect those points with a straight line, and we must put arrowheads at each end of the line (see Figure 9.3.2).

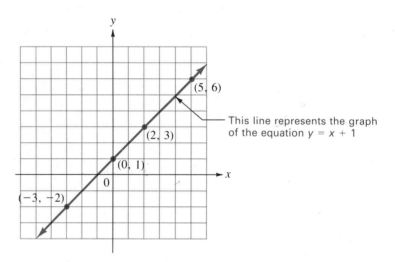

FIGURE 9.3.2.   ■

Intercepts   Students often ask which points to use to plot the graph of a straight line. We often start by choosing zeros for the x- and y-values. The points found by this method are called the x-intercept and the y-intercept.

The **x-intercept** of a line is the point where its graph meets the x-axis. Because the y-value of *every* point on the x-axis is zero, the y-coordinate of the x-intercept is zero. Therefore, to find the x-intercept, we let y equal zero and solve for x (Figure 9.3.3).

The **y-intercept** of a line is the point where its graph meets the y-axis. Because the x-value of *every* point on the y-axis is zero, the x-coordinate of the y-intercept is zero. Therefore, to find the y-intercept, we let x equal zero and solve for y (Figure 9.3.3).

Example 3   Graph $3x - 2y = -6$.

Solution We first let x equal zero and let y equal zero.

x	y
0	
	0

y-intercept: If $x =$ 0 , $3(0) - 2y = -6$

$$-2y = -6$$

$$y = \ 3$$

x	y
0	3
	0

Therefore, the *y*-intercept is (0, 3). (Because the *x*-coordinate of the *y*-intercept is *always* zero, we sometimes say, "The *y*-intercept is 3," naming only the *y*-coordinate of the point rather than the point itself.)

x-intercept: If $y =$ 0 , $3x - 2(0) = -6$

$$3x = -6$$

$$x = \ -2$$

x	y
0	3
−2	0

Therefore, the *x*-intercept is (−2, 0). (Because the *y*-coordinate of the *x*-intercept is *always* zero, we sometimes say "The *x*-intercept is −2," naming only the *x*-coordinate of the point rather than the point itself.)

Checkpoint: If $x =$ 2 , $3(2) - 2y = -6$

$$6 - 2y = -6$$

$$\underline{-6 \qquad \quad -6}$$

$$-2y = -12$$

$$y = 6$$

x	y
0	3
−2	0
2	6

Therefore, this checkpoint is (2, 6).

We graph these three points, draw the straight line through them, and put arrowheads at each end of the line (Figure 9.3.3).

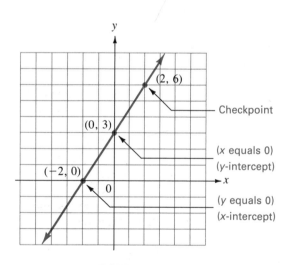

FIGURE 9.3.3

NOTE   Some students find it helpful to first solve the equation for *y* (using the method given in Section 8.10 for solving literal equations) and then let *x* have three different values. In Example 3, the equation becomes $y = \frac{3}{2}x + 3$ (verify this!), and the table of values remains unchanged.

447

Example 4   Graph $4x + 3y = 12$.
Solution

x-intercept: If $y = \boxed{0}$, $4x + 3(0) = 12$

$$4x = 12$$

$$x = \boxed{3}$$

x	y
3	0
0	4
6	−4

Therefore, the *x*-intercept is (3, 0).

y-intercept: If $x = \boxed{0}$, $4(0) + 3y = 12$

$$3y = 12$$

$$y = \boxed{4}$$

Therefore, the *y*-intercept is (0, 4).

Checkpoint: If $x = \boxed{6}$, $4(6) + 3y = 12$

$$24 + 3y = 12$$
$$\underline{-24 \qquad\qquad -24}$$
$$3y = -12$$

$$y = \boxed{-4}$$

Therefore, this checkpoint is (6, −4).

NOTE   For the checkpoint, if we let *x* be some multiple of the coefficient of *y* (in this case, we let *x* be a multiple of 3), we often eliminate fractions. You should verify that if we had let $x = 4$ (4 is *not* a multiple of the coefficient of *y*), we would have had $y = -\frac{4}{3}$. While it is possible to graph the point $\left(4, -\frac{4}{3}\right)$, it is *easier* to graph (6, −4). ☑

We plot the *x*-intercept (3, 0), the *y*-intercept (0, 4), and the checkpoint (6, −4); then we draw the straight line through these points and put arrowheads at each end of the line (Figure 9.3.4).

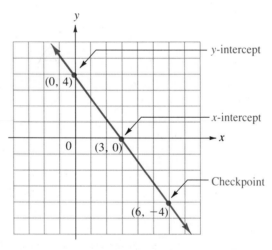

FIGURE 9.3.4 ■

Example 5   Graph $3x - 4y = 0$.
Solution

x-intercept: If $y = \boxed{0}$, $3x - 4(0) = 0$

$$3x = 0$$

$$x = \boxed{0}$$

Therefore, the *x*-intercept is (0, 0). Since the line goes through the origin, the *y*-intercept is also (0, 0).

We have found only one point on the line: (0, 0). Therefore, we must find another point on the line. To find another point, we must set either variable equal to a number and then solve the equation for the other variable. For example,

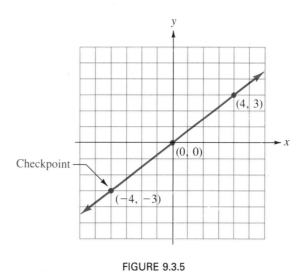

$$\text{if } y = \boxed{3}, \; 3x - 4(3) = 0$$

$$3x - 12 = 0$$

$$\underline{12 \quad 12}$$

$$3x \quad = 12$$

$$x = \boxed{4}$$

x	*y*
0	0
4	3
−4	−3

This gives the point (4, 3) on the line.

Checkpoint: If $x = \boxed{-4}$, $3(-4) - 4y = 0$

$$-12 - 4y = 0$$

$$\underline{12 \qquad\quad 12}$$

$$-4y = 12$$

$$y = \boxed{-3}$$

We graph the points (0, 0), (4, 3), and (−4, −3); then we draw the straight line through them and put arrowheads at each end of the line (Figure 9.3.5).

FIGURE 9.3.5

NOTE Sometimes the *x*- and *y*-intercepts are so close together that drawing the line through them accurately would be very difficult. In this case, we find another point on the line far enough away from the intercepts that an accurate line can be drawn easily. To find the other point, we set either variable equal to a number and then solve the equation for the other variable. ☑

In Example 6, we find three points on the line by letting *x* have *any* three values.

Example 6   Graph $3x - 5y = 15$.

Solution We substitute three values for x, and find the corresponding values for y.

Equation: $3x - 5y = 15$

If $x = \boxed{0}$, $3(0) - 5y = 15$

$$-5y = 15$$

$$y = \boxed{-3}$$

If $x = \boxed{2}$, $3(2) - 5y = 15$

$$6 - 5y = 15$$
$$\underline{-6 \qquad -6}$$
$$-5y = 9$$

$$y = -\frac{9}{5} = \boxed{-1\frac{4}{5}}$$

If $x = \boxed{5}$, $3(5) - 5y = 15$

$$15 - 5y = 15$$
$$\underline{-15 \qquad -15}$$
$$-5y = 0$$

$$y = \boxed{0}$$

x	y
0	-3
2	$-1\frac{4}{5}$
5	0

Now we graph the three points from the table of values, draw a straight line through them, and put arrowheads at each end of the line (Figure 9.3.6).

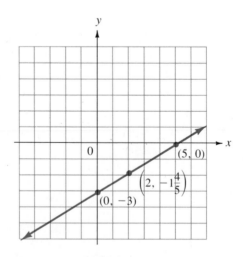

FIGURE 9.3.6 ■

Some equations of a line have only one variable. The graphs of such equations are either vertical or horizontal lines (see Examples 7 and 8).

Example 7   Graph $x = 3$.

Solution The equation $x = 3$ is equivalent to $0y + x = 3$.

If $y = \boxed{0}$, $0(0) + x = 3$

$0 + x = 3$

$x = \boxed{3}$

If $y = \boxed{5}$, $0(5) + x = 3$

$0 + x = 3$

$x = \boxed{3}$

If $y = \boxed{-2}$, $0(-2) + x = 3$

$0 + x = 3$

$x = \boxed{3}$

x	y
3	0
3	5
3	-2

We can see that no matter what value y has in this equation, x is always 3. Therefore, all the points with an x-value of 3 lie on a vertical line whose x-intercept is (3, 0) (Figure 9.3.7).

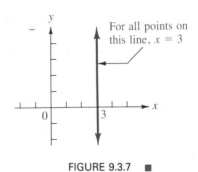

For all points on this line, $x = 3$

FIGURE 9.3.7 ■

The graph of $x = a$, where a is any real number, is always a *vertical line*.

Example 8   Graph $y + 4 = 0$.
Solution

$$y + 4 = 0$$
$$\underline{-4 \quad -4}$$
$$y = -4$$

In the equation $y + 4 = 0$, no matter what value x has, y is always -4. Therefore, all the points with a y-value of -4 lie on a horizontal line whose y-intercept is (0, -4) (Figure 9.3.8).

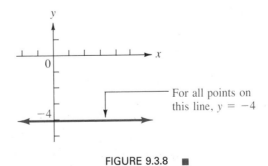

For all points on this line, $y = -4$

FIGURE 9.3.8 ■

The graph of $y = b$, where b is any real number, is always a *horizontal line*.
The methods used in these examples are summarized below.

TO GRAPH A STRAIGHT LINE

Method 1 (Intercept method)	*Method 2*
1. Find, if it exists, the x-intercept: Set $y = 0$; then solve for x. **2.** Find, if it exists, the y-intercept: Set $x = 0$; then solve for y. **3.** If both intercepts are (0, 0), an additional point must be found. **4.** Find one other point by letting x or y have any value. **5.** Draw a straight line through the three points and draw arrowheads at each end of the line.	**1.** In the equation, substitute a number for one of the variables and solve for the corresponding value of the other variable. Find *three* points in this way, and list them in a table of values. **2.** Graph the points listed in the table of values. **3.** Draw a straight line through the points and draw arrowheads at each end of the line.

NOTE   Before you use step 3 of method 1 or step 1 of method 2, you may find it convenient to solve the equation for one of the variables. ☑

Special cases:

The graph of $x = a$ is a vertical line that passes through $(a, 0)$.

The graph of $y = b$ is a horizontal line that passes through $(0, b)$.

EXERCISES 9.3

Set I   In Exercises 1–26, graph the line whose equation is given.

1. $x + y = 3$ **2.** $x + y = 5$ **3.** $2x - 3y = 6$

4. $3x - 4y = 12$ **5.** $y = 8$ **6.** $y = -3$

7. $x = -2$ **8.** $x = 9$ **9.** $x + 5 = 0$

10. $x + 1 = 0$ **11.** $y + 5 = 0$ **12.** $y + 2 = 0$

13. $3x = 5y + 15$ **14.** $2x = 5y + 10$ **15.** $x = -2y$

16. $x = -5y$ **17.** $y = -4x$ **18.** $y = -2x$

19. $4 - y = x$ **20.** $6 - y = x$ **21.** $y = x$

22. $y = -x$ **23.** $y = \frac{1}{2}x + 2$ **24.** $y = \frac{3}{4}x + 1$

25. $3x = 24 + 4y$ **26.** $9x = 18 + 2y$

In Exercises 27 and 28, (a) graph the two lines for each exercise on the same set of axes and (b) name the ordered pair corresponding to the point where the two lines cross.

27. $x - y = 5$ **28.** $3x - 4y = -12$

$\quad\ \ x + y = 1$ $\qquad\ \ 3x + y = 18$

Set II   In Exercises 1–26, graph the line whose equation is given.

1. $x + y = 1$ **2.** $x = \frac{2}{3}y + 4$ **3.** $4x - 3y = 12$

4. $x - y = 5$ **5.** $y = 1$ **6.** $y = 0$

7. $x = -4$ **8.** $x = 0$ **9.** $x + 7 = 0$

10. $x + y = -3$ **11.** $y + 8 = 0$ **12.** $3 - y = 2x$

13. $2x = 4y + 8$ **14.** $5y = x - 5$ **15.** $x = -3y$

16. $x + y = 0$ **17.** $y = -3x$ **18.** $x - y = 0$

19. $2 - y = x$ **20.** $5x - y = 0$ **21.** $y = x + 3$

22. $y = x + 5$ **23.** $y = \frac{2}{3}x + 1$ **24.** $y = \frac{2}{3}x - 2$

25. $y = \frac{2}{3}x$ **26.** $7x = 3y + 21$

In Exercises 27 and 28, (a) graph the two lines for each exercise on the same set of axes and (b) name the ordered pair corresponding to the point where the two lines cross.

27. $x - 2y = 4$ **28.** $2x + 3y = 12$
     $x + 2y = 2$ $2x = 3y$

9.4 The Slope of a Line

Subscripts   A small number written below and to the right of a variable is used to indicate a *particular value* of that variable. It is called a **subscript**. For example, 1 is a subscript in the following expression.

$$x_1$$

x_1 is read "x sub one."

x_2 ◄— A different subscript indicates a different *value* of that variable

x_2 is read "x sub two."

Example 1   Examples of subscripted variables:

a. y_1 and y_2 represent two different values of y.

b. P_1 and P_2 represent two different values of P.

c. $P_1(x_1, y_1)$ and $P_2(x_2, y_2)$ represent two different *points*. ■

The Slope of a Line
If we imagine a line as representing a hill, then the **slope** of the line is a measure of the steepness of the hill. To measure the slope of a line, we choose any two points on the line, $P_1(x_1, y_1)$ and $P_2(x_2, y_2)$ (Figure 9.4.1).

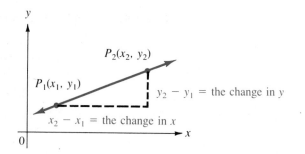

FIGURE 9.4.1

The letter m is used to represent the slope of a line. The slope is defined as follows:

THE SLOPE OF A LINE

If $P_1(x_1, y_1)$ and $P_2(x_2, y_2)$ are any two points on a nonvertical line, and if m represents the slope of the line, then

$$m = \frac{y_2 - y_1}{x_2 - x_1}$$

$$\text{or} \quad m = \frac{y_1 - y_2}{x_1 - x_2}$$

A WORD OF CAUTION   It is *incorrect* to write $\dfrac{x_2 - x_1}{y_2 - y_1}$, $\dfrac{x_1 - x_2}{y_1 - y_2}$, $\dfrac{y_2 - y_1}{x_1 - x_2}$, or

$\dfrac{y_1 - y_2}{x_2 - x_1}$ as the slope of a line.   ☑

Example 2   Find the slope of the line through the points $(-2, -4)$ and $(6, 0)$ (see Figure 9.4.2).
Solution

Let $P_1 = (-2, -4)$

and $P_2 = (6, 0)$

That is, $x_1 = -2$, $y_1 = -4$,
$x_2 = 6$, and $y_2 = 0$. Therefore,

$$m = \frac{y_2 - y_1}{x_2 - x_1}$$

$$= \frac{0 - (-4)}{6 - (-2)} = \frac{4}{8} = \frac{1}{2}$$

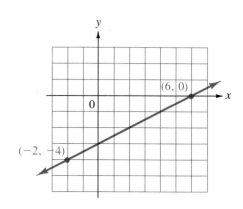

FIGURE 9.4.2

The slope is not changed if the points P_1 and P_2 are interchanged or if we use the equation

$$m = \frac{y_1 - y_2}{x_1 - x_2}$$

$$= \frac{-4 - 0}{-2 - 6} = \frac{-4}{-8} = \frac{1}{2} \quad \blacksquare$$

Notice that in Example 2 a point moving along the line in the positive x-direction *rises*, and the slope of the line is *positive*.

Example 3 Find the slope of the line through the points $A(-1, 3)$ and $B(2, -3)$ (see Figure 9.4.3).
Solution

$$m = \frac{y_2 - y_1}{x_2 - x_1}$$

$$= \frac{-3 - (3)}{2 - (-1)} = \frac{-6}{3} = -2$$

or

$$m = \frac{y_1 - y_2}{x_1 - x_2}$$

$$= \frac{3 - (-3)}{-1 - 2} = \frac{6}{-3} = -2$$

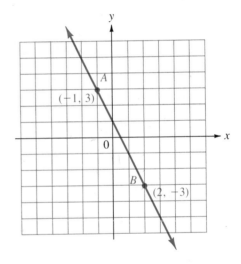

FIGURE 9.4.3 ■

Notice that in Example 3 a point moving along the line in the positive x-direction *falls*, and the slope of the line is *negative*.

Example 4 Find the slope of the line through the points $E(-4, -3)$ and $F(2, -3)$ (see Figure 9.4.4).
Solution

$$m = \frac{y_2 - y_1}{x_2 - x_1}$$

$$= \frac{(-3) - (-3)}{(2) - (-4)} = \frac{0}{6} = 0$$

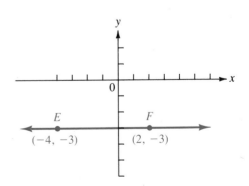

Notice that the line is horizontal. Whenever a line is horizontal, its slope is zero, and whenever the slope of a line is zero, the line is horizontal.

FIGURE 9.4.4 ■

Example 5 Find the slope of the line through the points $R(4, 5)$ and $S(4, -2)$ (see Figure 9.4.5).
Solution

$$m = \frac{y_2 - y_1}{x_2 - x_1}$$

$$= \frac{(-2) - (5)}{(4) - (4)} = \frac{-7}{0}$$

Not defined ⤴

Note that the line is vertical. Whenever a line is vertical, its slope does
not exist, and whenever the slope of a
line does not exist, the line is vertical.

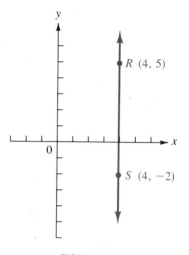

FIGURE 9.4.5 ∎

The facts about the slope of a line are summarized as follows:

> *The slope of a line is positive* if a point moving along the line in the positive
> x-direction (to the right) rises (Figure 9.4.2).
>
> *The slope of a line is negative* if a point moving along the line in the positive
> x-direction falls (Figure 9.4.3).
>
> *The slope is zero* if the line is horizontal (Figure 9.4.4).
>
> *The slope does not exist* if the line is vertical (Figure 9.4.5).

Given the equation of a line, we can find its slope by finding the coordinates of any
two points on the line (see Example 6).

Example 6 For each of the following equations, graph the line and find its slope.

a. $y = 3x$

Solution

x	y	
-2	-6	Point A
0	0	Point B
1	3	Point C

For the slope, if we use points A and B, we have

$$m = \frac{-6 - (0)}{-2 - (0)} = \frac{-6}{-2} = 3$$

If we use points B and C, we have

$$m = \frac{0 - (3)}{0 - (1)} = \frac{-3}{-1} = 3$$

b. $y = 3x + 2$
   Solution

x	y	
-2	-4	Point D
0	2	Point E
1	5	Point F

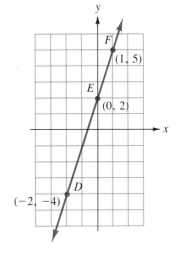

For the slope, if we use points D and E, we have

$$m = \frac{-4 - (2)}{-2 - (0)} = \frac{-6}{-2} = 3$$

If we use points E and F, we have

$$m = \frac{2 - (5)}{0 - (1)} = \frac{-3}{-1} = 3$$

c. $y = -\frac{1}{3}x$
   Solution

x	y	
-3	1	Point G
0	0	Point H
1	$-\frac{1}{3}$	Point I

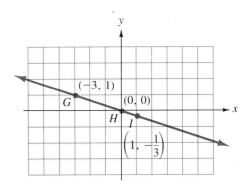

For the slope, if we use points G and H, we have

$$m = \frac{1 - (0)}{-3 - (0)} = \frac{1}{-3} = -\frac{1}{3}$$

If we use points H and I, we have

$$m = \frac{0 - \left(-\frac{1}{3}\right)}{0 - (1)} = \frac{\frac{1}{3}}{-1} = -\frac{1}{3}$$

You can verify that, for each of the lines, if any other points on the lines are used, the slope remains unchanged.

d. Graph the lines from Examples 6a and 6b on the same axes. Note that the slopes of both lines were 3.

Solution

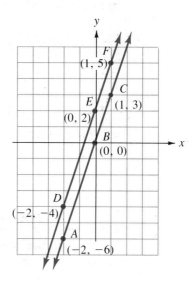

e. Graph the lines from Examples 6a and 6c on the same axes. Note that the slope of the line from Example 6a was 3 and the slope of the line from Example 6c was $-\frac{1}{3}$.

Solution

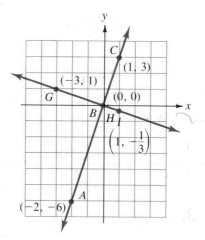

Parallel Lines   Two or more lines in the same plane are **parallel** if they will not meet even if they are extended indefinitely.

Perpendicular Lines   Two lines are **perpendicular** if they intersect in a right (90°) angle.

The following rules about the slopes of parallel and perpendicular lines are stated here without proof.

Given two nonvertical lines, let the slope of one line be m_1 and the slope of the other line be m_2. Then

If the two nonvertical lines are parallel, they have the same slope; that is, $m_1 = m_2$.

If $m_1 = m_2$, the two lines are parallel.

If the two nonvertical lines are perpendicular, $m_1 m_2 = -1$, or $m_1 = -\dfrac{1}{m_2}$.

If $m_1 m_2 = -1$, the two lines are perpendicular.

A vertical line is parallel to any other vertical line and perpendicular to any horizontal line.

Notice that in Examples 6a and 6b the lines were *parallel* to each other and the *slopes* of the two lines were equal. Notice also that the lines in Examples 6a and 6c appear to be perpendicular to each other and the *product* of their slopes is $(3)\left(-\frac{1}{3}\right) = -1$.

EXERCISES 9.4

Set I   In Exercises 1–14, find the slope of the line through the given pair of points.

1. $(-2, 1)$ and $(7, 3)$

2. $(-1, -3)$ and $(5, 2)$

3. $(-1, -1)$ and $(5, -3)$

4. $(-2, -1)$ and $(6, -5)$

5. $(-5, -3)$ and $(4, -3)$

6. $(-2, -4)$ and $(3, -4)$

7. $(-7, 5)$ and $(8, -3)$

8. $(12, -9)$ and $(-5, -4)$

9. $(-6, -15)$ and $(4, -5)$

10. $(-7, 8)$ and $(-4, 5)$

11. $(-2, 5)$ and $(-2, 8)$

12. $(6, -4)$ and $(6, -11)$

13. $(-4, 3)$ and $(-4, -2)$

14. $(-5, 2)$ and $(-5, 7)$

In Exercises 15–18, graph each set of lines on the same axes and find the slope of each line. Verify that the three lines in each set appear to be parallel to each other.

15. a. $y = \frac{1}{2}x$ b. $y = \frac{1}{2}x + 3$ c. $y = \frac{1}{2}x - 2$

16. a. $y = 2x$ b. $y = 2x + 1$ c. $y = 2x - 3$

17. a. $y = -3x$ b. $y = -3x + 4$ c. $y = -3x - 1$

18. a. $y = -\frac{3}{4}x$ b. $y = -\frac{3}{4}x + 2$ c. $y = -\frac{3}{4}x - 3$

In Exercises 19 and 20, graph each pair of lines on the same axes and find the slope of each line. Verify that the pairs of lines appear to be perpendicular to each other.

19. a. $y = -\frac{1}{3}x + 2$ b. $y = 3x + 2$

20. a. $y = \frac{1}{2}x - 3$ b. $y = -2x - 3$

Set II   In Exercises 1–14, find the slope of the line through the given pair of points.

1. $(-3, 5)$ and $(2, -1)$ **2.** $(4, 0)$ and $(7, 0)$

3. $(-3, 2)$ and $(6, 4)$ **4.** $(-2, -2)$ and $(4, -1)$

5. $(-4, -3)$ and $(6, -3)$ **6.** $(-10, 2)$ and $(-4, -8)$

7. $(-3, 8)$ and $(9, 12)$ **8.** $(-7, 3)$ and $(-7, -2)$

9. $(4, -2)$ and $(4, 3)$ **10.** $(3, -1)$ and $(0, 0)$

11. $(5, -2)$ and $(1, 1)$ **12.** $(-3, -1)$ and $(2, -1)$

13. $(-3, 7)$ and $(-3, -5)$ **14.** $(8, -2)$ and $(-2, 8)$

In Exercises 15–18, graph each set of lines on the same axes and find the slope of each line. Verify that the three lines in each set appear to be parallel to each other.

15. a. $y = \frac{1}{4}x$ b. $y = \frac{1}{4}x + 2$ c. $y = \frac{1}{4}x - 3$

16. a. $y = -4x$ b. $y = -4x + 3$ c. $y = -4x - 1$

17. a. $y = -x$ b. $y = -x + 2$ c. $y = -x - 3$

18. a. $y = -\frac{2}{3}x$ b. $y = -\frac{2}{3}x + 4$ c. $y = -\frac{2}{3}x - 2$

In Exercises 19 and 20, graph each pair of lines on the same axes and find the slope of each line. Verify that the pairs of lines appear to be perpendicular to each other.

19. a. $y = -\frac{1}{4}x + 3$ b. $y = 4x + 3$

20. a. $y = \frac{2}{3}x - 1$ b. $y = -\frac{3}{2}x - 1$

9.5 Deriving Equations of Straight Lines

In Section 9.3, we discussed the graph of a straight line. In this section, we show how to write the *equation* of a line when certain facts about the line are known.

Three forms of the equation of a straight line are especially useful: the *general form*, the *point-slope form*, and the *slope-intercept form*.

General Form of the Equation of a Line

The general form of the equation of a line, *as used in this text*, is defined as follows:

GENERAL FORM OF THE EQUATION OF A LINE

$$Ax + By + C = 0$$

where A, B, and C are integers, $A \geq 0$, and A and B are not both 0.

A is the coefficient of x, and the x-term is written first on the left side of the equation; B is the coefficient of y, and the y-term is written next; C is the constant, and it is the last term on the left side. Zero must be the only term on the right side of the equation.

To write the equation of a straight line in the *general form*, multiply both sides of the equation by the LCD and move all terms to the left side of the equal sign. If A is then negative, multiply both sides of the equation by -1.

Example 1   Write $-\frac{2}{3}x + \frac{1}{2}y = 1$ in the general form.
Solution LCD $= 6$

$$\frac{6}{1}\left(-\frac{2}{3}x\right) + \frac{6}{1}\left(\frac{1}{2}y\right) = \frac{6}{1}\left(\frac{1}{1}\right)$$

$$-4x + 3y \quad = \quad 6$$
$$\underline{\qquad\qquad - 6 \quad -6}$$
$$-4x + 3y - 6 = \quad 0$$
$$4x - 3y + 6 = 0 \qquad \text{General form} \quad \blacksquare$$

Point-Slope Form of the Equation of a Line

Let $P_1(x_1, y_1)$ be a known point on a line whose slope is m. Let $P(x, y)$ represent *any* other point on that line. Then, using the definition of slope and simplifying the equation, we have

$$m = \frac{y - y_1}{x - x_1}$$

$$y - y_1 = m(x - x_1)$$

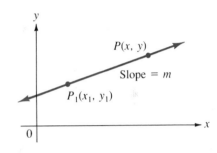

$A = m$ when $y = 1$

POINT-SLOPE FORM OF THE EQUATION OF A LINE

$$y - y_1 = m(x - x_1)$$

where $m =$ the slope of the line and $P_1(x_1, y_1)$ is a known point on the line.

When we are given the *slope* of a line and a *point* through which it passes, we use the **point-slope form** of the equation of a line to write the equation of the line (see Examples 2 and 3).

Example 2   Write the general form of the equation of the line that passes through $(2, -3)$ and has a slope of 4.
Solution Into the point-slope form of the equation of a line, we substitute 2 for x_1, -3 for y_1, and 4 for m.

$$y - y_1 = m(x - x_1) \qquad \text{Point-slope form}$$

$$y - (-3) = 4(x - 2) \qquad \text{Substituting}$$

$$y + 3 = 4x - 8 \qquad \text{Simplifying}$$
$$\underline{-4x \qquad + 8 \quad -4x + 8} \qquad \text{Adding } -4x + 8 \text{ to both sides}$$
$$-4x + y + 11 = 0$$

$$4x - y - 11 = 0 \qquad \text{General form} \quad \blacksquare$$

461

Example 3   Write the general form of the equation of the line that passes through $(-1, 4)$ and has a slope of $-\frac{2}{3}$.

Solution

$$y - y_1 = m(x - x_1) \qquad \text{Point-slope form}$$

$$y - 4 = -\frac{2}{3}[x - (-1)] \qquad \text{Substituting}$$

$$3y - 12 = -2(x + 1) \qquad \text{Multiplying both sides by 3}$$

$$3y - 12 = -2x - 2 \qquad \text{Using the distributive property}$$
$$\underline{2x \qquad + 2 \qquad 2x + 2} \qquad \text{Adding } 2x + 2 \text{ to both sides}$$
$$2x + 3y - 10 = \quad 0 \qquad \text{General form} \quad \blacksquare$$

Slope-Intercept Form of the Equation of a Line

Let $(0, b)$ be the *y-intercept* of a line whose slope is m. Then

$$y - y_1 = m(x - x_1)$$

$$y - b = m(x - 0)$$

$$y - b = mx$$
$$\underline{+ b \qquad\qquad + b}$$
$$y = m\ x + b$$

Slope $\rightarrow$ $\qquad$ $\llcorner$ *y*-intercept

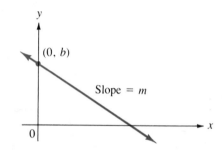

$(0, b)$

Slope $= m$

MEM + DISTRIBUTE
AX+C=BY

┌───┐
│ SLOPE-INTERCEPT FORM OF THE EQUATION OF A LINE │
│   │
│ $$y = mx + b$$ │
│   │
│ where $m =$ the slope of the line and $b =$ the *y*-intercept of the line.* │
└───┘

When we are given the *slope* of a line and its *y-intercept* and are asked to write the equation of the line, we can use the **slope-intercept form** of the equation (see Example 4). The point-slope form could also be used.

Example 4   Write the general form of the equation of the line that has a slope of $-\frac{1}{2}$ and a *y*-intercept of -3.

Solution Into the slope-intercept form, we substitute $-\frac{1}{2}$ for m and -3 for b.

$$y = mx + b \qquad \text{Slope-intercept form}$$

$$y = -\frac{1}{2}x + (-3) \qquad \text{Substituting}$$

$$2y = 2\left(-\frac{1}{2}x - 3\right) \qquad \text{Multiplying both sides by 2}$$

$$2y \qquad = -x - 6 \qquad \text{Using the distributive property}$$
$$\underline{x \qquad + 6 \qquad x + 6} \qquad \text{Adding } x + 6 \text{ to both sides}$$
$$x + 2y + 6 = 0 \qquad \text{General form} \quad \blacksquare$$

---

*Of course, when we say that the *y*-intercept is b, we really mean that the *y*-intercept is the point $(0, b)$.

We sometimes need to change an equation into the slope-intercept form. To do this, we use the method given in Section 8.10 to solve the equation for y.

Example 5 Write $2x + 3y + 6 = 0$ in the slope-intercept form, find the slope of the line, and find the y-intercept of the line.

Solution We solve the given equation for y.

$$2x + 3y + 6 = 0 \qquad \text{General form}$$
$$\underline{-2x \qquad -6 \quad -2x - 6} \qquad \text{Adding } -2x - 6 \text{ to both sides}$$
$$3y \qquad = -2x - 6$$
$$\frac{3y}{3} = \frac{-2x - 6}{3} \qquad \text{Dividing both sides by 3}$$
$$y = -\frac{2}{3}x - 2 \qquad \text{Slope-intercept form}$$

Slope $\uparrow$ $\qquad$ $\uparrow$ y-intercept

The slope is $-\frac{2}{3}$, and the y-intercept is -2. ∎

Example 6 Write (a) the equation of the line that passes through the point $(-1, 3)$ and is *parallel* to the line $2x + 3y + 6 = 0$; (b) the equation of the line that passes through the point $(-1, 3)$ and is *perpendicular* to the line $2x + 3y + 6 = 0$.

Solution We found in Example 5 that the slope of the line $2x + 3y + 6 = 0$ is $-\frac{2}{3}$.

a. We must write the equation of the line that passes through $(-1, 3)$ and whose slope *equals* $-\frac{2}{3}$, since the lines are to be parallel.

$$y - y_1 = m(x - x_1)$$
$$y - 3 = -\frac{2}{3}(x - [-1]) \qquad \text{Substituting 3 for } y_1, -\frac{2}{3} \text{ for } m, \text{ and } -1 \text{ for } x_1$$
$$3y - 9 = -2(x + 1) \qquad \text{Multiplying both sides by 3}$$
$$3y - 9 = -2x - 2 \qquad \text{Using the distributive property}$$
$$\underline{2x \qquad +2 \qquad 2x + 2} \qquad \text{Adding } 2x + 2 \text{ to both sides}$$
$$2x + 3y - 7 = 0 \qquad \text{General form}$$

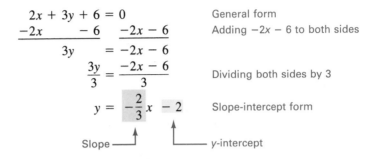

b. We must write the equation of the line that passes through $(-1, 3)$ and whose slope is $\frac{3}{2}$ $\left(\text{the } \textit{negative reciprocal} \text{ of } -\frac{2}{3}\right)$, since the lines are to be perpendicular.

$$y - y_1 = m(x - x_1)$$
$$y - 3 = \frac{3}{2}(x - [-1]) \qquad \text{Substituting}$$
$$2y - 6 = 3(x + 1) \qquad \text{Multiplying both sides by 2}$$
$$2y - 6 = 3x \qquad +3 \qquad \text{Using the distributive property}$$
$$\underline{-2y + 6 \qquad -2y + 6} \qquad \text{Adding } -2y + 6 \text{ to both sides}$$
$$0 = 3x - 2y + 9$$
$$3x - 2y + 9 = 0 \qquad \text{General form} ∎$$

Writing the Equation of a Line When Two Points on It Are Known

In Example 7, we show how to write the equation of a line if we're given two points on the line.

Example 7 Find the equation of the line passing through the points $(-5, 3)$ and $(-15, -9)$.

Solution

1. We find the slope from the two given points.

$$m = \frac{(-9) - (3)}{(-15) - (-5)} = \frac{-12}{-10} = \frac{6}{5}$$

2. We use this slope with *either* given point to find the equation of the line.

Using the point $(-5, 3)$	*Using the point* $(-15, -9)$
$y - y_1 = m(x - x_1)$	$y - y_1 = m(x - x_1)$
$y - 3 = \frac{6}{5}[x - (-5)]$	$y - (-9) = \frac{6}{5}[x - (-15)]$
$5y - 15 = 6(x + 5)$	$5(y + 9) = 6(x + 15)$
$5y - 15 = 6x \qquad + 30$	$5y + 45 = 6x \qquad + 90$
$\underline{-5y + 15 \qquad -5y + 15}$	$\underline{-5y - 45 \qquad -5y - 45}$
$0 = 6x - 5y + 45$	$0 = 6x - 5y + 45$

This result shows that the same equation is obtained no matter which of the two given points is used. The general form is $6x - 5y + 45 = 0$. ∎

The Equation of a Horizontal Line

Because a horizontal line has a slope, its equation can be written using the slope-intercept form or the point-slope form, *or* the following statement can be memorized:

> The equation of a horizontal line is of the form $y = b$ or $y - b = 0$, where b is the y-coordinate of every point on the line.

Example 8 Find the equation of the horizontal line that passes through $(-2, 4)$.

Solution

Method 1 The slope of every horizontal line is 0. Therefore,

$$y - 4 = 0(x - [-2]) \qquad \text{Point-slope form}$$
$$y - 4 = 0 \qquad \text{General form}$$

Method 2 The y-coordinate of the given point is 4. Therefore, the equation of the line must be

$$y \quad = \quad 4$$
$$\underline{-4 \qquad -4} \qquad \text{Adding } -4 \text{ to both sides}$$
or $\qquad y - 4 = \quad 0 \qquad \text{General form} \quad ∎$

The Equation of a Vertical Line

Because a vertical line has no slope, we cannot use the point-slope or the slope-intercept form when we write its equation. The only way to write the equation of a vertical line is to memorize the following statement:

> The equation of a vertical line is of the form $x = a$ or $x - a = 0$, where a is the x-coordinate of every point on the line.

Example 9 Write the equation of the vertical line that passes through $(-2, 4)$. *any #*

Solution Because the x-coordinate of the given point is -2, the equation of the line must be

$$
\begin{aligned}
x &= -2 \\
\underline{ 2} & \underline{2} \qquad \text{Adding 2 to both sides} \\
x + 2 &= 0 \qquad \text{General form} \quad \blacksquare
\end{aligned}
$$

We summarize here the important forms for the equation of a line:

FORMS OF THE EQUATION OF A LINE

The general form:	$Ax + By + C = 0,$	where A, B, and C are integers, $A \geq 0$, and A and B are not both 0.
The point-slope form:	$y - y_1 = m(x - x_1),$	where m is the slope and $P_1(x_1, y_1)$ is a known point on the line.
The slope-intercept form:	$y = mx + b,$	where m is the slope and b is the y-intercept.
The equation of a horizontal line:	$y = b,$	where b is the y-coordinate of every point on the line.
The equation of a vertical line:	$x = a,$	where a is the x-coordinate of every point on the line.

EXERCISES 9.5

Set I In Exercises 1–6, write each equation in the general form.

1. $3x = 2y - 4$

2. $2x = 3y + 7$

3. $y = -\frac{3}{4}x - 2$

4. $y = -\frac{3}{5}x - 4$

5. $2(3x + y) = 5(x - y) + 4$

6. $3(2x - y) = 2(x + 3y) - 5$

In Exercises 7–10, write the general form of the equation of the line through the given point and with the indicated slope.

7. $(3, 4)$, $m = \frac{1}{2}$

8. $(5, 6)$, $m = \frac{1}{3}$

9. $(-1, -2)$, $m = -\frac{2}{3}$

10. $(-2, -3)$, $m = -\frac{5}{4}$

check

In Exercises 11–14, write the general form of the equation of the line with the indicated slope and y-intercept.

11. $m = \frac{3}{4}$, y-intercept $= -3$ **12.** $m = \frac{2}{7}$, y-intercept $= -2$

13. $m = -\frac{2}{5}$, y-intercept $= \frac{1}{2}$ **14.** $m = -\frac{5}{3}$, y-intercept $= \frac{3}{4}$

In Exercises 15–18, find the general form of the equation of the line that passes through the given points.

Find Slope
then geral form

15. $(4, -1)$ and $(2, 4)$ **16.** $(5, -2)$ and $(3, 1)$

17. $(0, 0)$ and $(3, 4)$ **18.** $(0, 0)$ and $(-2, -5)$

In Exercises 19–26, (a) write the given equation in the slope-intercept form, (b) give the slope of the line, (c) give the y-intercept of the line, (d) write the equation of the line that passes through the point $(2, -4)$ and is parallel to the given line, and (e) write the equation of the line that passes through the point $(-3, 5)$ and is perpendicular to the given line.

19. $3x + 4y + 12 = 0$ **20.** $2x + 5y + 10 = 0$

21. $2x + 3y - 9 = 0$ **22.** $2x + 5y - 15 = 0$

23. $5x - 3y = 30$ **24.** $4x - 7y = 28$

25. $x + 2y - 6 = 0$ **26.** $x + 5y - 5 = 0$

27. Write the equation of the horizontal line that passes through the point $(-3, 5)$.

28. Write the equation of the horizontal line that passes through the point $(-2, -3)$.

29. Write the equation of the vertical line that passes through the point $(-3, 5)$.

30. Write the equation of the vertical line that passes through the point $(-2, -3)$.

Parallel, slope stay the same
Perpendicular, invert Slope & Sign

Set II   In Exercises 1–6, write each equation in the general form.

1. $5x = 4y - 7$ **2.** $3 - 2y = 5x$

3. $y = -\frac{4}{5}x - 3$ **4.** $\frac{2}{3}x = 3y - \frac{1}{6}$

5. $3(2x - y) = 4(x + y) - 6$ **6.** $x = 4$

In Exercises 7–10, write the general form of the equation of the line through the given point and with the indicated slope.

7. $(-6, 3)$, $m = \frac{1}{2}$ **8.** $(5, -7)$, $m = -\frac{3}{4}$

9. $(3, -5)$, $m = -\frac{3}{5}$ **10.** $(0, 0)$, $m = 0$

In Exercises 11–14, write the general form of the equation of the line with the indicated slope and y-intercept.

11. $m = -\frac{2}{3}$, y-intercept $= -4$ **12.** $m = \frac{5}{4}$, y-intercept $= -3$

13. $m = -\frac{1}{3}$, y-intercept $= \frac{2}{5}$ **14.** $m = 0$, y-intercept $= 0$

In Exercises 15–18, find the general form of the equation of the line that passes through the given points.

15. $(-3, 4)$ and $(5, -2)$ **16.** $(5, -3)$ and $(-2, -4)$

17. $(3, -5)$ and $(-3, 2)$ **18.** $(4, 6)$ and $(-3, 6)$

In Exercises 19–26, (a) write the given equation in the slope-intercept form, (b) give the slope of the line, (c) give the y-intercept of the line, (d) write the equation of the line that passes through the point $(5, -3)$ and is parallel to the given line, and (e) write the equation of the line that passes through the point $(-2, 7)$ and is perpendicular to the given line.

19. $5x + 3y + 15 = 0$ **20.** $\frac{3}{4}x - 2y - 3 = 0$

21. $4x + 2y - 7 = 0$ **22.** $x = \frac{1}{2}y + 3$

23. $3x - 7y = 21$ **24.** $\frac{1}{2}x - 3y = 6$

25. $x + 3y + 6 = 0$ **26.** $x + 3y - 3 = 0$

27. Write the equation of the horizontal line that passes through the point $(-4, -2)$.

28. Write the equation of the horizontal line that passes through the point $(2, -5)$.

29. Write the equation of the vertical line that passes through the point $(-4, -2)$.

30. Write the equation of the vertical line that passes through the point $(2, -5)$.

9.6 Graphing Curves

We mentioned in Section 9.3 that there are infinitely many points whose coordinates satisfy an equation in one or two variables. This is true whether the equation is a first-degree equation or a higher-degree equation. In this section, we graph equations such as $y = ax^2 + bx + c$ (a second-degree equation) and $y = ax^3 + bx^2 + cx + d$ (a third-degree equation). The graph of such an equation will be a smooth *curve*, not a straight line. The graph must contain, within the limits of the size of the graph, the set of *all* the ordered pairs, and only those ordered pairs, that satisfy the given equation.

The *x-intercepts* of a curve are the points at which the curve crosses the x-axis. The curves graphed in this book sometimes have no x-intercepts, sometimes have one, and sometimes have more than one. We find the x-intercepts by setting y equal to zero and trying to solve the resulting equation for x. (We won't always be able to solve the equation; this doesn't necessarily mean there are no x-intercepts.) If the x-intercepts are $(a, 0)$ and $(c, 0)$, we commonly say, "The x-intercepts are a and c."

The *y-intercepts* of a curve are the points at which the curve intersects the y-axis. We find the y-intercepts by setting x equal to zero and solving the resulting equation for y. The graphs of higher-degree equations *given in this book* will always have exactly one y-intercept.

TO GRAPH AN EQUATION THAT IS IN THE FORM
$$y = ax^2 + bx + c \quad \text{or} \quad y = ax^3 + bx^2 + cx + d$$

In this book, you will be given some of the values to use for x*.

1. Find the y-intercept, and find the x-intercepts if possible.

2. Use the equation to find the y-values that correspond to the other given x-values.

3. Graph the ordered pairs found in steps 1 and 2.

4. Join the points graphed in step 3 *from left to right* with a smooth curve. Use arrowheads to indicate that the graph extends beyond the curve just drawn.

---

*In higher-level courses, you will be given suggestions to help you decide what values to choose for x.

For easy graphing, list the ordered pairs found in steps 1 and 2 in a table of values, *being sure to list the x-values in order of size.*

Example 1   Find the x- and y-intercepts for $y = x^2 - x - 2$. In addition, find the y-values that correspond to x-values of $-2, \frac{1}{2}, 1$, and 3. Graph the curve.
Solution
x-intercepts: To find the x-intercepts, we let $y = 0$ and solve for x.

$$0 = x^2 - x - 2$$
$$0 = (x - 2)(x + 1)$$

$$
\begin{array}{c|c}
x - 2 = 0 & x + 1 = 0 \\
\underline{2\quad 2} & \underline{-1\quad -1} \\
x = 2 & x = -1
\end{array}
$$

Therefore, the x-intercepts are 2 and -1.
y-intercept: To find the y-intercept, we let $x = 0$ and solve for y.

$$y = 0^2 - 0 - 2 = -2$$

Therefore, the y-intercept is -2.

We use the given equation to find the y-values that correspond to x-values of $-2, \frac{1}{2}, 1$, and 3.

Notice that in the table of values the x-values are listed *in order of size.*

If $x = -2$, $y = (-2)^2 - (-2) - 2 = 4 + 2 - 2 = 4$

An x-intercept

The y-intercept

If $x = \frac{1}{2}$, $y = \left(\frac{1}{2}\right)^2 - \left(\frac{1}{2}\right) - 2 = \frac{1}{4} - \frac{1}{2} - 2 = -2\frac{1}{4}$

If $x = 1$, $y = (1)^2 - (1) - 2 = 1 - 1 - 2 = -2$

An x-intercept

If $x = 3$, $y = (3)^2 - (3) - 2 = 9 - 3 - 2 = 4$

x	y
-2	4
-1	0
0	-2
$\frac{1}{2}$	$-2\frac{1}{4}$
1	-2
2	0
3	4

In Figure 9.6.1, we graph these points and draw a smooth curve through them. In drawing the smooth curve, start with the point in the table of values with the smallest x-value.

Draw to the point having the next larger x-value. Continue in this way through all the points. The graph of the equation $y = x^2 - x - 2$ is called a **parabola**.

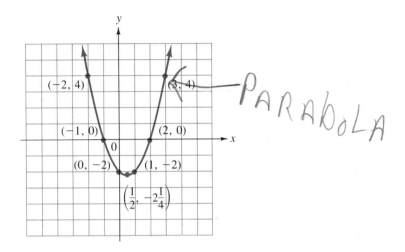

FIGURE 9.6.1 ∎

The graph of any equation that can be expressed in the form $y = ax^2 + bx + c$ is a parabola. A parabola always has the general shape of the graph in Figure 9.6.1, although the parabola might open downward or sideways rather than upward.

Example 2   Try to find the x-intercepts and find the y-intercept for $y = x^2 - 2x - 1$. In addition, find the y-values that correspond to x-values of $-2, -1, 1, 2, 3,$ and 4. Graph the curve.
Solution This is a parabola, because the equation is in the form

$$y = ax^2 + bx + c$$

x-intercepts: We cannot factor $x^2 - 2x - 1$. Therefore, we cannot find the x-intercepts.
y-intercept: If $x = 0$, $y = (0)^2 - 2(0) - 1 = -1$. Therefore, the y-intercept is -1.

We make a table of values, listing the x-values in order of size.

If $x = -2$, $y = (-2)^2 - 2(-2) - 1 = 7$

If $x = -1$, $y = (-1)^2 - 2(-1) - 1 = 2$

The y-intercept

If $x = 1$, $y = (1)^2 - 2(1) - 1 = -2$

If $x = 2$, $y = (2)^2 - 2(2) - 1 = -1$

If $x = 3$, $y = (3)^2 - 2(3) - 1 = 2$

If $x = 4$, $y = (4)^2 - 2(4) - 1 = 7$

x	y
-2	7
-1	2
0	-1
1	-2
2	-1
3	2
4	7

In Figure 9.6.2, we graph these points and draw a smooth curve through them.

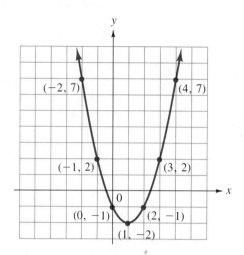

FIGURE 9.6.2

We see from the graph that there *were* x-intercepts even though we couldn't find them algebraically. ■

Example 3

Find the x and y-intercepts for $y = x^3 - 4x$. In addition, find the y-values that correspond to x-values of -3, -1, 1, and 3. Graph the curve.

Solution

x-intercepts: If $y = 0$, we have

$$0 = x^3 - 4x$$

$$0 = x(x^2 - 4)$$

$$0 = x(x - 2)(x + 2)$$

$x = 0$	$x - 2 = 0$	$x + 2 = 0$
	$x = 2$	$x = -2$

Therefore, the x-intercepts are 0, 2, and -2.

y-intercept: If $x = 0$, $y = (0)^3 - 4(0) = 0$. Therefore, the y-intercept is 0.

We make a table of values, listing the x-values in order of size.

x	y
-3	-15
-2	0
-1	3
0	0
1	-3
2	0
3	15

If $x = -3$, $y = (-3)^3 - 4(-3) = -27 + 12 = -15$

An x-intercept

If $x = -1$, $y = (-1)^3 - 4(-1) = -1 + 4 = 3$

The y-intercept *and* an x-intercept

If $x = 1$, $y = (1)^3 - 4(1) = 1 - 4 = -3$

An x-intercept

If $x = 3$, $y = (3)^3 - 4(3) = 27 - 12 = 15$

In Figure 9.6.3, we graph these points and draw a smooth curve through them.

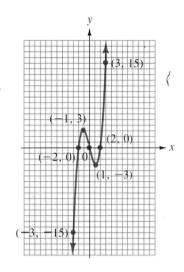

FIGURE 9.6.3 ∎

EXERCISES 9.6

Set I   In Exercises 1–4, complete the table of values for each equation; then draw its graph. Name the x- and y-intercepts.

1. $y = x^2$

x	y
-2	
-1	
0	
1	
2	

2. $y = \dfrac{x^2}{4}$

x	y
-3	
-2	
-1	
0	
1	
2	
3	

3. $y = x^2 - 2x$

x	y
-2	
-1	
0	
1	
2	
3	
4	

4. $y = 3x - x^2$

x	y
-2	
-1	
0	
1	
2	
3	
4	

5. Find the x-intercepts, if possible, for $y = x^2 + 2x - 2$, and find the y-intercept. In addition, find the y-values that correspond to x-values of $-4, -3, -2, -1, 1$, and 2. Graph the curve.

6. Find the x-intercepts, if possible, for $y = x^2 + 2x + 2$, and find the y-intercept. In addition, find the y-values that correspond to x-values of $-4, -3, -2, -1, 1$, and 2. Graph the curve.

7. Find the x-intercepts, if possible, for $y = 2x - x^2$, and find the y-intercept. In addition, find the y-values that correspond to x-values of $-2, -1, 1, 3$, and 4. Graph the curve.

8. Find the *x*-intercepts, if possible, for $y = 2x + x^2$, and find the *y*-intercept. In addition, find the *y*-values that correspond to *x*-values of -3, -1, and 1. Graph the curve.

9. Find the *y*-intercept for $y = x^3$. In addition, find the *y*-values that correspond to *x*-values of -2, -1, 1, and 2. Graph the curve.

10. Find the *y*-intercept for $y = x^3 - 3x + 4$. In addition, find the *y*-values that correspond to *x*-values of -2, -1, 1, and 2. Graph the curve.

Set II   In Exercises 1–4, complete the table of values for each equation; then draw its graph. Name the *x*- and *y*-intercepts.

1. $y = \dfrac{x^2}{2}$

x	y
−3	
−2	
−1	
0	
1	
2	
3	

2. $y = x^2 + 4x$

x	y
−5	
−4	
−3	
−2	
−1	
0	
1	

3. $y = x^2 - 4$

x	y
−3	
−2	
−1	
0	
1	
2	
3	

4. $y = 4 - x^2$

x	y
−3	
−2	
−1	
0	
1	
2	
3	

5. Find the *x*-intercepts, if possible, for $y = x^2 - 4x + 2$, and find the *y*-intercept. In addition, find the *y*-values that correspond to *x*-values of -1, 1, 2, 3, 4, and 5. Graph the curve.

6. Find the *x*-intercepts, if possible, for $y = x^2 - 4x - 2$, and find the *y*-intercept. In addition, find the *y*-values that correspond to *x*-values of -1, 1, 2, 3, 4, and 5. Graph the curve.

7. Find the x-intercepts, if possible, for $y = 5x - x^2$, and find the y-intercept. In addition, find the y-values that correspond to x-values of -1, 1, 2, 3, 4, and 6. Graph the curve.

8. Find the x-intercepts, if possible, for $y = -2x - x^2$, and find the y-intercept. In addition, find the y-values that correspond to x-values of -3, -1, and 1. Graph the curve.

9. Find the y-intercept for $y = -x^3$. In addition, find the y-values that correspond to x-values of -2, -1, 1, and 2. Graph the curve.

10. Find the y-intercept for $y = 1 - 2x - x^3$. In addition, find the y-values that correspond to x-values of -2, -1, 1, and 2. Graph the curve.

9.7 Graphing First-Degree Inequalities in Two Variables

In Section 3.6, we solved inequalities in one variable and graphed the solution sets on the real number line. In this section, we graph solution sets of first-degree inequalities in one or two variables *in the plane* of the rectangular coordinate system. The graph must contain, within the limits of the size of the graph, the set of *all* the ordered pairs, and only those ordered pairs, that satisfy the given inequality.

Half-planes   Any line in a plane separates that plane into two half-planes. For example, in Figure 9.7.1, the line AB separates the plane into the two half-planes shown.

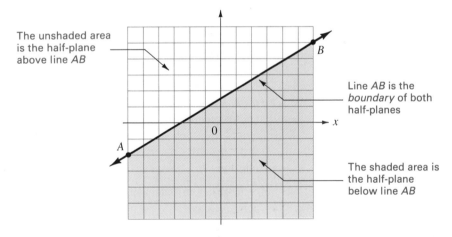

FIGURE 9.7.1

The following statement must be memorized; it will not be proved.

Any first-degree inequality in one or two variables has a graph that is a half-plane.

The equation of the boundary line of the half-plane is obtained by replacing the inequality sign with an equal sign.

How to Determine Whether the Boundary is a Dashed or Solid Line

[handwritten: Solid line when equality is included ≥ ≤]

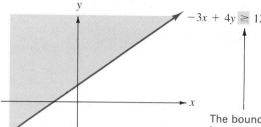

$-3x + 4y \geq 12$

The boundary is a *solid* line when the inequality is ≥ or ≤ (that is, when equality is included); in this case, the boundary line is part of the solution

[handwritten: dashed line when equality not included > <]

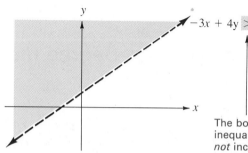

$-3x + 4y > 12$

The boundary is a *dashed* line when the inequality is > or < (that is, when equality is *not* included); in this case, the boundary line is *not* part of the solution

How to Determine the Correct Half-plane

1. *If the boundary does not go through the origin*, substitute the coordinates of the origin (0, 0) into the inequality.

   [handwritten: 0, 0 - True ½ plane] If the resulting statement is *true*, the solution is the half-plane containing (0, 0).

   [handwritten: false, not ½ plane] If the resulting statement is *false*, the solution is the half-plane *not* containing (0, 0).

2. *If the boundary goes through the origin*, select a point *not* on the boundary. Substitute the coordinates of this point into the inequality.

   If the resulting statement is *true*, the solution is the half-plane containing the point selected.

   If the resulting statement is *false*, the solution is the half-plane *not* containing the point selected.

[handwritten: 1. find point not on boundary 2. Substitute into inequality 3. If statement true, ½ plane 4. If statement false, other]

Graphing a First-Degree Inequality

TO GRAPH A FIRST-DEGREE INEQUALITY (IN THE PLANE)

1. The boundary line is *solid* if equality is included (≥, ≤).
   The boundary line is *dashed* if equality is *not* included (>, <).

2. Graph the boundary line.

3. Select and shade the correct half-plane.

Example 1   Graph $2x - 3y < 6$.

Solution Change $<$ to $=$ to find the boundary line.

Boundary line: $2x - 3y = 6$

Step 1. The boundary is a *dashed* line because the equality is *not* included.

$$2x - 3y < 6$$

Step 2. We graph the boundary line $2x - 3y = 6$ (Figure 9.7.2).

x-intercept: If $y = 0$, $2x - 3(0) = 6$

$$2x = 6$$

$$x = 3$$

y-intercept: If $x = 0$, $2(0) - 3y = 6$

$$-3y = 6$$

$$y = -2$$

x	y
3	0
0	-2

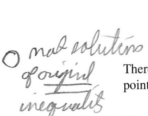

Therefore, the boundary line goes through $(3, 0)$ and $(0, -2)$. We'll graph these points as *hollow* dots, since they are *not* solutions of the original inequality.

Step 3. We select the correct half-plane. The solution of the inequality is only one of the two half-planes determined by the boundary line. We substitute the coordinates of the origin $(0, 0)$ into the inequality:

$$2x - 3y < 6$$

$$2(0) - 3(0) < 6$$

$$0 < 6 \quad \text{True}$$

Therefore, the half-plane containing the origin is the solution. The solution is the shaded area in Figure 9.7.2.

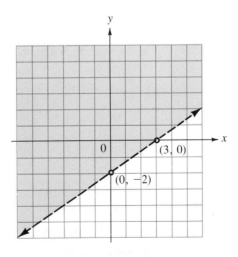

FIGURE 9.7.2 ■

Example 2   Graph $3x + 4y \leq -12$.
Solution

Step 1. The boundary is a *solid* line because the equality is included.

$$3x + 4y \leq -12$$

Step 2. We graph the boundary line $3x + 4y = -12$ (Figure 9.7.3).

x-intercept: If $y = 0$, $3x + 4(0) = -12$

$$3x = -12$$

$$x = -4$$

y-intercept: If $x = 0$, $3(0) + 4y = -12$

$$4y = -12$$

$$y = -3$$

x	y
−4	0
0	−3

Therefore, the boundary line goes through $(-4, 0)$ and $(0, -3)$. We'll graph these points as *solid* dots, since they *are* solutions of the original inequality.

Step 3. To select the correct half-plane, we substitute the coordinates of the origin $(0, 0)$ into the inequality:

$$3x + 4y \leq -12$$

$$3(0) + 4(0) \leq -12$$

$$0 \leq -12 \quad \text{False}$$

Therefore, the solution is the half-plane *not* containing $(0, 0)$. The solution is the shaded area in Figure 9.7.3.

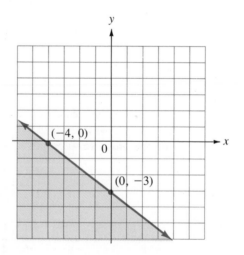

FIGURE 9.7.3 ■

Some inequalities contain only one variable. Such inequalities have graphs whose boundaries are either *vertical* or *horizontal* lines.

Example 3   Graph $x + 4 < 0$.
Solution

Step 1. The boundary is a *dashed* line because the equality is *not* included.

$$x + 4 \ < \ 0$$

Step 2. We graph the boundary line $x + 4 = 0$ or $x = -4$ (Figure 9.7.4).

Step 3. To select the correct half-plane, we substitute the coordinates of the origin (0, 0) into the inequality:

$$x + 4 < 0$$
$$(0) + 4 < 0$$
$$4 < 0 \qquad \text{False}$$

Therefore, the solution is the half-plane *not* containing (0, 0). The solution is the shaded area in Figure 9.7.4.

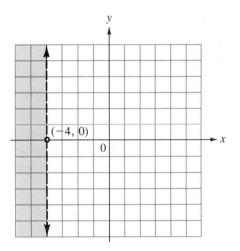

FIGURE 9.7.4 ■

In Section 3.6, we discussed how to graph the solution set of an inequality such as $x + 4 < 0$ on a *single number line*. Of course, if $x + 4 < 0$, then $x < -4$, and on the number line, we have the following graph:

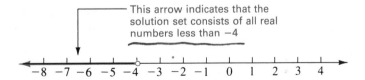

Example 3 of this section shows that the graph of the solution set of this same inequality, $x + 4 < 0$, is an entire half-plane when it is graphed in the rectangular coordinate system (see Figure 9.7.4).

Example 4   Graph $2y - 5x \geq 0$.
Solution

Step 1. The boundary is a *solid* line because the equality is included.

$$2y - 5x \geq 0$$

Step 2. We graph the boundary line $2y - 5x = 0$ (Figure 9.7.5).

x-intercept: If $y = 0$, $2(0) - 5x = 0$

$$-5x = 0$$

$$x = 0$$

y-intercept: If $x = 0$, $2y - 5(0) = 0$

$$2y = 0$$

$$y = 0$$

Therefore, the boundary line passes through the origin $(0, 0)$.

Find another point on the line:

If $x = 2$, $2y - 5(2) = 0$

$$2y - 10 = 0$$

$$2y = 10$$

$$y = 5$$

The point $(2, 5)$ is on the line.

x	y
0	0
0	0
2	5

Step 3. To select the correct half-plane, we substitute the coordinates of $(1, 0)$, a point *not* on the boundary, into the inequality:

$$2y - 5x \geq 0$$

$$2(0) - 5(1) \geq 0$$

$$-5 \geq 0 \qquad \text{False}$$

Therefore, the solution is the half-plane *not* containing $(1, 0)$ (Figure 9.7.5).

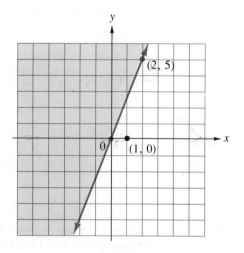

FIGURE 9.7.5 ■

EXERCISES 9.7

Set I   In the plane, graph the solution set for each of the following inequalities.

1. $x + 2y < 4$ **2.** $3x + y < 6$ **3.** $2x - 3y > 6$ **4.** $5x - 2y > 10$

5. $x \geq -2$ **6.** $x \leq 3$ **7.** $y \leq 4$ **8.** $y \geq -3$

9. $3y - 4x \geq 12$ **10.** $2y - 3x \geq 6$ **11.** $3x + 2y \geq -6$ **12.** $2x + y \geq -4$

13. $3x - 4y \geq 10$ **14.** $2x - 3y \leq 11$

15. $2(x + 1) + 3 \leq 3(2x - 1)$ **16.** $4(2x - 1) + 6 \leq 3(4x + 2)$

17. $4(2y + 3) - 1 \leq 5(2y - 1)$ **18.** $5(y - 1) + 4 \leq 2(3y + 2)$

19. $\dfrac{x}{5} - \dfrac{y}{3} > 1$ **20.** $\dfrac{x}{2} - \dfrac{y}{2} < 1$

21. $\dfrac{y}{4} - \dfrac{x}{5} > 1$ **22.** $\dfrac{y}{3} - \dfrac{x}{2} > 1$

23. $\dfrac{2x + y}{3} - \dfrac{x - y}{2} \geq \dfrac{5}{6}$ **24.** $\dfrac{4x - 3y}{5} - \dfrac{2x - y}{2} \geq \dfrac{2}{5}$

Set II   In the plane, graph the solution set for each of the following inequalities.

1. $4x + 2y < 8$ **2.** $2x - 7y \geq 14$ **3.** $5x + 2y > 10$ **4.** $x + y > 0$

5. $x \geq -4$ **6.** $y \geq -4$ **7.** $y \leq 1$ **8.** $x \leq 1$

9. $2y - 5x \geq 10$ **10.** $x - y > 0$ **11.** $x + 3y \geq -3$ **12.** $2x + 7y < 14$

13. $5x - 4y > 12$ **14.** $y \leq -2$

15. $4(x + 2) - 3 \leq 2(x - 3)$ **16.** $x > 0$

17. $3(y - 4) + 4 \leq 2(3y + 2)$ **18.** $3x - 2(2x - 7) \leq 2(3 + x) - 4$

19. $\dfrac{x}{5} - \dfrac{y}{2} > 1$ **20.** $\dfrac{x}{4} - \dfrac{y}{2} \geq 1$

21. $\dfrac{y}{3} - \dfrac{x}{2} > 1$ **22.** $\dfrac{y}{2} + \dfrac{x}{4} < 1$

23. $\dfrac{3x + y}{2} - \dfrac{x + y}{3} \geq \dfrac{7}{6}$ **24.** $\dfrac{x - y}{2} - \dfrac{x - 4y}{7} \leq \dfrac{5}{14}$

9.8 Review: 9.1–9.7

**Ordered Pairs
9.1**

An ordered pair of numbers is used to represent a point in the plane.

x-coordinate —
Abscissa
Horizontal coordinate (a , b) Vertical coordinate
First coordinate   y-coordinate
                          Ordinate
                          Second coordinate

**Linear Equation
in Two Variables
9.2**

An equation that can be expressed in the form $Ax + By + C = 0$ is a **linear equation in two variables**. An ordered pair (a, b) is a solution of the equation $Ax + By + C = 0$ if we get a true statement when we substitute a for x and b for y in that equation.

Graphing Straight Lines
9.3

Method 1 (Intercept method)	*Method 2*
1. Find, if it exists, the *x*-intercept: Set $y = 0$; then solve for *x*.	1. In the equation, substitute a number for one of the variables and solve for the corresponding value of the other variable. Find *three* points in this way, and list them in a table of values.
2. Find, if it exists, the *y*-intercept: Set $x = 0$; then solve for *y*.	
3. If both intercepts are (0, 0), an additional point must be found.	2. Graph the points listed in the table of values.
4. Find one other point by letting *x* or *y* have any value.	3. Draw a straight line through the points and put arrowheads at each end of the line.
5. Draw a straight line through the three points and put arrowheads at each end of the line.	

Special cases:

The graph of $x = a$ is a vertical line that passes through $(a, 0)$

The graph of $y = b$ is a horizontal line that passes through $(0, b)$.

Slope
9.4

The slope of a line through the points $P_1(x_1, y_1)$ and $P_2(x_2, y_2)$ is

$$m = \frac{y_2 - y_1}{x_2 - x_1} \quad \text{or} \quad m = \frac{y_1 - y_2}{x_1 - x_2}$$

The slope of a horizontal line is 0.

The slope of a vertical line *does not exist.*

Parallel lines have the same slope.

The product of the slopes of two **perpendicular** lines is -1.

Equations of Lines
9.5

1. *General form:* $Ax + By + C = 0$
   where *A*, *B*, and *C* are integers and *A* and *B* are not both 0.

2. *Point-slope form:* $y - y_1 = m(x - x_1)$
   where $m =$ the slope and $P_1(x_1, y_1)$ is a known point on the line.

3. *Slope-intercept form:* $y = mx + b$
   where $m =$ the slope and $b =$ the *y*-intercept.

4. The equation of any *horizontal line* is $y = b$, where *b* is the *y*-coordinate of every point on the line.

5. The equation of any *vertical line* is $x = a$, where *a* is the *x*-coordinate of every point on the line.

To Graph a Curve
9.6

1. Find the *y*-intercept and find the *x*-intercepts if possible.

2. Use the equation to find the *y*-values corresponding to several (given) *x*-values.

3. Graph the ordered pairs found in steps 1 and 2.

4. Join the points graphed in step 3 *from left to right* with a smooth curve, and put arrowheads at the ends of the curve just drawn.

To Graph a First-Degree
Inequality in the Plane
9.7

Find the equation of the boundary line by replacing the inequality symbol with an equal sign.

1. The boundary line is *solid* if equality is included ($\geq$, $\leq$).

   The boundary line is *dashed* if equality is *not* included ($>$, $<$).

2. Graph the boundary line.

3. Select and shade the correct half-plane.

Review Exercises 9.8 Set I

1. Draw the four-sided figure with sides that are straight lines and with vertices at the following points: $A(-4, -5)$, $B(2, -4)$, $C(5, 1)$, $D(-2, 3)$.

2. Given the point $B(-2, 4)$, do the following:

   a. Find the *y*-coordinate of *B*.   b. Find the abscissa of *B*.

   c. Find the *x*-coordinate of *B*.   d. Graph *B*.

In Exercises 3–9, graph each line.

3. $x - y = 5$ 4. $y - x = 3$ 5. $x = -2$ 6. $y = -3$

7. $4y - 5x = 20$ 8. $5y - 4x = 20$ 9. $x + 2y = 0$

In Exercises 10–12, find the slope of the line through the given pair of points.

10. $(4, -1)$ and $(4, 3)$ 11. $(2, -6)$ and $(-3, 5)$ 12. $(3, -5)$ and $(-2, 4)$

In Exercises 13–17, write the general form of the equation of the given line.

13. The line through $(3, -5)$ with slope $m = \frac{2}{3}$.

14. The line with a slope of $-\frac{1}{3}$ and a *y*-intercept of 9.

15. The line that passes through the points $(-3, 4)$ and $(1, -2)$.

16. The horizontal line through $(-2, 5)$.

17. The vertical line through $(-2, 5)$.

In Exercises 18–20, complete the table of values and graph the curve.

18. $y = x^2 - 3x$

x	y
-2	
-1	
0	
1	
2	
3	
4	

19. $y = x^3 - 3x$

x	y
-3	
-2	
-1	
0	
1	
2	
3	

20. $y = x^3 + 3x^2 - 2$

x	y
-4	
-3	
-2	
-1	
0	
1	
2	

In Exercises 21–25, graph the solution set for each of the inequalities.

21. $3x - 7y < 21$ **22.** $x + 4y \le 0$ **23.** $y > 3$

24. $\dfrac{x}{2} - \dfrac{y}{6} \ge 1$ **25.** $\dfrac{y}{3} - \dfrac{x}{4} \le 1$

Review Exercises 9.8 Set II

1. Draw the four-sided figure with sides that are straight lines and with vertices at the following points: $A(4, -5)$, $B(-2, 4)$, $C(3, 2)$, $D(5, -1)$.

ANSWERS

1.

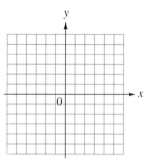

2. Given the point $P(2, -4)$, do the following.

a. Find the x-coordinate of P.

b. Find the ordinate of P.

c. Graph P.

2a. _____

b. _____

c.

In Exercises 3–9, graph each line.

3. $x - y = -4$

3.

4. $y + 2 = 0$

4.

5. $x + 3y = 6$

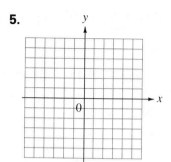

6. $x = \frac{3}{5}y$

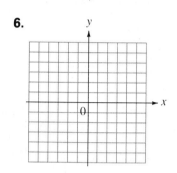

7. $2x - 3y = 7$

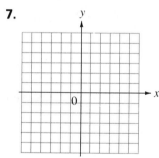

8. $y - 2(x - 3) = 0$

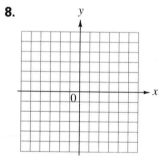

9. $x = 4$

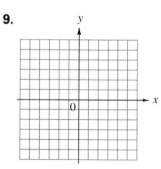

In Exercises 10–12, find the slope of the line through the given pair of points.

10. $(4, -3)$ and $(2, -3)$ **11.** $(-13, -10)$ and $(-8, 7)$

10. _____

11. _____

12. $(-5, 3)$ and $(-5, 0)$

12. _____

13. _____

In Exercises 13–17, write the general form of the equation of the given line.

13. The line through $(4, -3)$ with slope $m = -\frac{2}{5}$.

14. _____

15. _____

14. The line with a slope of $-\frac{4}{3}$ and a y-intercept of 5.

16. _____

17. _____

15. The line that passes through the points $(-4, 3)$ and $(2, -1)$.

16. The horizontal line through $(6, -5)$.

18.

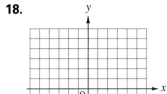

17. The vertical line through $(6, -5)$.

In Exercises 18–20, complete the table of values and graph the curve.

18. $y = x^2 + 1$

x	y
-3	
-2	
-1	
0	
1	
2	
3	

19. $y = 1 - x^2$

x	y
-3	
-2	
-1	
0	
1	
2	
3	

20. $y = 2x - x^3$

x	y
-2	
-1	
0	
1	
2	

19.

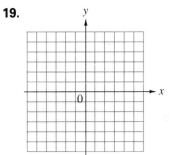

20.

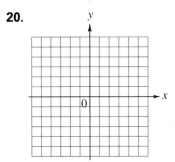

In Exercises 21–25, graph the solution set for each of the inequalities.

21. $2y - 5x \leq 11$

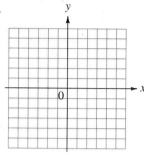

22. $2x - 3y \leq 0$

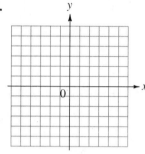

23. $x - 3 < 0$

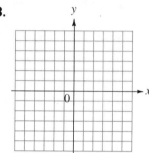

24. $\dfrac{x}{5} - \dfrac{y}{4} > 1$

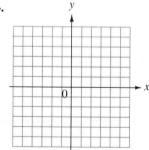

25. $\dfrac{y}{3} - \dfrac{x}{2} \geq 1$

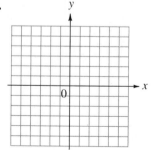

Chapter 9 Diagnostic Test

The purpose of this test is to see how well you understand graphing. We recommend that you work this diagnostic test *before* your instructor tests you on this chapter. Allow yourself about 50 minutes.

Complete solutions for all the problems on this test, together with section references, are given in the answer section in the back of the book. For the problems you do incorrectly, study the sections referred to.

1. a. Draw the triangle with vertices at the following points: $A(4, -3)$, $B(2, 0)$, $C(-3, 2)$.

 b. Name the ordered pairs corresponding to points D, E, F, and G in the graph below.

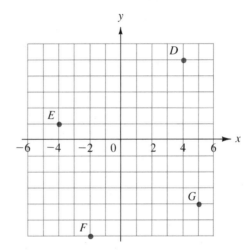

2. Determine whether the given ordered pairs are solutions of the equation $5x - 3y = 15$.

 a. $(0, 0)$ b. $(3, 0)$ c. $(0, 3)$ d. $(0, -5)$ e. $(6, 5)$

In Problems 3 and 4, graph each line.

3. a. $y = -2$ b. $3x + y = 3$

4. a. $2x - y = 4$ b. $4x + 9y = 18$

5. Find the slopes of the lines through the given points.

 a. $(5, 2)$ and $(-3, 1)$ b. $(-3, 2)$ and $(-3, 0)$ c. $(2, -4)$ and $(0, -4)$

6. Write the equation of each given line.

 a. The vertical line that passes through the point $(-3, 4)$.

 b. The horizontal line that passes through the point $(-3, 4)$.

7. Write the equation of the line with a slope of 5 and a y-intercept of 3.

8. Given the points $A(-3, 2)$ and $B(1, -4)$, find the following.

 a. The slope of the line through A and B.

 b. The general form of the equation of the line through A and B.

9. Graph the curve $y = x^2 - x - 6$, using integral values of x from -3 to 4.

10. Graph $3x - 2y \le 6$ in the plane.

Cumulative Review Exercises: Chapters 1–9

1. Find A, using the given formula and the values of the variables given with the formula.

$$A = \frac{h}{2}(B + b) \qquad h = 5, B = 11, b = 7$$

2. Simplify. Write your answer using only positive exponents.

$$\left(\frac{35a}{15a^{-2}b}\right)^{-2}$$

In Exercises 3–7, perform the indicated operations and simplify.

3. $(6z^2 - z - 7) \div (3z - 5)$

4. $\dfrac{2x - 1}{x - 2} - \dfrac{1}{x + 1}$

5. $\dfrac{b + 1}{a^2 + ab} \div \dfrac{b^2 - b - 2}{a^3 - ab^2}$

6. $\dfrac{x^2 - \dfrac{4}{y^2}}{x + \dfrac{2}{y}}$

7. $\dfrac{3}{x^2 - 7x + 6} - \dfrac{4}{2x^2 + x - 3}$

8. Solve for x: $\dfrac{2x - 4}{3} = \dfrac{4 - 6x}{5}$

9. Solve for x: $\dfrac{2x}{9} - \dfrac{11}{3} = \dfrac{5x}{6}$

10. Solve for k: $C = \dfrac{5}{2h} - \dfrac{h}{k}$

11. Graph $y = 3x - 2$.

12. Graph $y = x^2 - 4x$, using integral values of x from -1 to 5.

13. Write the equation of the straight line that passes through $(7, 1)$ and $(2, -4)$.

In Exercises 14–22, set up the problem algebraically, solve, and check. Be sure to state what your variables represent.

14. One number is 4 more than another. Their product is equal to $-\frac{7}{4}$. Find the two numbers.

15. A 50-lb mixture of Delicious and Jonathan apples costs $44.58. If the Delicious apples cost 98¢ a pound and the Jonathan apples cost 85¢ a pound, how many pounds of each kind are there?

16. The base of a triangle is 3 cm more than its altitude. Its area is 44 sq. cm. Find the lengths of the altitude and the base.

17. The sum of a number and its reciprocal is $\frac{53}{14}$. Find the number.

18. Find two consecutive integers whose product is 5 more than their sum.

19. How many cubic centimeters of water must be added to 10 cc of a 17% solution of a disinfectant to reduce it to a 0.2% solution?

20. Charles types 473 words in 8.6 min. What is his rate of typing?

21. A camera shop buys eighteen lenses. If it had bought fifteen lenses of higher quality, it would have paid $23 more per lens for the same total cost. Find the cost of each type of lens.

22. A hardware store buys twelve saws. If it had bought eight saws of higher quality, it would have paid $2.75 more per saw for the same total cost. Find the cost of each type of saw.

Critical Thinking

Each of the following problems has an error. Can you find it?

1. Simplify $\dfrac{4}{x} + \dfrac{2}{x + 2}$.

LCD $= x(x + 2)$

$$\overset{1}{\cancel{x}}(x + 2)\dfrac{4}{\underset{1}{\cancel{x}}} + x(\overset{1}{\cancel{x + 2}})\dfrac{2}{\underset{1}{\cancel{x + 2}}} = 4x + 8 + 2x = 6x + 8$$

2. Simplify $\dfrac{1 - \dfrac{1}{x^2}}{1 - \dfrac{1}{x}}$.

$$\dfrac{x^2\left(1 - \dfrac{1}{x^2}\right)}{x\left(1 - \dfrac{1}{x}\right)} = \dfrac{x^2 - 1}{x - 1}$$

$$= \dfrac{(x + 1)(\overset{1}{\cancel{x - 1}})}{\underset{1}{\cancel{x - 1}}}$$

$$= x + 1$$

3. Solve $x^2 - 3x = 4$.

$$x^2 - 3x = 4$$

$$x(x - 3) = 4$$

$$x = 4 \quad \bigg| \quad x - 3 = 4$$

$$x = 7$$

4. Solve $4x - 2(x - 3) < 3(x + 3)$

$$4x - 2(x - 3) < 3(x + 3)$$

$$4x - 2x + 6 < 3x + 9$$

$$2x + 6 < 3x + 9$$

$$-x + 6 < 9$$

$$-x < 3$$

$$x < -3$$

10 Systems of Equations

In previous chapters, we showed how to solve a single equation for a single variable. In this chapter, we show how to solve systems of two linear equations in two variables. Systems that have *more* than two equations and variables are not discussed in this book.

10.1 Basic Definitions

A **system of two equations** in (the same) two variables is a set of two equations considered together, or simultaneously. The **solution**, if one exists, of a system of two equations in two variables is the set of ordered pairs that satisfies *both* equations.

Example 1

Examples of systems of two equations in two variables and their solutions:

a. $\begin{cases} x - y = 6 \\ x + y = 2 \end{cases}$ is a system of two equations in x and y.

   $(4, -2)$ is a solution of the system, because when we substitute 4 for x and -2 for y into the first equation, we have

$$4 - (-2) = 6 \qquad \text{True}$$

   and when we substitute 4 for x and -2 for y in the second equation, we have

$$4 + (-2) = 2 \qquad \text{True}$$

   This can be stated as "$(4, -2)$ satisfies the system $\begin{cases} x - y = 6 \\ x + y = 2 \end{cases}$."

b. $\begin{cases} x + y = 4 \\ 2x - y = 2 \end{cases}$ is a system of two equations in x and y.

   $(2, 2)$ is a solution of the system, because when we substitute 2 for x and 2 for y in the first equation, we have

$$2 + 2 = 4 \qquad \text{True}$$

   and substituting 2 for x and 2 for y in the second equation gives

$$2(2) - 2 = 2 \qquad \text{True} \quad \blacksquare$$

Example 2

Determine whether $(0, 3)$ is a solution of the system $\begin{cases} x + y = 3 \\ 3x - y = 5 \end{cases}$.

Solution

$$0 + 3 = 3 \qquad \text{Substituting 0 for } x \text{ and 3 for } y \text{ in the first equation}$$
$$3 = 3 \qquad \text{True}$$
$$3(0) - 3 = 5 \qquad \text{Substituting 0 for } x \text{ and 3 for } y \text{ in the second equation}$$
$$-3 = 5 \qquad \text{False}$$

Because a false statement results when we substitute the ordered pair into the second equation, $(0, 3)$ does not satisfy *both* equations. Therefore, $(0, 3)$ is not a solution of the given system of equations. $\blacksquare$

EXERCISES 10.1

Set I

Determine whether each ordered pair is a solution of the given system of equations.

1. $\begin{cases} 3x + y = 8 \\ 2x - y = 2 \end{cases}$ 　　a. $(0, 8)$ 　　b. $(1, 0)$ 　　c. $(2, 2)$

2. $\begin{cases} 4x - y = 3 \\ 3x + y = 11 \end{cases}$ a. $(0, -3)$ b. $(2, 5)$ c. $(1, 1)$

3. $\begin{cases} 2x + 3y = 6 \\ 4x + 6y = 12 \end{cases}$ a. $(3, 2)$ b. $(0, 2)$ c. $(3, 0)$

4. $\begin{cases} 3x - 2y = 6 \\ 6x - 4y = 12 \end{cases}$ a. $(2, 0)$ b. $(0, -3)$ c. $(-2, -6)$

5. $\begin{cases} x + y = 2 \\ x + y = 4 \end{cases}$ a. $(0, 4)$ b. $(0, 2)$ c. $(4, 0)$

6. $\begin{cases} 2x - y = 4 \\ 2x - y = 2 \end{cases}$ a. $(0, -4)$ b. $(2, 0)$ c. $(1, 0)$

Set II   Determine whether each ordered pair is a solution of the given system of equations.

1. $\begin{cases} 2x + y = 6 \\ 3x - y = -1 \end{cases}$ a. $(0, 6)$ b. $(1, 4)$ c. $(3, 0)$

2. $\begin{cases} 4x - 2y = 8 \\ 2x - y = 4 \end{cases}$ a. $(0, -4)$ b. $(2, 0)$ c. $(6, 8)$

3. $\begin{cases} x + 3y = 6 \\ 2x + 6y = 12 \end{cases}$ a. $(6, 0)$ b. $(0, 2)$ c. $(1, 3)$

4. $\begin{cases} 3x - y = 3 \\ 6x - 2y = 3 \end{cases}$ a. $(1, 0)$ b. $(0, -3)$ c. $\left(\frac{1}{2}, 0\right)$

5. $\begin{cases} 4x + y = 4 \\ 8x + 2y = 4 \end{cases}$ a. $(0, 4)$ b. $(1, 0)$ c. $\left(\frac{1}{2}, 0\right)$

6. $\begin{cases} 2x + y = 6 \\ 2x - y = 2 \end{cases}$ a. $(0, 6)$ b. $(2, 2)$ c. $(1, 0)$

10.2 Graphical Method for Solving a Linear System

The graphical method for solving a system of equations is not an exact method of solution, but it can sometimes be used successfully in solving such systems. Recall from Chapter 9 that the graph of a linear equation in two variables is a straight line.

TO SOLVE A SYSTEM OF LINEAR EQUATIONS BY THE GRAPHICAL METHOD

1. Graph each straight line on the same set of axes.

2. Three outcomes are possible:
   a. The lines intersect in exactly one point (Figure 10.2.1). The solution is the ordered pair that represents the point of intersection.
   b. The lines will never cross, even if they are extended indefinitely (Figure 10.2.2). Such lines are called *parallel lines*. There is no solution for the system of equations.
   c. Both equations have the same line for their graph (Figure 10.2.3). Infinitely many ordered pairs are solutions; any ordered pair that represents a point on the line is a solution.

3. If the lines graphed in step 1 intersect in exactly one point, check the coordinates of that point in *both* equations.

Example 1   Solve the system $\begin{Bmatrix} x + 2y = 8 \\ x - y = 2 \end{Bmatrix}$ graphically.

Solution We draw the graph of each line on the same set of axes (Figure 10.2.1).

Line 1: $x + 2y = 8$

y-intercept: If $x = 0$, then $y = 4$.

x-intercept: If $y = 0$, then $x = 8$.

Checkpoint: If $x = 2$, then $y = 3$.

x	y
0	4
8	0
2	3

Line 2: $x - y = 2$

y-intercept: If $x = 0$, then $y = -2$.

x-intercept: If $y = 0$, then $x = 2$.

Checkpoint: If $x = -2$, then $y = -4$.

x	y
0	-2
2	0
-2	-4

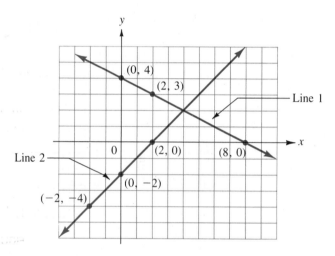

FIGURE 10.2.1

The lines *appear* to intersect at the point (4, 2). We check that solution by substituting 4 for *x* and 2 for *y* in *both* equations.

(1) $x + 2y = 8$ (2) $x - y = 2$

$4 + 2(2) \overset{?}{=} 8$ $4 - 2 \overset{?}{=} 2$

$8 = 8$ True $2 = 2$ True

Therefore, the solution of the system is (4, 2).

The coordinates of any point on line 1 satisfy the equation of line 1. The coordinates of any point on line 2 satisfy the equation of line 2. The only point that lies on *both* lines is (4, 2). Therefore, it is the only point whose coordinates satisfy *both* equations. ∎

When a system of equations has a solution, it is called a **consistent system**. When each equation in the system has a different graph, the system is called an **independent system**. Therefore, the system in Example 1 is a consistent, independent system.

Example 2   Solve the system $\begin{Bmatrix} 3x + 2y = 6 \\ 6x + 4y = 24 \end{Bmatrix}$ graphically.

Solution We draw the graph of each line on the same set of axes (Figure 10.2.2).

Line 1: $3x + 2y = 6$

y-intercept: If $x = 0$, then $y = 3$.

x-intercept: If $y = 0$, then $x = 2$.

Checkpoint: If $x = -2$, then $y = 6$.

x	y
0	3
2	0
-2	6

Line 2: $6x + 4y = 24$

y-intercept: If $x = 0$, then $y = 6$.

x-intercept: If $y = 0$, then $x = 4$.

Checkpoint: If $x = 2$, then $y = 3$.

x	y
0	6
4	0
2	3

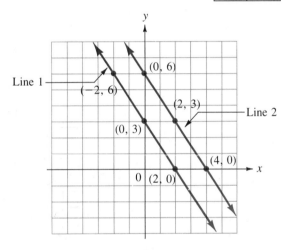

FIGURE 10.2.2

The lines will never meet; they are parallel. (You should verify that their slopes are equal.) There is *no solution* for the system of equations. ■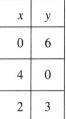

A system (such as that in Example 2) that has no solution is called an **inconsistent system**. Since each equation in Example 2 has a different graph, the system is independent.

Example 3   Solve the system $\begin{Bmatrix} x - 2y = -4 \\ 3x - 6y = -12 \end{Bmatrix}$ graphically.

Solution We draw the graph of each line on the same set of axes (Figure 10.2.3).

Line 1: $x - 2y = -4$

y-intercept: If $x = 0$, then $y = 2$.

x-intercept: If $y = 0$, then $x = -4$.

Checkpoint: If $x = 4$, then $y = 4$.

x	y
0	2
-4	0
4	4

Line 2: $3x - 6y = -12$

y-intercept: If $x = 0$, then $y = 2$.

x-intercept: If $y = 0$, then $x = -4$.

Checkpoint: If $x = 2$, then $y = 3$.

x	y
0	2
−4	0
2	3

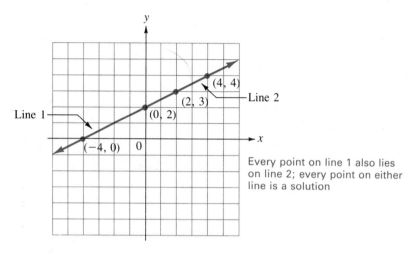

FIGURE 10.2.3

Since both lines go through (0, 2) and (−4, 0), they must be the same line (Figure 10.2.3). There are infinitely many solutions for this system of equations; four of them are (0, 2), (−4, 0), (4, 4), and (2, 3). ∎

When both equations in the system have the same graph, the system is called a **dependent system**. Therefore, the system of equations in Example 3 is a consistent, dependent system.

Example 4 illustrates the drawbacks of the graphical method and shows the need for the algebraic methods of solving systems of equations given in Sections 10.3 and 10.4. Nevertheless, we feel that the graphical method helps you understand what it means to *solve* a system of equations, and especially what it means when a system has no solution or has many solutions.

Example 4   Try to solve the system $\begin{Bmatrix} -x + y = 3 \\ x + 2y = -2 \end{Bmatrix}$ graphically.

Solution We draw the graph of each line on the same set of axes (Figure 10.2.4).

Line 1: $-x + y = 3$

y-intercept: If $x = 0$, then $y = 3$.

x-intercept: If $y = 0$, then $x = -3$.

Checkpoint: If $y = 2$, then $x = -1$.

x	y
0	3
−3	0
−1	2

Line 2: $x + 2y = -2$

y-intercept: If $x = 0$, then $y = -1$.

x-intercept: If $y = 0$, then $x = -2$.

Checkpoint: If $y = 1$, then $x = -4$.

x	y
0	-1
-2	0
-4	1

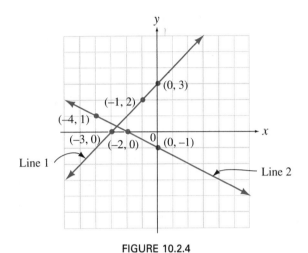

FIGURE 10.2.4

It's difficult to name the coordinates of the point of intersection of the two lines. One guess might be $\left(-2\frac{1}{2}, \frac{1}{2}\right)$; however, this ordered pair does *not* check in the second equation (verify this). We'll find in Example 1 of Section 10.3 that the correct solution is $\left(-2\frac{2}{3}, \frac{1}{3}\right)$. (Verify that this ordered pair does satisfy *both* equations.) ∎

EXERCISES 10.2

Set I Find the solution of each system of equations graphically. If the lines are parallel, write "No solution." If both equations have the same line for their graph, write "Many solutions."

1. $\begin{cases} 2x + y = 6 \\ 2x - y = -2 \end{cases}$ **2.** $\begin{cases} 2x - y = -4 \\ x + y = 1 \end{cases}$ **3.** $\begin{cases} x - 2y = -6 \\ 4x + 3y = 20 \end{cases}$

4. $\begin{cases} x - 3y = 6 \\ 4x + 3y = 9 \end{cases}$ **5.** $\begin{cases} x + 2y = 0 \\ x - 2y = -2 \end{cases}$ **6.** $\begin{cases} 2x + y = 0 \\ 2x - y = -6 \end{cases}$

7. $\begin{cases} 3x - 2y = -9 \\ x + y = 2 \end{cases}$ **8.** $\begin{cases} x + 2y = -4 \\ 2x - y = -3 \end{cases}$ **9.** $\begin{cases} 10x - 4y = 20 \\ 6y - 15x = -30 \end{cases}$

10. $\begin{cases} 12x - 9y = 36 \\ 6y - 8x = -24 \end{cases}$ **11.** $\begin{cases} 2x - 4y = -2 \\ -3x + 6y = 12 \end{cases}$ **12.** $\begin{cases} 4x - 2y = 8 \\ -6x + 3y = 6 \end{cases}$

Set II Find the solution of each system of equations graphically. If the lines are parallel, write "No solution." If both equations have the same line for their graph, write "Many solutions."

1. $\begin{cases} 2x + y = -4 \\ x - y = -5 \end{cases}$ **2.** $\begin{cases} 3x + 2y = 0 \\ 3x - 2y = 12 \end{cases}$ **3.** $\begin{cases} 3x - 4y = -3 \\ 3x + 2y = 6 \end{cases}$

4. $\begin{cases} 3x + 2y = 3 \\ 5x - 4y = 16 \end{cases}$ **5.** $\begin{cases} x + y = 1 \\ x - y = -3 \end{cases}$ **6.** $\begin{cases} x - 4y = 4 \\ 8y - 2x = -8 \end{cases}$

7. $\begin{cases} 4x - 3y = 6 \\ x - y = 1 \end{cases}$ **8.** $\begin{cases} 3x - y = 6 \\ y - 3x = 3 \end{cases}$ **9.** $\begin{cases} 4x - 6y = 12 \\ 9y - 6x = -18 \end{cases}$

10. $\begin{cases} 4x + 3y = 3 \\ x + y = -1 \end{cases}$ **11.** $\begin{cases} x - 2y = 10 \\ 2y - x = 2 \end{cases}$ **12.** $\begin{cases} x - 4y = 2 \\ x + 2y = -4 \end{cases}$

10.3 Addition Method for Solving a Linear System

The graphical method for solving a system of equations has two disadvantages: (1) It is slow, and (2) it is not an exact method of solution. The method we discuss in this section has neither of these disadvantages. We need one definition before we proceed.

EQUIVALENT SYSTEMS OF EQUATIONS

Two systems of equations are **equivalent** if the systems both have the same solution set.

When we solve a system of two equations in x and y algebraically, we must try to find an equivalent system in which one of the equations is in the form $x = a$ and the other is in the form $y = b$. That is, we try to find an equivalent system in which each equation contains only one variable.

The operations in the following box always give us a system of equations equivalent to the system we started with and are used in the addition method for solving a system of equations.

OPERATIONS THAT YIELD EQUIVALENT SYSTEMS OF EQUATIONS

1. We can interchange the equations; that is, either one can be written first.

2. We can rewrite either or both equations using the multiplication property of equality.

3. We can add the two equations together. That is, we can use the following property of equality:

$$\text{If } a = b \text{ and } c = d, \text{ then } a + c = b + d.$$

We will demonstrate in several examples how to use these properties in order to *eliminate one of the variables*, and we'll show how doing so helps us solve the system of equations. In addition to using these properties, we can, of course, simplify either equation by removing grouping symbols, by combining like terms, and by using *any* of the properties of equality on the equation.

The system in Example 1 is the system we tried to solve graphically in Example 4 of Section 10.2. In solving it algebraically, we don't need to use the first two properties listed in the box above.

Example 1   Solve the system $\left\{\begin{array}{l} -x + y = 3 \\ x + 2y = -2 \end{array}\right\}$ by the addition method.

Solution We first want to eliminate one of the variables. In this case, if we simply add the two equations together, the x's will be eliminated.

$$
\begin{array}{r}
-x + y = 3 \\
x + 2y = -2 \\
\hline
3y = 1
\end{array}
$$ Adding the two equations together

$$\frac{3y}{3} = \frac{1}{3}$$ Dividing both sides by 3

$$y = \frac{1}{3}$$ This is the y-coordinate of the point of intersection

By eliminating the x's, we were able to solve for y. Now that we know what the y-coordinate of the point of intersection is, we find the x-coordinate of the point by substituting $y = \frac{1}{3}$ into *either* of the original equations (this time we'll substitute into the first equation) and solving for x:

$$-x + y = 3$$ The first equation

$$-x + \frac{1}{3} = 3$$ Substituting $\frac{1}{3}$ for y

$$x - \frac{1}{3} = -3$$ Multiplying both sides of the equation by -1

$$\frac{+\frac{1}{3} \quad +\frac{1}{3}}{x \quad = -2\frac{2}{3}}$$ Adding $\frac{1}{3}$ to both sides of the equation

We isolate x

The system $\left\{\begin{array}{l} x = -2\frac{2}{3} \\ y = \frac{1}{3} \end{array}\right\}$ is equivalent to the given system, because both systems have the same solution set. The solution is $\left(-2\frac{2}{3}, \frac{1}{3}\right)$. (You verified in Example 4 of Section 10.2 that this is the correct solution for the system of equations.) ∎

In Example 1, the coefficients of x were the negatives of each other; therefore, when we added the two equations together, the x's were eliminated. Often, however, we need to use the multiplication property of equality to *make* the coefficients of one of the variables become equal numerically but opposite in sign (that is, to become the *negatives* of each other). Then, when we add the equations, at least one of the variables will be eliminated because the sum of its coefficients will be zero.

The general method for solving a system of equations in two variables by the **addition method** is as follows:

TO SOLVE A SYSTEM OF EQUATIONS BY THE ADDITION METHOD

1. If necessary, multiply one or both equations by numbers that make the coefficients of one of the variables become the negatives of each other (see Examples 2 and 3). The smallest possible value for the *new* coefficients is the least common multiple (LCM) of both of the original coefficients (see Example 4). Write the equations one under the other *with like terms lined up*.

2. Add the equations from step 1 together. (This will eliminate at least one of the variables.)

3. Three outcomes are possible:

a. *One variable remains*. Solve the resulting equation for that variable. Then substitute this known value into either of the system's equations to solve for the other variable. There is one solution for the system. The system is consistent and independent.

b. *Both variables are eliminated and a false statement results*. There is no solution; the system is inconsistent and independent.

c. *Both variables are eliminated and a true statement results*. There are many solutions; the system is consistent and dependent.

4. If one solution is found in step 3, check it in both equations.

Solving Systems with Only One Solution by the Addition Method

Example 2   Solve the system $\begin{cases} x - 2y = -6 \\ 4x + 3y = 20 \end{cases}$.

Solution

Equation 1: $x - 2y = -6$
Equation 2: $4x + 3y = 20$

If both sides of Equation 1 are multiplied by -4, the coefficients of x will add to zero.

This symbol means both sides of Equation 1 are to be multiplied by -4

The coefficients of x add to zero

$$(-4) \; x - 2y = -6 \Rightarrow^* \quad -4x + 8y = +24$$
$$4x + 3y = 20 \Rightarrow \quad \underline{\;\;4x + 3y = \;\; 20\;\;}$$
$$11y = \;\; 44 \qquad \text{Adding the equations}$$
$$y = 4 \qquad \begin{array}{l}\text{This is the } y\text{-coordinate} \\ \text{of the point of intersection}\end{array}$$

---

*The symbol $\Rightarrow$, read "implies," means that the second statement is true if the first statement is true. For example,

$$x - 2y = -6 \Rightarrow -4x + 8y = +24$$

is read "$x - 2y = -6$ implies $-4x + 8y = +24$" and means that $-4x + 8y = +24$ is true if $x - 2y = -6$ is true.

The value $y = 4$ can be substituted into either Equation 1 or Equation 2 to find the value of x. It is usually easier to substitute into the equation that has the smaller coefficients. Substituting 4 for y in Equation 1, we have

$x - 2y = -6$	Equation 1
$x - 2(4) = -6$	Substituting 4 for y
$\begin{aligned} x - 8 &= -6 \\ +8 \quad & +8 \end{aligned}$	Adding 8 to both sides
$x = 2$	This is the x-coordinate of the point of intersection

If we substitute 4 for y in Equation 2, we will still get 2 as the value for x:

$4x + 3y = 20$	Equation 2
$4x + 3(4) = 20$	Substituting 4 for y
$\begin{aligned} 4x + 12 &= 20 \\ -12 \quad & -12 \end{aligned}$	Adding -12 to both sides
$4x = 8$	We now divide both sides by 4
$x = 2$	This is the x-coordinate of the point of intersection

The check, which is left to the student, will verify that the solution of the system is $(2, 4)$. ■

Example 3   Solve the system $\begin{cases} 3x + 4y = 6 \\ 2x + 3y = 5 \end{cases}$ using the addition method (a) eliminating the x's and (b) eliminating the y's.
Solution

a. Equation 1: $3x + 4y = 6$
   Equation 2: $2x + 3y = 5$

If we're to eliminate the x's, we must make the coefficients of x (now 3 and 2) be equal numerically. We can do this by simply multiplying both sides of Equation 1 by 2 (the coefficient of x in Equation 2) and multiplying both sides of Equation 2 by 3 (the coefficient of x in Equation 1). However, to make the new coefficients opposite in sign, we must make one of the multipliers negative. Therefore, let's multiply Equation 1 by 2 and Equation 2 by -3.

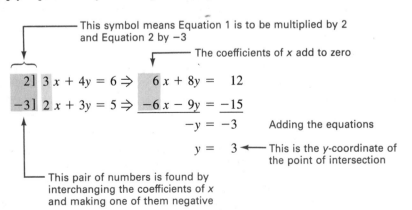

We substitute 3 for y in Equation 2:

$$2x + 3y = 5 \qquad \text{Equation 2}$$

$$2x + 3(3) = 5 \qquad \text{Substituting 3 for } y$$

$$2x + 9 = 5$$
$$\underline{-9 \quad -9} \qquad \text{Adding } -9 \text{ to both sides}$$
$$2x = -4$$

$$x = -2 \qquad \begin{array}{l}\text{This is the } x\text{-coordinate of}\\ \text{the point of intersection}\end{array}$$

Checking will verify that the solution of the system is $(-2, 3)$.

b. Equation 1: $3x + 4y = 6$
Equation 2: $2x + 3y = 5$

If we're to eliminate the y's, we must make the coefficients of y (now 4 and 3) be equal numerically but opposite in sign. We can do this by multiplying both sides of Equation 1 by 3 (the coefficient of y in Equation 2) and multiplying both sides of Equation 2 by -4 (the *negative* of the coefficient of y in Equation 1).

$$\begin{array}{l} 3\,]\;3x + 4y = 6 \Rightarrow 9x + 12y = 18 \\ -4\,]\;2x + 3y = 5 \Rightarrow \underline{-8x - 12y = -20} \\ \;x = -2 \end{array}$$

⌐ The coefficients of y add to zero

└ This pair of numbers is found by interchanging the coefficients of y and making one of them negative

The value $x = -2$ can be substituted into either Equation 1 or Equation 2 to find the value of y. Substituting into Equation 2, we have

$$2x + 3y = 5 \qquad \text{Equation 2}$$

$$2(-2) + 3y = 5 \qquad \text{Substituting } -2 \text{ for } x$$

$$-4 + 3y = 5$$
$$\underline{+4 \quad +4} \qquad \text{Adding 4 to both sides}$$
$$3y = 9$$

$$y = 3 \qquad \text{Dividing both sides by 3}$$

The solution, $(-2, 3)$, is, of course, the same solution we obtained in Example 3a. The check will not be shown. ∎

We can always use the method shown in Example 3 and find the pair of numbers to use as multipliers by interchanging the coefficients of one of the variables. Sometimes, however, smaller numbers can be used; the *smallest* possible value for the new coefficients is the *least common multiple* (LCM) of both of the original coefficients (see Example 4).

Example 4   Find the smallest numbers that can be used to eliminate (a) the x's and (b) the y's for the system $\begin{cases} 10x - 9y = 5 \\ 15x + 6y = 4 \end{cases}$. (Do not solve the system.)

Solution

a. The LCM of 10 and 15 (the coefficients of the x's) is 30. Therefore, we want the new coefficients to be 30 and -30. However, instead of determining the LCM beforehand, the work can be done as follows:

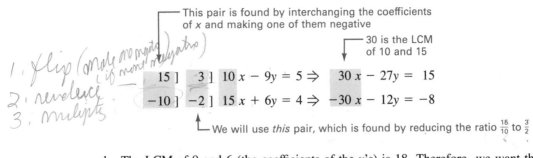

This pair is found by interchanging the coefficients of x and making one of them negative

30 is the LCM of 10 and 15

$$15\,]\quad 3\,]\quad 10x - 9y = 5 \Rightarrow \quad 30x - 27y = 15$$
$$-10\,]\quad -2\,]\quad 15x + 6y = 4 \Rightarrow \quad -30x - 12y = -8$$

We will use *this* pair, which is found by reducing the ratio $\frac{15}{10}$ to $\frac{3}{2}$

b. The LCM of 9 and 6 (the coefficients of the y's) is 18. Therefore, we want the new coefficients to be 18 and -18. Because the coefficients of y already have different signs, we do *not* need to make one of the multipliers negative. The work can be done as follows:

This pair is found by interchanging the coefficients of y

18 is the LCM of 9 and 6

$$6\,]\quad 2\,]\quad 10x - 9y = 5 \Rightarrow \quad 20x - 18y = 10$$
$$9\,]\quad 3\,]\quad 15x + 6y = 4 \Rightarrow \quad 45x + 18y = 12$$

We will use *this* pair, which is found by reducing the ratio $\frac{6}{9}$ to $\frac{2}{3}$ ∎

Example 5 Solve the system $\begin{cases} x - y = 6 \\ y + x = 2 \end{cases}$ by the addition method.

Solution In this example, the coefficients of the y's are already the negatives of each other. Therefore, we can omit step 1; when we add the equations, the y's will be eliminated. We must, however, be sure to line up like terms when we write the equations.

Equation 1: $x - y = 6$
Equation 2: $\underline{x + y = 2}$ Rearranging so like terms are lined up
$\qquad\quad 2x \qquad = 8$ Adding the equations
$\qquad\qquad\; x = 4$ Dividing both sides by 2

Substituting 4 for x in Equation 1, we have

$$x - y = \; 6 \qquad \text{Equation 1}$$
$$4 - y = \; 6 \qquad \text{Substituting 4 for } x$$
$$\underline{-4 \qquad\quad -4} \qquad \text{Adding } -4 \text{ to both sides}$$
$$-y = \; 2$$
$$y = -2 \qquad \text{Multiplying both sides by } -1$$

The system $\begin{cases} x = 4 \\ y = -2 \end{cases}$ is equivalent to the system $\begin{cases} x - y = 6 \\ y + x = 2 \end{cases}$, because both systems have the same solution set.

Check See Example 1a, Section 10.1.

Therefore, the solution is $(4, -2)$. ∎

Example 6   Solve the system $\left\{\begin{array}{l}3y + 2x = 4 \\ 5x + 6y = 11\end{array}\right\}$.

Solution We first arrange the equations so like terms are lined up.

$$5] \; 2x + 3y = 4 \Rightarrow \quad 10x + 15y = 20$$
$$-2] \; 5x + 6y = 11 \Rightarrow \quad \underline{-10x - 12y = -22}$$
$$3y = -2 \qquad \text{Adding the equations}$$
$$y = -\frac{2}{3} \qquad \text{Dividing both sides by 3}$$

Substituting $-\frac{2}{3}$ for y in $2x + 3y = 4$, we have

$$2x + 3y = 4$$
$$2x + \frac{\overset{1}{\cancel{3}}}{1}\left(\frac{-2}{\underset{1}{\cancel{3}}}\right) = 4$$
$$2x - 2 = 4$$
$$\underline{+2 \quad +2} \qquad \text{Adding 2 to both sides}$$
$$2x = 6$$
$$x = 3 \qquad \text{Dividing both sides by 2}$$

The check, which is left to the student, will confirm that the solution of the system is $\left(3, -\frac{2}{3}\right)$. ■

In all the examples so far, when we attempted to solve the system by the addition method, we had one variable left after we added the equations. Therefore, there was exactly one solution for each system. Each system was consistent and independent, and the lines representing the equations intersected at exactly one point.

Solving Systems with No Solution by the Addition Method

Sometimes when we attempt to solve a system of equations by the addition method, both variables drop out and we're left with a *false statement*. In this case, there is no solution for the system. The system is inconsistent, and the graphs of the equations are parallel lines.

Example 7   Solve the system $\left\{\begin{array}{l}4x - 2y = 5 \\ 2x - y = 2\end{array}\right\}$ by the addition method.

Solution

Equation 1: $-2]$ $-1] \, 4x - 2y = 5 \Rightarrow \quad -4x + 2y = -5$
Equation 2: $4]$ $2] \, 2x - y = 2 \Rightarrow \quad \underline{4x - 2y = 4}$
$$0 = -1 \qquad \text{False}$$

This is the pair we'll use

Because both variables were eliminated and we were left with a *false* statement, there is no solution for the system. The system is independent and inconsistent. ■

Solving Systems with More Than One Solution by the Addition Method

Sometimes when we attempt to solve a system of equations by the addition method, both variables drop out and we're left with a *true statement*. In this case, there are many solutions for the system; the graphs of the equations are lines that coincide.

Example 8   Solve the system $\begin{Bmatrix} 4x - 2y = 4 \\ 2x - y = 2 \end{Bmatrix}$ by the addition method.

Solution

Equation 1: $\boxed{-2}\ ] \quad \boxed{-1}\ ]\ 4x - 2y = 4 \Rightarrow \quad -4x + 2y = -4$

Equation 2: $\boxed{4}\ ] \quad \boxed{2}\ ]\ 2x - y = 2 \Rightarrow \quad \underline{4x - 2y = 4}$

$$0 = 0 \qquad \text{True}$$

This is the pair we'll use

Because both variables were eliminated and we were left with a *true* statement, there are many solutions for the system. Any ordered pair that satisfies Equation 1 or Equation 2 is a solution of the system. The system of equations is consistent and dependent. ■

EXERCISES 10.3

Set I   Find the solution of each system of equations by the addition method. Check your solutions. Write "Inconsistent" if no solution exists. Write "Dependent" if many solutions exist.

1. $\begin{Bmatrix} 2x - y = -4 \\ x + y = -2 \end{Bmatrix}$

2. $\begin{Bmatrix} 2x + y = 6 \\ x - y = 0 \end{Bmatrix}$

3. $\begin{Bmatrix} x - 2y = 10 \\ x + y = 4 \end{Bmatrix}$

4. $\begin{Bmatrix} x + 4y = 4 \\ x - 2y = -2 \end{Bmatrix}$

5. $\begin{Bmatrix} x - 3y = 6 \\ 4x + 3y = 9 \end{Bmatrix}$

6. $\begin{Bmatrix} 2x + 5y = 2 \\ 3x - 5y = 3 \end{Bmatrix}$

7. $\begin{Bmatrix} x + y = 2 \\ 3x - 2y = -9 \end{Bmatrix}$

8. $\begin{Bmatrix} x + 2y = -4 \\ 2x - y = -3 \end{Bmatrix}$

9. $\begin{Bmatrix} x + 2y = 0 \\ y - 2x = 0 \end{Bmatrix}$

10. $\begin{Bmatrix} x - 3y = 0 \\ 3y - x = 0 \end{Bmatrix}$

11. $\begin{Bmatrix} 4x + 3y = 2 \\ 3x + 5y = -4 \end{Bmatrix}$

12. $\begin{Bmatrix} 5x + 7y = 1 \\ 3x + 4y = 1 \end{Bmatrix}$

13. $\begin{Bmatrix} 6x - 10y = 6 \\ 9x - 15y = -4 \end{Bmatrix}$

14. $\begin{Bmatrix} 7x - 2y = 7 \\ 21x - 6y = 6 \end{Bmatrix}$

15. $\begin{Bmatrix} 3x - 5y = -2 \\ 10y - 6x = 4 \end{Bmatrix}$

16. $\begin{Bmatrix} 15x - 9y = -3 \\ 6y - 10x = 2 \end{Bmatrix}$

Set II   Find the solution of each system of equations by the addition method. Check your solutions. Write "Inconsistent" if no solution exists. Write "Dependent" if many solutions exist.

1. $\begin{Bmatrix} x - y = -5 \\ 3x + y = -3 \end{Bmatrix}$

2. $\begin{Bmatrix} x - 2y = 4 \\ -2x + 4y = 3 \end{Bmatrix}$

3. $\begin{Bmatrix} x + y = 5 \\ x - 3y = 3 \end{Bmatrix}$

4. $\begin{Bmatrix} x - 2y = 3 \\ 3x + 7y = -4 \end{Bmatrix}$

5. $\begin{Bmatrix} x - 2y = 6 \\ 3x - 2y = 12 \end{Bmatrix}$

6. $\begin{Bmatrix} 3x + 5y = 4 \\ 2x + 3y = 4 \end{Bmatrix}$

7. $\begin{Bmatrix} 2x + 6y = 2 \\ 3x + 9y = 3 \end{Bmatrix}$

8. $\begin{Bmatrix} 4x - 3y = 7 \\ 2x + 7y = 12 \end{Bmatrix}$

9. $\begin{Bmatrix} y - 5x = 0 \\ x - 5y = 0 \end{Bmatrix}$

10. $\begin{Bmatrix} 5x + 2y = 7 \\ 10x - 4y = 14 \end{Bmatrix}$

11. $\begin{Bmatrix} 7x + 2y = 14 \\ 2x - 3y = 4 \end{Bmatrix}$

12. $\begin{Bmatrix} 6x + 3y = 4 \\ y + 2x = 2 \end{Bmatrix}$

13. $\begin{Bmatrix} 2x - 5y = 4 \\ 6x - 15y = 8 \end{Bmatrix}$

14. $\begin{Bmatrix} 5x + 3y = -12 \\ 3y - 6x = 10 \end{Bmatrix}$

15. $\begin{Bmatrix} 6x + 4y = 2 \\ 3x - 2y = 3 \end{Bmatrix}$

16. $\begin{Bmatrix} 9x + 3y = 0 \\ 2x - 5y = 0 \end{Bmatrix}$

10.4 Substitution Method for Solving a Linear System

All linear systems of two equations in two variables can be solved by the addition method shown in Section 10.3. However, it is important that you also learn the **substitution method** of solution so that you can apply it later in solving more complicated systems.

TO SOLVE A LINEAR SYSTEM OF TWO EQUATIONS BY THE SUBSTITUTION METHOD

1. Solve one equation for one of the variables in terms of the other by using the method for solving literal equations given in Section 8.10.

2. Substitute the expression obtained in step 1 into the other equation. Simplify both sides of the resulting equation.

3. Three outcomes are possible:
   a. *One variable remains.* Solve the equation resulting from step 2 for its variable, and substitute that value into either equation to find the value of the other variable. There is one solution for the system. The system is consistent and independent.
   b. *Both variables are eliminated and a false statement results.* There is no solution; the system is inconsistent and independent.
   c. *Both variables are eliminated and a true statement results.* There are many solutions; the system is consistent and dependent.

4. If one solution was found in step 3, check it in both equations.

When students use the substitution method, they often worry about which equation to start with and which variable to solve for. Sometimes one of the equations is already solved for a variable (see Examples 1 and 4); if so, use that equation and variable. Sometimes in one of the equations the coefficient of one of the variables is 1 (see Examples 2 and 5); if so, solve that equation for the variable with a coefficient of 1. Otherwise, solve for the variable with the smallest coefficient (see Examples 3 and 6).

Example 1   Examples of systems with one equation already solved for one variable:

a. $\begin{cases} x + y = 4 \\ y = x + 2 \end{cases}$

　　　　└── Already solved for *y*

　　┌── Already solved for *x*

b. $\begin{cases} x = -3 - y \\ 2x + 5y = 0 \end{cases}$ ■

Example 2   Examples of systems in which one variable has a coefficient of 1:

┌── *y* has a coefficient of 1

a. $\begin{cases} 2x + y = 5 \\ 3x - 2y = 18 \end{cases} \Rightarrow y = 5 - 2x$

b. $\begin{Bmatrix} 2x + 6y = 3 \\ x - 4y = 2 \end{Bmatrix} \Rightarrow x = 4y + 2$

↑
x has a coefficient of 1   ∎

Example 3 Examples of choosing the variable with the smallest coefficient:

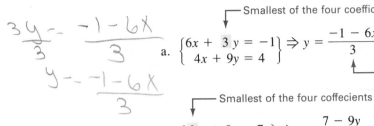

a.   ┌─ Smallest of the four coefficients
$\begin{Bmatrix} 6x + 3\,y = -1 \\ 4x + 9y = 4 \end{Bmatrix} \Rightarrow y = \dfrac{-1 - 6x}{3}$ We solved for y by using the method given in Section 8.10
└─ Smallest possible denominator

b.   ┌─ Smallest of the four coffecients
$\begin{Bmatrix} 3\,x + 9y = 7 \\ 12x - 8y = 15 \end{Bmatrix} \Rightarrow x = \dfrac{7 - 9y}{3}$ We solved for x by using the method given in Section 8.10
└─ Smallest possible denominator   ∎

Solving Systems with One Solution by the Substitution Method

Example 4 Solve the system $\begin{Bmatrix} x + y = 4 \\ y = x + 2 \end{Bmatrix}$ by the substitution method.

Solution

Equation 1: $x + y = 4$
Equation 2: $y = x + 2$

Step 1. Equation 2 has already been solved for y: $y = x + 2$.

Step 2. Substitute $x + 2$ for y in Equation 1:

$$x + y \qquad = 4$$
$$x + (x + 2) = 4$$

Step 3. $2x + 2 = 4$ Both sides are simplified; the variable x remains
$$\dfrac{-2 \qquad -2}{2x \quad = \quad 2}$$ Adding -2 to both sides

$x = 1$ Dividing both sides by 2

To find y, substitute 1 for x in Equation 2:

$$y = x + 2$$
$$y = 1 + 2$$
$$y = 3$$

Step 4. (1) $x + y = 4$ (2) $y = x + 2$
$1 + 3 \overset{?}{=} 4$ $3 \overset{?}{=} 1 + 2$
$4 = 4$ True $3 = 3$ True

Therefore, (1, 3) is the solution for this system.   ∎

Example 5   Solve the system $\left\{ \begin{array}{l} 2x + y = 5 \\ 3x - 2y = 18 \end{array} \right\}$ by the substitution method.

Solution

Equation 1: $2x + y = 5$
Equation 2: $3x - 2y = 18$

Step 1. The coefficient of y in Equation 1 is 1; therefore, solve Equation 1 for y:

$$\begin{array}{rl} 2x + y &= 5 \\ -2x \qquad\quad &\quad -2x \\ \hline y &= 5 - 2x \end{array}$$ Adding $-2x$ to both sides

Step 2. Substitute $5 - 2x$ for y in Equation 2:

$$3x - 2y = 18$$
$$3x - 2(5 - 2x) = 18$$
$$3x - 10 + 4x = 18$$

Step 3. $\begin{array}{rl} 7x - 10 = & 18 \\ + 10 & +10 \\ \hline 7x \quad = & 28 \end{array}$ Both sides are simplified; the variable x remains
                                                                                                              Adding 10 to both sides

$$x = 4$$ Dividing both sides by 7

To find y, substitute 4 for x in $y = 5 - 2x$.

$$y = 5 - 2x$$
$$y = 5 - 2(4)$$
$$y = 5 - 8$$
$$y = -3$$

The check, which is left to the student, will verify that $(4, -3)$ is the solution for this system. ∎

⌐Smallest of the four coefficients

Example 6   Solve the system $\left\{ \begin{array}{l} 6x + 3y = -1 \\ 4x + 9y = 4 \end{array} \right\}$ by the substitution method.

Solution

Equation 1: $6x + 3y = -1$
Equation 2: $4x + 9y = 4$

Step 1. No variable has a coefficient of 1. Solve Equation 1 for y because it has the smallest coefficient.

$$\begin{array}{rl} 6x + 3y &= -1 \\ -6x \qquad\quad &\quad -6x \\ \hline 3y &= -1 - 6x \end{array}$$ Adding $-6x$ to both sides

$$y = \frac{-1 - 6x}{3}$$ Dividing both sides by 3

Step 2. Substitute $\dfrac{-1-6x}{3}$ for y in Equation 2:

$$4x + 9y = 4$$

$$4x + \overset{3}{\cancel{9}}\left(\dfrac{-1-6x}{\underset{1}{\cancel{3}}}\right) = 4$$

$$4x + 3(-1-6x) = 4$$

$$
\begin{array}{rl}
4x - 3 - 18x = & 4 \\
\underline{+3 \qquad\qquad} & \underline{+3} \\
-14x = & 7
\end{array}
$$
Adding 3 to both sides

$$x = \dfrac{7}{-14} = -\dfrac{1}{2} \qquad \text{Dividing both sides by } -14$$

Step 3. Substitute $-\dfrac{1}{2}$ for x in $y = \dfrac{-1-6x}{3}$:

$$y = \dfrac{-1 - \overset{3}{\cancel{6}}\left(-\dfrac{1}{\underset{1}{\cancel{2}}}\right)}{3} = \dfrac{-1 - 3(-1)}{3} = \dfrac{-1+3}{3} = \dfrac{2}{3}$$

The check, which is left to the student, will confirm that $\left(-\frac{1}{2}, \frac{2}{3}\right)$ is the solution for this system. ∎

Solving Systems with No Solution by the Substitution Method

Example 7 shows how to identify systems that have no solution when you are using the substitution method.

Coefficient is 1

Example 7 Solve the system $\begin{Bmatrix} x + 3y = 4 \\ 2x + 6y = 4 \end{Bmatrix}$ using the substitution method.

Solution

Equation 1: $x + 3y = 4$
Equation 2: $2x + 6y = 4$

Step 1. Solve Equation 1 for x:

$$x + 3y = 4 \Rightarrow x = 4 - 3y \qquad \text{Adding } -3y \text{ to both sides}$$

Step 2. Substitute $4 - 3y$ for x in Equation 2:

$$2x + 6y = 4$$

$$2(4 - 3y) + 6y = 4$$

$$8 - 6y + 6y = 4$$

Step 3: $\qquad\qquad\qquad\qquad\qquad\qquad 8 = 4 \qquad \text{A false statement}$

Because both variables were eliminated and we were left with a *false* statement, there is *no solution* for this system of equations. ∎

Solving Systems with More Than One Solution by the Substitution Method

Example 8 shows how to identify systems that have more than one solution when you are using the substitution method.

Example 8   Solve the system $\begin{cases} 6x + 3y = 6 \\ 2x + y = 2 \end{cases}$ using the substitution method.

Coefficient is 1

Solution

Equation 1: $6x + 3y = 6$
Equation 2: $2x + y = 2$

Step 1. Solve Equation 2 for y:

$$2x + y = 2 \Rightarrow y = 2 - 2x \qquad \text{Adding } -2x \text{ to both sides}$$

Step 2. Substitute $2 - 2x$ for y in Equation 1:

$$6x + 3y = 6$$

$$6x + 3(2 - 2x) = 6$$

$$6x + 6 - 6x = 6$$

Step 3. $6 = 6$ A true statement

Because both variables were eliminated and we were left with a *true* statement, this system of equations has many solutions. Any ordered pair that satisfies Equation 1 also satisfies Equation 2. ∎

EXERCISES 10.4

Set I   Find the solution of each system of equations. Write "Inconsistent" if no solution exists. Write "Dependent" if many solutions exist. In Exercises 1–10, use the substitution method.

1. $\begin{cases} 2x - 3y = 1 \\ x = y + 2 \end{cases}$ 2. $\begin{cases} y = 2x + 3 \\ 3x + 2y = 20 \end{cases}$ 3. $\begin{cases} 3x + 4y = 2 \\ y = x - 3 \end{cases}$

4. $\begin{cases} 2x + 3y = 11 \\ x = y - 2 \end{cases}$ 5. $\begin{cases} 4x + y = 2 \\ 7x + 3y = 1 \end{cases}$ 6. $\begin{cases} 5x + 7y = 1 \\ x + 4y = -5 \end{cases}$

7. $\begin{cases} 4x - y = 3 \\ 8x - 2y = 6 \end{cases}$ 8. $\begin{cases} x - 3y = 2 \\ 3x - 9y = 6 \end{cases}$ 9. $\begin{cases} x + 3 = 0 \\ 3x - 2y = 6 \end{cases}$

10. $\begin{cases} y - 4 = 0 \\ 3y - 5x = 15 \end{cases}$

In Exercises 11–18, use any convenient method.

11. $\begin{cases} 8x + 4y = 7 \\ 3x + 6y = 6 \end{cases}$ 12. $\begin{cases} 5x - 4y = 2 \\ 15x + 12y = 12 \end{cases}$ 13. $\begin{cases} 3x - 2y = 8 \\ 2y - 3x = 4 \end{cases}$

14. $\begin{cases} 4x - 5y = 15 \\ 5y - 4x = 10 \end{cases}$ 15. $\begin{cases} 8x + 5y = 2 \\ 7x + 4y = 1 \end{cases}$ 16. $\begin{cases} 4x - 9y = 7 \\ 3x - 8y = 4 \end{cases}$

17. $\begin{cases} 4x + 4y = 3 \\ 6x + 12y = -6 \end{cases}$ **18.** $\begin{cases} 4x + 9y = -11 \\ 10x + 6y = 11 \end{cases}$

Set II   Find the solution of each system of equations. Write "Inconsistent" if no solution exists. Write "Dependent" if many solutions exist. In Exercises 1–10, use the substitution method.

1. $\begin{cases} x - y = 1 \\ y = 2x - 3 \end{cases}$ **2.** $\begin{cases} x + 3y = 4 \\ x = 2 - y \end{cases}$ **3.** $\begin{cases} x + 2y = -1 \\ x = 5 + y \end{cases}$

4. $\begin{cases} x + y = 1 \\ y = x + 7 \end{cases}$ **5.** $\begin{cases} x - 4y = 9 \\ 3x + 8y = 7 \end{cases}$ **6.** $\begin{cases} 3x + 4y = 18 \\ 5x - y = 7 \end{cases}$

7. $\begin{cases} 5x + y = 6 \\ 10x + 2y = 12 \end{cases}$ **8.** $\begin{cases} 2x + 5y = -5 \\ 3y + x = -2 \end{cases}$ **9.** $\begin{cases} 5x + y = 0 \\ 3x + 2y = 7 \end{cases}$

10. $\begin{cases} x + y = 4 \\ 2x - y = 2 \end{cases}$

In Exercises 11–18, use any convenient method.

11. $\begin{cases} 7x + 5y = -4 \\ 4x + 3y = -2 \end{cases}$ **12.** $\begin{cases} 4x + 7y = 9 \\ 6x + 5y = -3 \end{cases}$ **13.** $\begin{cases} 2x - y = 3 \\ y = 2x + 1 \end{cases}$

14. $\begin{cases} 3x + 6y = 9 \\ 4x + 8y = 12 \end{cases}$ **15.** $\begin{cases} 8x + 3y = 4 \\ 2x - 3y = 6 \end{cases}$ **16.** $\begin{cases} x + 3y = 0 \\ x - 2y = 10 \end{cases}$

17. $\begin{cases} 3x + 2y = 7 \\ 6x - 4y = 7 \end{cases}$ **18.** $\begin{cases} 7x - 2y = 5 \\ 5x + 3y = 8 \end{cases}$

10.5 Using Systems of Equations to Solve Word Problems

In solving word problems involving more than one unknown, it is sometimes difficult to represent each unknown in terms of a single variable. In this section, we eliminate that difficulty by using a different variable for each unknown. When we use two variables, we must find two equations to represent the given facts; we then use a system of equations to solve the problem.

TO SOLVE A WORD PROBLEM USING A SYSTEM OF EQUATIONS

Read   First read the problem very carefully.

Think   Determine what *type* of problem it is, if possible. Determine what is unknown.

Sketch   Draw a sketch *with labels*, if possible.

Step 1   Represent each unknown number by a *different* variable.

Reread   Reread the word problem, breaking it up into small pieces.

> **Step 2** Translate each English phrase into an algebraic expression; fit these expressions together into *two different equations*.
>
> **Step 3** Solve the *system* of equations for *both* variables, using one of the following:
>   a. Addition method (Section 10.3)
>   b. Substitution method (Section 10.4)
>   c. Graphical method (Section 10.2)
>
> **Step 4** Be sure you've answered *all* the questions asked.
>
> **Step 5** Check the solutions *in the word statements*.
>
> **Step 6** State your results clearly.

In Example 1, the word problem is solved on the left by using a single variable and a single equation. On the right, it is solved by using two variables and a system of equations. (We will not show "Step 1," "Step 2," and so on in this example.)

Example 1 The sum of two numbers is 20. Their difference is 6. What are the numbers?
Solution

Using one variable

Let $\quad x$ = larger number

$20 - x$ = smaller number

Their difference	is	6

$$x - (20 - x) = \quad 6$$
$$x - 20 + x = \quad 6$$
$$\underline{+ 20 \qquad +20}$$
$$2x = \quad 26$$

Larger number $\qquad x = 13$

Smaller number $\quad 20 - x = 7$

The difficulty in using the one-variable method to solve this problem is that some students cannot decide whether to represent the second unknown number by $x - 20$ or by $20 - x$.

Using two variables

Let x = larger number

$\qquad y$ = smaller number

The sum of two numbers	is	20

(1) $\qquad x + y \qquad = \quad 20$

Their difference	is	6

(2) $\qquad x - y \qquad = \quad 6$

Using the addition method:

$$\begin{aligned}(1)\ & \left\{ \begin{aligned} x + y &= 20 \\ x - y &= 6 \end{aligned} \right\} \\ (2)\ & \\ 2x \quad &= 26 \quad \text{Adding (1) and (2)} \\ x &= 13 \quad \text{Larger number}\end{aligned}$$

Substituting 13 for x in Equation 1, we have

$$x + y = 20$$
$$13 + y = 20$$
$$y = 7 \qquad \text{Smaller number}$$

Check

$$13 + 7 = 20 \quad \text{The sum of the numbers is 20.}$$
$$13 - 7 = 6 \quad \text{The difference of the numbers is 6.}$$

Therefore, the numbers are 13 and 7. ∎

512

Example 2 Kevin spent $3.76 on eighteen stamps, buying only 18¢ and 22¢ stamps. How many of each kind did he buy?

Solution

Step 1. Let x = the number of 18¢ stamps

y = the number of 22¢ stamps

Reread	Number of 18¢ stamps	+	number of 22¢ stamps	=	total number of stamps

Step 2. x + y =   18

Reread	Value of 18¢ stamps	+	value of 22¢ stamps	=	total value of stamps

Step 2. $18x$ + $22y$ =   376   Value in cents

Step 3. Using the addition method:

$$(1) \begin{cases} x + y = 18 \\ (2) \ 18x + 22y = 376 \end{cases}$$

$$\begin{array}{rl} -18x - 18y = -324 & \text{Multiplying Equation 1 by } -18 \\ \underline{18x + 22y = \ \ 376} & \text{Equation 2} \\ 4y = \ \ 52 & \text{Adding the equations} \\ y = \ \ 13 & \text{The number of 22¢ stamps} \end{array}$$

Substituting 13 for y into Equation 1, we have

$$\begin{array}{rl} x + y = & 18 \\ x + 13 = & 18 \\ \underline{-13} & \underline{-13} \end{array}$$

Step 4. x =   5   The number of 18¢ stamps

Step 5. *Check* $13 + 5 = 18$ There are eighteen stamps altogether.

$13(\$0.22) + 5(\$0.18) = \$2.86 + \$0.90 = \$3.76$ The total value of the stamps is $3.76.

Step 6. Therefore, Kevin bought thirteen 22¢ stamps and five 18¢ stamps. ∎

Example 3 A fraction has the value $\frac{2}{3}$. If 3 is added to the numerator and 2 is added to the denominator, the resulting fraction has the value $\frac{3}{4}$. Find the original fraction.

Solution

Step 1. Let x = the numerator of the original fraction

y = the denominator of the original fraction

Step 2.   (1) $\begin{cases} \dfrac{x}{y} = \dfrac{2}{3} \\[2mm] \dfrac{x+3}{y+2} = \dfrac{3}{4} \end{cases}$ The value of the original fraction is $\frac{2}{3}$

(2)   The value of the new fraction is $\frac{3}{4}$

$\dfrac{x}{y} = \dfrac{2}{3}$

2. $\dfrac{x+3}{y+2} = \dfrac{3}{4}$

1. $x = \dfrac{2}{3}y$

2. $\dfrac{2}{3}y + 3 = \dfrac{3}{4}(y+2)$

3. $\dfrac{2}{3}y + 3 = \dfrac{3}{4}y + \dfrac{3}{2}$

$\dfrac{3}{2} = \dfrac{1}{12}y$

$18 = y$

1) $3x = 2y$
2) $4(x+3) = 3(y+2)$

L.1 $3x - 2y = 0$
L.2 $4x + 12 = 3y + 6$
$= 4x - 3y = -6$

L.1 $3\,|\,3x - 2y = 0$
-2) $4x - 3y = -6$
$9x - 6y = 0$
$-8x + 6y = 12$
$x = 12$

$3(12) - 2y = 0$
$36 - 2y = 0$
$-2y = -36$
$\dfrac{-2y}{-2} = \dfrac{-36}{-2}$
$y = 18$

Before we can solve the system of equations, we must clear fractions and line up the like terms:

Step 3.

(1) $\qquad 3x = 2y \qquad$ Multiplying both sides by the LCD or cross-multiplying

(2) $4(x + 3) = 3(y + 2) \qquad$ Multiplying both sides by the LCD or cross-multiplying

(1) $\qquad 3x = 2y \qquad \Rightarrow 3x - 2y = 0 \qquad$ Verify this

(2) $4(x + 3) = 3(y + 2) \Rightarrow 4x + 12 = 3y + 6 \Rightarrow 4x - 3y = -6 \qquad$ Verify this

Therefore, the system is

$$
\begin{array}{ll}
(1) & \left\{\begin{array}{l} 3x - 2y = 0 \\ 4x - 3y = -6 \end{array}\right\}
\end{array}
$$

We now solve the system:

$$
\begin{aligned}
(1) \quad & 3]\; 3x - 2y = 0 \;\Rightarrow\; 9x - 6y = 0 \\
(2) \quad & -2]\; 4x - 3y = -6 \Rightarrow \underline{-8x + 6y = 12} \\
& \hphantom{-2]\; 4x - 3y = -6 \Rightarrow} x \hphantom{+ 6y} = 12
\end{aligned}
$$

Substituting 12 for x in Equation 1, we have

$$
\begin{aligned}
3x - 2y &= 0 \\
3(12) - 2y &= 0 \\
36 - 2y &= 0 \\
\underline{-36 \hphantom{- 2y}} &\;\; \underline{-36} \\
-2y &= -36 \\
y &= 18
\end{aligned}
$$

Step 4. The *original fraction* is apparently $\dfrac{12}{18}$.

Step 5. *Check* $\dfrac{12}{18} = \dfrac{2}{3}$ The value of the original fraction is $\dfrac{2}{3}$.

$\dfrac{12 + 3}{18 + 2} = \dfrac{15}{20} = \dfrac{3}{4}$ If 3 is added to the numerator and 2 is added to the denominator, the value of the resulting fraction is $\dfrac{3}{4}$.

Step 6. Therefore, the original fraction *is* $\dfrac{12}{18}$. ∎

Some word problems should be solved using one unknown, while others are best solved using a system of equations.

HOW TO CHOOSE WHICH METHOD TO USE FOR SOLVING A WORD PROBLEM

1. Read the problem completely and determine how many unknown numbers there are.

2. If there is only one unknown number, use the one-variable method.

3. If there is more than one unknown number, try to represent all the unknowns in terms of one variable. If this is too difficult, represent each unknown number by a different variable and then solve using a system of equations.

Systems that have more than two equations or more than two variables are not discussed in this book.

EXERCISES 10.5

Set I   Set up each of the following problems algebraically *using two variables*, solve, and check. Be sure to state what your variables represent.

1. The sum of two numbers is 80. Their difference is 12. What are the numbers?

2. The sum of two numbers is 90. Their difference is 4. What are the numbers?

3. The sum of two angles is 90°. Their difference is 60°. What are the angles?

4. The sum of two angles is 180°. Their difference is 32°. What are the angles?

5. Find two numbers such that twice the smaller plus 3 times the larger is 34, and 5 times the smaller minus twice the larger is 9.

6. Find two numbers such that 5 times the larger plus 3 times the smaller is 47, and 4 times the larger minus twice the smaller is 20.

7. Jason paid $16.88 for a 6-lb mixture of granola and dried apple chunks. If the granola cost $2.10 per pound and the dried apple chunks cost $4.24 per pound, how many pounds of each did he buy?

8. A 100-lb mixture of two different grades of coffee costs $351.50. If grade A costs $3.80 per pound and grade B costs $3.30 per pound, how many pounds of each grade were used?

9. Don spent $4.48 for twenty-two stamps. If he bought only 22¢ and 18¢ stamps, how many of each kind did he buy?

10. Sue spent $12.38 for fifty stamps. If she bought only 22¢ and 45¢ stamps, how many of each type did she buy?

11. The length of a rectangle is 1 ft 6 in. longer than its width. Its perimeter is 19 ft. Find its dimensions.

12. The length of a rectangle is 2 ft 6 in. longer than its width. Its perimeter is 25 ft. Find its dimensions.

13. A fraction has the value $\frac{2}{3}$. If 4 is added to the numerator and the denominator is decreased by 2, the resulting fraction has the value $\frac{6}{7}$. What is the original fraction?

14. A fraction has the value $\frac{3}{4}$. If 4 is added to its numerator and 8 is subtracted from its denominator, the value of the resulting fraction is 1. What is the original fraction?

15. A boat takes 7 hr to travel 252 mi with the current and takes 9 hr to travel the same distance against the same current. Find the average speed of the boat in still water and the average speed of the current.

16. A pilot takes 5 hr to fly 450 mi against the wind and only 3 hr to return with the wind. Find the average speed of the plane in still air and the average speed of the wind.

17. A tie and a pin cost $1.10. The tie costs $1 more than the pin. What is the cost of each?

18. A number of birds are resting on two limbs of a tree. One limb is above the other. A bird on the lower limb says to the birds on the upper limb, "If one of you will come down here, we will have an equal number on each limb." A bird from above replies, "If one of you will come up here we will have twice as many up here as you will have down there." How many birds are sitting on each limb?

Set II Set up each of the following problems algebraically *using two variables*, solve, and check. Be sure to state what your variables represent.

1. The sum of two numbers is 42. Their difference is 12. What are the numbers?

2. Half the sum of two numbers is 15. Half their difference is 8. Find the numbers.

3. The sum of two angles is 90°. Their difference is 16°. Find the angles.

4. A 20-lb mixture of Product A and Product B costs $19.75. If Product A costs 85¢ per pound and Product B costs $1.40 per pound, find the number of pounds of each.

5. Find two numbers such that twice the smaller plus 4 times the larger is 66, and 6 times the smaller minus 3 times the larger is 3.

6. Beatrice has seventeen coins with a total value of $3.05. If these coins are all nickels and quarters, how many of each kind does she have?

7. A 20-lb mixture of almonds and hazelnuts costs $53.00. If almonds cost $2.50 per pound and hazelnuts cost $3.00 per pound, find the number of pounds of each.

8. A fraction has the value $\frac{1}{2}$. If 2 is added to the numerator and 1 is subtracted from the denominator, the value of the resulting fraction is $\frac{4}{7}$. Find the original fraction.

9. Rachelle spent $4.55 for fifteen stamps. If she bought only 25¢ and 35¢ stamps, how many of each kind did she buy?

10. A class received $233 for selling 200 tickets to the school play. If student tickets cost $1 each and nonstudent tickets cost $2 each, how many nonstudent tickets were sold?

11. The length of a rectangle is 3 ft 6 in. longer than its width. Its perimeter is 13 ft. Find its dimensions.

12. Jill spent $59.50 for fourteen tickets to a movie. Some were for children and some for adults. If children's tickets cost $3.50 and adults' tickets cost $5.25, how many of each kind did she buy?

13. A fraction has the value $\frac{1}{2}$. If 4 is added to the numerator and 2 is subtracted from the denominator, the resulting fraction has the value $\frac{2}{3}$. What is the original fraction?

14. Several families went to a school play together. They spent $10.60 for eight tickets. If adults' tickets cost $1.95 and children's tickets cost 95¢, how many of each kind of ticket were bought?

15. Wing takes 6 hr to ride his bicycle 36 mi against the wind and only 2 hr to return with the wind. Find the average riding speed in still air and the average speed of the wind.

16. A 100-lb mixture of Product A and Product B costs $22. If Product A costs 20¢ per pound and Product B costs 25¢ per pound, how many pounds of each kind are there in the mixture?

17. A bracelet and a pin cost $11. One costs $10 more than the other. What is the cost of each?

18. A pilot takes 7 hr to fly 875 mi against the wind and only 5 hr to return with the wind. Find the average speed of the plane in still air and the average speed of the wind.

10.6 Review: 10.1–10.5

Solution of a System of Equations
10.1

A solution of a system of two equations in two unknowns is an ordered pair that satisfies both equations.

Solving a System of Equations
10.2, 10.3, 10.4

In solving a system of equations, there are three possibilities:

1. There is only one solution.
   a. Graphical method: The lines intersect at one point.
   b. Algebraic methods:
      Addition   } The equations can be solved for
      Substitution   } a single ordered pair.

2. There is no solution.
   a. Graphical method: The lines are parallel.
   b. Algebraic methods:
      Addition   } Both variables drop out and
      Substitution   } a false statement results.

3. There are many solutions.
   a. Graphical method: Both equations have the same line for a graph.
   b. Algebraic methods:
      Addition   } Both variables drop out and
      Substitution   } a true statement results.

To Solve a Word Problem Using a System of Equations
10.5

Read   First read the problem very carefully.

Think   Determine what type of problem it is, if possible. Determine what is unknown.

Sketch   Draw a sketch *with labels*, if possible.

Step 1   Represent each unknown number by a *different* variable.

Reread   Reread the word problem, breaking it up into small pieces.

Step 2   Translate each English phrase into an algebraic expression; fit these expressions together into *two different equations*.

Step 3 Solve the *system* of equations for *both* variables, using one of the following:
    a. Addition method (Section 10.3)
    b. Substitution method (Section 10.4)
    c. Graphical method (Section 10.2)

Step 4 Be sure you've answered *all* the questions asked.

Step 5 Check the solutions *in the word statements*.

Step 6 State your results clearly.

Review Exercises 10.6 Set I

In Exercises 1–3, find the solution of each system graphically. Write "Inconsistent" if no solution exists. Write "Dependent" if many solutions exist.

1. $\begin{cases} x + y = 6 \\ x - y = 4 \end{cases}$

2. $\begin{cases} x + y = 5 \\ x - y = -3 \end{cases}$

3. $\begin{cases} 3x - y = 6 \\ 6x - 2y = 12 \end{cases}$

In Exercises 4–6, find the solution of each system, using the addition method. Write "Inconsistent" if no solution exists. Write "Dependent" if many solutions exist.

4. $\begin{cases} 4x + 5y = 22 \\ 3x + y = 11 \end{cases}$

5. $\begin{cases} 4x - 8y = 4 \\ 3x - 6y = 3 \end{cases}$

6. $\begin{cases} 4x - 7y = 28 \\ 7y - 4x = 20 \end{cases}$

In Exercises 7–9, find the solution of each system, using the substitution method. Write "Inconsistent" if no solution exists. Write "Dependent" if many solutions exist.

7. $\begin{cases} x = y + 2 \\ 4x - 5y = 3 \end{cases}$

8. $\begin{cases} 7x - 3y = 1 \\ y = x + 5 \end{cases}$

9. $\begin{cases} x + y = 4 \\ 2x - y = 2 \end{cases}$

In Exercise 10, solve the system by any convenient method. Write "Inconsistent" if no solution exists. Write "Dependent" if many solutions exist.

10. $\begin{cases} 5x - 4y = -7 \\ -6x + 8y = 2 \end{cases}$

In Exercises 11–14, set up each problem algebraically *using two variables*, solve, and check. Be sure to state what your variables represent.

11. The sum of two numbers is 84. Their difference is 22. What are the numbers?

12. A fraction has the value $\frac{4}{5}$. If 3 is subtracted from the numerator and 5 is added to the denominator, the value of the resulting fraction is $\frac{3}{5}$. What is the original fraction?

13. An office manager paid $107.75 for twenty boxes of paper for the copy machine and the laser printer. If paper for the laser printer costs $7.50 per box and paper for the copy machine costs $4.25 per box, how many boxes of each kind were purchased?

14. A boat takes 7 hr to travel 231 mi with the current and 11 hr to travel the same distance against the same current. Find the average speed of the boat in still water and the average speed of the current.

Review Exercises 10.6 Set II

In Exercises 1 and 2, find the solution of each system graphically. Write "Inconsistent" if no solution exists. Write "Dependent" if many solutions exist.

ANSWERS

1. $\begin{cases} 3x + y = -9 \\ 3x - 2y = 0 \end{cases}$

1. _____

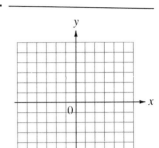

2. $\begin{cases} 2x - 3y = 3 \\ 3y - 2x = 6 \end{cases}$

2. _____

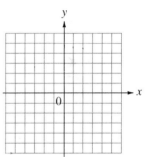

In Exercises 3–5, find the solution of each system using the addition method. Write "Inconsistent" if no solution exists. Write "Dependent" if many solutions exist.

3. _____

3. $\begin{cases} 3x - 2y = 10 \\ 5x + 4y = 24 \end{cases}$

4. $\begin{cases} 6x + 4y = -1 \\ 4x + 6y = -9 \end{cases}$

4. _____

5. _____

6. _____

5. $\begin{cases} 6x + 4y = 13 \\ 8x + 10y = 1 \end{cases}$

In Exercises 6 and 7, solve each system using the substitution method. Write "Inconsistent" if no solution exists. Write "Dependent" if many solutions exist.

6. $\begin{cases} x - 3y = 15 \\ 5x + 7y = -13 \end{cases}$

7. $\begin{cases} 4x - 6y = 2 \\ 6x - 9y = 3 \end{cases}$

In Exercises 8–10, solve each system by any convenient method. Write "Inconsistent" if no solution exists. Write "Dependent" if many solutions exist.

8. $\begin{cases} 5x + 4y = 2 \\ 2x + 5y = 11 \end{cases}$ **9.** $\begin{cases} 3x - 5y = 2 \\ 2x + 7y = 22 \end{cases}$

10. $\begin{cases} 5x - 7y = 17 \\ 4x - 5y = 13 \end{cases}$

In Exercises 11–14, set up each problem algebraically *using two variables*, solve, and check. Be sure to state what your variables represent.

11. The sum of two numbers is 95. Their difference is 19. What are the numbers?

12. The width of a rectangle is 5 m shorter than its length. Its perimeter is 98 m. Find its dimensions.

13. An office manager paid $730 for a total of eighty rolls of two different kinds of tape. If one kind costs $10 per roll and the other kind costs $8 per roll, how many rolls of each kind were bought?

14. A fraction has the value $\frac{1}{3}$. If 5 is added to the numerator and 5 is subtracted from the denominator, the value of the resulting fraction is $\frac{1}{2}$. What is the original fraction?

Chapter 10 Diagnostic Test

The purpose of this test is to see how well you understand systems of equations. We recommend that you work this diagnostic test *before* your instructor tests you on this chapter. Allow yourself about 50 minutes.

Complete solutions for all the problems on this test, together with section references, are given in the answer section in the back of the book. For problems you do incorrectly, study the sections referred to.

In Problems 1–7, write "Inconsistent" if no solution exists, and write "Dependent" if many solutions exist.

1. Solve graphically: $\begin{Bmatrix} 2x + 3y = 6 \\ x + 3y = 0 \end{Bmatrix}$

2. Solve by addition: $\begin{Bmatrix} x + 3y = 5 \\ -x + 4y = 2 \end{Bmatrix}$

3. Solve by addition: $\begin{Bmatrix} 4x + 3y = 5 \\ 5x - 2y = 12 \end{Bmatrix}$

4. Solve by substitution: $\begin{Bmatrix} 2x - 5y = -20 \\ x = y - 7 \end{Bmatrix}$

5. Solve by any convenient method: $\begin{Bmatrix} 3x + 4y = -1 \\ 2x - 5y = -16 \end{Bmatrix}$

6. Solve by any convenient method: $\begin{Bmatrix} 5x + y = 10 \\ 10x + 2y = 9 \end{Bmatrix}$

7. Solve by any convenient method: $\begin{Bmatrix} 6x - 9y = 3 \\ 15y - 10x = -5 \end{Bmatrix}$

In Problems 8–10, set up each problem algebraically *using two variables*, solve, and check. Be sure to state what your variables represent.

8. The sum of two numbers is 13. Their difference is 37. What are the numbers?

9. A pilot takes 11 hr to fly 1,650 mi against the wind and only 7.5 hr to return with the wind. Find the average speed of the plane in still air and the average speed of the wind.

10. The length of a rectangle is 3 cm more than its width. Its perimeter is 98 cm. Find the dimensions of the rectangle.

Cumulative Review Exercises: Chapters 1–10

1. Solve for x: $\dfrac{2 + 5x}{6x - 1} = \dfrac{3}{4}$

2. Solve for x: $\dfrac{3x - 1}{4} + 6 = \dfrac{5 - x}{5}$

3. Add and simplify:

$$\frac{1 + a}{2 - 3a} + \frac{2 + a}{4a}$$

4. Graph the equation $5x - 7y = 15$.

5. Draw the graph of $4y = x^2$. Use the following values for x in your table of values: $-4, -2, -1, 0, 1, 2, 4$.

6. Solve the inequality $2(x - 3) \leq 4$ and graph its solution set on the number line.

7. Graph the inequality $3x - 4y > 12$ in the plane.

8. Write the general form of the equation of the line that passes through the points $A(-3, 2)$ and $B(4, -1)$.

In Exercises 9–12, solve each system of equations by any convenient method. Write "Inconsistent" if no solution exists. Write "Dependent" if many solutions exist.

9. $\begin{cases} 2x - 3y = 3 \\ 3y - 2x = 5 \end{cases}$

10. $\begin{cases} 8x - 4y = 12 \\ 6x - 3y = 9 \end{cases}$

11. $\begin{cases} x + 5y = 11 \\ 3x + 4y = 11 \end{cases}$

12. $\begin{cases} 6x + 8y = 15 \\ 4x + 2y = -5 \end{cases}$

In Exercises 13 and 14, factor the polynomial completely, or write "Not factorable."

13. $5x^2 - 2x - 13$

14. $7x^2 - 3x - 11$

In Exercises 15–18, set up each problem algebraically, solve, and check. Be sure to state what your variables represent.

15. A drugstore buys eighteen cameras. If it had bought ten cameras of a higher quality, it would have paid $48 more per camera for the same total expenditure. Find the price of each type of camera.

16. Mr. McAllister invested part of $27,300 at 14% and the remainder at 21%. His total yearly income from these investments is $4,648. How much is invested at each rate?

In Exercises 17 and 18, *use two variables*.

17. The sum of two numbers is 6. Their difference is 40. What are the numbers?

18. The width of a rectangle is 5 m less than its length. Its perimeter is 58 m. Find the dimensions of the rectangle.

11 Radicals

Roots of numbers were introduced in Section 1.10, and square roots of algebraic terms were introduced in Section 7.2. In this chapter, we discuss the simplification of square roots and operations involving square roots. We also solve radical equations and some word problems that lead to such equations.

11.1 Review of Square Roots

We learned in Section 1.10 that the *square root* of a number p is a number that, when squared, gives p. We also learned that every positive real number has both a positive and a negative square root and that the *positive* square root is called the *principal square root.* Recall that the symbol $\sqrt{p}$ represents the principal square root of p and that p is called the *radicand* (Figure 11.1.1).

$$\sqrt{p} \qquad \text{Read "The square root of } p\text{"}$$

Radical sign ———————————————— Radicand

FIGURE 11.1.1

Example 1   Examples of identifying the radicand:

a. $\sqrt{7}$ 7 is the radicand.

b. $\sqrt{2x}$ $2x$ is the radicand.

c. $\sqrt{\dfrac{3x}{2y}}$ $\dfrac{3x}{2y}$ is the radicand. ∎

Just as the number 4 has two square roots, 2 and -2, the algebraic expression x^2 has two square roots, x and $-x$. Since x itself can be either a positive or a negative number, we don't know which is the principal square root. However, we do know that $|x|$ must be positive (or zero); therefore, $|x|$ is the principal square root of x^2; in symbols, $\sqrt{x^2} = |x|$. However, in the remainder of this chapter, we assume that all variables represent positive numbers. For this reason, the absolute value symbol need not be used, and we will write $\sqrt{x^2} = x$.

The Square of a Square Root   As stated in Rule 11.1, when you square the square root of a number, you get that number.

RULE 11.1   THE SQUARE OF A SQUARE ROOT

$$(\sqrt{a})^2 = a$$

Example 2   Examples of squaring the square root of a number:

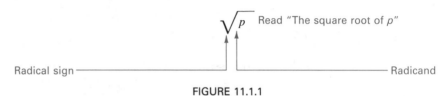

a. $(\sqrt{25})^2 = (5)^2 = 25$; therefore, $(\sqrt{25})^2 = 25$.

b. $(\sqrt{16})^2 = (4)^2 = 16$; therefore, $(\sqrt{16})^2 = 16$.

Replacing $\sqrt{121}$ with 11

c. $(\sqrt{121})^2 = (11)^2 = 121$; therefore, $(\sqrt{121})^2 = 121$. ∎

Perfect Square When a factor has an even exponent or is the square of an integer, it can be called a **perfect square**. In Example 2, all the radicands were perfect squares. However, Rule 11.1 applies even when a is not a perfect square.

Example 3 Examples of simplifying expressions by using the rule $(\sqrt{a})^2 = a$:

a. $(\sqrt{7})^2 = 7$ b. $(\sqrt{13})^2 = 13$

c. $(\sqrt{x})^2 = x$ d. $(\sqrt{cd^4})^2 = cd^4$

e. $(\sqrt{x + 3})^2 = x + 3$ ∎

EXERCISES 11.1

Set I In Exercises 1–6, identify the radicand.

1. $\sqrt{17}$ 2. $\sqrt{23}$ 3. $\sqrt{4xy^2}$

4. $\sqrt{9a^2b}$ 5. $\sqrt{x + 1}$ 6. $\sqrt{m - n}$

In Exercises 7–14, simplify each expression.

7. $(\sqrt{9})^2$ 8. $(\sqrt{100})^2$ 9. $(\sqrt{15})^2$

10. $(\sqrt{29})^2$ 11. $(\sqrt{3a^2})^2$ 12. $(\sqrt{7y^2})^2$

13. $(\sqrt{x + y})^2$ 14. $(\sqrt{u + v})^2$

Set II In Exercises 1–6, identify the radicand.

1. $\sqrt{35}$ 2. $\sqrt{82}$ 3. $\sqrt{25ab^2}$

4. $\sqrt{49c^2d}$ 5. $\sqrt{y + 5}$ 6. $\sqrt{s + t}$

In Exercises 7–14, simplify each expression.

7. $(\sqrt{49})^2$ 8. $(\sqrt{225})^2$ 9. $(\sqrt{31})^2$ 10. $(\sqrt{53})^2$

11. $(\sqrt{5r^2})^2$ 12. $(\sqrt{2m^2})^2$ 13. $(\sqrt{e + 4})^2$ 14. $(\sqrt{c + d})^2$

11.2 Simplifying Square Roots

We will consider square roots of two kinds:

1. Square roots that do not involve fractions

2. Square roots involving fractions

11.2A Square Roots That Do Not Involve Fractions

A number of conditions must be satisfied for a square root to be in *simplest radical form*. The *first* of these conditions is that no factor of the radicand can have an exponent greater than 1 when the radicand is expressed in prime factored, exponential form.

No EXPONANTS

In simplifying radicals, we may make use of the following property:

(No Negative Numbers) (handwritten)

RULE 11.2

$\sqrt{ab} = \sqrt{a}\sqrt{b}$, if $a \geq 0$ and $b \geq 0$

$$\text{The square root of a product} = \text{the product of the square roots}$$

Even exponent Divide By 2 (handwritten)

When the radicand contains a factor with an *even* exponent, we can remove that factor from the radical by dividing its exponent by 2, as was done in Section 7.2 (see Example 1).

Example 1   Examples of simplifying radicals:

a. $\sqrt{5^2x^2y^4} = 5^{2\div2}x^{2\div2}y^{4\div2} = 5xy^2$ *(handwritten)* $\sqrt{5^{(2)}x^{(2)}y^{(4)}} = 5xy^2$

b. $\sqrt{3^4z^6} = 3^{4\div2}z^{6\div2} = 3^2z^3$

c. $\sqrt{a^2} = a^{2\div2} = a^1 = a$ ∎

NOTE   We saw in Rule 11.1 that $(\sqrt{a})^2 = a$, and we see from Example 1c that $\sqrt{a^2} = a$. Since $(\sqrt{a})^2$ and $\sqrt{a^2}$ both equal the same number (a), they must equal each other. That is, $(\sqrt{a})^2 = \sqrt{a^2} = a$. ☑

When any of the factors of the radicand have *odd* exponents, we can use the following procedure:

TO SIMPLIFY THE PRINCIPAL SQUARE ROOT OF A PRODUCT

1. Express the radicand in prime factored, exponential form.

2. Find the square root of each factor as follows:
   a. If the exponent of a factor is 1, that factor must remain under the radical sign.
   b. If the exponent of a factor is an even number, remove that factor from the radical by dividing its exponent by 2.
   c. If the exponent of a factor is an odd number, write that factor as the product of two factors—one factor with an even exponent, and the other factor with an exponent of 1. Then remove from the radical the factor that has an even exponent by dividing its exponent by 2.

3. The simplified expression is the product of all the factors found in step 2.

Example 2   Examples of simplifying radicals:

a. $\sqrt{3^5} = \sqrt{3^4 \cdot 3^1}$ ← $3^5 = 3^4 \cdot 3^1$

$= \sqrt{3^4}\sqrt{3}$ The factor 3^4 is the highest power of 3 whose exponent (4) is exactly divisible by 2

$= 3^2\sqrt{3}$

$= 9\sqrt{3}$

b. $\sqrt{2^7} = \sqrt{2^6 \cdot 2^1}$ ◄─── $2^7 = 2^6 \cdot 2^1$

$\quad\quad = \sqrt{2^6}\sqrt{2}$ └── The factor 2^6 is the highest power of 2 whose exponent (6) is exactly divisible by 2

$\quad\quad = 2^3\sqrt{2}$

$\quad\quad = 8\sqrt{2}$

c. $\sqrt{48} = \sqrt{2^4 \cdot 3}$ ◄─── Prime factored form of 48

$\quad\quad = \sqrt{2^4}\sqrt{3}$

$\quad\quad = 2^2\sqrt{3}$

$\quad\quad = 4\sqrt{3}$

2	48
2	24
2	12
2	6
	3

$48 = 2^4 \cdot 3$

d. $\sqrt{75} = \sqrt{3 \cdot 5^2}$ ◄─── Prime factored form of 75

$\quad\quad = \sqrt{3}\sqrt{5^2}$

$\quad\quad = \sqrt{3}(5)$

$\quad\quad = 5\sqrt{3}$

5	75
5	15
	3

$75 = 3 \cdot 5^2$

e. $\sqrt{360} = \sqrt{2^3 \cdot 3^2 \cdot 5}$

$\quad\quad = \sqrt{2^2 \cdot 2 \cdot 3^2 \cdot 5}$

$\quad\quad = 2 \cdot 3\sqrt{2 \cdot 5}$

$\quad\quad = 6\sqrt{10}$

2	360
2	180
2	90
3	45
3	15
	5

$360 = 2^3 \cdot 3^2 \cdot 5$ ■

If you see that a radicand contains a factor that is a perfect square, you can sometimes simplify the square root "by inspection" (see Example 3).

Example 3   Examples of simplifying radicals by inspection:

a. $\sqrt{12} = \sqrt{4 \cdot 3}$ 4 is a factor of 12 and is a perfect square

$\quad\quad = \sqrt{4} \cdot \sqrt{3}$

$\quad\quad = 2 \cdot \sqrt{3}$

$\quad\quad = 2\sqrt{3}$

b. $\sqrt{50} = \sqrt{25 \cdot 2}$ 25 is a factor of 50 and is a perfect square

$\quad\quad = \sqrt{25} \cdot \sqrt{2}$

$\quad\quad = 5\sqrt{2}$

c. $\sqrt{600} = \sqrt{100 \cdot 6}$ 100 is a factor of 600 and is a perfect square

$\quad\quad = \sqrt{100}\sqrt{6}$

$\quad\quad = 10\sqrt{6}$ ■

When the radicand contains variables, we follow the rules given in the box on page 526 for simplifying the radicals (see Example 4).

Example 4   Examples of simplifying radicals when the radicand contains variables:

a. $\sqrt{x^7} = \sqrt{x^6 \cdot x^1}$ ← $x^7 = x^6 \cdot x^1$

$= \sqrt{x^6}\sqrt{x}$

$= x^3\sqrt{x}$

— The factor x^6 is the highest power of x whose exponent (6) is exactly divisible by 2

b. $\sqrt{12x^4y^3} = \sqrt{2^2 \cdot 3 \cdot x^4 \cdot y^2 \cdot y}$

$= \sqrt{2^2x^4y^2(3y)}$ ← Rearranging factors

$= 2x^2y\sqrt{3y}$

$$\begin{array}{r|l} 2 & 12 \\ 2 & 6 \\ & 3 \end{array}$$ $12 = 2^2 \cdot 3$

c. $\sqrt{24a^5b^7} = \sqrt{2^2 \cdot 2 \cdot 3 \cdot a^4 \cdot a \cdot b^6 \cdot b}$

$= \sqrt{2^2a^4b^6(2 \cdot 3ab)}$ ← Rearranging factors

$= 2a^2b^3\sqrt{2 \cdot 3ab}$

$= 2a^2b^3\sqrt{6ab}$ ∎

$$\begin{array}{r|l} 2 & 24 \\ 2 & 12 \\ 2 & 6 \\ & 3 \end{array}$$ $24 = 2^3 \cdot 3$

EXERCISES 11.2A

Set I   Simplify each of the square roots.

1. $\sqrt{25x^2}$ 2. $\sqrt{100y^2}$ 3. $\sqrt{81s^4}$ 4. $\sqrt{64t^6}$

5. $\sqrt{4x^2}$ 6. $\sqrt{9y^2}$ 7. $\sqrt{16z^4}$ 8. $\sqrt{25b^6}$

9. $\sqrt{98}$ 10. $\sqrt{20}$ 11. $\sqrt{18}$ 12. $\sqrt{45}$

13. $\sqrt{8}$ 14. $\sqrt{32}$ 15. $\sqrt{x^3}$ 16. $\sqrt{y^5}$

17. $\sqrt{m^7}$ 18. $\sqrt{n^9}$ 19. $\sqrt{24}$ 20. $\sqrt{54}$

21. $\sqrt{32x^2y^3}$ 22. $\sqrt{98ab^3}$ 23. $\sqrt{8s^4t^5}$ 24. $\sqrt{44x^3y^6}$

25. $\sqrt{18a^2b^3}$ 26. $\sqrt{27c^5d^2}$ 27. $\sqrt{40x^6y^2}$ 28. $\sqrt{135a^6b^8}$

29. $\sqrt{60h^5k^4}$ 30. $\sqrt{90m^7n^6}$ 31. $\sqrt{280y^5z^6}$

32. $\sqrt{270g^8h^9}$ 33. $\sqrt{500x^7y^9}$ 34. $\sqrt{216x^{11}y^{13}}$

Set II   Simplify each of the square roots.

1. $\sqrt{36z^2}$ 2. $\sqrt{120a^2}$ 3. $\sqrt{25w^2}$ 4. $\sqrt{64u^6}$

5. $\sqrt{16t^2}$ 6. $\sqrt{12b^2}$ 7. $\sqrt{25a^4}$ 8. $\sqrt{84y^8}$

9. $\sqrt{75}$ 10. $\sqrt{72}$ 11. $\sqrt{125}$ 12. $\sqrt{400}$

13. $\sqrt{432}$ 14. $\sqrt{200}$ 15. $\sqrt{a^5}$ 16. $\sqrt{u^7}$

17. $\sqrt{z^5}$ 18. $\sqrt{a^9}$ 19. $\sqrt{28}$ 20. $\sqrt{96u^6}$

21. $\sqrt{12a^2b^5}$ 22. $\sqrt{8x^5y^3}$ 23. $\sqrt{18c^6d^7}$ 24. $\sqrt{50s^3t^5}$

25. $\sqrt{84c^2d^4}$ 26. $\sqrt{125xy^2}$ 27. $\sqrt{28ab^5}$ 28. $\sqrt{150x^3y}$

29. $\sqrt{80a^7b^4}$ 30. $\sqrt{121x^4y^2}$ 31. $\sqrt{88s^6t}$

32. $\sqrt{40p^3q^4}$ 33. $\sqrt{54a^3b^4}$ 34. $\sqrt{250x^5y^2z}$

11.2B Square Roots Involving Fractions

Another basic rule for working with square roots follows:

RULE 11.3

$$\sqrt{\frac{a}{b}} = \frac{\sqrt{a}}{\sqrt{b}}, \text{ if } a \geq 0, b > 0$$

The square root of a quotient	=	the quotient of the square roots

---

A *second* condition that must be satisfied for a square root to be in *simplest radical form* is that the radicand cannot be a fraction. Rule 11.3 can be used in simplifying radicals in which the radicand is a fraction (see Example 5).

Example 5 Examples of simplifying the principal square root of a fraction (assume $y \neq 0$ and $k \neq 0$):

a. $\sqrt{\frac{4}{9}} = \frac{\sqrt{4}}{\sqrt{9}} = \frac{2}{3}$

b. $\sqrt{\frac{25}{36}} = \frac{\sqrt{25}}{\sqrt{36}} = \frac{5}{6}$

c. $\sqrt{\frac{x^4}{y^6}} = \frac{\sqrt{x^4}}{\sqrt{y^6}} = \frac{x^2}{y^3}$

d. $\sqrt{\frac{50h^2}{2k^4}} = \sqrt{\frac{\overset{25}{\cancel{50}}h^2}{\underset{1}{\cancel{2}}k^4}} = \frac{\sqrt{25h^2}}{\sqrt{k^4}} = \frac{5h}{k^2}$

          └── We simplify the fraction first

e. $\sqrt{\frac{3x^2}{4}} = \frac{\sqrt{3x^2}}{\sqrt{4}} = \frac{x\sqrt{3}}{2}$ ∎

In Example 5, all the denominators were (or became, after the fraction was simplified) perfect squares.

Rationalizing the Denominator

If the denominator is *not* a perfect square, we must multiply it by a number that makes it a perfect square. The denominator then becomes a *rational* number; therefore, this procedure is called **rationalizing the denominator**.

In Examples 6a and 6c, we show how to rationalize the denominator by extending the radical sign and multiplying the radicand by a fraction that has a value of 1; that fraction is placed *under* the radical sign. In Example 6b, we show how to rationalize the denominator by using the fundamental property of rational expressions: We extend the radical sign and multiply the numerator and denominator of the radicand by the same number. Either method is acceptable.

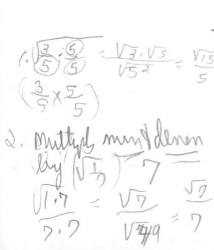

Example 6   Examples of rationalizing denominators:

Extending the bar of the radical sign to include $\frac{5}{5}$

a. $\sqrt{\dfrac{4}{5}} = \sqrt{\dfrac{4}{5} \cdot \dfrac{5}{5}} = \sqrt{\dfrac{4 \cdot 5}{5 \cdot 5}} = \dfrac{\sqrt{4 \cdot 5}}{\sqrt{5^2}} = \dfrac{\sqrt{4}\sqrt{5}}{5} = \dfrac{2\sqrt{5}}{5}$

Multiplying $\frac{4}{5}$ by $\frac{5}{5}$ in order to make the new denominator 5^2, a perfect square

b. $\sqrt{\dfrac{1}{7}} = \sqrt{\dfrac{1 \cdot 7}{7 \cdot 7}} = \dfrac{\sqrt{7}}{\sqrt{7^2}} = \dfrac{\sqrt{7}}{7}$

Multiplying both the numerator and the denominator of $\frac{1}{7}$ by 7 to make the denominator 7^2, a perfect square

c. $\sqrt{\dfrac{3}{20}} = \sqrt{\dfrac{3}{20} \cdot \dfrac{5}{5}} = \sqrt{\dfrac{3 \cdot 5}{20 \cdot 5}} = \dfrac{\sqrt{15}}{\sqrt{100}} = \dfrac{\sqrt{15}}{10}$

Note: 20·5
NOT 20·20 →
(need to make perfect
square)
↓
(20·20 LONG WAY)

Multiplying $\frac{3}{20}$ by $\frac{5}{5}$ in order to make the new denominator 100, a perfect square

This problem could also have been done as follows:

$$\sqrt{\dfrac{3}{20}} = \sqrt{\dfrac{3}{20} \cdot \dfrac{20}{20}} = \sqrt{\dfrac{3 \cdot 2^2 \cdot 5}{20^2}} = \dfrac{\sqrt{2^2}\sqrt{3 \cdot 5}}{20} = \dfrac{\overset{1}{\cancel{2}}\sqrt{15}}{\underset{10}{\cancel{20}}} = \dfrac{\sqrt{15}}{10}$$

However, the arithmetic is easier the first way (that is, by multiplying numerator and denominator by the *smallest* number that will make the new denominator a perfect square). ∎

A *third* condition that must be satisfied for an algebraic expression to be in *simplest radical form* is that no denominator can contain a square root.

In Example 7a, we show how to rationalize a denominator that contains a square root by multiplying the given fraction by a fraction that equals 1. In Examples 7b–7d, we rationalize the denominator by using the fundamental property of rational expressions; that is, we multiply both numerator and denominator by the same number. Either method is acceptable.

Example 7   Examples of rationalizing denominators (assume $x > 0$):

Multiply num & den
by number equal to one
Exam $\left(\frac{5}{5}\right)\left(\frac{4}{4}\right)$

↑ by √ root of denom

a. $\dfrac{2}{\sqrt{5}} = \dfrac{2}{\sqrt{5}} \cdot \dfrac{\sqrt{5}}{\sqrt{5}} = \dfrac{2\sqrt{5}}{\sqrt{5}\sqrt{5}} = \dfrac{2\sqrt{5}}{(\sqrt{5})^2} = \dfrac{2\sqrt{5}}{5}$

The denominator is now a rational number

Multiplying by $\dfrac{\sqrt{5}}{\sqrt{5}}$ does not change the value of the fraction, since $\dfrac{\sqrt{5}}{\sqrt{5}} = 1$; the product of the two denominators will be a rational number

The denominator is *not* a rational number

b. $\dfrac{5}{\sqrt{6}} = \dfrac{5 \cdot \sqrt{6}}{\sqrt{6} \cdot \sqrt{6}} = \dfrac{5\sqrt{6}}{(\sqrt{6})^2} = \dfrac{5\sqrt{6}}{6}$

The denominator is now a rational number

Multiplying both numerator and denominator by $\sqrt{6}$ does not change the value of the fraction, and $\sqrt{6} \cdot \sqrt{6}$ will be a rational number

Cancel

c. $\dfrac{6}{\sqrt{3}} = \dfrac{6 \cdot \sqrt{3}}{\sqrt{3} \cdot \sqrt{3}} = \dfrac{6\sqrt{3}}{(\sqrt{3})^2} = \dfrac{\overset{2}{\cancel{6}}\sqrt{3}}{\underset{1}{\cancel{3}}} = 2\sqrt{3}$

— The denominator is now a rational number

— Multiplying both numerator and denominator by $\sqrt{3}$
does not change the value of the fraction, and $\sqrt{3} \cdot \sqrt{3}$
will be a rational number

d. $\dfrac{3xy}{\sqrt{x}} = \dfrac{3xy \cdot \sqrt{x}}{\sqrt{x} \cdot \sqrt{x}} = \dfrac{3xy\sqrt{x}}{(\sqrt{x})^2} = \dfrac{3xy\sqrt{x}}{x} = 3y\sqrt{x}$

— The denominator is now a rational number

— Multiplying both numerator and denominator by $\sqrt{x}$
does not change the value of the fraction, and $\sqrt{x} \cdot \sqrt{x}$
will be a rational number ■

We summarize here the rules we've given for simplifying a radical:

THE SIMPLIFIED FORM OF AN EXPRESSION THAT HAS SQUARE ROOTS

1. When the radicand is in prime factored, exponential form, no factor of the radicand has an exponent greater than 1.

2. No radicand contains a fraction.

3. No denominator contains a square root.

EXERCISES 11.2B

Set I   Simplify each of the following expressions; assume that all variables in denominators are nonzero.

1. $\sqrt{\dfrac{9}{25}}$ 2. $\sqrt{\dfrac{36}{49}}$ 3. $\sqrt{\dfrac{16}{25}}$ 4. $\sqrt{\dfrac{81}{100}}$

5. $\sqrt{\dfrac{y^4}{x^2}}$ 6. $\sqrt{\dfrac{x^6}{v^8}}$ 7. $\sqrt{\dfrac{4x^2}{9}}$ 8. $\sqrt{\dfrac{16a^2}{49}}$

 *Reduce* $\sqrt{\dfrac{1}{4k^2}}$

9. $\boxed{\sqrt{\dfrac{2}{8k^2}}}$ 10. $\sqrt{\dfrac{2m^2}{18}}$ 11. $\sqrt{\dfrac{4x^3y}{xy^3}}$ 12. $\sqrt{\dfrac{x^5y}{9xy^3}}$

13. $\sqrt{\dfrac{1}{5}}$ 14. $\sqrt{\dfrac{1}{3}}$ 15. $\sqrt{\dfrac{7}{13}}$ 16. $\sqrt{\dfrac{5}{17}}$

17. $\sqrt{\dfrac{1}{18}}$ 18. $\sqrt{\dfrac{1}{50}}$ 19. $\sqrt{\dfrac{7}{27}}$ 20. $\sqrt{\dfrac{5}{8}}$

21. $\dfrac{3}{\sqrt{7}}$ 22. $\dfrac{2}{\sqrt{3}}$ 23. $\dfrac{10}{\sqrt{5}}$ 24. $\dfrac{14}{\sqrt{2}}$

25. $\sqrt{\dfrac{m^2}{3}}$ 26. $\sqrt{\dfrac{k^2}{5}}$ 27. $\sqrt{\dfrac{x^5z^4}{36y^2}}$ 28. $\sqrt{\dfrac{a^3c^6}{25b^2}}$

29. $\sqrt{\dfrac{3a^2b}{4b^3}}$ 30. $\sqrt{\dfrac{10uv^2}{8u}}$ 31. $\sqrt{\dfrac{b^2c^4}{16d^3}}$ 32. $\sqrt{\dfrac{h^4k^8}{49p^5}}$

33. $\sqrt{\dfrac{8m^2n}{2n^2}}$ 34. $\sqrt{\dfrac{18xy^2}{2x^2}}$

Set II Simplify each of the following expressions; assume that all variables in denominators are nonzero.

1. $\sqrt{\dfrac{16}{49}}$ 2. $\sqrt{\dfrac{64}{81}}$ 3. $\sqrt{\dfrac{25}{64}}$ 4. $\sqrt{\dfrac{81}{121}}$

5. $\sqrt{\dfrac{a^2}{b^6}}$ 6. $\sqrt{\dfrac{c^8}{d^4}}$ 7. $\sqrt{\dfrac{36m^2}{25}}$ 8. $\sqrt{\dfrac{16a^4}{49}}$

9. $\sqrt{\dfrac{3}{27x^2}}$ 10. $\sqrt{\dfrac{18c^2}{2d^4}}$ 11. $\sqrt{\dfrac{h^3k^3}{16hk^5}}$ 12. $\sqrt{\dfrac{50a^4b}{2b^3}}$

13. $\sqrt{\dfrac{1}{11}}$ 14. $\sqrt{\dfrac{1}{2}}$ 15. $\sqrt{\dfrac{2}{19}}$ 16. $\sqrt{\dfrac{2}{18}}$

17. $\sqrt{\dfrac{1}{45}}$ 18. $\sqrt{\dfrac{1}{32}}$ 19. $\sqrt{\dfrac{2}{75}}$ 20. $\sqrt{\dfrac{3}{28}}$

21. $\dfrac{5}{\sqrt{10}}$ 22. $\dfrac{3}{\sqrt{75}}$ 23. $\dfrac{6}{\sqrt{15}}$ 24. $\dfrac{7}{\sqrt{28}}$

25. $\sqrt{\dfrac{e^2}{7}}$ 26. $\sqrt{\dfrac{f^2}{2}}$ 27. $\sqrt{\dfrac{r^3s^6}{9t^2}}$ 28. $\sqrt{\dfrac{50a}{2a^5b^4}}$

29. $\sqrt{\dfrac{15zw^4}{18z}}$ 30. $\sqrt{\dfrac{12ab^2}{5a^5b}}$ 31. $\sqrt{\dfrac{d^4e^6}{100f^3}}$ 32. $\sqrt{\dfrac{125x^3y^4}{3xy^2}}$

33. $\sqrt{\dfrac{45tu^2}{5t^2}}$ 34. $\sqrt{\dfrac{7cd^3}{5c^2d}}$

11.3 Products and Quotients of Square Roots

11.3A Multiplying Square Roots

In Section 11.2A, we used Rule 11.2 in the following direction:

$$\overrightarrow{\sqrt{ab}} = \sqrt{a}\sqrt{b}$$
$$\sqrt{4 \cdot 3} = \sqrt{4}\sqrt{3} \qquad \text{Using Rule 11.2 to simplify the square root of a product}$$
$$= 2\sqrt{3}$$

In this section, we use Rule 11.2 in the opposite way:

$$\overrightarrow{\sqrt{a}\sqrt{b}} = \sqrt{ab}$$
$$\sqrt{2}\sqrt{8} = \sqrt{2 \cdot 8} \qquad \text{Using Rule 11.2 to find the product of square roots}$$
$$= \sqrt{16} = 4$$

In words,

> A product of square roots = the square root of the product
>
> $$\sqrt{a}\sqrt{b} = \sqrt{ab}$$

A WORD OF CAUTION   While a product of square roots equals the square root of the product, a *sum* of square roots does *not* equal the square root of the sum. Students often make the following error:

$$\sout{\sqrt{a} + \sqrt{b} = \sqrt{a + b}}$$

However, this is incorrect. To see that it is incorrect, let $a = 9$ and $b = 16$. Then

$$\sqrt{a} + \sqrt{b} = \sqrt{9} + \sqrt{16} = 3 + 4 = 7$$
$$\sqrt{a + b} = \sqrt{9 + 16} = \sqrt{25} = 5$$

Since $7 \neq 5$, $\sqrt{9} + \sqrt{16} \neq \sqrt{9 + 16}$ and, in general, $\sqrt{a} + \sqrt{b} \neq \sqrt{a + b}$.   ☑

Example 1   Examples of multiplying square roots:

$$\underset{\substack{\downarrow}}{\text{Product of}\atop\text{square roots}} = \underset{\substack{\downarrow}}{\text{square root}\atop\text{of product}}$$

a. $\sqrt{2}\sqrt{2} = \sqrt{2 \cdot 2} = \sqrt{2^2} = 2$

b. $\sqrt{2}\sqrt{8} = \sqrt{2 \cdot 8} = \sqrt{16} = 4$

c. $\sqrt{4x}\sqrt{x} = \sqrt{4x \cdot x} = \sqrt{4x^2} = 2x$

d. $\sqrt{3y}\sqrt{12y^3} = \sqrt{3y \cdot 12y^3} = \sqrt{36y^4} = 6y^2$

e. $\sqrt{3}\sqrt{6}\sqrt{2} = \sqrt{3 \cdot 6 \cdot 2} = \sqrt{36} = 6$

f. $\sqrt{5z}\sqrt{10}\sqrt{2z^3} = \sqrt{5z \cdot 10 \cdot 2z^3} = \sqrt{100z^4} = 10z^2$ ■

Multiplying a Square Root by Itself

$$\sqrt{a}\sqrt{a} = (\sqrt{a})^2 \qquad \text{Definition of exponents}$$
$$(\sqrt{a})^2 = a \qquad \text{Using Rule 11.1}$$

Therefore, $\sqrt{a}\sqrt{a} = a$.

RULE 11.4   MULTIPLYING A SQUARE ROOT BY ITSELF

$$\sqrt{a}\sqrt{a} = a, \text{ if } a \geq 0$$

Also by Rule 11.2, $\sqrt{ab} = \sqrt{a}\sqrt{b}$.

$$\sqrt{aa} = \sqrt{a}\sqrt{a} \qquad \text{Substituting } a \text{ for } b \text{ in Rule 11.2}$$
$$\sqrt{a^2} = a \qquad \text{Rewriting } aa \text{ as } a^2 \text{ and, on the right, using Rule 11.4}$$

Therefore, we can now reword Rule 11.1: $(\sqrt{a})^2 = \sqrt{a}\sqrt{a} = \sqrt{a^2} = a$.

Example 2   Examples of multiplying a square root by itself:

a. $\sqrt{16}\sqrt{16} = 16$

b. $\sqrt{7}\sqrt{7} = 7$

c. $\sqrt{2x}\sqrt{2x} = 2x$ ■

A *fourth* condition that must be satisfied for an algebraic expression to be in *simplest radical form* is that no term can have more than one radical sign in it.

Example 3   Examples of finding and simplifying products of radicals:

a. $\sqrt{3x}\sqrt{6x^2} = \sqrt{3x \cdot 6x^2} = \sqrt{18x^3}$ Multiplying the square roots

$$\left. \begin{array}{l} = \sqrt{2 \cdot 9 \cdot x^2 \cdot x} \\ \\ = 3x\sqrt{2x} \end{array} \right\} \text{Simplifying}$$

b. $\sqrt{3w^2}\sqrt{2}\sqrt{8w^3} = \sqrt{3w^2 \cdot 2 \cdot 8w^3} = \sqrt{48w^5}$ Multiplying the square roots

$$\left. \begin{array}{l} = \sqrt{3 \cdot 16 \cdot w^4 \cdot w} \\ \\ = 4w^2\sqrt{3w} \end{array} \right\} \text{Simplifying}$$

c. $5\sqrt{2y^3} \cdot 3\sqrt{6y^4} = (5 \cdot 3)\sqrt{2y^3 \cdot 6y^4} = 15\sqrt{12y^7}$ Multiplying the square roots

$$\left. \begin{array}{l} = 15\sqrt{3 \cdot 4 \cdot y^6 \cdot y} \\ \\ = 15 \cdot 2 \cdot y^3\sqrt{3y} \\ \\ = 30y^3\sqrt{3y} \end{array} \right\} \text{Simplifying} \quad \blacksquare$$

EXERCISES 11.3A

Set I   Find the following products and simplify the results.

1. $\sqrt{3}\sqrt{3}$ 2. $\sqrt{7}\sqrt{7}$ 3. $\sqrt{4}\sqrt{4}$

4. $\sqrt{9}\sqrt{9}$ 5. $\sqrt{3}\sqrt{12}$ 6. $\sqrt{2}\sqrt{32}$

7. $\sqrt{9x}\sqrt{x}$ 8. $\sqrt{25y}\sqrt{y}$ 9. $\sqrt{5}\sqrt{10}\sqrt{2}$

10. $\sqrt{6}\sqrt{12}\sqrt{2}$ 11. $\sqrt{8}\sqrt{18}\sqrt{20}$ 12. $\sqrt{12}\sqrt{75}\sqrt{28}$

13. $\sqrt{5x}\sqrt{10x^2}$ 14. $\sqrt{11y^2}\sqrt{33y}$ 15. $\sqrt{3x}\sqrt{2x^2y}\sqrt{30y}$

16. $\sqrt{7a^2}\sqrt{14ab^2}\sqrt{10b}$ 17. $\sqrt{5ab^2}\sqrt{20ab}$ 18. $\sqrt{3x^2y}\sqrt{27xy}$

19. $\sqrt{2a}\sqrt{6}\sqrt{3a}$ 20. $\sqrt{2}\sqrt{h^3}\sqrt{8h}$

21. $5\sqrt{2x} \cdot \sqrt{8x^3}(2\sqrt{3x^5})$ 22. $4\sqrt{2M^3} \cdot \sqrt{3M}(3\sqrt{12M^3})$

23. $8\sqrt{5x} \cdot \sqrt{15x^3}(3\sqrt{6x})$ 24. $6\sqrt{3a^3}(5\sqrt{45a^2})(\sqrt{5a})$

Set II   Find the following products and simplify the results.

1. $\sqrt{11}\sqrt{11}$ 2. $\sqrt{6}\sqrt{6}$ 3. $\sqrt{8}\sqrt{8}$

4. $\sqrt{3}\sqrt{3}$ 5. $\sqrt{18}\sqrt{2}$ 6. $\sqrt{4x}\sqrt{4x}$

7. $\sqrt{16a}\sqrt{a}$ 8. $\sqrt{49y}\sqrt{2y}$ 9. $\sqrt{3}\sqrt{6}\sqrt{2}$

10. $\sqrt{6x}\sqrt{2x}\sqrt{8x}$ 11. $\sqrt{15}\sqrt{12}\sqrt{35}$ 12. $\sqrt{35}\sqrt{42}\sqrt{3}$

13. $\sqrt{3b}\sqrt{15b^2}$ 14. $\sqrt{13x^2}\sqrt{39x}$ 15. $\sqrt{5a}\sqrt{2a^2b}\sqrt{20b}$

16. $\sqrt{17x^2}\sqrt{34xy}\sqrt{6y}$ 17. $\sqrt{18x^3}\sqrt{2xy^2}$ 18. $\sqrt{30ab^5}\sqrt{3a^2b}$

19. $\sqrt{12mn}\sqrt{mn^3}\sqrt{3}$ 20. $\sqrt{b^3}\sqrt{32}\sqrt{2b}$

21. $6\sqrt{5z^3} \cdot \sqrt{2z}(4\sqrt{8z^3})$ 22. $2\sqrt{6x^5} \cdot \sqrt{3x}(5\sqrt{20x^3})$

23. $7\sqrt{3y} \cdot \sqrt{8y^3}(5\sqrt{2y})$ 24. $6\sqrt{5x^2}(5\sqrt{6x^5})(\sqrt{7x})$

11.3B Dividing Square Roots

In Section 11.2B, we used Rule 11.3 in the following direction:

$$\overrightarrow{\sqrt{\frac{a}{b}} = \frac{\sqrt{a}}{\sqrt{b}}} \quad (b \neq 0) \qquad \text{Using Rule 11.3 to find the square root of a quotient}$$

$$\sqrt{\frac{4}{9}} = \frac{\sqrt{4}}{\sqrt{9}} = \frac{2}{3}$$

In this section, we use Rule 11.3 in the opposite way:

$$\overrightarrow{\frac{\sqrt{a}}{\sqrt{b}} = \sqrt{\frac{a}{b}}} \quad (b \neq 0) \qquad \text{Using Rule 11.3 to find the quotient of square roots}$$

$$\frac{\sqrt{50}}{\sqrt{2}} = \sqrt{\frac{50}{2}} = \sqrt{25} = 5$$

In words,

$$\boxed{\begin{array}{c} \text{A quotient of square roots} = \text{the square root of the quotient} \\[2mm] \dfrac{\sqrt{a}}{\sqrt{b}} = \sqrt{\dfrac{a}{b}} \quad (b \neq 0) \end{array}}$$

Example 4   Examples of dividing square roots (assume $a > 0$, $x > 0$, $y > 0$):

$$\underset{\substack{\text{Quotient of} \\ \text{square roots}}}{} = \underset{\substack{\text{square root} \\ \text{of quotient}}}{}$$

a. $\dfrac{\sqrt{8}}{\sqrt{2}} = \sqrt{\dfrac{8}{2}} = \sqrt{4} = 2$

b. $\dfrac{\sqrt{a^5}}{\sqrt{a^3}} = \sqrt{\dfrac{a^5}{a^3}} = \sqrt{a^2} = a$

c. $\dfrac{\sqrt{27x}}{\sqrt{3x^3}} = \sqrt{\dfrac{27x}{3x^3}} = \sqrt{\dfrac{9}{x^2}} = \dfrac{3}{x}$

d. $\dfrac{\sqrt{28xy^3}}{\sqrt{7xy}} = \sqrt{\dfrac{28xy^3}{7xy}} = \sqrt{4y^2} = 2y$ ∎

In Example 5, we rationalize the denominator after the division has been performed.

Example 5   Examples of finding and simplifying quotients of radicals (assume $x > 0$):

a. $\dfrac{\sqrt{5x}}{\sqrt{10x^2}} = \sqrt{\dfrac{5x}{10x^2}} = \sqrt{\dfrac{1}{2x}} = \dfrac{\sqrt{1}}{\sqrt{2x}} = \dfrac{1}{\sqrt{2x}} \cdot \dfrac{\sqrt{2x}}{\sqrt{2x}} = \dfrac{\sqrt{2x}}{2x} \quad {\scriptstyle \sqrt{2x}\,\sqrt{2x}\,=\,2x \atop \text{by Rule 11.4}}$

b. $\dfrac{6\sqrt{25x^3}}{5\sqrt{3x}} = \dfrac{6}{5}\sqrt{\dfrac{25x^3}{3x}} = \dfrac{6}{5} \cdot \dfrac{\sqrt{25x^2}}{\sqrt{3}} = \dfrac{6}{\cancel{5}} \cdot \dfrac{\cancel{5}x}{\sqrt{3}} \cdot \dfrac{\sqrt{3}}{\sqrt{3}} = \dfrac{\cancel{6}x\sqrt{3}}{\cancel{3}} = 2x\sqrt{3}$ ∎

EXERCISES 11.3B

Set I   Simplify each of the following expressions; assume that all variables in denominators are nonzero.

1. $\dfrac{\sqrt{20}}{\sqrt{5}}$ 2. $\dfrac{\sqrt{7}}{\sqrt{28}}$ 3. $\dfrac{\sqrt{32}}{\sqrt{2}}$ 4. $\dfrac{\sqrt{98}}{\sqrt{2}}$

5. $\dfrac{\sqrt{4}}{\sqrt{5}}$ 6. $\dfrac{\sqrt{9}}{\sqrt{7}}$ 7. $\dfrac{\sqrt{15x}}{\sqrt{5x}}$ 8. $\dfrac{\sqrt{18y}}{\sqrt{3y}}$

9. $\dfrac{\sqrt{75a^3}}{\sqrt{5a}}$ 10. $\dfrac{\sqrt{24b^5}}{\sqrt{8b}}$ 11. $\dfrac{\sqrt{35s^4}}{\sqrt{10s}}$ 12. $\dfrac{\sqrt{42t^2}}{\sqrt{12t}}$

13. $\dfrac{\sqrt{8xy^3}}{\sqrt{12x^2y}}$ 14. $\dfrac{\sqrt{10a^3b}}{\sqrt{15ab^2}}$ 15. $\dfrac{\sqrt{72x^3y^2}}{\sqrt{2xy^2}}$ 16. $\dfrac{\sqrt{27x^2y^3}}{\sqrt{3x^2y}}$

17. $\dfrac{\sqrt{x^4y}}{\sqrt{5y}}$ 18. $\dfrac{\sqrt{m^6n}}{\sqrt{3n}}$ 19. $\dfrac{4\sqrt{45m^3}}{3\sqrt{10m}}$ 20. $\dfrac{6\sqrt{400x^4}}{5\sqrt{6x}}$

21. $\dfrac{5\sqrt{35a^5}}{7\sqrt{5a}}$ 22. $\dfrac{8\sqrt{27x^7}}{3\sqrt{6x}}$

Set II   Simplify each of the following expressions; assume that all variables in denominators are nonzero.

1. $\dfrac{\sqrt{27}}{\sqrt{3}}$ 2. $\dfrac{\sqrt{2}}{\sqrt{50}}$ 3. $\dfrac{\sqrt{75}}{\sqrt{3}}$ 4. $\dfrac{\sqrt{7}}{\sqrt{28}}$

5. $\dfrac{\sqrt{25}}{\sqrt{2}}$ 6. $\dfrac{\sqrt{36}}{\sqrt{11}}$ 7. $\dfrac{\sqrt{35x}}{\sqrt{5x}}$ 8. $\dfrac{\sqrt{18m}}{\sqrt{6m}}$

9. $\dfrac{\sqrt{45x^3}}{\sqrt{5x}}$ 10. $\dfrac{\sqrt{40y^4}}{\sqrt{5y}}$ 11. $\dfrac{\sqrt{63a^4}}{\sqrt{14a}}$ 12. $\dfrac{\sqrt{38b^3}}{\sqrt{57b}}$

13. $\dfrac{\sqrt{9ab^3}}{\sqrt{12a^2b}}$ 14. $\dfrac{\sqrt{9x^3y}}{\sqrt{15xy^2}}$ 15. $\dfrac{\sqrt{75a^5b^3}}{\sqrt{3a^3b^3}}$ 16. $\dfrac{\sqrt{18x^3y}}{\sqrt{5xy^5}}$

17. $\dfrac{\sqrt{x^7y}}{\sqrt{3x}}$ 18. $\dfrac{\sqrt{c^3d^4}}{\sqrt{7c^2d}}$ 19. $\dfrac{9\sqrt{20a^6}}{2\sqrt{15a^3}}$ 20. $\dfrac{12\sqrt{15b^3c}}{7\sqrt{3bc^5}}$

21. $\dfrac{9\sqrt{36x^5}}{4\sqrt{18x}}$ 22. $\dfrac{6\sqrt{32y^3}}{5\sqrt{6y}}$

11.4 Sums and Differences of Square Roots

Like square roots are square roots that have the same radicand.

Example 1   Examples of like square roots:

a. $3\sqrt{5}$, $2\sqrt{5}$, $-7\sqrt{5}$

b. $2\sqrt{x}$, $-9\sqrt{x}$, $11\sqrt{x}$ ■

Unlike square roots are square roots that have different radicands.

Example 2 Examples of unlike square roots:

a. $2\sqrt{15}$, $-6\sqrt{11}$, $8\sqrt{24}$

b. $5\sqrt{y}$, $3\sqrt{x}$, $-4\sqrt{13}$ ∎

Combining Like Square Roots

The addition of like square roots is very much like the addition of like terms; it is an application of the distributive property. For example,

$$3\sqrt{7} + 2\sqrt{7} = (3 + 2)\sqrt{7} = 5\sqrt{7}$$

Therefore, we add like square roots by adding their coefficients and then multiplying that sum by the square root.

Example 3 Examples of combining like radicals:

a. $5\sqrt{2} + 3\sqrt{2} = $ (5 + 3)$\sqrt{2}$ $= 8\sqrt{2}$

                         — This step need not be shown

b. $6\sqrt{3} - 4\sqrt{3} = $ (6 − 4)$\sqrt{3}$ $= 2\sqrt{3}$

                         — This step need not be shown

c. $7\sqrt{x} - 3\sqrt{x} = $ (7 − 3)$\sqrt{x}$ $= 4\sqrt{x}$

                         — This step need not be shown

                         — This step need not be shown

d. $\dfrac{3}{2}\sqrt{5} + \dfrac{\sqrt{5}}{2} = \left(\dfrac{3}{2} + \dfrac{1}{2}\right)\sqrt{5} = 2\sqrt{5}$

           $\dfrac{\sqrt{5}}{2} = \dfrac{1}{2}\sqrt{5}$, because $\dfrac{1}{2}\sqrt{5} = \dfrac{1}{2} \cdot \dfrac{\sqrt{5}}{1} = \dfrac{\sqrt{5}}{2}$ ∎

Combining Unlike Square Roots

When two or more *unlike* radicals are connected with *addition* or *subtraction* symbols, we usually cannot express the sum or difference with just one radical sign. That is, most unlike radicals cannot be combined. However, some *can* be combined. Sometimes terms that were not like radicals *become* like radicals after they have been expressed in simplest radical form; if this is the case, they can be combined (see Example 4).

TO COMBINE UNLIKE RADICALS

1. Express each term in simplest radical form.

2. Combine any terms that have like radicals by adding their coefficients and multiplying that sum by the like radical.

Example 4 Examples of simplifying and combining radicals:

a. $\quad \sqrt{8} + \sqrt{18}$

$\quad = \sqrt{4 \cdot 2} + \sqrt{9 \cdot 2} = 2\sqrt{2} + 3\sqrt{2} = 5\sqrt{2}$ $2\sqrt{2}$ and $3\sqrt{2}$ are like radicals

b. $\sqrt{12} - \sqrt{27} + 5\sqrt{3}$

$= \sqrt{4 \cdot 3} - \sqrt{9 \cdot 3} + 5\sqrt{3} = 2\sqrt{3} - 3\sqrt{3} + 5\sqrt{3} = 4\sqrt{3}$

c. $\sqrt{20} - 2\sqrt{45} - \sqrt{15}$

$= \sqrt{4 \cdot 5} - 2\sqrt{9 \cdot 5} - \sqrt{3 \cdot 5}$ *already lowest*

$= 2\sqrt{5} - 2 \cdot 3\sqrt{5} - \sqrt{15}$ $2\sqrt{5}$ and $-2 \cdot 3\sqrt{5}$ are like radicals

$= 2\sqrt{5} - 6\sqrt{5} - \sqrt{15} = -4\sqrt{5} - \sqrt{15}$

d. $2\sqrt{\dfrac{1}{2}} - 6\sqrt{\dfrac{1}{8}} - 10\sqrt{\dfrac{4}{5}}$

$= 2\sqrt{\dfrac{1 \cdot 2}{2 \cdot 2}} - 6\sqrt{\dfrac{1 \cdot 2}{8 \cdot 2}} - 10\sqrt{\dfrac{4 \cdot 5}{5 \cdot 5}}$ Rationalizing the denominators

$= \dfrac{2}{1} \cdot \dfrac{\sqrt{2}}{\sqrt{4}} - \dfrac{6}{1} \cdot \dfrac{\sqrt{2}}{\sqrt{16}} - \dfrac{10}{1} \cdot \dfrac{\sqrt{4}\sqrt{5}}{\sqrt{25}}$

$= \dfrac{\overset{1}{\cancel{2}}}{1} \cdot \dfrac{\sqrt{2}}{\cancel{2}} - \dfrac{\overset{3}{\cancel{6}}\sqrt{2}}{\underset{2}{\cancel{4}}} - \dfrac{\overset{2}{\cancel{10}}}{1} \cdot \dfrac{2\sqrt{5}}{\underset{1}{\cancel{5}}}$

$= \sqrt{2} - \dfrac{3}{2}\sqrt{2} - 4\sqrt{5}$ $\sqrt{2}$ and $-\frac{3}{2}\sqrt{2}$ are like radicals

$= \left(1 - \dfrac{3}{2}\right)\sqrt{2} - 4\sqrt{5} = -\dfrac{1}{2}\sqrt{2} - 4\sqrt{5}$ ∎

Although we cannot add the unlike square roots in the expression $-\frac{1}{2}\sqrt{2} - 4\sqrt{5}$, we can find the *approximate* value by using a calculator or Table I, found inside the back cover (see Example 5).

Example 5   Approximate $-\dfrac{1}{2}\sqrt{2} - 4\sqrt{5}$.

Using Table I

$$\sqrt{2} \approx 1.414$$
$$\sqrt{5} \approx 2.236$$

Therefore,

$$-\dfrac{1}{2}\sqrt{2} - 4\sqrt{5} \approx -\dfrac{1}{2}(1.414) - 4(2.236)$$

$$= -0.707 - 8.944$$

$$= -9.651$$

Using a calculator

$$\sqrt{2} \approx 1.414213562$$
$$\sqrt{5} \approx 2.236067977$$

Therefore,

$$-\frac{1}{2}\sqrt{2} - 4\sqrt{5} \approx -\frac{1}{2}(1.414213562) - 4(2.236067977)$$

$$= -0.707106781 - 8.944271908$$

$$= -9.651378689$$

$$\approx \boxed{-9.651} \qquad \text{Rounded off to three decimal places}$$

Notice that when the calculator answer is rounded off to the same number of decimal places as the answer obtained using Table I, we get the same number, -9.651. However, answers sometimes differ slightly because of the rounding off process. ■

Example 6   Find the approximate value of (a) $\sqrt{5} + \sqrt{7}$ and (b) $\sqrt{12}$. Determine whether $\sqrt{5} + \sqrt{7}$ equals $\sqrt{5+7}$, or $\sqrt{12}$.

Solutions

a. Using a calculator and rounding off each approximation to three decimal places, we have

$$\sqrt{5} + \sqrt{7} \approx 2.236 + 2.646 = 4.882$$

b. $\sqrt{12} \approx 3.464$

Because $4.882 \neq 3.464$, $\sqrt{5} + \sqrt{7} \neq \sqrt{12}$. ■

A WORD OF CAUTION   Remember that, in general,

$$\sqrt{a} + \sqrt{b} \neq \sqrt{a+b}$$

We saw in Example 6 that $\sqrt{5} + \sqrt{7} \neq \sqrt{5+7}$, and we saw in the Word of Caution in Section 11.3A that $\sqrt{9} + \sqrt{16} \neq \sqrt{9+16}$. ☑

A *fifth* condition that must be satisfied for an algebraic expression to be in *simplest radical form* is that all like radicals must be combined.

EXERCISES 11.4

Set I   In Exercises 1–32, simplify each expression.

1. $2\sqrt{3} + 5\sqrt{3}$
2. $4\sqrt{2} + 3\sqrt{2}$
3. $3\sqrt{x} - \sqrt{x}$

4. $5\sqrt{a} - \sqrt{a}$
5. $\frac{3}{2}\sqrt{2} - \frac{\sqrt{2}}{2}$
6. $\frac{4}{3}\sqrt{3} - \frac{\sqrt{3}}{3}$

7. $5 \cdot 8\sqrt{5} + \sqrt{5}$
8. $3 \cdot 4\sqrt{7} + \sqrt{7}$
9. $\sqrt{25} + \sqrt{5}$

10. $\sqrt{16} + \sqrt{6}$
11. $\sqrt{50} + \sqrt{50}$
12. $\sqrt{32} + \sqrt{32}$

13. $\sqrt{68} - \sqrt{17}$
14. $\sqrt{52} - \sqrt{13}$
15. $\sqrt{45} - \sqrt{28}$

16. $\sqrt{90} - \sqrt{54}$
17. $2\sqrt{3} + \sqrt{12}$
18. $3\sqrt{2} + \sqrt{8}$

19. $2\sqrt{50} - \sqrt{32}$
20. $3\sqrt{24} - \sqrt{54}$
21. $3\sqrt{32} - \sqrt{8}$

22. $4\sqrt{27} - 3\sqrt{12}$
23. $\sqrt{\frac{1}{2}} + \sqrt{8}$
24. $\sqrt{\frac{1}{3}} + \sqrt{12}$

25. $\sqrt{24} - \sqrt{\dfrac{2}{3}}$　　　　**26.** $\sqrt{45} - \sqrt{\dfrac{4}{5}}$　　　　**27.** $10\sqrt{\dfrac{3}{5}} + \sqrt{60}$

28. $8\sqrt{\dfrac{3}{16}} + \sqrt{48}$　　　　**29.** $\sqrt{\dfrac{25}{2}} - \dfrac{3}{\sqrt{2}}$　　　　**30.** $5\sqrt{\dfrac{1}{5}} - \sqrt{\dfrac{9}{20}}$

31. $3\sqrt{\dfrac{1}{6}} + \sqrt{12} - 5\sqrt{\dfrac{3}{2}}$　　　　**32.** $3\sqrt{\dfrac{5}{2}} + \sqrt{20} - 5\sqrt{\dfrac{1}{10}}$

In Exercises 33 and 34, approximate each expression using Table I or a calculator. Round off answers to three decimal places.

33. $\dfrac{1}{3}\sqrt{7} + 2\sqrt{3}$　　　　　　　　**34.** $3\sqrt{6} + \dfrac{1}{4}\sqrt{5}$

Set II　In Exercises 1–32, simplify each expression.

1. $3\sqrt{7} + 2\sqrt{7}$　　　　**2.** $8\sqrt{14} + 7\sqrt{14}$　　　　**3.** $4\sqrt{m} - \sqrt{m}$

4. $12\sqrt{c} - \sqrt{c}$　　　　**5.** $\dfrac{2}{3}\sqrt{5} - \dfrac{\sqrt{5}}{3}$　　　　**6.** $\dfrac{3}{2}\sqrt{6} - \dfrac{\sqrt{6}}{2}$

7. $3 \cdot 6\sqrt{3} + \sqrt{3}$　　　　**8.** $5 \cdot 8\sqrt{11} + \sqrt{11}$　　　　**9.** $\sqrt{9} + \sqrt{17}$

10. $\sqrt{25} + \sqrt{26}$　　　　**11.** $\sqrt{18} + \sqrt{18}$　　　　**12.** $\sqrt{48} + \sqrt{48}$

13. $\sqrt{44} - \sqrt{11}$　　　　**14.** $\sqrt{117} - \sqrt{13}$　　　　**15.** $\sqrt{52} - \sqrt{44}$

16. $\sqrt{160} - \sqrt{63}$　　　　**17.** $5\sqrt{2} + \sqrt{18}$　　　　**18.** $7\sqrt{7} + \sqrt{28}$

19. $\sqrt{45} - 3\sqrt{20}$　　　　**20.** $\sqrt{48} - 5\sqrt{27}$　　　　**21.** $4\sqrt{28} - \sqrt{63}$

22. $8\sqrt{45} - 3\sqrt{20}$　　　　**23.** $\sqrt{20} + \sqrt{\dfrac{1}{5}}$　　　　**24.** $\sqrt{18} + \sqrt{\dfrac{1}{2}}$

25. $\sqrt{75} - \sqrt{\dfrac{4}{3}}$　　　　**26.** $\sqrt{\dfrac{1}{3}} - \sqrt{108}$　　　　**27.** $6\sqrt{\dfrac{5}{9}} + \sqrt{45}$

28. $5\sqrt{8} + 3\sqrt{\dfrac{1}{2}}$　　　　**29.** $\sqrt{\dfrac{16}{3}} - \dfrac{6}{\sqrt{3}}$　　　　**30.** $\dfrac{4}{\sqrt{5}} - \sqrt{\dfrac{36}{5}}$

31. $10\sqrt{\dfrac{1}{15}} + \sqrt{60} - 8\sqrt{\dfrac{3}{5}}$　　　　**32.** $7\sqrt{40} - 3\sqrt{\dfrac{5}{2}} + 8\sqrt{\dfrac{1}{10}}$

In Exercises 33 and 34, approximate each expression using Table I or a calculator. Round off answers to three decimal places.

33. $\dfrac{1}{4}\sqrt{13} + 3\sqrt{6}$　　　　　　　　**34.** $5\sqrt{5} + \dfrac{1}{5}\sqrt{7}$

11.5 Products and Quotients Involving More Than One Term

Example 1 illustrates the use of the distributive property when radicals are involved.

Example 1 $\sqrt{2}(3\sqrt{2} - 5)$
Solution

distribute

$$\sqrt{2}(3\sqrt{2} - 5) = \sqrt{2} \cdot (3\sqrt{2}) - \sqrt{2}(5)$$
$$= 3\sqrt{2 \cdot 2} - 5\sqrt{2}$$
$$= 3 \cdot 2 - 5\sqrt{2}$$
$$= 6 - 5\sqrt{2} \quad \blacksquare$$

The FOIL method can be used when we multiply two expressions that contain square roots and have two terms (see Example 2).

Example 2 $(2\sqrt{3} - 5)(4\sqrt{3} - 6)$
Solution

FOIL

$(2\sqrt{3} \cdot 4\sqrt{3}) + (2\sqrt{3} \cdot (-6))$
$+ ((-5) \cdot 4\sqrt{3}) + (-5 \cdot (-6))$
$= 24 + (-12\sqrt{3}) + (-20\sqrt{3}) + 30$
$= 54 - 32\sqrt{3}$

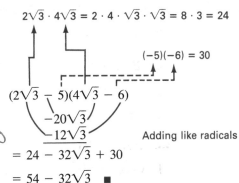

$$2\sqrt{3} \cdot 4\sqrt{3} = 2 \cdot 4 \cdot \sqrt{3} \cdot \sqrt{3} = 8 \cdot 3 = 24$$

$(-5)(-6) = 30$

$(2\sqrt{3} - 5)(4\sqrt{3} - 6)$

$-20\sqrt{3}$

$-12\sqrt{3}$ Adding like radicals

$$= 24 - 32\sqrt{3} + 30$$
$$= 54 - 32\sqrt{3} \quad \blacksquare$$

Example 3 illustrates squaring an expression that contains square roots and has two terms.

Example 3 $(3\sqrt{5} - 7x)^2$
Solution Using the formula $(a - b)^2 = a^2 - 2ab + b^2$, we have

$a^2 - 2ab + b^2$

$$(3\sqrt{5} - 7x)^2 = (3\sqrt{5})^2 - 2(3\sqrt{5})(7x) + (7x)^2$$
$$= 3^2(\sqrt{5})^2 - 42x\sqrt{5} + 7^2x^2$$
$$= 9 \cdot 5 - 42x\sqrt{5} + 49x^2$$
$$= 45 - 42x\sqrt{5} + 49x^2$$

    We put x before $\sqrt{5}$ so it will be clear that x is *not* under the radical sign

We obtain the same answer if we write $(3\sqrt{5} - 7x)^2$ as $(3\sqrt{5} - 7x)(3\sqrt{5} - 7x)$ and use the FOIL method. $\blacksquare$

In a division problem, when the dividend contains more than one term but the divisor contains only one term, we divide *each term* of the dividend by the divisor (see Example 4).

Example 4 $\dfrac{3\sqrt{14} - \sqrt{8}}{\sqrt{2}}$
Solution

$$\frac{3\sqrt{14} - \sqrt{8}}{\sqrt{2}} = \frac{3\sqrt{14}}{\sqrt{2}} - \frac{\sqrt{8}}{\sqrt{2}} = 3\sqrt{\frac{14}{2}} - \sqrt{\frac{8}{2}} = 3\sqrt{7} - \sqrt{4} = 3\sqrt{7} - 2 \quad \blacksquare$$

Rationalizing a Denominator That Contains Two Terms

In order to discuss division problems in which the divisor contains square roots and has two terms, we need a new definition.

CONJUGATE

The **conjugate** of the algebraic expression $a + b$ is the algebraic expression $a - b.$

---

opposite

Example 5   Examples of conjugates of algebraic expressions that contain square roots:

a. The conjugate of $1 - \sqrt{2}$ is $1 + \sqrt{2}$.

b. The conjugate of $2\sqrt{3} + 5$ is $2\sqrt{3} - 5$.

c. The conjugate of $\sqrt{x} - \sqrt{y}$ is $\sqrt{x} + \sqrt{y}$. ■

Example 6   Find the product of $1 - \sqrt{2}$ and its conjugate.
Solution The conjugate of $1 - \sqrt{2}$ is $1 + \sqrt{2}$. Therefore, the product is

$$(1 - \sqrt{2})(1 + \sqrt{2}) = 1^2 - (\sqrt{2})^2 = 1 - 2 = -1$$

Notice that -1 is a *rational number*. ■

The product of an algebraic expression that contains square roots and its conjugate is *always* a rational number. Because of this fact, the following procedure should be used when a denominator contains square roots and has two terms:

TO RATIONALIZE A DENOMINATOR THAT CONTAINS SQUARE ROOTS AND HAS TWO TERMS

Multiply the numerator and the denominator by the conjugate of the denominator.

(Multiplying both numerator and denominator by the same number is, of course, equivalent to multiplying the given fraction by a fraction whose value is 1.)

Example 7   Examples of rationalizing denominators that contain square roots and have two terms:

a. $\dfrac{2}{1 + \sqrt{3}} = \dfrac{2}{(1 + \sqrt{3})} \dfrac{(1 - \sqrt{3})}{(1 - \sqrt{3})} = \dfrac{2(1 - \sqrt{3})}{(1)^2 - (\sqrt{3})^2} = \dfrac{2(1 - \sqrt{3})}{1 - 3} = \dfrac{2(1 - \sqrt{3})}{-2}$

$= \dfrac{2(1 - \sqrt{3})}{-2}$ Dividing both numerator and denominator by -2

$= \sqrt{3} - 1$

We multiply both numerator and denominator by $1 - \sqrt{3}$ (the conjugate of the denominator $1 + \sqrt{3}$)

b. $\dfrac{6}{\sqrt{5} - \sqrt{3}} = \dfrac{6}{(\sqrt{5} - \sqrt{3})} \dfrac{(\sqrt{5} + \sqrt{3})}{(\sqrt{5} + \sqrt{3})} = \dfrac{6(\sqrt{5} + \sqrt{3})}{(\sqrt{5})^2 - (\sqrt{3})^2} = \dfrac{6(\sqrt{5} + \sqrt{3})}{5 - 3}$

$= \dfrac{\overset{3}{\cancel{6}}(\sqrt{5} + \sqrt{3})}{\underset{1}{\cancel{2}}} = 3\sqrt{5} + 3\sqrt{3}$

We multiply both numerator and denominator by $\sqrt{5} + \sqrt{3}$ (the conjugate of the denominator $\sqrt{5} - \sqrt{3}$) ∎

EXERCISES 11.5

Set I   In Exercises 1–18, simplify each expression.

1. $\sqrt{2}(\sqrt{2} + 1)$ **2.** $\sqrt{3}(\sqrt{3} + 1)$ **3.** $\sqrt{3}(2\sqrt{3} + 1)$ $6\oplus\sqrt{3}$

4. $\sqrt{5}(3\sqrt{5} + 1)$ **5.** $\sqrt{x}(\sqrt{x} - 3)$ **6.** $\sqrt{y}(4 - \sqrt{y})$ $\underline{not}$ $6\sqrt{3}$

7. $(\sqrt{7} + 2)(\sqrt{7} + 3)$ **8.** $(\sqrt{3} + 2)(\sqrt{3} + 4)$

9. $(\sqrt{8} - 3\sqrt{2})(\sqrt{8} + 2\sqrt{5})$ **10.** $(\sqrt{2} + 4\sqrt{3})(\sqrt{12} - 2\sqrt{3})$

11. $(\sqrt{2} + \sqrt{14})^2$ **12.** $(\sqrt{5} + \sqrt{11})^2$

13. $(\sqrt{2x} + 3)^2$ **14.** $(\sqrt{7x} + 4)^2$

15. $\dfrac{\sqrt{8} + \sqrt{18}}{\sqrt{2}}$ **16.** $\dfrac{\sqrt{12} + \sqrt{27}}{\sqrt{3}}$

17. $\dfrac{\sqrt{20} + 5\sqrt{10}}{\sqrt{5}}$ $= 2 + 5\sqrt{2}$ NOT $7\sqrt{2}$ **18.** $\dfrac{2\sqrt{6} + \sqrt{14}}{\sqrt{6}}$

In Exercises 19 and 20, write the conjugate for each expression.

19. a. $2 + \sqrt{3}$ b. $2\sqrt{5} - 7$

20. a. $3\sqrt{2} - 5$ b. $\sqrt{7} + 4$

In Exercises 21–28, rationalize the denominators and simplify.

21. $\dfrac{3}{\sqrt{2} - 1}$ $\dfrac{3\sqrt{2} + 1}{(1)} = 3\sqrt{2} + 1$ **22.** $\dfrac{5}{\sqrt{2} - 1}$ **23.** $\dfrac{6}{\sqrt{3} - \sqrt{2}}$ $\left(6(\sqrt{3} + \sqrt{2})\right)$

24. $\dfrac{8}{\sqrt{2} - \sqrt{3}}$ **25.** $\dfrac{6}{\sqrt{5}\oplus\sqrt{2}}$ $6\sqrt{5}\ominus\sqrt{2})$ **26.** $\dfrac{4}{\sqrt{7} + \sqrt{5}}$

27. $\dfrac{x - 4}{\sqrt{x} + 2}$ **28.** $\dfrac{y - 9}{\sqrt{y} - 3}$ $(y \neq 9)$

 In Exercises 29–32, approximate each expression using Table I or a calculator. Round off answers to two decimal places.

29. $\dfrac{1}{3}\sqrt{7} - 2\sqrt{3}$ **30.** $3\sqrt{6} - \dfrac{1}{4}\sqrt{5}$

31. $\dfrac{3 + 2\sqrt{11}}{6}$ **32.** $\dfrac{3 - 2\sqrt{11}}{6}$

Set II   In Exercises 1–18, simplify each expression.

1. $\sqrt{5}(\sqrt{5} + 1)$ **2.** $\sqrt{7}(\sqrt{7} - 5)$ **3.** $\sqrt{2}(3\sqrt{2} + 1)$

4. $\sqrt{5}(8\sqrt{5} - \sqrt{3})$ **5.** $\sqrt{z}(\sqrt{3} - \sqrt{z})$ **6.** $\sqrt{8}(\sqrt{5} + \sqrt{2})$

7. $(\sqrt{6} + 5)(\sqrt{6} + 2)$

8. $(\sqrt{11} - 2)(\sqrt{11} - 3)$

9. $(\sqrt{2} - 5\sqrt{8})(\sqrt{2} + 4\sqrt{8})$

10. $(\sqrt{7} + 3\sqrt{2})(3\sqrt{7} + 2\sqrt{2})$

11. $(\sqrt{7} + \sqrt{18})^2$

12. $(\sqrt{13} + \sqrt{12})^2$

13. $(3\sqrt{y} + 5)^2$

14. $(5\sqrt{z} + 2)^2$

15. $\dfrac{\sqrt{15} + \sqrt{10}}{\sqrt{5}}$

16. $\dfrac{\sqrt{21} - \sqrt{14}}{\sqrt{3}}$

17. $\dfrac{5\sqrt{32} + \sqrt{24}}{\sqrt{8}}$

18. $\dfrac{8\sqrt{48} - \sqrt{16}}{\sqrt{3}}$

In Exercises 19 and 20, write the conjugate for each expression.

19. a. $5 - \sqrt{7}$ b. $3\sqrt{11} - 2$

20. a. $-2 + \sqrt{3}$ b. $-8 + \sqrt{5}$

In Exercises 21–28, rationalize the denominators and simplify.

21. $\dfrac{10}{1 + \sqrt{2}}$

22. $\dfrac{8}{2 + \sqrt{12}}$

23. $\dfrac{13}{\sqrt{6} - \sqrt{5}}$

24. $\dfrac{21}{\sqrt{7} + \sqrt{3}}$

25. $\dfrac{15}{\sqrt{3} + \sqrt{8}}$

26. $\dfrac{12}{\sqrt{5} - 2}$

27. $\dfrac{z - 25}{\sqrt{z} + 5}$

28. $\dfrac{m^2 - 16}{2 - \sqrt{m}}$ $(m \neq 4)$

 In Exercises 29–32, approximate each expression using Table I or a calculator. Round off answers to two decimal places.

29. $\dfrac{-5 + 2\sqrt{5}}{7}$

30. $\dfrac{-1 - 5\sqrt{3}}{3}$

31. $\dfrac{5 + 6\sqrt{7}}{11}$

32. $\dfrac{\sqrt{2} + \sqrt{3}}{5}$

11.6   Radical Equations

A **radical equation** is an equation in which the variable appears in a radicand. In this text, we will consider only radical equations that have square roots.

Example 1   Examples of radical equations:

a. $\sqrt{x} = 7$

b. $\sqrt{x + 2} = 3$

c. $\sqrt{2x - 3} = \sqrt{x + 7} - 2$ ∎

The following property of real numbers is used in solving radical equations:

RULE 11.5

If two numbers are equal, then their squares are equal.

$$\text{If} \quad a = b$$
$$\text{then} \quad a^2 = b^2$$

We can remove square root signs from equations by using Rule 11.5; that is, we square both sides of the equation. However, we sometimes introduce *extraneous roots* (see Section 8.9) when we do this. Therefore, all apparent solutions to the equation must be checked in the original equation.

TO SOLVE A RADICAL EQUATION

1. Arrange the terms so that one term with a radical is by itself on one side of the equation.

2. Square both sides of the equation.

3. Combine like terms.

4. If a radical still remains, repeat steps 1, 2, and 3.

5. Solve the resulting equation for the variable.

6. Check apparent solutions in the original equation.

Example 2 Solve $\sqrt{x} = 7$.

Solution

$$\sqrt{x} = 7$$
$$(\sqrt{x})^2 = (7)^2 \longleftarrow \text{Squaring both sides}$$
$$x = 49$$

Check

$$\sqrt{x} = 7$$
$$\sqrt{49} \stackrel{?}{=} 7$$
$$7 = 7 \quad \blacksquare$$

Example 3 Solve $\sqrt{3x - 7} = \sqrt{x + 5}$.

Solution

$$\sqrt{3x - 7} = \sqrt{x + 5}$$
$$(\sqrt{3x - 7})^2 = (\sqrt{x + 5})^2 \longleftarrow \text{Squaring both sides}$$

$$\begin{array}{rcl} 3x - 7 &=& x + 5 \\ -x + 7 &=& -x + 7 \\ \hline 2x &=& 12 \end{array}$$

Adding $-x + 7$ to both sides

$$\frac{2x}{2} = \frac{12}{2}$$

Dividing both sides by 2

$$x = 6$$

Check

$$\sqrt{3x - 7} = \sqrt{x + 5}$$
$$\sqrt{3(6) - 7} \stackrel{?}{=} \sqrt{(6) + 5}$$
$$\sqrt{18 - 7} \stackrel{?}{=} \sqrt{11}$$
$$\sqrt{11} = \sqrt{11}$$

The check confirms that the solution is 6. $\quad \blacksquare$

Example 4   Solve $\sqrt{x+2} = 3$.

Solution   *Check*

$$\sqrt{x+2} = 3$$
$$(\sqrt{x+2})^2 = (3)^2$$
$$x + 2 = 9$$
$$\underline{-2 \quad -2}$$
$$x = 7$$

$$\sqrt{x+2} = 3$$
$$\sqrt{7+2} \overset{?}{=} 3$$
$$\sqrt{9} \overset{?}{=} 3$$
$$3 = 3$$

The check confirms that the solution is 7. ∎

A WORD OF CAUTION   If one side of the equation has *more than one term*, squaring each *term* is not the same as squaring both sides of the equation. That is, students often think

$$\text{if} \quad a = b + c$$
$$\text{then} \quad a^2 = b^2 + c^2$$

However, if $a = b + c$, $a^2 \neq b^2 + c^2$. To verify this with numbers, suppose that $a = 7$, $b = 2$, and $c = 5$.

$$7 = 2 + 5 \quad \text{A true statement}$$

But
$$7^2 \neq 2^2 + 5^2$$

since $7^2 = 49$ and $2^2 + 5^2 = 4 + 25 = 29$. It *is* true that if $a = b + c$, $a^2 = (b + c)^2$. To verify this with the same numbers:

$$7^2 = (2 + 5)^2$$

since $(2 + 5) = 7$.   ☑

In Examples 5 and 6 we must be sure to square the *entire* right side of the equation.

Example 5   Solve $\sqrt{2x+1} = x - 1$.

Solution

$$\sqrt{2x+1} = x - 1$$
$$(\sqrt{2x+1})^2 = (x-1)^2 \qquad \text{Squaring both sides}$$

When squaring $(x - 1)$, do not forget this middle term

$$2x + 1 = x^2 - 2x + 1$$
$$\underline{-2x - 1 \qquad -2x - 1} \qquad \text{Adding } -2x - 1 \text{ to both sides}$$
$$0 = x^2 - 4x$$

$$0 = x(x - 4) \qquad \text{Factoring the right side}$$

$x = 0$ $\qquad$ $x - 4 = 0$ $\qquad$ Setting each factor equal to 0
$\qquad\qquad$ $\underline{+4 \quad 4}$ $\qquad$ Adding 4 to both sides
$\qquad\qquad$ $x = 4$

Check for x = 0

$$\sqrt{2x + 1} = x - 1$$

$$\sqrt{2(0) + 1} \stackrel{?}{=} (0) - 1$$

$$\sqrt{1} \stackrel{?}{=} -1 \quad \text{The symbol } \sqrt{1} \text{ always stands for}$$
the *principal* square root of 1, which is 1 (*not* −1)

$$1 = -1 \quad \text{False}$$

Therefore, 0 is not a solution of $\sqrt{2x + 1} = x - 1$ because it does not satisfy the equation.

Check for x = 4

$$\sqrt{2x + 1} = x - 1$$

$$\sqrt{2(4) + 1} \stackrel{?}{=} (4) - 1$$

$$\sqrt{9} \stackrel{?}{=} 3$$

$$3 = 3 \quad \text{True}$$

Therefore, 4 is a solution because it does satisfy the equation. ∎

Example 6   Solve $\sqrt{2x - 3} = \sqrt{x + 7} - 2$.

Solution

$$\sqrt{2x - 3} = \sqrt{x + 7} - 2$$

$$(\sqrt{2x - 3})^2 = (\sqrt{x + 7} - 2)^2 \qquad \text{Squaring both sides}$$

When squaring $(\sqrt{x + 7} - 2)$, do not forget this middle term
↓

$$2x - 3 = x + 7 \;\boxed{-4\sqrt{x + 7}}\; + 4$$

$$2x - 3 = x + 11 - 4\sqrt{x + 7} \qquad \text{We must repeat steps 1, 2, and 3}$$
$$\underline{-x - 11 \qquad -x - 11} \qquad \text{Adding } -x - 11 \text{ to both sides}$$
$$x - 14 = \qquad\qquad -4\sqrt{x + 7}$$

$$(x - 14)^2 = (-4\sqrt{x + 7})^2 \qquad \text{Squaring both sides again}$$

$$x^2 - 28x + 196 = 16(x + 7) \qquad \text{Simplifying}$$

$$x^2 - 28x + 196 = 16x + 112$$
$$\underline{-16x - 112 \qquad -16x - 112} \qquad \text{Adding } -16x - 112 \text{ to both sides}$$
$$x^2 - 44x + 84 = \qquad 0$$

$$(x - 2)(x - 42) = 0 \qquad \text{Factoring the left side}$$

$$\begin{array}{c|c} x - 2 = 0 & x - 42 = 0 \quad \text{Setting each factor equal to 0} \\ \underline{+2 \quad +2} & \underline{+42 \quad +42} \\ x = 2 & x = 42 \end{array}$$

Check for x = 2

$$\sqrt{2x - 3} = \sqrt{x + 7} - 2$$

$$\sqrt{2(2) - 3} \stackrel{?}{=} \sqrt{(2) + 7} - 2$$

$$\sqrt{4 - 3} \stackrel{?}{=} \sqrt{9} - 2$$

$$\sqrt{1} \stackrel{?}{=} 3 - 2$$

$$1 = 1 \quad \text{True}$$

Therefore, 2 is a solution because it does satisfy the equation.

Check for x = 42

$$\sqrt{2x - 3} = \sqrt{x + 7} - 2$$
$$\sqrt{2(42) - 3} \stackrel{?}{=} \sqrt{(42) + 7} - 2$$
$$\sqrt{84 - 3} \stackrel{?}{=} \sqrt{49} - 2$$
$$\sqrt{81} \stackrel{?}{=} 7 - 2$$
$$9 = 5 \qquad \text{False}$$

Therefore, 42 is not a solution because it does not satisfy the equation. ■

Word Problems Involving Radical Equations

Many applications of algebra in business, engineering, the sciences, and so forth involve radical equations.

Example 7 Find the amount of power, P, consumed if an appliance has a resistance of 16 ohms and draws 5 amps (amperes) of current, using the following formula from electricity:

$$I = \sqrt{\frac{P}{R}}$$

Current = $\dfrac{Power}{Resistance}$
(amps)

where I, the current, is measured in amps (amperes); P, the power, is measured in watts; and R, the resistance, is measured in ohms.

Solution We must find the power, P, when $R = 16$ and $I = 5$.

$$I = \sqrt{\frac{P}{R}}$$

$$5 = \sqrt{\frac{P}{16}}$$

$$(5)^2 = \left(\sqrt{\frac{P}{16}}\right)^2 \qquad \text{Squaring both sides}$$

$$25 = \frac{P}{16}$$

$$P = 400 \qquad \text{Multiplying both sides by 16}$$

Check

$$I = \sqrt{\frac{P}{R}}$$

$$5 \stackrel{?}{=} \sqrt{\frac{400}{16}}$$

$$5 \stackrel{?}{=} \sqrt{25}$$

$$5 = 5 \qquad \text{True}$$

Therefore, 400 watts of power is consumed. ■

EXERCISES 11.6

Set I In Exercises 1–22, solve each equation.

1. $\sqrt{x} = 5$
2. $\sqrt{x} = 4$
3. $\sqrt{2x} = 4$
4. $\sqrt{3x} = 6$
5. $\sqrt{x - 3} = 2$
6. $\sqrt{x + 4} = 6$
7. $\sqrt{2x + 1} = 9$
8. $\sqrt{5x - 4} = 4$
9. $\sqrt{3x + 1} = 5$
10. $\sqrt{7x + 8} = 6$
11. $\sqrt{x + 1} = \sqrt{2x - 7}$
12. $\sqrt{3x - 2} = \sqrt{x + 4}$
13. $\sqrt{3x - 2} = x$
14. $\sqrt{5x - 6} = x$
15. $\sqrt{4x - 1} = 2x$
16. $\sqrt{6x - 1} = 3x$
17. $\sqrt{x - 3} + 5 = x$
18. $\sqrt{4x + 5} + 5 = 2x$
19. $\sqrt{2x + 2} = 1 + \sqrt{x + 2}$
20. $\sqrt{3x + 1} = 1 + \sqrt{x + 4}$
21. $\sqrt{3x - 5} = \sqrt{x + 6} - 1$
22. $\sqrt{5x - 1} = \sqrt{x + 14} - 1$

In Exercises 23–26, solve each problem for the indicated variable, using the formula $I = \sqrt{\dfrac{P}{R}}$, as given in Example 7.

23. Find the amount of power, P, consumed if an appliance has a resistance of 9 ohms and draws 7 amps of current.

24. Find the amount of power, P, consumed if an appliance has a resistance of 25 ohms and draws 2 amps of current.

25. Find R, the resistance in ohms, for an electrical system that consumes 500 watts and draws 5 amps of current.

26. Find R, the resistance in ohms, for an electrical system that consumes 600 watts and draws 5 amps of current.

In Exercises 27 and 28, use this formula from statistics: $\sigma = \sqrt{npq}$, where σ (*sigma*) is the *standard deviation*, n is the number of trials, p is the probability of success, and q is the probability of failure.

27. Find the probability of success, p, if the standard deviation is 3, the probability of failure is $\frac{3}{4}$, and the number of trials is 48.

28. Find the probability of success, p, if the standard deviation is $3\frac{1}{5}$, the probability of failure is $\frac{4}{5}$, and the number of trials is 64.

In Exercises 29 and 30, use this formula from geometry: $c = \sqrt{a^2 + b^2}$, where c is the length of the longest side of a right triangle and a and b are the lengths of the other two sides.

29. Find the length of one side of a right triangle if the length of the longest side is 5 and the length of one of the other sides is 3.

30. Find the length of one side of a right triangle if the length of the longest side is 13 and the length of one of the other sides is 5.

Set II   In Exercises 1–22, solve each equation.

1. $\sqrt{x} = 8$ **2.** $\sqrt{x} = 11$

3. $\sqrt{5x} = 10$ **4.** $\sqrt{3x} = 12$

5. $\sqrt{x - 6} = 3$ **6.** $\sqrt{x - 5} = 2$

7. $\sqrt{6x + 1} = 5$ **8.** $\sqrt{3x - 2} = 5$

9. $\sqrt{9x - 5} = 7$ **10.** $\sqrt{5x + 6} = 6$

11. $\sqrt{9 - 2x} = \sqrt{5x - 12}$ **12.** $\sqrt{6x - 1} = \sqrt{4x - 5}$

13. $\sqrt{3x + 10} = x$ **14.** $x = \sqrt{8x - 15}$

15. $\sqrt{3x + 2} = 3x$ **16.** $2x = \sqrt{40 - 12x}$

17. $\sqrt{x - 6} + 8 = x$ **18.** $1 + \sqrt{3x - 3} = x$

19. $\sqrt{3x + 1} = 2 + \sqrt{x + 1}$ **20.** $\sqrt{2x + 7} = 1 + \sqrt{x + 3}$

21. $\sqrt{2x - 1} = 1 + \sqrt{x - 1}$ **22.** $\sqrt{2x - 5} = \sqrt{x + 9} - 1$

In Exercises 23–26, solve each problem for the indicated variable, using the formula $I = \sqrt{\dfrac{P}{R}}$, as given in Example 7.

23. Find the amount of power, P, consumed if an appliance has a resistance of 12 ohms and draws 5 amps of current.

24. Find the amount of power, P, consumed if an appliance has a resistance of 20 ohms and draws 6 amps of current.

25. Find R, the resistance in ohms, for an electrical system that consumes 700 watts and draws 5 amps of current.

26. Find R, the resistance in ohms, for an electrical system that consumes 400 watts and draws 4 amps of current.

In Exercises 27 and 28, use this formula from statistics; $\sigma = \sqrt{npq}$, where σ (*sigma*) is the *standard deviation*, n is the number of trials, p is the probability of success, and q is the probability of failure.

27. Find the probability of success, p, if the standard deviation is $2\frac{2}{3}$, the probability of failure is $\frac{2}{3}$, and the number of trials is 32.

28. Find the probability of success, p, if the standard deviation is 5, the probability of failure is $\frac{5}{6}$, and the number of trials is 180.

In Exercises 29 and 30, use this formula from geometry: $c = \sqrt{a^2 + b^2}$, where c is the length of the longest side of a right triangle and a and b are the lengths of the other two sides.

29. Find the length of one side of a right triangle if the length of the longest side is 17 and the length of one of the other sides is 8.

30. Find the length of one side of a right triangle if the length of the longest side is 10 and the length of one of the other sides is 6.

11.7 Review: 11.1–11.6

Square Roots
11.1

The **square root** of a number p is a number that gives p when squared. A positive real number p has two square roots, a positive root called the **principal square root**, written $\sqrt{p}$, and a negative square root, written $-\sqrt{p}$.

Square Root Rules
11.1–11.3

In this text it is understood that all variables in radicands represent positive numbers.

Rule 11.1 $(\sqrt{a})^2 = \sqrt{a^2} = a$

Rule 11.2 $\sqrt{ab} = \sqrt{a}\sqrt{b}$

Rule 11.3 $\sqrt{\dfrac{a}{b}} = \dfrac{\sqrt{a}}{\sqrt{b}}$ $(b \neq 0)$

Rule 11.4 $\sqrt{a}\sqrt{a} = a$

To Simplify the Square Root of a Product
11.2

1. Express the radicand in prime factored form.

2. Find the square root of each factor as follows:
   a. If the exponent of a factor is 1, that factor must remain under the radical sign.
   b. If the exponent of a factor is an even number, remove that factor from the radical by dividing its exponent by 2.
   c. If the exponent of a factor is an odd number, write that factor as the product of two factors—one factor with an even exponent, and the other factor with an exponent of 1. Then remove from the radical the factor that has an even exponent by dividing its exponent by 2.

3. The simplified expression is the product of all the factors found in step 2.

The Simplified Form of an Expression That Has Square Roots
11.2
11.3
11.4

1. When the radicand is in prime factored, exponential form, no factor of the radicand has an exponent greater than 1.

2. No radicand contains a fraction.

3. No denominator contains a square root.

4. No term contains more than one radical sign.

5. All like radicals are combined.

To Multiply Square Roots
11.3, 11.5

Use Rule 11.2: $\sqrt{a}\sqrt{b} = \sqrt{ab}$. Simplify the result. If one of the factors has more than one term, use the distributive property. If both factors have two terms, the FOIL method can be used.

To Divide Square Roots
11.3

Use Rule 11.3: $\dfrac{\sqrt{a}}{\sqrt{b}} = \sqrt{\dfrac{a}{b}}$. You may, however, need to rationalize the denominator.

Conjugates
11.5

The **conjugate** of $a + b$ is $a - b$.

To Rationalize a Denominator
11.3
11.5

A. With one term: Multiply numerator and denominator by a number that makes the denominator a perfect square.

B. With two terms: Multiply numerator and denominator by the conjugate of the denominator.

To Combine Square Roots
11.4

A. Like radicals: Add their coefficients and multiply that sum by the like radical.

B. Unlike radicals:

    1. Express each term in simplest radical form.

    2. Combine any terms that have like radicals by adding their coefficients and multiplying that sum by the like radical.

To Solve a Radical Equation
11.6

1. Arrange the terms so that one term with a radical is by itself on one side of the equation.

2. Square both sides of the equation.

3. Combine like terms.

4. If a radical still remains, repeat steps 1, 2, and 3.

5. Solve the resulting equation for the variable.

6. Check the solutions in the original equation. There may be extraneous roots.

Review Exercises 11.7 Set I

1. Identify the radicand in each expression.

    a. $\sqrt{9x}$ b. $\sqrt{\dfrac{1}{2}}$ c. $\sqrt{x-5}$

In Exercises 2–26, simplify each expression.

2. $\sqrt{100}$ **3.** $(\sqrt{3})^2$ **4.** $(\sqrt{a+b})^2$

5. $\sqrt{x^3}$ **6.** $\sqrt{36a^4b^2}$ **7.** $\sqrt{a^3b^3}$

8. $\sqrt{11}\sqrt{11}$ **9.** $\sqrt{2}\sqrt{32}$ **10.** $\sqrt{50}$

11. $\sqrt{180}$ **12.** $4\sqrt{2}-\sqrt{2}$ **13.** $\sqrt{18}-\sqrt{8}$

14. $\sqrt{27}-\sqrt{12}$ **15.** $\dfrac{1}{\sqrt{5}}$ **16.** $\sqrt{\dfrac{1}{7}}$

17. $\dfrac{6}{\sqrt{3}}$ **18.** $\sqrt{\dfrac{9}{2}}$ **19.** $\sqrt{8}-\sqrt{\dfrac{1}{2}}$

20. $\sqrt{54}-\sqrt{\dfrac{2}{3}}$ **21.** $\sqrt{2}(\sqrt{8}+\sqrt{18})$ **22.** $(5+\sqrt{2})(5-\sqrt{2})$

23. $(2\sqrt{3}+1)^2$ **24.** $(3\sqrt{2}-1)^2$ **25.** $\dfrac{8}{\sqrt{3}-2}$

26. $\dfrac{6}{2+\sqrt{5}}$

In Exercises 27–34, solve each equation.

27. $\sqrt{x}=4$ **28.** $\sqrt{3a}=6$

29. $\sqrt{2x-1}=5$ **30.** $\sqrt{5a-4}=\sqrt{3a+2}$

31. $\sqrt{7x - 6} = x$

32. $\sqrt{9x - 8} = x$

33. $\sqrt{2x + 7} = \sqrt{x} + 2$

34. $\sqrt{3x + 1} = \sqrt{x + 8} - 1$

In Exercise 35, use the formula $c = \sqrt{a^2 + b^2}$ (where c is the length of the longest side of a right triangle).

35. Find the length of one side of a right triangle if the length of the longest side is 13 and the length of one of the other sides is 12.

Review Exercises 11.7 Set II

1. Identify the radicand in each expression.

a. $\sqrt{4-z}$ 　　　　　 b. $\sqrt{\dfrac{2}{3}}$ 　　　　　 c. $\sqrt{5a}$

In Exercises 2–26, simplify each expression.

2. $\sqrt{64}$ 　　　　　　　　　 **3.** $(\sqrt{11})^2$

4. $(\sqrt{m+n})^2$ 　　　　　 **5.** $\sqrt{w^7}$

6. $\sqrt{25t^4s^2}$ 　　　　　 **7.** $\sqrt{u^3v^5}$

8. $\sqrt{6}\sqrt{6}$ 　　　　　 **9.** $\sqrt{18}\sqrt{2}$

10. $\sqrt{68}$ 　　　　　　　 **11.** $\sqrt{175}$

12. $5\sqrt{6}-\sqrt{6}$ 　　　　 **13.** $\sqrt{45}-\sqrt{20}$

14. $\sqrt{63}-\sqrt{28}$ 　　　 **15.** $\dfrac{1}{\sqrt{13}}$

16. $\sqrt{\dfrac{1}{17}}$ 　　　　　 **17.** $\dfrac{12}{\sqrt{6}}$

18. $\sqrt{\dfrac{16}{5}}$ 　　　　　 **19.** $\sqrt{60}-\sqrt{\dfrac{3}{5}}$

ANSWERS

1a. _____

b. _____

c. _____

2. _____

3. _____

4. _____

5. _____

6. _____

7. _____

8. _____

9. _____

10. _____

11. _____

12. _____

13. _____

14. _____

15. _____

16. _____

17. _____

18. _____

19. _____

20. $\sqrt{24} - \sqrt{\dfrac{2}{3}}$

21. $\sqrt{5}(\sqrt{5} + \sqrt{20})$

22. $(\sqrt{6} - 2)(\sqrt{6} + 2)$

23. $(3 - 2\sqrt{7})^2$

24. $(\sqrt{2} - 5)^2$

25. $\dfrac{3}{1 - \sqrt{2}}$

26. $\dfrac{12}{\sqrt{5} + 3}$

In Exercises 27–34, solve each equation.

27. $\sqrt{z} = 5$

28. $\sqrt{5h} = 10$

29. $\sqrt{4x - 3} = 5$

30. $\sqrt{7 - 3m} = \sqrt{m + 3}$

31. $x = \sqrt{x + 20}$

32. $\sqrt{4x - 7} = \sqrt{x} + 1$

33. $\sqrt{5x - 4} = 2 + \sqrt{x}$

34. $\sqrt{2x + 5} = \sqrt{x + 14} - 1$

20. _____

21. _____

22. _____

23. _____

24. _____

25. _____

26. _____

27. _____

28. _____

29. _____

30. _____

31. _____

32. _____

33. _____

34. _____

35. _____

In Exercise 35, use the formula $c = \sqrt{a^2 + b^2}$ (where c is the length of the longest side of a right triangle).

35. Find the length of one side of a right triangle if the length of the longest side is 17 and the length of one of the other sides is 15.

Chapter 11 Diagnostic Test

The purpose of this test is to see how well you understand operations with radicals. We recommend that you work this diagnostic test *before* your instructor tests you on this chapter. Allow yourself about 50 minutes.

Complete solutions for all the problems on this test, together with section references, are given in the answer section in the back of the book. For the problems you do incorrectly, study the sections referred to.

1. Identify the radicand in each of the following:

   a. $\sqrt{3x^3y}$

   b. $\sqrt{2x + 5}$

In Problems 2–22, simplify the expression; assume that all variables in denominators are nonzero.

2. $\sqrt{16z^2}$

3. $\sqrt{50}$

4. $\sqrt{x^6y^5}$

5. $\sqrt{\dfrac{3}{14}}$

6. $\sqrt{\dfrac{48}{3n^2}}$

7. $\sqrt{60}$

8. $\sqrt{13}\sqrt{13}$

9. $\sqrt{3}\sqrt{75y^2}$

10. $\sqrt{5}(2\sqrt{5} - 3)$

11. $(\sqrt{13} + \sqrt{3})(\sqrt{13} - \sqrt{3})$

12. $(4\sqrt{x} + 3)(3\sqrt{x} + 4)$

13. $(8 + \sqrt{10})^2$

14. $\dfrac{\sqrt{5}}{\sqrt{80}}$

15. $\dfrac{\sqrt{3}}{\sqrt{6}}$

16. $\dfrac{\sqrt{18} - \sqrt{36}}{\sqrt{3}}$

17. $\dfrac{\sqrt{x^4y}}{\sqrt{3y}}$

18. $\dfrac{6}{1 + \sqrt{5}}$

19. $8\sqrt{t} - \sqrt{t}$

20. $7\sqrt{3} - \sqrt{27}$

21. $\sqrt{54} + 4\sqrt{24}$

22. $\sqrt{45} + 6\sqrt{\dfrac{4}{5}}$

In Problems 23–25, solve each equation.

23. $\sqrt{8x} = 12$

24. $\sqrt{5x + 11} = 6$

25. $\sqrt{6x - 5} = x$

Cumulative Review Exercises: Chapters 1–11

1. Simplify. Write your answer using only positive exponents.

$$\left(\frac{30x^4y^{-3}}{12y^{-1}}\right)^{-2}$$

2. Simplify the following expression. Then factor if possible.

$$(2h - 6)^2 - 2[-4h(3 - h)]$$

3. Divide and simplify.

$$\frac{x^2 + x - 2}{x^2 - 1} \div \frac{x^2 - 2x - 8}{x^2 - 4x}$$

4. Solve the inequality $3(x - 2) \le 6$ and graph its solution set on the number line.

5. Solve for x: $\dfrac{2x - 3}{2} = \dfrac{5x + 4}{6} - \dfrac{5}{3}$

6. Subtract: $\dfrac{2x}{x + 2} - \dfrac{5}{x - 1}$

7. Solve graphically:
$$\begin{cases} 4x - 3y = -9 \\ 3y + x = -6 \end{cases}$$

8. Solve by addition:
$$\begin{cases} 3x + 2y = 8 \\ 2x - y = 17 \end{cases}$$

9. Solve by substitution: $\begin{cases} 2x - 5y = 2 \\ 3x - 7y = 2 \end{cases}$

10. What is the multiplicative inverse of $-\frac{4}{7}$?

In Exercises 11–15, simplify each expression.

11. $(\sqrt{2x})^2$

12. $(\sqrt{11} + \sqrt{14})^2$

13. $\sqrt{5}(3 + \sqrt{80})$

14. $\sqrt{175} - \sqrt{28}$

15. $\dfrac{8}{\sqrt{6} - 2}$

16. Solve the equation $\sqrt{3x + 2} = 4$.

In Exercises 17–20, set up each problem algebraically, solve, and check. Be sure to state what your variables represent.

17. The sum of two numbers is $\frac{1}{2}$. Their difference is $7\frac{1}{2}$. Find the two numbers.

18. Farah takes 20 min to drive to work in the morning, but she takes 36 min to return home over the same route during the evening rush hour. If her average morning speed is 16 mph faster than her average evening speed, how far is it from her apartment to her office?

19. Sylvia has thirty-two coins with a total value of $2.45. If all the coins are nickels and dimes, how many of each does she have?

20. The sum of the first two of three consecutive odd integers added to the sum of the last two is 140. Find the integers.

12 Quadratic Equations

In this chapter, we discuss several methods for solving quadratic equations. We have already solved quadratic equations by factoring in Sections 7.8, 8.9C, and 11.6.

12.1 The General Form for Quadratic Equations

A **quadratic equation** is a polynomial equation whose highest-degree term is a second-degree term.

Example 1

Examples of quadratic equations:

a. $8x^2 + 2x - 4 = 0$ b. $x^2 + 1 = 0$

c. $\frac{1}{3}x - 3x^2 = \frac{4}{5}$ d. $3x^2 = 2x$ ∎

The **general form** of a quadratic equation, as used in this text, is as follows:

THE GENERAL FORM OF A QUADRATIC EQUATION

$$ax^2 + bx + c = 0 \quad \text{or} \quad 0 = ax^2 + bx + c$$

where a, b, and c are integers and $a > 0$.

Notice that a is the coefficient of x^2, b is the coefficient of x, and c is the constant.

TO CHANGE A QUADRATIC EQUATION INTO THE GENERAL FORM

1. Remove fractions by multiplying each term by the LCD.

2. Remove grouping symbols.

3. Collect and combine like terms.

4. Arrange all nonzero terms in descending powers on one side of the equal sign, leaving only zero on the other side.

Example 2

Change each of the following quadratic equations into the general form and identify a, b, and c.

a. $8x = 3 - 2x^2$

Solution

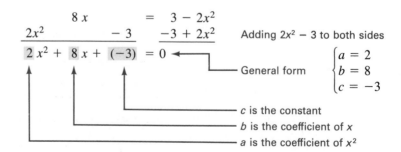

b. $3x^2 = 5$

Solution We add -5 to both sides.

$$
\begin{array}{rcl}
3x^2 & = & 5 \\
\underline{-5} & & \underline{-5} \\
3x^2 - 5 & = & 0 \longleftarrow \text{General form}
\end{array}
\qquad
\begin{cases}
a = 3 \\
b = 0 \\
c = -5
\end{cases}
$$

The coefficient of x is an understood 0.

c. $8x = 7x^2$

Solution We add $-8x$ to both sides so that a will be positive.

$$
\begin{array}{rcl}
8x & = & 7x^2 \\
\underline{-8x} & & \underline{-8x} \\
0 & = & 7x^2 - 8x \longleftarrow \text{General form}
\end{array}
\qquad
\begin{cases}
a = 7 \\
b = -8 \\
c = 0
\end{cases}
$$

The constant is an understood 0.

d. $\dfrac{1}{2}x^2 - 3x = \dfrac{2}{3}$

Solution

$$6\left(\frac{1}{2}x^2 - \overset{6}{3x}\right) = 6\left(\frac{2}{3}\right) \qquad \text{Multiplying both sides by the LCD, 6}$$

$$
\begin{array}{rcl}
3x^2 - 18x & = & 4 \\
\underline{-4} & & \underline{-4} \qquad \text{Adding } -4 \text{ to both sides} \\
3x^2 - 18x - 4 & = & 0 \longleftarrow \text{General form}
\end{array}
\qquad
\begin{cases}
a = 3 \\
b = -18 \\
c = -4
\end{cases}
$$

e. $x(x - 2) = 5$

Solution We remove parentheses first.

$$
\begin{array}{rcll}
x(x - 2) & = & 5 & \\
x^2 - 2x & = & 5 & \text{Using the distributive property} \\
\underline{-5} & & \underline{-5} & \text{Adding } -5 \text{ to both sides} \\
x^2 - 2x - 5 & = & 0 \longleftarrow \text{General form}
\end{array}
\qquad
\begin{cases}
a = 1 \\
b = -2 \\
c = -5
\end{cases}
$$

f. $(x + 3)(x - 2) = 2x + 1$

Solution

$$
\begin{array}{rcll}
(x + 3)(x - 2) & = & 2x + 1 & \\
x^2 + x - 6 & = & 2x + 1 & \text{Using the FOIL method to} \\
& & & \text{remove the parentheses on the left} \\
\underline{-2x - 1} & & \underline{-2x - 1} & \text{Adding } -2x - 1 \text{ to both sides} \\
x^2 - x - 7 & = & 0 \longleftarrow \text{General form}
\end{array}
\qquad
\begin{cases}
a = 1 \\
b = -1 \\
c = -7 \quad \blacksquare
\end{cases}
$$

EXERCISES 12.1

Set I   Write each of the following quadratic equations in the general form and identify a, b, and c.

1. $2x^2 = 5x + 3$

2. $3x^2 = 4 - 2x$

3. $6x^2 = x$

4. $2x - 3x^2 = 0$

5. $\dfrac{3x}{2} + 5 = x^2$

6. $4 - x^2 = \dfrac{2x}{3}$

7. $x^2 - \dfrac{5x}{4} + \dfrac{2}{3} = 0$

8. $2x^2 + \dfrac{3x}{5} = \dfrac{1}{3}$

9. $x(x - 3) = 4$

10. $2x(x + 1) = 12$

11. $3x(x + 1) = (x + 1)(x + 2)$

12. $(x + 1)(x - 3) = 4x(x - 1)$

Set II   Write each of the folowing quadratic equations in the general form and identify a, b, and c.

1. $5x = 4x^2 + 1$

2. $3 = 5x - x^2$

3. $5x^2 = x$

4. $(x - 2)(2x + 4) = 0$

5. $\dfrac{4x}{3} + 2 = x^2$

6. $7x^2 = 3$

7. $3x^2 + \dfrac{x}{4} = \dfrac{1}{5}$

8. $\left(\dfrac{x}{2} - 3\right)\left(x + \dfrac{1}{4}\right) = 2$

9. $(x - 1)(3x) = 7$

10. $2 = 3x^2$

11. $3x(x + 1) = (x + 2)(x - 3)$

12. $(4x - 1)(x + 2) = x(x - 4)$

12.2 Solving Quadratic Equations by Factoring

We will review the method of solving quadratic equations by factoring first used in Section 7.8.

TO SOLVE A QUADRATIC EQUATION BY FACTORING

1. Arrange the equation in the general form (Section 12.1):

$$ax^2 + bx + c = 0$$

2. Factor the polynomial.

3. Set each factor equal to zero and solve for the variable.

4. Check apparent solutions in the original equation.

Example 1   Solve $x^2 - 2x - 8 = 0$.
Solution

$$x^2 - 2x - 8 = 0 \qquad \text{The equation is already in the general form}$$

$$(x + 2)(x - 4) = 0 \qquad \text{Factoring the left side}$$

$$
\begin{array}{c|c}
\begin{aligned}
x + 2 &= 0 \\
-2 \quad &-2 \\
\hline
x \quad\;\; &= -2
\end{aligned}
&
\begin{aligned}
x - 4 &= 0 \\
+4 \quad &+4 \\
\hline
x \quad\;\; &= 4
\end{aligned}
\end{array}
\qquad \text{Setting each factor equal to zero}
$$

Check for $x = -2$	*Check for $x = 4$*
$x^2 - 2x - 8 = 0$	$x^2 - 2x - 8 = 0$
$(-2)^2 + 2(-2) - 8 \overset{?}{=} 0$	$(4)^2 - 2(4) - 8 \overset{?}{=} 0$
$4 + 4 - 8 \overset{?}{=} 0$	$16 - 8 - 8 \overset{?}{=} 0$
$0 = 0$ True	$0 = 0$ True

Therefore, -2 and 4 are solutions. ■

We can also solve equations of this type by getting all terms to the *right* side, leaving only zero on the *left* side (see Example 2).

Example 2   Solve $5 - 2x^2 = 3x$.
Solution

$$
\begin{aligned}
5 - 2x^2 &= 3x \\
-5 + 2x^2 \quad\;\; &\quad -5 + 2x^2 \qquad \text{Adding } -5 + 2x^2 \text{ to both sides}\\
\hline
0 &= 3x - 5 + 2x^2
\end{aligned}
$$

$$0 = 2x^2 + 3x - 5 \qquad \text{The coefficient of } x^2 \text{ is } positive$$

$$0 = (x - 1)(2x + 5) \qquad \text{Factoring the right side}$$

$$
\begin{array}{c|c}
\begin{aligned}
x - 1 &= 0 \\
+1 \quad &+1 \\
\hline
x \quad\;\; &= 1
\end{aligned}
&
\begin{aligned}
2x + 5 &= 0 \\
-5 \quad &-5 \\
\hline
2x \quad\;\; &= -5 \\
x &= -\frac{5}{2}
\end{aligned}
\end{array}
\qquad \text{Setting each factor equal to zero}
$$

Checking confirms that the solutions are 1 and $-\dfrac{5}{2}$. ■

Example 3   Solve $x^2 = \dfrac{3 - 5x}{2}$.

Solution We can treat this like a proportion. There are no excluded values.

$$\frac{x^2}{1} = \frac{3 - 5x}{2}$$

$$2 \cdot x^2 = 1 \cdot (3 - 5x) \qquad \text{Cross-multiplying}$$

$$
\begin{aligned}
2x^2 \qquad\;\; &= 3 - 5x \\
+ 5x - 3 \quad &\; -3 + 5x \qquad \text{Adding } -3 + 5x \text{ to both sides} \\
\hline
2x^2 + 5x - 3 &= 0
\end{aligned}
$$

$$(2x - 1)(x + 3) = 0 \qquad \text{Factoring the left side}$$

$$\begin{array}{rcl} 2x - 1 &=& 0 \\ +1 & & +1 \\ \hline 2x &=& 1 \end{array} \qquad \begin{array}{rcl} x + 3 &=& 0 \\ -3 & & -3 \\ \hline x &=& -3 \end{array}$$

Setting each factor equal to zero

$$x = \frac{1}{2}$$

Checking confirms that the solutions are $\frac{1}{2}$ and -3. ∎

Example 4 Solve $\dfrac{x - 1}{x - 3} = \dfrac{12}{x + 1}$.

Solution This is a proportion. Excluded values are 3 and -1.

$$\frac{x - 1}{x - 3} = \frac{12}{x + 1}$$

$$(x - 1)(x + 1) = 12(x - 3) \qquad \text{Cross-multiplying}$$

$$\begin{array}{rcl} x^2 \qquad - 1 &=& 12x - 36 \\ -12x + 36 & & -12x + 36 \\ \hline x^2 - 12x + 35 &=& 0 \end{array}$$

Adding $-12x + 36$ to both sides

$$(x - 7)(x - 5) = 0 \qquad \text{Factoring the left side}$$

$$\begin{array}{rcl} x - 7 &=& 0 \\ +7 & & +7 \\ \hline x &=& 7 \end{array} \qquad \begin{array}{rcl} x - 5 &=& 0 \\ +5 & & +5 \\ \hline x &=& 5 \end{array}$$

Setting each factor equal to zero

Checking confirms that the solutions are 7 and 5. ∎

Example 5 Solve $(6x + 2)(x - 4) = 2 - 11x$.

Solution

$$(6x + 2)(x - 4) = 2 - 11x$$

$$\begin{array}{rcl} 6x^2 - 22x - 8 &=& 2 - 11x \\ +11x - 2 & & -2 + 11x \\ \hline 6x^2 - 11x - 10 &=& 0 \end{array}$$

Using the FOIL method to remove the parentheses on the left

Adding $-2 + 11x$ to both sides

$$(3x + 2)(2x - 5) = 0 \qquad \text{Factoring the left side}$$

$$\begin{array}{rcl} 3x + 2 &=& 0 \\ -2 & & -2 \\ \hline 3x &=& -2 \end{array} \qquad \begin{array}{rcl} 2x - 5 &=& 0 \\ +5 & & +5 \\ \hline 2x &=& 5 \end{array}$$

Setting each factor equal to zero

$$\frac{3x}{3} = \frac{-2}{3} \qquad \qquad \frac{2x}{2} = \frac{5}{2}$$

$$x = -\frac{2}{3} \qquad \qquad x = \frac{5}{2}$$

Checking shows that the solutions are $-\dfrac{2}{3}$ and $\dfrac{5}{2}$. ∎

Example 6   Michelle bicycled from her home to a beach and back, a total distance of 120 mi. Her average speed returning was 3 mph slower than her average speed going to the beach. If her total bicycling time was 9 hr, what was her average speed going to the beach?

Solution

Let x = average speed going to the beach (in mph)

     $x - 3$ = average speed returning from the beach (in mph)

The distance each way is 60 mi. We solve the distance-rate-time formula $d = rt$ for t:

$$t = \frac{d}{r}$$

We then use this formula to find the time going to the beach and the time returning from the beach:

$$\frac{60}{x} = \text{time going to the beach (in hours)}$$

$$\frac{60}{x - 3} = \text{time returning from the beach (in hours)}$$

The sum of the two times equals 9 hr:

$$\frac{60}{x} + \frac{60}{x - 3} = 9$$

$$x(x - 3)\left(\frac{60}{x} + \frac{60}{x - 3}\right) = x(x - 3)\,(9) \qquad \text{Multiplying both sides by the LCD}$$

$$\overset{1}{x}(x - 3)\left(\frac{60}{\underset{1}{x}}\right) + x(\overset{1}{x - 3})\left(\frac{60}{\underset{1}{x - 3}}\right) = x(x - 3)(9) \qquad \text{Simplifying}$$

$$60(x - 3) + 60x = 9x(x - 3)$$

$$60x - 180 + 60x = 9x^2 - 27x$$

$$120x - 180 = 9x^2 - 27x \qquad \text{Combining like terms}$$

$$\underline{-120x + 180 \qquad\qquad -120x + 180} \qquad \begin{array}{l}\text{Adding } -120x + 180 \\ \text{to both sides}\end{array}$$

$$0 = 9x^2 - 147x + 180$$

$$0 = 3(3x^2 - 49x + 60) \qquad \text{Factoring out the GCF}$$

$$0 = 3(3x - 4)(x - 15) \qquad \text{Completing the factoring}$$

$3 \neq 0$	$3x - 4 = \quad 0$	$x - 15 = \quad 0$	Setting each factor equal to zero
	$\underline{\quad + 4 \quad\quad +4}$	$\underline{\quad + 15 \quad\quad +15}$	
	$3x \quad\quad = \quad 4$	$x \quad\quad = \quad 15$	
	$x = \dfrac{4}{3}$	$x = 15$	The average speed going to the beach, in mph

Check for $x = \frac{4}{3}$ If $\frac{4}{3}$ is the speed going to the beach, then the speed returning is

$$x - 3 = \frac{4}{3} - 3 = -\frac{5}{3}$$

But this is not possible, since a speed cannot be negative.

Check for x = 15 If 15 is the speed going to the beach, then

$$x - 3 = 15 - 3 = 12 \qquad \text{The speed returning, in mph}$$

$$\frac{60}{x} = \frac{60}{15} = 4 \qquad \text{The time going to the beach, in hours}$$

$$\frac{60}{x - 3} = \frac{60}{12} = 5 \qquad \text{The time returning, in hours}$$

The sum of the two times is 9 hr. Therefore, Michelle's average speed going to the beach was 15 mph. ■

EXERCISES 12.2

Set I   In Exercises 1–22, solve each equation by factoring.

1. $x^2 + x - 6 = 0$ **2.** $x^2 - x - 6 = 0$

3. $x^2 + x = 12$ **4.** $x^2 + 2x = 15$

5. $2x^2 - x = 1$ **6.** $2x^2 + x = 1$

7. $\dfrac{x}{8} = \dfrac{2}{x}$ **8.** $\dfrac{x}{3} = \dfrac{12}{x}$

9. $\dfrac{x + 2}{3} = \dfrac{-1}{x - 2}$ **10.** $\dfrac{x + 3}{2} = \dfrac{-4}{x - 3}$

11. $x^2 + 9x + 8 = 0$ **12.** $x^2 + 7x + 6 = 0$

13. $2x^2 + 4x = 0$ **14.** $3x^2 - 9x = 0$

15. $x^2 = x + 2$ **16.** $x^2 = x + 6$

17. $\dfrac{x}{2} + \dfrac{2}{x} = \dfrac{5}{2}$ **18.** $\dfrac{x}{3} + \dfrac{2}{x} = \dfrac{7}{3}$

19. $\dfrac{x - 1}{4} + \dfrac{6}{x + 1} = 2$ **20.** $\dfrac{3x + 1}{5} + \dfrac{8}{3x - 1} = 3$

21. $2x^2 = \dfrac{2 - x}{3}$ **22.** $4x^2 = \dfrac{14x - 3}{2}$

In Exercises 23–28, set up each problem algebraically and solve. Be sure to state what your variables represent and to check your solutions.

23. The length of a rectangle is 5 more than its width. If its area is 24, find its dimensions.

24. The length of a rectangle is 4 more than its width. If its area is 77, find its dimensions.

25. If the product of two consecutive even integers is increased by 4, the result is 84. Find the integers.

26. If the product of two consecutive integers is decreased by 6, the result is 36. Find the integers.

27. Bruce drove from Los Angeles to the Mexican border and back to Los Angeles, a total distance of 240 mi. His average speed returning to Los Angeles was 20 mph faster than his average speed going to Mexico. If his total driving time was 5 hr, what was his average speed driving from Los Angeles to Mexico?

28. Nick drove from his home to San Diego and back, a total distance of 360 mi. His average speed returning to his home was 15 mph faster than his average speed going to San Diego. If his total driving time was 7 hr, what was his average speed driving from his home to San Diego?

Set II In Exercises 1–22, solve each equation by factoring.

1. $x^2 + x - 12 = 0$

2. $x^2 + 2x - 8 = 0$

3. $x^2 + 4x = 5$

4. $x^2 + 2x = 63$

5. $3x^2 - 2x = 1$

6. $2x^2 = 3 - 5x$

7. $\dfrac{x}{27} = \dfrac{3}{x}$

8. $\dfrac{x - 2}{5} = \dfrac{1}{x + 2}$

9. $\dfrac{x + 1}{1} = \dfrac{3}{x - 1}$

10. $2x^2 + 7x + 3 = 0$

11. $x^2 + 6x + 5 = 0$

12. $x^2 + x = 0$

13. $2x^2 - 6x = 0$

14. $x^2 + 2x = 15$

15. $x^2 = x + 12$

16. $x + \dfrac{1}{4} = \dfrac{3}{4x}$

17. $\dfrac{x}{2} - \dfrac{3}{x} = \dfrac{5}{4}$

18. $\dfrac{x + 3}{4} + \dfrac{3}{x + 3} = \dfrac{61}{28}$

19. $\dfrac{2x + 1}{5} + \dfrac{10}{2x + 1} = 3$

20. $\dfrac{x}{2} = \dfrac{19}{10} + \dfrac{3}{x}$

21. $3x^2 = \dfrac{11x + 10}{2}$

22. $2x^2 = \dfrac{1 - 5x}{3}$

In Exercises 23–28, set up each problem algebraically and solve. Be sure to state what your variables represent and to check your solutions.

23. The length of a rectangle is 5 more than its width. If its area is 36, find its dimensions.

24. The length of a rectangle is 3 times its width. If the numerical sum of its area and perimeter is 80, find its dimensions.

25. If the product of two consecutive even integers is increased by 8, the result is 16. Find the integers.

26. If the product of two consecutive odd integers is decreased by 3, the result is 12. Find the integers.

27. Leon drove from his home to Lake Shasta and back, a total distance of 600 mi. His average speed returning to his home was 10 mph faster than his average speed going to Lake Shasta. If his total driving time was 11 hr, what was his average speed driving from his home to Lake Shasta?

28. Ruth drove from Creston to Des Moines, a distance of 90 mi. Then she continued on from Des Moines to Omaha, a distance of 120 mi. Her average speed was 10 mph faster on the second part of the journey than on the first part. If the total driving time was 6 hr, what was her average speed on the first leg of the journey?

12.3 Incomplete Quadratic Equations

An **incomplete quadratic equation** is one in which b or c (or both) is zero. If a were zero, the equation would not be a quadratic equation. Therefore, in these problems, the coefficient a cannot be zero.

Example 1   Examples of incomplete quadratic equations:

a. $12x^2 + 5 = 0$ $(b = 0)$

b. $7x^2 - 2x = 0$ $(c = 0)$

c. $3x^2 = 0$ $(b = 0 \text{ and } c = 0)$ ∎

We don't need any special rules or methods for solving an incomplete quadratic equation when $c = 0$ (see Examples 2 and 3).

Example 2   Solve $12x^2 = 3x$.
Solution

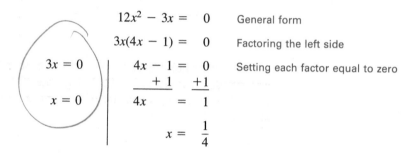

$$12x^2 - 3x = 0 \qquad \text{General form}$$

$$3x(4x - 1) = 0 \qquad \text{Factoring the left side}$$

$3x = 0$	$4x - 1 = 0$	Setting each factor equal to zero
	$\underline{+1 \qquad +1}$	
$x = 0$	$4x \quad = 1$	
	$x = \dfrac{1}{4}$	

Checking confirms that the solutions are 0 and $\dfrac{1}{4}$.   ∎

A WORD OF CAUTION   In doing Example 2, a common mistake students make is to divide both sides of the equation by x.

$$\cancel{12x^2 = 3x}$$

$$12x = 3 \qquad \text{Dividing both sides by } x$$

$$x = \frac{1}{4} \qquad \text{Using this method, only the solution } \frac{1}{4} \text{ is found}$$

By dividing both sides of the equation by x, we lost the solution 0. Do not divide both sides of an equation by an expression containing the variable, because you may lose solutions.   ☑

Example 3   Solve $\frac{2}{5}x = 3x^2$.
Solution LCD = 5

$$\frac{5}{1} \cdot \frac{2}{5}x = 5 \cdot (3x^2) \qquad \text{Multiplying both sides by 5}$$

$$2x = 15x^2$$

$$\underline{-2x \qquad \qquad - 2x} \qquad \text{Adding } -2x \text{ to both sides}$$

$$0 = 15x^2 - 2x \qquad \text{General form}$$

$$0 = x(15x - 2) \qquad \text{Factoring the right side}$$

$$x = 0 \qquad \begin{aligned} 15x - 2 &= 0 \\ +2 \quad &+2 \\ \hline 15x \quad &= 2 \\ x &= \frac{2}{15} \end{aligned}$$

Checking will confirm that the solutions are 0 and $\frac{2}{15}$. ∎

Some incomplete quadratic equations in which $b = 0$ can be solved by factoring. Other such equations can be solved by using the following rule:

RULE 12.1

If $x^2 = p$, then $x = \sqrt{p}$ or $x = -\sqrt{p}$.

We can combine both answers by using the $\pm$ symbol (read ''plus or minus'') that was introduced in Section 1.12. Using Rule 12.1, an incomplete quadratic equation in which $b = 0$ (that is, an equation of the form $ax^2 + c = 0$) can be solved by the following method:

TO SOLVE A QUADRATIC EQUATION WHEN $b = 0$

1. Arrange the equation so the second-degree term is on one side of the equal sign and the constant term is on the other side.

2. Divide both sides by a (the coefficient of the second-degree term).

3. Use Rule 12.1: If $x^2 = p$, then $x = \pm\sqrt{p}$.

4. Express the solutions in simplest form. When the radicand is *positive* or *zero*, the square roots are *real numbers*; when it is *negative*, the square roots are *not* real numbers.

5. Check all real solutions by substituting them into the original equation.

In Example 4, we demonstrate the above method and also show that the given equation can be solved by factoring.

Example 4 Solve $x^2 - 4 = 0$.
Solution $x^2 - 4 = 0$ is in the general form, and $b = 0$.

Step 1. $x^2 = 4$

Step 3. $x = \pm\sqrt{4}$ Using Rule 12.1

Step 4. $x = \pm\, 2$

This equation can also be solved by factoring.

$$x^2 - 4 = 0$$

$$(x + 2)(x - 2) = 0$$

$$
\begin{array}{c|c}
\begin{aligned}
x + 2 &= 0 \\
-2 \quad & -2 \\
\hline
x \quad\;\; &= -2
\end{aligned}
&
\begin{aligned}
x - 2 &= 0 \\
+2 \quad & +2 \\
\hline
x \quad\;\; &= 2
\end{aligned}
\end{array}
$$

Notice that we do obtain both 2 and -2 as solutions when we solve by factoring. The checking is left to the student. ∎

Example 5 can be solved by factoring; however, we show only the method described in the box on page 569.

Example 5   Solve $4x^2 - 9 = 0$.
Solution

Step 1.
$$
\begin{aligned}
4x^2 - 9 &= 0 \\
+9 \quad & +9 \\
\hline
4x^2 \quad\;\; &= 9
\end{aligned}
$$
Adding 9 to both sides

Step 2.
$$x^2 = \frac{9}{4}$$
Dividing both sides by 4

Step 3.
$$x = \pm\sqrt{\frac{9}{4}}$$
Using Rule 12.1

Step 4.
$$x = \pm\frac{3}{2}$$

Checking shows that the solutions are $\frac{3}{2}$ and $-\frac{3}{2}$. ∎

Examples 6 and 7 *cannot* be solved by factoring over the integers.

Example 6   Solve $3x^2 - 5 = 0$.
Solution

Step 1.
$$
\begin{aligned}
3x^2 - 5 &= 0 \\
+5 \quad & +5 \\
\hline
3x^2 \quad\;\; &= 5
\end{aligned}
$$
Adding 5 to both sides

Step 2.
$$x^2 = \frac{5}{3}$$
Dividing both sides by 3

Step 3.
$$x = \pm\sqrt{\frac{5}{3}} = \pm\frac{\sqrt{5}}{\sqrt{3}} \cdot \frac{\sqrt{3}}{\sqrt{3}}$$
← Rationalizing the denominator

Step 4.
$$x = \pm\frac{\sqrt{15}}{3}$$

We can use Table I or a calculator to approximate $\sqrt{15}$ and find decimal approximations to the answers.

$$x = \pm\frac{\sqrt{15}}{3} \approx \pm\frac{3.873}{3} = \pm1.291 \approx \pm1.29$$ Rounded off to two decimal places

Step 5.   *Check for* $x = +\dfrac{\sqrt{15}}{3}$ *Check for* $x = -\dfrac{\sqrt{15}}{3}$

$$3x^2 - 5 = 0 \qquad\qquad 3x^2 - 5 = 0$$

$$3\left(\frac{\sqrt{15}}{3}\right)^2 - 5 \stackrel{?}{=} 0 \qquad 3\left(-\frac{\sqrt{15}}{3}\right)^2 - 5 \stackrel{?}{=} 0$$

$$3\left(\frac{15}{9}\right) - 5 \stackrel{?}{=} 0 \qquad\qquad 3\left(\frac{15}{9}\right) - 5 \stackrel{?}{=} 0$$

$$5 - 5 \stackrel{?}{=} 0 \qquad\qquad\qquad 5 - 5 \stackrel{?}{=} 0$$

$$0 = 0 \qquad\qquad\qquad\qquad 0 = 0$$

The solutions are $\dfrac{\sqrt{15}}{3}$ and $-\dfrac{\sqrt{15}}{3}$, or approximately 1.29 and -1.29. ∎

In Example 7, the solutions *do exist*. However, they are not real numbers; they are *imaginary numbers*, and imaginary numbers are not discussed in this text.

Example 7   Solve $x^2 + 25 = 0$.
Solution

$$\begin{array}{rcr} x^2 + 25 = & & 0 \\ -25 & & -25 \\ \hline \end{array}$$

Step 1. $x^2 \qquad = -25$

Step 3. $x = \boxed{\pm\sqrt{-25}}$ The solutions are not real numbers because the radicand is negative

Because the radicand is negative, our answer is "The solutions are not real numbers." ∎

EXERCISES 12.3

Set I   Solve and check each equation, or write "The solutions are not real numbers."

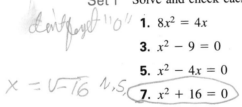

1. $8x^2 = 4x$

2. $15x^2 = 5x$

3. $x^2 - 9 = 0$

4. $x^2 - 36 = 0$

5. $x^2 - 4x = 0$

6. $x^2 - 16x = 0$

7. $x^2 + 16 = 0$

8. $x^2 + 1 = 0$

9. $x^2 - 4 = 0$

10. $x^2 - 16 = 0$

11. $5x^2 = 4$

12. $3x^2 = 25$

13. $8 - 2x^2 = 0$

14. $27 - 3x^2 = 0$

15. $x(x + 3) = 3x - 4$

16. $x(x + 5) = 5x - 9$

17. $2(x + 3) = 6 + x(x + 2)$

18. $5(2x - 3) - x(2 - x) = 8(x - 1) - 7$

19. $2x(3x - 4) = 2(3 - 4x)$

20. $5x(2x - 3) = 3(4 - 5x)$

Set II   Solve and check each equation, or write "The solutions are not real numbers."

1. $8x^2 = 12x$

2. $3x = 6x^2$

3. $x^2 - 25 = 0$

4. $x^2 = 100$

5. $x^2 - 49x = 0$

6. $x^2 = 9x$

7. $x^2 + 81 = 0$

8. $x^2 + 4x = 0$

9. $x^2 - 121 = 0$

10. $5x^2 = 25$

11. $3x^2 = 7$

12. $x^2 - 25x = 0$

13. $64 - 4x^2 = 0$

14. $x(x + 7) = 7x - 1$

15. $x(x + 1) = x - 9$

16. $x^2 + 3(x - 1) = -3$

17. $3(x + 4) = 12 + x(x + 2)$

18. $x(x + 3) - 7 = 3(x - 1)$

19. $4x(5x - 2) = 2(40 - 4x)$

20. $5x(2x - 3) = 3(10 - 5x)$

12.4 The Quadratic Formula

The methods we have shown in previous sections can be used to solve only *some* quadratic equations. The methods we show in this section can be used to solve *all* quadratic equations.

In Example 1, we use a method called **completing the square** to solve a quadratic equation.

Example 1   Solve $x^2 - 4x + 1 = 0$.

Solution We begin by moving the constant to the right side, and next we "complete the square" on the left side.

$$x^2 - 4x + 1 = 0$$

$$x^2 - 4x = -1 \qquad \text{Adding } -1 \text{ to both sides}$$

Find $\frac{1}{2}$ of -4: $\frac{1}{2}(-4) = -2$

Then $(-2)^2 = 4$

$$x^2 - 4x + 4 = -1 + 4 \qquad \text{Adding } 4 \text{ to both sides makes the left side a perfect square}$$

$$(x - 2)^2 = 3 \qquad \text{Factoring the left side}$$

$$x - 2 = \pm\sqrt{3} \qquad \text{Using Rule 12.1}$$

$$x = 2 \pm \sqrt{3} \qquad \text{Adding 2 to both sides}$$

Check for $x = 2 + \sqrt{3}$.

$$x^2 \quad - \quad 4x \quad + 1 = 0$$

$$(2 + \sqrt{3})^2 - 4(2 + \sqrt{3}) + 1 \overset{?}{=} 0$$

$$4 + 4\sqrt{3} + 3 - 8 - 4\sqrt{3} + 1 \overset{?}{=} 0$$

$$0 = 0$$

We leave the check for $x = 2 - \sqrt{3}$ to the student. The solutions are $2 + \sqrt{3}$ and $2 - \sqrt{3}$. ∎

The method of completing the square can be used to solve *any* quadratic equation. We now use it to solve the general form of the quadratic equation, and, in this way, we derive the **quadratic formula**.

Derivation of the Quadratic Formula

$$ax^2 + bx + c = 0 \qquad \text{General form}$$

$$ax^2 + bx = -c \qquad \text{Adding } -c \text{ to both sides}$$

$$x^2 + \frac{b}{a}x = -\frac{c}{a} \qquad \text{Dividing both sides by } a$$

Find $\frac{1}{2}$ of $\frac{b}{a}$: $\frac{1}{2}\left(\frac{b}{a}\right) = \frac{b}{2a}$

Then $\left(\frac{b}{2a}\right)^2 = \frac{b^2}{4a^2}$

$$x^2 + \frac{b}{a}x + \frac{b^2}{4a^2} = \frac{b^2}{4a^2} - \frac{c}{a} = \frac{b^2}{4a^2} - \frac{4ac}{4a^2} \qquad \text{Adding } \frac{b^2}{4a^2} \text{ to both sides to make the left side a perfect square}$$

$$\left(x + \frac{b}{2a}\right)^2 = \frac{b^2 - 4ac}{4a^2} \qquad \text{Factoring the left side and adding the fractions on the right side}$$

$$x + \frac{b}{2a} = \pm\sqrt{\frac{b^2 - 4ac}{4a^2}} \qquad \text{Using Rule 12.1}$$

$$x + \frac{b}{2a} = \pm\frac{\sqrt{b^2 - 4ac}}{\sqrt{4a^2}} = \pm\frac{\sqrt{b^2 - 4ac}}{2a} \qquad \text{Simplifying the radicals}$$

$$x = -\frac{b}{2a} \pm \frac{\sqrt{b^2 - 4ac}}{2a} \qquad \text{Adding } -\frac{b}{2a} \text{ to both sides}$$

Therefore, $\qquad x = \dfrac{-b \pm \sqrt{b^2 - 4ac}}{2a} \qquad$ The *quadratic formula*

The procedure for using the quadratic formula can be summarized as follows:

TO SOLVE AN EQUATION BY USING THE QUADRATIC FORMULA

1. Arrange the equation in the general form (Section 12.1):

$$ax^2 + bx + c = 0$$

2. Substitute the values of a, b, and c into the *quadratic formula*:

$$x = \frac{-b \pm \sqrt{b^2 - 4ac}}{2a} \qquad (a \neq 0)$$

3. Simplify the apparent solutions.

4. Check the apparent solutions in the original equation.

A WORD OF CAUTION   Be sure that your fraction bar and the bar of your radical sign are long enough.

$$x = \frac{-b}{2a} \pm \sqrt{b^2 - 4ac} \quad \text{is incorrect,}$$

$$x = \frac{-b \pm \sqrt{b^2 - 4ac}}{2a} \quad \text{is incorrect,}$$

and

$$x = \frac{-b \pm \sqrt{b^2} - 4ac}{2a} \quad \text{is incorrect.} \quad \boxed{\checkmark}$$

Example 2   Solve $x^2 - 5x + 6 = 0$ by using the quadratic formula.

Solution

Substitute the values $\begin{cases} a = 1 \\ b = -5 \\ c = 6 \end{cases}$ into the formula $x = \dfrac{-b \pm \sqrt{b^2 - 4ac}}{2a}$.

$$x = \frac{-(-5) \pm \sqrt{(-5)^2 - 4(1)(6)}}{2(1)}$$

$$= \frac{5 \pm \sqrt{25 - 24}}{2} = \frac{5 \pm \sqrt{1}}{2}$$

$$= \frac{5 \pm 1}{2} = \begin{cases} \dfrac{5 + 1}{2} = \dfrac{6}{2} = 3 \\[2mm] \dfrac{5 - 1}{2} = \dfrac{4}{2} = 2 \end{cases}$$

We write $\dfrac{5 \pm 1}{2}$ as two separate fractions

We obtain the same answers if we solve the equation by factoring.

$$x^2 - 5x + 6 = 0$$

$$(x - 2)(x - 3) = 0$$

$x - 2 =$	0		$x - 3 =$	0
$+ 2$	$+2$		$+ 3$	$+3$
x $=$	2		x $=$	3

Checking shows that the solutions are 2 and 3.   ∎

Solving a quadratic equation by factoring is ordinarily faster than using the formula. Therefore, first check to see if the equation can be solved by factoring. If it cannot, use the formula. The equations in Examples 3–6 cannot be solved by factoring.

Example 3   Solve $x^2 - 6x - 3 = 0$.

Solution

Substitute the values $\begin{cases} a = 1 \\ b = -6 \\ c = -3 \end{cases}$ into the formula $x = \dfrac{-b \pm \sqrt{b^2 - 4ac}}{2a}$.

$$x = \frac{-(-6) \pm \sqrt{(-6)^2 - 4(1)(-3)}}{2(1)}$$

$$= \frac{6 \pm \sqrt{36 + 12}}{2} = \frac{6 \pm \sqrt{48}}{2}$$

$$= \frac{6 \pm 4\sqrt{3}}{2} = \frac{\overset{1}{\cancel{2}}(3 \pm 2\sqrt{3})}{\underset{1}{\cancel{2}}} = 3 \pm 2\sqrt{3}$$

Checking confirms that the solutions are $3 + 2\sqrt{3}$ and $3 - 2\sqrt{3}$. ∎

Example 4   Solve $\frac{1}{4}x^2 = 1 - x$.

Solution LCD = 4

$$\frac{4}{1} \cdot \frac{1}{4}x^2 = 4 \cdot (1 - x) \qquad \text{First, change the equation to the general form}$$

$$x^2 = 4 - 4x$$

$$x^2 + 4x - 4 = 0 \qquad\qquad \text{General form}$$

Substitute the values $\begin{cases} a = 1 \\ b = 4 \\ c = -4 \end{cases}$ into $x = \dfrac{-b \pm \sqrt{b^2 - 4ac}}{2a}$.

$$x = \frac{-(4) \pm \sqrt{(4)^2 - 4(1)(-4)}}{2(1)}$$

$$= \frac{-4 \pm \sqrt{16 + 16}}{2} = \frac{-4 \pm \sqrt{32}}{2}$$

$$= \frac{-4 \pm 4\sqrt{2}}{2} = \frac{\overset{1}{\cancel{2}}(-2 \pm 2\sqrt{2})}{\underset{1}{\cancel{2}}} = -2 \pm 2\sqrt{2}$$

Checking verifies that the solutions are $-2 + 2\sqrt{2}$ and $-2 - 2\sqrt{2}$. ∎

Example 5   Solve $4x^2 - 5x + 2 = 0$.

Solution

Substitute the values $\begin{cases} a = 4 \\ b = -5 \\ c = 2 \end{cases}$ into $x = \dfrac{-b \pm \sqrt{b^2 - 4ac}}{2a}$.

$$x = \frac{-(-5) \pm \sqrt{(-5)^2 - 4(4)(2)}}{2(4)}$$

$$= \frac{5 \pm \sqrt{25 - 32}}{8} = \frac{5 \pm \sqrt{-7}}{8} \quad \leftarrow \text{The solutions are not real numbers because the radicand is negative}$$

There are no real solutions for the equation. ∎

In some practical applications, we need decimal approximations to the exact answers. In such cases, we can use a calculator or Table I to approximate the square roots; we can then express the answers as decimal approximations (see Example 6).

Example 6 Solve $x^2 - 5x + 3 = 0$. Round off the answers to two decimal places.
Solution

Substitute the values $\begin{cases} a = 1 \\ b = -5 \\ c = 3 \end{cases}$ into $x = \dfrac{-b \pm \sqrt{b^2 - 4ac}}{2a}$.

$$x = \frac{-(-5) \pm \sqrt{(-5)^2 - 4(1)(3)}}{2(1)}$$

$$= \frac{5 \pm \sqrt{25 - 12}}{2} = \frac{5 \pm \sqrt{13}}{2} \qquad (\sqrt{13} \approx 3.606 \text{ from Table I})$$

$$\approx \frac{5 \pm 3.606}{2} = \begin{cases} \dfrac{5 + 3.606}{2} = \dfrac{8.606}{2} = 4.303 \approx 4.30 \\[2ex] \dfrac{5 - 3.606}{2} = \dfrac{1.394}{2} = 0.697 \approx 0.70 \end{cases}$$

Check We suggest that you use a calculator to verify that the solutions are approximately 4.30 and 0.70. The solutions will not check exactly, since they are approximations and not exact solutions. ∎

EXERCISES 12.4

Set I In Exercises 1–16, use the quadratic formula or the method of completing the square to solve each equation. If the radicand is negative, write "The solutions are not real numbers." Express all other solutions in simplest radical form.

1. $3x^2 - x - 2 = 0$

2. $2x^2 + 3x - 2 = 0$

3. $x^2 - 4x + 1 = 0$

4. $x^2 - 4x - 1 = 0$

5. $\dfrac{x}{2} + \dfrac{2}{x} = \dfrac{5}{2}$

6. $\dfrac{x}{3} + \dfrac{2}{x} = \dfrac{7}{3}$

7. $2x^2 = 8x - 5$

8. $3x^2 = 6x - 2$

9. $\dfrac{1}{x} + \dfrac{x}{x - 1} = 3$

10. $\dfrac{2}{x} - \dfrac{x}{x + 1} = 4$

11. $x^2 + x + 1 = 0$

12. $x^2 + x + 2 = 0$

13. $4x^2 + 4x = 1$

14. $9x^2 - 6x = 2$

15. $3x^2 + 2x + 1 = 0$

16. $4x^2 + 3x + 2 = 0$

In Exercises 17 and 18, set up each problem algebraically and solve. Express all answers in simplest radical form.

17. A number less its reciprocal is $\frac{2}{3}$. What is the number?

18. If a number is subtracted from twice its reciprocal, the result is $\frac{5}{4}$. What is the number?

In Exercises 19–24, find the solutions and express them in simplest radical form. Also use a calculator or Table I to approximate the solutions, rounding them off to two decimal places. (In Exercises 21–24, set up each problem algebraically.)

19. $4x^2 = 1 - 4x$

20. $9x^2 = 2 + 6x$

21. The length of a rectangle is 2 more than its width. If its area is 2, find its dimensions.

22. The length of a rectangle is 4 more than its width. If its area is 6, find its dimensions.

23. The perimeter of a square is numerically 6 less than its area. Find the length of a side.

24. The area of a square is numerically 2 more than its perimeter. Find the length of a side.

Set II   In Exercises 1–16, use the quadratic formula or the method of completing the square to solve each equation. If the radicand is negative, write "The solutions are not real numbers." Express all other solutions in simplest radical form.

1. $3x^2 - 8x - 3 = 0$

2. $x^2 - 3x + 3 = 0$

3. $x^2 - 4x + 2 = 0$

4. $2x^2 = 2x - 5$

5. $\dfrac{x}{3} + \dfrac{1}{x} = \dfrac{7}{6}$

6. $\dfrac{x}{2} + \dfrac{1}{x} = \dfrac{33}{8}$

7. $3x^2 + 4 = 2x$

8. $5x^2 + 3x + 1 = 0$

9. $\dfrac{2}{x} = \dfrac{3x}{x + 2} - 1$

10. $\dfrac{3}{x} = \dfrac{x}{x + 2} + 1$

11. $x^2 + 3x + 5 = 0$

12. $x^2 - 3x + 4 = 0$

13. $9x^2 + 9x = 1$

14. $5x(x + 5) = 1$

15. $3x^2 + 4 = 2x$

16. $x^2 = \dfrac{3 - 5x}{2}$

In Exercises 17 and 18, set up each problem algebraically and solve. Express all answers in simplest radical form.

17. A number less its reciprocal is $\frac{1}{4}$. What is the number?

18. A number is equal to the sum of its reciprocal and $\frac{2}{5}$. What is the number?

In Exercises 19–24, find the solutions and express them in simplest radical form. Also use a calculator or Table I to approximate the solutions, rounding them off to two decimal places. (In Exercises 21–24, set up each problem algebraically.)

19. $5x^2 + 4x = 3$

20. $3x^2 = 5 + 6x$

21. The length of a rectangle is 2 more than its width. If its area is 6, find its dimensions.

22. The length of a rectangle is 5 more than its width. If its area is 3, find its dimensions.

23. The perimeter of a square is numerically 3 more than its area. Find the length of a side.

24. The area of a square is numerically 3 more than its perimeter. Find the length of a side.

12.5 The Pythagorean Theorem

Right Triangles   A triangle that has a right angle (a 90° angle) is called a **right triangle**. The side opposite the right angle of a right triangle is called the **hypotenuse**; it is always the longest side of the triangle. The other two sides are often called the **legs** of the right triangle.

The diagonal of a rectangle divides the rectangle into two right triangles. The parts of a right triangle and a rectangle and one of its diagonals are shown in Figure 12.5.1.

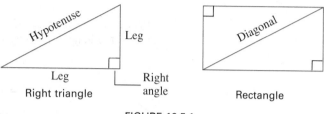

FIGURE 12.5.1

The Pythagorean theorem, which follows, applies only to right triangles.

THE PYTHAGOREAN THEOREM

The square of the hypotenuse of a right triangle is equal to the sum of the squares of the two legs.

$$c^2 = a^2 + b^2$$

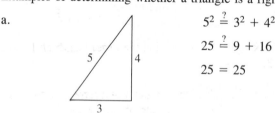

The Pythagorean theorem can be used to determine whether or not a given triangle is a right triangle; to do this, we use the rule as follows: If the square of the longest side of a triangle equals the sum of the squares of the two shorter sides, the triangle is a right triangle.

Example 1   Examples of determining whether a triangle is a right triangle:

a.

5   4

3

$$5^2 \overset{?}{=} 3^2 + 4^2$$

$$25 \overset{?}{=} 9 + 16$$

$$25 = 25$$

Therefore, the given triangle *is* a right triangle.

b.

$\sqrt{29}$

2

5

$$(\sqrt{29})^2 \overset{?}{=} 5^2 + 2^2$$

$$29 \overset{?}{=} 25 + 4$$

$$29 = 29$$

Therefore, the given triangle *is* a right triangle.

c.

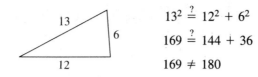

$$13^2 \stackrel{?}{=} 12^2 + 6^2$$

$$169 \stackrel{?}{=} 144 + 36$$

$$169 \neq 180$$

Therefore, this triangle *is not* a right triangle. ∎

The Pythagorean theorem can also be used to find one side of a right triangle when the other two sides are known. In such problems, we can consider *either* leg to be a, but the hypotenuse *must* be c.

Example 2   Find the length of the hypotenuse of a right triangle whose legs are 8 cm and 6 cm.
Solution Let x = the length of the hypotenuse.

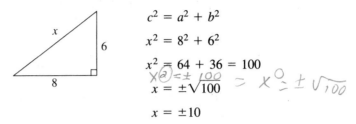

$$c^2 = a^2 + b^2$$

$$x^2 = 8^2 + 6^2$$

$$x^2 = 64 + 36 = 100$$

$$x = \pm\sqrt{100}$$

$$x = \pm 10$$

NOTE   Checking will show that both 10 and -10 are solutions of the equation. However, -10 cannot be a solution of the *geometric problem*, because we consider lengths in geometric figures to be positive numbers. For this reason, in the problems of this section we will consider only the positive (*principal*) square root.   ☑

Therefore, the length of the hypotenuse is 10 cm. ∎

While lengths in geometric figures cannot be negative, they *can* be irrational. In fact, in a square whose sides measure 1 in., the *exact* length of the diagonal is $\sqrt{2}$ in. The *decimal approximation* of this length is 1.41 in.

Example 3   Use the Pythagorean theorem to find x in the given right triangle.
Solution

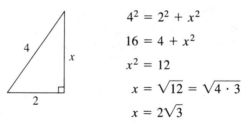

$$4^2 = 2^2 + x^2$$

$$16 = 4 + x^2$$

$$x^2 = 12$$

$$x = \sqrt{12} = \sqrt{4 \cdot 3}$$

$$x = 2\sqrt{3}$$

Checking will confirm that the *exact* length of the side is $2\sqrt{3}$; the *approximate* length is 3.46. ∎

Example 4   Find the length of the diagonal of a rectangle that has a length of 6 ft and a width of 4 ft.
Solution Let x = the length of the diagonal.

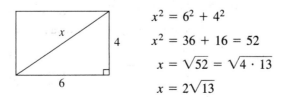

$$x^2 = 6^2 + 4^2$$

$$x^2 = 36 + 16 = 52$$

$$x = \sqrt{52} = \sqrt{4 \cdot 13}$$

$$x = 2\sqrt{13}$$

Checking will show that the length of the diagonal is $2\sqrt{13}$ ft. ∎

Example 5   The length of a rectangle is 2 m more than its width. If the length of its diagonal is 10 m, find the dimensions of the rectangle.

Solution

Let x = width

$x + 2$ = length

$(10)^2 = $ ▨$(x + 2)^2$▨ $+ (x)^2$

$100 = $ ▨$x^2 + 4x + 4$▨ $+ x^2$

$100 = 2x^2 + 4x + \quad 4$ Combining like terms

$\underline{-100 \qquad\qquad\quad - 100}$ Adding -100 to both sides

$0 = 2x^2 + 4x - \quad 96$

$0 = 2(x^2 + 2x - 48)$ Factoring out the GCF

$0 = 2(x + 8)(x - 6)$ Completing the factoring

$2 \neq 0$ | $x + 8 = \quad 0$ | $x - 6 = \quad 0$
| $\underline{\quad -8 \quad -8}$ | $\underline{\quad +6 \quad +6}$
| $x \qquad = -8$ | $x \qquad = \quad 6$ Width
| Not a length | $x + 2 = \quad 8$ Length

The check is left to the student. The rectangle is 6 m wide and 8 m long. ■

EXERCISES 12.5

Set I   In Exercises 1–14, use the Pythagorean theorem to find x in the given right triangle. If a solution is not rational, express it in simplest radical form.

1.
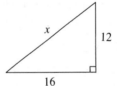

2.
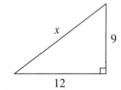

3.
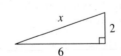

4.

5.
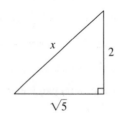

6.
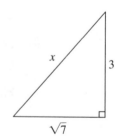

7.
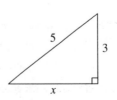

8.
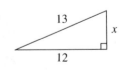

9.

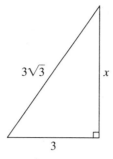

$3\sqrt{3}$ x

3

10.

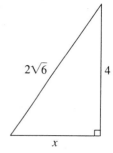

$2\sqrt{6}$ 4

x

11.

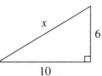

x 6

10

12.

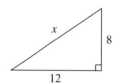

x 8

12

13.

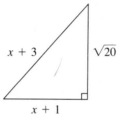

$x + 3$ $\sqrt{20}$

$x + 1$

14.

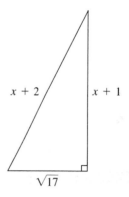

$x + 2$ $x + 1$

$\sqrt{17}$

In Exercises 15–22, set up each problem algebraically and solve. If the answers are not rational numbers, express them in simplest radical form.

15. Find the width of a rectangle with a diagonal of 25 cm and a length of 24 cm.

16. Find the width of a rectangle with a diagonal of 41 in. and a length of 40 in.

17. One leg of a right triangle is 4 m less than twice the other leg. If its hypotenuse is 10 m, how long are the two legs?

18. The length of a rectangle is 3 yd more than its width. If the length of its diagonal is 15 yd, find the dimensions of the rectangle.

19. Find the length of the diagonal of a square whose side is $\sqrt{6}$ in.

20. Find the length of the diagonal of a square whose side is $\sqrt{14}$ m.

21. Find the length of one side of a square with a diagonal of $4\sqrt{2}$ cm.

22. Find the length of one side of a square with a diagonal of $5\sqrt{2}$ in.

In Exercises 23 and 24, set up each problem algebraically and solve. Use a calculator or Table I and round off answers to two decimal places.

23. Find the length of the diagonal of a square whose side is 41.6 cm.

24. Find the width of a rectangle with a diagonal of 4.53 m and a length of 3.67 m.

Set II   In Exercises 1–14, use the Pythagorean theorem to find x in the given right triangle. If a solution is not rational, express it in simplest radical form.

1.
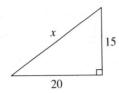

2.
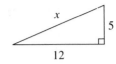

3.
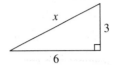

4.
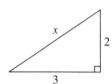

5.
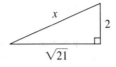

6.
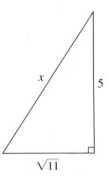

7.
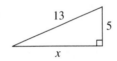

8.
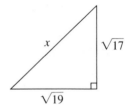

9.
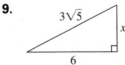

10.
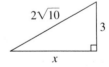

11.
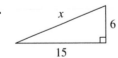

12.
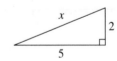

13.
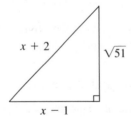

14.
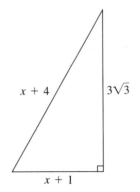

In Exercises 15–22, set up each problem algebraically and solve. If the answers are not rational numbers, express them in simplest radical form.

15. Find the width of a rectangle with a diagonal of 17 cm and a length of 15 cm.

16. Find the width of a rectangle with a diagonal of $10\sqrt{3}$ ft and a length of 15 ft.

17. One leg of a right triangle is 2 cm less than twice the other leg. If the hypotenuse is 5 cm, how long are the two legs?

18. The width of a rectangle is 4 cm less than its length. If the length of the diagonal is $4\sqrt{5}$ cm, find the dimensions of the rectangle.

19. Find the length of the diagonal of a square whose side is $\sqrt{3}$ cm.

20. Find the length of the diagonal of a square whose side is $\sqrt{15}$ ft.

21. Find the length of one side of a square with a diagonal of $7\sqrt{2}$ cm.

22. Find the length of one side of a square with a diagonal of 24 in.

In Exercises 23 and 24, set up each problem algebraically and solve. Use a calculator or Table I and round off answers to two decimal places.

23. Find the length of the diagonal of a square whose side is 7.13 in.

24. Find the width of a rectangle with a diagonal of 7.81 cm and a length of 2.68 cm.

12.6 Review: 12.1–12.5

Quadratic Equations
12.1

A **quadratic equation** is a polynomial equation whose highest-degree term is a second-degree term.

The **general form** of a quadratic equation as used in this text is

$$ax^2 + bx + c = 0$$

where a, b, and c are integers, and $a > 0$.

Methods of Solving
Quadratic Equations
12.2

Factoring

1. Arrange in the general form.

2. Factor the polynomial.

3. Set each factor equal to zero and solve for the variable.

4. Check the apparent solutions in the original equation.

12.3

Incomplete quadratic

a. *When $c = 0$:* Find the greatest common factor (GCF); then solve by factoring.

b. *When $b = 0$:* 1. Write the equation as $ax^2 = -c$.

2. Divide both sides by a.

3. Use Rule 12.1: If $x^2 = p$, then $x = \pm\sqrt{p}$. Simplify.

Check the apparent solutions in the original equation.

12.4 Quadratic Formula

1. Arrange in the general form.

2. Substitute the values of a, b, and c into the quadratic formula

$$x = \frac{-b \pm \sqrt{b^2 - 4ac}}{2a}$$

3. Simplify the apparent solutions.

4. Check the apparent solutions in the original equation.

The Pythagorean Theorem
12.5   The square of the hypotenuse of a right triangle is equal to the sum of the squares of the two legs.

$$c^2 = a^2 + b^2$$

Review Exercises 12.6 Set I

In Exercises 1–14, solve each equation by any convenient method, or write "The solutions are not real numbers." Express real answers that are not rational in simplest radical form.

1. $x^2 + x = 6$

2. $x^2 - 49x = 0$

3. $x^2 - 2x - 4 = 0$

4. $x^2 - 4x + 1 = 0$

5. $x^2 = 5x$

6. $\dfrac{3x}{5} = \dfrac{5}{12x}$

7. $\dfrac{x + 2}{3} = \dfrac{1}{x - 2} + \dfrac{2}{3}$

8. $x^2 + 144 = 0$

9. $5(x + 2) = x(x + 5)$

10. $3(x + 4) = x(x + 3)$

11. $\dfrac{2}{x} + \dfrac{x}{x + 1} = 5$

12. $\dfrac{3}{x} - \dfrac{x}{x + 2} = 2$

13. $3x^2 + 2x + 1 = 0$

14. $(2x - 1)(3x + 5) = x(x + 7) + 4$

15. Use the Pythagorean theorem to solve for x.

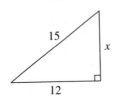

In Exercises 16–18, set up each problem algebraically and solve. Express any answer that is not rational in simplest radical form.

16. One leg of a right triangle is 7 in. more than the other. The hypotenuse is 13 in. Find the length of the longer leg.

17. Find the length of the diagonal of a square whose side is $\sqrt{10}$ m.

18. Find the length of the diagonal of a square whose side is 8 yd.

In Exercises 19 and 20, set up each problem algebraically and solve. Express each answer in simplest radical form, and use a calculator to approximate each answer, rounding off to two decimal places.

19. A rectangle is 8 in. long and 4 in. wide. Find the length of its diagonal.

20. The width of a rectangle is 6 ft less than its length. Its area is 6 sq. ft. Find its dimensions.

Review Exercises 12.6 Set II

In Exercises 1–14, solve each equation by any convenient method, or write "The solutions are not real numbers." Express real answers that are not rational in simplest radical form.

1. $x^2 + x = 12$

2. $x^2 - 36 = 0$

1. _____

2. _____

3. _____

3. $x^2 - 2x - 2 = 0$

4. $x^2 = 11x$

4. _____

5. _____

6. _____

5. $x^2 + 1 = 0$

6. $x^2 + x + 1 = 0$

7. _____

8. _____

7. $\dfrac{2x}{3} = \dfrac{3}{8x}$

8. $\dfrac{x + 2}{4} = \dfrac{1}{x - 2} + 1$

9. _____

10. _____

11. _____

9. $2(x + 1) = x(x - 4)$

10. $\dfrac{3}{x} + \dfrac{x}{x - 1} = 2$

12. _____

13. _____

11. $3x^2 - x = 0$

12. $4x^2 = 2x - 1$

14. _____

15. _____

13. $x^2 + 2x + 5 = 0$

14. $x^2 + 2x - 5 = 0$

15. Use the Pythagorean theorem to solve for x.

In Exercises 16–18, set up each problem algebraically and solve. Express any answer that is not rational in simplest radical form.

16. The length of the diagonal of a square is $\sqrt{32}$ yd. What is the length of a side?

17. Find the width of a rectangle with a diagonal of 13 in. and a length of 12 in.

18. A rectangle is 9 ft long and 3 ft wide. Find the length of its diagonal.

In Exercises 19 and 20, set up each problem algebraically and solve. Express each answer in simplest radical form, and use a calculator to approximate each answer, rounding off to two decimal places.

19. Find the length of the diagonal of a rectangle with a width of 8 cm and a length of 10 cm.

20. One leg of a right triangle is 2 m longer than the other. The hypotenuse is $2\sqrt{2}$ m. Find the length of each leg.

16. _____

17. _____

18. _____

19. _____

20. _____

_____

Chapter 12 Diagnostic Test

The purpose of this test is to see how well you understand quadratic equations. We recommend that you work this diagnostic test *before* your instructor tests you on this chapter. Allow yourself about 50 minutes.

Complete solutions for all the problems on this test, together with section references, are given in the answer section in the back of the book. For the problems you do incorrectly, study the sections referred to.

In Problem 1, write each quadratic equation in the general form, and identify a, b, and c in each equation.

1. a. $5 + 3x^2 = 7x$ b. $3(x + 5) = 3(4 - x^2)$

In Problems 2–7, solve each equation, or write "The solutions are not real numbers." Use any convenient method. If necessary, use the quadratic formula, which is as follows:

$$\text{If } ax^2 + bx + c = 0, \text{ then } x = \frac{-b \pm \sqrt{b^2 - 4ac}}{2a}.$$

2. $x^2 = 8x + 15$ **3.** $5x^2 = 17x$

4. $x^2 + 2x + 6 = 0$ **5.** $6x^2 = 54$

6. $3(x + 1) = x^2$ **7.** $x(x + 3) = x$

8. Use the Pythagorean theorem to find x in the right triangle.

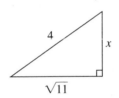

In Problems 9 and 10, set up each problem algebraically and solve. If answers are not rational numbers, express them in simplest radical form.

9. Find the length of the diagonal of a square whose side is 7 m.

10. The length of a rectangle is 6 ft more than its width. Its area is 78 sq. ft. Find its length and its width.

Cumulative Review Exercises: Chapters 1–12

1. Is division associative?

2. Is subtraction commutative?

3. What is the additive identity?

4. Simplify. Write your answer using only positive exponents.

$$\left(\frac{18s^{-1}t^{-3}}{12s}\right)^{-2}$$

In Exercises 5–14, perform the indicated operation, or write "Not defined." Express all answers in simplest form.

5. $73 \div 0$

6. $(2\sqrt{3} - 3)(4\sqrt{3} + 5)$

7. $(4x^2 + 3x - 2)(5x - 3)$

8. $(\sqrt{21} - \sqrt{5})^2$

9. $\dfrac{x^2 + 3x}{2x^2 + 7x + 5} \div \dfrac{x^2 - 9}{x^2 - 2x - 3}$

10. $(x^2 - 14x - 5) \div (x - 3)$ (Use long division)

11. $(x^2 + 2x - 8) \div \dfrac{x^2 - 4}{x^2 + 2x}$

12. $\dfrac{x + 3}{x - 3} - \dfrac{x^2 + 9}{x^2 - 9}$

13. $\dfrac{x^2 + 3x - 2}{x^2 + 5x + 6} - \dfrac{x - 1}{2x^2 + 7x + 3}$

14. $5\sqrt{18} - \sqrt{\dfrac{9}{2}}$

In Exercises 15–25, solve for all variables, or write "Identity," or write "No solution."

15. $7 - (2x + 3) = 3x - 2 - (4x - 3)$

16. $\begin{cases} 2x + 3y = 11 \\ 4x + 9y = 4 \end{cases}$

17. $\begin{cases} 3x + 4y = 3 \\ 2x - 5y = 25 \end{cases}$

18. $\sqrt{6x + 13} = 3$

19. $\sqrt{5x + 3} = 10x$

20. $\dfrac{3x - 5}{4} + 2 = \dfrac{3 + 4x}{3}$

21. $12x - 3 \le 7x - 13$

22. $x^2 + 5x = 14$

23. $3 + 4x^2 - 12x = 0$

24. $x(3x - 1) = 2x + 6$

25. $7x^2 - 1 = 0$

26. Rationalize the denominator and simplify: $\dfrac{14}{3 - \sqrt{2}}$

In Exercises 27–30, factor completely, or write "Not factorable."

27. $3x^2 - 12x - 15$

28. $5x^2 - 45$

29. $6x^2 - 12x + 5$

30. $7x^2 - 13x - 2$

In Exercises 31–34, set up each problem algebraically, solve, and check. Be sure to state what your variables represent.

31. On a scale drawing, $\frac{1}{4}$ in. represents 1 ft. What will be the size of the drawing of a room that is 12 ft by 18 ft?

32. Find two consecutive integers whose product is 11 more than their sum.

33. A 5-lb mixture of beef back ribs and beef stew meat costs $7.35. If the stew meat is $2.10 per pound and the back ribs are $1.20 per pound, how many pounds of each kind are there?

34. One leg of a right triangle is 7 cm longer than the other leg. The length of the hypotenuse is 13 cm. Find the lengths of the two legs.

Critical Thinking

Each of the following problems has an error. Can you find it?

1. Simplify $\sqrt{18} + 3\sqrt{32}$.

$$\sqrt{18} + 3\sqrt{32} = \sqrt{9 \cdot 2} + 3\sqrt{16 \cdot 2}$$
$$= 3\sqrt{2} + 7\sqrt{2}$$
$$= 10\sqrt{2}$$

2. Simplify $\dfrac{b}{a} - \dfrac{a-b}{a+b}$.

$$\frac{b}{a} - \frac{a-b}{a+b} = \frac{b}{a} \cdot \boxed{\frac{a+b}{a+b}} - \frac{a-b}{a+b} \cdot \boxed{\frac{a}{a}}$$

$$= \frac{ab + b^2}{a(a+b)} - \frac{a^2 - ab}{a(a+b)}$$

$$= \frac{ab + b^2 - a^2 - ab}{a(a+b)}$$

$$= \frac{b^2 - a^2}{a(a+b)}$$

$$= \frac{\overset{1}{\cancel{(b+a)}}(b-a)}{a\underset{1}{\cancel{(a+b)}}}$$

$$= \frac{b-a}{a}$$

3. Solve $\sqrt{3x - 2} = \sqrt{x} + 2$.

$$\sqrt{3x - 2} = \sqrt{x} + 2$$
$$(\sqrt{3x - 2})^2 = (\sqrt{x} + 2)^2$$
$$3x - 2 = x + 4$$
$$2x - 2 = 4$$
$$2x = 6$$
$$x = 3$$

Problem 4 has two errors. Can you find both?

4. Solve $2x^2 - 4x - 1 = 0$.
Using the quadratic formula,

$$\left. \begin{array}{l} a = 2 \\ b = -4 \\ c = -1 \end{array} \right\} \quad x = \frac{-(-4) \pm \sqrt{(-4)^2 - 4(2)(-1)}}{2(2)}$$

$$= \frac{4 \pm \sqrt{16 - 8}}{4}$$

$$= \frac{4 \pm \sqrt{8}}{4} = \frac{\overset{1}{\cancel{4}} \pm 2\sqrt{2}}{\underset{1}{\cancel{4}}}$$

$$= \pm 2\sqrt{2}$$

Appendix A

Sets

Ideas in all branches of mathematics, such as arithmetic, algebra, geometry, calculus, and statistics, can be explained in terms of sets. For this reason, you will find a basic understanding of sets helpful. We have already used sets in the text and in writing answers and solutions for the exercises.

The sets of numbers mentioned in this appendix are defined in Chapter 1, and a few of the definitions that follow also appear in Chapter 1.

A.1 Basic Definitions

Set   A **set** is a collection of objects or things.

Example 1   Examples of sets:

a. The first five letters of our alphabet

b. The set of whole numbers between 3 and 8

c. The set of students in your math class ■

Elements of a Set   The objects or things that make up a set are called its **elements** (or **members**).

Roster Method of Representing a Set   A **roster** is a list of the members of a group. The method of representing a set by listing its elements is called the **roster method**.

Whenever the roster method is used, it is customary to put the elements inside a pair of braces { }. We never use parentheses for sets. It is customary, though not necessary, to arrange numbers and letters in sets in numerical and alphabetical order (to make reading easier) and to represent elements of a set with lowercase letters.

Example 2   Examples showing the elements of sets and the roster method:

a. Set $\{a, b, c\}$ has elements a, b, and c.

b. Set $\{1, 5, 12, 23\}$ has elements 1, 5, 12, and 23.

c. Set $\{1, 2, 3, \ldots\}$ has elements 1, 2, 3, and so on. The three dots to the right of the 3 indicate that the numbers go on forever, following the pattern that has been established. ■

Naming a Set   A set is usually named by a capital letter such as A, N, W, and so on. The expression "$P = \{1, 5, 7\}$" is read "P is the set whose elements are 1, 5, and 7."

The Symbol $\in$ If we wish to show that a number or object is a member of a given set, we use the symbol $\in$, which is read "is an element of." Thus, the expression $2 \in A$ is read "2 is an element of set A" (or "2 is a member of set A"). If $A = \{2, 3, 4\}$, we can say $2 \in A$, $3 \in A$, and $4 \in A$. If we wish to show that a number or object is *not* a member of a given set, we use the symbol $\notin$, which is read "is not an element of" or "is not a member of." If $A = \{2, 3, 4\}$, then $5 \notin A$ (read "5 is not an element of set A"). To help you remember this symbol, notice that $\in$ looks like the first letter of the word *element*.

Example 3   Examples showing the use of $\in$ and $\notin$:

a. If $B = \{5, 9\}$, then $5 \in B$, $9 \in B$, $3 \notin B$, and $1 \notin B$.

b. If $C = \{b, e, g, m\}$, then $g \in C$, $e \in C$, and $k \notin C$.

c. If $D = \{\text{Abe, Helen, John}\}$, then Helen $\in D$ but Mary $\notin D$. ■

Cardinal Number of a Set   The **cardinal number** of a set is the number of elements in that set. The symbol $n(A)$ is read "the cardinal number of set A" and means the number of elements in set A. When a set is represented by the roster method, its cardinal number is found by counting its elements.

Example 4   Examples of the cardinal number of a set:

a. If $A = \{5, 8, 6, 9\}$, then $n(A) = 4$.

b. If $H = \{$Ed, Mabel$\}$, then $n(H) = 2$.

c. If $Q = \{a, h, l, s, t, v\}$, then $n(Q) = 6$.

d. If $E = \{2, 4, 3, 2\}$, then $n(E) = 3$. The cardinal number is 3 instead of 4 because the set has only 3 *different* elements. ∎

Equal Sets   Two sets are equal if they both have exactly the same members.

Example 5   Examples of equal sets and unequal sets:

a. $\{1, 5, 7\} = \{5, 1, 7\}$. Notice that both sets have exactly the same elements, even though they are not listed in the same order.

b. $\{1, 5, 5, 5\} = \{5, 1\}$. Notice that both sets have exactly the same elements. It is not necessary to write the same element more than once when writing the roster of a set.

c. $\{7, 8, 11\} \neq \{7, 11\}$. These sets are *not* equal because they do not both have exactly the same elements. ∎

The Empty Set (or Null Set)   If we think of the braces $\{\ \}$ as a basket, then the set $A = \{1, 5, 7\}$ has three elements in its basket: 1, 5, and 7. Set $B = \{1, 5\}$ has two elements in its basket, and set $C = \{5\}$ has only one element. Set $D = \{\ \}$ is empty, having no elements in its basket. A set with no elements is called the **empty set** (or **null set**). We use the symbols $\{\ \}$ or $\varnothing$ to represent the empty set. Whenever either symbol appears, read it as "the empty set."

Example 6   Examples of the empty set:

a. The set of all people in this classroom who are 10 ft tall $= \varnothing$.

b. The set of all digits greater than $10 = \{\ \}$. ∎

A WORD OF CAUTION $\{0\}$ is *not* the correct symbol for the empty set, and $\{\varnothing\}$ is *not* the correct symbol for the empty set. Use either the symbol $\{\ \}$ or the symbol $\varnothing$. Do not combine the two symbols. ☑

Universal Set   A **universal set** is a set containing all the elements being considered in a particular problem. We use the symbol U to represent the universal set under consideration.

Example 7   Examples of universal sets:

a. Suppose we are going to consider sets of digits. Then $U = \{0, 1, 2, 3, 4, 5, 6, 7, 8, 9\}$, since U contains *all* the digits.

b. If we are going to consider sets of whole numbers, then $U = \{0, 1, 2, 3, \ldots\}$, since U contains *all* the whole numbers.

c. Suppose we are going to consider sets of football players at East Los Angeles College in 1978. Then U is the set of all football players at East Los Angeles College in 1978.

d. Suppose we are going to consider different sets of cats. Then U is the set of *all* cats. ■

Notice that there can be different universal sets. We might want to consider the set of all students in this class as the universal set if we were going to deal only with class members. On the other hand, we might use the set of all fifty of the United States as our universal set if we were going to consider only sets of states.

Finite Set   If, in counting the elements of a set, the counting comes to an end, the set is called a **finite set**. This means that the number of elements in a finite set must be a particular whole number (which we call its cardinal number).

Example 8   Examples of finite sets:

a. $A = \{5, 9, 10, 13\}$; $n(A) = 4$.

b. D = the set of digits; $n(D) = 10$.

c. C = the set of whole numbers less than 100; $n(C) = 100$.

d. $\varnothing = \{\ \}$; $n(\varnothing) = 0$. ■

Infinite Set   If, in counting the elements of a set, the counting never comes to an end, the set is called an **infinite set**. A set is infinite if it is not finite. The cardinal number of an infinite set is not a whole number.

Example 9   Examples of infinite sets:

a. $N = \{1, 2, 3, \ldots\}$

b. $W = \{0, 1, 2, 3, \ldots\}$

c. The set of all fractions ■

EXERCISES A.1

1. Can the collection $\{\square, X, =, 5\}$ be called a set?

2. Are $\{2, 7\}$ and $\{7, 2\}$ equal sets?

3. Are $\{2, 2, 7, 7\}$ and $\{7, 2\}$ equal sets?

4. Write the set of digits < 3 (see Section 1.1 for the meaning of $<$ and $>$).

5. Write the set of digits > 9.

6. Write the set of whole numbers < 3.

7. Write the set of whole numbers > 9.

8. Write the set of whole numbers > 4 and < 6.

9. Write the set of whole numbers > 4 and < 5.

10. Write the set of whole numbers < 4 and < 5.

11. Write all the elements of the set $\{2, a, 3\}$.

12. Write all the elements in the set of digits.

13. Write the cardinal number of each of the following sets.

    a. $\{1, 1, 3, 5, 5, 5\}$

    b. $\{0\}$

    c. $\{a, b, g, x\}$

    d. The set of whole numbers < 8

    e. $\varnothing$

14. The empty set has no elements. Therefore, the statement $\varnothing = \{\ \ \}$ is true. The statement $\varnothing = \{0\}$ is not true. Explain.

15. State which of the following sets are finite and which are infinite.

    a. The set of digits

    b. The set of whole numbers

    c. The set of days in the week

    d. The set of books in the East Los Angeles College library

16. State which of the following statements are true and which are false.

    a. If $A = \{5, 11, 19\}$, then $11 \in A$.

    b. If $B = \{x, y, z, w\}$, then $a \notin B$.

    c. If $C = \{$Ann, Bill, Charles$\}$, then Dan $\in C$.

    d. $0 \in \varnothing$.

17. Given the universal set $U = \{7, 12, 15, 20, 23\}$, which of the following statements are true and which are false?

    a. $12 \in U$ b. $8 \in U$ c. $15 \notin U$ d. $11 \notin U$

18. Given the universal set $U = \{5, 7, 11, 14, 20\}$, which of the following statements are true and which are false?

    a. $12 \in U$ b. $5 \in U$ c. $15 \notin U$ d. $11 \notin U$

A.2 Subsets

Subset   A set A is called a **subset** of set B if every member of A is also a member of B. "A is a subset of B" is written $A \subseteq B$.

Example 1   Examples of subsets:

a. $A = \{3, 5\}$ is a subset of $B = \{3, 5, 7\}$ because every member of A is also a member of B. Therefore, $A \subseteq B$.

b. $P = \{a, c, g, f\}$ is a subset of $Q = \{d, f, a, g, h, c\}$ because every member of P is also a member of Q. Therefore, $P \subseteq Q$.

c. $X = \{$Joe, Betty$\}$ is a subset of $Y = \{$Mary, Betty, Jack, Joe$\}$ because every member of X is also a member of Y. Therefore, $X \subseteq Y$.

d. $D = \{4, 7\}$ is *not* a subset of $E = \{7, 8, 5\}$ because $4 \in D$, but $4 \notin E$. Therefore, $D \nsubseteq E$, read "D is not a subset of E."

e. $K = \{d, f, m\}$ is *not* a subset of $L = \{f, m\}$ because $d \in K$, but $d \notin L$. Therefore, $K \nsubseteq L$, read "K is not a subset of L."   ■

Proper Subset *A* is called a **proper subset** of *B* if *A* is a subset of *B* *and* there is at least one member of *B* that is not a member of *A*. "*A* is a proper subset of *B*" is written $A \subset B$. The empty set is a proper subset of every set except itself.

Example 2 Examples of proper subsets:

a. $A = \{3, 5\}$ is a proper subset of $B = \{3, 5, 7\}$ because *A* is a subset of *B* and $7 \in B$, but $7 \notin A$. Therefore, $A \subset B$.

b. $X = \{\text{Joe, Betty}\}$ is a proper subset of $Y = \{\text{Mary, Betty, Jack, Joe}\}$ because *X* is a subset of *Y* and Mary $\in Y$, but Mary $\notin X$. Therefore, $X \subset Y$. ■

Improper Subset *A* is called an **improper subset** of *B* if *A* is a subset of *B* *and* there is no member of *B* that is not also a member of *A*.

Example 3 Examples of improper subsets:

a. $A = \{10, 12\}$ is an improper subset of $B = \{12, 10\}$, because *A* is a subset of *B* *and* no member of *B* is not also a member of *A*. Note that $A = B$. This means that *A* is an improper subset of *A*. In fact, *any set is an improper subset of itself*.

b. *D* is an improper subset of *D*. ■

By considering set $A = \{3, 5, 7\}$, we find that all its subsets are $\varnothing$, $\{3\}$, $\{5\}$, $\{7\}$, $\{3, 5\}$, $\{3, 7\}$, $\{5, 7\}$, and $\{3, 5, 7\}$. Of these eight subsets, only $A = \{3, 5, 7\}$ is an improper subset of *A*; the seven others are proper subsets of *A*.

Example 4 Examples of finding all the subsets of a set:

a. All the subsets of $\{6, 8\}$ are $\varnothing$, $\{6\}$, $\{8\}$, and $\{6, 8\}$. Of these four subsets, only $\{6, 8\}$ is an improper subset.

b. All the subsets of $\{a\}$ are $\varnothing$ and $\{a\}$. Of these two subsets, only $\{a\}$ is an improper subset. ■

EXERCISES A.2

1. $M = \{1, 2, 3, 4, 5\}$. State whether each of the following sets is a proper subset, an improper subset, or not a subset of *M*.

   a. $A = \{3, 5\}$

   b. $B = \{0, 1, 7\}$

   c. $\varnothing$

   d. $C = \{2, 4, 1, 3, 5\}$

2. $P = \{x, z, w, r, t, y\}$. State whether each of the following sets is a proper subset, an improper subset, or not a subset of *P*.

   a. $D = \{x, y, r\}$

   b. P

   c. $\varnothing$

   d. $E = \{x, y, z, s, t\}$

3. Write all the subsets of $\{R, G, Y\}$.

4. Write all the subsets of $\{\square, \triangle\}$.

A.3 Union and Intersection of Sets

Venn Diagrams   A useful tool for helping us understand set concepts is the Venn diagram. A simple Venn diagram is shown in Figure A.3.1.

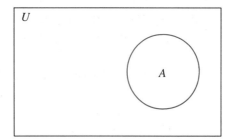

FIGURE A.3.1   Venn Diagram

It is customary to represent the universal set U by a rectangle and the other sets by circles enclosed within the U-rectangle. Just as we enclose all members of a particular set within braces in conventional set notation, in Venn diagrams we think of all members of a set A as being enclosed in a circle marked by the A. Using the same reasoning, we see that all elements in the universe under consideration are enclosed in the U-rectangle.

Union of Sets   The **union** of sets A and B, written $A \cup B$, is the set that contains all the elements of A as well as all the elements of B.

In Figure A.3.2, $A \cup B$ is represented by the shaded area. In terms of set notation, suppose $A = \{b, c, g\}$ and $B = \{1, 2, 5, 7\}$. Then $A \cup B = \{b, c, g, 1, 2, 5, 7\}$.

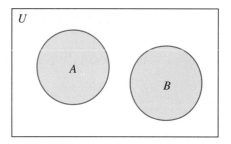

FIGURE A.3.2 $A \cup B$

Example 1   Examples of the union of sets:

a. If $C = \{x, y\}$ and $D = \{4,\}$, then $C \cup D = \{x, y, 4\}$.

b. If $E = \{$Mary, Helen$\}$ and $G = \{$high, low$\}$, then $E \cup G = \{$Mary, Helen, high, low$\}$.

c. If $H = \{4, 8, 26\}$ and $K = \{15, 17\}$, then $H \cup K = \{4, 8, 26, 15, 17\}$.

d. If $S = \{2, 5, 9\}$ and $T = \{9, 2, 7\}$, then $S \cup T = \{2, 5, 7, 9\}$.

e. If $P = \{a, 4, 3, c\}$ and $Q = \{3, k, a\}$, then $P \cup Q = \{a, 4, 3, c, k\}$.

f. If $A = \{2, 3, 4\}$ and $B = \{\ \}$, then $A \cup B = \{2, 3, 4\} = A$.   ■

Intersection of Sets   The **intersection** of sets C and D, written $C \cap D$, is the set that contains only elements in both C and D. Consider $C = \{g, f, m\}$ and $D = \{g, m, t, z\}$. Then $C \cap D = \{g, m\}$, because g and m are the only elements in both C and D. In the

Venn diagram (Figure A.3.3), the shaded area represents $C \cap D$ because that area lies in both circles.

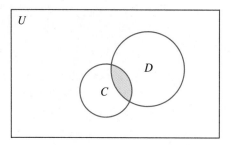

FIGURE A.3.3 $C \cap D$

To be certain that you can distinguish between the union and intersection of sets, two Venn diagrams (Figures A.3.4 and A.3.5) are shown for the same sets, P and Q.

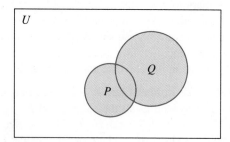

FIGURE A.3.4 Shaded Area Is $P \cup Q$ FIGURE A.3.5 Shaded Area Is $P \cap Q$

Example 2   Examples of the intersection and the union of sets:

a. If $A = \{$John, Henry$\}$ and $B = \{$Bill, Henry, Tom$\}$, then $A \cap B = \{$Henry$\}$ and $A \cup B = \{$John, Henry, Bill, Tom$\}$.

b. If $G = \{1, 5, 7\}$ and $H = \{2, 5, 6, 7, 8\}$, then $G \cap H = \{5, 7\}$ and $G \cup H = \{1, 5, 7, 2, 6, 8\}$.

c. If $K = \{a, 6, b\}$ and $M = \{c, 5, 4\}$, then $K \cap M = \{\ \ \}$ and $K \cup M = \{a, 6, b, c, 5, 4\}$. ∎

Disjoint Sets   **Disjoint sets** are sets whose intersection is the empty set. A Venn diagram showing disjoint sets is Figure A.3.2, in which sets A and B are disjoint.

Example 3   Examples of disjoint sets and of sets that are not disjoint:

a. If $A = \{1, 2\}$ and $B = \{3, 4\}$, then $A \cap B = \varnothing$. Therefore, A and B are disjoint sets.

b. If $P = \{a, b, c\}$ and $Q = \{2, 8\}$, then $P \cap Q = \varnothing$. Therefore, P and Q are disjoint sets.

c. If $R = \{5, 7, 9\}$ and $T = \{9, 10, 12\}$, then $R \cap T = \{9\} \neq \varnothing$. Therefore, R and T are *not* disjoint sets. ∎

EXERCISES A.3   **1.** Write the union and intersection of each pair of given sets.

     a. $\{1, 5, 7\}$, $\{2, 4\}$

     b. $\{a, b\}$, $\{x, y, z, a\}$

   c. { }, {k, 2}

   d. {river, boat}, {boat, streams, down}

2. Given that A = {1, 3, 5, 7}, B = {2, 4, 6}, C = {1, 2, 3, 4}, and D = {5, 6, 7}, find the following.

   a. $A \cup B$ b. $C \cup D$ c. $A \cap B$ d. $B \cap D$

3. Given that A = {Bob, John}, B = {Charles, Tom, Bob}, C = {Tom, John, Dick}, and D = {Ray, Bob}, find any two sets that are disjoint.

4. Given the Venn diagram below, write each of the following sets in roster notation. (The lowercase letters within a particular circle in the diagram are elements of that set.)

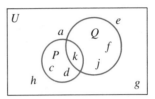

   a. P b. Q c. $P \cup Q$ d. $P \cap Q$ e. U

5. Given: X = {2, 5, 6, 11} and Y = {7, 5, 11, 13}.

   a. Find $X \cap Y$.   b. Find $Y \cap X$.   c. Is $X \cap Y = Y \cap X$?

6. Given: K = {a, 4, 7, b}, L = {m, 4, 6, b}, and M = {n, 4, 7, t}.

   a. Write $K \cap L$ in roster notation.   b. Find $n(K \cap L)$.

   c. Write $L \cup M$ in roster notation.   d. Find $n(L \cup M)$.

7. Shade in $A \cup B$ in the Venn diagram given here.

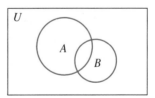

8. Shade in $P \cap Q$.

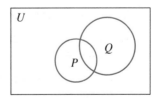

9. Write the name of the set representing the shaded area.

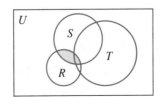

10. Write the name of the set representing the shaded area.

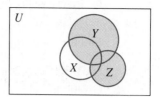

Appendix B

Brief Review of Arithmetic

This brief review is given as an aid for students who need to refresh their knowledge of arithmetic. For a more complete treatment, see Johnston, Willis, and Lazaris, *Essential Arithmetic, 6th ed.* (Belmont, Calif.: Wadsworth Publishing Company, 1991).

Whole Numbers

All whole numbers can be considered as decimals.

$$15 = 15.$$

Decimal point

Whole number ⌐ ⌐ Decimal

Therefore, all operations with whole numbers can be done in the same way as operations with decimals.

Decimals

Decimal Places   The number of decimal places in a number is the number of digits written to the right of the decimal point.

75.14 (two decimal places)   1.086 (three decimal places)

Approximately Equal   The symbol ≈ (read "approximately equal to") is used to show that two numbers are approximately equal to each other.

$78 \approx 80$ Read "78 is approximately equal to 80"

$\pi \approx 3.14$ Read "π is approximately equal to 3.14"

$0.99 \approx 1$ Read "0.99 is approximately equal to 1"

Rounding Off

$876. \approx 880$ Rounded to tens

$44.62 \approx 44.6$ Rounded to one decimal place (or to tenths)

$0.16504 \approx 0.17$ Rounded to two decimal places (or to hundredths)

Addition and Subtraction   To add or subtract decimals, write the numbers with their decimal points in the same vertical line and with like decimal places aligned.

Add: 17.6 + 2.65 + 439 + 0.015

```
   17.6
    2.65
  439.
+   0.015
  459.265
```

Subtract 26.54 from 518.3

```
  518.30
 − 26.54
  491.76
```

Multiplication   The number of decimal places in a product is the sum of the number of decimal places in the numbers being multiplied.

$$
\begin{array}{r}
56.75 \quad \text{(two decimal places)} \\
\times\ 3.8 \quad \text{(one decimal place)} \\
\hline
45\ 400 \\
170\ 25 \\
\hline
215.650 \quad \text{(three decimal places)}
\end{array}
$$

Division   When the quotient is to be rounded off to tenths, we carry out the division to hundredths and then round off.

Multiplying a Decimal by a Power of 10

$$23.4 \times 100 = 23.40 = 2{,}340 \qquad 1.56 \times 10^3 = 1.560 = 1{,}560$$

Dividing a Decimal by a Power of 10   Move the decimal point to the *left* as many places as the number of zeros in the power of ten.

$$\frac{72.8}{10} = 72.8 \div 10 = 7.28 \qquad \frac{875}{10^2} = 875. \div 100 = 8.75$$

Order of Operations

A. If there are any grouping symbols in the expression, that part of the expression within a set of grouping symbols is evaluated first.

B. The evaluation then proceeds in this order:

First: Powers and roots are done.

Next: Multiplication and division are done in order from left to right.

Last: Addition and subtraction are done in order from left to right.

Examples of using the correct order of operations:

a. $12 \div 3 \times 4$

$\quad\quad 4 \quad \times 4 = 16$

b. $4 \times 3^2 \div 6 - 2$

$\quad\quad 4 \times 9 \div 6 - 2$

$\quad\quad\quad 36 \ \div 6 - 2$

$\quad\quad\quad\quad 6 \ - 2 = 4$

c. $5(8 - 2) + 6(4)$

    $5(6)\quad + 6(4)$

      $30\quad + \;24 = 54$

d. $5\sqrt{16} \div 4 - 2$

    $5(4)\;\; \div 4 - 2$

     $20\;\; \div 4 - 2$

       $5\;\; - 2 = 3$

Fractions

Raising a Fraction to Higher Terms   Multiply both numerator and denominator by the same number (not zero).

$$\frac{2}{3} = \frac{2 \cdot 2}{3 \cdot 2} = \frac{4}{6}$$

The fractions $\frac{2}{3}$ and $\frac{4}{6}$ are called **equivalent fractions**.

Reducing a Fraction to Lower Terms   Divide both numerator and denominator by a number that is a divisor of both.

$$\frac{\overset{3}{\cancel{15}}}{\underset{15}{\cancel{75}}} = \frac{\overset{1}{\cancel{3}}}{\underset{5}{\cancel{15}}} = \frac{1}{5} \qquad \begin{cases} \dfrac{15}{75} \text{ and } \dfrac{1}{5} \text{ are } \textit{equivalent fractions.} \\[2mm] \dfrac{1}{5} \text{ is } \dfrac{15}{75} \text{ reduced to } \textit{lowest terms.} \end{cases}$$

A fraction is in *lowest* terms when 1 is the only number that is a divisor of both the numerator and the denominator.

Changing an Improper Fraction into a Mixed Number   Divide the numerator by the denominator and express the remainder as a fraction.

$$\frac{17}{3} = 3\overline{)17}^{\;5\ \text{R2}} = 5\frac{2}{3}$$

Changing a Mixed Number into an Improper Fraction

$$\frac{\text{whole number} \times \text{denominator} + \text{numerator}}{\text{denominator}}$$

$$2\frac{3}{5} = \frac{2 \times 5 + 3}{5} = \frac{10 + 3}{5} = \frac{13}{5}$$

Adding Fractions

$$\begin{aligned} \frac{3}{8} &= \frac{9}{24} \\[2mm] +\frac{5}{6} &= \frac{20}{24} \\[1mm] \hline \\[-2mm] \frac{29}{24} &= 1\frac{5}{24} \end{aligned} \qquad \begin{aligned} 8 &= 2^3 \\[2mm] 6 &= 2 \cdot 3 \\[2mm] \text{LCD} &= 2^3 \cdot 3 \\[2mm] &= 8 \cdot 3 \\[2mm] &= 24 \end{aligned}$$

Subtracting Mixed Numbers

$$13\frac{1}{2} = \quad 13\frac{3}{6} = 12 + 1 + \frac{3}{6} = 12 + \frac{6}{6} + \frac{3}{6} = \quad 12\frac{9}{6}$$

$$-\ 9\frac{2}{3} = -\ 9\frac{4}{6} \qquad\qquad\qquad\qquad\qquad = -\ 9\frac{4}{6}$$

$$\overline{\qquad\qquad\qquad\qquad\qquad\qquad\qquad\qquad 3\frac{5}{6}}$$

Multiplying Fractions

$$\frac{\overset{2}{\cancel{6}}}{\underset{1}{\cancel{7}}} \times \frac{\overset{2}{\cancel{14}}}{\underset{5}{\cancel{15}}} = \frac{4}{5}$$

Dividing Fractions

$$\frac{5}{8} \div \left(3\frac{3}{4}\right) = \frac{5}{8} \div \frac{15}{4} = \frac{\overset{1}{\cancel{5}}}{\underset{2}{\cancel{8}}} \times \frac{\overset{1}{\cancel{4}}}{\underset{3}{\cancel{15}}} = \frac{1}{6}$$

Simplifying a Complex Fraction Divide the numerator of the complex fraction by the denominator.

$$\frac{\dfrac{5}{12}}{\dfrac{15}{16}} = \frac{5}{12} \div \frac{15}{16} = \frac{\overset{1}{\cancel{5}}}{\underset{3}{\cancel{12}}} \times \frac{\overset{4}{\cancel{16}}}{\underset{3}{\cancel{15}}} = \frac{4}{9}$$

$$\frac{3 - \dfrac{1}{5}}{\dfrac{1}{3} + \dfrac{1}{5}} = \frac{\dfrac{15}{5} - \dfrac{1}{5}}{\dfrac{5}{15} + \dfrac{3}{15}} = \frac{\dfrac{14}{5}}{\dfrac{8}{15}} = \frac{14}{5} \div \frac{8}{15} = \frac{\overset{7}{\cancel{14}}}{\underset{1}{\cancel{5}}} \times \frac{\overset{3}{\cancel{15}}}{\underset{4}{\cancel{8}}} = \frac{21}{4} = 5\frac{1}{4}$$

Changing a Fraction to a Decimal Divide the numerator by the denominator.

$$\frac{5}{12} = 12\overline{)5.000}$$

$$\begin{array}{r} 0.416 \approx 0.42 \\ \hline 5.000 \\ 4\ 8 \\ \hline 20 \\ 12 \\ \hline 80 \\ 72 \\ \hline 8 \end{array}$$

Rounded to two decimal places
When the answer is to be rounded off to two decimal places, we need three decimal places in the dividend

Changing a Decimal to a Fraction

$$0.48 = 48 \text{ hundredths} = \frac{48}{100} = \frac{12}{25}$$

Percents

The Meaning of a Percent *Percent* means hundredths. For example, 5% of a quantity is $\frac{5}{100}$, or 0.05, of that quantity.

Changing a Decimal to a Percent Move the decimal point two places to the *right* and attach the percent symbol to the right of the number.

$$0.13 = 13\% \qquad 0.003 = 0.3\% \qquad 2.5 = 250\%$$

Changing a Percent to a Decimal Move the decimal point two places to the *left* and drop the percent symbol.

$$83\% = 0.83 \qquad 3\% = 0.03 \qquad 115\% = 1.15$$

Changing a Fraction to a Percent First change the fraction to a decimal (rounding off, if necessary); then move the decimal point two places to the right and attach the percent symbol to the right of the number.

$$\frac{1}{4} = 0.25 = 25\%$$

$$\frac{5}{12} \approx 0.4167 = 41.67\%$$

In this example, the decimal was rounded off to four decimal places, and the percent was rounded off to two decimal places.

Changing a Percent to a Fraction Write the number with a denominator of 100 and drop the percent symbol. Reduce the resulting fraction if possible.

$$32\% = \frac{32}{100} = \frac{8}{25} \qquad 105\% = \frac{105}{100} = \frac{21}{20} \qquad 0.07\% = \frac{0.07}{100} = \frac{7}{10,000}$$

EXERCISES B.1 In Exercises 1–4, reduce each fraction to lowest terms.

1. $\dfrac{6}{9}$ 2. $\dfrac{12}{20}$ 3. $\dfrac{49}{24}$ 4. $\dfrac{210}{240}$

In Exercises 5–8, change each improper fraction into a mixed number and reduce if possible.

5. $\dfrac{5}{2}$ 6. $\dfrac{11}{8}$ 7. $\dfrac{47}{25}$ 8. $\dfrac{78}{33}$

In Exercises 9–12, change each mixed number into an improper fraction.

9. $3\dfrac{7}{8}$ 10. $2\dfrac{7}{9}$ 11. $2\dfrac{5}{6}$ 12. $8\dfrac{3}{5}$

In Exercises 13–38, perform the indicated operations.

13. $\dfrac{2}{3} + \dfrac{1}{4}$ 14. $\dfrac{2}{7} + \dfrac{3}{14}$ 15. $\dfrac{7}{10} - \dfrac{2}{5}$ 16. $\dfrac{3}{4} - \dfrac{2}{5}$

17. $\dfrac{3}{16} \times \dfrac{20}{9}$ 18. $\dfrac{6}{35} \times \dfrac{15}{8}$ 19. $\dfrac{4}{3} \div \dfrac{8}{9}$ 20. $\dfrac{5}{2} \div \dfrac{5}{8}$

21. $2\dfrac{2}{3} + 1\dfrac{3}{5}$ 22. $3\dfrac{1}{4} + 4\dfrac{1}{2}$ 23. $5\dfrac{4}{5} - 3\dfrac{7}{10}$ 24. $4\dfrac{5}{6} - 2\dfrac{2}{3}$

25. $4\dfrac{1}{5} \times 2\dfrac{1}{7}$ 26. $2\dfrac{1}{3} \times 4\dfrac{1}{2}$ 27. $2\dfrac{2}{5} \div 1\dfrac{1}{15}$ 28. $2\dfrac{1}{4} \div 3\dfrac{3}{8}$

29. $\dfrac{\frac{5}{8}}{\frac{5}{6}}$

30. $\dfrac{\frac{3}{5}}{\frac{6}{5}}$

31. Add $34.5 + 1.74 + 18 + 0.016$.

32. Add $74.1 + 19 + 1.55 + 0.095$.

33. Subtract 34.67 from 356.4.

34. Subtract 81.94 from 418.5.

35. 100×7.45

36. $1{,}000 \times 3.54$

37. $\dfrac{46.8}{100}$

38. $\dfrac{89.5}{1{,}000}$

In Exercises 39–42, perform the indicated operations. Then round off your answer to the indicated place.

39. 70.9×94.78 (one decimal place)

40. 40.8×68.59 (one decimal place)

41. $6.007 \div 7.25$ (two decimal places)

42. $8.009 \div 4.67$ (two decimal places)

43. Change $5\frac{3}{4}$ to a decimal.

44. Change $4\frac{3}{5}$ to a decimal.

45. Change 0.65 to a fraction.

46. Change 0.25 to a fraction.

47. Change 5.9 to a mixed number.

48. Change 4.3 to a mixed number.

In Exercises 49–52, change each number to a percent.

49. 4.7

50. 0.001

51. 0.258

52. 12

In Exercises 53–56, change each percent to a decimal.

53. 3.5%

54. 0.02%

55. 157%

56. 17.8%

In Exercises 57–60, change each fraction to a percent. In Exercises 59 and 60, round off to two decimal places.

57. $\dfrac{1}{8}$

58. $\dfrac{3}{5}$

59. $\dfrac{7}{12}$

60. $\dfrac{1}{7}$

In Exercises 61–64, change each percent to a fraction reduced to lowest terms.

61. 35%

62. 80%

63. 12%

64. 18%

Appendix C

Functions

C.1 Definitions

We live in a world of functions. If we are paid at a fixed hourly rate, our weekly salary is a function of the number of hours we work each week. At any given time, the cost of sending a letter first class is fixed; therefore, the cost of mailing several letters (of equal weights) at once is a function of the number of letters being mailed.

Function   A **function** is a set of ordered pairs in which no two of the ordered pairs have the *same first coordinate* and *different second coordinates*. A function can also be thought of as a rule that assigns one and only one value to y for each value of x. Because the value of y depends on the value of x, we often call y the **dependent variable** and x the **independent variable**.

A function can be described in several different ways: by a table of values, by a set of ordered pairs, by a written statement, by an equation, or by a graph.

Domain   The **domain** of a function is the set of all the values that x can have so that y is a real number; it is the set of all the *first coordinates*, or x-values, of the ordered pairs.

Range   The **range** of a function is the set of all the values that y can have; it is the set of all the *second coordinates*, or y-values, of the ordered pairs. It is often more difficult to find the range than the domain of a function.

Example 1   Examples of functions and their domains:

a. $\{(3, 7), (-2, 7), (0, -5)\}$

(Recall from Section 1.1 that we enclose the elements of a set within braces.) This is a set of three ordered pairs; they all have different first coordinates. Therefore, it is a function. The domain of $\{(\,3\,, 7), (\,-2\,, 7), (\,0\,, -5)\}$ is the set $\{3, -2, 0\}$.

b. $y = x + 1$

For any value of x we choose, there will be one and only one corresponding value for y. Therefore, this is a function. Because we can choose *any* value for x, the domain of this function is the set of all real numbers.

c. $y = \dfrac{3}{x - 1}$

Excluded values were introduced in Section 8.1. For the fraction $\dfrac{3}{x - 1}$, 1 is an excluded value. Therefore, the domain is the set of all real numbers except 1. For any value of x from the domain we choose, there will be one and only one corresponding value for y. Therefore, it is a function.

d. $y = \sqrt{x + 4}$

We learned in Section 1.10A that square roots of negative numbers are not real numbers. Therefore, if y is to be a real number, the radicand $x + 4$ must be greater than or equal to zero. We must then solve the inequality $x + 4 \geq 0$.

$$
\begin{array}{rl}
x + 4 \geq & 0 \\
\underline{-4 \quad -4} & \qquad \text{Adding } -4 \text{ to both sides} \\
x \geq & -4
\end{array}
$$

Therefore, the domain is the set of all real numbers greater than or equal to -4. If we let x equal any number greater than or equal to -4, we get one and only one corresponding value for y. Therefore, this is a function.   ■

The rules for finding the domain of a function when an *equation* for the function has been given are as follows:

The domain of the function will be the set of all real numbers unless:

1. The domain is restricted by some statement accompanying the equation.

2. There are variables in a denominator or variables with negative exponents.

3. Variables occur under a radical sign when the index of the radical is an even number.

The Graph of a Function   The graph of a function is simply the graph of all the ordered pairs in the function.

Example 2   Find the domain of each of the following functions and graph each function.

a. {(2, 4), (−1, 3), (0, 3), (3, 1)}
   Solution The domain of {(2 , 4), (−1 , 3), (0 , 3), (3 , 1)} is {2, −1, 0, 3}.

   The graph is as follows:

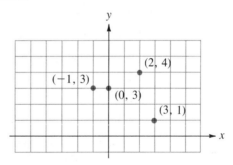

b. $y = x + 1, x \leq 2$
   Solution In this example, there *are* restrictions on the domain, namely, $x \leq 2$. Therefore, the domain is the set of all real numbers less than or equal to 2.

   For the graph, we make a table of values, taking care to choose *x*-values only from the domain.

x	y
−2	−1
0	1
2	3

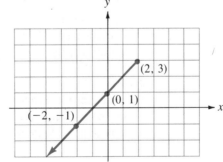

We then plot those points and connect them with a straight line that does not extend to the right of the point where $x = 2$. ■

NOTE   It is possible to graph *any* set of ordered pairs, whether it is a function or not.

The following statement is true because of the way in which a function is defined:

No vertical line can meet the graph of a function in more than one point.

Example 3   Examples of graphs of sets of ordered pairs. (These sets of ordered pairs are *not* functions.)

a. $\{(\,3\,, -2), (4, 1), (\,3\,, 3)\}$

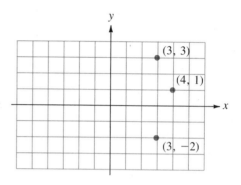

The set is *not* a function because the ordered pairs $(\,3\,, -2)$ and $(\,3\,, 3)$ have two different second elements but the same first element (3). The vertical line $x = 3$ would meet the graph in more than one point.

b. $x = 1$

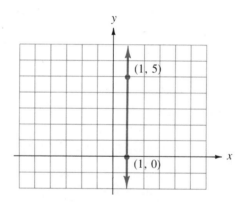

$x = 1$ is *not* a function because there are ordered pairs that satisfy the equation $x = 1$ that have the same first element but different second elements. (1, 5) and (1, 0) are two such pairs. ∎

Example 4   Determine from the graph whether each of the following is a function.

a.

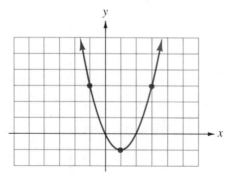

Solution This *is* the graph of a function, because no vertical line can meet the graph in more than one point.

b.

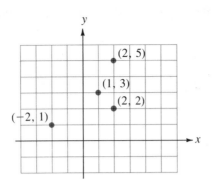

Solution This is *not* the graph of a function, because a vertical line through (2, 5) would also pass through (2, 2).

c.

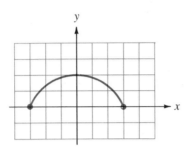

Solution This *is* the graph of a function because no vertical line can meet the graph in more than one point.

∎

EXERCISES C.1   In Exercises 1–6, determine whether each set of ordered pairs is a function, and find the domain of each of the functions.

1. {(4, −1), (2, 5), (3, 0), (0, 4)}

2. {(6, 3), (−2, 5), (4, −1), (0, 0)}

3. {(2, −5), (3, 4), (2, 0), (7, −4), (−1, 3)}

4. {(3, 7), (−1, 4), (5, −2), (3, 0), (−8, 2)}

5. {(−8, −2), (3, −4), (6, −2), (9, −4)}

6. {(−3, −7), (−1, −7), (0, −2), (3, 0)}

In Exercises 7–14, find the domain of each function.

7. $y = \dfrac{7}{x - 12}$

8. $y = \dfrac{5}{x - 2}$

9. $y = 3x + 4$

10. $y = 2x + 5$

11. $y = \sqrt{x - 5}$

12. $y = \sqrt{x - 10}$

13. $y = x + 3, x \geq 0$

14. $y = x + 7, x \geq -3$

15. Which of the following are graphs of functions?

a.   b.   c.

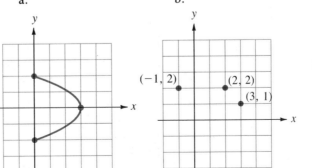

16. Which of the following are graphs of functions?

a.

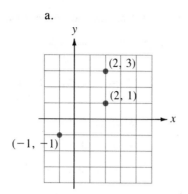

b.

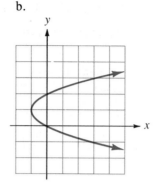

c.

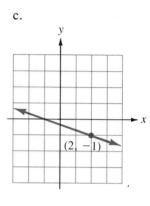

C.2 Functional Notation

Functional Notation   When we have a rule that assigns one and only one value to y for each value of x, we say that y is a function of x. This can be written $y = f(x)$, which is read "y equals f of x."

A WORD OF CAUTION $f(x)$ does *not* mean f times x. It is simply a different way of writing y when y is a function of x.   ☑

The equations $y = x + 1$ and $f(x) = x + 1$ mean exactly the same thing, and we often combine the equations in the following manner:

$$y = f(x) = x + 1$$

Evaluating a Function   To *evaluate* a function means to determine what value $f(x)$ or y has for a particular x-value. The notation $f(a)$ (read "f of a") means that the function $f(x)$ is to be evaluated at $x = a$.

Example 1   If $f(x) = 3x + 2$, find $f(4)$.

Solution   Since $f(x) = 3x + 2$,

$$f(4) = 3(4) + 2 \qquad \text{Substituting 4 for } x$$
$$f(4) = 12 + 2$$
$$f(4) = 14$$

The statement $f(4) = 14$ means that $y = 14$ when $x = 4$.   ∎

Example 2   If $f(x) = \dfrac{3x^2 + 1}{x - 2}$, find $f(1)$, $f(-3)$, and $f(0)$.

Solution

$$f(1) = \frac{3(1)^2 + 1}{1 - 2} = \frac{3 + 1}{-1} = \frac{4}{-1} = -4$$

$$f(-3) = \frac{3(-3)^2 + 1}{-3 - 2} = \frac{3(9) + 1}{-5} = \frac{27 + 1}{-5} = -\frac{28}{5}$$

$$f(0) = \frac{3(0)^2 + 1}{0 - 2} = \frac{0 + 1}{-2} = \frac{1}{-2} = -\frac{1}{2} \quad ∎$$

Letters other than f can be used to name functions.

Variables other than x can be used for the independent variable.

Variables other than y can be used for the dependent variable.

Example 3 Examples in which letters other than f, x, and y are used to represent functions and variables:

a. $A = F(r) = \pi r^2$ A is a function of r.

b. $s = g(t) = 4t - 1$ s is a function of t.

c. $V = h(s) = s^3$ V is a function of s. ∎

EXERCISES C.2 Evaluate the functions.

1. If $f(x) = (x + 2)^2$, find the following.

  a. $f(0)$ b. $f(2)$ c. $f(-3)$ d. $f(1)$

2. If $f(x) = (2x + 3)^2$, find the following.

  a. $f(0)$ b. $f(1)$ c. $f(-2)$ d. $f(2)$

3. If $g(x) = x^2 + 4x + 4$, find the following.

  a. $g(0)$ b. $g(2)$ c. $g(-3)$ d. $g(1)$

4. If $h(x) = 4x^2 + 12x + 9$, find the following.

  a. $h(0)$ b. $h(1)$ c. $h(-2)$ d. $h(2)$

5. If $F(x) = x^2 + 4$, find the following.

  a. $F(0)$ b. $F(2)$ c. $F(-3)$ d. $F(1)$

6. If $G(x) = 4x^2 + 9$, find the following.

  a. $G(0)$ b. $G(1)$ c. $G(-2)$ d. $G(2)$

7. If $f(x) = 3x^2 + x - 1$, find the following.

  a. $f(0)$ b. $f(-2)$ c. $f(5)$ d. $f(1)$

8. If $f(x) = 2x^2 + 3x - 5$, find the following.

  a. $f(0)$ b. $f(-3)$ c. $f(4)$ d. $f(1)$

9. If $f(x) = x^3 - 1$, find the following.

  a. $f(0)$ b. $f(-1)$ c. $f(2)$ d. $f(1)$

10. If $f(x) = x^3 + 1$, find the following.

  a. $f(0)$ b. $f(-2)$ c. $f(1)$ d. $f(-1)$

11. If $g(t) = \dfrac{4t^2 + 5t - 1}{t + 2}$, find the following.

  a. $g(0)$ b. $g(-1)$ c. $g(2)$ d. $g(4)$

12. If $h(t) = \dfrac{-2t^2 - 3t + 1}{t + 3}$, find the following.

  a. $h(0)$ b. $h(-2)$ c. $h(1)$ d. $h(-1)$

Answers

To Set I Exercises, Diagnostic Tests, and
Cumulative Review Exercises

Exercises 1.1 (page 6)

1. 5   **2.** 5   **3.** 1   **4.** 0   **5.** 9   **6.** 99   **7.** 10

8. 100   **9.** >   **10.** <   **11.** >   **12.** <   **13.** Yes

14. Yes   **15.** Yes   **16.** Yes   **17.** No   **18.** No

19. No   **20.** Yes   **21.** Three   **22.** Four

Exercises 1.2 (page 10)

1. Negative seventy-five (or minus seventy-five)

2. Negative forty-nine (or minus forty-nine)   **3.** -54 **4.** -109

5. -2, because -2 is to the right of -4 on the number line   **6.** 0

7. -5, because -5 is to the right of -10 on the number line

8. -15 **9.** -1 **10.** Yes   **11.** Yes   **12.** Yes

13. Yes   **14.** Yes   **15.** Yes   **16.** There is none

17. 1, 2, 3, 4   **18.** 7, 8, 9   **19.** 0, 2, 4   **20.** 1, 3, 5, 7

21. >, because 0 is to the right of -3 on the number line   **22.** >

23. <, because -5 is to the left of 2 on the number line   **24.** <

25. >, because -2 is to the right of -10 on the number line

26. <   **27.** -62 **28.** $-45°F$

Exercises 1.3 (page 17)

1. Since the signs are the same, add the absolute values: $4 + 5 = 9$. Sum has sign of both numbers: $+$. $4 + (5) = 9$.

2. 8

3. Since the signs are the same, add the absolute values: $3 + 4 = 7$. Sum has sign of both numbers: $-$. $-3 + (-4) = -7$.

4. -8

5. Since the signs are different, subtract the absolute values: $6 - 5 = 1$. Sum has sign of number with larger absolute value: $-$. Therefore, $-6 + (5) = -1$.

6. -5

7. Since the signs are different, subtract the absolute values: $7 - 3 = 4$. Sum has sign of number with larger absolute value: $+$. Therefore, $7 + (-3) = 4$.

8. 5

9. Since the signs are the same, add the absolute values: $8 + 4 = 12$. Sum has sign of both numbers: $-$. $-8 + (-4) = -12$.

10. -11

11. Since the signs are different, subtract the absolute values: $9 - 3 = 6$. Sum has sign of number with larger absolute value: $-$. Therefore, $3 + (-9) = -6$.

12. -4 **13.** $-5 + 0 = -5$, because 0 is the additive identity.

14. -17

15. Since the signs are different, subtract the absolute values: $9 - 7 = 2$. Sum has sign of number with larger absolute value: $+$. Therefore, $-7 + (9) = 2$.

16. 3

17. Since the signs are the same, add the absolute values: $2 + 11 = 13$. Sum has sign of both numbers: $-$. $-2 + (-11) = -13$.

18. -9

19. Since the signs are different, subtract the absolute values: $15 - 5 = 10$. Sum has sign of number with larger absolute value: $-$. Therefore, $5 + (-15) = -10$.

20. -8

21. Since the signs are different, subtract the absolute values: $9 - 8 = 1$. Sum has sign of number with larger absolute value: $+$. Therefore, $-8 + (9) = 1$.

22. 6

23. Since the signs are different, subtract the absolute values: $4 - 4 = 0$. Zero is neither positive nor negative. The sum $-4 + (4) = 0$.

24. 0

25. Since the signs are the same, add the absolute values: $27 + 13 = 40$. Sum has sign of both numbers: $-$. $-27 + (-13) = -40$.

26. -54

27. Since the signs are different, subtract the absolute values: $121 - 80 = 41$. Sum has sign of number with larger absolute value: $+$. Therefore, $-80 + 121 = 41$.

28. 65

29. Since the signs are different, subtract the absolute values: $105 - 73 = 32$. Sum has sign of number with larger absolute value: $+$. Therefore, $105 + (-73) = 32$.

30. 105

31. Since the signs are the same, add the absolute values. Sum has sign of both numbers: $-$.

$$\begin{array}{r} -1\frac{1}{2} = -1\frac{5}{10} \\ +\left(-3\frac{2}{5}\right) = -3\frac{4}{10} \\ \hline -4\frac{9}{10} \end{array} \Big\} \text{Add}$$

32. $-7\frac{3}{4}$

33. Since the signs are different, subtract the absolute values. Sum has sign of number with larger absolute value: $+$.

$$\begin{array}{r} 4\frac{5}{6} = 4\frac{5}{6} \\ +\left(-1\frac{1}{3}\right) = -1\frac{2}{6} \\ \hline 3\frac{3}{6} = 3\frac{1}{2} \end{array} \Big\} \text{Subtract}$$

34. $4\frac{5}{8}$

35. Since the signs are different, subtract the absolute values. Sum has sign of number with larger absolute value: $+$.

$$\begin{array}{r} 5\frac{1}{4} = 5\frac{3}{12} = 4\frac{15}{12} \\ +\left(-2\frac{1}{3}\right) = -2\frac{4}{12} = -2\frac{4}{12} \\ \hline 2\frac{11}{12} \end{array} \Big\} \text{Subtract}$$

36. $4\frac{7}{10}$

37. Since the signs are different, subtract the absolute values: $6.075 - 3.146 = 2.929$. Sum has sign of number with larger absolute value: $+$. Therefore, the sum is 2.929.

38. 88.373

39. Since the signs are different, subtract the absolute values: $5.2 - 2.345 = 2.855$. Sum has sign of number with larger absolute value: $+$. Therefore, $5.2 + (-2.345) = 2.855$.

40. 0.775

41. $|-73| = 73$. The absolute value of a number is never negative.

42. 55   **43.** $|48| = 48$. Therefore, $-|48| = -48$.   **44.** -26

45. $|-6| = 6$. Therefore, $|-6| + (-2) = 6 + (-2)$. Since the signs are different, subtract the absolute values: $6 - 2 = 4$. Sum has sign of number with larger absolute value: $+$. Therefore, $|-6| + (-2) = 6 + (-2) = 4$.

46. 3

47. $|-17| = 17$. Therefore, $-8 + |-17| = -8 + 17$. Since the signs are different, subtract the absolute values: $17 - 8 = 9$. Sum has sign of number with larger absolute value: $+$. Therefore, $-8 + |-17| = -8 + 17 = 9$.

48. 21

49. $|23| = 23$. Therefore, $-9 + |23| = -9 + 23$. Since the signs are different, subtract the absolute values: $23 - 9 = 14$. Sum has sign of number with larger absolute value: $+$. Therefore, $-9 + |23| = -9 + 23 = 14$.

50. 31

51. We must add $53°$ to $-35°$. Since the signs are different, subtract the absolute values: $53 - 35 = 18$. Sum has sign of number with larger absolute value: $+$. Therefore, the sum is $18°$ F.

52. $17°$ F

Exercises 1.4 (page 21)

1. 6   **2.** $-\frac{2}{3}$ **3.** $-3 - (-2) = -3 + (+2) = -1$

4. -1 **5.** $-6 - (2) = -6 + (-2) = -8$ **6.** -13

7. $9 - (-5) = 9 + (+5) = 14$ **8.** 10

9. $2 - (-7) = 2 + (+7) = 9$ **10.** 8

11. $-5 - (-9) = -5 + (+9) = 4$ **12.** 6

13. $-4 - (3) = -4 + (-3) = -7$ **14.** -15

15. $6 - (11) = 6 + (-11) = -5$ **16.** -4

17. $-9 - (-4) = -9 + (+4) = -5$ **18.** -3

19. $4 - (-7) = 4 + (+7) = 11$ **20.** 17

21. $-15 - (11) = -15 + (-11) = -26$ **22.** -40

23. $0 - 7 = -7$ **24.** -15 **25.** $0 - (-9) = -(-9) = 9$

26. 25   **27.** $16 - 0 = 16$ **28.** 10

29. $156 - (-97) = 156 + (+97) = 253$ **30.** 373

31. $-354 - (-286) = -354 + (+286) = -68$ **32.** -109

33. $-7 - (-2.009) = -7.000 + (+2.009) = -4.991$ **34.** -8.11

35. $-6\frac{1}{2} - \left(-3\frac{2}{3}\right) = -6\frac{1}{2} + \left(+3\frac{2}{3}\right) = -6\frac{3}{6} + 3\frac{4}{6} = -5\frac{9}{6} + 3\frac{4}{6} = -2\frac{5}{6}$

36. $-5\frac{1}{2}$

37. $6\frac{1}{3} - \left(+8\frac{1}{4}\right) = 6\frac{1}{3} + \left(-8\frac{1}{4}\right) = 6\frac{4}{12} + \left(-8\frac{3}{12}\right) = 6\frac{4}{12} + \left(-7\frac{15}{12}\right) = -1\frac{11}{12}$

38. $2\frac{7}{10}$ **39.** $(+5) - (-2) = (+5) + (+2) = 7$

40. -5

41. $\left(-5\frac{1}{4}\right) - \left(2\frac{1}{2}\right) = \left(-5\frac{1}{4}\right) + \left(-2\frac{1}{2}\right) = \left(-5\frac{1}{4}\right) + \left(-2\frac{2}{4}\right) = -7\frac{3}{4}$

42. $-10\frac{1}{2}$ **43.** $-8 - (5) = -8 + (-5) = -13$ **44.** -9

45. $7 - 12 = 7 + (-12) = -5$ **46.** -10

47. $0 - 4 = 0 + (-4) = -4$ **48.** -8

49. $\$473.29 - \$238.43 = \$473.29 + (-\$238.43) = \$234.86$

50. $\$578.89$

51. $42 - (-7) = 42 + (+7) = 49$. Therefore, the rise in temperature was $49°$ F.

52. $43.1°$ F   **53.** $-141 - 68 = -141 + (-68) = -209$ ft

54. 9,800 ft

55. $29,028 - (-36,198) = 29,028 + (+36,198) = 65,226$ ft (This is 12.35 miles.)

56. 3,997 ft

Exercises 1.5 (page 27)

1. $3(-2) = -(3 \times 2) = -6$ **2.** -24

3. $-5(2) = -(5 \times 2) = -10$ **4.** -35

5. $-8(-2) = +(8 \times 2) = 16$ **6.** 42

7. $8(-4) = -(8 \times 4) = -32$ **8.** -45

9. $-7(9) = -(7 \times 9) = -63$ **10.** -48

11. $-10(-10) = +(10 \times 10) = 100$ **12.** 81

13. $8(-7) = -(8 \times 7) = -56$ **14.** -72

15. $-26(10) = -(26 \times 10) = -260$ **16.** -132

17. $-20(-10) = +(20 \times 10) = 200$ **18.** 600

19. $75(-15) = -(75 \times 15) = -1,125$ **20.** $-1,118$

21. $-30(5) = -(30 \times 5) = -150$ **22.** -300

23. $-7(-20) = +(7 \times 20) = 140$ **24.** 360

25. $\left(-5\frac{1}{2}\right)(0) = 0$ **26.** 0

27. $\left(6\frac{3}{4}\right)\left(-8\frac{1}{4}\right) = \left(\frac{27}{4}\right)\left(-\frac{33}{4}\right) = -\frac{891}{16}$ or $-55\frac{11}{16}$

28. $-\frac{1,826}{27}$ or $-67\frac{17}{27}$

29. $\left(-3\frac{3}{5}\right)\left(-8\frac{1}{5}\right) = \left(-\frac{18}{5}\right)\left(-\frac{41}{5}\right) = \frac{738}{25}$ or $29\frac{13}{25}$

30. $\frac{684}{35}$ or $19\frac{19}{35}$ **31.** $-3.5(-1.4) = +(3.5 \times 1.4) = 4.90$

32. 7.52   **33.** $2.74(-100) = -(2.74 \times 100) = -274$

34. -304 **35.** $\left(2\frac{1}{3}\right)\left(-3\frac{1}{2}\right) = \left(\frac{7}{3}\right)\left(-\frac{7}{2}\right) = -\frac{49}{6} = -8\frac{1}{6}$

36. $13\frac{13}{20}$ **37.** $0\left(-\frac{2}{3}\right) = 0$ **38.** 0   **39.** 12   **40.** $-\frac{5}{7}$

41. $-\frac{1}{3}$ **42.** $-\frac{1}{2}$ **43.** No; their product is -1, not 1.

44. No; their product is -1, not 1.

Exercises 1.6 (page 32)

1. 2. Positive because numbers have same sign.   **2.** 3

3. -4. Negative because numbers have different signs.   **4.** -2

5. -5 **6.** -2 **7.** 2   **8.** 5   **9.** -5 **10.** -6

11. -4 **12.** -5 **13.** 3   **14.** 3   **15.** -3 **16.** -4

17. 9   **18.** 7   **19.** -15 **20.** -2.5 **21.** -3

22. -7 **23.** -3 **24.** -3 **25.** Not defined

26. Not defined   **27.** Not defined   **28.** Not defined   **29.** 0

30. 0   **31.** Not defined   **32.** Not defined

33. $\frac{-15}{6} = \frac{-5}{2} = -2\frac{1}{2}$ **34.** $-2\frac{1}{4}$

35. $\frac{7.5}{-0.5} = \frac{75}{-5} = \frac{15}{-1} = -15$ **36.** -5

37. $\frac{-6.3}{-0.9} = \frac{-63}{-9} = 7$ **38.** 8   **39.** $\frac{-367}{100} = -3.67$

40. -4.86 **41.** $-\frac{3}{8} \div \frac{2}{5} = -\frac{3}{8} \cdot \frac{5}{2} = -\frac{15}{16}$ **42.** 4

43. -5.6 **44.** 6.1

Review Exercises 1.7 (page 35)

1. 8, 9   **2.** 10   **3.** -9 **4.** -1

5a. $<$, because -3 is to the left of 8 on the number line

b. $>$, because 5 is to the right of -2 on the number line

6a. $>$ **b.** $>$ **7.** 1   **8.** -4

9. Since the signs are different, subtract the absolute values: $3 - 2 = 1$. Sum has same sign as number with larger absolute value: $+$. Therefore, sum is 1.

10. -1 **11.** 3. Positive because numbers have same sign.

12. 2   **13.** $-5 - (-3) = -5 + (+3) = -2$ **14.** -5

15. $+5 - |-2| = +5 - (+2) = +5 + (-2) = 3$ **16.** 5

17. 12. Positive because numbers have same sign.   **18.** 20

19. $-7 - (3) = -7 + (-3) = -10$ **20.** -6

21. $4 + (-12) = -8$ **22.** -2 **23.** -120 **24.** -90

25. -8. Negative because numbers have different signs.

26. $-2\frac{1}{3}$ **27.** $9 - (-4) = 9 + (+4) = 13$ **28.** 11

29. $\left(-2\frac{2}{3}\right)\left(2\frac{1}{2}\right) = \left(-\frac{8}{3}\right)\left(\frac{5}{2}\right) = -\left(\frac{\overset{4}{\cancel{8}}}{3} \times \frac{5}{\underset{1}{\cancel{2}}}\right) = -\frac{20}{3} = -6\frac{2}{3}$

30. $4\frac{3}{10}$ **31.** -3 **32.** -3

33. Since the signs are the same, find sum of the absolute values: $10 + 2 = 12$. Sum is $-$. Therefore, the sum is -12.

34. -11 **35.** -24. Negative because numbers have opposite signs.

36. -35 **37.** 5. Positive because both numbers have the same sign.

38. 8   **39.** 0. Zero divided by any number other than zero is zero.

40. 0   **41.** $-10 - (-6) = -10 + (+6) = -4$ **42.** 3

Sections 1.1–1.7 Diagnostic Test (page 39)

Following each problem number is the textbook section number (in parentheses) where that kind of problem is discussed.

1. (1.1)   False; zero is not a natural number.   **2.** (1.2)   True

3. (1.3)   False   **4.** (1.2)   False   **5.** (1.3)   False

6. (1.2)   True   **7.** (1.1)   False   **8.** (1.2)   False

9. (1.2)   True   **10.** (1.5)   False   **11.** (1.2) $<$

12. (1.2) $>$ **13.** (1.3) $|-7.6| = 7.6$

14. (1.3) $-|-56| = -(+56) = -56$ (Notice that $|-56| = +56$.)

15. (1.4) $3 - (+17) = 3 + (-17) = -14$

16. (1.4) $-12 - (-5) = -12 + (+5) = -7$

17. (1.4) $6 - (-2) = 6 + (+2) = 8$ **18.** (1.4) $-\frac{5}{8}$

19. (1.4)   6   **20.** (1.6) $-\frac{1}{6}$

21. (1.3) $8 + (-26) = -18$ (The signs were different; we subtracted absolute values.)

22. (1.3) $-13 + (-5) = -18$ (The signs were the same; we added absolute values.)

23. (1.3) $-21 + (-5) = -26$ (The signs were the same; we added absolute values.)

24. (1.3) $-\frac{2}{5} + \frac{3}{10} = -\frac{4}{10} + \frac{3}{10} = -\frac{1}{10}$ (The signs were different; we subtracted absolute values.)

25. (1.3) $6.16 + (-8.3) = -2.14$ (The signs were different; we subtracted absolute values.)

26. (1.3) $-3\frac{1}{2} + 2\frac{1}{8} = -3\frac{4}{8} + 2\frac{1}{8} = -1\frac{3}{8}$ (The signs were different; we subtracted absolute values.)

27. (1.4) $-8 - (-3) = -8 + (+3) = -5$

28. (1.4) $6 - (-12) = 6 + (+12) = 18$

29. (1.4) $-4 - 1 = -4 + (-1) = -5$

30. (1.4) $8 - 37 = 8 + (-37) = -29$

31. (1.4) $-5\frac{2}{3} - \left(-2\frac{8}{9}\right) = -5\frac{6}{9} + \left(+2\frac{8}{9}\right) = -4\frac{15}{9} + 2\frac{8}{9} = -2\frac{7}{9}$

32. (1.4) $3\frac{1}{3} - 8\frac{1}{6} = 3\frac{2}{6} + \left(-8\frac{1}{6}\right) = 3\frac{2}{6} + \left(-7\frac{7}{6}\right) = -4\frac{5}{6}$

33. (1.4) $-2.325 - (-6.3) = -2.325 + (+6.3) = 3.975$

34. (1.4) $0 - 12 = -12$ **35.** (1.5) $0(-12) = 0$

36. (1.5) $-4(-1) = 4$ **37.** (1.5) $12(-6) = -72$

38. (1.5) $-16(0) = 0$ **39.** (1.5) $-\frac{5}{6}\left(-\frac{3}{10}\right) = \frac{1}{4}$

40. (1.5) $-1\frac{1}{3}\left(1\frac{3}{4}\right) = -\frac{4}{3}\left(\frac{7}{4}\right) = -\frac{7}{3}$, or $-2\frac{1}{3}$

41. (1.5) $(6.32)(-0.1) = -0.632$

42. (1.6) $8 \div (-2) = -4$ **43.** (1.6) $\frac{17}{0}$ is not defined

44. (1.6) $\frac{-24}{10} = -2.4$ **45.** (1.6) $-36 \div (-12) = 3$

46. (1.6) $\frac{18}{-24} = -\frac{3}{4}$ **47.** (1.6) $0 \div (-2) = 0$

48. (1.6) $\frac{0}{0}$ is not defined   **49.** (1.6) $-54 \div 9 = -6$

50. (1.6) $\frac{-4.9}{-7} = 0.7$

Exercises 1.8 (page 45)

1. True. Commutative property of addition.

2. True. Commutative property of addition.

3. True. Associative property of addition.

4. True. Associative property of addition.   **5.** False.

6. False.   **7.** True. Associative property of multiplication.

8. True. Associative property of multiplication.   **9.** False.

10. False.   **11.** True. Commutative property of multiplication.

12. True. Commutative property of multiplication.

13. True. Commutative property of addition.

14. True. Commutative property of addition.

15. True. Commutative property of addition.

16. True. Commutative property of addition.   **17.** False.

18. True. Distributive property.

19. True. Commutative property of addition.

20. True. Commutative property of addition.

21. True. 1 is the multiplicative identity.

22. True. 0 is the additive identity.   **23.** True. Additive inverse.

24. True. Additive inverse.   **25.** True. Distributive property.

26. False.   **27.** False.   **28.** False.

29. True. Associative and commutative properties of addition.

30. True. Commutative and associative properties of multiplication.

31. False.   **32.** False.

33. True. Associative property of multiplication.

34. True. Associative property of multiplication.

35. True. Commutative property of addition.

36. True. Commutative property of addition.   **37.** $3 + (-7)$

38. $-5 + 12$ **39.** $[4(-3)](6)$ **40.** $(-8 \cdot 4)(-2)$

41. $(3 \cdot 5) - (3 \cdot 8)$ **42.** $(6 \cdot 15) + (6 \cdot 2)$ **43.** $3(-4)$

44. $-3(6)$ **45.** $[4 + (-3)] + 6$ **46.** $-3 + [4 + (-2)]$

47. $5 + (-2) + 4 + (-8) + (-5)$
$= (5 + 4) + ([-2] + [-8] + [-5]) = 9 + [-15] = -6$

48. -11 **49.** $8 + (-11) = -3$ **50.** -12

51. $(-5)(-4)(-2) = -(5 \times 4 \times 2) = -40$ **52.** -48

53. $(2)(-5)(-9) = +(2 \times 5 \times 9) = 90$ **54.** 24

55. $(-2)(-3)(-5)(-4) = +(2 \times 3 \times 5 \times 4) = 120$ **56.** 56

Exercises 1.9 (page 50)

1. $3^3 = 3 \cdot 3 \cdot 3 = 27$ **2.** 16 **3.** $(-5)^2 = (-5)(-5) = 25$

4. -216 **5.** $7^2 = 7 \cdot 7 = 49$ **6.** 81 **7.** $0^3 = 0$

8. 0 **9.** $(-10)^1 = -10$ **10.** 100

11. $10^3 = 10 \cdot 10 \cdot 10 = 1{,}000$ **12.** 10,000

13. $(-10)^5 = (-10)(-10)(-10)(-10)(-10) = -100{,}000$

14. 1,000,000 **15.** $2^1 = 2$ **16.** 32

17. $(-2)^6 = (-2)(-2)(-2)(-2)(-2)(-2) = 64$ **18.** -128

19. $2^8 = 2 \cdot 2 \cdot 2 \cdot 2 \cdot 2 \cdot 2 \cdot 2 \cdot 2 = 256$ **20.** 625

21. $40^3 = 40 \cdot 40 \cdot 40 = 64{,}000$ **22.** 0

23. $(-12)^3 = (-12)(-12)(-12) = -1{,}728$ **24.** 225

25. $(-1)^5 = (-1)(-1)(-1)(-1)(-1) = -1$ **26.** -1

27. $-2^2 = -(2 \cdot 2) = -4$ **28.** -9 **29.** $(-1)^{99} = -1$

30. 1 **31.** $-[(-1)^5] = -(-1) = 1$ **32.** -1

33. $-[(-9)^2] = -(81) = -81$ **34.** -25 **35.** $0^8 = 0$

36. 0 **37.** 161.29 **38.** 237.16 **39.** 0.024336

40. 0.007569

Exercises 1.10A (page 52)

1. 4, because $4^2 = 16$ **2.** 5

3. $\sqrt{4} = 2$; therefore, $-(\sqrt{4}) = -(2) = -2$ **4.** -3

5. 9, because $9^2 = 81$ **6.** 6 **7.** 10, because $10^2 = 100$

8. 12 **9.** $\sqrt{81} = 9$; therefore, $-(\sqrt{81}) = -(9) = -9$

10. -11 **11.** Not a real number **12.** Not a real number

13. Not a real number **14.** Not a real number

Exercises 1.10B (page 53)

1. Try 20: $20^2 = 400$, and $400 < 529$; therefore, 20 is too small. Try 23: $23^2 = 529$; therefore, $\sqrt{529} = 23$.

2. 19

3. Try 25: $25^2 = 625$, and $625 > 441$; therefore, 25 is too large. Try 22: $22^2 = 484$, and $484 > 441$; therefore, 22 is too large. Try 21: $21^2 = 441$; therefore, $\sqrt{441} = 21$.

4. 25

5. Try 18: $18^2 = 324$, and $324 > 289$; therefore, 18 is too large. Try 17: $17^2 = 289$; therefore, $\sqrt{289} = 17$.

6. 18

7. Try 29: $29^2 = 841$, and $841 > 729$; therefore, 29 is too large. Try 27: $27^2 = 729$; therefore, $\sqrt{729} = 27$.

8. 36

Exercises 1.10C (page 55)

1. 3.606 **2.** 4.243 **3.** 6.083 **4.** 7.071 **5.** 8.888

6. 7.746 **7.** 9.274 **8.** 9.592 **9.** 683 **10.** 821

11. 522 **12.** 299

Exercises 1.10D (page 57)

1. By trial, we find: $2^3 = 2 \cdot 2 \cdot 2 = 8$; $3^3 = 3 \cdot 3 \cdot 3 = 27$; $4^3 = 4 \cdot 4 \cdot 4 = 64$. Therefore, $\sqrt[3]{64} = 4$.

2. 3 **3.** 3 (See the trial in the solution for Exercise 1.)

4. 5 **5.** Because $\sqrt[3]{27} = 3$, $-(\sqrt[3]{27}) = -(3) = -3$.

6. -5 **7.** $-\sqrt[4]{1} = -(\sqrt[4]{1}) = -(1) = -1$ **8.** 2

9. -5, because $(-5)^3 = -125$ **10.** -2

11. Not a real number **12.** Not a real number

13. -10, because $(-10)^3 = -1{,}000$ **14.** -4 **15.** -1

16. -2 **17.** 3, because $3^6 = 729$ **18.** 6

19. Not a real number **20.** Not a real number

21. The rational numbers are $\frac{1}{2}$, -12, and $0.\overline{26}$. The irrational numbers are $\sqrt[3]{13}$ and $0.196732468\ldots$. The real numbers are $\sqrt[3]{13}$, $\frac{1}{2}$, -12, $0.\overline{26}$, and $0.196732468\ldots$. One number is not real; it is $\sqrt{-15}$.

22. The rational numbers are 18, $\frac{11}{32}$, and $0.\overline{37}$. The irrational numbers are $\sqrt[3]{12}$ and $0.67249713\ldots$. The real numbers are $0.67249713\ldots$, 18, $\frac{11}{32}$, $0.\overline{37}$, and $\sqrt[3]{12}$. One number is not real; it is $\sqrt{-27}$.

Exercises 1.11 (page 64)

1. $12 - 8 - 6 = 4 - 6 = -2$ **2.** 2

3. $17 - 11 + 13 - 9 = 6 + 13 - 9 = 19 - 9 = 10$ **4.** 12

5. $7 + 2 \cdot 4 = 7 + 8 = 15$ **6.** 28

7. $9 - 3 \cdot 2 = 9 - 6 = 3$ **8.** -10

9. $10 \div 2 \cdot 5 = 5 \cdot 5 = 25$ **10.** 60

11. $12 \div 6 \div 2 = 2 \div 2 = 1$ **12.** $\frac{1}{3}$

13. $(-12) \div 2 \cdot (-3) = (-6) \cdot (-3) = 18$ **14.** -36

15. $8 \cdot 5^2 = 8 \cdot 25 = 200$ **16.** 96

17. $(-485)^2 \cdot 0 \cdot (-5)^2 = 0 \cdot (-5)^2 = 0$ **18.** 0

19. $12 \cdot 4 + 16 \div 8 = 48 + 16 \div 8 = 48 + 2 = 50$ **20.** 15

21. $28 \div 4 \cdot 2(6) = 7 \cdot 2(6) = 14(6) = 84$ **22.** -48

23. $(-2)^2 + (-4)(5) - (-3)^2 = 4 + (-4)(5) - 9$
$= 4 + (-20) - (9) = -16 + (-9) = -25$

24. -3

25. $2 \cdot 3 + 3^2 - 4 \cdot 2 = 2 \cdot 3 + 9 - 4 \cdot 2 = 6 + 9 - 8$
$= 15 + (-8) = 7$

26. 624

27. $(10^2)\sqrt{16} + 5(4) - 80 = (100)(4) + 5(4) - 80$
$= 400 + 20 - 80 = 420 - 80 = 340$

28. 39

29. $2 \cdot (-6) \div 3 \cdot (8 - 4) = 2 \cdot (-6) \div 3 \cdot (4) = -12 \div 3 \cdot 4 = -4 \cdot 4 = -16$

30. -50 **31.** $24 - [(-6) + 18] = 24 - [12] = 12$ **32.** 11

33. $[12 - (-19)] - 16 = [31] - 16 = 15$ **34.** 6

35. $[11 - (5 + 8)] - 24 = [11 - (13)] - 24 = [-2] - 24 = -26$

36. -25

37. $20 - [5 - (7 - 10)] = 20 - [5 - (-3)] = 20 - [8] = 12$

38. 3 **39.** $\dfrac{7 + (-12)}{8 - 3} = \dfrac{-5}{5} = -1$ **40.** -4

41. $15 - \{4 - [2 - 3(6 - 4)]\} = 15 - \{4 - [2 - 3(2)]\}$
$= 15 - \{4 - [2 - 6]\} = 15 - \{4 - [-4]\} = 15 - \{8\} = 7$

42. 30

43. $32 \div (-2)^3 - 5\left\{7 - \dfrac{6 - 2}{5}\right\} = 32 \div (-2)^3 - 5\left\{7 - \dfrac{4}{5}\right\}$

$= 32 \div (-2)^3 - \overset{1}{\cancel{5}}\left\{\dfrac{31}{\underset{1}{\cancel{5}}}\right\} = 32 \div (-8) - 31 = (-4) - 31 = -35$

44. -16 **45.** $\sqrt{3^2 + 4^2} = \sqrt{9 + 16} = \sqrt{25} = 5$ **46.** 12

47. $\sqrt{16.3^2 - 8.35^2} = \sqrt{265.69 - 69.7225} = \sqrt{195.9675} \approx 13.999$

48. ≈ 45.400

49. $(1.5)^2 \div (-2.5) + \sqrt{35} \approx 2.25 \div (-2.5) + 5.916$
$= -0.900 + 5.916 = 5.016$

50. ≈ -55.152

51. $18.91 - [64.3 - (8.6^2 + 14.2)]$
$= 18.91 - [64.3 - (73.96 + 14.2)] = 18.91 - [64.3 - 88.16]$
$= 18.91 - [-23.86] = 18.91 + (+23.86) = 42.77$

52. ≈ 8.396

Exercises 1.12 (page 69)

1. $\pm 1, \pm 2, \pm 4$ **2.** $\pm 1, \pm 3, \pm 9$ **3.** $\pm 1, \pm 2, \pm 5, \pm 10$

4. $\pm 1, \pm 2, \pm 7, \pm 14$ **5.** $\pm 1, \pm 3, \pm 5, \pm 15$

6. $\pm 1, \pm 2, \pm 4, \pm 8, \pm 16$ **7.** $\pm 1, \pm 2, \pm 3, \pm 6, \pm 9, \pm 18$

8. $\pm 1, \pm 2, \pm 4, \pm 5, \pm 10, \pm 20$ **9.** $\pm 1, \pm 3, \pm 7, \pm 21$

10. $\pm 1, \pm 2, \pm 11, \pm 22$ **11.** $\pm 1, \pm 3, \pm 9, \pm 27$

12. $\pm 1, \pm 2, \pm 4, \pm 7, \pm 14, \pm 28$ **13.** $\pm 1, \pm 3, \pm 11, \pm 33$

14. $\pm 1, \pm 2, \pm 17, \pm 34$ **15.** $\pm 1, \pm 2, \pm 4, \pm 11, \pm 22, \pm 44$

16. $\pm 1, \pm 3, \pm 5, \pm 9, \pm 15, \pm 45$ **17.** Prime; 1, 5

18. Composite; 1, 2, 4, 8 **19.** Prime; 1, 13

20. Composite; 1, 3, 5, 15 **21.** Composite; 1, 2, 3, 4, 6, 12

22. Prime; 1, 11 **23.** Composite; 1, 3, 7, 21

24. Prime; 1, 23 **25.** Composite; 1, 5, 11, 55

26. Prime; 1, 41 **27.** Composite; 1, 7, 49 **28.** Prime; 1, 31

29. Composite; 1, 3, 17, 51

30. Composite; 1, 2, 3, 6, 7, 14, 21, 42

31. Composite; 1, 3, 37, 111 **32.** Prime, 1, 101

33. $2\underline{|14}$ **34.** $3 \cdot 5$ **35.** $3\underline{|21}$
    7 7
$14 = 2 \cdot 7$ $21 = 3 \cdot 7$

36. $2 \cdot 11$ **37.** $2\underline{|26}$
    13
$26 = 2 \cdot 13$

38. 3^3 **39.** 29 is prime **40.** 31 is prime

41. $2\underline{|32}$ **42.** $3 \cdot 11$ **43.** $2\underline{|34}$ **44.** $5 \cdot 7$
$2\underline{|16}$ 17
$2\underline{|8}$ $34 = 2 \cdot 17$
$2\underline{|4}$
2
$32 = 2^5$

45. $2\underline{|84}$ **46.** $3 \cdot 5^2$ **47.** $2\underline{|144}$ **48.** $2^2 \cdot 3^2 \cdot 5$
$2\underline{|42}$ $2\underline{|72}$
$3\underline{|21}$ $2\underline{|36}$
7 $2\underline{|18}$
$84 = 2^2 \cdot 3 \cdot 7$ $3\underline{|9}$
         3
      $144 = 2^4 \cdot 3^2$

49. The prime numbers greater than 17 and less than 37 are 19, 23, 29, and 31. The only one of these numbers that yields a remainder of 1 when divided by 5 is 31.

50. 37

Exercises 1.13 (page 71)

1. Enough information has been given; addition would be used.

2. Not enough information has been given; we need to know how many pencils Eric bought.

3. Not enough information has been given; we need to know how many members there are in the math department.

4. Not enough information has been given; we need to know how much money Jeffrey borrowed.

5. Enough information has been given; multiplication would be used.

6. Enough information has been given; subtraction would be used.

Review Exercises 1.14 (page 73)

1a. True. Associative property of multiplication.
  b. True. Commutative property of addition.
  c. False.
  d. True. Associative property of addition.
  e. True. Commutative property of multiplication.
  f. True. Commutative property of addition.
  g. True. Distributive property.

2. 23 **3.** 0 **4.** 1 **5.** 0 **6.** 16

7. $-5^2 = -5 \cdot 5 = -25$ **8.** 11

9. $-\sqrt[4]{16} = -(\sqrt[4]{16}) = -(2) = -2$ **10.** Not a real number

11. -2, because $(-2)^3 = -8$ **12.** 2

13. $16^2 = 16 \cdot 16 = 256$

14. The rational numbers are -31, $\frac{7}{12}$, and $4.\overline{15}$. The irrational numbers are $-\sqrt{5}$, $5.3892531\ldots$, and $\sqrt[3]{71}$. The real numbers are $-\sqrt{5}$, -31, $5.3892531\ldots$, $\frac{7}{12}$, $4.\overline{15}$, and $\sqrt[3]{71}$. There are no numbers that are not real.

15. 5.568 **16.** -13 **17.** $11 - 7 \cdot 3 = 11 - 21 = -10$
18. 12 **19.** $15 \div 5 \cdot 3 = 3 \cdot 3 = 9$ **20.** 58
21. $6 - [8 - (3 - 4)] = 6 - [8 - (-1)] = 6 - [9] = -3$

22. -3 **23.** $\dfrac{6 + (-14)}{3 - 7} = \dfrac{-8}{-4} = 2$ **24.** 6

25. The prime numbers greater than 19 and less than 31 are 23 and 29. The only one of these numbers that yields a remainder of 2 when divided by 7 is 23.

26a. The prime factorization of 270 is $2 \cdot 3^3 \cdot 5$.
  b. The integral factors of 270 are $\pm 1, \pm 2, \pm 3, \pm 5, \pm 6, \pm 9, \pm 10, \pm 15, \pm 18, \pm 27, \pm 30, \pm 45, \pm 54, \pm 90, \pm 135,$ and ± 270.

27a. $8(-7) + 8(3)$ **b.** $[3 \cdot (-4)] \cdot (6)$ **c.** $-4 + (-3)$

28. Not enough information has been given; we need to know how many books cost $3.95 and how many cost $5.95.

Chapter 1 Diagnostic Test (page 77)

Following each problem number is the textbook section number (in parentheses) where that kind of problem is discussed.

1a. (1.1) True **b.** (1.8) False **c.** (1.10) True
  d. (1.8) False

2a. (1.1) True **b.** (1.11) False **c.** (1.2) True
  d. (1.8) False

3a. (1.10) False **b.** (1.8) True **c.** (1.5) True
  d. (1.3) False

4. (1.10) The rational numbers are -63, $0.\overline{38}$, and $\frac{2}{23}$. The irrational numbers are $-\sqrt[3]{19}$, and $0.25375913\ldots$. The real numbers are $-\sqrt[3]{19}$, -63, $0.\overline{38}$, $\frac{2}{23}$, and $0.25375913\ldots$. One number is not real; it is $\sqrt{-121}$.

5. (1.4) **a.** $5 - 24 = 5 + (-24) = -19$
      b. $-17 - 8 = -17 + (-8) = -25$

6. (1.3) **a.** $4 + (-18) = -14$ **b.** $15 + (-8.62) = 6.38$

7. (1.5) **a.** $-6(0) = 0$ **b.** $-\dfrac{\overset{1}{\cancel{2}}}{3}\left(-\dfrac{1}{\underset{2}{\cancel{4}}}\right) = \dfrac{1}{6}$

8. (1.4) **a.** $-10 - 5 = -10 + (-5) = -15$
   b. $3 - 22 = 3 + (-22) = -19$

9. (1.9) **a.** $-7^2 = -(7)(7) = -49$ **b.** $(-7)^2 = (-7)(-7) = 49$

10. (1.6) **a.** $3 \div 0$ is not defined   **b.** $\dfrac{15}{-21} = -\dfrac{5}{7}$

11. (1.6) **a.** $-\dfrac{5}{8} \div \left(-\dfrac{1}{2}\right) = -\dfrac{5}{8} \cdot \left(-\dfrac{2}{1}\right) = \dfrac{5}{4}$, or $1\dfrac{1}{4}$
   b. $\dfrac{0}{4.7} = 0$

12. (1.10) **a.** $\sqrt{49} = 7$ **b.** $\sqrt[3]{64} = 4$

13. (1.4) **a.** $3 - (-12) = 3 + (+12) = 15$
   b. $0 - 6 = -6$

14. (1.11) **a.** $2 \cdot 5^2 = 2 \cdot 25 = 50$
   b. $\sqrt{17^2 - 15^2} = \sqrt{289 - 225} = \sqrt{64} = 8$

15. (1.11) $23 - 19 - 3 + 11 = 4 - 3 + 11 = 1 + 11 = 12$

16. (1.11) $\left(-\dfrac{\overset{2}{\cancel{6}}}{}\right)(-5)(-2)\left(-\dfrac{1}{\underset{1}{\cancel{3}}}\right) = 20$

17. (1.11) $54 \div 9 \cdot 6 = (54 \div 9) \cdot (6) = 6 \cdot 6 = 36$

18. (1.11) $8 + 9 \cdot 7 = 8 + 63 = 71$

19. (1.11) $6 \cdot 4^2 - 4 = 6 \cdot 16 - 4 = 96 - 4 = 92$

20. (1.11) $\dfrac{7 - 15}{-2 + 6} = \dfrac{-8}{4} = -2$

21. (1.11) $5\sqrt{36} - 6(-5) = 5(6) - 6(-5) = 30 + 30 = 60$

22. (1.11) $(2^3 - 8)(8^2 - 9^2) = (8 - 8)(64 - 81) = 0(-17) = 0$

23. (1.11) $\{-10 - [5 + 2(4 - 7)]\} - 3 = \{-10 - [5 + 2(-3)]\} -$
   $3 = \{-10 - [5 + (-6)]\} - 3 = \{-10 - [-1]\} - 3$
   $= \{-10 + [+1]\} - 3 = -9 - 3 = -9 + (-3) = -12$

24. (1.12) **a.** The integral factors of 54 are ±1, ±2, ±3, ±6, ±9,
   ±18, ±27, and ±54.
   b. $54 = 2 \cdot 3^3$

25. (1.13) Enough information has been given; subtraction would be
   used.

Exercises 2.1 (page 83)

1a. 2, 4   **b.** x, y **2a.** 7, 3   **b.** s, t
3a. 7, -8, 2   **b.** u, v **4a.** 3, -5, -2 **b.** x, y
5a. Factor   **b.** Term   **c.** Factor   **d.** Factor
6a. Term   **b.** Factor   **c.** Factor   **d.** Factor
7a. Term   **b.** Term   **c.** Factor   **d.** Term
8a. Factor   **b.** Factor   **c.** Term   **d.** Term
9a. One term   **b.** no second term
10a. One term   **b.** no second term
11a. Three terms   **b.** second term: $-5F$
12a. Three terms   **b.** second term: $-2T$
13a. Three terms   **b.** second term: $\dfrac{2x + y}{3xy}$
14a. Three terms   **b.** second term: $\dfrac{5x - y}{7xy}$
15a. Two terms   **b.** second term: $-6u(2u + v^2)$
16a. Two terms   **b.** second term: $-2E(8E + F^2)$
17a. One term   **b.** no second term
18a. One term   **b.** no second term
19. 3   **20.** 4   **21.** 1   **22.** 1   **23.** -1 **24.** -1
25. $\dfrac{4}{5}$ **26.** $\dfrac{1}{2}$ **27.** No   **28.** No   **29.** Yes   **30.** Yes
31. Yes   **32.** Yes

Exercises 2.2A (page 88)

1. $5a + 30$ **2.** $4x + 40$ **3.** $7x + 7y$ **4.** $5m + 5n$
5. $3m - 12$ **6.** $3a - 15$ **7.** $4x - 4y$ **8.** $9m - 9n$
9. $6a + ax$ **10.** $7b + by$
11. $(-2)(x) + (-2)(-3) = -2x + 6$ **12.** $-3x + 15$
13. $6x - 24$ **14.** $-15 + 10x$
15. $-3(x) + (-3)(-2y) + (-3)(2) = -3x + 6y - 6$
16. $-2x + 6y - 8$

Exercises 2.2B (page 91)

1. $(15 - 3)x = 12x$ **2.** $3a$ **3.** $(5 - 12)a = -7a$
4. $-14x$ **5.** $(2 - 5 + 6)a = 3a$ **6.** $4y$
7. $(5 - 8 + 1)x = -2x$ **8.** $-a$
9. $3x - 3x + 2y = (3 - 3)x + 2y = 0x + 2y = 2y$ **10.** $4a$
11. $(4 + 1 - 10)y = -5y$ **12.** $-3x$
13. $(3 - 5 + 2)mn = (0)mn = 0$ **14.** 0
15. $2xy - 5xy + xy = (2 - 5 + 1)xy = -2xy$ **16.** $4mn$
17. $(8 - 2)x^2y = 6x^2y$ **18.** $7ab^2$ **19.** $(1 - 3)a^2b = -2a^2b$
20. $-4x^2y^2$ **21.** $5ab - 2ab + 2c = (5 - 2)ab + 2c = 3ab + 2c$
22. $3xy - 3z$ **23.** $(5 - 2 - 4)xyz^2 = -xyz^2$ **24.** $2a^2bc$
25. $(5 - 2)u + 10v = 3u + 10v$ **26.** $4w + 5v$
27. $(8 - 4)x - 2y = 4x - 2y$ **28.** $11x - 8y$
29. $(7 - 4)x^2y - 2xy^2 = 3x^2y - 2xy^2$ **30.** $2xy^2 - 5x^2y$
31. $(5x^2 - 2x^2) + (-3x + 8x) + (7 - 9) = 3x^2 + 5x - 2$
32. $-2y^2 + 2y + 1$
33. $(12.67 + 9.08 - 6.73)\text{sec} = 15.02$ sec   **34.** 83.0 ft

Exercises 2.2C (page 94)

1. $8 + a - b$ **2.** $7 + m - n$
3. $5 - 1(x - y) = 5 - x + y$ **4.** $6 - a + b$
5. $12 - 3m + 3n$ **6.** $14 - 5x + 5y$ **7.** $R - S - 8$
8. $x - y - 2$ **9.** $-8R + 8S$ **10.** $-2x + 2y$
11. $-15xz + 10yz$ **12.** $-10ac + 14bc$ **13.** $3x - 2y - 5z$
14. $5a - 7b - 2c$ **15.** $10 - 2a + 2b$ **16.** $12 - 5x + 10y$
17. $2x - 2y + 3$ **18.** $4a - 8b + 5$ **19.** $2a - 3x + 3y$
20. $5x - 2a + 2b$
21. $a - [x - 1(b - c)] = a - 1[x - b + c] = a - x + b - c$
22. $a - x - b + c$
23. $4 - 2[a - 3x + 3y] = 4 - 2a + 6x - 6y$
24. $8 - 3x + 18a - 6b$
25. $7 + 6[x - 3a + 3b] = 7 + 6x - 18a + 18b$
26. $3 + 5y - 40x + 20a$
27. $12 - 4\{a - 1(b - 7c)\} = 12 - 4\{a - b + 7c\}$
   $= 12 - 4a + 4b - 28c$
28. $15 - 7x + 7y + 21z$ **29.** $3a - 6x - 2y + 6b$
30. $4x - 2b - 3y + 15a$
31. $-10[-2x + 6y + a] - b = 20x - 60y - 10a - b$
32. $30x - 15y - 5a - c$
33. $(a - b) - \{[x - 1(3 - y)] - R\} = (a - b) - \{[x - 3 + y] - R\}$
   $= (a - b) - 1\{x - 3 + y - R\} = a - b - x + 3 - y + R$

34. $x - y - a + c - 5 + b$

35. $\underline{a} + 5b + 15c + \underline{5a} = 6a + 5b + 15c$

36. $10x + 9y + 63z$

37. $9y - 2\{y - \underline{4y} - 4)\} = 9y - 2(-3y - 4) = \underline{9y} + \underline{6y} + 8$
   $= 15y + 8$

38. $28x + 60$

39. $4x - 1(5 - 2y + 8x) = \underline{4x} - 5 + 2y - \underline{8x} = 2y - 5 - 4x$

40. $1 - 2a - 3b - 4c$

41. $6x^2 + 5 - 1(3x + 4 - x^2) = \underline{6x^2} + 5 - 3x - 4 + \underline{x^2}$
   $= 7x^2 - 3x + 1$

42. $-3a^2 + 5a - 3$

Exercises 2.3 (page 98)

1. $ab = 3(-5) = -15$ **2.** -28

3. $a + b = 3 + (-5) = -2$ **4.** -3

5. $3b = 3(-5) = -15$ **6.** -75

7. $9 - 6b = 9 - 6(-5) = 9 + 30 = 39$ **8.** -27

9. $b^2 = (-5)^2 = (-5)(-5) = 25$ **10.** -49

11. $2a - 3b = 2(3) - 3(-5) = 6 + 15 = 21$ **12.** 26

13. $x - y - 2b = (4) - (-7) - 2(-5) = 4 + 7 + 10 = 21$

14. 29

15. $3b - ab + xy = 3(-5) - (3)(-5) + (4)(-7)$
   $= -15 + 15 - 28 = -28$

16. -27 **17.** $x^2 - y^2 = (4)^2 - (-7)^2 = 16 - 49 = -33$

18. 24

19. $4 + a(x + y) = 4 + (3)[(4) + (-7)] = 4 + 3[-3] = 4 - 9 = -5$

20. 15

21. $2(a - b) - 3c = 2[(3) - (-5)] - 3(-1) = 2[8] - 3(-1)$
   $= 16 + 3 = 19$

22. 17

23. $3x^2 - 10x + 5 = 3(4)^2 - 10(4) + 5 = 3(16) - 10(4) + 5$
   $= 48 - 40 + 5 = 13$

24. 156

25. $a^2 - 2ab + b^2 = (3)^2 - 2(3)(-5) + (-5)^2 = 9 - 2(3)(-5) + 25$
   $= 9 + 30 + 25 = 64$

26. 121 **27.** $\dfrac{3x}{y + b} = \dfrac{3(4)}{(-7) + (-5)} = \dfrac{12}{-12} = -1$ **28.** 3

29. $\dfrac{E + F}{EF} = \dfrac{(-1) + (3)}{(-1)(3)} = \dfrac{2}{-3} = -\dfrac{2}{3}$ **30.** $-\dfrac{9}{20}$

31. $\dfrac{(1 + G)^2 - 1}{H} = \dfrac{[1 + (-5)]^2 - 1}{-4} = \dfrac{[-4]^2 - 1}{-4}$
   $= \dfrac{16 - 1}{-4} = \dfrac{15}{-4} = -3\dfrac{3}{4}$

32. $\dfrac{1}{3}$

33. $2E - [F - (3K - H)] = 2(-1) - [(3) - \{3(0) - (-4)\}]$
   $= 2(-1) - [(3) - \{0 + 4\}] = 2(-1) - [3 - \{4\}]$
   $= 2(-1) - [-1] = -2 + 1 = -1$

34. 1

35. $G - \sqrt{G^2 - 4EH} = (-5) - \sqrt{(-5)^2 - 4(-1)(-4)}$
   $= (-5) - \sqrt{25 - 4(-1)(-4)} = (-5) - \sqrt{25 - 16}$
   $= (-5) - \sqrt{9} = -5 - 3 = -8$

36. -8

37. $\dfrac{\sqrt{2H - 5G}}{0.2F^2} = \dfrac{\sqrt{2(-4) - 5(-5)}}{0.2(3)^2} = \dfrac{\sqrt{-8 + 25}}{0.2(9)}$
   $= \dfrac{\sqrt{17}}{1.8} \approx \dfrac{4.123}{1.8} \approx 2.291$

38. 20

Exercises 2.4 (page 101)

1. $A = \frac{1}{2}bh$
   $A = \dfrac{1}{\underset{1}{\overset{}{2}}}\left(\dfrac{15}{1}\right)\left(\dfrac{\overset{7}{14}}{1}\right)$
   $A = 105$

2. 486

3. $I = \dfrac{E}{R}$
   $I = \dfrac{110}{22}$
   $I = 5$

4. $6\frac{2}{3}$

5. $I = prt$
   $I = 600(0.09)(4.5)$
   $I = 243$

6. 140

7. $F = \frac{9}{5}C + 32$
   $F = \frac{9}{5}(25) + 32$
   $F = 77$

8. -13

9. $A = \pi r^2$
   $A \approx 3.14(10)^2$
   $A \approx 314$

10. $\approx 1,256$

11. $s = \frac{1}{2}gt^2$
   $s = \frac{1}{2}(32)(3)^2$
   $s = \frac{1}{2}(32)(9)$
   $s = 144$

12. 400

13. $V = C - Crt$
   $V = 500 - (500)(0.1)(2)$
   $V = 500 - 100$
   $V = 400$

14. 600

15. $\sigma = \sqrt{npq}$
   $\sigma = \sqrt{(100)(0.9)(0.1)}$
   $\sigma = \sqrt{9}$
   $\sigma = 3$

16. 4

17. $C = \frac{5}{9}(F - 32)$
   $C = \frac{5}{9}(-4 - 32)$
   $C = \frac{5}{9}(-36)$
   $C = -20$

18. -15

19. $C = \dfrac{a}{a + 12} \cdot A$
   $C = \dfrac{6}{6 + 12}(30)$
   $C = \dfrac{6}{18}(30)$
   $C = 10$

20. 12

21. $V = \frac{4}{3}\pi r^3$
   $V \approx \frac{4}{3}(3.14)(3)^3$
   $V \approx \frac{4}{3}(3.14)(27)$
   $V \approx 113.04$

22. ≈ 904.32

23. $A = P(1 + i)^n$
   $A = 1,000(1 + 0.06)^2$
   $A = 1,000(1.1236)$
   $A = 1,123.60$

24. $\approx 2,519.42$

Exercises 2.5 (page 105)

1. $F = \frac{9}{5}C + 32$; $F = \frac{9}{5}(100) + 32 = 180 + 32 = 212$; $212°$ F

2. $1,920

3. $A = P(1 + rt)$; $A = (2,000)[1 + (0.07)(5)] = 2,000(1 + 0.35)$
$= 2,000(1.35) = 2,700$; $2,700

4. ≈ 904 cu. in.

5. $s = \frac{1}{2}gt^2$; $s = \frac{1}{2}(32)(3)^2 = 16(9) = 144$; 144 ft　　**6.** $7\frac{1}{2}$ sq. ft

Review Exercises 2.6 (page 108)

1a. 3 terms　　**2a.** 2 terms　　**3a.** 1 term
　b. $2ab$　　　　**b.** $-2(x^2 + y^2)$　　**b.** No second term

4a. 2 terms　　　**5.** Term　　**6.** Factor
　b. $-2(x^3 + 1)$

7. $41x$　　**8.** $-54xy^2$　　**9.** $-9ab - 3a + 5b$

10. $2x - y - 3$　　**11.** $24x - 8y$　　**12.** $-5a - 5b$

13. $-6s + 4t$　　**14.** $3x - 2t - 2$

15. $7 - 3[x - 8 - 4x] = 7 - 3[-3x - 8] = 7 + 9x + 24$
$= 9x + 31$

16. $36x + 21$

17. $3x - y + z = 3(-2) - (3) + (-4) = -6 - 3 - 4 = -13$

18. 25

19. $5x^2 - 3x + 10 = 5(-2)^2 - 3(-2) + 10 = 5(4) - 3(-2) + 10$
$= 20 + 6 + 10 = 36$

20. -17

21. $x - 2[y - x(y + z)] = -2 - 2[3 - (-2)(3 + (-4))]$
$= -2 - 2[3 - (-2)(-1)] = -2 - 2[3 - 2] = -2 - 2(1)$
$= -2 - 2 = -4$

22. $6\frac{5}{6}$　　**23.** $C = \dfrac{a}{a + 12}(A)$　　**24.** 110

$C = \dfrac{8}{8 + 12}(35)$

$C = \frac{8}{20}\left(\frac{35}{1}\right)$

$C = 14$

25. $C = \frac{5}{9}(F - 32)$　　**26.** 665.50　　**27.** $\sigma = \sqrt{npq}$
$C = \frac{5}{9}\left(15\frac{1}{2} - 32\right)$　　　　　　　　　$\sigma = \sqrt{(400)(0.5)(0.5)}$
$C = \frac{5}{9}\left(-\frac{33}{2}\right)$　　　　　　　　　　$\sigma = \sqrt{100}$
$C = -9\frac{1}{6}$　　　　　　　　　　　　$\sigma = 10$

28. 1,100

29. $A = P(1 + rt)$; $A = (2,500)[1 + (0.06)(3)] = 2,500(1 + 0.18)$
$= 2,500(1.18) = 2,950$; $2,950

30. 10

Chapter 2 Diagnostic Test (page 111)

Following each problem number is the textbook section number (in parentheses) where that kind of problem is discussed.

1. (2.1)　**a.** Factor　　**b.** Term　　**c.** Term　　**d.** Term
　　　e. Factor

2. (2.1)　**a.** Three terms　　**b.** Second term: $\dfrac{7x + 1}{3}$　　**c.** 6

3. (2.1)　**a.** Four terms　　**b.** Second term: $3x$　　**c.** -1

4. (2.2)　$5 - (x - y) = 5 - 1(x - y) = 5 - x + y$

5. (2.2C)　$-2[-4(3c - d) + a] - b = -2[-12c + 4d + a] - b$
$= 24c - 8d - 2a - b$

6. (2.2B)　$4x - 3x + 5x = (4 - 3 + 5)x = 6x$

7. (2.2B)　$2a - 5b - 7 - 3b + 4 - 5a = -3a - 8b - 3$

8. (2.2)　$5x - 3(y - x) = 5x - 3y + 3x = 8x - 3y$

9. (2.2)　$(x - 4)(-5) = -5x + 20$

10. (2.2)　$(x - 4) - 5 = x - 4 - 5 = x - 9$

11. (2.2)　$3 + 5[3x - 1(2y - x)] = 3 + 5[3x - 2y + x]$
$= 3 + 5[4x - 2y] = 3 + 20x - 10y$

12. (2.3)　$3c - by + cx = 3(-2) - (-7)(5) + (-2)(-6)$
$= -6 - (-35) + 12 = -6 + (+35) + 12$
$= 29 + 12 = 41$

13. (2.3)　$4x - [a - (3c - b)] = 4(-6) - \{(-4) - (3[-2] - [-7])\}$
$= 4(-6) - \{-4 - (-6 + 7)\} = 4(-6) - \{-4 - (+1)\}$
$= 4(-6) - \{-4 + (-1)\} = 4(-6) - \{-5\}$
$= -24 + \{+5\} = -19$

14. (2.3)　$x^2 + 2xy + y^2 = (-6)^2 + 2(-6)(5) + (5)^2$
$= 36 + (-60) + 25 = -24 + 25 = 1$

15. (2.3)　$(x + y)^2 = ([-6] + [5])^2 = (-1)^2 = 1$

16. (2.4)　$C = \frac{5}{9}(F - 32)$　　**17.** (2.4)　$A = \pi r^2$
　　　$C = \frac{5}{9}[(68) - 32]$　　　　　　$A \approx 3.14(3)^2$
　　　$C = \frac{5}{9}(36)$　　　　　　　　$A \approx 3.14(9)$
　　　$C = 20$　　　　　　　　　　$A \approx 28.26$

18. (2.4)　$A = P(1 + rt)$
　　　$A = 500[1 + (0.1)(3.5)]$
　　　$A = 500(1.35)$
　　　$A = 675$

19. (2.4)　$V = C - Crt$
　　　$V = 800 - 800(0.06)(10)$
　　　$V = 800 - 480$
　　　$V = 320$

20. (2.5)　$V = \frac{1}{3}\pi r^2 h$
　　　$V \approx \frac{1}{3}(3.14)(4\text{ ft})^2(3\text{ ft})$
　　　$V \approx \frac{1}{3}(3.14)(16\text{ sq. ft})(3\text{ ft})$
　　　$V \approx 50.24$ cu. ft

Cumulative Review Exercises: Chapters 1 and 2 (page 112)

1. 32　　**2.** -21　　**3.** $(-5) - (-11) = (-5) + (+11) = 6$

4. -4　　**5.** 0　　**6.** -8　　**7.** -17　　**8.** 4　　**9.** 4

10. 0　　**11.** $(25)^2 = (25)(25) = 625$　　**12.** -45

13. $-2^4 = -(2^4) = -16$　　**14.** -4

15. $\dfrac{-12}{0}$ is not defined; we cannot divide by zero.　　**16.** 2

17. 8　　**18.** 0　　**19.** $8 - 17 = 8 + (-17) = -9$　　**20.** -29

21. $24 \div 12 \cdot 2 = 2 \cdot 2 = 4$　　**22.** 24

23. $8 - 6 \cdot 5 = 8 - 30 = -22$　　**24.** 1

25. $17 - 9 - 2 = 8 - 2 = 6$　　**26.** 38　　**27.** 0

28. $C = \frac{5}{9}(F - 32)$; $F = 21\frac{1}{2} = \frac{43}{2}$
　　　$C = \frac{5}{9}\left(\frac{43}{2} - \frac{64}{2}\right)$
　　　$C = \frac{5}{9}\left(-\frac{21}{2}\right)$
　　　$C = -\frac{35}{6}$ or $-5\frac{5}{6}$

29. True　　**30.** True　　**31.** True　　**32.** False　　**33.** True

34. True　　**35.** False　　**36.** True　　**37.** False　　**38.** False

39. $5x - 1(z - 2x) = 5x - z + 2x = 7x - z$

40. $-3x - 3y + 20$

Exercises 3.1 (page 115)

1. $(-4) + 2 \neq 3$. Therefore, -4 is not a solution of $x + 2 = 3$.

2. No   **3.** $(4) + 1 = 5$. Therefore, 4 is a solution of $x + 1 = 5$.

4. Yes   **5.** $2 + (-3) = -1$. Therefore, -3 is in the solution set.

6. Yes   **7.** $5 + (3) \neq 1$. Therefore, 3 is not in the solution set.

8. No

Exercises 3.2 (page 119)

1.
$$
\begin{aligned}
x + 5 &= 8 \\
-5 &\quad -5 \\
x &= 3
\end{aligned}
$$
The solution is 3.

Check: $x + 5 = 8$
$$3 + 5 \overset{?}{=} 8$$
$$8 = 8$$

2. 5

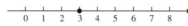

3.
$$
\begin{aligned}
x - 3 &= 4 \\
+3 &\quad +3 \\
x &= 7
\end{aligned}
$$
The solution is 7.

Check: $x - 3 = 4$
$$7 - 3 \overset{?}{=} 4$$
$$4 = 4$$

4. 9

5.
$$
\begin{aligned}
3 + x &= -4 \\
-3 &\quad -3 \\
x &= -7
\end{aligned}
$$
The solution is -7.

Check: $3 + x = -4$
$$3 + (-7) \overset{?}{=} -4$$
$$-4 = -4$$

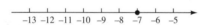

6. -7

7.
$$
\begin{aligned}
x + 4 &= 21 \\
-4 &\quad -4 \\
x &= 17
\end{aligned}
$$
The solution is 17.

Check: $x + 4 = 21$
$$17 + 4 \overset{?}{=} 21$$
$$21 = 21$$

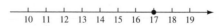

8. 9

9.
$$
\begin{aligned}
x - 35 &= 7 \\
+35 &\quad +35 \\
x &= 42
\end{aligned}
$$
The solution is 42.

Check: $x - 35 = 7$
$$42 - 35 \overset{?}{=} 7$$
$$7 = 7$$

10. 51

11.
$$
\begin{aligned}
9 &= x + 5 \\
-5 &\quad -5 \\
4 &= x \\
x &= 4
\end{aligned}
$$

Check: $9 = x + 5$
$$9 \overset{?}{=} 4 + 5$$
$$9 = 9$$
The solution is 4.

12. 3

13.
$$
\begin{aligned}
12 &= x - 11 \\
+11 &\quad +11 \\
23 &= x \\
x &= 23
\end{aligned}
$$

Check: $12 = x - 11$
$$12 \overset{?}{=} 23 - 11$$
$$12 = 12$$
The solution is 23.

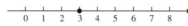

14. 29

15.
$$
\begin{aligned}
-17 + x &= 28 \\
+17 &\quad +17 \\
x &= 45
\end{aligned}
$$
The solution is 45.

Check: $-17 + x = 28$
$$-17 + 45 \overset{?}{=} 28$$
$$28 = 28$$

16. 47

17.
$$
\begin{aligned}
-28 &= -15 + x \\
+15 &\quad +15 \\
-13 &= x \\
x &= -13
\end{aligned}
$$

Check: $-28 = -15 + x$
$$-28 \overset{?}{=} -15 + (-13)$$
$$-28 = -28$$
The solution is -13.

18. -29

19.
$$
\begin{aligned}
x + \tfrac{1}{2} &= 2\tfrac{1}{2} \\
-\tfrac{1}{2} &\quad -\tfrac{1}{2} \\
x &= 2
\end{aligned}
$$
The solution is 2.

Check: $x + \tfrac{1}{2} = 2\tfrac{1}{2}$
$$2 + \tfrac{1}{2} \overset{?}{=} 2\tfrac{1}{2}$$
$$2\tfrac{1}{2} = 2\tfrac{1}{2}$$

20. 5

21.
$$
\begin{aligned}
5.6 + x &= 2.8 \\
-5.6 &\quad -5.6 \\
x &= -2.8
\end{aligned}
$$
The solution is -2.8.

Check: $5.6 + x = 2.8$
$$5.6 + (-2.8) \overset{?}{=} 2.8$$
$$2.8 = 2.8$$

22. -0.08

23.

$$7.84 = x - 3.98$$
$$\underline{+\ 3.98 \qquad +\ 3.98}$$
$$11.82 = x$$
$$x = 11.82$$

Check: $7.84 = x - 3.98$
$$7.84 \overset{?}{=} 11.82 - 3.98$$
$$7.84 = 7.84$$
The solution is 11.82.

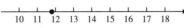

24. 7.07

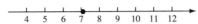

Exercises 3.3A (page 124)

1. $2x = 8$

$$\frac{\overset{1}{\cancel{2}}x}{\cancel{2}} = \frac{8}{2}$$
$$x = 4$$

Check: $2x = 8$
$$2(4) \overset{?}{=} 8$$
$$8 = 8$$
The solution is 4.

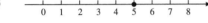

2. 5

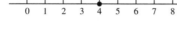

3. $21 = 7x$

$$\frac{21}{7} = \frac{\overset{1}{\cancel{7}}x}{\cancel{7}}$$
$$3 = x$$
$$x = 3$$

Check: $21 = 7x$
$$21 \overset{?}{=} 7(3)$$
$$21 = 21$$
The solution is 3.

4. 7

5. $11x = 33$

$$\frac{\overset{1}{\cancel{11}}x}{\cancel{11}} = \frac{33}{11}$$
$$x = 3$$

Check: $11x = 33$
$$11(3) \overset{?}{=} 33$$
$$33 = 33$$
The solution is 3.

6. 4

7. $\dfrac{x}{3} = 4$

$$\overset{1}{\cancel{3}}\left(\frac{x}{\cancel{3}}\right) = 3(4)$$
$$x = 12$$

Check: $\dfrac{x}{3} = 4$
$$\frac{12}{3} \overset{?}{=} 4$$
$$4 = 4$$
The solution is 12.

8. 15

9. $\dfrac{x}{5} = -2$

$$\overset{1}{\cancel{5}}\left(\frac{x}{\cancel{5}}\right) = 5(-2)$$
$$x = -10$$

Check: $\dfrac{x}{5} = -2$
$$\frac{-10}{5} \overset{?}{=} -2$$
$$-2 = -2$$
The solution is -10.

10. -24

11. $4 = \dfrac{x}{7}$

$$7(4) = \overset{1}{\cancel{7}}\left(\frac{x}{\cancel{7}}\right)$$
$$28 = x$$
$$x = 28$$

Check: $4 = \dfrac{x}{7}$
$$4 \overset{?}{=} \frac{28}{7}$$
$$4 = 4$$
The solution is 28.

12. 24

The checks will not be shown for Exercises 13–16.

13. $-13 = \dfrac{x}{9}$

$$9(-13) = \overset{1}{\cancel{9}}\left(\frac{x}{\cancel{9}}\right)$$
$$-117 = x$$
$$x = -117 \qquad \text{The solution is } -117.$$

14. -120

15. $\dfrac{x}{10} = 3.14$

$$\overset{1}{\cancel{10}}\left(\frac{x}{\cancel{10}}\right) = 10(3.14) \qquad \text{The solution is 31.4.}$$
$$x = 31.4$$

16. 39

Exercises 3.3B (page 129)

1. $4x + 1 = 9$

$$\underline{\qquad -1 \qquad -1}$$
$$4x \quad = 8$$
$$\frac{\overset{1}{\cancel{4}}x}{\cancel{4}} = \frac{8}{4}$$
$$x = 2$$

Check: $4x + 1 = 9$
$$4(2) + 1 \overset{?}{=} 9$$
$$8 + 1 \overset{?}{=} 9$$
$$9 = 9$$
The solution is 2.

2. 2

3. $6x - 2 = 10$

$$\underline{\qquad +2 \qquad +2}$$
$$6x \quad = 12$$
$$\frac{\overset{1}{\cancel{6}}x}{\cancel{6}} = \frac{12}{6}$$
$$x = 2$$

Check: $6x - 2 = 10$
$$6(2) - 2 \overset{?}{=} 10$$
$$12 - 2 \overset{?}{=} 10$$
$$10 = 10$$
The solution is 2.

4. 1

5. $2x - 15 = 11$

$\quad\quad \underline{+\ 15 \quad +15}$

$\quad\quad 2x \quad\quad = \quad 26$

$\quad\quad \dfrac{\overset{1}{\cancel{2}}x}{\underset{1}{\cancel{2}}} = \dfrac{26}{2}$

$\quad\quad x = 13$

Check: $\quad 2x - 15 = 11$

$\quad\quad 2(13) - 15 \overset{?}{=} 11$

$\quad\quad 26 - 15 \overset{?}{=} 11$

$\quad\quad 11 = 11$

The solution is 13.

[number line: 10 11 12 13 14 15 16 17 18, point at 13]

6. 6 [number line: 0 1 2 3 4 5 6 7 8, point at 6]

The checks will not be shown for Exercises 7–38.

7. $4x + 2 = -14$

$\quad\quad \underline{-\ 2 \quad\quad -2}$

$\quad\quad 4x \quad\quad = -16$

$\quad\quad \dfrac{\overset{1}{\cancel{4}}x}{\underset{1}{\cancel{4}}} = \dfrac{-16}{4}$

$\quad\quad x = -4$

The solution is -4.

[number line: −6 −5 −4 −3 −2 −1 0 1 2, point at −4]

8. -3 [number line: −6 −5 −4 −3 −2 −1 0 1 2, point at −3]

9. $14 = 9x - 13$

$\quad\quad \underline{+13 \quad\quad +\ 13}$

$\quad\quad 27 = 9x$

$\quad\quad \dfrac{27}{9} = \dfrac{\overset{1}{\cancel{9}}x}{\underset{1}{\cancel{9}}}$

$\quad\quad 3 = x$

$\quad\quad x = 3$

The solution is 3.

[number line: 0 1 2 3 4 5 6 7 8, point at 3]

10. 5 [number line: 0 1 2 3 4 5 6 7 8, point at 5]

11. $12x + 17 = 65$

$\quad\quad \underline{-\ 17 \quad\quad -17}$

$\quad\quad 12x \quad\quad = 48$

$\quad\quad \dfrac{\overset{1}{\cancel{12}}x}{\underset{1}{\cancel{12}}} = \dfrac{48}{12}$

$\quad\quad x = 4$

The solution is 4.

[number line: 0 1 2 3 4 5 6 7 8, point at 4]

12. 2 [number line: 0 1 2 3 4 5 6 7 8, point at 2]

13. $8x - 23 = 31$

$\quad\quad \underline{+\ 23 \quad\quad +23}$

$\quad\quad 8x \quad\quad = 54$

$\quad\quad \dfrac{\overset{1}{\cancel{8}}x}{\underset{1}{\cancel{8}}} = \dfrac{54}{8}$

$\quad\quad x = 6\frac{3}{4}$

The solution is $6\frac{3}{4}$.

[number line: 0 1 2 3 4 5 6 7 8, point at 6¾]

14. $10\frac{1}{3}$ [number line: 9 10 11 12 13 14 15 16 17 18, point at 10⅓]

15. $\quad 14 - 4x = -28$

$\quad\quad \underline{-14 \quad\quad\quad -14}$

$\quad\quad -4x = -42$

$\quad\quad \dfrac{\overset{1}{\cancel{-4}}x}{\underset{1}{\cancel{-4}}} = \dfrac{-42}{-4}$

$\quad\quad x = 10\frac{1}{2}$

The solution is $10\frac{1}{2}$.

[number line: 9 10 11 12 13 14 15 16 17 18, point at 10½]

16. $10\frac{1}{3}$ [number line: 9 10 11 12 13 14 15 16 17 18, point at 10⅓]

17. $\quad\quad 8 = 25 - 3x$

$\quad\quad \underline{-25 \quad\quad -25}$

$\quad\quad -17 = \quad\quad -3x$

$\quad\quad \dfrac{-17}{-3} = \dfrac{\overset{1}{\cancel{-3}}x}{\underset{1}{\cancel{-3}}}$

$\quad\quad 5\frac{2}{3} = x$

$\quad\quad x = 5\frac{2}{3}$

The solution is $5\frac{2}{3}$.

[number line: 0 1 2 3 4 5 6 7 8, point at 5⅔]

18. $8\frac{1}{2}$ [number line: 4 5 6 7 8 9 10 11 12, point at 8½]

19. $\quad -73 = 24x + 31$

$\quad\quad \underline{-\ 31 \quad\quad\quad -31}$

$\quad\quad -104 = 24x$

$\quad\quad \dfrac{-104}{24} = \dfrac{\overset{1}{\cancel{24}}x}{\underset{1}{\cancel{24}}}$

$\quad\quad -4\frac{1}{3} = x$

$\quad\quad x = -4\frac{1}{3}$

The solution is $-4\frac{1}{3}$.

[number line: −6 −5 −4 −3 −2 −1 0 1 2, point at −4⅓]

20. $-2\frac{1}{2}$ [number line: −6 −5 −4 −3 −2 −1 0 1 2, point at −2½]

21. $18x - 4.8 = \quad 6$

$\quad\quad \underline{+\ 4.8 \quad +\ 4.8}$

$\quad\quad 18x \quad\quad = \quad 10.8$

$\quad\quad \dfrac{\overset{1}{\cancel{18}}x}{\underset{1}{\cancel{18}}} = \dfrac{10.8}{18}$

$\quad\quad x = 0.6$

The solution is 0.6.

[number line: −6 −5 −4 −3 −2 −1 0 1 2, point at 0.6]

22. $1.03\frac{1}{3}$ [number line: −6 −5 −4 −3 −2 −1 0 1 2, point near 1]

23. $2.5x - 3.8 = -7.9$

$\quad\quad \underline{+\ 3.8 \quad +3.8}$

$\quad\quad 2.5x \quad\quad = -4.1$

$\quad\quad \dfrac{\overset{1}{\cancel{2.5}}x}{\underset{1}{\cancel{2.5}}} = \dfrac{-4.1}{2.5}$

$\quad\quad x = -1.64$

The solution is -1.64.

[number line: −6 −5 −4 −3 −2 −1 0 1 2, point at −1.64]

24. ≈ -0.0667 [number line: −6 −5 −4 −3 −2 −1 0 1 2, point near 0]

25. $\dfrac{x}{4} + 6 = \quad 9$

$\quad\quad \underline{-\ 6 \quad\quad -6}$

$\quad\quad \dfrac{x}{4} \quad\quad = \quad 3$

$\quad\quad \overset{1}{\cancel{4}}\left(\dfrac{x}{\underset{1}{\cancel{4}}}\right) = 4(3)$

$\quad\quad x = 12$

The solution is 12.

[number line: 10 11 12 13 14 15 16 17 18, point at 12]

26. 25 [number line: 20 21 22 23 24 25 26 27 28, point at 25]

27. $\dfrac{x}{10} - 5 = 13$

$\phantom{\dfrac{x}{10}}\ \ \underline{+5 \qquad +5}$

$\dfrac{x}{10} = 18$

$\overset{1}{\cancel{10}}\left(\dfrac{x}{\cancel{10}}\right) = 10(18)$ The solution is 180.

$x = 180$

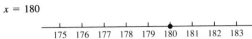

175 176 177 178 179 **180** 181 182 183

28. 320

315 316 317 318 319 **320** 321 322 323

29. $-14 = \dfrac{x}{6} - 7$

$\underline{+7 \qquad +7}$

$-7 = \dfrac{x}{6}$

$6(-7) = \overset{1}{\cancel{6}}\left(\dfrac{x}{\cancel{6}}\right)$

$-42 = x$

$x = -42$ The solution is -42.

$-50\ -49\ -48\ -47\ -46\ -45\ -44\ -43\ \mathbf{-42}\ -41$

30. -88

$-92\ -91\ -90\ -89\ \mathbf{-88}\ -87\ -86\ -85\ -84$

31. $7 = \dfrac{2x}{5} + 3$

$\underline{-3 \qquad\ -3}$

$4 = \dfrac{2x}{5}$

$5(4) = \overset{1}{\cancel{5}}\left(\dfrac{2x}{\cancel{5}}\right)$

$20 = 2x$

$\dfrac{20}{2} = \dfrac{\overset{1}{\cancel{2}}x}{\cancel{2}}$ The solution is 10.

$10 = x$

$x = 10$

4 5 6 7 8 9 **10** 11 12

32. 4

0 1 2 3 **4** 5 6 7 8

33. $4 - \dfrac{7x}{5} = 11$

$\underline{-4 \qquad\ -4}$

$-\dfrac{7x}{5} = 7$

$\overset{1}{\cancel{5}}\left(\dfrac{-7x}{\cancel{5}}\right) = 5(7)$

$-7x = 35$

$\dfrac{\overset{1}{\cancel{-7}}x}{\cancel{-7}} = \dfrac{35}{-7}$ The solution is -5.

$x = -5$

$-6\ \mathbf{-5}\ -4\ -3\ -2\ -1\ 0\ 1\ 2$

34. -45

$-50\ -49\ -48\ -47\ -46\ \mathbf{-45}\ -44\ -43\ -42$

35. $-24 + \dfrac{5x}{8} = 41$

$\underline{+24 \qquad\quad +24}$

$\dfrac{5x}{8} = 65$

$\overset{1}{\cancel{8}}\left(\dfrac{5x}{\cancel{8}}\right) = 8(65)$

$5x = 8(65)$

$\dfrac{5x}{5} = \dfrac{8\overset{13}{\cancel{(65)}}}{\cancel{5}}$

$x = 8(13) = 104$ The solution is 104.

101 102 103 **104** 105 106 107

36. 20

16 17 18 19 **20** 21 22 23 24

37. $41 = 25 - \dfrac{4x}{5}$

$\underline{-25 \qquad -25}$

$16 = -\dfrac{4x}{5}$

$5(16) = \overset{1}{\cancel{5}}\left(\dfrac{-4x}{\cancel{5}}\right)$

$5(16) = -4x$

$\dfrac{5(16)}{-4} = \dfrac{\overset{1}{\cancel{-4}}x}{\cancel{-4}}$

$-20 = x$ The solution is -20.

$x = -20$

$-23\ -22\ -21\ \mathbf{-20}\ -19\ -18\ -17$

38. -35

$-40\ -39\ -38\ -37\ -36\ \mathbf{-35}\ -34\ -33\ -32$

Exercises 3.4A (page 133)

(The checks will not be shown.)

1. $3x + 11 = 14x$

$\underline{-3x \qquad\ -3x}$

$11 = 11x$

$\dfrac{11}{11} = \dfrac{\overset{1}{\cancel{11}}x}{\cancel{11}}$ The solution is 1.

$1 = x$

$x = 1$

0 **1** 2 3 4

2. 1

0 **1** 2 3 4 5 6 7 8

3. $9x - 7 = 2x$

$\underline{-2x \qquad\ -2x}$

$7x - 7 = 0$

$\underline{+7 \qquad\ +7}$

$7x = 7$

$\dfrac{\overset{1}{\cancel{7}}x}{\cancel{7}} = \dfrac{7}{7}$ The solution is 1.

$x = 1$

0 **1** 2 3 4

4. 1

0 **1** 2 3 4 5 6 7 8

5.
$$
\begin{aligned}
2x - 7 &= x \\
-x &\quad -x \\
\hline
x - 7 &= 0 \\
+7 &\quad +7 \\
\hline
x &= 7
\end{aligned}
$$
The solution is 7.

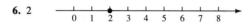

6. 2

7.
$$
\begin{aligned}
5x &= 3x - 4 \\
-3x &\quad -3x \\
\hline
2x &= -4 \\
\frac{\cancel{2}x}{\cancel{2}} &= \frac{-4}{2} \\
x &= -2
\end{aligned}
$$
The solution is −2.

8. −3

9.
$$
\begin{aligned}
9 - 2x &= x \\
+2x &\quad +2x \\
\hline
9 &= 3x \\
\frac{9}{3} &= \frac{\cancel{3}x}{\cancel{3}} \\
3 &= x \\
x &= 3
\end{aligned}
$$
The solution is 3.

10. 1

11.
$$
\begin{aligned}
3x - 4 &= 2x + 5 \\
-2x &\quad -2x \\
\hline
x - 4 &= 5 \\
+4 &= +4 \\
\hline
x &\quad\;\; 9
\end{aligned}
$$
The solution is 9.

12. 6

13.
$$
\begin{aligned}
6x + 7 &= 3 + 8x \\
-6x &\quad -6x \\
\hline
7 &= 3 + 2x \\
-3 &\quad -3 \\
\hline
4 &= 2x \\
\frac{4}{2} &= \frac{\cancel{2}x}{\cancel{2}} \\
2 &= x \\
x &= 2
\end{aligned}
$$
The solution is 2.

14. −7

15.
$$
\begin{aligned}
7x - 8 &= 8 - 9x \\
+9x &\quad +9x \\
\hline
16x - 8 &= 8 \\
+8 &\quad +8 \\
\hline
16x &= 16 \\
\frac{\cancel{16}x}{\cancel{16}} &= \frac{16}{16} \\
x &= 1
\end{aligned}
$$
The solution is 1.

16. 1

17.
$$
\begin{aligned}
3x - 7 - x &= 15 - 2x - 6 \\
2x - 7 &= 9 - 2x \\
+2x &\quad +2x \\
\hline
4x - 7 &= 9 \\
+7 &\quad +7 \\
\hline
4x &= 16 \\
\frac{\cancel{4}x}{\cancel{4}} &= \frac{16}{4} \\
x &= 4
\end{aligned}
$$
The solution is 4.

18. −3

19.
$$
\begin{aligned}
8x - 13 + 3x &= 12 + 5x - 7 \\
11x - 13 &= 5 + 5x \\
-5x &\quad -5x \\
\hline
6x - 13 &= 5 \\
+13 &\quad +13 \\
\hline
6x &= 18 \\
\frac{\cancel{6}x}{\cancel{6}} &= \frac{18}{6} \\
x &= 3
\end{aligned}
$$
The solution is 3.

20. 2

21.
$$
\begin{aligned}
7 - 9x - 12 &= 3x + 5 - 8x \\
-9x - 5 &= -5x + 5 \\
+9x &\quad +9x \\
\hline
-5 &= 4x + 5 \\
-5 &\quad -5 \\
\hline
-10 &= 4x \\
\frac{-10}{4} &= \frac{\cancel{4}x}{\cancel{4}} \\
-\tfrac{5}{2} &= x \\
x &= -2\tfrac{1}{2}
\end{aligned}
$$
The solution is $-2\tfrac{1}{2}$.

22. $-1\tfrac{1}{3}$

23.
$$
\begin{aligned}
7.84 - 1.15x &= 2.45 \\
-7.84 &= -7.84 \\
\hline
-1.15x &= -5.39 \\
\frac{\cancel{-1.15}x}{\cancel{-1.15}} &= \frac{-5.39}{-1.15} \\
x &\approx 4.69
\end{aligned}
$$
The solution is about 4.69.

24. ≈ 0.662

Exercises 3.4B (page 136)

(The checks will not be shown.)

1. $5x - 3(2 + 3x) = 6$

$5x - 6 - 9x = 6$

$-4x = 12$

$\dfrac{\cancel{-4}x}{\cancel{-4}} = \dfrac{12}{-4}$ The solution is -3.

$x = -3$

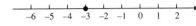

2. -18

3. $6x + 2(3 - 8x) = -14$

$6x + 6 - 16x = -14$

$-10x = -20$

$\dfrac{\cancel{-10}x}{\cancel{-10}} = \dfrac{-20}{-10}$ The solution is 2.

$x = 2$

4. 2

5. $7x + 5 = 3(3x + 5)$

$7x + 5 = 9x + 15$

$\underline{-7x - 15 \quad -7x - 15}$

$-10 = 2x$

$\dfrac{-10}{2} = \dfrac{\cancel{2}x}{\cancel{2}}$ The solution is -5.

$x = -5$

6. -2

7. $9 - 4x = 5(9 - 8x)$

$9 - 4x = 45 - 40x$

$\underline{-9 + 40x \quad -9 + 40x}$

$36x = 36$

$\dfrac{\cancel{36}x}{\cancel{36}} = \dfrac{36}{36}$ The solution is 1.

$x = 1$

8. 2

9. $3y - 2(2y - 7) = 2(3 + y) - 4$

$3y - 4y + 14 = 6 + 2y - 4$

$-y + 14 = 2 + 2y$

$\underline{-2y - 14 \quad -14 - 2y}$

$-3y = -12$

$\dfrac{\cancel{-3}y}{\cancel{-3}} = \dfrac{-12}{-3}$ The solution is 4.

$y = 4$

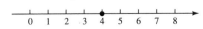

10. 1

11. $6(3 - 4x) + 12 = 10x - 2(5 - 3x)$

$18 - 24x + 12 = 10x - 10 + 6x$

$30 - 24x = 16x - 10$

$\underline{+10 + 24x \quad +24x + 10}$

$40 = 40x$

$\dfrac{40}{40} = \dfrac{\cancel{40}x}{\cancel{40}}$ The solution is 1.

$x = 1$

12. 1

13. $2(3x - 6) - 3(5x + 4) = 5(7x - 8)$

$6x - 12 - 15x - 12 = 35x - 40$

$-9x - 24 = 35x - 40$

$\underline{+9x + 40 \quad +9x + 40}$

$16 = 44x$

$\dfrac{\cancel{16}}{\cancel{44}} = \dfrac{\cancel{44}x}{\cancel{44}}$

$x = \dfrac{4}{11}$ The solution is $\dfrac{4}{11}$.

14. $\dfrac{1}{18}$

15. $6(5 - 4h) = 3(4h - 2) - 7(6 + 8h)$

$30 - 24h = 12h - 6 - 42 - 56h$

$30 - 24h = -48 - 44h$

$\underline{-30 + 44h \quad -30 + 44h}$

$20h = -78$

$\dfrac{\cancel{20}h}{\cancel{20}} = \dfrac{-78}{20}$ The solution is $-3\dfrac{9}{10}$.

$h = -3\dfrac{9}{10}$

16. $8\dfrac{1}{2}$

17. $2[3 - 5(x - 4)] = 10 - 5x$

$2[3 - 5x + 20] = 10 - 5x$

$2[23 - 5x] = 10 - 5x$

$46 - 10x = 10 - 5x$

$\underline{-10 + 10x \quad -10 + 10x}$

$36 = 5x$

$\dfrac{36}{5} = \dfrac{\cancel{5}x}{\cancel{5}}$ The solution is $7\dfrac{1}{5}$.

$x = 7\dfrac{1}{5}$

18. 16

19. $3[2h - 6] = 2\{2(3 - h) - 5\}$

$6h - 18 = 2\{6 - 2h - 5\}$

$6h - 18 = 2\{1 - 2h\}$

$6h - 18 = \quad 2 - 4h$

$\underline{+4h + 18 \quad +18 + 4h}$

$10h \quad = \quad 20$

$\dfrac{\overset{1}{\cancel{10}h}}{\cancel{10}_{1}} = \dfrac{20}{10}$ \qquad The solution is 2.

$h = 2$

(number line: points marked 0 1 2 3 4 5 6 7 8, dot at 2)

20. $\frac{7}{10}$

(number line: −4 −3 −2 −1 0 1 2 3 4, dot between 0 and 1)

21. $5(3 - 2x) - 10 = 4x + [-(2x - 5) + 15]$

$15 - 10x - 10 = 4x + [-2x + 5 + 15]$

$15 - 10x - 10 = 4x - 2x + 20$

$5 - 10x = \quad 2x + 20$

$\underline{-20 + 10x \quad +10x - 20}$

$-15 \quad = \quad 12x$

$-\dfrac{15}{12} = \dfrac{\overset{1}{\cancel{12}x}}{\cancel{12}_{1}}$

$x = -1\frac{1}{4}$ \qquad The solution is $-1\frac{1}{4}$.

(number line: −4 −3 −2 −1 0 1 2 3 4, dot between −2 and −1)

22. -1

(number line: −4 −3 −2 −1 0 1 2 3 4, dot at −1)

23. $9 - 3(2x - 7) - 9x = 5x - 2[6x - (4 - x) - 20]$

$9 - 6x + 21 - 9x = 5x - 2[6x - 4 + x - 20]$

$9 - 6x + 21 - 9x = 5x - 2[7x - 24]$

$9 - 6x + 21 - 9x = 5x - 14x + 48$

$30 - 15x = -9x + 48$

$\underline{-30 + 9x \quad +9x - 30}$

$-6x = \quad 18$

$\dfrac{\overset{1}{\cancel{-6}x}}{\cancel{-6}_{1}} = \dfrac{18}{-6}$

$x = -3$ \qquad The solution is -3.

(number line: −6 −5 −4 −3 −2 −1 0 1 2, dot at −3)

24. 21

(number line: 18 19 20 21 22 23 24 25 26, dot at 21)

25. $-2\{5 - [6 - 3(4 - x)] - 2x\} = 13 - [-(2x - 1)]$

$-2\{5 - [6 - 12 + 3x] - 2x\} = 13 - [-2x + 1]$

$-2\{5 - 6 + 12 - 3x - 2x\} = 13 + 2x - 1$

$-2\{11 - 5x\} = 12 + 2x$

$-22 + 10x = \quad 12 + 2x$

$\underline{+22 - 2x \quad +22 - 2x}$

$8x = \quad 34$

$\dfrac{\overset{1}{\cancel{8}x}}{\cancel{8}_{1}} = \dfrac{34}{8}$

$x = 4\frac{1}{4}$ \qquad The solution is $4\frac{1}{4}$.

(number line: 0 1 2 3 4 5 6 7 8, dot between 4 and 5)

26. 10

(number line: 7 8 9 10 11 12 13 14 15, dot at 10)

27. $5.073x - 2.937(8.622 + 7.153x) = 6.208$

$5.073x - 25.322814 - 21.008361x = 6.208$

$-25.322814 - 15.935361x = \quad 6.208$

$\underline{+25.322814 \quad\quad\quad\quad +25.322814}$

$-15.935361x = 31.530814$

$\dfrac{\overset{1}{\cancel{-15.935361}x}}{\cancel{-15.935361}_{1}} = \dfrac{31.530814}{-15.935361}$

$x \approx -1.979$

The solution is about -1.979.

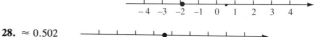

(number line: −4 −3 −2 −1 0 1 2 3 4, dot between −2 and −1)

28. ≈ 0.502

(number line: −4 −3 −2 −1 0 1 2 3 4, dot between 0 and 1)

29. $8.23x - 4.07(6.75x - 5.59) = 3.84(9.18 - x) - 2.67$

$8.23x - 27.4725x + 22.7513 = 35.2512 - 3.84x - 2.67$

$-19.2425x + 22.7513 = \quad 32.5812 - 3.84x$

$\underline{+3.84 \; x - 22.7513 \quad -22.7513 + 3.84x}$

$-15.4025x \quad = \quad 9.8299$

$\dfrac{\overset{1}{\cancel{-15.4025}x}}{\cancel{-15.4025}_{1}} = \dfrac{9.8299}{-15.4025}$

$x \approx -0.63820159$

$x \approx -0.638$

The solution is about -0.638.

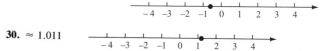

(number line: −4 −3 −2 −1 0 1 2 3 4, dot between −1 and 0)

30. ≈ 1.011

(number line: −4 −3 −2 −1 0 1 2 3 4, dot at 1)

Exercises 3.5 (page 140)

1. $x + 3 = \quad 8$

$\underline{\quad - 3 \quad -3}$

$x \quad = \quad 5$

Conditional; the solution is 5.

2. -2; Conditional

3. $2x + 5 = 7 + 2x$

$\underline{-2x \quad\quad - 2x}$

$5 = 7$ \qquad False

No solution

4. No solution

5. $6 + 4x = \quad 4x + 6$

$\underline{\quad - 4x \quad -4x}$

$6 \quad = \quad 6$ \quad True

Identity

6. Identity

7. $5x - 2(4 - x) = 6$

$5x - 8 + 2x = 6$

$7x - 8 = \quad 6$

$\underline{\quad + 8 \quad +8}$

$7x \quad = \quad 14$

$\dfrac{\overset{1}{\cancel{7}x}}{\cancel{7}_{1}} = \dfrac{14}{7}$

$x = 2$

Conditional; the solution is 2.

8. 2; Conditional

9. $6x - 3(5 + 2x) = -15$
$6x - 15 - 6x = -15$
$-15 = -15$ True
Identity

10. Identity

11. $4x - 2(6 + 2x) = -15$
$4x - 12 - 4x = -15$
$-12 = -15$ False
No solution

12. No solution

13. $7(2 - 5x) - 32 = 10x - 3(6 + 15x)$
$14 - 35x - 32 = 10x - 18 - 45x$
$-35x - 18 = -35x - 18$
$\underline{+35x \qquad\quad +35x}$
$-18 = \qquad -18$ True
Identity

14. $\frac{34}{41}$; Conditional

15. $2(2x - 5) - 3(4 - x) = 7x - 20$
$4x - 10 - 12 + 3x = 7x - 20$
$7x - 22 = 7x - 20$
$\underline{-7x \qquad\quad -7x}$
$-22 = \qquad -20$ False
No solution

16. Identity

17. $2[3 - 4(5 - x)] = 2(3x - 11)$
$2[3 - 20 + 4x] = 6x - 22$
$6 - 40 + 8x = 6x - 22$
$-34 + 8x = 6x - 22$
$\underline{+34 - 6x \quad -6x + 34}$
$2x = 12$
$\dfrac{\overset{1}{\cancel{2}}x}{\underset{1}{\cancel{2}}} = \dfrac{12}{2}$
$x = 6$
Conditional; the solution is 6.

18. No solution

19. $460.2x - 23.6(19.5x - 51.4) = 1213.04$
$460.2x - 460.2x + 1213.04 = 1213.04$
$1213.04 = 1213.04$ True
Identity

20. ≈ 2.415; Conditional

Exercises 3.6 (page 147)

1. $x - 5 < 2$
$\underline{+ 5 \quad +5}$
$x < 7$

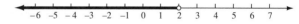

2. $x < 11$

3. $5x + 4 \le 19$
$\underline{-4 \quad -4}$
$5x \le 15$
$\dfrac{\overset{1}{\cancel{5}}x}{\underset{1}{\cancel{5}}} \le \dfrac{15}{5}$
$x \le 3$

4. $x \le 3$

5. $6x + 7 > 3 + 8x$
$\underline{-6x - 3 \quad -3 - 6x}$
$4 > 2x$
$\dfrac{4}{2} > \dfrac{\overset{1}{\cancel{2}}x}{\underset{1}{\cancel{2}}}$
$2 > x$
$x < 2$

6. $x > -7$

7. $2x - 9 > 3(x - 2)$
$2x - 9 > 3x - 6$
$\underline{-3x + 9 \quad -3x + 9}$
$-x > 3$
$\dfrac{-x}{-1} < \dfrac{3}{-1}$
$x < -3$

8. $x < -3$

9. $6(3 - 4x) + 12 \ge 10x - 2(5 - 3x)$
$18 - 24x + 12 \ge 10x - 10 + 6x$
$30 - 24x \ge 16x - 10$
$\underline{10 + 24x \quad 24x + 10}$
$40 \ge 40x$
$\dfrac{40}{40} \ge \dfrac{\overset{1}{\cancel{40}}x}{\underset{1}{\cancel{40}}}$
$x \le 1$

10. $x \le 1$

11. $4(6 - 2x) \ne 5x - 2$
$24 - 8x \ne 5x - 2$
$\underline{2 + 8x \quad 8x + 2}$
$26 \ne 13x$
$\dfrac{26}{13} \ne \dfrac{\overset{1}{\cancel{13}}x}{\underset{1}{\cancel{13}}}$
$x \ne 2$

12. $x \ne 4$

13.
$$2[3 - 5(x - 4)] < 10 - 5x$$
$$2[3 - 5x + 20] < 10 - 5x$$
$$2[23 - 5x] < 10 - 5x$$
$$46 - 10x < 10 - 5x$$
$$\underline{-10 + 10x \qquad -10 + 10x}$$
$$36 < 5x$$
$$\frac{36}{5} < \frac{\cancel{5}^{1}x}{\cancel{5}_{1}}$$
$$7\tfrac{1}{5} < x$$
$$x > 7\tfrac{1}{5}$$

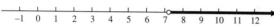

14. $x > 16$

15.
$$7(x - 5) - 4x > x - 8$$
$$7x - 35 - 4x > x - 8$$
$$3x - 35 > x - 8$$
$$\underline{-x + 35 \qquad -x + 35}$$
$$2x > 27$$
$$\frac{\cancel{2}^{1}x}{\cancel{2}_{1}} > \frac{27}{2}$$
$$x > \tfrac{27}{2} \text{ or } 13\tfrac{1}{2}$$

16. $x \neq 1$

17.
$$3x - 5(x + 2) \leq 4x + 8$$
$$3x - 5x - 10 \leq 4x + 8$$
$$-2x - 10 \leq 4x + 8$$
$$\underline{+2x - 8 \qquad 2x - 8}$$
$$-18 \leq 6x$$
$$\frac{-18}{6} \leq \frac{\cancel{6}^{1}x}{\cancel{6}_{1}}$$
$$x \geq -3$$

18. $x < 1$

19.
$$12.85x - 15.49 \geq 22.06(9.66x - 12.74)$$
$$12.85x - 15.49 \geq 213.0996x - 281.0444$$
$$\underline{-12.85x + 281.0444 \qquad -12.85x + 281.0444}$$
$$265.5544 \geq 200.2496x$$
$$\frac{265.5544}{200.2496} \geq \frac{\cancel{200.2496}^{1}x}{\cancel{200.2496}_{1}}$$
$$x \leq 1.326 \quad \text{(approx.)}$$

20. $x > -8.410 \quad \text{(approx.)}$

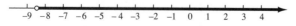

Review Exercises 3.7 (page 150)

The checks will not be shown.

1.
$$3x - 5 = 4$$
$$\underline{+ 5 \qquad +5}$$
$$3x = 9$$
$$\frac{\cancel{3}^{1}x}{\cancel{3}_{1}} = \frac{9}{3} \qquad \text{The solution is 3.}$$
$$x = 3$$

2. 2

3.
$$2 = 20 - 9x$$
$$\underline{-20 \qquad -20}$$
$$-18 = -9x$$
$$\frac{-18}{-9} = \frac{\cancel{-9}^{1}x}{\cancel{-9}_{1}}$$
$$2 = x \qquad \text{The solution is 2.}$$
$$x = 2$$

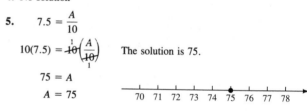

4. No solution

5.
$$7.5 = \frac{A}{10}$$
$$10(7.5) = \cancel{10}^{1}\left(\frac{A}{\cancel{10}_{1}}\right) \qquad \text{The solution is 75.}$$
$$75 = A$$
$$A = 75$$

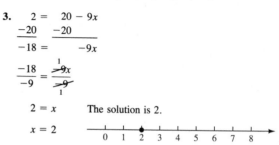

6. 180

7.
$$7 - 2(M - 4) = 5$$
$$7 - 2M + 8 = 5$$
$$-2M + 15 = 5$$
$$\underline{- 15 \qquad -15}$$
$$-2M = -10$$
$$\frac{\cancel{-2}^{1}M}{\cancel{-2}_{1}} = \frac{-10}{-2} \qquad \text{The solution is 5.}$$
$$M = 5$$

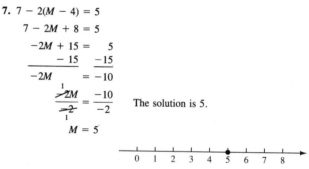

8. Identity

9.
$$6R - 8 = 6(2 - 3R)$$
$$6R - 8 = 12 - 18R$$
$$\underline{+18R + 8 \qquad 8 + 18R}$$
$$24R = 20$$
$$\frac{\cancel{24}^{1}R}{\cancel{24}_{1}} = \frac{20}{24} \qquad \text{The solution is } \tfrac{5}{6}.$$
$$R = \tfrac{5}{6}$$

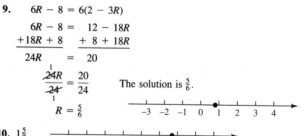

10. $1\tfrac{5}{7}$

11. $56T - 18 = 7(8T - 4)$

$56T - 18 = 56T - 28$

$\underline{-56T \qquad -56T}$

$-18 = \qquad -28$ False

No solution

12. No solution

13. $15(4 - 5V) = 16(4 - 6V) + 10$

$60 - 75V = 64 - 96V + 10$

$60 - 75V = 74 - 96V$

$\underline{-60 + 96V \qquad -60 + 96V}$

$21V = 14$

$\dfrac{\overset{1}{\cancel{21}}V}{\underset{1}{\cancel{21}}} = \dfrac{14}{21}$

The solution is $\frac{2}{3}$.

$V = \frac{2}{3}$

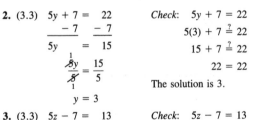

number line from −4 to 4, point at 1

14. Identity

15. $5x - 7(4 - 2x) + 8 = 10 - 9(11 - x)$

$5x - 28 + 14x + 8 = 10 - 99 + 9x$

$19x - 20 = -89 + 9x$

$\underline{-9x + 20 \qquad +20 - 9x}$

$10x \qquad = -69$

$\dfrac{\overset{1}{\cancel{10}}x}{\underset{1}{\cancel{10}}} = \dfrac{-69}{10}$

$x = -6.9$ The solution is -6.9.

number line from −10 to −2, point at −7

16. -8.3 number line from −10 to −2, point at −8

17. $2[-7y - 3(5 - 4y) + 10] = 10y - 12$

$2[-7y - 15 + 12y + 10] = 10y - 12$

$2[5y - 5] = 10y - 12$

$10y - 10 = 10y - 12$

$\underline{-10y \qquad -10y}$

$-10 = \qquad -12$ False

No solution

18. No solution

19. $4[-24 - 6(3x - 5) + 22x] = 0$

$4[-24 - 18x + 30 + 22x] = 0$

$4[6 + 4x] = 0$

$24 + 16x = \quad 0$

$\underline{-24 \qquad -24}$

$16x = -24$

$\dfrac{\overset{1}{\cancel{16}}x}{\underset{1}{\cancel{16}}} = \dfrac{-24}{16}$

$x = -1\frac{1}{2}$

The solution is $-1\frac{1}{2}$.   number line from −4 to 4, point at −1½

20. 1   number line from −4 to 4, point at 1

21. $x + 7 > \quad 2$

$\underline{-7 \qquad -7}$

$x \qquad > -5$

number line from −6 to 6, open circle at −5

22. $x > -4$

number line from −6 to 6, open circle at −4

23. $x - 3 \le -8$

$\underline{+3 \qquad +3}$

$x \qquad \le -5$

number line from −8 to 3, point at −5

24. $x \le -4$

number line from −6 to 1, point at −4

25. $2(x - 4) - 5 \ge 7 + 3(2x - 1)$

$2x - 8 - 5 \ge 7 + 6x - 3$

$2x - 13 \ge \quad 4 + 6x$

$\underline{-6x + 13 \qquad +13 - 6x}$

$-4x \qquad \ge \quad 17$

$\dfrac{\overset{1}{\cancel{-4}}x}{\underset{1}{\cancel{-4}}} \le \dfrac{17}{-4}$

$x \le -4\frac{1}{4}$

number line from −8 to 3, point at −4¼

26. $x \le \frac{1}{3}$

number line from −5 to 5, point at ⅓

Chapter 3 Diagnostic Test (page 153)

Following each problem number is the textbook section number (in parentheses) where that kind of problem is discussed.

1. (3.2) $x - 5 = \quad 3$

$\underline{+5 \qquad +5}$

$x \quad = \quad 8$

The solution is 8.

Check: $x - 5 = 3$

$8 - 5 \overset{?}{=} 3$

$3 = 3$

number line from 5 to 13, point at 8

2. (3.3) $5y + 7 = \quad 22$

$\underline{-7 \qquad -7}$

$5y \quad = \quad 15$

$\dfrac{\overset{1}{\cancel{5}}y}{\underset{1}{\cancel{5}}} = \dfrac{15}{5}$

$y = 3$

Check: $5y + 7 = 22$

$5(3) + 7 \overset{?}{=} 22$

$15 + 7 \overset{?}{=} 22$

$22 = 22$

The solution is 3.

3. (3.3) $5z - 7 = \quad 13$

$\underline{+7 \qquad +7}$

$5z \quad = \quad 20$

$\dfrac{\overset{1}{\cancel{5}}z}{\underset{1}{\cancel{5}}} = \dfrac{20}{5}$

$z = 4$

Check: $5z - 7 = 13$

$5(4) - 7 \overset{?}{=} 13$

$20 - 7 \overset{?}{=} 13$

$13 = 13$

The solution is 4.

4. (3.4) $14 + 3x = \quad 7 - 4x$

$\underline{-14 + 4x \qquad -14 + 4x}$

$7x = \quad -7$

$\dfrac{\overset{1}{\cancel{7}}x}{\underset{1}{\cancel{7}}} = \dfrac{-7}{7}$

$x = -1$

Check:

$14 + 3x = 7 - 4x$

$14 + 3(-1) \overset{?}{=} 7 - 4(-1)$

$14 - 3 \overset{?}{=} 7 + 4$

$11 = 11$

The solution is -1.

5. (3.5)
$$7x - 4 = 3x + 4(x - 1)$$
$$7x - 4 = 3x + 4x - 4$$
$$7x - 4 = 7x - 4$$
$$\underline{-7x \qquad -7x}$$
$$-4 = -4 \quad \textit{True}$$

Identity (The variable dropped out and we were left with a true statement.)

6. (3.3)
$$8 = 4y - 1$$
$$\underline{+1 \qquad +1}$$
$$9 = 4y$$
$$\frac{9}{4} = \frac{\cancel{4}y}{\cancel{4}}$$
$$y = \frac{9}{4}$$

Check: $8 = 4y - 1$
$$8 \stackrel{?}{=} \cancel{4}\left(\frac{9}{\cancel{4}}\right) - 1$$
$$8 \stackrel{?}{=} 9 - 1$$
$$8 = 8$$
The solution is $\frac{9}{4}$.

7. (3.5)
$$3x - 4 = x + 2(x - 1)$$
$$3x - 4 = x + 2x - 2$$
$$3x - 4 = 3x - 2$$
$$\underline{-3x \qquad -3x}$$
$$-4 = -2 \quad \textit{False}$$

No solution (The variable dropped out and we were left with a false statement.)

8. (3.3)
$$17 - 3z = -1$$
$$\underline{-17 \qquad -17}$$
$$-3z = -18$$
$$\frac{\cancel{-3}z}{\cancel{-3}} = \frac{-18}{-3}$$
$$z = 6$$

Check: $17 - 3z = -1$
$$17 - 3(6) \stackrel{?}{=} -1$$
$$17 - 18 \stackrel{?}{=} -1$$
$$-1 = -1$$
The solution is 6.

9. (3.3)
$$\frac{x}{6} = 5.1$$
$$\cancel{6}\left(\frac{x}{\cancel{6}}\right) = (5.1)(6)$$
$$x = 30.6$$

Check: $\frac{x}{6} = 5.1$
$$\frac{30.6}{6} \stackrel{?}{=} 5.1$$
$$5.1 = 5.1$$
The solution is 30.6.

10. (3.3)
$$-6 = \frac{w}{7}$$
$$7(-6) = \left(\frac{w}{\cancel{7}}\right)\cancel{7}$$
$$w = -42$$

Check: $-6 = \frac{w}{7}$
$$-6 \stackrel{?}{=} \frac{-42}{7}$$
$$-6 = -6$$
The solution is -42.

11. (3.3)
$$\frac{x}{4} - 5 = 3$$
$$\underline{+5 \qquad +5}$$
$$\frac{x}{4} = 8$$
$$\cancel{4}\left(\frac{x}{\cancel{4}}\right) = (8)(4)$$
$$x = 32$$

Check: $\frac{x}{4} - 5 = 3$
$$\frac{32}{4} - 5 \stackrel{?}{=} 3$$
$$8 - 5 \stackrel{?}{=} 3$$
$$3 = 3$$
The solution is 32.

12. (3.4)
$$6x + 1 = 17 - 2x$$
$$\underline{+2x - 1 \qquad -1 + 2x}$$
$$8x = 16$$
$$\frac{\cancel{8}x}{\cancel{8}} = \frac{16}{8}$$
$$x = 2$$

Check: $6x + 1 = 17 - 2x$
$$6(2) + 1 \stackrel{?}{=} 17 - 2(2)$$
$$12 + 1 \stackrel{?}{=} 17 - 4$$
$$13 = 13$$
The solution is 2.

13. (3.4)
$$3z - 21 + 5z = 4 - 6z + 17$$
$$8z - 21 = 21 - 6z$$
$$\underline{+6z + 21 \qquad +21 + 6z}$$
$$14z = 42$$
$$\frac{\cancel{14}z}{\cancel{14}} = \frac{42}{14}$$
$$z = 3$$

Check: $3z - 21 + 5z = 4 - 6z + 17$
$$3(3) - 21 + 5(3) \stackrel{?}{=} 4 - 6(3) + 17$$
$$9 - 21 + 15 \stackrel{?}{=} 4 - 18 + 17$$
$$3 = 3 \qquad \text{The solution is 3.}$$

14. (3.4)
$$5k - 9(7 - 2k) = 6$$
$$5k - 63 + 18k = 6$$
$$23k - 63 = 6$$
$$\underline{+ 63 \qquad +63}$$
$$23k = 69$$
$$\frac{\cancel{23}k}{\cancel{23}} = \frac{69}{23}$$
$$k = 3$$

Check: $5k - 9(7 - 2k) = 6$
$$5(3) - 9(7 - 2[3]) \stackrel{?}{=} 6$$
$$15 - 9(7 - 6) \stackrel{?}{=} 6$$
$$15 - 9(1) \stackrel{?}{=} 6$$
$$15 - 9 \stackrel{?}{=} 6$$
$$6 = 6$$
The solution is 3.

15. (3.5)
$$10x - 2(5x - 7) = 14$$
$$10x - 10x + 14 = 14$$
$$14 = 14 \quad \textit{True}$$

Identity (The variable dropped out and we were left with a true statement.)

16. (3.4)
$$2y - 4(3y - 2) = 5(6 + y) - 7$$
$$2y - 12y + 8 = 30 + 5y - 7$$
$$-10y + 8 = 23 + 5y$$
$$\underline{+10y - 23 \qquad -23 + 10y}$$
$$-15 = 15y$$
$$\frac{-15}{15} = \frac{\cancel{15}y}{\cancel{15}}$$
$$y = -1$$

Check: $2y - 4(3y - 2) = 5(6 + y) - 7$
$$2(-1) - 4(3[-1] - 2) \stackrel{?}{=} 5(6 + [-1]) - 7$$
$$-2 - 4(-3 - 2) \stackrel{?}{=} 5(5) - 7$$
$$-2 - 4(-5) \stackrel{?}{=} 25 - 7$$
$$-2 + 20 \stackrel{?}{=} 25 - 7$$
$$18 = 18 \qquad \text{The solution is } -1.$$

17. (3.5)
$$7(3z + 4) = 14 + 3(7z - 1)$$
$$21z + 28 = 14 + 21z - 3$$
$$21z + 28 = 11 + 21z$$
$$\underline{-21z - 28 \qquad -28 - 21z}$$
$$0 = -17 \quad \textit{False}$$

No solution (The variable dropped out and we were left with a false statement.)

18. (3.4) $3[7 - 6(x - 2)] = -3 + 2x$

$$3[7 - 6x + 12] = -3 + 2x$$

$$3[19 - 6x] = -3 + 2x$$

$$57 - 18x = -3 + 2x$$

$$\underline{+3 + 18x \quad\quad +3 + 18x}$$

$$60 \quad = \quad 20x$$

$$\frac{60}{20} = \frac{\overset{1}{\cancel{20}}x}{\underset{1}{\cancel{20}}}$$

$$x = 3$$

Check: $3[7 - 6(x - 2)] = -3 + 2x$

$$3[7 - 6(3 - 2)] \overset{?}{=} -3 + 2[3]$$

$$3[7 - 6(1)] \overset{?}{=} -3 + 6$$

$$3[7 - 6] \overset{?}{=} 3$$

$$3[1] \overset{?}{=} 3$$

$$3 = 3 \quad\quad \text{The solution is 3.}$$

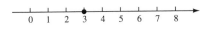

19. (3.6) $4x + 5 > -3$

$$\underline{\quad - 5 \quad\quad -5}$$

$$4x \quad\quad > -8$$

$$\frac{\overset{1}{\cancel{4}}x}{\underset{1}{\cancel{4}}} > \frac{-8}{4}$$

$$x > -2$$

20. (3.6) $5x - 2 \le 10 - x$

$$5x - 2 \le 10 - x$$

$$\underline{+ x + 2 \quad\quad + 2 + x}$$

$$6x \quad \le \quad 12$$

$$\frac{\overset{1}{\cancel{6}}x}{\underset{1}{\cancel{6}}} \le \frac{12}{6}$$

$$x \le 2$$

Cumulative Review Exercises: Chapters 1–3 (page 154)

1. $\dfrac{-7}{0}$ Not defined   **2.** 10   **3.** $7\sqrt{16} - 5(-4)$

$$= 7(4) - 5(-4)$$

$$= 28 + 20 = 48$$

4. 37   **5.** $V = \frac{4}{3}\pi r^3$ **6.** 3,630

$$V \approx \frac{\overset{}{4}}{\underset{1}{\cancel{3}}}(3.14)(\cancel{9})(9)(9)$$

$$V \approx 4(3.14)(3)(9)(9)$$

$$V \approx 3052.08$$

7. $15x - [9y - 1(7x - 10y)]$

$$= 15x - 1[9y - 7x + 10y]$$

$$= \underline{15x} - \underline{9y} + \underline{7x} - \underline{10y}$$

$$= 22x - 19y$$

8. $-14a - 57b$

9. $8 - 3[2x - 1(1 - 4x)] = 8 - 3[2x - 1 + 4x] = 8 - 3[6x - 1]$

$$= 8 - 18x + 3 = 11 - 18x$$

10. $-13x + 88$

11. $2x + 1 = \quad 2x + 7$ **12.** $x = 3$

$$\underline{-2x \quad\quad -2x}$$

$$1 = \quad\quad 7 \quad \text{False}$$

No solution

13. $10 - 4(2 - 3x) = 2 + 12x$ **14.** $N = 4$

$$10 - 8 + 12x = 2 + 12x$$

$$2 + 12x = 2 + 12x$$

$$\underline{\quad - 12x \quad\quad - 12x}$$

$$2 \quad = 2 \quad\quad \text{True}$$

Identity

15. $\dfrac{C}{7} - 15 = \quad 13$

$$\underline{\quad + 15 \quad\quad +15}$$

$$\frac{C}{7} \quad = \quad 28$$

$$\overset{1}{\cancel{7}}\left(\frac{C}{\underset{1}{\cancel{7}}}\right) = 28(7)$$

$$C = 196 \quad\quad \text{The solution is 196.}$$

16. $W = \frac{5}{3}$ or $1\frac{2}{3}$

17. $9 - 3(x - 2) = 3(5 - x)$

$$9 - 3x + 6 = 15 - 3x$$

$$15 - 3x = 15 - 3x$$

$$\underline{\quad + 3x \quad\quad + 3x}$$

$$15 \quad = 15 \quad\quad \text{True}$$

Identity

18. $x < 4$

19. $3 - x \ge \quad 4$

$$\underline{-4 + x \quad\quad -4 + x}$$

$$-1 \quad \ge \quad x$$

$$x \le -1$$

20. $x \ge -2$ **21.** True   **22.** False   **23.** False

24. True   **25.** False

Exercises 4.1A (page 159)

1. $x + 10$ **2.** $A + B$ **3.** $A - 5$ **4.** $B - C$ **5.** $6z$

6. AB **7.** $x - 7$ **8.** $9 + A$ **9.** $3x - 4$ **10.** $6 + 2x$

11. $x - uv$ **12.** $PQ - x$ **13.** $5x^2$ **14.** $10x^3$

15. $(A + B)^2$ **16.** $\left(\dfrac{A}{B}\right)^2$ **17.** $\dfrac{x + 7}{y}$ **18.** $\dfrac{T}{x + 9}$

19. $x(y - 6)$ **20.** $A(3 + B)$

Exercises 4.1B (page 164)

1. Let S represent Fred's salary. Then $S + 75$ is the algebraic expression for "Fred's salary plus seventy-five dollars."

2. $S - 42$, where S is Jaime's salary

3. Let N represent the number of children. Then $N - 2$ is the algebraic expression for "two less than the number of children in Mr. Moore's family."

4. $N + 2$, where $N =$ the number of players on Jerry's team

5. Let x represent Joyce's age. Then $4x$ is the algebraic expression for "four times Joyce's age."

6. $\dfrac{x}{4}$ or $\dfrac{1}{4}x$, where x is Rene's age

7. Let C represent the cost of a record (in cents). Then $20C + 89$ is the algebraic expression for "twenty times the cost of a record, increased by eighty-nine cents."

8. $5C - 17$, where C is the cost of the pen (in cents)

9. Let C represent the cost of a hamburger. Then $\dfrac{1}{5}C$ or $\dfrac{C}{5}$ is the algebraic expression for "one-fifth the cost of a hamburger."

10. $\dfrac{\ell}{8}$, where ℓ is the length of the building

11. Let x = speed of car. Then $5x + 100$ is the algebraic expression for "five times the speed of the car plus one hundred miles per hour."

12. $2x - 40$, where x is the speed of the car

13. Let x = the unknown number. Then x^2 = the square of the unknown, and $5x^2 - 10$ is the algebraic expression that represents "ten less than five times the square of the unknown."

14. $4x^3 + 8$, where x is the unknown number

15. *First method:* Let w = width
$w + 12$ = length

Second method: Let ℓ = length
$\ell - 12$ = width

16. *First method:* Let b = base
$b - 7$ = altitude

Second method: Let a = altitude
$a + 7$ = base

17. Let x = the unknown number. Then $50 - \dfrac{7}{x}$ represents the statement in the problem.

18. $15 + \dfrac{x}{9}$, where x is the unknown number

19. Let ℓ = length of a rectangle. Then $2\ell + 11$ represents the statement in the problem.

20. $d - 1$, where d is the diameter

21. If W = Walter's weight, then $320 - W$ = Carlos's weight.
If C = Carlos's weight, then $320 - C$ = Walter's weight.

22. *First method:* Let T = Teresa's weight
$224 - T$ = Lucy's weight

Second method: Let L = Lucy's weight
$224 - L$ = Teresa's weight

23. Let w = weight in kilograms. Then $2.2w$ represents the statement in the problem.

24. $0.62D$, where D is the distance in kilometers

25. Let x = the unknown number. Then $\dfrac{8 + x}{x^2}$ represents the statement in the problem.

26. $\dfrac{x^2 + 11}{x}$, where x is the unknown number

27. Let C = Celsius temperature. Then $32 + \dfrac{9}{5}C$ represents the statement in the problem.

28. $\dfrac{5}{9}(F - 32)$, where F is the Fahrenheit temperature

29. Let x = the smallest integer
$x + 1$ = the next integer
$x + 2$ = the largest integer
The sum is $x + (x + 1) + (x + 2)$.

30. $x + (x + 1)$, where x = the smaller integer and $x + 1$ = the larger integer

31. Let x = first odd integer
$x + 2$ = second odd integer
$x + 4$ = third odd integer
$x + 6$ = fourth odd integer
The sum is $x + (x + 2) + (x + 4) + (x + 6)$.

32. $x + (x + 2) + (x + 4)$, where x = first even integer, $x + 2$ = second even integer, and $x + 4$ = third even integer

33. Let r = radius
$6 + r$

34. $5 + r$, where r = radius

35. Let r = radius
$2r$ = height

36. $4h$ is radius, where h = height

37. Let h = height   or   Let r = radius
$h - 4$ = radius $r + 4$ = height
Volume = $\pi(h - 4)^2 h$ Volume = $\pi r^2(r + 4)$

38. Area = $\frac{1}{2}(h - 5)h$, where or Area = $\frac{1}{2}b(b + 5)$, where
h = altitude and b = base and
$h - 5$ = base $b + 5$ = altitude

39. Let r = radius
$4r$ = height
Volume = $\frac{1}{3}\pi r^2(4r)$

40. Volume = $\pi(6h)^2 h$, where or Volume = $\pi r^2\left(\dfrac{r}{6}\right)$, where
h = height and   radius = r
$6h$ = radius   height = $\dfrac{r}{6}$

Exercises 4.2 (page 168)

1. Let x = unknown number.
$13 + 2x = 25$

2. Let x = unknown number.
$25 + 3x = 34$

3. Let x = unknown number.
$5x - 8 = 22$

4. Let x = unknown number.
$4x - 5 = 15$

5. Let x = unknown number.
$7 - x = x + 1$

6. Let x = unknown number.
$6 + x = 12 - x$

7. Let x = unknown number.
$\frac{1}{5}x = 4$

8. Let x = unknown number.
$\dfrac{x}{12} = 6$

9. Let x = unknown number.
$\frac{1}{2}x - 4 = 6$

10. Let x = unknown number.
$\frac{1}{3}x - 5 = 4$

11. Let x = unknown number.
$2(5 + x) = 26$

12. Let x = unknown number.
$4(9 + x) = 18$

13. Let x = unknown number.
$(x + x)(3) = 24$

14. Let x = unknown number.
$5(x + x) = 40$

15. Let x = length of shorter piece of rope (in meters)
$x + 13$ = length of longer piece (in meters)
$x + (x + 13) = 75$

16. Let x = length of shorter piece of wire (in meters)
$x + 8$ = length of longer piece (in meters)
$x + (x + 8) = 46$

17. Let x = length of shorter piece of wire (in centimeters)
$2x$ = length of longer piece of wire (in centimeters)
$x + 2x = 72$

18. Let x = length of shorter piece of wire (in centimeters)
$3x$ = length of longer piece of wire (in centimeters)
$x + 3x = 72$

19. Let $\quad x$ = number of cans of pears
$\quad\quad x + 8$ = number of cans of peaches
$\quad x + (x + 8) = 42$

20. Let $\quad x$ = number of cans of corn
$\quad\quad x + 11$ = number of cans of peas
$\quad x + (x + 11) = 49$

21. Let $\quad x$ = first integer
$\quad\quad x + 1$ = second integer
$\quad\quad x + 2$ = third integer
$\quad\quad x + 3$ = fourth integer
$\quad x + (x + 1) + (x + 2) + (x + 3) = 106$

22. Let $\quad x$ = smallest integer
$\quad\quad x + 1$ = next integer
$\quad\quad x + 2$ = third integer
$\quad x + (x + 1) + (x + 2) = -72$

23. Let $\quad w$ = width (in cm) $\quad$ or $\quad$ Let $\quad \ell$ = length (in cm)
$\quad\quad w + 4$ = length (in cm) $\quad\quad\quad \ell - 4$ = width (in cm)
$\quad 2w + 2(w + 4) = 36 \quad\quad\quad 2\ell + 2(\ell - 4) = 36$

24. Let $\quad w$ = width (in feet) $\quad$ or $\quad$ Let $\quad \ell$ = length (in feet)
$\quad\quad w + 6$ = length (in feet) $\quad\quad\quad \ell - 6$ = width (in feet)
$\quad 2w + 2(w + 6) = 64 \quad\quad\quad 2\ell + 2(\ell - 6) = 64$

Exercises 4.3 (page 174)

(The checks will not be shown.)

1. Let x = unknown number. $\quad\quad$ **2.** 3

$$13 + 2x = 25$$
$$\underline{-13 \quad\quad -13}$$
$$2x = 12$$
$$\frac{\overset{1}{\cancel{2}}x}{\underset{1}{\cancel{2}}} = \frac{12}{2}$$
$$x = 6 \quad\quad \text{The number is 6.}$$

3. Let x = unknown number. $\quad\quad$ **4.** 5

$$5x - 8 = 22$$
$$\underline{+ 8 \quad\quad + 8}$$
$$5x = 30$$
$$\frac{\overset{1}{\cancel{5}}x}{\underset{1}{\cancel{5}}} = \frac{30}{5}$$
$$x = 6 \quad\quad \text{The number is 6.}$$

5. Let x = unknown number. $\quad\quad$ **6.** 3

$$7 - x = x + 1$$
$$\underline{-1 + x \quad\quad +x - 1}$$
$$6 = 2x$$
$$\frac{6}{2} = \frac{\overset{1}{\cancel{2}}x}{\underset{1}{\cancel{2}}}$$
$$x = 3 \quad\quad \text{The number is 3.}$$

7. Let x = unknown number. $\quad\quad$ **8.** 27

$$\tfrac{1}{2}x - 4 = 6$$
$$\underline{+ 4 \quad\quad + 4}$$
$$\tfrac{1}{2}x = 10$$
$$\overset{1}{\cancel{2}}\left(\frac{1}{\underset{1}{\cancel{2}}}x\right) = 2(10)$$
$$x = 20 \quad\quad \text{The number is 20.}$$

9. Let x = unknown number. $\quad\quad$ **10.** -4

$$2(5 + x) = 26$$
$$10 + 2x = 26$$
$$\underline{-10 \quad\quad -10}$$
$$2x = 16$$
$$\frac{\overset{1}{\cancel{2}}x}{\underset{1}{\cancel{2}}} = \frac{16}{2}$$
$$x = 8 \quad\quad \text{The number is 8.}$$

The checks will not be shown.

11. Let $\quad x$ = length of shorter piece of wire (in cm)
$\quad\quad x + 10$ = length of longer piece (in cm)

$$x + (x + 10) = 36$$
$$x + x + 10 = 36$$
$$2x + 10 = 36$$
$$\underline{- 10 \quad\quad -10}$$
$$2x = 26$$
$$\frac{\overset{1}{\cancel{2}}x}{\underset{1}{\cancel{2}}} = \frac{26}{2}$$
$$x = 13$$
$$x + 10 = 13 + 10 = 23$$

The shorter piece is 13 cm long and the longer one is 23 cm long.

12. 4 yd and 6 yd

13. Let $\quad x$ = number of packages of apricots
$\quad\quad x + 7$ = number of packages of apples

$$x + (x + 7) = 15$$
$$x + x + 7 = 15$$
$$2x + 7 = 15$$
$$\underline{- 7 \quad\quad - 7}$$
$$2x = 8$$
$$\frac{\overset{1}{\cancel{2}}x}{\underset{1}{\cancel{2}}} = \frac{8}{2}$$
$$x = 4$$
$$x + 7 = 4 + 7 = 11$$

Rebecca buys 4 packages of apricots and 11 packages of apples.

14. 5 skeins of green yarn; 8 skeins of blue yarn

15. $A = \tfrac{1}{2}bh$

$$A = \frac{1}{\underset{1}{\cancel{2}}}(\overset{13}{\cancel{26}} \text{ cm})(13 \text{ cm}) = 169 \text{ sq. cm}$$

16. $V = 30$ cu. m; $\quad S = 62$ sq. m

17. $V = \tfrac{4}{3}\pi r^3 \quad\quad\quad S = 4\pi r^2$

$$V \approx \tfrac{4}{3}(3.14)(6 \text{ yd})^3 \quad S \approx 4(3.14)(6 \text{ yd})^2$$
$$V \approx 904.32 \text{ cu. yd} \quad S \approx 452.16 \text{ sq. yd}$$

18. $A \approx 50.24$ sq. ft
$\quad\quad C \approx 25.12$ ft

19. Let x = first odd integer
$x + 2$ = second one
$x + 4$ = third one

$x + (x + 2) + (x + 4) = 177$
$x + x + 2 + x + 4 = 177$
$3x + 6 = 177$
$\underline{\quad -6 \quad -6}$
$3x = 171$

$\dfrac{\overset{1}{\cancel{3}}x}{\underset{1}{\cancel{3}}} = \dfrac{171}{3}$

$x = 57$, first odd integer

$x + 2 = 57 + 2 = 59$, second odd integer
$x + 4 = 57 + 4 = 61$, third odd integer

20. -50, -48, and -46

21. Let x = first integer **22.** 14, 15, and 16
$x + 1$ = second integer
$x + 2$ = third integer

$4[x + (x + 2)] = 140 + (x + 1)$
$4[x + x + 2] = 140 + x + 1$
$4[2x + 2] = 141 + x$
$8x + 8 = 141 + x$
$\underline{-x - 8 \quad -8 - x}$
$7x = 133$

$\dfrac{\overset{1}{\cancel{7}}x}{\underset{1}{\cancel{7}}} = \dfrac{133}{7}$

$x = 19$, first integer

$x + 1 = 19 + 1 = 20$, second integer
$x + 2 = 19 + 2 = 21$, third integer

23. Let x = number of degrees in third angle. **24.** $125°$
$62 + 47 + x = 180$

$109 + x = 180$
$\underline{-109 \qquad -109}$
$x = 71$

The third angle measures $71°$.

25. Let w = width of rectangle (in ft)
$2w$ = length (in ft)

$2w + 2(2w) = 102$
$2w + 4w = 102$
$6w = 102$

$\dfrac{\overset{1}{\cancel{6}}w}{\underset{1}{\cancel{6}}} = \dfrac{102}{6}$

$w = 17$
$2w = 2(17) = 34$

The rectangle is 34 ft long and 17 ft wide.

26. Length = 64 cm; width = 16 cm

27. Let x = length of rectangle (in ft). **28.** 12 yd
$2x + 2(5) = 44$

$2x + 10 = 44$
$\underline{\quad -10 \quad -10}$
$2x = 34$

$\dfrac{\overset{1}{\cancel{2}}x}{\underset{1}{\cancel{2}}} = \dfrac{34}{2}$

$x = 17$ The rectangle is 17 ft long.

29. Let x = unknown number. **30.** 5
$2(4 + x) + x = x + 10$
$8 + 2x + x = x + 10$
$8 + 3x = x + 10$
$\underline{-8 - x \qquad -x - 8}$
$2x = 2$

$\dfrac{\overset{1}{\cancel{2}}x}{\underset{1}{\cancel{2}}} = \dfrac{2}{2}$

$x = 1$ The number is 1.

31. Let x = unknown number. **32.** $\frac{10}{11}$
$3(8 + 2x) = 4(3x + 8)$
$24 + 6x = 12x + 32$
$\underline{-24 - 12x \quad -12x - 24}$
$-6x = 8$

$\dfrac{\overset{1}{\cancel{-6}}x}{\underset{1}{\cancel{-6}}} = \dfrac{8}{-6}$

$x = -1\frac{1}{3}$ The number is $-1\frac{1}{3}$.

33. Let x = unknown number. **34.** $-2\frac{1}{2}$
$10x - 3(4 + x) = 5(9 + 2x)$
$10x - 12 - 3x = 45 + 10x$
$7x - 12 = 45 + 10x$
$\underline{-10x + 12 \quad +12 - 10x}$
$-3x = 57$

$\dfrac{\overset{1}{\cancel{-3}}x}{\underset{1}{\cancel{-3}}} = \dfrac{57}{-3}$

$x = -19$ The number is -19.

35. Let x = unknown number.
$8.66x - 5.75(6.94 + x) = 4.69(8.55 + 3.48x)$
$8.66x - 39.905 - 5.75x = 40.0995 + 16.3212x$
$2.91x - 39.905 = 40.0995 + 16.3212x$
$\underline{-2.91x - 40.0995 \quad -40.0995 - 2.91 x}$
$-80.0045 = 13.4112x$

$\dfrac{-80.0045}{13.4112} = \dfrac{\cancel{13.4112}x}{\cancel{13.4112}}$

$x \approx -5.97$

The number is about -5.97.

36. ≈ -32.59

37. Let x = unknown number.
$x + 18 \geq 5$
$\underline{\quad -18 \quad -18}$
$x \geq -13$

The unknown number can be any number greater than or equal to -13.

38. The unknown number can be any number greater than or equal to 35.

39. Let x = unknown length.

$$\begin{array}{rr} 37 + x < & 80 \\ -37 & -37 \\ \hline x < & 43 \end{array}$$

The second piece must be less than 43 m.

40. Any length greater than 17 m

Exercises 4.4A (page 180)

1. 7(5 cents) = 35 cents   **2.** $2.75

3. 9(50 cents) = 450 cents = $4.50   **4.** $1.20

5. x(25 cents) = 25x cents   **6.** 5y cents

7. 7($1.50) + 5($0.75) = $10.50 + $3.75 = $14.25

8. $21.25   **9.** x($3.50) + y($1.90) = (3.50$x$ + 1.90y) dollars

10. (2.75x + 1.50y) dollars

11. x(25 cents) + y(6 cents) = (25x + 6y) cents

12. (4x + 2y) cents

Exercises 4.4B (page 182)

(The checks will not be shown.)

1. Let D = number of dimes   **2.** 5 nickels, 6 dimes

$13 - D$ = number of nickels

$$10D + 5(13 - D) = 95$$
$$10D + 65 - 5D = 95$$
$$\begin{array}{rr} 5D + 65 = & 95 \\ -65 & -65 \\ \hline 5D & = & 30 \end{array}$$
$$\frac{\overset{1}{\cancel{5}}D}{\underset{1}{\cancel{5}}} = \frac{30}{5}$$
$$D = 6$$
$$13 - D = 13 - 6 = 7$$

Bill has 6 dimes and 7 nickels.

3. Let N = number of nickels   **4.** 10 nickels, 8 quarters

$12 - N$ = number of quarters

$$5N + 25(12 - N) = 220$$
$$5N + 300 - 25N = 220$$
$$\begin{array}{rr} 300 - 20N = & 220 \\ -300 & = -300 \\ \hline -20N & = -80 \end{array}$$
$$\frac{\overset{1}{\cancel{-20}}N}{\underset{1}{\cancel{-20}}} = \frac{-80}{-20}$$
$$N = 4$$
$$12 - N = 12 - 4 = 8$$

Jennifer has 4 nickels and 8 quarters.

5. Let x = number of nickels

$x + 4$ = number of quarters

$3x$ = number of dimes

$$5x + 25(x + 4) + 10(3x) = 400$$
$$5x + 25x + 100 + 30x = 400$$
$$\begin{array}{rr} 60x + 100 = & 400 \\ -100 & -100 \\ \hline 60x & = & 300 \end{array}$$
$$\frac{\overset{1}{\cancel{60}}x}{\underset{1}{\cancel{60}}} = \frac{300}{60}$$
$$x = 5$$
$$x + 4 = 5 + 4 = 9$$
$$3x = 3(5) = 15$$

Derek has 5 nickels, 9 quarters, and 15 dimes.

6. 2 nickels, 9 dimes, 18 quarters

7. Let x = number of quarters

$x - 3$ = number of dimes

$2x - 3$ = number of nickels

$$25x + 10(x - 3) + 5(2x - 3) = 225$$
$$25x + 10x - 30 + 10x - 15 = 225$$
$$\begin{array}{rr} 45x - 45 = & 225 \\ + 45 & +45 \\ \hline 45x & = & 270 \end{array}$$
$$\frac{\overset{1}{\cancel{45}}x}{\underset{1}{\cancel{45}}} = \frac{270}{45}$$
$$x = 6$$
$$x - 3 = 6 - 3 = 3$$
$$2x - 3 = 12 - 3 = 9$$

Michael has 6 quarters, 3 dimes, and 9 nickels.

8. 7 pennies, 12 nickels, 19 dimes

9. Let x = number of box seats sold

$5x$ = number of balcony seats sold

This leaves $1{,}080 - x - 5x = 1{,}080 - 6x$ for the number of orchestra seats.

$$30x + 12(5x) + 21(1{,}080 - 6x) = 19{,}800$$
$$30x + 60x + 22{,}680 - 126x = 19{,}800$$
$$\begin{array}{rr} -36x + 22{,}680 = & 19{,}800 \\ - 22{,}680 & -22{,}680 \\ \hline -36x & = -2{,}880 \end{array}$$
$$\frac{\overset{1}{\cancel{-36}}x}{\underset{1}{\cancel{-36}}} = \frac{-2{,}880}{-36}$$
$$x = 80$$
$$5x = 5(80) = 400$$
$$1{,}080 - 6x = 1{,}080 - 6(80) = 600$$

There were 80 box seat tickets, 400 balcony seat tickets, and 600 orchestra seat tickets sold.

10. 4,400 box seats, 17,600 reserved seats, 35,200 general admission seats

11. Let x = number of 12¢ stamps
$2x$ = number of 10¢ stamps

This leaves $60 - x - 2x = 60 - 3x$ for the number of 2¢ stamps.

$$12x + 10(2x) + 2(60 - 3x) = 380$$
$$12x + 20x + 120 - 6x = 380$$
$$26x + 120 = 380$$
$$\underline{ - 120 \quad -120}$$
$$26x = 260$$
$$\frac{\cancel{26}^{1}x}{\cancel{26}_{1}} = \frac{260}{26}$$
$$x = 10 \quad 12\text{¢ stamps}$$
$$2x = 2(10) = 20 \quad 10\text{¢ stamps}$$
$$60 - 3x = 60 - 3(10) = 30 \quad 2\text{¢ stamps}$$

Christy bought 10 12¢ stamps, 20 10¢ stamps, and 30 2¢ stamps.

12. Ten 8¢ stamps, thirty 6¢ stamps, sixty 12¢ stamps.

Exercises 4.5 (page 189)

(The checks will not be shown.)

1. Let $4x$ = larger number
$3x$ = smaller number

$$3x + 4x = 35$$
$$7x = 35$$
$$\frac{\cancel{7}^{1}x}{\cancel{7}_{1}} = \frac{35}{7}$$
$$x = 5$$

Smaller number = $3x = 3(5) = 15$
Larger number = $4x = 4(5) = 20$

2. 91, 39

3. Let $4x$ = width
$9x$ = length

$$2(9x) + 2(4x) = 78$$
$$18x + 8x = 78$$
$$26x = 78$$
$$\frac{\cancel{26}^{1}x}{\cancel{26}_{1}} = \frac{78}{26}$$
$$x = 3$$

Length = $9x = 9(3) = 27$
Width = $4x = 4(3) = 12$

4. 28, 8

5. Let $4x$ = amount spent for food
$5x$ = amount spent for rent
x = amount spent for clothing

$$4x + 5x + x = 850$$
$$10x = 850$$
$$\frac{\cancel{10}^{1}x}{\cancel{10}_{1}} = \frac{850}{10}$$
$$x = \$85, \text{ amount for clothing}$$
$$4x = 4(\$85) = \$340, \text{ amount for food}$$
$$5x = 5(\$85) = \$425, \text{ amount for rent}$$

6. $5,400 for tuition, $5,400 for housing, $1,350 for food

7. Let $6x$ = one part
$5x$ = other part

$$6x + 5x = 88$$
$$11x = 88$$
$$\frac{\cancel{11}^{1}x}{\cancel{11}_{1}} = \frac{88}{11}$$
$$x = 8$$
$$6x = 6(8) = 48$$
$$5x = 5(8) = 40$$

The numbers are 48 and 40.

8. 22, 77

9. Let $4x$ = shortest side
$5x$ = next side
$6x$ = longest side

$$4x + 5x + 6x = 90$$
$$15x = 90$$
$$\frac{\cancel{15}^{1}x}{\cancel{15}_{1}} = \frac{90}{15}$$
$$x = 6$$
$$4x = 4(6) = 24$$
$$5x = 5(6) = 30$$
$$6x = 6(6) = 36$$

The lengths of the sides are 24 ft, 30 ft, and 36 ft.

10. 20 yd, 24 yd, 28 yd

11. Let $2x$ = amount received by first nephew
$3x$ = amount received by second nephew
$4x$ = amount received by third nephew

$$2x + 3x + 4x = \$27,000$$
$$9x = \$27,000$$
$$\frac{\cancel{9}^{1}x}{\cancel{9}_{1}} = \frac{\$27,000}{9}$$
$$x = \$3,000$$
$$2x = 2(\$3,000) = \$6,000$$
$$3x = 3(\$3,000) = \$9,000$$
$$4x = 4(\$3,000) = \$12,000$$

The first nephew received $6,000, the second $9,000, and the third $12,000.

12. $75,000; $15,000; $60,000

13. Let $4x$ = cu. yd of gravel
$2\frac{1}{2}x$ = cu. yd of sand
$1x$ = cu. yd of cement

$$4x = 5$$
$$\frac{\cancel{4}^{1}x}{\cancel{4}_{1}} = \frac{5}{4}$$
$$x = \frac{5}{4} = 1\frac{1}{4}$$
$$2\frac{1}{2}x = \frac{5}{2}\left(\frac{5}{4}\right) = \frac{25}{8} = 3\frac{1}{8}$$

Mr. Mora used $3\frac{1}{8}$ cu. yd of sand and $1\frac{1}{4}$ cu. yd of cement.

14. 519 seafood dinners

15. Let $7x$ = length

$4x$ = width

$2(7x) + 2(4x) < 88$

$14x + 8x < 88$

$22x < 88$

$\dfrac{\overset{1}{\cancel{22}}x}{\underset{1}{\cancel{22}}} < \dfrac{88}{22}$

$x < 4$

$4x < 16$

The width must be less than 16.

16. The length must be greater than 24.

17. Let $2x$ = length of shortest side

$4x$ = length of next side

$7x$ = length of longest side

$2x + 4x + 7x > 117$

$13x > 117$

$\dfrac{\overset{1}{\cancel{13}}x}{\underset{1}{\cancel{13}}} > \dfrac{117}{3}$

$x > 9$

$7x > 63$

The longest side must be greater than 63.

18. The length of the shortest side must be less than 24.

19. $\dfrac{360 \text{ miles}}{7 \text{ hours}}$ **20.** $\dfrac{35 \text{ miles}}{4 \text{ hours}}$ **21.** $\dfrac{3 \text{ afghans}}{25 \text{ days}}$ **22.** $\dfrac{500 \text{ words}}{11 \text{ minutes}}$

23. Let x = number of square yards   **24.** 36 hours

$x \cancel{\text{ sq yd}}\left(23\dfrac{\text{dollars}}{\cancel{\text{sq yd}}}\right) = 805 \text{ dollars}$

$23x = 805$

$\dfrac{\overset{1}{\cancel{23}}x}{\underset{1}{\cancel{23}}} = \dfrac{805}{23}$

$x = 35$, number of square yards

25. Let x = number of gallons of gasoline   **26.** 12 rolls

$x \cancel{\text{ gallons}}\left(23\dfrac{\text{miles}}{\cancel{\text{gallon}}}\right) = 368 \text{ miles}$

$23x = 368$

$\dfrac{\overset{1}{\cancel{23}}x}{\underset{1}{\cancel{23}}} = \dfrac{368}{23}$

$x = 16$

Mr. Lee will use 16 gal of gasoline.

Exercises 4.6 (page 195)

1. Let x = number.   **2.** 80

$0.30x = 15$

$x = \dfrac{15}{0.30} = 50$

3. Let x = percent.   **4.** $146\frac{2}{3}\%$

$250x = 115$

$x = \frac{115}{250} = 0.46 = 46\%$

5. Let x = unknown number.   **6.** 29.25

$x = 0.25(40) = 10$

7. Let x = number.   **8.** 800

$0.15x = 127.5$

$x = \dfrac{127.5}{0.15} = 850$

9. Let x = percent.   **10.** 200%

$8x = 17$

$x = \dfrac{17}{8} = 2.125 = 212.5\%$

11. Let x = unknown number.   **12.** 42.63

$x = 0.63(48) = 30.24$

13. Let x = number.   **14.** 250

$1.25x = 750$

$x = \dfrac{750}{1.25} = 600$

15. Let x = percent.   **16.** $\approx 247.8\%$

$16x = 23$

$x = \dfrac{23}{16} = 1.4375 = 143.75\%$

17. Let x = unknown number.   **18.** 27

$x = 2.00(12) = 24$

19. Let x = number.   **20.** $36.45

$0.15x = 37.5$

$x = 250$

21. Let x = number.

$0.66\tfrac{2}{3}x = 42 \quad \left(0.66\tfrac{2}{3} = \dfrac{66\frac{2}{3}}{100} = \dfrac{\frac{200}{3}}{100} = \tfrac{2}{3}\right)$

$\tfrac{2}{3}x = 42$

$\tfrac{3}{2}\left(\tfrac{2}{3}\right)x = \tfrac{3}{2}(42)$

$x = 63$

22. 216   **23.** Sixty-eight is 80% of what number?

Let x = number.

$0.80x = 68$

$x = \dfrac{68}{0.80} = 85$

The team has played 85 games.

24. 32,200   **25.** Seven is what percent of 42?

Let x = percent.

$42x = 7$

$x = \tfrac{7}{42} = 0.16\tfrac{2}{3} = 16\tfrac{2}{3}\%$

26. $57.50   **27.** Fifty-four is what percent of 210?

Let x = percent.

$210x = 54$

$x = \tfrac{54}{210} \approx 0.257 = 25.7\%$

28. 11%   **29.** The markup is 35% of $125.

Let x = number.

$x = 0.35(\$125) = \$43.75$ (Markup)

Selling price = Cost + Markup

Selling price = $125 + $43.75 = $168.75

30. $86.40

31. Total weight of steers = $15 \times 1{,}027 = 15{,}405$ lb
Total weight of heifers = $18 \times 956 = 17{,}208$ lb
If 3% of the weight is lost in shipping, then 97% of the weight remains at time of sale.

Let x = amount of money from sale of steers.

$x = 0.97 \times 15{,}405 \times 0.84 = 12{,}551.9940$

Let y = amount of money from sale of heifers.

$y = 0.97 \times 17{,}208 \times 0.78 = 13{,}019.5728$

$x + y = 25{,}571.5668 \approx 25{,}571.57$

Amount of check = \$25,571.57

32. $\approx 38.6\%$

Exercises 4.7 (page 202)

(The checks will not be shown.)

1. Let x = number of hours.

$$\left(51 \frac{\text{miles}}{\text{hour}}\right)(x \text{ hours}) = 408 \text{ miles}$$

$$51x = 408$$

$$\frac{\overset{1}{\cancel{51}}x}{\underset{1}{\cancel{51}}} = \frac{408}{51}$$

$x = 8$, number of hours for Robbie to drive 408 mi.

2. 9 hours

3. Let x = speed with no traffic
$x - 10$ = speed with traffic

$$4x = 5(x - 10)$$

$$4x = 5x - 50$$

$$\frac{50 - 4x \quad -4x + 50}{50 \quad = \quad x}$$

a. Jennie's average speed is 50 mph, with no traffic

b. The distance is $\left(50 \frac{\text{miles}}{\text{hour}}\right)(4 \text{ hours}) = 200$ miles

4a. 60 mph **b.** 300 mi

5. Let x = Mr. Robinson's speed
$x + 9$ = Mr. Reid's speed

$$5x = 4(x + 9)$$

$$5x = 4x + 36$$

$$\frac{-4x \quad -4x}{}$$

a. $x = 36$ Mr. Robinson's speed is 36 mph.

b. $x + 9 = 45$ Mr. Reid's speed is 45 mph.

c. The distance is $\left(\frac{36 \text{ miles}}{\text{hour}}\right)(5 \text{ hours}) = 180$ mi.

6a. 45 mph **b.** 54 mph **c.** 270 mi

7. Let x = speed of boat in still water **8.** 24 mph
$x - 3$ = speed upstream
$x + 3$ = speed downstream

$$4(x - 3) = 3(x + 3)$$

$$4x - 12 = 3x + 9$$

$$\frac{-3x + 12 \quad -3x + 12}{x \quad = \quad 21}$$

The speed of the boat was 21 mph in still water.

9. Let x = speed of boat in still water **10.** 24 mph
$x - 4$ = speed upstream
$x + 4$ = speed downstream

$$4(x - 4) + 6 = 3(x + 4)$$

$$4x - 16 + 6 = 3x + 12$$

$$\frac{4x - 10 \quad 3x + 12}{-3x + 10 \quad -3x + 10}$$

$$\frac{}{x \quad = \quad 22}$$

The speed of the boat was 22 mph in still water.

11. Let t = number of hours driven on Saturday
$17 - t$ = number of hours driven the next day
$54t$ = distance driven on Saturday
$48(17 - t)$ = distance driven the next day

$$54t = 48(17 - t) \quad \text{(The distances are equal.)}$$

$$54t = 816 - 48t$$

$$\frac{48t \qquad + 48t}{102t = 816}$$

$$\frac{\overset{1}{\cancel{102}}t}{\underset{1}{\cancel{102}}} = \frac{816}{102}$$

a. $t = 8$ It took Matthew 8 hr to get to his friend's house.

b. The distance is $\left(\frac{54 \text{ miles}}{\text{hour}}\right)(8 \text{ hours}) = 432$ mi

12a. 3 hours **b.** 48 miles

Exercises 4.8 (page 208)

1. Let x = number of pounds of apple chunks
$10 - x$ = number of pounds of granola

$$4.2x + 2.2(10 - x) = 3(10)$$

$$42x + 22(10 - x) = 30(10)$$

$$42x + 220 - 22x = 300$$

$$\frac{20x + 220 = 300}{- 220 \quad -220}$$

$$\frac{20x \quad = \quad 80}{}$$

$$\frac{\overset{1}{\cancel{20}}x}{\underset{1}{\cancel{20}}} = \frac{80}{20}$$

$$x = 4$$

$$10 - x = 6$$

The grocer should use 4 lb of apple chunks and 6 lb of granola.

2. 23 lb of macadamia nuts and 27 lb of peanuts

3. Let x = number of pounds of apples at 95¢ per pound
$30 - x$ = number of pounds of apples at 65¢ per pound

$$95x + 65(30 - x) = 78(30)$$

$$95x + 1{,}950 - 65x = 2{,}340$$

$$\frac{30x + 1{,}950 = 2{,}340}{- 1{,}950 \quad -1{,}950}$$

$$\frac{30x \quad = \quad 390}{}$$

$$\frac{\overset{1}{\cancel{30}}x}{\underset{1}{\cancel{30}}} = \frac{390}{30}$$

$$x = 13$$

$$30 - x = 17$$

Alice should use 13 lb of the more expensive apples and 17 lb of the cheaper ones.

4. 35 lb of English toffee, 25 lb of peanut brittle

5. Let $\quad x$ = number of pounds of gumdrops
$\quad 27 - x$ = number of pounds of caramels

$$2.8x + 2.2(27 - x) = 66$$
$$28x + 22(27 - x) = 660$$
$$28x + 594 - 22x = 660$$
$$6x + 594 = 660$$
$$\underline{\quad - 594 \quad -594}$$
$$6x \quad = \quad 66$$
$$\frac{\overset{1}{\cancel{6}x}}{\cancel{6}} = \frac{66}{6}$$
$$x = 11$$
$$27 - x = 16$$

Margie should use 11 lb of gumdrops and 16 lb of caramels.

6. 23 lb of Brand C and 17 lb of Brand D

7. Let x = number of pounds of candy.
(There will be $50 + x$ pounds of the mixture.)

$$3x + 2.4(50) = 2.5(50 + x)$$
$$30x + 24(50) = 25(50 + x)$$
$$30x + 1,200 = 1,250 + 25x$$
$$\underline{-25x - 1,200 \quad -1,200 - 25x}$$
$$5x \quad = \quad 50$$
$$\frac{\overset{1}{\cancel{5}x}}{\cancel{5}} = \frac{50}{5}$$
$$x = 10$$

Dorothy should use 10 lb of candy.

8. 20 lb of cashews

Exercises 4.9 (page 212)

1. What is 20% of 16 ml?   **2.** 0.4 ℓ
Let x = unknown number.
$$x = (0.20)(16 \text{ ml}) = 3.2 \text{ ml}$$

3. Forty-three is what percent of 500?   **4.** 5.4%
Let x = percent.
$$500x = 43$$
$$x = \tfrac{43}{500} = 0.086 = 8.6\%$$

5. Twenty-four is 40% of what number?   **6.** 50 ml
Let x = unknown number.
$$0.40x = 24$$
$$x = \frac{24}{0.40} = 60$$

There are 60 ml of solution.

7. Let x = number of cc of 20% solution.   **8.** 20 pt
$$0.50(100) + 0.20(x) = 0.25(x + 100)$$
$$50(100) + 20x = 25(x + 100)$$
$$5,000 + 20x = 25x + 2,500$$
$$\underline{-2,500 - 20x \quad -20x - 2,500}$$
$$2,500 = 5x$$
$$\frac{2,500}{5} = \frac{\overset{1}{\cancel{5}x}}{\cancel{5}}$$
$$x = 500$$

500 cc of the 20% solution should be added.

9. Let x = number of milliliters of water.   **10.** 120 ml
(There will be $[500 + x]$ ml altogether.)

$$0.25(x + 500) = 0.40(500)$$
$$25(x + 500) = 40(500)$$
$$25x + 12,500 = 20,000$$
$$\underline{\quad - 12,500 \quad -12,500}$$
$$25x \quad = \quad 7,500$$
$$\frac{\overset{1}{\cancel{25}x}}{\cancel{25}} = \frac{7,500}{25}$$
$$x = 300$$

300 ml of water should be added.

11. Let x = number of liters of pure alcohol.   **12.** 5 ℓ
(There will be $[10 - x]$ ℓ altogether.)

$$1x + 0.2(10) = 0.5(10 + x)$$
$$10x + 2(10) = 5(10 + x)$$
$$10x + 20 = 50 + 5x$$
$$\underline{- 5x - 20 \quad -20 - 5x}$$
$$5x \quad = \quad 30$$
$$\frac{\overset{1}{\cancel{5}x}}{\cancel{5}} = \frac{30}{5}$$
$$x = 6 \qquad 6 \; \ell \text{ of pure alcohol should be added.}$$

Exercises 4.10 (page 214)

(The checks will not be shown.)

1. Use the formula $I = prt$, with $I = 756$, $r = 0.07$, and $t = 3$.
Let p = the amount invested originally.

$$I = prt$$
$$756 = p(0.07)(3)$$
$$756 = 0.21p$$
$$\frac{756}{0.21} = \frac{\overset{1}{\cancel{0.21}p}}{\cancel{0.21}}$$
$$p = 3,600; \text{ Jeffrey had invested \$3,600 originally.}$$

2. The club has 200 members.

3. Use the formula $V = \pi r^2 h$, with $V = 100\pi$ and $r = 5$.
Let h = the height.

$$V = \pi r^2 h$$
$$100\pi = \pi(5^2)h$$
$$\frac{100\pi}{25\pi} = \frac{\overset{1}{\cancel{25}\pi h}}{\cancel{25}\pi}$$
$$h = 4; \text{ the height of the cylinder is 4 in.}$$

4. 720 cu. in.

5. Let $\quad x$ = amount invested at 5% interest
$\quad 5,000 - x$ = amount invested at 6% interest

$$0.05x + 0.06(5,000 - x) = 272$$
$$5x + 6(5,000 - x) = 27,200$$
$$5x + 30,000 - 6x = 27,200$$
$$30,000 - x = 27,200$$
$$\underline{-30,000 \qquad -30,000}$$
$$-x = -2,800$$
$$x = 2,800; \text{ \$2,800 was invested at 5% interest.}$$

6. 120 mi

Review Exercises 4.11 (page 215)

(The checks will not be shown.)

1. Let $6x$ = one number
$7x$ = other number

$6x + 7x = 52$

$13x = 52$

$\dfrac{\overset{1}{\cancel{13}}x}{\underset{1}{\cancel{13}}} = \dfrac{52}{13}$

$x = 4$

$6x = 6(4) = 24$, one number

$7x = 7(4) = 28$, other number

2. $1,102.50

3. Let x = number.

$0.31x = 77.5$

$x = \dfrac{77.5}{0.31} = 250$

4. 37.5%

5. The amount of increase is 5% of $380.

Let x = increase.

$x = 0.05(\$380) = \$19$

The new rent = $380 + $19 = $399.

6. 9 oz lead and 15 oz tin

7. Let $3x$ = smallest side
$4x$ = next side
$5x$ = longest side

$3x + 4x + 5x = 108$

$12x = 108$

$x = 9$

$3x = 3(9) = 27$, shortest side

$4x = 4(9) = 36$, next side

$5x = 5(9) = 45$, longest side

8. 15 dimes; seven 50¢ pieces

9. Let x = number of 10¢ stamps
$2x$ = number of 12¢ stamps
$2(2x)$ = number of 2¢ stamps

$10x + 12(2x) + 2(4x) = 210$

$10x + 24x + 8x = 210$

$42x = 210$

$\dfrac{\overset{1}{\cancel{42}}x}{\underset{1}{\cancel{42}}} = \dfrac{210}{42}$

$x = 5$, number of 10¢ stamps

$2x = 2(5) = 10$, number of 12¢ stamps

$4x = 4(5) = 20$, number of 2¢ stamps

10. 1,000 box seats; 5,500 reserved seats; 22,000 general admission seats

11. We first find 24% of 1,800.

Let x = that number.

$x = 0.24(1,800) = 432$

The number of minority students needed is $432 - 256 = 176$.

12. 5 mph

13. Let x = Mrs. Koontz's average speed
$x + 5$ = Mrs. Fowler's average speed

$\left(\dfrac{x \text{ miles}}{\cancel{\text{hour}}}\right)(10 \, \cancel{\text{hours}}) = \left(\dfrac{[x + 5] \text{ miles}}{\cancel{\text{hour}}}\right)(9 \, \cancel{\text{hours}})$ (The distances are equal.)

$10x = 9(x + 5)$

$10x = 9x + 45$

$\underline{-9x \quad -9x}$

a. $x = 45$ Mrs. Koontz's average speed was 45 mph.

b. $x + 5 = 45 + 5 = 50$ Mrs. Fowler's average speed was 50 mph.

c. The distance is $\left(\dfrac{45 \text{ miles}}{\cancel{\text{hour}}}\right)(10 \, \cancel{\text{hours}}) = 450$ miles.

14. 12 lb at 85¢ per pound and 18 lb at 95¢ per pound

15. Let x = number of cc of water added.

$0.25(500) + 0 = 0.05(500 + x)$

$25(500) + 0 = 5(500 + x)$

$12,500 = 2,500 + 5x$

$\underline{-2,500 \quad -2,500}$

$10,000 = 5x$

$\dfrac{10,000}{5} = \dfrac{\overset{1}{\cancel{5}}x}{\underset{1}{\cancel{5}}}$

$x = 2,000$

2,000 cc of water must be added.

Chapter 4 Diagnostic Test (page 219)

Following each problem number is the textbook section number (in parentheses) where that kind of problem is discussed.

1. (4.6) Twenty-eight is 40% of what number?

Let x = unknown number.

$0.40x = 28$

$x = \dfrac{28}{0.40} = 70$ ml

Check: 40% of 70 ml = (0.40)(70 ml)
$= 28$ ml

2. (4.6) Seventeen is what percent of 20?

Let x = percent.

$20x = 17$

$x = \frac{17}{20} = 0.85 = 85\%$

Check: 85% of 20 = (0.85)(20)
$= 17$

3. (4.3) Let x = unknown number.

$16 + 3x = 37$

$\underline{-16 \qquad -16}$

$3x = 21$

$\dfrac{\overset{1}{\cancel{3}}x}{\underset{1}{\cancel{3}}} = \dfrac{21}{3}$

$x = 7$

Check: $16 + 3x = 37$

$17 + 3(7) \overset{?}{=} 37$

$16 + 21 \overset{?}{=} 37$

$37 = 37$

The number is 7.

4. (4.4) Let $\quad x =$ number of nickels
$\qquad 25 - x =$ number of dimes

Then amount of money in nickels $= 5x$
and amount of money in dimes $= 10(25 - x)$

$$5x + 10(25 - x) = 165$$
$$5x + 250 - 10x = 165$$
$$-5x + 250 = \quad 165$$
$$\underline{\quad - 250 \quad -250}$$
$$-5x \quad = -85$$

$$\frac{\overset{1}{\cancel{-5}x}}{\underset{1}{\cancel{-5}}} = \frac{-85}{-5}$$

$\qquad x = 17$, number of nickels

Check: nickels $17(5¢) = \quad 85¢$
dimes $\;\; 8(10¢) = \quad 80¢$
$\qquad\qquad\qquad\qquad \overline{165¢}$

$25 - x = 25 - 17 = 8$, number of dimes

5. (4.8) Let $\quad x =$ number of pounds of apricots
$\qquad 50 - x =$ number of pounds of granola

$$2.70x + 2.20(50 - x) = 2.34(50)$$
$$270x + 220(50 - x) = 234(50)$$
$$270x + 11{,}000 - 220x = 11{,}700$$
$$50x + 11{,}000 = \quad 11{,}700$$
$$\underline{\quad - 11{,}000 \quad -11{,}000}$$
$$50x \quad = \quad 700$$

$$\frac{\overset{1}{\cancel{50}x}}{\underset{1}{\cancel{50}}} = \frac{700}{50}$$

$\qquad x = 14$

$50 - x = 50 - 14 = 36$

Check: $\$2.70(14) + \$2.20(36) \overset{?}{=} \$2.34(50)$
$\qquad\qquad \$37.80 + \$79.20 \overset{?}{=} \$117$
$\qquad\qquad\qquad\qquad \$117 = \$117$

14 lb of apricots and 36 lb of granola should be used.

6. (4.7) Let $\quad x =$ Kevin's average speed
$\qquad x + 9 =$ Jason's average speed
$\qquad 6x =$ Kevin's distance
$\qquad 5(x + 9) =$ Jason's distance

$6x = 5(x + 9)$ (The distances are equal.)
$$6x = \quad 5x + 45$$
$$\underline{-5x \quad -5x}$$

a. $\qquad x = \qquad 45$, Kevin's average speed (in mph)

b. $\qquad x + 9 = 45 + 9 = 54$, Jason's average speed (in mph)

c. The distance is $\left(\dfrac{45 \text{ miles}}{\cancel{\text{hour}}}\right)(6 \; \cancel{\text{hours}}) = 270$ mi

Check: $45\dfrac{\text{mi}}{\cancel{\text{hr}}}(6 \; \cancel{\text{hr}}) \overset{?}{=} 54\dfrac{\text{mi}}{\cancel{\text{hr}}}(5 \; \cancel{\text{hr}})$

$\qquad\qquad 270 \text{ mi} = 270 \text{ mi}$

7. (4.8) Let $x =$ number of pounds of nuts to be added.

$$3.50x + 8.00(30) = 4.85(x + 30)$$
$$350x + 800(30) = 485(x + 30)$$
$$350x + 24{,}000 = \quad 485x + 14{,}550$$
$$\underline{-350x - 14{,}550 \quad -350x - 14{,}550}$$
$$9{,}450 = \quad 135x$$

$$\frac{9{,}450}{135} = \frac{\overset{1}{\cancel{135}x}}{\underset{1}{\cancel{135}}}$$

$\qquad x = 70$

(The mixture will contain 70 lb + 30 lb, or 100 lb.)

Check: $3.50\dfrac{\text{dollars}}{\cancel{\text{lb}}}(70 \; \cancel{\text{lb}}) + 8\dfrac{\text{dollars}}{\cancel{\text{lb}}}(30 \; \cancel{\text{lb}}) \overset{?}{=} 4.85\dfrac{\text{dollars}}{\cancel{\text{lb}}}(100 \; \cancel{\text{lb}})$

$\qquad\qquad\qquad \$245 + \$240 \overset{?}{=} \$485$
$\qquad\qquad\qquad\qquad \$485 = \$485$

70 lb of nuts should be added.

8. (4.9) Let $\quad x =$ number of milliliters of 30% solution
$\qquad 200 - x =$ number of milliliters of 80% solution

$$0.30x + 0.80(200 - x) = 0.45(200)$$
$$30x + 80(200 - x) = 45(200)$$
$$30x + 16{,}000 - 80x = 9{,}000$$
$$16{,}000 - 50x = \quad 9{,}000$$
$$\underline{-9{,}000 + 50x \quad -9{,}000 + 50x}$$
$$7{,}000 = \quad 50x$$

$$\frac{7{,}000}{50} = \frac{\overset{1}{\cancel{50}x}}{\underset{1}{\cancel{50}}}$$

$\qquad x = 140$

$200 - x = 60$

Check: $0.30(140) + 0.80(60) \overset{?}{=} 0.45(200)$
$\qquad\qquad\qquad 42 + 48 \overset{?}{=} 90$
$\qquad\qquad\qquad\qquad 90 = 90$

140 ml of the 30% solution and 60 ml of the 80% solution should be used.

9. (4.5) Let $7x =$ length of smallest side (in meters)
$\qquad 9x =$ length of next side
$\qquad 11x =$ length of longest side

$$7x + 9x + 11x = 135$$
$$27x = 135$$

$$\frac{\overset{1}{\cancel{27}x}}{\underset{1}{\cancel{27}}} = \frac{135}{27}$$

$\qquad x = 5$
$\qquad 7x = 7(5) = 35$
$\qquad 9x = 9(5) = 45$
$\qquad 11x = 11(5) = 55$

Check: $35 \text{ m} + 45 \text{ m} + 55 \text{ m} \overset{?}{=} 135 \text{ m}$
$\qquad\qquad\qquad\qquad 135 \text{ m} = 135 \text{ m}$

The lengths are 35 m, 45 m, and 55 m.

10. (4.6) We must first find the markup, which is 30% of $240.

Let $x =$ markup.

$x = 0.30(\$240) = \$72$

The selling price is $240 + $72 = $312.

Cumulative Review Exercises: Chapters 1–4 (page 220)

1. Not defined **2.** 98

3. $46 - 2\{4 - [3(5 - 8) - 10]\} = 46 - 2\{4 - [3(-3) - 10]\}$
$= 46 - 2\{4 - [-9 - 10]\} = 46 - 2\{4 - [-19]\} = 46 - 2\{23\}$
$= 46 - 46 = 0$

4. 45 **5.** $S = 4\pi r^2$ **6.** 11
$S \approx 4(3.14)5^2$
$S \approx 4(3.14)25$
$S \approx 4(25)3.14$
$S \approx 100(3.14)$
$S \approx 314$

7. $8 + 12\{3 - 2(x + 4)\} = 8 + 12\{3 - 2x - 8\}$
$= 8 + 12\{-5 - 2x\} = 8 - 60 - 24x = -52 - 24x$

8. $-13x + 10$

9.
$$32 - 4x = 15$$
$$\underline{-15 + 4x} \quad \underline{-15 + 4x}$$
$$17 = 4x$$
$$\frac{17}{4} = \frac{\overset{1}{\cancel{4}}x}{\underset{1}{\cancel{4}}}$$
$x = \frac{17}{4}$ or $4\frac{1}{4}$ The solution is $4\frac{1}{4}$.

10. -33

11.
$$-12 + 12w = 3 - 8w$$
$$\underline{+12 + 8w} \quad \underline{+12 + 8w}$$
$$20w = 15$$
$$\frac{\overset{1}{\cancel{20}}w}{\underset{1}{\cancel{20}}} = \frac{\overset{3}{\cancel{15}}}{\underset{4}{\cancel{20}}}$$
$w = \frac{3}{4}$ The solution is $\frac{3}{4}$.

12. $y = -3$

13.
$$6z - 22 = 2[8 - 20z + 4]$$
$$6z - 22 = 2[12 - 20z]$$
$$6z - 22 = 24 - 40z$$
$$\underline{40z + 22} \quad \underline{22 + 40z}$$
$$46z = 46$$
$$\frac{\overset{1}{\cancel{46}}z}{\underset{1}{\cancel{46}}} = \frac{46}{46}$$
$z = 1$ The solution is 1.

14. Yes **15.** Yes **16.** Yes **17.** 1

(For 18–20, the checks will not be shown.)

18. 88 and 72

19. Let x = number of nickels
$15 - x$ = number of quarters
$$5x + 25(15 - x) = 175$$
$$5x + 375 - 25x = 175$$
$$375 - 20x = 175$$
$$\underline{-375} \quad \underline{-375}$$
$$-20x = -200$$
$$\frac{\overset{1}{\cancel{-20}}x}{\underset{1}{\cancel{-20}}} = \frac{-200}{-20}$$
$$x = 10$$
$15 - x = 15 - 10 = 5$
Leona has 10 nickels and 5 quarters.

20. 60 mph

Exercises 5.1 (page 223)

1. $x^{3+4} = x^7$ **2.** x^{11} **3.** $y^{1+3} = y^4$ **4.** z^5

5. $m^{2+1} = m^3$ **6.** a^4 **7.** $10^{2+3} = 10^5$ **8.** 10^7

9. $5^{1+4+5} = 5^{10}$ **10.** 3^6 **11.** $x^{1+3+4} = x^8$ **12.** y^9

13. x^2y^5 **14.** a^3b^2 **15.** $3^2 \cdot 5^3$ or $1,125$

16. $2^3 \cdot 3^2$ or 72 **17.** $a^4 + a^2$ (cannot be rewritten)

18. $x^3 + x^4$ **19.** a^{x+w} **20.** x^{a+b}

21. $x^y y^x$ (cannot be combined) **22.** $a^b b^a$ **23.** $x^{2+5}y^3 = x^7y^3$

24. z^7w^2 **25.** $a^{2+5}b^3 = a^7b^3$ **26.** $x^{12}y$

27. $s^7 + s^4$ (cannot be rewritten) **28.** $t^3 + t^8$

29. $7^{3+5} = 7^8$ **30.** 6^{19}

Exercises 5.2 (page 225)

1. $-(2 \cdot 4)(a \cdot a^2) = -8a^3$ **2.** $-15x^4$

3. $+(5 \cdot 6)(h^2h^3) = 30h^5$ **4.** $48k^4$

5. $(-5x^3)(-5x^3) = +(5 \cdot 5)(x^3x^3) = 25x^6$ **6.** $49y^8$

7. $+(2 \cdot 4 \cdot 3)(a^3aa^4) = 24a^8$ **8.** $-48b^6$

9. $+(9 \cdot 2)(mm^5m^2) = 18m^8$ **10.** $28n^7$

11. $-(5 \cdot 7)x^2y = -35x^2y$ **12.** $-12x^3y$

13. $+(6 \cdot 4)(m^3m)(n^2n^2) = 24m^4n^4$ **14.** $-40h^6k^4$

15. $-(2 \cdot 3)(x^{10} \cdot x^{12})(y^2 \cdot y^7) = -6x^{22}y^9$ **16.** $10a^{12}b^{15}$

17. $(3xy^2)(3xy^2) = (3 \cdot 3)(x \cdot x)(y^2 \cdot y^2) = 9x^2y^4$ **18.** $16x^4y^6$

19. $-(5)(x^4)(y^5 \cdot y^4)(z \cdot z^7) = -5x^4y^9z^8$ **20.** $-21E^2F^{11}G^{18}$

21. $-(2^3 \cdot 2^2)(R \cdot R^5)S^2T^4 = -32R^6S^2T^4$ **22.** $-243x^9yz^5$

23. $+(5 \cdot 4)(c^2 \cdot c^5)(d \cdot d)(e^3 \cdot e^2) = 20c^7d^2e^5$ **24.** $24m^4n^7r^5$

25. $-(2 \cdot 3 \cdot 7)(x^2 \cdot x)(y^2 \cdot y)(z \cdot z) = -42x^3y^3z^2$ **26.** $420x^4y^3z^5$

27. $-(3 \cdot 5)(x \cdot x^2 \cdot x^3)(y \cdot y^2 \cdot y^3) = -15x^6y^6$ **28.** $-6x^2y^2z^2$

29. $-(2 \cdot 5)(a \cdot a)(b \cdot b)(c \cdot c) = -10a^2b^2c^2$ **30.** $6x^3y^3z^3$

31. $(5 \cdot 2)(x^2 \cdot x)(y \cdot y)(z^3 \cdot z) = 10x^3y^2z^4$ **32.** $6a^5b^7$

33. $-(5 \cdot 7)(x^2 \cdot x^5 \cdot x)(y^2 \cdot y^3 \cdot y)(z \cdot z \cdot z^5) = -35x^8y^6z^7$

34. $-32R^9S^7T^{13}$

35. $-(3 \cdot 7)(h^2 \cdot h^4)(k^1 \cdot k^5)(m^3 \cdot m^1) = -21h^6k^6m^4$

36. $-48k^3m^3n^3$

Exercises 5.3 (page 228)

1. $(-3)(2x^2) + (-3)(-4x) + (-3)(5) = -6x^2 + 12x - 15$

2. $-15x^2 + 10x + 35$

3. $(4x)(3x^2) + (4x)(-6) = 12x^3 - 24x$ **4.** $15x^3 - 30x$

5. $(-2x)(5x^2) + (-2x)(3x) + (-2x)(-4) = -10x^3 - 6x^2 + 8x$

6. $-8x^3 + 20x^2 - 12x$

7. $(y^2)(7) + (-4y)(7) + (3)(7) = 7y^2 - 28y + 21$

8. $63 - 7z + 14z^2$

9. $(2x^2)(4x) + (-3x)(4x) + (5)(4x) = 8x^3 - 12x^2 + 20x$

10. $15w^3 + 10w^2 - 40w$ **11.** $(x)(xy) + (x)(-3) = x^2y - 3x$

12. $a^2b - 4a$ **13.** $(3a)(ab) + (3a)(-2a^2) = 3a^2b - 6a^3$

14. $12x^2 - 8xy^2$

15. $(-2x)(-3y) + (4x^2y)(-3y) = 6xy - 12x^2y^2$ **16.** $6az - 4a^2z^2$

17. $(-2xy)(x^2y) + (-2xy)(-y^2x) + (-2xy)(-y) + (-2xy)(-5)$
$= -2x^3y^2 + 2x^2y^3 + 2xy^2 + 10xy$

18. $-24ab + 3a^3b + 3ab^3 - 3a^2b^2$

19. $(3x^3)(-2xy) + (-2x^2y)(-2xy) + (y^3)(-2xy)$
$= -6x^4y + 4x^3y^2 - 2xy^4$

20. $-8yz^4 + 2y^2z^3 + 2y^4z$

21. $(2xy^2z)(-5xz^3) + (-7x^2z^2)(-5xz^3) = -10x^2y^2z^4 + 35x^3z^5$

22. $-9a^3bc^4 + 12a^2b^3c^3$

23. $(5x^2y^3z)(-4xz^2) + (-2xz^3)(-4xz^2) + (y^4)(-4xz^2)$
$= -20x^3y^3z^3 + 8x^2z^5 - 4xy^4z^2$

24. $-12x^3y^3z^2 + 6xy^3z^3 + 8x^2y^2z^4$

25. $6 - 3(4) + (-3)(-3z) + (-3)(-2z^3) = \underline{6} - \underline{12} + 9z + 6z^3$
$= -6 + 9z + 6z^3$

26. $9 - 12x + 8x^2$

27. $5x - 2x(3x^2) + (-2x)(7x) + (-2x)(-3)$
$= \underline{5x} - 6x^3 - 14x^2 + \underline{6x} = 11x - 14x^2 - 6x^3$

28. $35y - 18y^2$ **29.** $3x^2 - \underline{4x} + 2 - \underline{5x} = 3x^2 - 9x + 2$

30. $14y^2 + 5y - 4$

31. $(3x^2)(-5x) - 4x(-5x) + 2(-5x) = -15x^3 + 20x^2 - 10x$

32. $42y^3 + 6y^2 - 12y$ **33.** $(3)(2)x^3xy = 6x^4y$ **34.** $30x^3z^2$

35. $4x^2(2x^2) + (4x^2)(-3x) + (4x^2)(7) + (-4x)(x^3) + (-4x)(2x) + (-4x)(-5) = \underline{8x^4} - 12x^3 + \underline{28x^2} - \underline{4x^4} - \underline{8x^2} + 20x$
$= 4x^4 - 12x^3 + 20x^2 + 20x$

36. $-15y^4 + 3y^3 + 9y^2 + 3y$

37. $8x^2(7x^2) + 8x^2(-3x) + 8x^2(1) + (-5x)(6x^2) + (-5x)(-8x) + (-5x)(9) = 56x^4 - \underline{24x^3} + \underline{8x^2} - \underline{30x^3} + 40x^2 - 45x$
$= 56x^4 - 54x^3 + 48x^2 - 45x$

38. $42a^5 + 18a^4 + 33a^3 - 15a^2$

39. $3x - 5[8 - 2(4x) + (-2)(-y)] = 3x - 5[8 - 8x + 2y]$
$= 3x - 5(8) + (-5)(-8x) + (-5)(2y) = \underline{3x} - 40 + \underline{40x} - 10y$
$= 43x - 40 - 10y$

40. $14x - 12 - 24y$

41. $3 - 2[7x^2 + (-3x)(8x) + (-3x)(2)] = 3 - 2(7x^2 - 24x^2 - 6x)$
$= 3 - 2(-17x^2 - 6x) = 3 + (-2)(-17x^2) + (-2)(-6x)$
$= 3 + 34x^2 + 12x$

42. $7 - 4y^2 - 32y$

43. $8x - 2x[6 - 3(2x) + (-3)(-1)] = 8x - 2x[6 - 6x + 3]$
$= 8x - 2x[9 - 6x] = 8x + (-2x)(9) + (-2x)(-6x)$
$= \underline{8x} - \underline{18x} + 12x^2 = -10x + 12x^2$

44. $7y + 12y^2$

45. $7x + 4x(3y) + 4x(-5) + 4x(8x) = \underline{7x} + 12xy - \underline{20x} + 32x^2$
$= 32x^2 + 12xy - 13x$

46. $65y + 32yz - 72y^2$

Exercises 5.4A (page 231)

1. y^{10} **2.** N^{12} **3.** x^{16} **4.** z^{28} **5.** 2^8 **6.** 3^8

7. x^5y^5 **8.** a^4b^4 **9.** 2^6c^6 **10.** 3^4x^4 **11.** x^{28}

12. v^{24} **13.** 10^6 **14.** 10^{14} **15.** $(-6)^2 = 36$ **16.** 144

17. $2(-27) = -54$ **18.** -24 **19.** $-4 \cdot 25 = -100$

20. -48

Exercises 5.4B (page 235)

1. $x^{7-2} = x^5$ **2.** y^2 **3.** $x^4 - x^2$ (cannot be rewritten)

4. $s^8 - s^3$ **5.** $a^{5-1} = a^4$ **6.** b^6 **7.** $10^{11-1} = 10^{10}$

8. 5^5 **9.** $\dfrac{x^8}{y^4}$ (cannot be rewritten)   **10.** $\dfrac{a^4}{b^3}$

11. $\dfrac{\overset{3}{\cancel{6}}}{\underset{1}{\cancel{2}}} \cdot \dfrac{x^2}{x} = \dfrac{3}{1} \cdot x^{2-1} = 3x$ **12.** $3y^2$

13. $\dfrac{a^3}{b^2}$ cannot be simplified because the bases are different.

14. $\dfrac{x^5}{y^3}$ **15.** $\dfrac{10x^4}{5x^3} = \dfrac{\overset{2}{\cancel{10}}}{\underset{1}{\cancel{5}}} \cdot \dfrac{x^4}{x^3} = \dfrac{2}{1}x^{4-3} = 2x$ **16.** $\dfrac{5y^3}{3}$

17. $\dfrac{12h^4k^3}{8h^2k} = \dfrac{12}{8} \cdot \dfrac{h^4}{h^2} \cdot \dfrac{k^3}{k} = \dfrac{3}{2}h^{4-2}k^{3-1} = \dfrac{3h^2k^2}{2}$ **18.** $\dfrac{4a^4b}{3}$

19. $x^{5a-3a} = x^{2a}$ **20.** M^{4x}

21. $\dfrac{a^4 - b^3}{a^2}$ or $a^2 - \dfrac{b^3}{a^2}$ **22.** $\dfrac{x^6 + y^4}{y^2}$ or $\dfrac{x^6}{y^2} + y^2$

23. $-\dfrac{\overset{1}{\cancel{5}}}{\underset{2}{\cancel{10}}} \cdot \dfrac{x^8}{x^2} = -\dfrac{1}{2} \cdot x^{8-2} = -\dfrac{1}{2}x^6$ or $-\dfrac{x^6}{2}$ **24.** $-\dfrac{y^5}{3}$

25. $\dfrac{s^7}{t^7}$ **26.** $\dfrac{x^9}{y^9}$ **27.** $\dfrac{2^4}{x^4}$ **28.** $\dfrac{3^2}{z^2}$ **29.** $\dfrac{x^6}{2^6}$ **30.** $\dfrac{c^3}{5^3}$

31. $a^{2\cdot2}b^{3\cdot2} = a^4b^6$ **32.** $x^{12}y^{15}$ **33.** $2^{1\cdot2}z^{3\cdot2} = 2^2z^6$

34. 3^3w^6 **35.** $\dfrac{x^{1\cdot2}y^{4\cdot2}}{z^{2\cdot2}} = \dfrac{x^2y^8}{z^4}$ **36.** $\dfrac{a^9b^3}{c^6}$

37. $\dfrac{5^{1\cdot4}y^{3\cdot4}}{2^{1\cdot4}x^{2\cdot4}} = \dfrac{5^4y^{12}}{2^4x^8}$ **38.** $\dfrac{6^3b^{12}}{7^3c^6}$ **39.** $(x^2y^5)^3 = x^{2\cdot3}y^{5\cdot3} = x^6y^{15}$

40. a^6b^{15} **41.** $(-2x^3y^4)^3 = (-2)^3x^{3\cdot3}y^{4\cdot3} = -8x^9y^{12}$

42. $-8s^6t^9$ **43.** $\dfrac{(-4)^2}{-4^2} = \dfrac{16}{-16} = -1$ **44.** -1

Exercises 5.5 (page 243)

1. $\dfrac{1}{x^4}$ **2.** $\dfrac{1}{y^7}$ **3.** a^4 **4.** b^5

5. $\dfrac{r^{-4}}{1} \cdot \dfrac{s}{1} \cdot \dfrac{t^{-2}}{1} = \dfrac{1}{r^4} \cdot \dfrac{s}{1} \cdot \dfrac{1}{t^2} = \dfrac{s}{r^4t^2}$ **6.** $\dfrac{t}{r^5s^3}$

7. $\dfrac{1}{(xy)^2} = \dfrac{1}{x^2y^2}$ **8.** $\dfrac{1}{a^4b^4}$ **9.** $\dfrac{h^2}{1} \cdot \dfrac{1}{k^{-4}} = \dfrac{h^2}{1} \cdot \dfrac{k^4}{1} = h^2k^4$

10. m^3n^2 **11.** $\dfrac{x^{-4}}{1} \cdot \dfrac{1}{y} = \dfrac{1}{x^4} \cdot \dfrac{1}{y} = \dfrac{1}{x^4y}$ **12.** $\dfrac{1}{a^5b}$

13. $\dfrac{a}{1} \cdot \dfrac{b^{-2}}{1} \cdot \dfrac{c^0}{1} = \dfrac{a}{1} \cdot \dfrac{1}{b^2} \cdot \dfrac{1}{1} = \dfrac{a}{b^2}$ **14.** $\dfrac{z}{x^3}$

15. $x^{-3+4} = x^1 = x$ **16.** y^4 **17.** $10^{3-2} = 10^1 = 10$

18. $\frac{1}{2}$ **19.** $x^{2(-4)} = x^{-8} = \dfrac{1}{x^8}$ **20.** $\dfrac{1}{z^6}$

21. $a^{-2(3)} = a^{-6} = \dfrac{1}{a^6}$ **22.** $\dfrac{1}{b^{10}}$ **23.** $y^{(-2)-(5)} = y^{-7} = \dfrac{1}{y^7}$

24. $\dfrac{1}{z^4}$ **25.** $10^{2-(-5)} = 10^7$ **26.** 2^5 **27.** $\left(\dfrac{y}{x}\right)^3 = \dfrac{y^3}{x^3}$

28. $\dfrac{t^2}{s^2}$ **29.** $\dfrac{x^{-2}}{1} = x^{-2}$ **30.** y^{-3}

31. $\dfrac{h}{1} \cdot \dfrac{1}{k} = \dfrac{h}{1} \cdot \dfrac{k^{-1}}{1} = hk^{-1}$ **32.** mn^{-1}

33. $\dfrac{x^2}{1} \cdot \dfrac{1}{y} \cdot \dfrac{1}{z^5} = \dfrac{x^2}{1} \cdot \dfrac{y^{-1}}{1} \cdot \dfrac{z^{-5}}{1} = x^2y^{-1}z^{-5}$ **34.** $a^3b^{-2}c^{-1}$

35. $10^{4-2} = 10^2 = 100$ **36.** 3 **37.** $\dfrac{1}{10^4} = \dfrac{1}{10,000}$ **38.** $\frac{1}{8}$

39. $1 \cdot 49 = 49$ **40.** 64 **41.** $\dfrac{10^0}{10^2} = \dfrac{1}{100}$ **42.** 25

43. $10^{-3+2-5} = 10^{-6} = \dfrac{1}{10^6} = \dfrac{1}{1,000,000}$ **44.** $\frac{1}{8}$

45. $10^{2(-1)} = 10^{-2} = \dfrac{1}{10^2} = \dfrac{1}{100}$ **46.** $\frac{1}{64}$

47. $\dfrac{1}{(-5)^3} = -\dfrac{1}{125}$ **48.** $-\frac{1}{64}$ **49.** $\dfrac{1}{(-12)^2} = \dfrac{1}{144}$

50. $\frac{1}{169}$ **51.** $\dfrac{a^3}{1} \cdot \dfrac{b^0}{1} \cdot \dfrac{1}{c^{-2}} = \dfrac{a^3}{1} \cdot \dfrac{1}{1} \cdot \dfrac{c^2}{1} = a^3c^2$ **52.** e^2f^3

53. $\dfrac{p^4}{1} \cdot \dfrac{r^{-1}}{1} \cdot \dfrac{1}{t^{-2}} = \dfrac{p^4}{1} \cdot \dfrac{1}{r^1} \cdot \dfrac{t^2}{1} = \dfrac{p^4t^2}{r}$ **54.** $\dfrac{u^5w^3}{v^2}$

55. $\dfrac{\overset{2}{\cancel{8}}}{\underset{3}{\cancel{12}}} \cdot \dfrac{x^{-3}}{x} = \dfrac{2}{3} \cdot \dfrac{x^{-3-1}}{1} = \dfrac{2}{3} \cdot \dfrac{x^{-4}}{1} = \dfrac{2}{3} \cdot \dfrac{1}{x^4} = \dfrac{2}{3x^4}$ **56.** $\dfrac{3}{2y^3}$

57. $\dfrac{\overset{4}{\cancel{20}}}{\underset{7}{\cancel{35}}} \cdot \dfrac{h^{-2}}{h^{-4}} = \dfrac{4}{7} \cdot \dfrac{h^{(-2)-(-4)}}{1} = \dfrac{4}{7} \cdot \dfrac{h^2}{1} = \dfrac{4h^2}{7}$ **58.** $\dfrac{5k^3}{4}$

59. $\dfrac{\overset{3}{\cancel{15}}}{\underset{1}{\cancel{5}}} \cdot \dfrac{m^0}{m^{-3}} \cdot \dfrac{n^{-2}}{n^4} = \dfrac{3}{1} \cdot \dfrac{m^{0-(-3)}}{1} \cdot \dfrac{n^{-2-4}}{1}$
$= \dfrac{3}{1} \cdot \dfrac{m^3}{1} \cdot \dfrac{n^{-6}}{1} = \dfrac{3}{1} \cdot \dfrac{m^3}{1} \cdot \dfrac{1}{n^6} = \dfrac{3m^3}{n^6}$

60. $\dfrac{7x^2y}{6}$ **61.** $x^{3m-m} = x^{2m}$ **62.** y^{3n}

63. $x^{3b(-2)} = x^{-6b} = \dfrac{1}{x^{6b}}$ **64.** $\dfrac{1}{y^{6a}}$ **65.** $x^{2a-(-5a)} = x^{7a}$

66. a^{8x} **67.** $m^{(-2)4}n^{1 \cdot 4} = m^{-8}n^4 = \dfrac{n^4}{m^8}$ **68.** $\dfrac{r^5}{p^{15}}$

69. $x^{(-2)(-4)}y^{3(-4)} = x^8y^{-12} = \dfrac{x^8}{y^{12}}$ **70.** $\dfrac{w^6}{z^8}$

71. $\dfrac{M^{(-2)4}}{N^{3 \cdot 4}} = \dfrac{M^{-8}}{N^{12}} = \dfrac{1}{M^8N^{12}}$ **72.** $R^{15}S^{12}$

73. $(a^2b^{5-4})^2 = a^{2 \cdot 2}b^{1 \cdot 2} = a^4b^2$ **74.** x^3y^6

75. $(m^{1-3}n^{-1})^{-2} = m^{(-2)(-2)}n^{(-1)(-2)} = m^4n^2$ **76.** a^3b^6

77. $(x^{4+1}y^2)^{-1} = x^{5(-1)}y^{2(-1)} = x^{-5}y^{-2} = \dfrac{1}{x^5y^2}$ **78.** $\dfrac{1}{x^5y^4}$

79. $k^{(-4)(-2)} = k^8$ **80.** z^{10} **81.** 1 **82.** 1

83. $\dfrac{1}{(-3x)^2} = \dfrac{1}{9x^2}$ **84.** $\dfrac{1}{25y^2}$ **85.** $-3(x^{-2}) = -3\left(\dfrac{1}{x^2}\right) = -\dfrac{3}{x^2}$

86. $-\dfrac{5}{y^2}$ **87.** $\dfrac{1}{(-2z)^3} = \dfrac{1}{-8z^3}$ or $-\dfrac{1}{8z^3}$ **88.** $-\dfrac{1}{32a^5}$

89. $-2(z^{-3}) = -2\left(\dfrac{1}{z^3}\right) = -\dfrac{2}{z^3}$ **90.** $-\dfrac{2}{a^5}$

91. $\left(\dfrac{1}{3mn^3}\right)^4 = \dfrac{1^4}{3^4m^4n^{3 \cdot 4}} = \dfrac{1}{81m^4n^{12}}$ **92.** $\dfrac{1}{16x^2y^4}$

93. $\left(\dfrac{9y^{-2}}{6x^{-3}}\right)^3 = \left(\dfrac{3x^3}{2y^2}\right)^3 = \dfrac{3^3x^{3 \cdot 3}}{2^3y^{2 \cdot 3}} = \dfrac{27x^9}{8y^6}$ **94.** $\dfrac{64a^6}{27b^9}$

Exercises 5.6 (page 249)

1. 8,060 **2.** 31,400 **3.** 0.00132 **4.** 0.00082

5. 5.26 **6.** 9.11 **7.** 3.53×10^4 **8.** 8.25×10^5

9. 3.12×10^{-3} **10.** 1.45×10^{-4} **11.** 8.97×10^0

12. 2.497×10^0 **13.** 8.15×10^{-1} **14.** 2.74×10^{-1}

15. 2×10^{-4} **16.** 6×10^{-3} **17.** 4.5×10^1 **18.** 1.2×10^1

19. $\dfrac{(5 \times 10^6) \times (3 \times 10^{-6})}{(6 \times 10^{-4}) \times (2 \times 10^4)} = \dfrac{(5 \times 3) \times (10^6 \times 10^{-6})}{(6 \times 2) \times (10^{-4} \times 10^4)} = \dfrac{\overset{5}{\cancel{15}} \times 10^0}{\underset{4}{\cancel{12}} \times 10^0}$
$= \dfrac{5}{4} = 1.25 \times 10^0$

20. 0.175, or 1.75×10^{-1} **21.** 5.418×10^{11}

22. 3×10^{-10} **23.** 9×10^{-4} **24.** 1.5×10^{-3}

25. $\left(6.02 \times 10^{23}\ \dfrac{\text{molecules}}{\cancel{\text{mole}}}\right)(700\ \cancel{\text{moles}}) = 4,214 \times 10^{23}$ molecules
$= (4.214 \times 10^3) \times 10^{23}$ molecules $= 4.214 \times 10^{26}$ molecules

26. \$9,120,000,000

27. $\dfrac{4,400,000,000\text{ miles}}{12\ \cancel{\text{years}}} \times \dfrac{1\ \cancel{\text{year}}}{365\ \cancel{\text{days}}} \times \dfrac{1\ \cancel{\text{day}}}{24\text{ hours}} \approx 41,857\ \dfrac{\text{mi}}{\text{hr}}$

28. $1.607\ 04 \times 10^{10}$ mi or $16,070,400,000$ mi

Review Exercises 5.7 (page 252)

1. $m^{2+3} = m^5$ **2.** y^8 **3.** $2^{1+2} = 2^3$ **4.** x^{7y}

5. 10^{1+y} **6.** a^2 **7.** $5x(x^2) + 5x(7) = 5x^3 + 35x$

8. $-10x^3y - 6x^2 + 2x$

9. $(-5)(-1)(-10)(-2)(ee^4e)(f^2f^7f)(g^3g^2) = 100e^6f^{10}g^5$

10. $27x^3$ **11.** $x^{2 \cdot 4}y^{3 \cdot 4} = x^8y^{12}$ **12.** n^{15}

13. $p^{-3 \cdot 5} = p^{-15} = \dfrac{1}{p^{15}}$ **14.** $\dfrac{1}{16c^4}$ **15.** 1 **16.** $\dfrac{t^3}{s^{12}}$

17. $4x^8$ **18.** $-\dfrac{125}{a^{12}}$ **19.** $\dfrac{1}{(-10)^3} = -\dfrac{1}{1,000}$ **20.** r^2

21. $\dfrac{1}{x^5x^4} = \dfrac{1}{x^9}$ **22.** $\dfrac{c^4}{d^4}$ **23.** $\dfrac{1 \cdot m^3}{1} = m^3$

24. $x^4 + x^2$ (cannot be simplified) **25.** $\dfrac{1}{n^0n^6} = \dfrac{1}{n^6}$

26. $\dfrac{x^{10}y^{15}}{32z^{20}}$ **27.** $\left(\dfrac{b^3c^0}{a^{-4}}\right)^5 = (b^3a^4)^5 = a^{20}b^{15}$ **28.** $\dfrac{1}{x^8}$

29. $\left(\dfrac{t^3}{s^5r^6}\right)^4 = \dfrac{t^{12}}{r^{24}s^{20}}$ **30.** 1 **31.** $5^{-2}a^{-6}b^8 = \dfrac{b^8}{5^2a^6}$ or $\dfrac{b^8}{25a^6}$

32. $-63x^{12}y^8$

33. $2x(3x^2 - x) - 1(3x^2 - 4) = 6x^3 - 2x^2 - 3x^2 + 4$
$= 6x^3 - 5x^2 + 4$

34. $21m^3n^3 - 14m^2n^2$ **35.** a^3b^{-2} **36.** m^2n^3

37. $10^{-2}u^{-4}v^3w^5$ **38.** $\frac{1}{16}$ **39.** $10^{-4} = \dfrac{1}{10^4}$ or $\dfrac{1}{10,000}$

40. 8 **41.** $1 \cdot 9 = 9$ **42.** -1 **43.** 4.53×10^4

44. 3.156×10^{-2}

Chapter 5 Diagnostic Test (page 255)

Following each problem number is the textbook section number (in parentheses) where that kind of problem is discussed.

1. (5.2) $(-3xy)(5x^3y)(-2xy^4) = (-3)(5)(-2)(xx^3x)(yyy^4) = 30x^5y^6$
(Because two factors—an even number—were negative, the answer is positive.)

2. (5.3) $2xy^2(x^2 - 3y - 4) = 2xy^2(x^2) + 2xy^2(-3y) + 2xy^2(-4)$
$= 2x^3y^2 - 6xy^3 - 8xy^2$

3. (5.1) $x^3 \cdot x^4 = x^{3+4} = x^7$ **4.** (5.4) $(x^2)^3 = x^{2 \cdot 3} = x^6$

5. (5.4) $\dfrac{x^5}{x^2} = x^{5-2} = x^3$ **6.** (5.5) $x^{-4} = \dfrac{1}{x^4}$

7. (5.5) $x^2y^{-3} = \dfrac{x^2}{1} \cdot \dfrac{1}{y^3} = \dfrac{x^2}{y^3}$

8. (5.5) $\dfrac{a^{-3}}{b} = \dfrac{a^{-3}}{1} \cdot \dfrac{1}{b} = \dfrac{1}{a^3} \cdot \dfrac{1}{b} = \dfrac{1}{a^3b}$

9. (5.4) $\dfrac{x^{5a}}{x^{3a}} = x^{5a-3a} = x^{2a}$ **10.** (5.5) $(4^{3x})^0 = 1$

11. (5.4) $(x^2y^4)^3 = x^{2 \cdot 3}y^{4 \cdot 3} = x^6y^{12}$

12. (5.5) $(a^{-3}b)^2 = a^{-3\cdot2}b^{1\cdot2} = a^{-6}b^2 = \dfrac{1}{a^6} \cdot \dfrac{b^2}{1} = \dfrac{b^2}{a^6}$

13. (5.4) $\left(\dfrac{p^3}{q^2}\right)^2 = \dfrac{p^{3\cdot2}}{q^{2\cdot2}} = \dfrac{p^6}{q^4}$

14. (5.5) $\left(\dfrac{x}{y^2}\right)^{-3} = \dfrac{x^{1(-3)}}{y^{2(-3)}} = \dfrac{x^{-3}}{y^{-6}} = \dfrac{1}{x^3} \cdot \dfrac{y^6}{1} = \dfrac{y^6}{x^3}$

15. (5.5) $\left(\dfrac{\overset{2}{\cancel{4}}x^{-2}}{\underset{1}{\cancel{2}}x^{-3}}\right)^{-1} = (2x^{-2-(-3)})^{-1} = (2^1 x^1)^{-1}$

$= 2^{1(-1)}x^{1(-1)} = 2^{-1}x^{-1} = \dfrac{1}{2^1} \cdot \dfrac{1}{x^1} = \dfrac{1}{2x}$

16. (5.3) $3h(2k^2 - 5h) - h(2h - 3k^2) = 6hk^2 - 15h^2 - 2h^2 + 3hk^2$
$= 9hk^2 - 17h^2$

17. (5.3) $x(x^2 + 2x + 4) - 2(x^2 + 2x + 4)$
$= x^3 + 2x^2 + 4x - 2x^2 - 4x - 8 = x^3 - 8$

18. (5.5) $\dfrac{a^3}{b} = a^3b^{-1}$ **19.** (5.1) $2^3 \cdot 2^2 = 2^{3+2} = 2^5 = 32$

20. (5.5) $10^{-4} \cdot 10^2 = 10^{-2} = \dfrac{1}{10^2} = \dfrac{1}{100}$

21. (5.5) $5^{-2} = \dfrac{1}{5^2} = \dfrac{1}{25}$

22. (5.5) $(2^{-3})^2 = 2^{(-3)(2)} = 2^{-6} = \dfrac{1}{2^6} = \dfrac{1}{64}$

23. (5.5) $\dfrac{10^{-3}}{10^{-4}} = 10^{-3-(-4)} = 10^1 = 10$

24. (5.5) $(5^0)^2 = 1^2 = 1$

25. (5.6) **a.** 1.326×10^0 **b.** 5.27×10^{-1}

Cumulative Review Exercises: Chapters 1–5 (page 256)

1. False   **2.** True   **3.** False   **4.** False   **5.** False

6. False   **7.** $\dfrac{-9}{3} = -3$ **8.** 12

9. $6(-3) - 4(6) = -18 - 24 = -42$ **10.** -28 **11.** 0

12. -25 **13.** $A = P(1 + rt)$
$A = 1{,}200[1 + 0.15(4)]$
$A = 1{,}200[1.6]$
$A = 1{,}920$

14. $x = -4$ **15.** $(-5)(4)(pp^3) = -20p^4$ **16.** $200h^9j^{10}k^5$

17. $7 - 12xy + 12 = 19 - 12xy$ **18.** $-26x^2y^2 + 9x^2y^3$

19. Let $\quad x =$ number of adults' tickets
$\quad 9 - x =$ number of children's tickets

$2.50x + 1.25(9 - x) = 16.25$
$250x + 125(9 - x) = 1{,}625$
$250x + 1{,}125 - 125x = 1{,}625$

$$\begin{array}{rcr} 125x + 1{,}125 & = & 1{,}625 \\ -1{,}125 & & -1{,}125 \\ \hline 125x & = & 500 \end{array}$$

$\dfrac{\overset{1}{\cancel{125}}x}{\underset{1}{\cancel{125}}} = \dfrac{500}{125}$

$x = 4$

$9 - x = 9 - 4 = 5$

There were 4 adults' tickets and 5 children's tickets purchased.

20. -2

Exercises 6.1 (page 260)

1a. First degree   **2a.** First degree
  b. Second degree   **b.** Third degree

3. Not a polynomial   **4.** Not a polynomial

5a. Third degree   **6a.** Third degree
  b. Sixth degree   **b.** Third degree

7a. Second degree   **8a.** Second degree
  b. Fourth degree   **b.** Third degree

9. Not a polynomial   **10.** Not a polynomial

11. Not a polynomial   **12.** Not a polynomial

13a. Second degree   **14a.** Third degree
  b. Second degree   **b.** Third degree

15. Not a polynomial   **16.** Not a polynomial

17. $8x^5 + 7x^3 - 4x - 5$ Leading coefficient is 8

18. $-3y^5 - 2y^3 + 4y^2 + 10$ Leading coefficient is -3

19. $xy^3 + 8xy^2 - 4x^2y$ Leading coefficient is 1

20. $x^4y^2 + 3x^3y - 3xy^3$ Leading coefficient is 1

Exercises 6.2 (page 264)

1. $\underline{2m^2} - \underset{uu}{m} + \underline{\underline{4}} + \underline{3m^2} + \underset{u}{\underline{m}} - \underline{\underline{5}} = 5m^2 - 1$

2. $11n^2 + 2n + 3$

3. $2x^3 \,\underline{-4} + 4x^2 \,\underline{+8x} \,\underline{-9x} \,\underline{+7} = 2x^3 + 4x^2 - x + 3$

4. $9z^2 + 9$

5. $(3x^2 + 4x - 10) + (-5x^2 + 3x - 7)$
$= \underline{3x^2} + \underset{u}{\underline{4x}} - \underline{\underline{10}} - \underline{5x^2} + \underset{u}{\underline{3x}} - \underline{\underline{7}} = -2x^2 + 7x - 17$

6. $-a^2 - 7a + 14$

7. $(8b^2 + 2b - 14) - (-5b^2 + 4b + 8)$
$= (8b^2 + 2b - 14) + (+5b^2 - 4b - 8)$
$= 8b^2 + 2b - 14 + 5b^2 - 4b - 8 = 13b^2 - 2b - 22$

or $\begin{array}{l} 8b^2 + 2b - 14 \\ -(-5b^2 + 4b + 8) \end{array} \Rightarrow \begin{array}{l} 8b^2 + 2b - 14 \\ +(+5b^2 - 4b - 8) \\ \hline 13b^2 - 2b - 22 \end{array}$

8. $19c^2 + 5c + 1$

9. $\underline{6a} - \underline{5a^2} + \underset{u}{\underline{\underline{6}}} + \underline{4a^2} + \underset{u}{\underline{\underline{6}}} - \underline{3a} = -a^2 + 3a + 12$

10. $11b^2 + 3$

11. $(4a^2 + 6 - 3a) - (5a + 3a^2 - 4)$
$= (4a^2 + 6 - 3a) + (-5a - 3a^2 + 4)$
$= 4a^2 + 6 - 3a - 5a - 3a^2 + 4 = a^2 - 8a + 10$

or $\begin{array}{l} 4a^2 - 3a + 6 \\ -(3a^2 + 5a - 4) \end{array} \Rightarrow \begin{array}{l} 4a^2 - 3a + 6 \\ +(-3a^2 - 5a + 4) \\ \hline a^2 - 8a + 10 \end{array}$

12. $2b^2 - 9b + 15$ **13.** $17a^3 + 8a^2 - 2a$

14. $-20b^4 + b^3 + 5b^2 - 1$ **15.** $12x^2y^3 - 9xy^2$

16. $9a^2b + 2ab^2 - 10ab$

17. $\begin{array}{l} 15x^3 - 4x^2 \quad\quad + 12 \\ -(8x^3 \quad\quad + 9x - 5) \end{array} \Rightarrow \begin{array}{l} 15x^3 - 4x^2 \quad\quad + 12 \\ +(-8x^3 \quad\quad - 9x + 5) \\ \hline 7x^3 - 4x^2 - 9x + 17 \end{array}$

18. $-7y^3 - 28y^2 + 19y - 24$

19. $\begin{array}{l} 10a^2b - 6ab + 5ab^2 \\ -(3a^2b + 6ab - 7ab^2) \end{array} \Rightarrow \begin{array}{l} 10a^2b - 6ab + 5ab^2 \\ +(-3a^2b - 6ab + 7ab^2) \\ \hline 7a^2b - 12ab + 12ab^2 \end{array}$

20. $22m^3n^2 - 4m^2n^2 - 9mn$

21. $\underline{7m^8} - 4m^4 + 4m^4 + \underset{u}{\underline{\underline{m^5}}} + \underline{8m^8} - \underset{u}{\underline{\underline{m^5}}} = 15m^8$

22. $-h^6 + 17h$

23. $\underline{6r^3t} + \underline{14r^2t} - 11 + \underline{19} - \underline{8r^2t} + \underline{r^3t} + \underline{8} - \underline{6r^2t} = 7r^3t + 16$

24. $4m^2n^2 + 11$

25. $(7x^2y^2 - 3x^2y + xy + 7) + (-3x^2y^2 + 5xy - 4 - 7x^2y)$
$= \underline{7x^2y^2} - \underline{3x^2y} + \underline{xy} + \underline{7} - \underline{3x^2y^2} + \underline{5xy} - \underline{4} - \underline{7x^2y}$
$= 4x^2y^2 - 10x^2y + 6xy + 3$

26. $9x^2y^2 - 2x^2y - 4xy - 13$

27. $(x^2 + 4) - [(x^2 - 5) + (-3x^2 - 1)]$
$= (x^2 + 4) - [x^2 - 5 - 3x^2 - 1]$
$= (x^2 + 4) + [-x^2 + 5 + 3x^2 + 1]$
$= \underline{x^2} + \underline{4} - \underline{x^2} + \underline{5} + \underline{3x^2} + \underline{1} = 3x^2 + 10$

28. $6x^2 - 7$

29. $[(5x^2 - 2x + 1) + (-4x^2 + 6x - 8)] - (2x^2 - 4x + 3)$
$= [5x^2 - 2x + 1 - 4x^2 + 6x - 8] + (-2x^2 + 4x - 3)$
$= \underline{5x^2} - \underline{2x} + \underline{1} - \underline{4x^2} + \underline{6x} - \underline{8} - \underline{2x^2} + \underline{4x} - \underline{3}$
$= -x^2 + 8x - 10$

30. $-4y$

31. $[(5 + xy^2 + x^3y) + (-6 - 3xy^2 + 4x^3y)] - [(x^3y + 3xy^2 - 4) + (2x^3y - xy^2 + 5)] = [\underline{5} + \underline{xy^2} + \underline{x^3y} - \underline{6} - \underline{3xy^2} + \underline{4x^3y}]$
$- [\underline{x^3y} + \underline{3xy^2} - \underline{4} + \underline{2x^3y} - \underline{xy^2} + \underline{5}]$
$= [5x^3y - 2xy^2 - 1] - [3x^3y + 2xy^2 + 1]$
$= [5x^3y - 2xy^2 - 1] + [-3x^3y - 2xy^2 - 1]$
$= \underline{5x^3y} - \underline{2xy^2} - \underline{1} - \underline{3x^3y} - \underline{2xy^2} - \underline{1} = 2x^3y - 4xy^2 - 2$

32. $6m^2n + 6$

33. $\underline{7.239x^2} - \underline{4.028x} + \underline{6.205} - \underline{2.846x^2} + \underline{8.096x} + \underline{5.307}$
$= 4.393x^2 + 4.068x + 11.512$

34. $37.528x^2 + 6.15x - 52.64$

Exercises 6.3A (page 268)

1. $x - 3x - 3y = -2x - 3y$ **2.** $-y - 2z$

3. $2a - 4a + 4b = -2a + 4b$ **4.** $c + 2d$

5. $u^3 + \underline{2u^2} + \underline{4u} - \underline{2u^2} - \underline{4u} - 8 = u^3 - 8$ **6.** $x^3 + 27$

7. $x^4 + \underline{x^2y^2} - \underline{x^2y^2} - y^4 = x^4 - y^4$ **8.** $w^4 - 16$

9. $\underline{6x^3} - \underline{10x^2} + \underline{2x} - \underline{8x^3} + \underline{12x^2} + \underline{20x} = -2x^3 + 2x^2 + 22x$

10. $6x^3 - 4x^2 - 11x$

11. $-3a + 6b + 2a - 6b = -a$ **12.** $2m - 2n$

13. $-10x + 15y - 10x - 50y = -20x - 35y$ **14.** $-36s + 60t$

15. $-30x^2y^2$ **16.** $-30abc^2$ **17.** $6x^2y - 15xy^2$

18. $10abc - 6ac^2$ **19.** $-15xy^2 - 10xy$ **20.** $-6ac^2 - 15bc$

21. $3xy + 2x - 5y$ **22.** $2ac + 5b - 3c$

23. $\underline{3x^3y^3} - \underline{x^2y^2} - \underline{8x^2y^2} + \underline{2x^3y^3} = 5x^3y^3 - 9x^2y^2$

24. $-4a^2b^2 + 2a^2b^3$

25. $6h^3 - 2hk - hk + 3k^4 = 6h^3 - 3hk + 3k^4$ **26.** $11xy^2 - 14x^2$

27. $6f^3 - 12fg - 2fg + g^3 = 6f^3 - 14fg + g^3$ **28.** $3ab^2 - 7ab$

29. $2x - [3a + 4x - 5a] = 2x - [-2a + 4x] = 2x + 2a - 4x$
$= 2a - 2x$

30. $-c - y$

31. $5x + [-2x + 10 + 7] = 5x - 2x + 10 + 7 = 3x + 17$

32. $x + 9$

33. $25 - 2[3g - 10g + 35] = 25 - 2[-7g + 35] = 25 + 14g - 70$
$= 14g - 45$

34. $66h - 200$

35. $-2\{-3[20 + 15z - 2z] + 30z\} = -2\{-60 - 45z + 6z + 30z\}$
$= -2\{-60 - 9z\} = 120 + 18z$

36. $138z + 90$ **37.** $4x(9x^2) = 36x^3$ **38.** $50y^3$

39. $-45x^2 + 10xy$ **40.** $-27cd + 63d^2$

41. $(4x^2y^4)(3y) = 12x^2y^5$ **42.** $45a^4b^3$

43. $9x - 2y - 5x = 4x - 2y$ **44.** $3c - 16d$

Exercises 6.3B (page 272)

1. $-56x^4y^3$ **2.** $-24a^5b^5$ **3.** $x^2 + x - 6$

4. $a^2 - a - 12$ **5.** $6x^3 + 14x$ **6.** $12y^4 - 6y$

7. $y^2 - y - 72$ **8.** $z^2 + 7z - 30$

9. $15x^3 + 5x^2y - 6xy - 2y^2$ **10.** $2s^3 + 3s^2t - 8st - 12t^2$

11. $10a^3 - 2ab^2$ **12.** $21c^3 - 3cd^2$ **13.** $x^2 + 5x + 4$

14. $x^2 + 4x + 3$ **15.** $a^2 + 7a + 10$ **16.** $a^2 + 8a + 7$

17. $m^2 - 2m - 8$ **18.** $n^2 + 4n - 21$ **19.** $y^2 - 3y - 108$

20. $z^2 - 16z + 55$ **21.** $-abx - aby$ **22.** $abx - aby$

23. $ax - bx + ay - by$ **24.** $ax + bx - ay - by$ **25.** $16x^2$

26. $9x^2$ **27.** $(4 + x)(4 + x) = 16 + 8x + x^2$

28. $9 + 6x + x^2$ **29.** $(b - 4)(b - 4) = b^2 - 8b + 16$

30. $b^2 - 12b + 36$ **31.** $3x^2 + 7x + 2$ **32.** $2x^2 + 7x + 6$

33. $8x^2 + 10x - 12$ **34.** $6x^2 + 7x - 5$

35. $20x^2 - 38x + 12$ **36.** $15x^2 - 32x + 16$

37. $(2x + 5)(2x + 5) = 4x^2 + 20x + 25$

38. $9x^2 + 24x + 16$ **39.** $6x^3 + 3x^2 - 4x - 2$

40. $24a^3 - 16a^2 + 3a - 2$ **41.** $8x^2 + 26xy - 7y^2$

42. $12x^2 + 7xy - 10y^2$ **43.** $49x^2 - 140xy + 100y^2$

44. $16u^2 - 72uv + 81v^2$

45. $(3a + 2b)(3a + 2b) = 9a^2 + 12ab + 4b^2$

46. $4x^2 - 24xy + 36y^2$ **47.** $16c^2 - 9d^2$ **48.** $25e^2 - 4f^2$

Exercises 6.3C (page 276)

1.
$$
\begin{array}{r}
2x^2 + x - 1 \\
x - 3 \\
\hline
-6x^2 - 3x + 3 \\
2x^3 + x^2 - x \\
\hline
2x^3 - 5x^2 - 4x + 3
\end{array}
$$

2. $3x^3 - 5x^2 - 3x + 2$ **3.** $6x^3y^2 - 2x^2y^3 + 8xy^4$

4. $12x^2y^3 + 3x^3y^2 - 9x^4y$

5.
$$
\begin{array}{r}
x^2 + x + 1 \\
x^2 + x + 1 \\
\hline
x^2 + x + 1 \\
x^3 + x^2 + x \\
x^4 + x^3 + x^2 \\
\hline
x^4 + 2x^3 + 3x^2 + 2x + 1
\end{array}
$$

6. $x^4 - 2x^3 - x^2 + 2x + 1$ **7.** $4z^3 - 16z^2 + 64z$

8. $-5a^3 - 25a^2 - 125a$

9.
$$
\begin{array}{r}
z^2 - 4z + 16 \\
z + 4 \\
\hline
4z^2 - 16z + 64 \\
z^3 - 4z^2 + 16z \\
\hline
z^3 \qquad\quad + 64
\end{array}
$$

10. $a^3 - 125$

11.
$$
\begin{array}{r}
-3z^3 + z^2 - 5z + 4 \\
-z + 4 \\
\hline
-12z^3 + 4z^2 - 20z + 16 \\
3z^4 - z^3 + 5z^2 - 4z \\
\hline
3z^4 - 13z^3 + 9z^2 - 24z + 16
\end{array}
$$

12. $v^4 - 4v^3 + 5v + 6$

13.
$$
\begin{array}{r}
2x^2 - x + 4 \\
x^2 - 5x - 3 \\
\hline
-6x^2 + 3x - 12 \\
-10x^3 + 5x^2 - 20x \\
2x^4 - x^3 + 4x^2 \\
\hline
2x^4 - 11x^3 + 3x^2 - 17x - 12
\end{array}
$$

14. $12a^4 - 14a^3 + 18a^2 + 19a - 15$

15.
$$
\begin{array}{r}
3y^2 - 2y + 7 \\
4y^2 + 8y - 3 \\
\hline
-9y^2 + 6y - 21 \\
24y^3 - 16y^2 + 56y \\
12y^4 - 8y^3 + 28y^2 \\
\hline
12y^4 + 16y^3 + 3y^2 + 62y - 21
\end{array}
$$

16. $6x^4 - 17x^3 + 47x^2 + 21x - 36$

17.
$$
\begin{array}{r}
5x - 2 \\
5x - 2 \\
\hline
-10x + 4 \\
25x^2 - 10x \\
\hline
25x^2 - 20x + 4
\end{array}
$$

18. $4x^2 - 20x + 25$

19. $(x + y)^2 = (x + y)(x + y)$
$$= x^2 + 2xy + y^2 \quad \text{and}$$
$(x - y)^2 = (x - y)(x - y)$
$$= x^2 - 2xy + y^2$$

Then $(x + y)^2(x - y)^2 = (x^2 + 2xy + y^2)(x^2 - 2xy + y^2)$
$$
\begin{array}{r}
x^2 + 2xy + y^2 \\
x^2 - 2xy + y^2 \\
\hline
x^2y^2 + 2xy^3 + y^4 \\
-2x^3y - 4x^2y^2 - 2xy^3 \\
x^4 + 2x^3y + x^2y^2 \\
\hline
x^4 \qquad - 2x^2y^2 \qquad + y^4
\end{array}
$$

20. $x^4 - 8x^2 + 16$

21. $(x + 2)^3 = \underbrace{(x + 2)(x + 2)}(x + 2)$

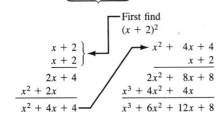

First find
$(x + 2)^2$

$$
\begin{array}{r}
x + 2 \\
x + 2 \\
\hline
2x + 4 \\
x^2 + 2x \\
\hline
x^2 + 4x + 4
\end{array}
\qquad
\begin{array}{r}
x^2 + 4x + 4 \\
x + 2 \\
\hline
2x^2 + 8x + 8 \\
x^3 + 4x^2 + 4x \\
\hline
x^3 + 6x^2 + 12x + 8
\end{array}
$$

22. $x^3 + 9x^2 + 27x + 27$

23. $(x^2 + 2x - 3)^2 = (x^2 + 2x - 3)(x^2 + 2x - 3)$
$$
\begin{array}{r}
x^2 + 2x - 3 \\
x^2 + 2x - 3 \\
\hline
-3x^2 - 6x + 9 \\
+ 2x^3 + 4x^2 - 6x \\
x^4 + 2x^3 - 3x^2 \\
\hline
x^4 + 4x^3 - 2x^2 - 12x + 9
\end{array}
$$

24. $y^4 - 8y^3 + 6y^2 + 40y + 25$

25. $(x + 3)^2 = (x + 3)(x + 3) = x^2 + 6x + 9$

Then $(x + 3)^4 = (x + 3)^2(x + 3)^2$
$$= (x^2 + 6x + 9)(x^2 + 6x + 9)$$
$$
\begin{array}{r}
x^2 + 6x + 9 \\
x^2 + 6x + 9 \\
\hline
9x^2 + 54x + 81 \\
6x^3 + 36x^2 + 54x \\
x^4 + 6x^3 + 9x^2 \\
\hline
x^4 + 12x^3 + 54x^2 + 108x + 81
\end{array}
$$

26. $x^4 + 8x^3 + 24x^2 + 32x + 16$

Exercises 6.4A (page 278)

1. $(x)^2 - (3)^2 = x^2 - 9$ **2.** $z^2 - 16$

3. $(w)^2 - (6)^2 = w^2 - 36$ **4.** $y^2 - 25$

5. $(5a)^2 - (4)^2 = 25a^2 - 16$ **6.** $36a^2 - 25$

7. $(2u)^2 - (5v)^2 = 4u^2 - 25v^2$ **8.** $9m^2 - 49n^2$

9. $(4b)^2 - (9c)^2 = 16b^2 - 81c^2$ **10.** $49a^2 - 64b^2$

11. $(2x^2)^2 - (9)^2 = 4x^4 - 81$ **12.** $100y^4 - 9$

13. $(1)^2 - (8z^3)^2 = 1 - 64z^6$ **14.** $81v^8 - 1$

15. $(5xy)^2 - (z)^2 = 25x^2y^2 - z^2$ **16.** $100a^2b^2 - c^2$

17. $(7mn)^2 - (2rs)^2 = 49m^2n^2 - 4r^2s^2$ **18.** $64h^2k^2 - 25e^2f^2$

Exercises 6.4B (page 280)

1. $(x)^2 - 2(x)(1) + (1)^2 = x^2 - 2x + 1$ **2.** $x^2 - 10x + 25$

3. $(x)^2 + 2(x)(3) + (3)^2 = x^2 + 6x + 9$ **4.** $x^2 + 8x + 16$

5. $(4x)^2 - 2(4x)(1) + (1)^2 = 16x^2 - 8x + 1$

6. $49x^2 - 14x + 1$

7. $(12x)^2 + 2(12x)(1) + 1^2 = 144x^2 + 24x + 1$

8. $121x^2 + 22x + 1$

9. $(2s)^2 + 2(2s)(4t) + (4t)^2 = 4s^2 + 16st + 16t^2$

10. $9u^2 + 42uv + 49v^2$

11. $(5x)^2 - 2(5x)(3y) + (3y)^2 = 25x^2 - 30xy + 9y^2$

12. $16x^2 - 56xy + 49y^2$

13. $(3x)^2 + 2(3x)(2z) + (2z)^2 = 9x^2 + 12xz + 4z^2$

14. $4x^2 + 28xs + 49s^2$

Exercises 6.5A (page 281)

1. $\dfrac{3x}{3} + \dfrac{6}{3} = x + 2$ **2.** $2x + 3$ **3.** $\dfrac{4}{4} + \dfrac{8x}{4} = 1 + 2x$

4. $1 - 2x$ **5.** $\dfrac{6x}{2} + \dfrac{-8y}{2} = 3x - 4y$ **6.** $x - 2y$

7. $\dfrac{2x^2}{x} + \dfrac{3x}{x} = 2x + 3$ **8.** $4y - 3$

9. $\dfrac{15x^3}{5x^2} + \dfrac{-5x^2}{5x^2} = 3x - 1$ **10.** $2y^2 - 1$

11. $\dfrac{3a^2b}{ab} + \dfrac{-ab}{ab} = 3a - 1$ **12.** $5n - 1$

13. $\dfrac{8x^7}{4x^2} + \dfrac{4x^5}{4x^2} + \dfrac{-12x^3}{4x^2} = 2x^5 + x^3 - 3x$ **14.** $y^4 + 3y^2 - 2y$

15. $\dfrac{5x^5}{-5x^2} + \dfrac{-4x^3}{-5x^2} + \dfrac{10x^2}{-5x^2} = -x^3 + \frac{4}{5}x - 2$

16. $-y^2 + \frac{5}{7}y - 2$ **17.** $\dfrac{-15x^2y^2z^2}{-5xyz} + \dfrac{-30xyz}{-5xyz} = 3xyz + 6$

18. $3abc + 2$ **19.** $\dfrac{13x^4y^2}{13x^2y^2} + \dfrac{-26x^2y^3}{13x^2y^2} + \dfrac{39x^2y^2}{13x^2y^2} = x^2 - 2y + 3$

20. $3n^3 - 5m - 2$

Exercises 6.5B (page 287)

NOTE: Sign changes for the subtractions are not shown here.

1.
$$
\begin{array}{r}
x + 3 \\
x + 2 \overline{) x^2 + 5x + 6} \\
\underline{x^2 + 2x} \\
3x + 6 \\
\underline{3x + 6} \\
0
\end{array}
$$

2. $x + 2$

3.
$$
\begin{array}{r}
x + 3 \\
x - 4 \overline{) x^2 - x - 12} \\
\underline{x^2 - 4x} \\
3x - 12 \\
\underline{3x - 12} \\
0
\end{array}
$$

4. $x - 4$

5.
$$
\begin{array}{r}
2x + 3 \\
3x - 2 \overline{) 6x^2 + 5x - 6} \\
\underline{6x^2 - 4x} \\
9x - 6 \\
\underline{9x - 6} \\
0
\end{array}
$$

6. $4x + 5$

7.
$$
\begin{array}{r}
3v + 8 \quad \text{R } 66 \\
5v - 7 \overline{) 15v^2 + 19v + 10} \\
\underline{15v^2 - 21v} \\
40v + 10 \\
\underline{40v - 56} \\
66
\end{array}
$$

8. $5v - 7$ R 52

9.
$$
\begin{array}{r}
3x + 5 \\
2x - 3 \overline{) 6x^2 + x - 15} \\
\underline{6x^2 - 9x} \\
10x - 15 \\
\underline{10x - 15} \\
0
\end{array}
$$

10. $4x - 2$

11.
$$
\begin{array}{r}
4x^2 + 8x + 8 \quad \text{R } -6 \\
-x + 2 \overline{) -4x^3 + 0x^2 + 8x + 10} \\
\underline{-4x^3 + 8x^2} \\
-8x^2 + 8x \\
\underline{-8x^2 + 16x} \\
-8x + 10 \\
\underline{-8x + 16} \\
-6
\end{array}
$$

12. $x^2 + 3x - 3$ R -6

13.
$$
\begin{array}{r}
3a - 2b \quad \text{R } 7b^2 \\
2a + 3b \overline{) 6a^2 + 5ab + b^2} \\
\underline{6a^2 + 9ab} \\
-4ab + b^2 \\
\underline{-4ab - 6b^2} \\
7b^2
\end{array}
$$

14. $2a + 3b$ R $5b^2$

15.
$$
\begin{array}{r}
a^2 + 2a + 4 \\
a - 2 \overline{) a^3 + 0a^2 + 0a - 8} \\
\underline{a^3 - 2a^2} \\
2a^2 + 0a \\
\underline{2a^2 - 4a} \\
4a - 8 \\
\underline{4a - 8} \\
0
\end{array}
$$

16. $c^2 + 3c + 9$

17.
$$
\begin{array}{r}
x^2 + 4x + 8 \quad \text{R } 17 \\
x - 4 \overline{) x^3 + 0x^2 - 8x - 15} \\
\underline{x^3 - 4x^2} \\
4x^2 - 8x \\
\underline{4x^2 - 16x} \\
8x - 15 \\
\underline{8x - 32} \\
17
\end{array}
$$

18. $2x^2 - 3x + 6$ R -3

19.
$$
\begin{array}{r}
x^2 + x - 1 \quad \text{R } 3 \\
x^2 + x - 1 \overline{) x^4 + 2x^3 - x^2 - 2x + 4} \\
\underline{x^4 + x^3 - x^2} \\
x^3 + 0x^2 - 2x \\
\underline{x^3 + x^2 - x} \\
-x^2 - x + 4 \\
\underline{-x^2 - x + 1} \\
3
\end{array}
$$

20. $x^2 - x + 1$ R 6

21.
$$
\begin{array}{r}
x^2 + x + 1 \quad \text{R } -8 \\
x^2 + 2x + 3 \overline{) x^4 + 3x^3 + 6x^2 + 5x - 5} \\
\underline{x^4 + 2x^3 + 3x^2} \\
x^3 + 3x^2 + 5x \\
\underline{x^3 + 2x^2 + 3x} \\
x^2 + 2x - 5 \\
\underline{x^2 + 2x + 3} \\
-8
\end{array}
$$

22. $x^2 + 2x + 3$ R 2

Exercises 6.6 (page 288)

1. If x is the smaller integer, the larger one is $x + 2$.
$x(x + 2) = x^2 + 2x$

2. $3x + 6$

3. The length is $4x$ and the height is $(3 + x)$. The volume is
$4x(x)(3 + x) = 4x^2(3 + x) = 12x^2 + 4x^3$

4. $(3x - 4)(4 + x)(x) = 3x^3 + 8x^2 - 16x$

5. The length of the photograph is $(x + 2)$, so the area of the photograph is $x(x + 2)$. The width of the mat is $x + 2 + 2$ (or $x + 4$), and the length of the mat is $(x + 2) + 3 + 3$ (or $x + 8$). The *required* area is

$(x + 4)(x + 8) - x(x + 2) = x^2 + 12x + 32 - x^2 - 2x$
$= 10x + 32$

6. $8x + 28$

7. The height will be x, and the width and length will both be $(10 - 2x)$. The volume is

$x(10 - 2x)(10 - 2x) = x(100 - 40x + 4x^2)$
$= 100x - 40x^2 + 4x^3$

8. $168x - 52x^2 + 4x^3$

Review Exercises 6.7 (page 291)

1a. Third degree   **b.** Third degree

2. Not a polynomial   **3.** Not a polynomial

4a. Fourth degree   **b.** First degree

5a. $3x^4 + 7x^2 + x - 6$ **b.** 3

6a. $y^4 + 3xy^2 + 3x^2y + x^3$ **b.** 1

7. $5x^2y + 3xy^2 - 4y^3 + 2xy^2 + 4y^3 + 3x^2y = 8x^2y + 5xy^2$

8. $-18x^4y^4$ **9.** $9xy^2(x^2y) + 9xy^2(-2xy) = 9x^3y^3 - 18x^2y^3$

10. $5a^2b + 9ab^2 - 2b^2 + 4a - 8$ **11.** $\dfrac{8x}{2} + \dfrac{2}{2} = 4x + 1$

12. $-20x^3y^3 + 12xy^3$ **13.** $x^2 + 2x - 35$ **14.** $x^3 - 8$

15.
$$
\begin{array}{r}
3x + 0 \ \text{R } 10 \\
2x - 3\overline{)6x^2 - 9x + 10} \\
\underline{6x^2 - 9x} \\
+ 10
\end{array}
$$
16. $x^2 - 16x + 64$

17. $(x^2 + 3 - 5x) - (4x^3 - x + 4x^2 - 1)$
$= (x^2 + 3 - 5x) + (-4x^3 + x - 4x^2 + 1)$
$= \underline{x^2} + \underline{\underline{3}} - 5x - 4x^3 + x - \underline{4x^2} + \underline{\underline{1}} = -4x^3 - 3x^2 - 4x + 4$

18. $a^2 - 1$

19. $(2m^2 - 5) - [(7 - m^2) + (-4m^2 + 3)]$
$= (2m^2 - 5) - [\underline{7} - m^2 - 4m^2 + \underline{3}] = (2m^2 - 5) - [10 - 5m^2]$
$= (2m^2 - 5) + [-10 + 5m^2] = \underline{2m^2} - 5 - 10 + \underline{5m^2}$
$= 7m^2 - 15$

20. $-14x^2 + 22x - 2$ **21.** $12x^3y^4 - 15x^2y^3 - 30xy^2$

22. $x^3 - 6x^2 + 12x - 8$ **23.** $9a^2 - 25$

24. $9 + 6x + 6y + x^2 + 2xy + y^2$ **25.**
$$
\begin{array}{r}
-3y + 2 \\
3y + 2\overline{)-9y^2 + 0y + 4} \\
\underline{-9y^2 - 6y} \\
6y + 4 \\
\underline{6y + 4}
\end{array}
$$

26. $3ab^2 - \frac{4}{5}b + 2$ **27.** $-3xyz - z^2$ **28.** $3xy$

29.
$$
\begin{array}{l}
7a^2 - 3ab + 5 - b^2 \qquad 7a^2 - 3ab + 5 - b^2 \\
\underline{-(9a^2 + 2ab - 8 + b^2)} \Rightarrow \underline{+(-9a^2 - 2ab + 8 - b^2)} \\
 -2a^2 - 5ab + 13 - 2b^2
\end{array}
$$

30. $6x^3 - 7x^2 + x - 4$

31. If x is the smallest integer, the second integer is $x + 2$ and the third is $x + 4$.

$x(x + 2)(x + 4) = x(x^2 + 6x + 8) = x^3 + 6x^2 + 8x$

32. $9\pi x^3$, where the height is x

Chapter 6 Diagnostic Test (page 297)

Following each problem number is the textbook section number (in parentheses) where that kind of problem is discussed.

1. (6.1) **a.** Second degree   **b.** Third degree   **c.** -4
d. -4 **e.** 0

2. (6.2) **a.**
$$
\begin{array}{r}
-7x^3 + 4x^2 + 3 \\
3x^3 + 6x - 5 \\
\underline{7x^2 - 4x + 8} \\
-4x^3 + 11x^2 + 2x + 6
\end{array}
$$

  b. $(6xy^2 - 5xy) + (17xy - 7x^2y) + (3xy^2 - y^3)$
$= \underline{6xy^2} - \underline{\underline{5xy}} + \underline{\underline{17xy}} - 7x^2y + \underline{3xy^2} - y^3$
$= 9xy^2 + 12xy - 7x^2y - y^3$

3. (6.2) **a.** $(-3x^2 - 6x + 9) - (8 - 2x + 5x^2)$
$= (-3x^2 - 6x + 9) + (-8 + 2x - 5x^2)$
$= -\underline{3x^2} - \underline{\underline{6x}} + \underline{9} - \underline{8} + \underline{\underline{2x}} - \underline{5x^2} = -8x^2 - 4x + 1$

  b.
$$
\begin{array}{l}
-6a^3 + 5a^2 + 4 \qquad -6a^3 + 5a^2 + 4 \\
\underline{-(4a^3 + 6a - 7)} \Rightarrow \underline{+(-4a^3 - 6a + 7)} \\
 -10a^3 + 5a^2 - 6a + 11
\end{array}
$$

4. (6.3) **a.** $(8x - 3) - (10x - 5) + (9 - 5x)$
$= (8x - 3) + (-10x + 5) + (9 - 5x)$
$= \underline{8x} - \underline{\underline{3}} - \underline{10x} + \underline{\underline{5}} + \underline{\underline{9}} - \underline{5x} = -7x + 11$

  b. $-6xy(3x^2 - 5xy^2 + 8y) = -18x^3y + 30x^2y^3 - 48xy^2$

5a. (6.3) $(9x - 7)(8x + 9) = 72x^2 + 25x - 63$

b. (6.4) $(3x - 8)^2 = (3x)^2 - 2(3x)(8) + (8)^2 = 9x^2 - 48x + 64$

6. (6.3)
$$
\begin{array}{r}
2x^2 - x - 4 \\
\underline{x^2 + 3x - 5} \\
-10x^2 + 5x + 20 \\
6x^3 - 3x^2 - 12x \\
\underline{2x^4 - x^3 - 4x^2} \\
2x^4 + 5x^3 - 17x^2 - 7x + 20
\end{array}
$$

7. (6.3) $(4abc^2)(3b)(-2a^2c) = (4)(3)(-2)(aa^2)(bb)(c^2c) = -24a^3b^2c^3$

8. (6.5) **a.** $\dfrac{8x^4 - 4x^3 + 12x^2}{4x^2} = \dfrac{8x^4}{4x^2} + \dfrac{-4x^3}{4x^2} + \dfrac{12x^2}{4x^2} = 2x^2 - x + 3$

  b.
$$
\begin{array}{r}
5x + 2 \ \text{R } 3 \\
3x - 1\overline{)15x^2 + x + 1} \\
\underline{-15x^2 - 5x} \\
6x + 1 \\
\underline{6x - 2} \\
3
\end{array}
$$

9. (6.5)
$$
\begin{array}{r}
x^2 - 2x - 8 \ \text{R } -20 \\
x - 4\overline{)x^3 - 6x^2 + 0x + 12} \\
\underline{x^3 - 4x^2} \\
-2x^2 + 0x \\
\underline{-2x^2 + 8x} \\
-8x + 12 \\
\underline{-8x + 32} \\
-20
\end{array}
$$

10. (6.6) Let $\qquad x$ = the width of the cardboard
$\qquad x + 5$ = the length of the cardboard
$\qquad x - 4$ = the width of the *box*
$\qquad (x + 5) - 4$ = the length of the *box* $\qquad (x + 5) - 4 = x + 1$
$\qquad 2$ = the height of the box

Volume $= (x + 1)(x - 4)(2) = (x^2 - 3x - 4)(2)$
$\qquad\qquad = 2x^2 - 6x - 8$

Cumulative Review Exercises: Chapters 1–6 (page 298)

1. $F = \frac{9}{5}C + 32$ **2.** -13

$F = \dfrac{9}{\cancel{5}}(\cancel{-20}) + 32$

$F = -36 + 32$

$F = -4$

3. $5[11 - 2(-3)] - 4(13 - 5) = 5[11 + 6] - 4(13 - 5)$
$= 5[17] - 4(8) = 85 - 32 = 53$

4. $6x^3 - 4x + 24$

5. $-25 \cdot 4 - 15 \div 3(5) = -25 \cdot 4 - 5(5) = -100 - 25 = -125$

6. -22 **7.**
$$
\begin{array}{r}
2\, \lfloor\underline{294} \\
3\, \lfloor\underline{147} \\
7\, \lfloor\underline{49} \\
7
\end{array}
$$
Therefore, $294 = 2 \cdot 3 \cdot 7^2$.

8. $6x^3 - 19x^2 + 16x - 15$ **9.** $(9x)^2 - (2)^2 = 81x^2 - 4$

10. $25x^2 - 30x + 9$

11.
$$\begin{array}{r} 5x^2 + 3x + 7 \quad \text{R } 6 \\ x - 1\overline{)5x^3 - 2x^2 + 4x - 1} \\ \underline{5x^3 - 5x^2} \\ 3x^2 + 4x \\ \underline{3x^2 - 3x} \\ 7x - 1 \\ \underline{7x - 7} \\ 6 \end{array}$$

12. 10^5 or $100{,}000$ **13.** $x^{3c-c} = x^{2c}$ **14.** $\dfrac{8a^6}{b^3}$

15. $\left(\dfrac{\overset{2}{6}y^{-1}}{\underset{1}{3}y^3}\right)^2 = (2y^{-1-3})^2 = (2y^{-4})^2 = 2^2 y^{-8} = \dfrac{4}{y^8}$ **16.** $\dfrac{x^4}{9}$

17. Twenty-two is what percent of 25? **18.** $175
Let x = percent.
$25x = 22$
$$\dfrac{25x}{25} = \dfrac{22}{25}$$
$x = \frac{22}{25} = 0.88 = 88\%$; Susan's score is 88%.

19. Let x = number of pounds of walnuts
$10 - x$ = number of pounds of almonds
$4.50x + 3.10(10 - x) = 3.52(10)$
$450x + 310(10 - x) = 3{,}520$
$450x + 3{,}100 - 310x = 3{,}520$
$$\begin{array}{r} 140x + 3{,}100 = 3{,}520 \\ -3{,}100 \quad -3{,}100 \\ \hline 140x = 420 \end{array}$$
$$\dfrac{140x}{140} = \dfrac{420}{140}$$
$x = 3$
$10 - x = 7$
Three pounds of walnuts and 7 pounds of almonds should be used.

20. 15, 25, and 35

Exercises 7.1A (page 305)

1. $4(3x + 2)$ **2.** $3(2x + 3)$ **3.** Not factorable
4. Not factorable **5.** $2(x + 4)$ **6.** $3(x + 3)$
7. $5(a - 2)$ **8.** $7(b - 2)$ **9.** $3(2y - 1)$ **10.** $5(3z - 1)$
11. $3x(3x + 1)$ **12.** $4y(2y - 1)$ **13.** $5a^2(2a - 5)$
14. $9b^2(3 - 2b^2)$ **15.** Not factorable **16.** Not factorable
17. $2ab(a + 2b)$ **18.** $3mn^2(1 + 2m)$ **19.** $6c^2d^2(2c - 3d)$
20. $15ab^3(1 - 3ab)$ **21.** $4x(x^2 - 3 - 6x)$
22. $6y(3 - y - 5y^2)$ **23.** Not factorable **24.** Not factorable
25. $2x(4x^2 - 3x + 1)$ **26.** $3y^2(3y^2 + 2y - 1)$
27. $8(3a^4 + a^2 - 5)$ **28.** $15(3b^3 - b^4 - 2)$
29. $14xy^3(-x^7y^6 + 3x^4y - 2)$ **30.** $7uv^5(-3u^6v^3 - 9 + 5u)$
31. Not factorable **32.** Not factorable
33. $11a^{10}b^4(-4a^4b^3 - 3b + 2a)$
34. $13e^8f^5(-2f + e^2f^3 - 3e^4)$ **35.** $2(9u^{10}v^5 + 12 - 7u^{10}v^6)$
36. $15(2a^3b^4 - 1 + 3a^8b^7)$ **37.** $6y^2(3x^3y^2 - 2z^3 - 8x^4y)$
38. $8m^3(4m^2n^7 - 3m^5p^9 - 5n^6)$

Exercises 7.1B (page 307)

1. $(s + t)(c + b)$ [GCF is $(s + t)$] **2.** $(b + c)(a + d)$
3. $(a - b)(x + 5)$ [GCF is $(a - b)$] **4.** $(s - t)(y + 7)$
5. $(u + v)(x - 3)$ [GCF is $(u + v)$] **6.** $(t + u)(s - 2)$
7. $(x - y)(8 - a)$ [GCF is $(x - y)$] **8.** $(a - b)(7 - c)$
9. $(s - t)(4 + u - v)$ [GCF is $(s - t)$]
10. $(x - y)(a + 5 - b)$
11. $3xy^2(a + b)(xy + 3)$ [GCF is $3xy^2(a + b)$]
12. $5ab(s + t^2)(2b^2 + 3a)$
13. $2u^3v^4(x - y)(2v + 3u)$ [GCF is $2u^3v^4(x - y)$]
14. $5x^2y^3(3a - 2b)(2y^2 + 3x^3)$

Exercises 7.2A (page 308)

1. 8 **2.** 9 **3.** $2x$ **4.** $3y$ **5.** $10a^4$ **6.** $7b^3$
7. m^2n **8.** u^5v^3 **9.** $x^{10/2}y^{4/2} = x^5y^2$ **10.** x^6y^4
11. $5a^{4/2}b^{2/2} = 5a^2b^1 = 5a^2b$ **12.** $10b^2c$
13. $6e^{8/2}f^{2/2} = 6e^4f^1 = 6e^4f$ **14.** $9h^6k^7$
15. $10a^{10/2}y^{2/2} = 10a^5y^1 = 10a^5y$ **16.** $11a^{12}b^2$
17. $3a^{4/2}b^{2/2}c^{6/2} = 3a^2b^1c^3 = 3a^2bc^3$ **18.** $12x^4yz^3$

Exercises 7.2B (page 311)

1. $(\sqrt{m^2} + \sqrt{n^2})(\sqrt{m^2} - \sqrt{n^2}) = (m + n)(m - n)$
2. $(u + v)(u - v)$
3. $(\sqrt{x^2} + \sqrt{9})(\sqrt{x^2} - \sqrt{9}) = (x + 3)(x - 3)$
4. $(x + 5)(x - 5)$
5. $(\sqrt{a^2} + \sqrt{1})(\sqrt{a^2} - \sqrt{1}) = (a + 1)(a - 1)$
6. $(1 + b)(1 - b)$
7. $(\sqrt{4c^2} + \sqrt{1})(\sqrt{4c^2} - \sqrt{1}) = (2c + 1)(2c - 1)$
8. $(4d + 1)(4d - 1)$
9. $(\sqrt{16x^2} + \sqrt{9y^2})(\sqrt{16x^2} - \sqrt{9y^2}) = (4x + 3y)(4x - 3y)$
10. $(5a + 2b)(5a - 2b)$ **11.** Not factorable
12. Not factorable **13.** $2x(2x^3 - 1)$ [GCF is $2x$]
14. $3a(3a^3 - 1)$ **15.** Not factorable **16.** Not factorable
17. $(\sqrt{49u^4} + \sqrt{36v^4})(\sqrt{49u^4} - \sqrt{36v^4}) = (7u^2 + 6v^2)(7u^2 - 6v^2)$
18. $(9m^3 + 10n^2)(9m^3 - 10n^2)$
19. $x^6 - a^4 = (\sqrt{x^6} + \sqrt{a^4})(\sqrt{x^6} - \sqrt{a^4}) = (x^3 + a^2)(x^3 - a^2)$
20. $(b + y^3)(b - y^3)$
21. $2x^2 - 18 = 2(x^2 - 9) = 2(x + 3)(x - 3)$
22. $3(x + 2)(x - 2)$ **23.** Not factorable **24.** Not factorable
25. $(\sqrt{a^2b^2} + \sqrt{c^2d^2})(\sqrt{a^2b^2} - \sqrt{c^2d^2}) = (ab + cd)(ab - cd)$
26. $(mn + rs)(mn - rs)$
27. $(\sqrt{49} + \sqrt{25w^2z^2})(\sqrt{49} - \sqrt{25w^2z^2}) = (7 + 5wz)(7 - 5wz)$
28. $(6 + 5uv)(6 - 5uv)$
29. $(\sqrt{4h^4k^4} + \sqrt{1})(\sqrt{4h^4k^4} - \sqrt{1}) = (2h^2k^2 + 1)(2h^2k^2 - 1)$
30. $(3x^2y^2 + 1)(3x^2y^2 - 1)$
31. $(\sqrt{81a^4b^6} + \sqrt{16m^2n^8})(\sqrt{81a^4b^6} - \sqrt{16m^2n^8})$
$= (9a^2b^3 + 4mn^4)(9a^2b^3 - 4mn^4)$
32. $(7c^4d^2 + 10e^3f)(7c^4d^2 - 10e^3f)$ **33.** $7x^2(7x^2y^2 - 1)$
34. $5y^2(5x^2y^2 - 1)$ **35.** $5x(x^2 - 9y^2) = 5x(x + 3y)(x - 3y)$
36. $11r(r + 2s)(r - 2s)$

Exercises 7.3A (page 318)

1. $(x + 2)(x + 4)$ **2.** $(x + 1)(x + 8)$ **3.** $(x + 1)(x + 4)$

4. $(x + 2)(x + 2)$ or $(x + 2)^2$ **5.** $(k + 1)(k + 6)$

6. $(k + 2)(k + 3)$ **7.** $u^2 + 7u + 10 = (u + 2)(u + 5)$

8. $(u + 1)(u + 10)$ **9.** Not factorable **10.** Not factorable

11. $(b - 2)(b - 7)$ **12.** $(b - 1)(b - 14)$

13. $(z - 4)(z - 5)$ **14.** $(z - 2)(z - 10)$

15. $x^2 - 11x + 18 = (x - 2)(x - 9)$ **16.** $(x - 3)(x - 6)$

17. $(x + 10)(x - 1)$ **18.** $(y - 5)(y + 2)$

19. $(z - 3)(z + 2)$ **20.** $(m + 6)(m - 1)$

21. $5x(x + 2)$ [The GCF is $5x$] **22.** $4y^2(2y + 1)$

23. $(x + 5)(x - 1)$ **24.** $(y + 7)(y - 1)$ **25.** Not factorable

26. Not factorable **27.** $z^3(z^2 + 9z - 10) = z^3(z + 10)(z - 1)$

28. $x^2(x + 8)(x - 1)$ **29.** Not factorable **30.** Not factorable

31. $(u^2 - 4)(u^2 + 16) = (u + 2)(u - 2)(u^2 + 16)$

32. $(v^2 + 2)(v^2 - 32)$ **33.** $(4 - v)(4 - v)$ or $(4 - v)^2$ or $(v - 4)^2$

34. $(2 - v)(8 - v)$ or $(v - 2)(v - 8)$ **35.** $(b + 4d)(b - 15d)$

36. $(c + 20x)(c - 3x)$ **37.** $(r + 3s)(r - 16s)$

38. $(s + 24t)(s - 2t)$ **39.** $x^2(x^2 + 2x - 35) = x^2(x + 7)(x - 5)$

40. $x^2(x + 8)(x - 6)$ **41.** $x(x^2 + 14x - 15) = x(x + 15)(x - 1)$

42. $x^2(x + 9)(x - 1)$ **43.** $3(x^2 + 2x - 8) = 3(x + 4)(x - 2)$

44. $5(x + 5)(x - 2)$ **45.** $4(x^2 - 4x + 3) = 4(x - 3)(x - 1)$

46. $2(x - 3)(x - 4)$ **47.** $x^2(x^2 + 6x + 1)$

48. $y^2(y^2 + 5y + 1)$ **49.** $(x^2 + 3)(x^2 - 3)$

50. $(a^2 + 5)(a^2 - 5)$

Exercises 7.3B (page 324)

1. $(x + 2)(3x + 1)$ **2.** $(x + 1)(3x + 2)$

3. $(x + 1)(5x + 2)$ **4.** $(x + 2)(5x + 1)$

5. $4x^2 + 7x + 3 = (x + 1)(4x + 3)$ **6.** $(x + 3)(4x + 1)$

7. Not factorable **8.** Not factorable **9.** $(a - 3)(5a - 1)$

10. $(m - 1)(5m - 3)$ **11.** $(b - 7)(3b - 1)$

12. $(u - 1)(3u - 7)$ **13.** $(z - 7)(5z - 1)$

14. $(z - 1)(5z - 7)$ **15.** $3n^2 + 14n - 5 = (n + 5)(3n - 1)$

16. $(k - 1)(5k + 7)$ **17.** $(3x + 7)(3x - 7)$

18. $(4y + 1)(4y - 1)$ **19.** $(x + 3y)(7x + 2y)$

20. $(a + 6b)(7a + b)$ **21.** $(h - k)(7h - 4k)$

22. $(h - 2k)(7h - 2k)$ **23.** Not factorable

24. Not factorable **25.** $(7x - 3)(7x - 3)$ or $(7x - 3)^2$

26. $(5x - 2)(5x - 2)$ or $(5x - 2)^2$

27. $3u(6u^2 + 13u - 5) = 3u(2u + 5)(3u - 1)$

28. $2y(2y - 1)(3y + 5)$ **29.** $4(x^2 + x + 1)$

30. $6(x^2 + x + 3)$ **31.** $7(x^2 - 7)$ **32.** $5(x^2 - 5)$

33. $(3 - v)(2 - 5v)$ **34.** $(1 - v)(6 - 5v)$

35. $3x^2 + 20x + 12 = (3x + 2)(x + 6)$ **36.** $(3x + 4)(x + 3)$

37. $5(9x^2 - 24x + 16) = 5(3x - 4)(3x - 4)$ or $5(3x - 4)^2$

38. $4(4x - 3)(4x - 3)$ or $4(4x - 3)^2$ **39.** $(2x - 3)(4x + 5)$

40. $(8x + 5)(x - 3)$ **41.** $(2y - 5)(3y - 2)$

42. $(2y - 1)(3y - 10)$ **43.** $(3a - 2)(3a + 10)$

44. $(3b + 5)(3b - 4)$ **45.** $(2e^2 - 5)(3e^2 + 4)$

46. $(5f^2 + 3)(2f^2 - 7)$ **47.** Not factorable **48.** Not factorable

Review Exercises 7.4 (page 325)

1. $4(2x - 1)$ **2.** Not factorable **3.** $(m + 2)(m - 2)$

4. Not factorable **5.** $(x + 3)(x + 7)$ **6.** Not factorable

7. $2u(u + 2)$ **8.** $3b(1 - 2b + 4b^2)$ **9.** $(z + 2)(z - 9)$

10. Not factorable **11.** $(4x - 1)(x - 6)$

12. $(2x + 5)(2x + 5)$ or $(2x + 5)^2$

13. $9(k^2 - 16) = 9(k + 4)(k - 4)$ **14.** Not factorable

15. $2(4 - a^2) = 2(2 + a)(2 - a)$ **16.** $2(5c - 2)(c - 4)$

17. $3uv(5u - 1)$ **18.** $5ab(b - 2a - 1)$

19. $(2x + 3y)(5x - 8y)$ **20.** $4x^2y^2(2x^3 - 3y^2 - 4)$

Exercises 7.5 (page 331)

1. $am + bm + an + bn = m(a + b) + n(a + b) = (a + b)(m + n)$

2. $(u + v)(c + d)$

3. $st + 4t + 3su + 12u = t(s + 4) + 3u(s + 4) = (s + 4)(t + 3u)$

4. $(c + 1)(5z + 7t)$

5. $3xr - 6yr + 4x - 8y = 3r(x - 2y) + 4(x - 2y) =$
$(x - 2y)(3r + 4)$

6. $(2s - 3t)(2m + 5n)$

7. $mx - nx - my + ny = x(m - n) - y(m - n) = (m - n)(x - y)$

8. $(h - k)(a - b)$

9. $xy + x - y - 1 = x(y + 1) - 1(y + 1) = (y + 1)(x - 1)$

10. $(a - 1)(d + 1)$

11. $3a^2 - 6ab + 2a - 4b = 3a(a - 2b) + 2(a - 2b)$
$= (a - 2b)(3a + 2)$

12. $(h - 3k)(2h + 5)$

13. $6e^2 - 2ef - 9e + 3f = 2e(3e - f) - 3(3e - f)$
$= (3e - f)(2e - 3)$

14. $(2m - n)(4m - 3)$

15. $h^2 - k^2 + 2h + 2k = (h + k)(h - k) + 2(h + k)$
$= (h + k)(h - k + 2)$

16. $(x + y)(x - y + 4)$

17. $x^3 + 3x^2 - 4x + 12 = x^2(x + 3) - 4(x - 3)$; not factorable

18. Not factorable

19. $a^3 - 2a^2 - 4a + 8 = a^2(a - 2) - 4(a - 2) = (a - 2)(a^2 - 4)$
$= (a - 2)(a + 2)(a - 2) = (a - 2)^2(a + 2)$

20. $(x - 3)(x + 3)(x - 3) = (x - 3)^2(x + 3)$

21. $10xy - 15y + 8x - 12 = 5y(2x - 3) + 4(2x - 3)$
$= (2x - 3)(5y + 4)$

22. $(7 + 3n)(5 - 6m)$

23. $a^2 - 4 + ab - 2b = (a + 2)(a - 2) + b(a - 2)$
$= (a - 2)(a + 2 + b)$

24. $(x - 5)(x + 5 - y)$

Exercises 7.6 (page 334)

1. MP $= 3 \cdot 2 = 6$

$6 = (-1)(-6) = (1)(6) \Rightarrow (1) + (6) = 7$
$ = (-2)(-3) = (2)(3)$

$\underbrace{3x^2 + 1x}_{} + \underbrace{6x + 2}_{}$
$= x(3x + 1) + 2(3x + 1)$
$= (3x + 1)(x + 2)$

2. $(x + 1)(3x + 2)$

3. MP $= 5 \cdot 2 = 10$

$10 = (-1)(-10 = (1)(10)$
$ = (-2)(-5) \; = (2)(5) \Rightarrow (2) + (5) = 7$

$\underbrace{5x^2 + 2x}_{} + \underbrace{5x + 2}_{}$
$= x(5x + 2) + 1(5x + 2)$
$= (5x + 2)(x + 1)$

4. $(x + 2)(5x + 1)$

5. $4x^2 + 7x + 3$

MP $= 4 \cdot 3 = 12$

$12 = (-1)(-12) = (1)(12)$
$ = (-2)(-6) \; = (2)(6)$
$ = (-3)(-4) \; = (3)(4) \Rightarrow (3) + (4) = 7$

$\underbrace{4x^2 + 3x}_{} + \underbrace{4x + 3}_{}$
$= x(4x + 3) + 1(4x + 3)$
$= (4x + 3)(x + 1)$

6. $(x + 3)(4x + 1)$

7. MP $= 5 \cdot 4 = 20$

$20 = (1)(20) = (-1)(-20)$
$ = (2)(10) = (-2)(-10)$ — None of the sums of these pairs is $+20$
$ = (4)(5) \; = (-4)(-5)$

Not factorable

8. Not factorable

9. MP $= 5 \cdot 3 = 15$

$15 = (1)(15) = (-1)(-15) \Rightarrow (-1) + (-15) = -16$
$ = (3)(5) \; = (-3)(-5)$

$\underbrace{5a^2 - 1a}_{} - \underbrace{15a + 3}_{}$
$= a(5a - 1) - 3(5a - 1)$
$= (5a - 1)(a - 3)$

10. $(m - 1)(5m - 3)$

11. MP $= 3 \cdot 7 = 21$

$21 = (1)(21) = (-1)(-21) \Rightarrow (-1) + (-21) = -22$
$ = (3)(7) \; = (-3)(-7)$

$\underbrace{3b^2 - 1b}_{} - \underbrace{21b + 7}_{}$
$= b(3b - 1) - 7(3b - 1)$
$= (3b - 1)(b - 7)$

12. $(u - 1)(3u - 7)$

13. MP $= 5 \cdot 7 = 35$

$35 = (1)(35) = (-1)(-35) \Rightarrow (-1) + (-35) = -36$
$ = (5)(7) \; = (-5)(-7)$

$\underbrace{5z^2 - 1z}_{} - \underbrace{35z + 7}_{}$
$= z(5z - 1) - 7(5z - 1)$
$= (5z - 1)(z - 7)$

14. $(z - 1)(5z - 7)$

15. $3n^2 + 14n - 5$

MP $= 3 \cdot (-5) = -15$

$-15 = (1)(-15) = (-1)(15) \Rightarrow (-1) + (15) = 14$
$ = (3)(-5) \; = (-3)(5)$

$\underbrace{3n^2 - 1n}_{} + \underbrace{15n - 5}_{}$
$= n(3n - 1) + 5(3n - 1)$
$= (3n - 1)(n + 5)$

16. $(k - 1)(5k + 7)$

17. (MP method need not be used.) $(3x + 7)(3x - 7)$

18. $(4y + 1)(4y - 1)$

19. MP $= 7 \cdot 6 = 42$

$42 = (-1)(-42) = (1)(42)$
$ = (-2)(-21) = (2)(21) \Rightarrow (2) + (21) = 23$
$ = (-3)(-14) = (3)(14)$
$ = (-6)(-7) \; = (6)(7)$

$\underbrace{7x^2 + 2xy}_{} + \underbrace{21xy + 6y^2}_{}$
$= x(7x + 2y) + 3y(7x + 2y)$
$= (7x + 2y)(x + 3y)$

20. $(a + 6b)(7a + b)$

21. MP $= 7 \cdot 4 = 28$

$28 = (1)(28) = (-1)(-28)$
$ = (2)(14) = (-2)(-14)$
$ = (4)(7) \; = (-4)(-7) \Rightarrow (-4) + (-7) = -11$

$\underbrace{7h^2 - 4hk}_{} - \underbrace{7hk + 4k^2}_{}$
$= h(7h - 4k) - k(7h - 4k)$
$= (7h - 4k)(h - k)$

22. $(h - 2k)(7h - 2k)$

23. MP $= 3(-6) = -18$

$-18 = (1)(-18) = (-1)(18)$ — None of the sums of these pairs is -18
$ = (2)(-9) \; = (-2)(9)$
$ = (3)(-6) \; = (-3)(6)$

Not factorable

24. Not factorable

25. MP $= 49 \cdot 9 = 441$

$441 = (1)(441) = (-1)(-441)$
$ = (3)(147) = (-3)(-147)$
$ = (7)(63) \; = (-7)(-63)$
$ = (9)(49) \; = (-9)(-49)$
$ = (21)(21) = (-21)(-21) \Rightarrow (-21) + (-21) = -42$

$49x^2 - 21x - 21x + 9$
$= 7x(7x - 3) - 3(7x - 3)$
$= (7x - 3)(7x - 3)$ or $(7x - 3)^2$

26. $(5x - 2)(5x - 2)$ or $(5x - 2)^2$

27. $3u(6u^2 + 13u - 5)$

To factor $6u^2 + 13u - 5$:
MP $= 6 \cdot (-5) = -30$

$-30 = (1)(-30) = (-1)(30)$
$ = (2)(-15) = (-2)(15) \Rightarrow (-2) + (15) = 13$
$ = (3)(-10) = (-3)(10)$
$ = (5)(-6) \; = (-5)(6)$

$6u^2 - 2u + 15u - 5$
$= 2u(3u - 1) + 5(3u - 1)$
$= (3u - 1)(2u + 5)$

The final answer is $3u(2u + 5)(3u - 1)$.

28. $2y(2y - 1)(3y + 5)$

29. $4(x^2 + x + 1)$
To factor $x^2 + x + 1$:
MP $= 1 \cdot 1 = 1$
$1 = (1)(1) = (-1)(-1)$
The sum of neither pair is 1. Therefore, $x^2 + x + 1$ is not factorable. The final answer is $4(x^2 + x + 1)$.

30. $6(x^2 + x + 3)$

Exercises 7.7 (page 335)

1. $2(x^2 - 4y^2) = 2(\sqrt{x^2} + \sqrt{4y^2})(\sqrt{x^2} - \sqrt{4y^2})$
$= 2(x + 2y)(x - 2y)$

2. $3(x + 3y)(x - 3y)$

3. $5(a^4 - 4b^2) = 5(\sqrt{a^4} + \sqrt{4b^2})(\sqrt{a^4} - \sqrt{4b^2})$
$= 5(a^2 + 2b)(a^2 - 2b)$

4. $6(m + 3n^2)(m - 3n^2)$

5. $(\sqrt{x^4} + \sqrt{y^4})(\sqrt{x^4} - \sqrt{y^4}) = (x^2 + y^2)(x^2 - y^2)$
$= (x^2 + y^2)(x + y)(x - y)$

6. $(a^2 + 4)(a + 2)(a - 2)$

7. $2(2v^2 + 7v - 4) = 2(v + 4)(2v - 1)$ **8.** $3(v - 5)(2v + 1)$

9. $4(2z^2 - 3z - 2) = 4(z - 2)(2z + 1)$ **10.** $3(2z - 3)(3z + 1)$

11. $4(x^2 - 25) = 4(\sqrt{x^2} + \sqrt{25})(\sqrt{x^2} - \sqrt{25}) = 4(x + 5)(x - 5)$

12. $9(x + 2)(x - 2)$ **13.** $2(6x^2 + 5x - 4) = 2(2x - 1)(3x + 4)$

14. $3(3x + 2)(5x - 4)$

15. $a(b^2 - 2b + 1) = a(b - 1)(b - 1) = a(b - 1)^2$

16. $a(u - 1)^2$

17. $(\sqrt{x^4} + \sqrt{81})(\sqrt{x^4} - \sqrt{81}) = (x^2 + 9)(x^2 - 9)$
$= (x^2 + 9)(x + 3)(x - 3)$

18. $(4y^4 + z^2)(2y^2 + z)(2y^2 - z)$ **19.** $16(x^2 + 1)$

20. $25(b^2 + 4)$ **21.** $2u(u^2 + uv - 6v^2) = 2u(u + 3v)(u - 2v)$

22. $3m(m + 3n)(m - 4n)$

23. $4h(2h^2 - 5hk + 3k^2) = 4h(2h - 3k)(h - k)$

24. $5k(h - 2k)(3h - k)$

25. $a^3b^2(a^2 - 4b^2) = a^3b^2(a + 2b)(a - 2b)$

26. $x^2y^2(y + 10x)(y - 10x)$

27. $2a(x^2 - 4a^2y^2) = 2a(\sqrt{x^2} + \sqrt{4a^2y^4})(\sqrt{x^2} - \sqrt{4a^2y^2})$
$= 2a(x + 2ay)(x - 2ay)$

28. $3b^2(x^2 + 2y)(x^2 - 2y)$

29. $4(3 + x) - x^2(3 + x) = (3 + x)(4 - x^2) = (3 + x)(2 + x)(2 - x)$

30. $(5 - z)(3 + z)(3 - z)$

31. $6my - \underline{4nz} + 15mz - \underline{5zn} = 6my - 9nz + 15mz$
$= 3(2my - 3nz + 5mz)$

32. $4(xy + mn)$

33. $(x^2 - 4)(x^2 - 4) = (x + 2)(x - 2)(x + 2)(x - 2)$
$= (x + 2)^2(x - 2)^2$

34. $(y + 3)^2(y - 3)^2$

35. $(x^4 + 1)(x^4 - 1) = (x^4 + 1)(x^2 + 1)(x^2 - 1)$
$= (x^4 + 1)(x^2 + 1)(x + 1)(x - 1)$

36. $(a^4 + b^4)(a^2 + b^2)(a + b)(a - b)$

37. $(x + a)(x - a) - 4(x - a) = (x - a)(x + a - 4)$

38. $(m - 5)(m + 5 + n)$

39. $6[ac + bc - ad - bd] = 6[c(a + b) - d(a + b)]$
$= 6(a + b)(c - d)$

40. $(2c + d)(5y - 3z)$

Exercises 7.8 (page 341)

(The checks will not be shown.)

1. $(x - 5)(x + 4) = 0$ **2.** $-7, 2$

$\begin{array}{c|c} x - 5 = 0 & x + 4 = 0 \\ x = 5 & x = -4 \end{array}$

The solutions are 5 and -4.

3. $3x(x - 4) = 0$ **4.** $0, -6$

$\begin{array}{c|c} 3x = 0 & x - 4 = 0 \\ x = 0 & x = 4 \end{array}$

The solutions are 0 and 4.

5. $(x + 10)(2x - 3) = 0$ **6.** $8, -\frac{2}{3}$

$\begin{array}{c|c} x + 10 = 0 & 2x - 3 = 0 \\ x = -10 & 2x = 3 \\ & x = \frac{3}{2} = 1\frac{1}{2} \end{array}$

The solutions are -10 and $1\frac{1}{2}$.

7. $(x + 1)(x + 8) = 0$ **8.** $-2, -4$

$\begin{array}{c|c} x + 1 = 0 & x + 8 = 0 \\ x = -1 & x = -8 \end{array}$

The solutions are -1 and -8.

9. $(x - 4)(x + 3) = 0$ **10.** $-4, 3$

$\begin{array}{c|c} x - 4 = 0 & x + 3 = 0 \\ x = 4 & x = -3 \end{array}$

The solutions are 4 and -3.

11. $\quad x^2 - 64 = 0$ **12.** $12, -12$
$(x + 8)(x - 8) = 0$

$\begin{array}{c|c} x + 8 = 0 & x - 8 = 0 \\ x = -8 & x = 8 \end{array}$

The solutions are -8 and 8.

13. $2x(3x - 5) = 0$ **14.** $0, \frac{7}{2}$

$\begin{array}{c|c} 2x = 0 & 3x - 5 = 0 \\ x = 0 & 3x = 5 \\ & x = \frac{5}{3} \end{array}$

The solutions are 0 and $\frac{5}{3}$.

15. $0 = 4w^2 - 24w$ **16.** $0, 4$
$0 = 4w(w - 6)$

$\begin{array}{c|c} 4w = 0 & w - 6 = 0 \\ w = 0 & w = 6 \end{array}$

The solutions are 0 and 6.

17. $\quad 5a^2 - 16a + 3 = 0$ **18.** $7, \frac{1}{3}$
$(1a - 3)(5a - 1) = 0$

$\begin{array}{c|c} a - 3 = 0 & 5a - 1 = 0 \\ a = 3 & 5a = 1 \\ & a = \frac{1}{5} \end{array}$

The solutions are 3 and $\frac{1}{5}$.

19. $\quad 3u^2 - 2u - 5 = 0$ **20.** $7, -\frac{1}{5}$
$(1u + 1)(3u - 5) = 0$

$\begin{array}{c|c} u + 1 = 0 & 3u - 5 = 0 \\ u = -1 & 3u = 5 \\ & u = \frac{5}{3} = 1\frac{2}{3} \end{array}$

The solutions are -1 and $1\frac{2}{3}$.

21. $x^2 - 5x + 6 = 2$
$x^2 - 5x + 4 = 0$
$(x - 1)(x - 4) = 0$

$x - 1 = 0$	$x - 4 = 0$
$x = 1$	$x = 4$

The solutions are 1 and 4.

22. 2, 6

23. $x^2 - 4x = 12$
$x^2 - 4x - 12 = 0$
$(x - 6)(x + 2) = 0$

$x - 6 = 0$	$x + 2 = 0$
$x = 6$	$x = -2$

The solutions are 6 and -2.

24. $-3, 5$

25. $4x(2x - 1)(3x + 7) = 0$

$4x = 0$	$2x - 1 = 0$	$3x + 7 = 0$
$x = 0$	$2x = 1$	$3x = -7$
	$x = \frac{1}{2}$	$x = -\frac{7}{3} = -2\frac{1}{3}$

The solutions are 0, $\frac{1}{2}$, and $-2\frac{1}{3}$.

26. $0, \frac{3}{4}, \frac{6}{7}$

27. $2x^3 + x^2 - 3x = 0$
$x(2x^2 + x - 3) = 0$
$x(1x - 1)(2x + 3) = 0$

$x = 0$	$x - 1 = 0$	$2x + 3 = 0$
	$x = 1$	$2x = -3$
		$x = -\frac{3}{2} = -1\frac{1}{2}$

The solutions are 0, 1, and $-1\frac{1}{2}$.

28. $0, -5, \frac{1}{2}$

29. $2a^2(a - 5) = 0$
$2aa(a - 5) = 0$

$2 \neq 0$	$a = 0$	$a = 0$	$a - 5 = 0$
		↑	$a = 5$

$a = 0$ need not be listed twice

The solutions are 0 and 5.

30. $0, 6$

Exercises 7.9 (page 346)

1. Let x = smaller number
$x + 5$ = larger number

$x(x + 5) = 14$
$x^2 + 5x = 14$
$x^2 + 5x - 14 = 0$
$(x + 7)(x - 2) = 0$

$x + 7 = 0$	$x - 2 = 0$
$x = -7$	$x = 2$
$x + 5 = -2$	$x + 5 = 7$

There are two answers: The numbers are 2 and 7, or the numbers are -7 and -2.

2. Two answers: The numbers are 3 and 9, or the numbers are -9 and -3.

3. Let x = one number
$12 - x$ = other number

$x(12 - x) = 35$
$12x - x^2 = 35$
$0 = x^2 - 12x + 35$
$0 = (x - 5)(x - 7)$

$x - 5 = 0$	$x - 7 = 0$
$x = 5$	$x = 7$
$12 - x = 7$	$12 - x = 5$

The numbers are 5 and 7.

4. 2 and -6

5. Let x = altitude (in inches)
$x + 3$ = base (in inches)
$\frac{1}{2}x(x + 3)$ = area

$\frac{1}{2}x(x + 3) = 20$
$(2)\left(\frac{1}{2}x\right)(x + 3) = 20(2)$
$x(x + 3) = 40$
$x^2 + 3x - 40 = 0$
$(x - 5)(x + 8) = 0$

$x - 5 = 0$	$x + 8 = 0$
$x = 5$	$x = -8$ Reject
$x + 3 = 8$	(A length cannot be negative.)

The altitude is 5 in. and the base is 8 in.

6. Altitude = 12 cm, base = 15 cm

7. Let x = the first integer
$x + 1$ = the second integer
$x + 2$ = the third integer

$x(x + 1) + (x + 1)(x + 2) = 8$
$x^2 + x + x^2 + 3x + 2 = 8$
$2x^2 + 4x - 6 = 0$
$2(x^2 + 2x - 3) = 0$
$2(x - 1)(x + 3) = 0$

$2 \neq 0$	$x - 1 = 0$	$x + 3 = 0$
	$x = 1$	$x = -3$
	$x + 1 = 2$	$x + 1 = -2$
	$x + 2 = 3$	$x + 2 = -1$

Two answers: The integers are 1, 2, and 3, or the integers are -3, -2, and -1.

8. The integers are 2, 3, and 4.

9. Let x = the first even integer
$x + 2$ = the second even integer
$x + 4$ = the third even integer

$2x(x + 2) = 16 + (x + 2)(x + 4)$
$2x^2 + 4x = 16 + x^2 + 6x + 8$
$x^2 - 2x - 24 = 0$
$(x + 4)(x - 6) = 0$

$x + 4 = 0$	$x - 6 = 0$
$x = -4$	$x = 6$
$x + 2 = -2$	$x + 2 = 8$
$x + 4 = 0$	$x + 4 = 10$

Two answers: The integers are -4, -2, and 0, or the integers are 6, 8, and 10.

10. Two answers: The integers are 5, 7, and 9, or the integers are -15, -13, and -11.

11. Let w = width
$w + 5$ = length

Area $= (w + 5)w$

$(w + 5)w = 84$
$w^2 + 5w = 84$
$w^2 + 5w - 84 = 0$
$(w + 12)(w - 7) = 0$

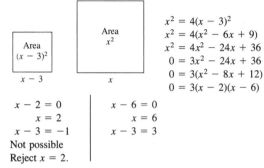
w
$w + 5$

$w + 12 = 0$	$w - 7 = 0$
$w = -12$	$w = 7$
Reject	$w + 5 = 12$

The width is 7 ft and the length is 12 ft.

12. Length = 7 ft, width = 4 ft

13. Let x = length of side of larger square
$x - 3$ = length of side of smaller square

Area $(x - 3)^2$

Area x^2

$x - 3$

x

$x^2 = 4(x - 3)^2$
$x^2 = 4(x^2 - 6x + 9)$
$x^2 = 4x^2 - 24x + 36$
$0 = 3x^2 - 24x + 36$
$0 = 3(x^2 - 8x + 12)$
$0 = 3(x - 2)(x - 6)$

$x - 2 = 0$	$x - 6 = 0$
$x = 2$	$x = 6$
$x - 3 = -1$	$x - 3 = 3$
Not possible	
Reject $x = 2$.	

The length of a side of the larger square is 6 cm and that of a side of the smaller square is 3 cm.

14. Side of smaller square = 2 ft; side of larger square = 6 ft

15.

$\ell - 4$

ℓ

Let ℓ = length
$\ell - 4$ = width
Area $= \ell(\ell - 4)$
Perimeter $= 2\ell + 2(\ell - 4)$

$\ell(\ell - 4) = 17 + 2\ell + 2(\ell - 4)$
$\ell^2 - 4\ell = 17 + 2\ell + 2\ell - 8$
$\ell^2 - 4\ell = 4\ell + 9$
$\ell^2 - 8\ell - 9 = 0$
$(\ell + 1)(\ell - 9) = 0$

$\ell + 1 = 0$	$\ell - 9 = 0$
$\ell = -1$	$\ell = 9$
Reject	$\ell - 4 = 5$

The length is 9 yd and the width is 5 yd.

16. Side = 8

17. Let x = altitude
$x + 3$ = base

Area $= \frac{1}{2}(x + 3)x$

$\frac{1}{2}(x + 3)x = 35$
$2\left(\frac{1}{2}\right)(x + 3)x = 2(35)$
$x^2 + 3x = 70$
$x^2 + 3x - 70 = 0$
$(x + 10)(x - 7) = 0$

$x + 10 = 0$	$x - 7 = 0$
$x = -10$	$x = 7$
Reject	$x + 3 = 10$
(A length cannot be negative.)	

The altitude is 7 in. and the base is 10 in.

18. Altitude = 4 m, base = 9 m

19. Let x = altitude (in inches)
$19 - x$ = base (in inches)
$\frac{1}{2}x(19 - x)$ = area

$\frac{1}{2}x(19 - x) = 42$

$(\cancel{2})\left(\frac{1}{\cancel{2}}x\right)(19 - x) = 42(2)$

$x(19 - x) = 84$
$19x - x^2 = 84$
$0 = x^2 - 19x + 84$
$0 = (x - 7)(x - 12)$

$x - 7 = 0$	$x - 12 = 0$
$x = 7$ (altitude)	$x = 12$ (altitude)
$19 - x = 12$ (base)	$19 - x = 7$ (base)

Two answers: The altitude is 7 in. and the base is 12 in., or the altitude is 12 in. and the base is 7 in.

20. Two answers: Altitude = 6 cm, base = 9 cm; or Base = 6 cm, altitude = 9 cm

21. Let x = height (in cm)
$x + 4$ = length (in cm)
$5x(x + 4)$ = volume

$5x(x + 4) = 225$
$5x^2 + 20x = 225$
$5x^2 + 20x - 225 = 0$
$5(x^2 + 4x - 45) = 0$
$5(x - 5)(x + 9) = 0$

$5 \neq 0$	$x - 5 = 0$	$x + 9 = 0$
	$x = 5$	$x = -9$ Reject
	$x + 4 = 9$	(A length cannot be negative.)

The height is 5 cm, and the length is 9 cm.

22. Width = 7 in., length = 9 in.

Review Exercises 7.10 (page 350)

1. $5(1 + 3a)$ **2.** $2(5 + n)(5 - n)$

3. $b(a + 2) - 1(a + 2) = (a + 2)(b - 1)$

4. $y(y + 2)(y + 8)$ **5.** Not factorable **6.** $3b(1 - 2b)$

7. $(5c - 2)(c - 4)$ **8.** $(m - 5)(n - 1)$

9. $5(x^2 - 7x - 30) = 5(x + 3)(x - 10)$ **10.** Not factorable

11. $x^2(x + 5) + 3(x + 5) = (x + 5)(x^2 + 3)$ **12.** $(x + 3)^2(x - 3)$

13. $(x + y)(x - y) + 1(x - y) = (x - y)([x + y] + 1)$
$= (x - y)(x + y + 1)$

14. $15(a + 2b)(a - b)$

15. $(x - 5)(x + 3) = 0$ **16.** 8, -3

$x - 5 = 0$	$x + 3 = 0$
$x = 5$	$x = -3$

The solutions are 5 and -3.

17. $m^2 - 3m - 18 = 0$ **18.** 6, -6
$(m - 6)(m + 3) = 0$

$m - 6 = 0$	$m + 3 = 0$
$m = 6$	$m = -3$

The solutions are 6 and -3.

19. $3z^2 - 12z = 0$ **20.** $\frac{5}{3}, \frac{9}{4}$
$3z(z - 4) = 0$

$3 \neq 0$	$z = 0$	$z - 4 = 0$
		$z = 4$

The solutions are 0 and 4.

21. $3x^2 + 13x - 10 = 0$
$(3x - 2)(x + 5) = 0$

$3x - 2 = 0$	$x + 5 = 0$
$3x = 2$	$x = -5$
$x = \frac{2}{3}$	

The solutions are $\frac{2}{3}$ and -5.

22. $\frac{1}{5}, -\frac{3}{2}$

23. $2u(u + 6)(u - 2) = 0$

$2 \neq 0$	$u = 0$	$u + 6 = 0$	$u - 2 = 0$
		$u = -6$	$u = 2$

The solutions are 0, -6, and 2.

24. $0, \frac{3}{2}, \frac{2}{3}$

25. Let $\quad x = $ one number
$x + 3 = $ other number

$x(x + 3) = 28$
$x^2 + 3x = 28$
$x^2 + 3x - 28 = 0$
$(x + 7)(x - 4) = 0$

$x + 7 = 0$	$x - 4 = 0$
$x = -7$	$x = 4$
$x + 3 = -4$	$x + 3 = 7$

Two answers: One number is -7 and the other is -4, or one number is 4 and the other is 7.

26. Small: 2 ft; large: 8 ft

27. Let $\quad x = $ width in meters
$x + 3 = $ length in meters

$x(x + 3) = 40$
$x^2 + 3x = 40$
$x^2 + 3x - 40 = 0$
$(x + 8)(x - 5) = 0$

$x + 8 = 0$	$x - 5 = 0$
$x = -8$ Reject	$x = 5$
(A length cannot be negative.)	$x + 3 = 8$

The width is 5 m and the length is 8 m.

28. Width: 4; length: 10

29. Let $\quad x = $ first integer
$x + 1 = $ second integer

$x(x + 1) = 1 + x + (x + 1)$
$x^2 + x = 2 + 2x$
$x^2 - x - 2 = 0$
$(x - 2)(x + 1) = 0$

$x - 2 = 0$	$x + 1 = 0$
$x = 2$	$x = -1$
$x + 1 = 3$	$x + 1 = 0$

Two answers: The integers are 2 and 3, or the integers are 0 and -1.

30. 1, 3, and 5

Chapter 7 Diagnostic Test (page 353)

Following each problem number is the textbook section number (in parentheses) where that problem is discussed.

1. (7.1) $\quad 8x + 12 = 4(2x + 3)$

2. (7.1) $\quad 5x^3 - 35x^2 = 5x^2(x - 7)$

3. (7.2) $\quad 25x^2 - 121y^2 = (5x + 11y)(5x - 11y)$

4. (7.1) $\quad 8x^2 + 7$ is not factorable

5. (7.1 and 7.2) $\quad 5a^2 - 180 = 5(a^2 - 36) = 5(a + 6)(a - 6)$

6. (7.3) $\quad z^2 + 9z + 8 = (z + 1)(z + 8)$

7. (7.3) $\quad m^2 + 5m - 6 = (m + 6)(m - 1)$

8. (7.3) $\quad 11x^2 - 18x + 7 = (11x - 7)(x - 1)$

9. (7.3) $\quad x^2 + 7x - 6$ is not factorable

10. (7.3) $\quad 4y^2 + 19y - 5 = (4y - 1)(y + 5)$

11. (7.5) $\quad 5n - mn - 5 + m = n(5 - m) - 1(5 - m)$
$= (5 - m)(n - 1)$

12. (7.3) $\quad 6h^2k - 8hk^2 + 2k^3 = 2k(3h^2 - 4hk + k^2)$
$= 2k(3h - k)(h - k)$

13. (7.8) $\quad 3x^2 - 12x = 0$
$3x(x - 4) = 0$

$3 \neq 0$	$x = 0$	$x - 4 = 0$
		$x = 4$

The solutions are 0 and 4.

14. (7.8) $\quad x^2 + 20 = 12x$
$x^2 - 12x + 20 = 0$
$(x - 10)(x - 2) = 0$

$x - 10 = 0$	$x - 2 = 0$
$x = 10$	$x = 2$

The solutions are 10 and 2.

15. (7.8) $\quad 3x^3 = x^2 + 10x$
$3x^3 - x^2 - 10x = 0$
$x(3x^2 - x - 10) = 0$
$x(3x + 5)(x - 2) = 0$

$x = 0$	$3x + 5 = 0$	$x - 2 = 0$
	$3x = -5$	$x = 2$
	$x = -\frac{5}{3}$	

The solutions are 0, $-\frac{5}{3}$, and 2.

16. (7.8) $\quad (x - 7)(x + 6) = -22$
$x^2 - x - 42 = -22$
$x^2 - x - 20 = 0$
$(x - 5)(x + 4) = 0$

$x - 5 = 0$	$x + 4 = 0$
$x = 5$	$x = -4$

The solutions are 5 and -4.

17. (7.8) $\quad 3x^2 = 75$
$3x^2 - 75 = 0$
$3(x^2 - 25) = 0$
$3(x + 5)(x - 5) = 0$

$3 \neq 0$	$x + 5 = 0$	$x - 5 = 0$
	$x = -5$	$x = 5$

The solutions are -5 and 5.

18. (7.8) $\quad 2x^2 - 5x = 3$
$2x^2 - 5x - 3 = 0$
$(2x + 1)(x - 3) = 0$

$2x + 1 = 0$	$x - 3 = 0$
$2x = -1$	$x = 3$
$x = -\frac{1}{2}$	

The solutions are $-\frac{1}{2}$ and 3.

19. (7.9) Let $\quad x =$ one odd integer
$x + 2 =$ next odd integer
$x + 4 =$ third odd integer

$$2x(x + 2) - x(x + 4) = 49$$
$$2x^2 + 4x - x^2 - 4x = 49$$
$$x^2 - 49 = 0$$
$$(x + 7)(x - 7) = 0$$

$x + 7 = 0$	$x - 7 = 0$
$x = -7$	$x = 7$
$x + 2 = -5$	$x + 2 = 9$
$x + 4 = -3$	$x + 4 = 11$

Two answers: The integers are -7, -5, and -3, or the integers are 7, 9, and 11.

20. (7.9) Let $\quad x =$ width (in ft)
$x + 3 =$ length (in ft)
$x(x + 3) =$ area

$$x(x + 3) = 28$$
$$x^2 + 3x - 28 = 0$$
$$(x - 4)(x + 7) = 0$$

$x - 4 = 0$	$x + 7 = 0$
$x = 4$	$x = -7$ Reject
$x + 3 = 7$	(A length cannot be negative.)

The width is 4 ft and the length is 7 ft.

Cumulative Review Exercises: Chapters 1–7 (page 354)

1. $24 \div 2\sqrt{16} - 3^2 \cdot 5 = 24 \div 2(4) - 9 \cdot 5 = 12(4) - 9 \cdot 5$
$= 48 - 45 = 3$

2. $\frac{12}{5}$ or $2\frac{2}{5}$

3. $C = \dfrac{a}{a + 12} \cdot A$

$C = \dfrac{8}{8 + 12} \cdot 35$

$C = \dfrac{\overset{2}{\cancel{8}}}{\underset{1}{\cancel{20}}} \cdot \dfrac{\overset{7}{\cancel{35}}}{1} = 14$

4. $\dfrac{y^6}{9z^4}$ **5.** $\left(\dfrac{\overset{3}{\cancel{15x^2}}}{\underset{2\ x}{\cancel{10x^3}}}\right)^3 = \left(\dfrac{3}{2x}\right)^3 = \dfrac{3^3}{2^3x^3} = \dfrac{27}{8x^3}$

6a. 5.73×10^7 **b.** 3.51×10^{-3}

7. $(2x^2 + 5x - 3) - 1(-4x^2 + 8x + 10) + (6x^2 + 3x - 8)$
$= \underline{2x^2} + \underline{\underline{5x}} - \underset{u}{3} + \underline{4x^2} - \underline{\underline{8x}} - \underset{u}{10} + \underline{6x^2} + \underline{\underline{3x}} - \underset{u}{8}$
$= 12x^2 - 21$

8. $3y^3 - 14y^2 + 13y - 20$ **9.** $\dfrac{12a^2}{3a} - \dfrac{3a}{3a} = 4a - 1$

10. $4x + 5$ R -2 or $4x + 5 - \dfrac{2}{2x - 3}$ **11.** $18x(2x - 1)$

12. $(1 + 6t)(1 - 6t)$ **13.** $(x - 3)(x - 5)$

14. Not factorable **15.** $(5k + 1)(k - 7)$ **16.** $(3n - 5)(n + 1)$

17. $5x^2 + 9x - 2 = 0$ **18.** $-\frac{11}{17}$
$(5x - 1)(x + 2) = 0$

$5x - 1 = 0$	$x + 2 = 0$
$5x = 1$	$x = -2$
$x = \frac{1}{5}$	

The solutions are $\frac{1}{5}$ and -2.

19. Let $x =$ number of pounds of cashews.
(There will be $20 + x$ lb in the mixture.)

$$7.50x + 3.50(20) = 5.90(20 + x)$$
$$75x + 35(20) = 59(20 + x)$$
$$75x + 700 = 1{,}180 + 59x$$
$$\underline{-59x - 700 \qquad - 700 - 59x}$$
$$16x = 480$$
$$x = 30$$

Therefore, 30 lb of cashews should be used.

20. The width is 10 m and the length is 15 m.

Exercises 8.1 (page 360)

1. 2, because 2 would make the denominator zero **2.** -3

3. None **4.** None

5. 0 and 2, because either of these would make the denominator zero

6. 0, 3

7. -1 and 2, because either of these would make the denominator zero

8. 3, -4

9. Yes, because if we multiply both x and $2y$ by 5, we get $5x$ and $10y$

10. Yes

11. No; we can't get the second rational expression from the first by multiplying both x and $2y$ by the same number

12. No

13. Yes, because if we multiply both $(x + 1)$ and $2(3x - 2)$ by 6, we get $6(x + 1)$ and $12(3x - 2)$

14. Yes

15. -5. The signs of the denominators are the same and the signs of the rational expressions are different; the signs of the numerators must be different.

16. -8

17. y. The signs of the rational expressions are the same and the signs of the numerators are different; the signs of the denominators must be different.

18. -2

19. $x - 2$. The signs of the rational expressions are the same and the signs of the denominators are different; the signs of the numerators must be different.

20. $y - 5$

21. -5. The signs of the rational expressions are the same and the signs of the numerators are different; the signs of the denominators must be different.

22. -7

23. $5 - x$. The signs of the numerators are the same and the signs of the rational expressions are different; the signs of the denominators must be different.

24. $4 - x$

25. $b - a$. The signs of the rational expressions are the same and the signs of the denominators are different; the signs of the numerators must be different.

26. $y - x$

Exercises 8.2 (page 365)

1. $\dfrac{\cancel{6} \cdot 3}{\cancel{6} \cdot 4} = \dfrac{3}{4}$ 2. $\dfrac{4}{7}$ 3. $\dfrac{\cancel{6}ab^2}{\cancel{3}ab} = 2b$ 4. $2m$

5. $\dfrac{\cancel{4}x^2y}{\cancel{2}xy} = 2x$ 6. $3x^2$ 7. $\dfrac{5(\cancel{x-2})}{x-2} = 5$ 8. 3

9. $\dfrac{7(\cancel{x-3})}{15x(\cancel{x-3})} = \dfrac{7}{15x}$ 10. $\dfrac{3}{y^2}$

11. $\dfrac{\cancel{6x^2}y}{\cancel{6xy}(5x^2-3y)} = \dfrac{x}{5x^2-3y}$ 12. $\dfrac{1}{2a^2-4b}$

13. $\dfrac{\cancel{4xy}(2x+3y)}{\cancel{6x^2y}(2x+3y)} = \dfrac{2}{3x}$ 14. $\dfrac{3s}{4t}$

15. $-\dfrac{5x-6}{6-5x} = +\dfrac{5x-6}{-(6-5x)} = \dfrac{5x-6}{5x-6} = 1$ 16. -1

17. $\dfrac{\cancel{5}x(x+6)}{\cancel{10}x(x-4)} = \dfrac{x+6}{2(x-4)}$ 18. $\dfrac{x}{3}$

19. $\dfrac{6}{4} = \dfrac{3}{2}$ Incorrect to cancel the 4's 20. 4

21. Cannot be reduced 22. Cannot be reduced

23. $\dfrac{(\cancel{x+1})(x-1)}{(\cancel{x+1})} = x-1$ 24. $x+2$

25. $\dfrac{(3x-2)(\cancel{2x+1})}{(5x-1)(\cancel{2x+1})} = \dfrac{3x-2}{5x-1}$ 26. $\dfrac{2x-3}{3x+2}$

27. $\dfrac{(\cancel{x+y})(x-y)}{(\cancel{x+y})(x+y)} = \dfrac{x-y}{x+y}$ 28. $\dfrac{a+3b}{a-3b}$

29. $\dfrac{(2y-3x)(y+2x)}{(3x-2y)(x+y)} = \dfrac{-(\cancel{3x-2y})(y+2x)}{(\cancel{3x-2y})(x+y)} = -\dfrac{2x+y}{x+y}$

30. $-\dfrac{3x+2y}{2x+3y}$

31. $\dfrac{2(4x^2-y^2)}{a(2x-y)+b(2x-y)} = \dfrac{2(2x+y)(\cancel{2x-y})}{(\cancel{2x-y})(a+b)} = \dfrac{2(2x+y)}{a+b}$

32. $\dfrac{3(x-2y)}{a+b}$ 33. $\dfrac{\cancel{8}-z}{\cancel{8}-z} = 1$ 34. 1

35. $\dfrac{(a-2b)(\cancel{a-b})}{(2a+b)(\cancel{a-b})} = \dfrac{a-2b}{2a+b}$ 36. $\dfrac{8}{3n+m}$

37. $\dfrac{(3+4x)(\cancel{3-4x})}{(4x-3)(\cancel{4x-3})} = -\dfrac{3+4x}{4x-3}$ or $\dfrac{3+4x}{3-4x}$

38. $-\dfrac{5+3x}{3x-5}$ or $\dfrac{5+3x}{5-3x}$ 39. $\dfrac{(5+3x)(\cancel{2-x})}{(2x+5)(\cancel{x-2})} = -\dfrac{5+3x}{2x+5}$

40. $-\dfrac{4-5x}{3x+4}$ or $\dfrac{5x-4}{3x+4}$

41. $\dfrac{3(6-x-x^2)}{6(x^2+x-6)} = \dfrac{(\cancel{3+x})(\cancel{2-x})}{2(\cancel{x+3})(\cancel{x-2})} = -\dfrac{1}{2}$ 42. $-\dfrac{1}{4}$

Exercises 8.3 (page 370)

1. $\dfrac{5}{6} \div \dfrac{5}{3} = \dfrac{5}{6} \cdot \dfrac{3}{5} = \dfrac{1}{2}$ 2. $\dfrac{3}{14}$ 3. $\dfrac{\cancel{4}a^3}{\cancel{5}b^2} \cdot \dfrac{\cancel{10}b}{\cancel{8}a^2} = \dfrac{a}{b}$ 4. $\dfrac{c}{d}$

5. $\dfrac{3x^2}{16} \div \dfrac{x}{8} = \dfrac{3x^2}{\cancel{16}} \cdot \dfrac{\cancel{8}}{x} = \dfrac{3x}{2}$ 6. $\dfrac{3y}{2}$

7. $\dfrac{\cancel{3}x^4y^2z}{\cancel{18}xy} \cdot \dfrac{\cancel{15}z}{x^3yz^2} = \dfrac{5x^4y^2z^2}{2x^4y^2z^2} = \dfrac{5}{2}$ 8. $\dfrac{9}{2}$

9. $\dfrac{x}{\cancel{x+2}} \cdot \dfrac{5(\cancel{x+2})}{x^3} = \dfrac{5}{x^2}$

10. $\dfrac{3}{b^3}$ 11. $\dfrac{\cancel{y-2}}{y} \cdot \dfrac{6}{3(\cancel{y-2})} = \dfrac{2}{y}$ 12. $\dfrac{2}{a}$

13. $\dfrac{b^3}{a+3} \div \dfrac{4b^2}{2a+6} = \dfrac{b^3}{\cancel{a+3}} \cdot \dfrac{\cancel{2}(\cancel{a+3})}{\cancel{4}b^2} = \dfrac{b}{2}$ 14. $\dfrac{m}{2}$

15. $\dfrac{5s-15}{30s} \div \dfrac{s-3}{45s^2} = \dfrac{\cancel{5}(\cancel{s-3})}{\cancel{30s}} \cdot \dfrac{\cancel{45}s^2}{\cancel{s-3}} = \dfrac{15s}{2}$ 16. $4n$

17. $\dfrac{a+4}{a-4} \div \dfrac{a^2+8a+16}{a^2-16} = \dfrac{\cancel{a+4}}{\cancel{a-4}} \cdot \dfrac{(\cancel{a+4})(\cancel{a-4})}{(\cancel{a+4})(\cancel{a+4})} = 1$ 18. 1

19. $\dfrac{5}{\cancel{z+4}} \cdot \dfrac{(\cancel{z+4})(\cancel{z-4})}{(z-4)(\cancel{z-4})} = \dfrac{5}{z-4}$ 20. $\dfrac{7}{x-3}$

21. $\dfrac{3(\cancel{a-b})}{\cancel{4}(\cancel{c+d})} \cdot \dfrac{\cancel{2}(\cancel{c+d})}{b-a} = -\dfrac{3}{2}$ or $-1\tfrac{1}{2}$ 22. $-\dfrac{7}{2}$ or $-3\tfrac{1}{2}$

23. $\dfrac{\cancel{4}(\cancel{x-2})}{\cancel{4}} \cdot \dfrac{\cancel{x+2}}{(\cancel{x+2})(\cancel{x-2})} = 1$ 24. 1

25. $\dfrac{4a+4b}{ab^2} \div \dfrac{3a+3b}{a^2b} = \dfrac{4(a+b)}{ab^2} \cdot \dfrac{a^2b}{3(a+b)} = \dfrac{4a}{3b}$ 26. $\dfrac{5y}{4x}$

27. $\dfrac{a^2-9b^2}{a^2-6ab+9b^2} \div \dfrac{a+3b}{a-3b} = \dfrac{(\cancel{a+3b})(\cancel{a-3b})}{(\cancel{a-3b})(\cancel{a-3b})} \cdot \dfrac{\cancel{a-3b}}{\cancel{a+3b}} = 1$

28. 1 29. $\dfrac{x-y}{9x+9y} \div \dfrac{x^2-y^2}{3(x^2+2xy+y^2)}$ 30. $\tfrac{1}{2}$

$= \dfrac{\cancel{x-y}}{\cancel{9}(\cancel{x+y})} \cdot \dfrac{\cancel{3}(\cancel{x+y})(\cancel{x+y})}{(\cancel{x+y})(\cancel{x-y})} = \dfrac{1}{3}$

31. $\dfrac{a^2-9b^2}{6(a^2-6ab+9b^2)} \div \dfrac{a+3b}{2a-6b}$ 32. $\tfrac{1}{2}$

$= \dfrac{(\cancel{a+3b})(\cancel{a-3b})}{\cancel{6}(\cancel{a-3b})(\cancel{a-3b})} \cdot \dfrac{\cancel{2}(\cancel{a-3b})}{\cancel{a+3b}} = \dfrac{1}{3}$

33. $\dfrac{2x^3y+2x^2y^2}{6x} \div \dfrac{x^2y^2-xy^3}{y-x} = \dfrac{\cancel{2}x^2y(x+y)}{\cancel{6}x} \cdot \dfrac{\cancel{y-x}}{xy^2(\cancel{x-y})} = -\dfrac{x+y}{3y}$

34. $-\dfrac{t+s}{2s^2t}$

Exercises 8.4 (page 375)

1. $\dfrac{7+2}{a} = \dfrac{9}{a}$ 2. $\dfrac{7}{b}$ 3. $\dfrac{6-2}{x-y} = \dfrac{4}{x-y}$ 4. $\dfrac{6}{m+n}$

5. $\dfrac{2+4}{3a} = \dfrac{\cancel{6}}{\cancel{3}a} = \dfrac{2}{a}$ 6. $\dfrac{2}{z}$ 7. $\dfrac{2y+2}{y+1} = \dfrac{2(\cancel{y+1})}{\cancel{y+1}} = 2$

8. 5 9. $\dfrac{3+x}{x+3} = 1$ 10. 1

11. $\dfrac{3x-12}{x-4} = \dfrac{3(\cancel{x-4})}{\cancel{x-4}} = 3$ 12. 7

13. $\dfrac{x - 3 - (x + 5)}{y - 2} = \dfrac{x - 3 - x - 5}{y - 2} = -\dfrac{8}{y - 2}$

14. $-\dfrac{7}{a - b}$

15. $\dfrac{a + 2 - (1 - a)}{2a + 1} = \dfrac{a + 2 - 1 + a}{2a + 1} = \dfrac{\overset{1}{\cancel{2a + 1}}}{\cancel{2a + 1}} = 1$ **16.** 1

17. $\dfrac{4x - 1 + 5x - 3}{2x + 3} = \dfrac{9x - 4}{2x + 3}$ **18.** $\dfrac{12x + 8}{3x - 2}$

19. $\dfrac{9 + 7x - 1(3x - 2)}{7x - 9} = \dfrac{9 + 7x - 3x + 2}{7x - 9} = \dfrac{4x + 11}{7x - 9}$

20. $\dfrac{9 - x}{5x - 4}$

21. $\dfrac{-x}{x - 2} + \dfrac{2}{x - 2} = \dfrac{-x + 2}{x - 2} = \dfrac{-1\overset{1}{\cancel{(x - 2)}}}{\underset{1}{\cancel{x - 2}}} = -1$ **22.** 1

23. $\dfrac{-15w}{1 - 5w} + \dfrac{3}{1 - 5w} = \dfrac{-15w + 3}{1 - 5w} = \dfrac{3\overset{1}{\cancel{(1 - 5w)}}}{\underset{1}{\cancel{1 - 5w}}} = 3$ **24.** 5

25. $\dfrac{7z}{8z - 4} + \dfrac{5z - 6}{8z - 4} = \dfrac{7z + 5z - 6}{8z - 4} = \dfrac{12z - 6}{8z - 4} = \dfrac{\overset{3}{\cancel{6}}\overset{1}{\cancel{(2z - 1)}}}{\underset{2}{\cancel{4}}\underset{1}{\cancel{(2z - 1)}}} = \dfrac{3}{2}$

26. $\dfrac{4}{3}$ **27.** $\dfrac{31 - 8x}{12 - 8x} + \dfrac{5 - 16x}{12 - 8x} = \dfrac{31 - 8x + 5 - 16x}{12 - 8x} = \dfrac{36 - 24x}{12 - 8x}$

$= \dfrac{\overset{3}{\cancel{12}}\overset{1}{\cancel{(3 - 2x)}}}{\underset{1}{\cancel{4}}\underset{1}{\cancel{(3 - 2x)}}} = 3$

28. 2

Exercises 8.5 (page 379)

1. (1) $2^2 \cdot 3$, 2^2 are the polynomials in factored form.   **2.** 9
(2) 2, 3 are all the different bases.
(3) 2^2, 3^1 are the highest powers of each base.
(4) LCM $= 2^2 \cdot 3^1 = 12$

3. (1) The polynomials are already factored.   **4.** $2y$
(2) 3, x are all the different bases.
(3) 3^1, x^1 are the highest powers of each base.
(4) LCM $= 3^1 \cdot x^1 = 3x$

5. (1) $5 \cdot x$, $2 \cdot x$ are the polynomials in factored form.   **6.** $12z$
(2) 2, 5, x are all the different bases.
(3) 2^1, 5^1, x^1 are the highest powers of each base.
(4) LCM $= 2^1 \cdot 5^1 \cdot x^1 = 10x$

7. (1) $3 \cdot 5 \cdot x^2$, $2^2 \cdot 3 \cdot x^3 \cdot y$ are the polynomials in factored form.
(2) 2, 3, 5, x, y are all the different bases.
(3) 2^2, 3^1, 5^1, x^3, y^1 are the highest powers of each base.
(4) LCM $= 2^2 \cdot 3^1 \cdot 5^1 \cdot x^3 \cdot y^1 = 60x^3y$

8. $144a^2b^2c$

9. (1) $2^2 \cdot 3 \cdot u^3 \cdot v^2$; $2 \cdot 3^2 \cdot u \cdot v^3$ are the denominators in factored form.
(2) 2, 3, u, v are all the different bases.
(3) 2^2, 3^2, u^3, v^3 are the highest powers of each base.
(4) LCD $= 2^2 \cdot 3^2 \cdot u^3 \cdot v^3 = 36u^3v^3$

10. $100x^3y^5$

11. (1) Denominators are already factored.
(2) a, $(a + 3)$ are the different bases.
(3) a^1, $(a + 3)^1$ are the highest powers of each base.
(4) LCD $= a(a + 3)$

12. $b(b - 5)$

13. (1) $2(x + 2)$, 2^2x are the denominators in factored form.
(2) 2, x, $(x + 2)$ are the different bases.
(3) 2^2, x^1, $(x + 2)^1$ are the highest powers of each base.
(4) LCD $= 4x(x + 2)$

14. $3x(x + 2)$

15. (1) $2^2 \cdot z^2$, $(z + 1)^2$, $z + 1$ are the factored denominators.
(2) 2, z, $(z + 1)$ are the different bases.
(3) 2^2, z^2, $(z + 1)^2$ are the highest powers of each base.
(4) LCD $= 4z^2(z + 1)^2$

16. $12x^3(x - 1)^2$

17. (1) $(x + 1)(x + 2)$, $(x + 1)^2$
(2) $(x + 1)$, $(x + 2)$
(3) $(x + 1)^2$, $(x + 2)^1$
(4) LCD $= (x + 1)^2(x + 2)$

18. $(x - 4)(x + 3)^2$

19. (1) $2^2 \cdot 3 \cdot x^2(x + 2)$, $(x - 2)^2$, $(x + 2)(x - 2)$
(2) 2, 3, x, $(x + 2)$, $(x - 2)$
(3) 2^2, 3^1, x^2, $(x + 2)^1$, $(x - 2)^2$
(4) LCD $= 12x^2(x + 2)(x - 2)^2$

20. $8y(y + 3)^2(y - 3)$

21. (1) $(2x - 1)(3x + 2)$; $x(3x + 2)^2$; $(2x - 1)^2$
(2) x; $(2x - 1)$; $(3x + 2)$
(3) x; $(2x - 1)^2$; $(3x + 2)^2$
(4) LCD $= x(2x - 1)^2(3x + 2)^2$

22. $x(2x + 3)^2(5x - 1)^2$

Exercises 8.6 (page 384)

1. LCD $= a^3$; $\dfrac{3}{a^2} \cdot \dfrac{a}{a} + \dfrac{2}{a^3} = \dfrac{3a}{a^3} + \dfrac{2}{a^3} = \dfrac{3a + 2}{a^3}$ **2.** $\dfrac{5u^2 + 4}{u^3}$

3. LCD $= 2x^2$; $\dfrac{1}{2} \cdot \dfrac{x^2}{x^2} + \dfrac{3}{x} \cdot \dfrac{2x}{2x} - \dfrac{5}{x^2} \cdot \dfrac{2}{2}$

$= \dfrac{x^2}{2x^2} + \dfrac{6x}{2x^2} - \dfrac{10}{2x^2} = \dfrac{x^2 + 6x - 10}{2x^2}$

4. $\dfrac{2y^2 - 3y + 12}{3y^2}$

5. LCD $= xy$; $\dfrac{2}{xy} - \dfrac{3}{y} \cdot \dfrac{x}{x} = \dfrac{2}{xy} - \dfrac{3x}{xy} = \dfrac{2 - 3x}{xy}$

6. $\dfrac{5 - 4b}{ab}$ **7.** LCD $= x$; $\dfrac{5}{1} \cdot \dfrac{x}{x} + \dfrac{2}{x} = \dfrac{5x + 2}{x}$ **8.** $\dfrac{3y + 4}{y}$

9. First reduce $\dfrac{9}{6y}$ to $\dfrac{3}{2y}$. LCD $= 4y^2$; $\dfrac{5}{4y^2} + \dfrac{3}{2y} \cdot \dfrac{2y}{2y} = \dfrac{5 + 6y}{4y^2}$

10. $\dfrac{8x + 3}{4x^2}$ **11.** LCD $= x$; $\dfrac{3x}{1} - \dfrac{3}{x} = \dfrac{3x}{1} \cdot \dfrac{x}{x} - \dfrac{3}{x} = \dfrac{3x^2 - 3}{x}$

12. $\dfrac{4y^2 - 5}{y}$

13. LCD $= a(a + 3)$; $\dfrac{3(a + 3)}{a(a + 3)} + \dfrac{a(a)}{(a + 3)(a)}$

$= \dfrac{3a + 9}{a(a + 3)} + \dfrac{a^2}{a(a + 3)} = \dfrac{3a + 9 + a^2}{a(a + 3)}$ or $\dfrac{a^2 + 3a + 9}{a(a + 3)}$

14. $\dfrac{b^2 + 5b - 25}{b(b - 5)}$

15. LCD $= 4x(x + 2)$; $\dfrac{x(2x)}{2(x + 2)(2x)} + \dfrac{-5(x + 2)}{4x(x + 2)}$

$= \dfrac{2x^2}{4x(x + 2)} + \dfrac{-5x - 10}{4x(x + 2)} = \dfrac{2x^2 - 5x - 10}{4x(x + 2)}$

16. $\dfrac{8 + 4x - 2x^2}{3x(x + 2)}$

17. LCD $= x(x - 2)$; $\dfrac{x}{1} \cdot \dfrac{x(x - 2)}{x(x - 2)} + \dfrac{2}{x} \cdot \dfrac{(x - 2)}{(x - 2)} + \dfrac{-3}{x - 2} \cdot \dfrac{x}{x}$

$= \dfrac{x^2(x - 2)}{x(x - 2)} + \dfrac{2(x - 2)}{x(x - 2)} + \dfrac{-3x}{x(x - 2)}$

$= \dfrac{x^3 - 2x^2 + 2x - 4 - 3x}{x(x - 2)} = \dfrac{x^3 - 2x^2 - x - 4}{x(x - 2)}$

18. $\dfrac{m^3 - 4m^2 + m - 12}{m(m - 4)}$

19. LCD $= b(a - b)$; $\dfrac{(a + b)(a - b)}{b} \cdot \dfrac{(a - b)}{(a - b)} + \dfrac{b}{(a - b)} \cdot \dfrac{b}{b}$

$= \dfrac{a^2 - b^2}{b(a - b)} + \dfrac{b^2}{b(a - b)} = \dfrac{a^2 - b^2 + b^2}{b(a - b)} = \dfrac{a^2}{b(a - b)}$

20. $\dfrac{y^2}{x(y - x)}$

21. LCD $= 6(e - 1)$; $\dfrac{3}{2(e - 1)} \cdot \dfrac{3}{3} + \dfrac{-2}{3(e - 1)} \cdot \dfrac{2}{2}$

$= \dfrac{9}{6(e - 1)} + \dfrac{-4}{6(e - 1)} = \dfrac{5}{6(e - 1)}$

22. $\dfrac{7}{15(f + 2)}$

23. Changing the signs of the last rational expression and of its denominator, we have

$\dfrac{3}{m - 2} + \dfrac{5}{m - 2} = \dfrac{8}{m - 2}$

24. $\dfrac{9}{n - 5}$

25. LCD $= (x - 1)(x + 1)$; $\dfrac{(x + 1)(x + 1)}{(x - 1)(x + 1)} + \dfrac{-(x - 1)(x - 1)}{(x + 1)\ (x - 1)}$

$= \dfrac{x^2 + 2x + 1}{(x - 1)(x + 1)} + \dfrac{-x^2 + 2x - 1}{(x - 1)(x + 1)} = \dfrac{4x}{x^2 - 1}$

26. $\dfrac{16x}{x^2 - 16}$

27. LCD $= xy(x - y)$

$\dfrac{y}{x(x - y)} \cdot \dfrac{y}{y} + \dfrac{-x}{y(x - y)} \cdot \dfrac{x}{x} = \dfrac{y^2 - x^2}{xy(x - y)}$

$= -\dfrac{x^2 - y^2}{xy(x - y)} = -\dfrac{(x + y)\cancel{(x - y)}}{xy\cancel{(x - y)}} = -\dfrac{x + y}{xy}$

28. $\dfrac{a + b}{ab}$

29. LCD $= (x - 3)(x + 3)$

$\dfrac{2x}{(x - 3)(x + 3)} + \dfrac{-2x}{(x + 3)(x - 3)} + \dfrac{36}{(x + 3)(x - 3)}$

$= \dfrac{2x^2 + 6x - 2x^2 + 6x + 36}{(x - 3)(x + 3)} = \dfrac{12x + 36}{(x - 3)(x + 3)}$

$= \dfrac{12\cancel{(x + 3)}}{(x - 3)\cancel{(x + 3)}} = \dfrac{12}{x - 3}$

30. $\dfrac{8}{4 - x}$

31. LCD $= (x + 2)^2$

$\dfrac{x}{x^2 + 4x + 4} + \dfrac{1}{(x + 2)(x + 2)} \,(x + 2) = \dfrac{x + x + 2}{(x + 2)^2} = \dfrac{2x + 2}{(x + 2)^2}$

32. $\dfrac{5 - 3x}{(x - 1)^2}$

33. LCD $= (x + 5)(x - 1)(2x + 3)$

$\dfrac{(2x + 3)(2x + 3)}{(x + 5)(x - 1)(2x + 3)} - \dfrac{(x + 5)(x + 5)}{(2x + 3)(x - 1)(x + 5)}$

$= \dfrac{(2x + 3)(2x + 3) - (x + 5)(x + 5)}{(x + 5)(x - 1)(2x + 3)}$

$= \dfrac{4x^2 + 12x + 9 - 1(x^2 + 10x + 25)}{(x + 5)(x - 1)(2x + 3)}$

$= \dfrac{4x^2 + 12x + 9 - x^2 - 10x - 25}{(x + 5)(x - 1)(2x + 3)} = \dfrac{3x^2 + 2x - 16}{(x + 5)(x - 1)(2x + 3)}$

34. $\dfrac{8x^2 - 2x - 45}{(x - 1)(x + 7)(3x + 2)}$

35. LCD $= (x + 4)^2(x - 4)$

$\dfrac{x^2(x - 4)}{(x + 4)^2(x - 4)} - \dfrac{(x + 1)(x + 4)}{(x + 4)(x - 4)(x + 4)}$

$= \dfrac{x^3 - 4x^2 - 1(x^2 + 5x + 4)}{(x + 4)^2(x - 4)} = \dfrac{x^3 - 4x^2 - x^2 - 5x - 4}{(x + 4)^2(x - 4)}$

$= \dfrac{x^3 - 5x^2 - 5x - 4}{(x + 4)^2(x - 4)}$

36. $\dfrac{x^3 - 4x^2 - 5x - 6}{(x + 3)^2(x - 3)}$

37. LCD $= x(2x + 5)(x - 1)(x + 1)$

$\dfrac{(x + 1)(x + 1)}{x(2x + 5)(x - 1)(x + 1)} - \dfrac{(x - 1)(x - 1)}{x(2x + 5)(x + 1)(x - 1)}$

$= \dfrac{x^2 + 2x + 1 - 1(x^2 - 2x + 1)}{x(2x + 5)(x - 1)(x + 1)}$

$= \dfrac{x^2 + 2x + 1 - x^2 + 2x - 1}{x(2x + 5)(x - 1)(x + 1)} = \dfrac{\overset{1}{\cancel{4x}}}{\underset{1}{\cancel{x}}(2x + 5)(x - 1)(x + 1)}$

$= \dfrac{4}{(2x + 5)(x - 1)(x + 1)}$

38. $\dfrac{4}{(3x + 4)(x - 1)(x + 1)}$

Exercises 8.7 (page 390)

1. $\dfrac{\frac{3}{4}}{\frac{5}{6}} = \dfrac{3}{4} \div \dfrac{5}{6} = \dfrac{3}{\cancel{4}} \cdot \dfrac{\overset{3}{\cancel{6}}}{5} = \dfrac{9}{10}$ **2.** $\dfrac{4}{5}$

3. $\dfrac{\frac{2}{3}}{\frac{4}{9}} = \dfrac{2}{3} \div \dfrac{4}{9} = \dfrac{\cancel{2}}{\cancel{3}} \cdot \dfrac{\overset{3}{\cancel{9}}}{\underset{2}{\cancel{4}}} = \dfrac{3}{2}$ or $1\frac{1}{2}$ **4.** $1\frac{1}{2}$

5. LCD of secondary denominators is 8.   **6.** $\dfrac{9}{7} = 1\frac{2}{7}$

$\dfrac{8}{8} \cdot \dfrac{\frac{3}{4} - \frac{1}{2}}{\frac{5}{8} + \frac{1}{4}} = \dfrac{8\left(\frac{3}{4}\right) + 8\left(-\frac{1}{2}\right)}{8\left(\frac{5}{8}\right) + 8\left(\frac{1}{4}\right)} = \dfrac{6 - 4}{5 + 2} = \dfrac{2}{7}$

7. LCD of secondary denominators is 20.

$\dfrac{20}{20} \cdot \dfrac{\frac{3}{5} + \frac{2}{1}}{\frac{2}{1} - \frac{3}{4}} = \dfrac{\overset{4}{\cancel{20}}\left(\frac{3}{\cancel{5}}\right)_1 + 20\left(\frac{2}{1}\right)}{20(2) + \overset{5}{\cancel{20}}\left(-\frac{3}{\cancel{4}}\right)_1} = \dfrac{12 + 40}{40 - 15} = \dfrac{52}{25} = 2\frac{2}{25}$

8. $1\frac{1}{82}$ **9.** $\dfrac{\frac{5x^3}{3y^4}}{\frac{10x}{9y}} = \dfrac{5x^3}{3y^4} \div \dfrac{10x}{9y} = \dfrac{\overset{x^2}{\cancel{5x^3}}}{\underset{1y^3}{\cancel{3y^4}}} \cdot \dfrac{\overset{3}{\cancel{9y}}}{\underset{2}{\cancel{10x}}}_1 = \dfrac{3x^2}{2y^3}$ **10.** $6ab$

11. $\dfrac{\dfrac{18cd^2}{5a^3b}}{\dfrac{12cd^2}{15ab^2}} = \dfrac{18cd^2}{5a^3b} \div \dfrac{12cd^2}{15ab^2} = \dfrac{\overset{3}{\cancel{18}}\overset{1}{\cancel{cd^2}}}{\underset{1}{\cancel{8}}\underset{a^2}{a^3}\underset{1}{b}} \cdot \dfrac{\overset{3}{\cancel{15}}\overset{1}{a}\overset{b}{\cancel{b^2}}}{\underset{2}{\cancel{12}}\underset{1}{\cancel{cd^2}}} = \dfrac{9b}{2a^2}$ **12.** $\dfrac{2xz^2}{y}$

13. $\dfrac{\dfrac{x+3}{5}}{\dfrac{2x+6}{10}} = \dfrac{x+3}{5} \div \dfrac{2x+6}{10} = \dfrac{\cancel{x+3}}{\underset{1}{\cancel{5}}} \cdot \dfrac{\overset{2}{\cancel{10}}}{\underset{1}{2(\underset{1}{\cancel{x+3}})}} = 1$ **14.** $1\frac{1}{2}$

15. $\dfrac{\dfrac{x+2}{2x}}{\dfrac{x+1}{4x^2}} = \dfrac{x+2}{2x} \div \dfrac{x+1}{4x^2} = \dfrac{x+2}{\underset{1}{\cancel{2x}}} \cdot \dfrac{\overset{2x}{\cancel{4x^2}}}{x+1} = \dfrac{2x^2+4x}{x+1}$

16. $\dfrac{3x-9}{x^2-9x}$ **17.** LCD $= b$

$\dfrac{b}{b} \cdot \dfrac{\left(\dfrac{a}{b}+1\right)}{\left(\dfrac{a}{b}-1\right)} = \dfrac{\cancel{b}\left(\dfrac{a}{\cancel{b}}\right)_1 + b(1)}{\cancel{b}\left(\dfrac{a}{\cancel{b}}\right)_1 - b(1)} = \dfrac{a+b}{a-b}$

18. $\dfrac{2y+x}{2y-x}$ **19.** LCD $= x$

$\dfrac{x}{x} \cdot \dfrac{\dfrac{1}{x}+x}{\dfrac{1}{x}-x} = \dfrac{x\left(\dfrac{1}{x}\right)+x(x)}{x\left(\dfrac{1}{x}\right)-x(x)} = \dfrac{1+x^2}{1-x^2}$

20. $\dfrac{a^2-4}{a^2+4}$ **21.** LCD $= d^2$

$\dfrac{d^2}{d^2} \cdot \dfrac{\left(\dfrac{c}{d}+2\right)}{\left(\dfrac{c^2}{d^2}-4\right)} = \dfrac{d^2\left(\dfrac{c}{d}\right)+d^2(2)}{d^2\left(\dfrac{c^2}{d^2}\right)-d^2(4)} = \dfrac{cd+2d^2}{c^2-4d^2}$

$= \dfrac{d(\overset{1}{\cancel{c+2d}})}{(c-2d)(\underset{1}{\cancel{c+2d}})} = \dfrac{d}{c-2d}$

22. $\dfrac{x+y}{y}$

23. LCD $= y$

$\dfrac{y}{y} \cdot \dfrac{x+\dfrac{x}{y}}{1+\dfrac{1}{y}} = \dfrac{y \cdot x + \cancel{y}\left(\dfrac{x}{\cancel{y}}\right)_1}{y \cdot 1 + \cancel{y}\dfrac{1}{\cancel{y}}} = \dfrac{yx+x}{y+1} = \dfrac{x(y+1)}{y+1} = x$

24. $\frac{1}{3}$ **25.** LCD $= x^2y^2$

$\dfrac{x^2y^2}{x^2y^2} \cdot \dfrac{\dfrac{1}{x^2}-\dfrac{1}{y^2}}{\dfrac{1}{x}+\dfrac{1}{y}} = \dfrac{\overset{1}{\cancel{x^2}}y^2\left(\dfrac{1}{\cancel{x^2}}\right)_1 - x^2\overset{1}{\cancel{y^2}}\left(\dfrac{1}{\cancel{y^2}}\right)_1}{\overset{x}{\cancel{x^2}}y^2\left(\dfrac{1}{\cancel{x}}\right)_1 + x^2\overset{y}{\cancel{y^2}}\left(\dfrac{1}{\cancel{y}}\right)_1}$

$= \dfrac{y^2-x^2}{xy^2+x^2y} = \dfrac{(\cancel{y+x})(y-x)}{xy(\underset{1}{\cancel{y+x}})} = \dfrac{y-x}{xy}$

26. $\dfrac{a+2}{2a}$

27. LCD $= x^2$

$\dfrac{x^2}{x^2} \cdot \dfrac{\dfrac{2}{x}-\dfrac{4}{x^2}}{\dfrac{1}{x}-\dfrac{2}{x^2}} = \dfrac{\overset{x}{\cancel{x^2}}\left(\dfrac{2}{\cancel{x}}\right)_1 - \overset{1}{\cancel{x^2}}\left(\dfrac{4}{\cancel{x^2}}\right)_1}{\overset{x}{\cancel{x^2}}\left(\dfrac{1}{\cancel{x}}\right)_1 - \overset{1}{\cancel{x^2}}\left(\dfrac{2}{\cancel{x^2}}\right)_1} = \dfrac{2x-4}{x-2} = \dfrac{2(\overset{1}{\cancel{x-2}})}{\underset{1}{\cancel{x-2}}} = 2$

28. $\frac{1}{4}$ **29.** LCD $= 3x(x+1)$

$\dfrac{3x(x+1)}{3x(x+1)} \cdot \dfrac{\dfrac{x}{x+1}+\dfrac{4}{3x}}{\dfrac{x}{x+1}-\dfrac{3}{x}}$

$= \dfrac{\dfrac{3x(\overset{1}{\cancel{x+1}})}{1}\left(\dfrac{x}{\cancel{x+1}}\right)_1 + \dfrac{\overset{1}{3\cancel{x}}(x+1)}{1}\left(\dfrac{4}{\cancel{3x}}\right)_1}{\dfrac{3x(\overset{1}{\cancel{x+1}})}{1}\left(\dfrac{x}{\cancel{x+1}}\right)_1 - \dfrac{3\overset{1}{\cancel{x}}(x+1)}{1}\left(\dfrac{3}{\cancel{x}}\right)_1}$

$= \dfrac{3x^2+4(x+1)}{3x^2-9(x+1)} = \dfrac{3x^2+4x+4}{3x^2-9x-9}$

30. $\dfrac{8x^2+4x+1}{20x+4}$

31. LCD $= 4x(x-6)$

$\dfrac{4x(x-6)}{4x(x-6)} \cdot \dfrac{\dfrac{1}{4x}+\dfrac{x}{x-6}}{\dfrac{x-1}{x}-\dfrac{3}{x-6}}$

$= \dfrac{\dfrac{\overset{1}{\cancel{4x}}(x-6)}{1}\left(\dfrac{1}{\cancel{4x}}\right)_1 + \dfrac{4x(\overset{1}{\cancel{x-6}})}{1}\left(\dfrac{x}{\cancel{x-6}}\right)_1}{\dfrac{4\overset{1}{\cancel{x}}(x-6)}{1}\left(\dfrac{x-1}{\cancel{x}}\right)_1 - \dfrac{4x(\overset{1}{\cancel{x-6}})}{1}\left(\dfrac{3}{\cancel{x-6}}\right)_1}$

$= \dfrac{x-6+4x^2}{4(x^2-7x+6)-12x} = \dfrac{4x^2+x-6}{4x^2-28x+24-12x}$

$= \dfrac{4x^2+x-6}{4x^2-40x+24}$

32. $\dfrac{2x^2-2+2y}{2y+x-1}$

Review Exercises 8.8 (page 393)

1. -4 must be excluded, because -4 makes the denominator zero.

2. 0

3. 3 and -3 must be excluded, because either makes the denominator zero.

4. $0, 1, -2$

5. Yes, because if we multiply both 1 and 2 by $(x+1)$, we get $x+1$ and $2x+2$.

6. No

7. $x-6$. The signs of the rational expressions are different and the signs of the denominators are different. Therefore, the signs of the numerators must be *the same*.

8. 2

9. -4. The signs of the rational expressions are the same and the signs of the numerators are different. Therefore, the signs of the denominators must be different.

10. $\dfrac{2b}{a^2}$ **11.** $\dfrac{\overset{1}{\cancel{2}}(1+2m)}{\underset{1}{\cancel{2}}} = 1+2m$ **12.** $1-2n$

13. $\dfrac{(\overset{1}{\cancel{a+2}})(a-2)}{\underset{1}{\cancel{a+2}}} = a-2$ **14.** $\dfrac{1}{x+2}$

15. $\dfrac{a-b}{a(x+y)-b(x+y)} = \dfrac{\overset{1}{\cancel{a-b}}}{(x+y)(\underset{1}{\cancel{a-b}})} = \dfrac{1}{x+y}$ **16.** $\dfrac{10}{k}$

17. LCD $= 2x$; $\dfrac{5(2x)}{1(2x)} - \dfrac{3}{2x} = \dfrac{10x - 3}{2x}$

18. 1 **19.** $\dfrac{-\overset{a}{\cancel{8a^2}}}{\underset{1}{\cancel{3b}}} \cdot \dfrac{\overset{3b}{\cancel{9b^2}}}{\underset{2}{\cancel{10a}}} = -\dfrac{3ab}{2}$

20. $-6x^2y^2$ **21.** $\dfrac{\overset{1}{\cancel{3(x+2)}}}{\underset{\underset{1}{2}}{\cancel{6}}} \cdot \dfrac{\overset{1}{\cancel{2x^2}}}{4\cancel{(x+2)}} = \dfrac{x^2}{4}$ **22.** $\dfrac{y - 17}{6}$

23. LCD $= (a - 1)(a + 2)$

$\dfrac{(a - 2)(a + 2)}{(a - 1)(a + 2)} + \dfrac{(a + 1)(a - 1)}{(a + 2)(a - 1)} = \dfrac{a^2 - 4 + a^2 - 1}{(a + 2)(a - 1)}$

$= \dfrac{2a^2 - 5}{(a + 2)(a - 1)}$

24. $-\dfrac{5}{(k - 2)(k - 3)}$

25. $\dfrac{2x^2 - 6x}{x + 2} \div \dfrac{x}{4x + 8} = \dfrac{2\overset{1}{\cancel{x}}(x - 3)}{\cancel{x + 2}} \cdot \dfrac{4\overset{1}{\cancel{(x + 2)}}}{\underset{1}{\cancel{x}}} = 8(x - 3)$

26. $8z$

27. LCD $= (2x + 3)(x + 2)(3x + 4)$

$\dfrac{(3x + 4)\,(3x + 4)}{(2x + 3)(x + 2)\,(3x + 4)} - \dfrac{(x + 2)\,(2x + 3)}{(3x + 4)(x + 2)\,(2x + 3)}$

$= \dfrac{9x^2 + 24x + 16 - 1(2x^2 + 7x + 6)}{(2x + 3)(x + 2)(3x + 4)}$

$= \dfrac{9x^2 + 24x + 16 - 2x^2 - 7x - 6}{(2x + 3)(x + 2)(3x + 4)} = \dfrac{7x^2 + 17x + 10}{(2x + 3)(x + 2)(3x + 4)}$

28. $\dfrac{-3x^3 + x^2 - 248x + 400}{15x(x + 4)^2(x - 4)}$

29. LCD of secondary denominators $= 9m^2$

$\dfrac{\dfrac{5k^2}{3m^2}}{\dfrac{10k}{9m}} = \dfrac{5k^2}{3m^2} \div \dfrac{10k}{9m} = \dfrac{\overset{k}{\cancel{5k^2}}}{\underset{m}{\cancel{3m^2}}} \cdot \dfrac{\overset{3}{\cancel{9m}}}{\underset{2}{\cancel{10k}}} = \dfrac{3k}{2m}$

30. $\dfrac{3b - a}{2b + a}$

31. LCD $= b^2$

$\dfrac{b^2}{b^2} \cdot \dfrac{\dfrac{a^2}{b^2} - 1}{\dfrac{a}{b} - 1} = \dfrac{\overset{1}{\cancel{b^2}}\left(\dfrac{a^2}{\cancel{b^2}}\right)_1 - b^2 \cdot 1}{\overset{b}{\cancel{b^2}}\left(\dfrac{a}{\cancel{b}}\right)_1 - b^2 \cdot 1} = \dfrac{a^2 - b^2}{ab - b^2}$

$= \dfrac{(a + b)\overset{1}{\cancel{(a - b)}}}{b\underset{1}{\cancel{(a - b)}}} = \dfrac{a + b}{b}$

32. $\dfrac{x + 4}{x}$

Sections 8.1–8.8 Diagnostic Test (page 397)

Following each problem number is the textbook section number (in parentheses) where that kind of problem is discussed.

1. (8.1) 5 must be excluded, because it makes the denominator zero.

2. (8.1) $x(x + 9) = 0$ will make the denominator zero. Therefore, 0 and -9 must be excluded.

3. (8.1) $3x$. The signs of the numerators are different and the signs of the rational expressions are the same. Therefore, the signs of the denominators must be different.

4. (8.1) -17. The signs of the denominators are different and the signs of the rational expressions are the same. Therefore, the signs of the numerators must be different.

5. (8.2) $\dfrac{5x + 1}{25x^2 + 10x + 1} = \dfrac{\overset{1}{\cancel{5x + 1}}}{(5x + 1)\underset{1}{\cancel{(5x + 1)}}} = \dfrac{1}{5x + 1}$

6. (8.2) $\dfrac{x^2 - 4y^2}{2x^2 + 3xy - 2y^2} = \dfrac{(x + 2y)(x - 2y)}{(2x - y)\underset{1}{\cancel{(x + 2y)}}} = \dfrac{x - 2y}{2x - y}$

7. (8.5) (1) $2 \cdot 3 \cdot x^3 \cdot y$, $3 \cdot 5 \cdot x^2 \cdot y^2$ are the polynomials in factored form.
(2) 2, 3, 5, x, y are all the different bases.
(3) 2^1, 3^1, 5^1, x^3, y^2 are the highest powers of each base.
(4) LCM $= 2^1 \cdot 3^1 \cdot 5^1 \cdot x^3 \cdot y^2 = 30x^3y^2$

8. (8.3) $\dfrac{x}{\underset{1}{\cancel{3y^2}}\overset{2}{}}\cdot\dfrac{\overset{2x}{\cancel{6xy}}}{5y^3} = \dfrac{2x^2}{5y^4}$

9. (8.3) $\dfrac{x}{x^2 + 6x + 8} \cdot \dfrac{3x + 12}{12x^2 + 24x} = \dfrac{\overset{1}{\cancel{x}}}{(x + 2)\cancel{(x + 4)}} \cdot \dfrac{\overset{1}{\cancel{3}}\overset{1}{\cancel{(x + 4)}}}{\underset{4}{\cancel{12}}x\underset{1}{(x + 2)}}$

$= \dfrac{1}{4(x + 2)^2}$

10. (8.3) $\dfrac{4}{x - 5} \div \dfrac{8}{x^2 - 5x} = \dfrac{4}{x - 5} \cdot \dfrac{x^2 - 5x}{8} = \dfrac{\overset{1}{\cancel{4}}}{\underset{1}{\cancel{x - 5}}} \cdot \dfrac{x\overset{1}{\cancel{(x - 5)}}}{\underset{2}{\cancel{8}}}$

$= \dfrac{x}{2}$

11. (8.3) $\dfrac{7y}{7y + y^2} \div \dfrac{1 - y^2}{7 + 8y + y^2} = \dfrac{7y}{7y + y^2} \cdot \dfrac{7 + 8y + y^2}{1 - y^2}$

$= \dfrac{7\overset{1}{\cancel{y}}}{\underset{1}{\cancel{y}}\underset{1}{\cancel{(7 + y)}}} \cdot \dfrac{\overset{1}{\cancel{(7 + y)}}(1 + y)}{(1 - y)\cancel{(1 + y)}} = \dfrac{7}{1 - y}$

12. (8.4) $\dfrac{8}{3y^2} - \dfrac{2}{3y^2} = \dfrac{8 - 2}{3y^2} = \dfrac{\overset{2}{\cancel{6}}}{\underset{1}{\cancel{3y^2}}} = \dfrac{2}{y^2}$

13. (8.4) $\dfrac{8x}{x + 7} + \dfrac{3}{x + 7} = \dfrac{8x + 3}{x + 7}$

14. (8.4) $\dfrac{8x}{4x - 1} - \dfrac{2}{4x - 1} = \dfrac{8x - 2}{4x - 1} = \dfrac{2\overset{1}{\cancel{(4x - 1)}}}{\underset{1}{\cancel{4x - 1}}} = \dfrac{2}{1} = 2$

15. (8.6) LCD $= (x - 1)(x + 3)$;

$\dfrac{3x\,(x + 3)}{(x - 1)\,(x + 3)} + \dfrac{4\,(x - 1)}{(x + 3)\,(x - 1)}$

$= \dfrac{3x^2 + 9x + 4x - 4}{(x + 3)(x - 1)} = \dfrac{3x^2 + 13x - 4}{(x + 3)(x - 1)}$

16. (8.6) LCD $= (x + 5)(x - 3)$

$\dfrac{(x - 3)\,(x - 3)}{(x + 5)\,(x - 3)} + \dfrac{(x + 5)\,(x + 5)}{(x + 5)\,(x - 3)}$

$= \dfrac{x^2 - 6x + 9 + x^2 + 10x + 25}{(x + 5)(x - 3)} = \dfrac{2x^2 + 4x + 34}{(x + 5)(x - 3)}$

17. (8.6) $x^2 + 7x + 12 = (x + 3)(x + 4)$;
$x^2 + 5x + 6 = (x + 2)(x + 3)$
LCD $= (x + 3)(x + 4)(x + 2)$

$\dfrac{(x + 14)\,(x + 2)}{(x + 3)(x + 4)\,(x + 2)} - \dfrac{(x + 5)\,(x + 4)}{(x + 2)(x + 3)\,(x + 4)}$

$= \dfrac{(x^2 + 16x + 28) - 1(x^2 + 9x + 20)}{(x + 3)(x + 4)(x + 2)}$

$= \dfrac{x^2 + 16x + 28 - x^2 - 9x - 20}{(x + 3)(x + 4)(x + 2)}$

$= \dfrac{7x + 8}{(x + 3)(x + 4)(x + 2)}$

18. (8.6) LCD $= (x - 4)(x + 4)$, since $x^2 - 16 = (x - 4)(x + 4)$

$$\frac{x\ (x + 4)}{(x - 4)\ (x + 4)} - \frac{x\ (x - 4)}{(x + 4)\ (x - 4)} - \frac{32}{(x + 4)(x - 4)}$$

$$= \frac{(x^2 + 4x) - 1(x^2 - 4x) - (32)}{(x - 4)(x + 4)}$$

$$= \frac{x^2 + 4x - x^2 + 4x - 32}{(x - 4)(x + 4)}$$

$$= \frac{8x - 32}{(x - 4)(x + 4)} = \frac{8(x - 4)}{(x - 4)(x + 4)} = \frac{8}{x + 4}$$

19. (8.7) $\dfrac{\dfrac{4x^3}{9y^4}}{\dfrac{2x}{27y^2}} = \dfrac{4x^3}{9y^4} \div \dfrac{2x}{27y^2} = \dfrac{4x^3}{9y^4} \cdot \dfrac{27y^2}{2x} = \dfrac{6x^2}{y^2}$

20. (8.7) LCD $= x^2$

$$\frac{\dfrac{6}{x} - \dfrac{4}{x^2}}{5 + \dfrac{2}{x}} = \frac{x^2}{x^2} \cdot \frac{\dfrac{6}{x} - \dfrac{4}{x^2}}{5 + \dfrac{2}{x}} = \frac{\dfrac{x^2}{1}\left(\dfrac{6}{x}\right) - \dfrac{x^2}{1}\left(\dfrac{4}{x^2}\right)}{\dfrac{x^2}{1}\left(\dfrac{5}{1}\right) + \dfrac{x^2}{1}\left(\dfrac{2}{x}\right)} = \frac{6x - 4}{5x^2 + 2x}$$

Exercises 8.9A (page 401)

(The checks usually will not be shown.)

1. LCD $= 12$. Multiply both sides of the equation by 12.   **2.** 15

$$(12)\frac{x}{3} + (12)\frac{x}{4} = (12)\,7$$
$$4x + 3x = 84$$
$$7x = 84$$
$$x = 12$$

The solution is 12.

3. LCD $= c$. Multiply both sides of the equation by c.   **4.** $1\frac{1}{2}$

$$\frac{10}{c}(c) = 2\,(c)$$
$$10 = 2c$$
$$5 = c, \text{ or } c = 5$$

The solution is 5.

5. LCD $= 10$; $(10)\dfrac{a}{2} + (10)\dfrac{-a}{5} = (10)\,6$ **6.** 63

$$5a - 2a = 60$$
$$3a = 60$$
$$a = 20$$

The solution is 20.

7. LCD $= 2x$; $\dfrac{9}{2x}(2x) = 3\,(2x)$ **8.** $\frac{2}{3}$

$$9 = 6x$$
$$\frac{9}{6} = x$$
$$x = \frac{3}{2} \text{ or } 1\frac{1}{2}$$

The solution is $1\frac{1}{2}$.

9. LCD $= 15$; $(15)\dfrac{M - 2}{5} + (15)\dfrac{M}{3} = (15)\dfrac{1}{5}$ **10.** $-\frac{5}{9}$

$$3(M - 2) + 5M = 3$$
$$3M - 6 + 5M = 3$$
$$8M = 9$$
$$M = \frac{9}{8} = 1\frac{1}{8}$$

The solution is $1\frac{1}{8}$.

11. LCD $= x(x + 4)$ **12.** 4

$$\frac{7}{x + 4}\ (x)(x + 4) = \frac{3}{x}\ (x)(x + 4)$$
$$7x = 3(x + 4)$$
$$7x = 3x + 12$$
$$4x = 12$$
$$x = 3$$

The solution is 3.

13. LCD $= x - 2$ **14.** 3

$$\frac{3x}{x - 2}\ (x - 2) = 5\ (x - 2)$$
$$3x = 5(x - 2)$$
$$3x = 5x - 10$$
$$-2x = -10$$
$$x = 5$$

The solution is 5.

15. LCD $= x - 2$

$$\frac{3x - 1}{x - 2}\ (x - 2) = 3\ (x - 2) + \frac{2x + 1}{x - 2}\ (x - 2)$$
$$3x - 1 = 3(x - 2) + (2x + 1)$$
$$3x - 1 = 3x - 6 + 2x + 1$$
$$3x - 1 = 5x - 5$$
$$4 = 2x$$
$$x = 2 \qquad 2 \text{ is an excluded value.}$$

No solution.
(If you try to check the apparent solution, you will have zeros in denominators.)

16. No solution

17. LCD $= (x^2 + 1)(1 + 2x)$ **18.** $\frac{2}{3}$

$$\frac{x}{x^2 + 1}\ (x^2 + 1)(1 + 2x) = \frac{2}{1 + 2x}\ (x^2 + 1)(1 + 2x)$$
$$x(1 + 2x) = 2(x^2 + 1)$$
$$x + 2x^2 = 2x^2 + 2$$
$$\underline{\quad -2x^2 \qquad -2x^2 \quad}$$
$$x = 2$$

The solution is 2.

19. LCD $= 12$ **20.** $1\frac{1}{9}$

$$(12)\frac{2x - 1}{3} + (12)\frac{3x}{4} = (12)\frac{5}{6}$$
$$4(2x - 1) + 3(3x) = 10$$
$$8x - 4 + 9x = 10$$
$$17x = 14$$
$$x = \frac{14}{17}$$

The solution is $\frac{14}{17}$.

21. LCD $= 10$ **22.** $7\frac{4}{11}$

$$(10)\frac{2(m - 3)}{5} + (10)\frac{(-3)(m + 2)}{2} = (10)\frac{7}{10}$$
$$4(m - 3) - 15(m + 2) = 7$$
$$4m - 12 - 15m - 30 = 7$$
$$-11m = 49$$
$$m = \frac{49}{-11} = -4\frac{5}{11}$$

The solution is $-4\frac{5}{11}$.

23. LCD $= x - 2$

$$\left(\frac{x-2}{1}\right)\frac{x}{x-2} = \left(\frac{x-2}{1}\right)\frac{2}{x-2} + \left(\frac{x-2}{1}\right)(5)$$
$$x = 2 + 5x - 10$$
$$8 = 4x$$
$$x = 2 \qquad 2 \text{ is an excluded value.}$$

No solution

(If you try to check the apparent solution, you will have zeros in denominators.)

24. No solution

Exercises 8.9B (page 407)

1. $\dfrac{x}{14} = \dfrac{-3}{7}$

$$7x = -3(14)$$
$$x = \frac{-3(14)}{7} = -6$$

The solution is -6.

2. -10

3. $\dfrac{x}{4} = \dfrac{2}{3}$

$$3x = 2(4) = 8$$
$$x = \frac{8}{3} = 2\frac{2}{3}$$

The solution is $2\frac{2}{3}$.

4. $7\frac{1}{2}$

5. $\dfrac{8}{x} = \dfrac{4}{5}$

$$4x = 8(5) = 40$$
$$x = 10$$

The solution is 10.

6. $2\frac{2}{3}$

7. $\dfrac{4}{7} = \dfrac{x}{21}$

$$7x = 4(21)$$
$$x = \frac{4(21)}{7} = 12$$

The solution is 12.

8. $11\frac{1}{4}$

9. $\dfrac{100}{x} = \dfrac{4\cancel{0}}{3\cancel{0}} = \dfrac{4}{3}$

$$4x = 300$$
$$x = 75$$

The solution is 75.

10. 24

11. $\dfrac{x+1}{x-1} = \dfrac{3}{2}$

$$2(x + 1) = 3(x - 1)$$
$$2x + 2 = 3x - 3$$
$$5 = x; \; x = 5$$

The solution is 5.

12. 4

13. $\dfrac{2x+7}{9} = \dfrac{2x+3}{5}$

$$5(2x + 7) = 9(2x + 3)$$
$$10x + 35 = 18x + 27$$
$$8 = 8x$$
$$x = 1$$

The solution is 1.

14. -2

15. $\dfrac{5x-10}{10} = \dfrac{3x-5}{7}$

$$7(5x - 10) = 10(3x - 5)$$
$$35x - 70 = 30x - 50$$
$$5x = 20$$
$$x = 4$$

The solution is 4.

16. $2\frac{2}{7}$

17. $\dfrac{\frac{3}{4}}{6} = \dfrac{P}{16}$

$$6P = \frac{3}{\cancel{4}} \cdot \frac{\cancel{16}^{\,4}}{1} = 12$$
$$\frac{\cancel{6}^{\,1}P}{\cancel{6}_{\,1}} = \frac{12}{6} = 2$$

The solution is 2.

18. $2\frac{1}{2}$

19. $\dfrac{A}{9} = \dfrac{3\frac{1}{3}}{5}$

$$5A = 9 \cdot 3\frac{1}{3}$$
$$5A = \frac{\cancel{9}^{\,3}}{1} \cdot \frac{10}{\cancel{3}_{\,1}} = 30$$
$$\frac{\cancel{5}^{\,1}A}{\cancel{5}_{\,1}} = \frac{30}{5} = 6$$

The solution is 6.

20. 1

21. $\dfrac{7.7}{B} = \dfrac{\overset{0.7}{\cancel{3.5}}}{\cancel{5}_{\,1}}$

$$0.7B = 7.7$$
$$\frac{\cancel{0.7}^{\,1}B}{\cancel{0.7}_{\,1}} = \frac{7.7}{0.7} = 11$$

The solution is 11.

22. 22.96

23. $\dfrac{P}{100} = \dfrac{\frac{3}{2}}{15}$

$$15P = \frac{\cancel{100}^{\,50}}{1} \cdot \frac{3}{\cancel{2}_{\,1}}$$
$$15P = 150$$
$$P = 10$$

The solution is 10.

24. 4

25. $\dfrac{12\frac{1}{2}}{100} = \dfrac{A}{48}$

$$100A = 12\frac{1}{2} \cdot 48$$
$$100A = \frac{25}{\cancel{2}_{\,1}} \cdot \frac{\cancel{48}^{\,24}}{1}$$
$$\frac{\cancel{100}^{\,1}A}{\cancel{100}_{\,1}} = \frac{25(24)}{\cancel{100}_{\,1}}^{\,6} = 6$$

The solution is 6.

26. 54

Exercises 8.9C (page 410)

(Checks will usually not be shown.)

1. LCD $= z$

$$(z)z + (\cancel{z})\frac{1}{\cancel{z}}^{\,1}_{\,1} = (\cancel{z})\frac{17}{\cancel{z}}^{\,1}_{\,1}$$
$$z^2 + 1 = 17$$
$$z^2 + 1 - 17 = 0$$
$$z^2 - 16 = 0$$
$$(z + 4)(z - 4) = 0$$

$z + 4 = 0$	$z - 4 = 0$
$z = -4$	$z = 4$

The solutions are -4 and 4.

2. $-3, 3$

3. LCD $= 2x^2$

$$\left(\frac{2x^2}{1}\right)\frac{2}{x} + \left(\frac{2x^2}{1}\right)\left(-\frac{2}{x^2}\right) = \left(\frac{2x^2}{1}\right)\frac{1}{2}$$
$$4x - 4 = x^2$$
$$0 = x^2 - 4x + 4$$
$$0 = (x - 2)(x - 2)$$

$x - 2 = 0$	$x - 2 = 2$
$x = 2$	$x = 2$

The solution is 2.

4. 2, 4

5. $\dfrac{x}{x+1} = \dfrac{4x}{3x+2}$

$4x(x+1) = x(3x+2)$
$4x^2 + 4x = 3x^2 + 2x$
$x^2 + 2x = 0$
$x(x+2) = 0$

$x = 0$ $\bigg|$ $x + 2 = 0$
$\qquad\qquad x = -2$

Check for $x = 0$:   *Check for $x = -2$:*

$\dfrac{(0)}{(0)+1} = \dfrac{4(0)}{3(0)+2}$ $\dfrac{(-2)}{(-2)+1} = \dfrac{4(-2)}{3(-2)+2}$

$\dfrac{0}{1} = \dfrac{0}{2}$ $\dfrac{-2}{-1} = \dfrac{-8}{-4}$

$\quad 0 = 0$ True $2 = 2$ True

The solutions are 0 and -2.

6. 0, 2

7. LCD $= x$

$\dfrac{5}{x}(x) - 1(x) = \dfrac{x+11}{x}(x)$

$\quad\; 5 - x = \; x + 11$
$\underline{-11 + x \quad\;\; x - 11}$
$\quad -6 \;\;\; = 2x$
$\qquad\quad x = -3$

The solution is -3.

8. -4

9. LCD $= 3(x+1)(x-1)$

$3(x+1)(x-1)\dfrac{1}{x+1} + 3(x+1)(x-1)\dfrac{2}{x+1}$
$\qquad\qquad\qquad = 3(x+1)(x-1)\dfrac{5}{3}$

$3(x+1) + 6(x-1) = 5(x^2 - 1)$
$3x + 3 + 6x - 6 = 5x^2 - 5$
$\qquad\qquad\quad 0 = 5x^2 - 9x - 2$
$\qquad\qquad\quad 0 = (5x+1)(x-2)$

$5x + 1 = 0$ $\bigg|$ $x - 2 = 0$
$\quad\; x = -\frac{1}{5}$ $\bigg|$ $\quad x = 2$

The solutions are $-\frac{1}{5}$ and 2.

10. $3, \frac{1}{21}$

11. LCD $= 4(2x+5)$

$\dfrac{4(2x+5)}{1} \cdot \dfrac{3}{2x+5} + \dfrac{4(2x+5)}{1} \cdot \dfrac{x}{4} = \dfrac{4(2x+5)}{1} \cdot \dfrac{3}{4}$

$4(3) + x(2x+5) = 3(2x+5)$
$12 + 2x^2 + 5x = 6x + 15$
$\quad\; 2x^2 - x - 3 = 0$
$\quad (1x+1)(2x-3) = 0$

$x + 1 = 0$ $\bigg|$ $2x - 3 = 0$
$\quad\; x = -1$ $\bigg|$ $\quad 2x = 3$
$\qquad\qquad\qquad\quad x = \frac{3}{2}$

The solutions are -1 and $\frac{3}{2}$.

12. $2, -\frac{19}{2}$

13. LCD $= 15x(x+1)$

$\dfrac{15x(x+1)}{1} \cdot \dfrac{4}{x+1} - \dfrac{15x(x+1)}{1} \cdot \dfrac{3}{x} = \dfrac{15x(x+1)}{1} \cdot \dfrac{1}{15}$

$60x - 45(x+1) = x^2 + x$
$60x - 45x - 45 = x^2 + x$
$\quad\; 15x - 45 = x^2 + x$
$\qquad\qquad 0 = x^2 - 14x + 45$
$\qquad\qquad 0 = (x-5)(x-9)$

$x - 5 = 0$ $\bigg|$ $x - 9 = 0$
$\quad\; x = 5$ $\bigg|$ $\quad x = 9$

The solutions are 5 and 9.

14. $-2, -3$

15. $x^2 - 9 = (x+3)(x-3)$; LCD $= 5(x+3)(x-3)$

$\dfrac{5(x+3)(x-3)}{1} \cdot \dfrac{6}{(x+3)(x-3)}$

$= \dfrac{5(x+3)(x-3)}{1} \cdot \dfrac{1}{x-3} - \dfrac{5(x+3)(x-3)}{1} \cdot \dfrac{1}{5}$

$30 = 5x + 15 - 1(x^2 - 9)$
$30 = 5x + 15 - x^2 + 9$
$30 = 5x + 24 - x^2$

$x^2 - 5x + 6 = 0$
$(x-2)(x-3) = 0$

$x - 2 = 0$ $\bigg|$ $x - 3 = 0$
$\quad\; x = 2$ $\bigg|$ $\quad x = 3$, but 3 cannot be a solution, since it is an
$\qquad\qquad\qquad\qquad$ excluded value.

The only solution is 2.

16. $\frac{7}{3}$

Exercises 8.10 (page 414)

1. $2x + y = 4$
$\quad 2x = 4 - y$
$\quad\;\; x = \dfrac{4-y}{2}$

2. $y = \dfrac{6-x}{3}$

3. $y - z = -8$
$\quad\; -z = -8 - y$
$\qquad z = 8 + y$

4. $n = m + 5$

5. $2x - y = -4$
$\quad\;\; -y = -2x - 4$
$\qquad y = 2x + 4$

6. $z = 3y + 5$

7. $2x - 3y = 6$
$\quad 2x = 3y + 6$
$\quad\;\; x = \dfrac{3y+6}{2}$

8. $x = \dfrac{2y+6}{3}$

9. $2(x - 3y) = x + 4$
$\quad 2x - 6y = x + 4$
$\qquad\qquad x = 6y + 4$

10. $x = \dfrac{2y+14}{7}$

11. $PV = k$
$\quad \dfrac{PV}{P} = \dfrac{k}{P}$
$\qquad V = \dfrac{k}{P}$

12. $R = \dfrac{E}{I}$

13. $I = prt$
$\quad \dfrac{I}{rt} = \dfrac{prt}{rt}$
$\quad\; p = \dfrac{I}{rt}$

14. $\ell = \dfrac{V}{wh}$

15.
$$p = 2\ell + 2w$$
$$p - 2w = 2\ell$$
$$\frac{p - 2w}{2} = \frac{\overset{1}{\cancel{2}}\ell}{\cancel{2}}$$
$$\ell = \frac{p - 2w}{2}$$

16. $w = \dfrac{P - 2\ell}{2}$

17.
$$y = mx + b$$
$$y - b = mx$$
$$\frac{y - b}{m} = \frac{\overset{1}{\cancel{m}}x}{\cancel{m}}$$
$$x = \frac{y - b}{m}$$

18. $t = \dfrac{V - k}{g}$

19.
$$\frac{S}{1} = \frac{a}{1 - r}$$
$$S(1 - r) = a$$
$$S - Sr = a$$
$$-Sr = a - S$$
$$\frac{-Sr}{-S} = \frac{a - S}{-S}$$
$$r = \frac{S - a}{S}$$

20. $R = \dfrac{E - Ir}{I}$

21. LCD = 9
$$9(C) = \frac{\overset{1}{\cancel{9}}}{1}\left(\frac{5(F - 32)}{\cancel{9}}\right)$$
$$9C = 5F - 160$$
$$9C + 160 = 5F$$
$$F = \frac{9C + 160}{5}$$

22. $B = \dfrac{2A - hb}{h}$

23.
$$L = a + (n - 1)d$$
$$L = a + nd - d$$
$$L - a + d = nd$$
$$\frac{L - a + d}{d} = \frac{nd}{\cancel{d}}\overset{1}{}$$
$$n = \frac{L - a + d}{d}$$

24. $h = \dfrac{A - 2\pi r^2}{2\pi r}$

25.
$$\frac{z}{1} = \frac{Rr}{R + r}$$
$$z(R + r) = 1 \cdot Rr$$
$$zR + zr = Rr$$
$$zr = Rr - zR$$
$$zr = R(r - z)$$
$$\frac{zr}{r - z} = \frac{R(\cancel{r - z})}{\cancel{r - z}}$$
$$R = \frac{zr}{r - z}$$

26. $b = \dfrac{ca}{a - c}$

27. LCD = Fuv
$$\frac{Fuv}{1}\left(\frac{1}{F}\right) = \frac{Fuv}{1}\left(\frac{1}{u}\right) + \frac{Fuv}{1}\left(\frac{1}{v}\right)$$
$$uv = Fv + Fu$$
$$uv - Fu = Fv$$
$$u(v - F) = Fv$$
$$\frac{u(v - F)}{v - F} = \frac{Fv}{v - F}$$
$$u = \frac{Fv}{v - F}$$

28. $a = \dfrac{bc}{b - c}$

Exercises 8.11 (page 420)

(The checks will not be shown.)

1. Let n = numerator
$n + 6$ = denominator

original fraction = $\dfrac{n}{n + 6}$

$$\frac{n + 4}{n + 6 - 4} = \frac{11}{10}$$
$$10(n + 4) = 11(n + 2)$$
$$10n + 40 = 11n + 22$$
$$18 = n$$

The original fraction = $\dfrac{n}{n + 6} = \dfrac{18}{18 + 6} = \dfrac{18}{24}$.

2. $\frac{16}{20}$

3. Let x = numerator
$2x$ = denominator

$$\frac{x + 5}{2x - 5} = \frac{4}{5}$$
$$(x + 5)(5) = 4(2x - 5)$$
$$5x + 25 = 8x - 20$$
$$\underline{-5x + 20 \quad -5x + 20}$$
$$45 = 3x$$
$$x = 15, \text{ numerator}$$
$$2x = 30, \text{ denominator}$$

The fraction is $\frac{15}{30}$.

4. $\frac{15}{45}$

5. Let x = the fraction
$\dfrac{1}{x}$ = its reciprocal

$$x + \frac{1}{x} = \frac{29}{10} \quad \text{LCD is } 10x$$
$$x(10x) + \frac{1}{x}(10x) = \frac{29}{10}(10x)$$
$$10x^2 + 10 = 29x$$
$$10x^2 - 29x + 10 = 0$$
$$(2x - 5)(5x - 2) = 0$$

$2x - 5 = 0$	$5x - 2 = 0$
$2x = 5$	$5x = 2$
$x = \frac{5}{2}$	$x = \frac{2}{5}$
$\frac{1}{x} = \frac{2}{5}$	$\frac{1}{x} = \frac{5}{2}$

Therefore, there are two answers: The number is $\frac{2}{5}$ and its reciprocal is $\frac{5}{2}$, or the number is $\frac{5}{2}$ and its reciprocal is $\frac{2}{5}$.

6. $\frac{3}{11}$, or $\frac{11}{3}$

7. Let w = width

$$\frac{1 \text{ in.}}{8 \text{ ft}} = \frac{2\frac{1}{2}\text{in.}}{w \text{ ft}} \Rightarrow \frac{1}{8} = \frac{\frac{5}{2}}{w}$$
$$w = 8\left(\frac{5}{2}\right) = 20$$

Let ℓ = length

$$\frac{1 \text{ in.}}{8 \text{ ft}} = \frac{3 \text{ in.}}{\ell \text{ ft}} \Rightarrow \frac{1}{8} = \frac{3}{\ell}$$
$$\ell = 8(3) = 24$$

Therefore, the room is 20 ft by 24 ft.

8. 26 ft by 34 ft

9. Let x = woman's weight on moon (in pounds). **10.** 78.4 lb

$$\frac{6 \text{ earth}}{1 \text{ moon}} = \frac{150 \text{ earth}}{x \text{ moon}}$$

$$\frac{6}{1} = \frac{150}{x}$$

$$6x = 150$$

$$x = \frac{150}{6} = 25$$

The woman would weigh 25 lb on the moon.

11. Let x = weight of lead bar. **12.** 7.5 lb

$$\frac{21 \text{ lead}}{5 \text{ aluminum}} = \frac{x \text{ lead}}{150 \text{ aluminum}}$$

$$\frac{21}{5} = \frac{x}{150}$$

$$5x = 21(150)$$

$$x = \frac{21(150)}{5} = 630$$

The lead bar would weigh 630 lb.

13. Let x = number of chains of fire line. **14.** $1,875

$$10 \text{ hr} - 6\tfrac{1}{2} \text{ hr} = 3\tfrac{1}{2} \text{ hr}$$

$$\frac{26 \text{ chains}}{6\frac{1}{2} \text{ hr}} = \frac{x \text{ chains}}{3\frac{1}{2} \text{ hr}}$$

$$6\tfrac{1}{2}x = 26\left(3\tfrac{1}{2}\right)$$

$$\frac{13}{2}x = \overset{13}{\cancel{26}}\left(\frac{7}{\underset{1}{\cancel{2}}}\right)$$

$$\frac{13x}{2} = \frac{91}{1}$$

$$13x = 2(91) = 182$$

$$\frac{13x}{13} = \frac{182}{13}$$

$$x = 14$$

The crew could build 14 chains of fire line.

15. Let x = amount received for 16 steers.

$$\frac{3,420 \text{ dollars}}{9 \text{ steers}} = \frac{x \text{ dollars}}{16 \text{ steers}}$$

$$9x = 16(3,420)$$

$$\frac{9x}{9} = \frac{54,720}{9}$$

$$x = 6,080$$

The producer will receive $6,080.

16. $45 per hog; 45¢ per kg

17. $\left(\text{Tricia's rate is } \dfrac{1 \text{ job}}{8 \text{ hr}}, \text{ or } \dfrac{1}{8} \dfrac{\text{job}}{\text{hr}}. \text{ Mike's rate is } \dfrac{1 \text{ job}}{10 \text{ hr}}, \text{ or } \dfrac{1}{10} \dfrac{\text{job}}{\text{hr}}.\right)$

Let x = number of hr for Tricia and Mike together to do the job.

They work for x hr

$$\tfrac{1}{8}x + \tfrac{1}{10}x = 1 \quad \text{LCD is } 40$$

$$\tfrac{1}{8}x(40) + \tfrac{1}{10}x(40) = 1(40)$$

$$5x + 4x = 40$$

$$9x = 40$$

$$x = \tfrac{40}{9} \text{ or } 4\tfrac{4}{9}$$

The job will take $4\tfrac{4}{9}$ hr.

18. $2\tfrac{11}{12}$ days **19.** Let x = numerator

$$42 - x = \text{denominator}$$

$$\frac{x}{42 - x} = \text{fraction}$$

$$\frac{x + 2}{(42 - x) + 6} = \frac{2}{3}$$

$$\frac{x + 2}{48 - x} = \frac{2}{3}$$

$$\begin{aligned} 3x + 6 &= 96 - 2x \\ \underline{2x - 6} &\quad \underline{- 6 + 2x} \\ 5x &= 90 \end{aligned}$$

$$x = 18, \text{ numerator}$$

$$42 - x = 24, \text{ denominator}$$

The fraction is $\tfrac{18}{24}$.

20. $\tfrac{17}{23}$

21. Ruth's rate is $\dfrac{220 \text{ pages}}{4 \text{ hour}}$, or $55 \dfrac{\text{pages}}{\text{hr}}$. **22.** 5 hr

Let t = time for Jerry to proofread 210 pages.

Then Jerry's rate is $\dfrac{210 \text{ pg}}{t \text{ hr}}$.

They read for 3 hr

$$\frac{210}{t}(3) + 55(3) = 291 \quad \text{Number of pages read}$$

$$\frac{630}{t}(t) + 165(t) = 291(t)$$

$$\begin{aligned} 630 + 165t &= 291t \\ \underline{- 165t} &\quad \underline{-165t} \\ 630 &= 126t \end{aligned}$$

$$t = 5$$

It takes Jerry 5 hr to proofread 210 pages.

23. Let x = Sandra's speed

$$\tfrac{3}{4}x = \text{Jim's speed}$$

They each ride 3 hr

$$x(3) + \tfrac{3}{4}x(3) = 63 \quad \text{LCD is } 4$$

$$3x(4) + \tfrac{9}{4}x(4) = 63(4)$$

$$12x + 9x = 252$$

$$21x = 252$$

$$x = 12$$

$$\tfrac{3}{4}x = 9$$

Sandra's speed is 12 mph and Jim's is 9 mph.

24. Rebecca's average speed is 16 mph, and Jill's is 14 mph.

25. Let x = speed of the boat in still water

$$x + 5 = \text{speed with current}$$

$$x - 5 = \text{speed against current}$$

Note: since $rt = d$, $t = \dfrac{d}{r}$ $\quad \dfrac{999}{x + 5} = \text{time going with current}$

$$\frac{729}{x - 5} = \text{time going against current}$$

The times are equal. $\quad \dfrac{999}{x + 5} = \dfrac{729}{x - 5}$

$$999(x - 5) = 729(x + 5)$$

$$\begin{aligned} 999x - 4995 &= 729x + 3645 \\ \underline{-729x + 4995} &\quad \underline{-729x + 4995} \\ 270x &= 8640 \end{aligned}$$

$$x = 32$$

The speed of the boat in still water is 32 mph.

26. 32 mph

27. Let $\quad x = $ larger number

$\qquad x - 12 = $ smaller number

$$\tfrac{1}{6}x - \tfrac{1}{5}(x - 12) = 2 \quad \text{LCD} = 30$$

$$\tfrac{1}{6}x(30) - \tfrac{1}{5}(x - 12)(30) = 2(30)$$

$$5x - 6(x - 12) = 60$$

$$5x - 6x + 72 = 60$$

$$-x = -12$$

$$x = 12$$

$$x - 12 = 0$$

The numbers are 12 and 0.

28. 48 and 56

Review Exercises 8.12 (page 425)

(The checks will not be shown.)

1. LCD $= m$

$$(m)\frac{6}{m} = (m)5$$

$$6 = 5m$$

$$\tfrac{6}{5} = m$$

$$m = 1\tfrac{1}{5}$$

The solution is $1\tfrac{1}{5}$.

2. 5

3. LCD $= 40$

$$(\cancel{40})\frac{z}{\cancel{5}} + (\cancel{40})\frac{-z}{\cancel{8}} = (40)3$$

$$8z - 5z = 120$$

$$3z = 120$$

$$z = 40$$

The solution is 40.

4. $\frac{11}{9}$

5. LCD $= 2z$

$$(2z)\frac{4}{2z} + (2z)\frac{2}{z} = (2z)1$$

$$4 + 4 = 2z$$

$$8 = 2z$$

$$z = 4$$

The solution is 4.

6. $-2, \frac{4}{5}$

7. LCD $= 18(2x - 1)$

$$\frac{7}{2x - 1}(18)(2x - 1) + \frac{1}{18}(18)(2x - 1) = \frac{x}{6}(18)(2x - 1)$$

$$126 + 2x - 1 = 3x(2x - 1)$$

$$125 + 2x = 6x^2 - 3x$$

$$0 = 6x^2 - 5x - 125$$

$$0 = (6x + 25)(x - 5)$$

$$6x + 25 = 0 \qquad \bigm| \qquad x - 5 = 0$$

$$6x = -25 \qquad \bigm| \qquad x = 5$$

$$x = -\tfrac{25}{6}$$

The solutions are $-\frac{25}{6}$ and 5.

8. $y = \dfrac{2x - 14}{7}$

9. LCD $= n$

$$(n)\frac{2m}{n} = (n)P$$

$$2m = nP$$

$$\frac{2m}{P} = n$$

$$\text{or} \quad n = \frac{2m}{P}$$

10. $B = \dfrac{3V}{h}$

11. $\dfrac{F - 32}{C} = \dfrac{9}{5}$

$$9C = 5(F - 32)$$

$$C = \tfrac{5}{9}(F - 32)$$

12. $\frac{17}{13}$

13. Let $x = $ speed returning from work (in mph).

$$\tfrac{1}{2}(x + 10) = \tfrac{3}{4}(x)$$

$$\tfrac{1}{2}x + 5 = \tfrac{3}{4}x$$

$$5 = \tfrac{1}{4}x$$

$$x = 20$$

$$\text{Distance} = (\text{rate}) \times (\text{time}) = \left(20\,\frac{\text{mi}}{\text{hr}}\right)\left(\frac{3}{4}\,\text{hr}\right) = 15 \text{ mi}$$

The distance is 15 mi.

14. Two answers: The number is $\frac{2}{3}$ and its reciprocal is $\frac{3}{2}$, or the number is $\frac{3}{2}$ and its reciprocal is $\frac{2}{3}$.

15. Let $x = $ number of hours for machine B to do the job.

Machine A's rate is $\dfrac{1 \text{ job}}{6 \text{ hr}}$. $\qquad$ Machine B's rate is $\dfrac{1 \text{ job}}{x \text{ hr}}$.

Both machines worked 4 hr

$$\frac{1 \text{ job}}{6 \text{ hr}}(4 \text{ hr}) + \frac{1 \text{ job}}{x \text{ hr}}(4 \text{ hr}) = 1 \text{ job} \quad \text{One job was completed}$$

$$\frac{1}{6}(4)(6x) + \frac{1}{x}(4)(6x) = 1(6x)$$

$$4x + 24 = 6x$$

$$24 = 2x$$

$$x = 12$$

It would take machine B 12 hr to do the job.

16. 810 mi

17. Let $x = $ number of yards in $\frac{1}{2}$ day.

$$\frac{2\frac{1}{2} \text{ yd}}{\frac{5}{8} \text{ day}} = \frac{x \text{ yd}}{\frac{1}{2} \text{ day}} \Rightarrow \frac{\frac{5}{2}}{\frac{5}{8}} = \frac{x}{\frac{1}{2}}$$

$$\tfrac{5}{8}x = \left(\tfrac{5}{2}\right)\left(\tfrac{1}{2}\right)$$

$$\tfrac{5}{8}x = \tfrac{5}{4}$$

$$\tfrac{5}{8}x(8) = \tfrac{5}{4}(8)$$

$$5x = 10$$

$$x = 2$$

Alan can weave 2 yd in half a day.

Chapter 8 Diagnostic Test (page 429)

Following each problem number is the textbook section number (in parentheses) where that kind of problem is discussed.

1. (8.1) **a.** -4 must be excluded, because it makes the denominator zero.

b. 0 and -4 must be excluded, because they make the denominator zero.

2. (8.1) **a.** 3. The signs of the rational expressions are different and the signs of the numerators are different. Therefore, the signs of the denominators must be *the same*.

b. -3. The signs of the denominators are different and the signs of the rational expressions are the same. Therefore, the signs of the numerators must be different.

3. (8.2) $\dfrac{x^2 - 9}{x^2 - 6x + 9} = \dfrac{(x + 3)\overset{1}{\cancel{(x - 3)}}}{(x - 3)\underset{1}{\cancel{(x - 3)}}} = \dfrac{x + 3}{x - 3}$

4. (8.2) $\dfrac{4x^2 - 23xy + 15y^2}{20x^2y - 15xy^2} = \dfrac{\overset{1}{\cancel{(4x - 3y)}}(x - 5y)}{5xy\cancel{(4x - 3y)}_1} = \dfrac{x - 5y}{5xy}$

5. (8.3) $\dfrac{a}{a^2 + 5a + 6} \cdot \dfrac{4a + 8}{6a^3 + 18a^2} = \dfrac{\overset{1}{\cancel{a}}}{\cancel{(a + 2)}_1(a + 3)} \cdot \dfrac{\overset{2}{\cancel{4}}\overset{1}{\cancel{(a + 2)}}}{\cancel{6a^2}_{3a}(a + 3)}$

$= \dfrac{2}{3a(a + 3)^2}$

6. (8.4) $\dfrac{6x}{x - 2} - \dfrac{3}{x - 2} = \dfrac{6x - 3}{x - 2}$ or $\dfrac{3(2x - 1)}{x - 2}$

7. (8.3) $\dfrac{2x}{x^2 - 9} \div \dfrac{4x^2}{x^2 - 6x + 9} = \dfrac{2x}{(x + 3)(x - 3)} \cdot \dfrac{(x - 3)(x - 3)}{4x^2}$

$= \dfrac{x - 3}{2x(x + 3)}$

8. (8.6) LCD $= b(b - 1)$

$\dfrac{b \cdot b}{(b - 1) \cdot b} - \dfrac{(b + 1)(b - 1)}{b(b - 1)} = \dfrac{b^2 - (b^2 - 1)}{b(b - 1)}$

$= \dfrac{b^2 - b^2 + 1}{b(b - 1)} = \dfrac{1}{b(b - 1)}$

9. (8.6) LCD $= (x - 5)^2(2x + 3)$

$\dfrac{(x + 3)(2x + 3)}{(x - 5)^2(2x + 3)} + \dfrac{x(x - 5)}{(x - 5)(2x + 3)(x - 5)}$

$= \dfrac{2x^2 + 9x + 9 + x^2 - 5x}{(x - 5)^2(2x + 3)} = \dfrac{3x^2 + 4x + 9}{(x - 5)^2(2x + 3)}$

10. (8.6) LCD $= (x + 5)(x - 5)$

$\dfrac{x(x - 5)}{(x + 5)(x - 5)} - \dfrac{x(x + 5)}{(x - 5)(x + 5)} - \dfrac{50}{(x + 5)(x - 5)}$

$= \dfrac{x^2 - 5x - 1(x^2 + 5x) - (50)}{(x + 5)(x - 5)}$

$= \dfrac{x^2 - 5x - x^2 - 5x - 50}{(x + 5)(x - 5)}$

$= \dfrac{-10x - 50}{(x + 5)(x - 5)} = \dfrac{-10(x + 5)}{(x + 5)(x - 5)} = \dfrac{-10}{x - 5}$

11. (8.7) $\dfrac{\frac{6x^4}{11y^2}}{\frac{9x}{22y^4}} = \dfrac{6x^4}{11y^2} \div \dfrac{9x}{22y^4} = \dfrac{6x^4}{11y^2} \cdot \dfrac{22y^4}{9x} = \dfrac{4x^3y^2}{3}$

12. (8.7) LCD $= x^2$; $\dfrac{x^2\left(\frac{6}{x} + \frac{15}{x^2}\right)}{x^2\left(2 + \frac{5}{x}\right)} = \dfrac{6x + 15}{2x^2 + 5x} = \dfrac{3(2x + 5)}{x(2x + 5)} = \dfrac{3}{x}$

13. (8.9) $\dfrac{x + 7}{3} = \dfrac{2x - 1}{4}$

$4(x + 7) = 3(2x - 1)$

$\begin{array}{r} 4x + 28 = 6x - 3 \\ \underline{-4x + 3 \quad -4x + 3} \\ 31 = 2x \end{array}$

$x = \frac{31}{2}$ (The check will not be shown.)

The solution is $\frac{31}{2}$.

14. (8.9) LCD is x; $x - \dfrac{4}{x} = 3$

$x(x) - \left(\dfrac{4}{x}\right)(x) = 3(x)$

$x^2 - 3x - 4 = 0$

$(x - 4)(x + 1) = 0$

$\begin{array}{c|c} x - 4 = 0 & x + 1 = 0 \\ x = 4 & x = -1 \end{array}$

Check for $x = 4$: *Check for $x = -1$:*

$x - \dfrac{4}{x} = 3 \qquad\qquad x - \dfrac{4}{x} = 3$

$4 - \dfrac{4}{4} \overset{?}{=} 3 \qquad\qquad -1 - \dfrac{4}{-1} \overset{?}{=} 3$

$4 - 1 \overset{?}{=} 3 \qquad\qquad -1 + 4 \overset{?}{=} 3$

$3 = 3 \qquad\qquad\qquad 3 = 3$

The solutions are 4 and -1.

15. (8.9) Since $4x + 2 = 2(2x + 1)$, the LCD is $6(2x + 1)$.

$6(2x + 1)\left(\dfrac{3}{2(2x + 1)}\right) + 6(2x + 1)\dfrac{(x + 1)}{6} = 1(6)(2x + 1)$

$9 + 2x^2 + 3x + 1 = 12x + 6$

$2x^2 + 3x + 10 = 12x + 6$

$2x^2 - 9x + 4 = 0$

$(x - 4)(2x - 1) = 0$

$\begin{array}{c|c} x - 4 = 0 & 2x - 1 = 0 \\ x = 4 & 2x = 1 \\ & x = \frac{1}{2} \end{array}$

Check for $x = 4$: *Check for $x = \frac{1}{2}$:*

$\dfrac{3}{2(2x + 1)} + \dfrac{x + 1}{6} = 1 \qquad \dfrac{3}{2(2x + 1)} + \dfrac{x + 1}{6} = 1$

$\dfrac{3}{2(2[4] + 1)} + \dfrac{[4] + 1}{6} \overset{?}{=} 1 \qquad \dfrac{3}{2[2(\frac{1}{2}) + 1]} + \dfrac{(\frac{1}{2}) + 1}{6} \overset{?}{=} 1$

$\dfrac{3}{2(8 + 1)} + \dfrac{5}{6} \overset{?}{=} 1 \qquad \dfrac{3}{2[1 + 1]} + \dfrac{\frac{3}{2}}{6} \overset{?}{=} 1$

$\dfrac{3}{2(9)} + \dfrac{5}{6} \overset{?}{=} 1 \qquad\qquad \dfrac{3}{2[2]} + \dfrac{1}{4} \overset{?}{=} 1$

$\dfrac{1}{6} + \dfrac{5}{6} \overset{?}{=} 1 \qquad\qquad\qquad \dfrac{3}{4} + \dfrac{1}{4} \overset{?}{=} 1$

$1 = 1 \qquad\qquad\qquad\qquad 1 = 1$

The solutions are 4 and $\frac{1}{2}$.

16. (8.9) $\dfrac{3x-5}{x} = \dfrac{x+1}{3}$

$3(3x-5) = x(x+1)$

$9x - 15 = x^2 + x$

$0 = x^2 - 8x + 15$

$0 = (x-3)(x-5)$

$x - 3 = 0 \quad | \quad x - 5 = 0$

$x = 3 \quad | \quad x = 5$

Check for $x = 3$:   *Check for* $x = 5$:

$\dfrac{3x-5}{x} = \dfrac{x+1}{3} \quad\quad \dfrac{3x-5}{x} = \dfrac{x+1}{3}$

$\dfrac{3(3)-5}{(3)} \overset{?}{=} \dfrac{(3)+1}{3} \quad\quad \dfrac{3(5)-5}{(5)} \overset{?}{=} \dfrac{(5)+1}{3}$

$\dfrac{9-5}{3} \overset{?}{=} \dfrac{4}{3} \quad\quad\quad \dfrac{15-5}{5} \overset{?}{=} \dfrac{6}{3}$

$\dfrac{4}{3} = \dfrac{4}{3} \quad\quad\quad\quad\quad \dfrac{10}{5} \overset{?}{=} \dfrac{6}{3}$

$\quad\quad\quad\quad\quad\quad\quad\quad\quad\quad 2 = 2$

The solutions are 3 and 5.

17. (8.10)   Solve for y:

$5x + 7y = 18$

$7y = 18 - 5x$

$y = \dfrac{18 - 5x}{7}$

18. (8.11)   Let $x = $ numerator

$x + 6 = $ denominator

$\dfrac{x-1}{(x+6)+7} = \dfrac{1}{3}$

$\dfrac{x-1}{x+13} = \dfrac{1}{3}$

$3(x-1) = 1(x+13)$

$3x - 3 = x + 13$

$\underline{-x + 3 \quad\quad -x + 3}$

$2x \quad = \quad\quad 16$

$x = 8$, numerator

$x + 6 = 14$, denominator

Check: $\dfrac{8-1}{14+7} = \dfrac{7}{21} = \dfrac{1}{3}$

The fraction is $\dfrac{8}{14}$.

19. (8.11)   Let $x = $ distance covered in $2\frac{2}{3}$ hr $\left(\text{or } \frac{8}{3} \text{ hr}\right)$.

$\dfrac{21 \text{ mi}}{\frac{2}{5} \text{ hr}} = \dfrac{x \text{ mi}}{\frac{8}{3} \text{ hr}}$

$\dfrac{2}{5}x = 21\left(\dfrac{8}{3}\right)$

$\dfrac{2}{5}x = 56 \quad\quad$ *Check:*   Rate is $\dfrac{21 \text{ mi}}{\frac{2}{5} \text{ hr}} = 52.5 \dfrac{\text{mi}}{\text{hr}}$

$(5)\left(\dfrac{2}{5}x\right) = 56(5)$

$2x = 280 \quad\quad\quad$ or   Rate is $\dfrac{140 \text{ mi}}{\frac{8}{3} \text{ hr}} = 52.5 \dfrac{\text{mi}}{\text{hr}}$

$x = 140$

Mr. Ames can drive 140 mi in $2\frac{2}{3}$ hr.

20. (8.11)   Let $t = $ the number of hours for Kenneth to type 111 pages.

Then Kenneth's rate is $\dfrac{111 \text{ pages}}{t \text{ hr}}$.

David's rate is $\dfrac{128 \text{ pages}}{4 \text{ hour}}$, or $32 \dfrac{\text{pages}}{\text{hr}}$.

They type for 5 hr

$\dfrac{111}{t}(5) + 32(5) = 345$ Number of pages typed

$\dfrac{555}{t}(t) + 160(t) = 345(t)$

$555 + 160t = 345t$

$\underline{\quad -160t \quad\quad -160t}$

$555 \quad\quad = \quad 185t$

$t = 3$

Check:

$\dfrac{111 \text{ pg}}{3 \text{ hr}} = 37 \dfrac{\text{pg}}{\text{hr}}$

$37 \dfrac{\text{pg}}{\text{hr}}(5 \text{ hr}) + 32 \dfrac{\text{pg}}{\text{hr}}(5 \text{ hr}) = 185 \text{ pg} + 160 \text{ pg} = 345 \text{ pg}$

It takes Kenneth 3 hr to type 111 pages.

Cumulative Review Exercises: Chapters 1–8 (page 431)

1. $10 - (3 \cdot 2 - 25) = 10 - (6 - 25) = 10 - (-19) = 29$

2. $V = \frac{10}{9}$ **3.** $\left(\dfrac{\overset{3}{\cancel{24}}y^{-3}}{\underset{1}{\cancel{8}}y^{-1}}\right)^{-2} = (3y^{-2})^{-2} = 3^{-2}y^4 = \dfrac{y^4}{3^2} = \dfrac{y^4}{9}$

4. $3x - 1$

5.

$$
\begin{array}{r}
5z + 7 \quad \text{R } 18 \\
3z - 2\overline{)15z^2 + 11z + 4} \\
\underline{15z^2 - 10z} \\
21z + 4 \\
\underline{21z - 14} \\
18
\end{array}
$$

6. x^3 **7.** $\dfrac{2x - (2x-1)}{x+1} = \dfrac{2x - 2x + 1}{x+1} = \dfrac{1}{x+1}$

8. $\dfrac{10x^3 + 3}{2x^3}$

9. LCD $= (x+4)(x+5)(2x-1)$ **10.** $\frac{140}{3}$

$\dfrac{(2x-1)(2x-1)}{(x+4)(x+5)(2x-1)} - \dfrac{(x+5)(x+5)}{(2x-1)(x+4)(x+5)}$

$= \dfrac{(4x^2 - 4x + 1) - (x^2 + 10x + 25)}{(x+4)(x+5)(2x-1)}$

$= \dfrac{4x^2 - 4x + 1 - x^2 - 10x - 25}{(x+4)(x+5)(2x-1)}$

$= \dfrac{3x^2 - 14x - 24}{(x+4)(x+5)(2x-1)}$

11. LCD $= 4x^2$

$(4x^2)\left(\dfrac{3}{x}\right) - (4x^2)\left(\dfrac{8}{x^2}\right) = \dfrac{1}{4}(4x^2)$

$12x - 32 = x^2$

$0 = x^2 - 12x + 32$

$0 = (x-4)(x-8)$

$x - 4 = 0 \quad | \quad x - 8 = 0$

$x = 4 \quad | \quad x = 8$

(The check is left to the student.) The solutions are 4 and 8.

12. $(x + 6)(x - 7)$ **13.** $3(w^2 - 16) = 3(w + 4)(w - 4)$

14. $(5a - 3b)(4a + b)$

15. $3x^2 - 6x - 2xy + 4y = 3x(x - 2) - 2y(x - 2)$
$= (x - 2)(3x - 2y)$

16. 16 and 17

17. Let x = one number
        $5 - x$ = other number

$x(5 - x) = -24$
$5x - x^2 = -24$
$0 = x^2 - 5x - 24$
$0 = (x - 8)(x + 3)$

$x - 8 = 0$	$x + 3 = 0$
$x = 8$	$x = -3$
$5 - x = -3$	$5 - x = 8$

(The check is left to the student.) The numbers are 8 and -3.

18. 5 lb of oranges at 99¢ per pound and 10 lb of oranges at 78¢ per pound

19. Let n = number of nickels
        $20 - n$ = number of dimes

$5n + 10(20 - n) = 165$
$5n + 200 - 10n = 165$
$-5n = -35$
$n = 7$
$20 - n = 13$

Manny has 7 nickels and 13 dimes.

20. 15 nickels, 6 dimes, 5 quarters

Exercises 9.1 (page 438)

1a. (3, 1): Start at the origin, move right 3 units, then up 1 unit.
b. (−4, −2): Start at the origin, move left 4 units, then down 2 units.
c. (0, 3): Start at the origin, move right 0 units, then up 3 units.
d. (5, −4): Start at the origin, move right 5 units, then down 4 units.
e. (4, 0): Start at the origin, move right 4 units, then up 0 units.
f. (−2, 4): Start at the origin, move left 2 units, then up 4 units.

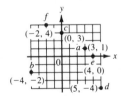

2.

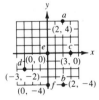

3a. (3, 0) Not in any quadrant
b. (0, 5) Not in any quadrant
c. (−5, 2) II
d. (−4, −3) III

4a. (4, 3) I
b. (−6, 0) Not in any quadrant
c. (3, −4) IV
d. (0, −3) Not in any quadrant

5a. 5 because A is 5 units to the right of the y-axis
b. −3 because C is 3 units to the left of the y-axis
c. 3 because E is 3 units to the right of the y-axis
d. 1 because F is 1 unit to the right of the y-axis

6a. 5   **b.** −2   **c.** 0   **d.** −5

7a. 1 because F is 1 unit to the right of the vertical axis.
b. 2 because C is 2 units above the horizontal axis

8a. 5   **b.** −4   **9.** 0

10. 0   **11.**

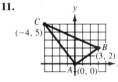

12.

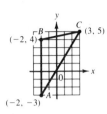

Exercises 9.2 (page 443)

1a. (0, 0) is not a solution, because $3(0) + 2(0) = 0 \neq 6$.
b. (0, 3) is a solution, because $3(0) + 2(3) = 0 + 6 = 6$.
c. (3, 0) is not a solution, because $3(3) + 2(0) = 9 \neq 6$.
d. (2, 0) is a solution, because $3(2) + 2(0) = 6 + 0 = 6$.

2a. No   **b.** No   **c.** Yes   **d.** Yes

3a. (0, 0) is a solution, because $3(0) + 4(0) = 0$.
b. (0, 3) is not a solution, because $3(0) + 4(3) = 0 + 12 \neq 0$.
c. (4, −3) is a solution, because $3(4) + 4(-3) = 12 - 12 = 0$.
d. (−4, 3) is a solution, because $3(-4) + 4(3) = -12 + 12 = 0$.

4a. Yes   **b.** No   **c.** No   **d.** Yes

5a. $4(0) - 3y = 12$ **b.** $4x - 3(0) = 12$
$-3y = 12$ $4x = 12$
$y = -4$ $x = 3$
$(0, -4)$ $(3, 0)$

c. $4(3) - 3y = 12$ **d.** $4x - 3(-4) = 12$
$12 - 3y = 12$ $4x + 12 = 12$
$-3y = 0$ $4x = 0$
$y = 0$ $x = 0$
$(3, 0)$ $(0, -4)$

6a. (0, 3)   **b.** (1, 0)   **c.** (3, −6)   **d.** (2, −3)

7a. $2(0) - 5y = 0$ **b.** $2x - 5(0) = 0$
$-5y = 0$ $2x = 0$
$y = 0$ $x = 0$
$(0, 0)$ $(0, 0)$

c. $2(2) - 5y = 0$ **d.** $2x - 5(2) = 0$
$4 - 5y = 0$ $2x - 10 = 0$
$-5y = -4$ $2x = 10$
$y = \frac{4}{5}$ $x = 5$
$\left(2, \frac{4}{5}\right)$ $(5, 2)$

8a. (0, 0)   **b.** (0, 0)   **c.** (3, −9)   **d.** (1, −3)

9a. $x - 5 = 0$ **b.** (5, 5)   **c.** (5, 2)   **d.** (5, −2)
$x = 5$
$(5, 0)$

10a. (0, −3)   **b.** (3, −3)   **c.** (−3, −3)   **d.** (5, −3)

Exercises 9.3 (page 452)

1. $x + y = 3$
$y = 3 - x$

If $x = 0$, $y = 3 - 0 = 3$
If $x = 1$, $y = 3 - 1 = 2$
If $x = 3$, $y = 3 - 3 = 0$

Table of values

x	y
0	3
1	2
3	0

2.

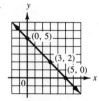

3. $2x - 3y = 6$
Intercepts:
Set $y = 0$. Then $x = 3$.
Set $x = 0$. Then $y = -2$.
This gives the points $(3, 0)$ and $(0, -2)$.
Checkpoint: Set $x = 2$. Then $y = -\frac{2}{3}$.

4.

5. x can be any number, but y is always 8.

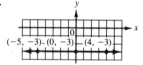

6.

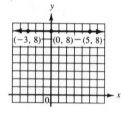

7. y can be any number, but x is always -2.

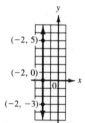

8.

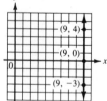

9. x is always -5. $x + 5 = 0 \Rightarrow x = -5$.

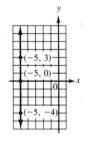

10.

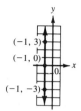

11. y must be -5.
$y + 5 = 0 \Rightarrow y = -5$.

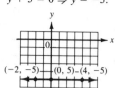

12.

13. $3x = 5y + 15$
Set $y = 0$. Then $x = 5$.
Set $x = 0$. Then $y = -3$.
This gives points $(5, 0)$ and $(0, -3)$.
Checkpoint: Set $x = 2$. Then $y = -1\frac{4}{5}$.

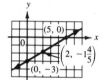

14.

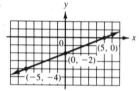

15. $x = -2y$
Set $y = 0$. Then $x = 0$.
Set $y = 2$. Then $x = -4$.
Checkpoint: Set $y = -1$.
Then $x = -2(-1) = 2$.

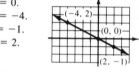

16.

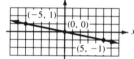

17. $y = -4x$
Set $x = 0$.
Then $y = 0$.
Set $x = 1$.
Then $y = -4(1) = -4$.
Checkpoint: Set $x = -1$.
Then $y = -4(-1) = 4$.

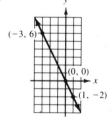

18.

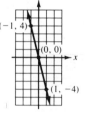

19. $4 - y = x$
Set $x = 0$.
Then $y = 4$.
Set $y = 0$.
Then $x = 4$.
Checkpoint: Set $x = 2$.
Then $y = 2$.

20.

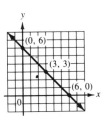

21. $y = x$

x	y
0	0
4	4
-3	-3

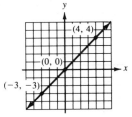

22.

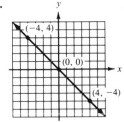

23. $y = \frac{1}{2}x + 2$
Set $x = 0$.
Then $y = 2$.
Set $y = 0$.
Then $x = -4$.
Checkpoint: Set $x = 4$.
Then $y = 4$.

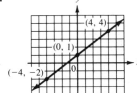

24.

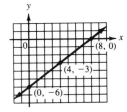

25. $3x = 24 + 4y$

x	y
0	-6
8	0
4	-3

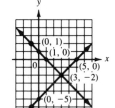

26.

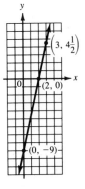

27. $x - y = 5$

a.

x	y
0	-5
5	0
-1	-6

$x + y = 1$

x	y
0	1
1	0
-1	2

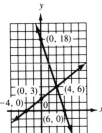

b. $(3, -2)$

28. Each line on the grid represents two units.

a.

b. $(4, 6)$

Exercises 9.4 (page 459)

1. $m = \dfrac{y_2 - y_1}{x_2 - x_1} = \dfrac{3 - 1}{7 - (-2)} = \dfrac{2}{9}$ **2.** $\dfrac{5}{6}$

3. $m = \dfrac{y_2 - y_1}{x_2 - x_1} = \dfrac{-3 - (-1)}{5 - (-1)} = \dfrac{-2}{6} = -\dfrac{1}{3}$ **4.** $-\dfrac{1}{2}$

5. $m = \dfrac{y_2 - y_1}{x_2 - x_1} = \dfrac{-3 - (-3)}{4 - (-5)} = \dfrac{0}{9} = 0$ **6.** 0

7. $m = \dfrac{y_2 - y_1}{x_2 - x_1} = \dfrac{(-3) - (5)}{(8) - (-7)} = \dfrac{-8}{15}$ **8.** $-\dfrac{5}{17}$

9. $m = \dfrac{y_2 - y_1}{x_2 - x_1} = \dfrac{(-5) - (-15)}{(4) - (-6)} = \dfrac{10}{10} = 1$ **10.** -1

11. $m = \dfrac{y_2 - y_1}{x_2 - x_1} = \dfrac{(8) - (5)}{(-2) - (-2)} = \dfrac{3}{0}$; slope does not exist

12. Slope does not exist

13. $m = \dfrac{y_2 - y_1}{x_2 - x_1} = \dfrac{-2 - 3}{-4 - (-4)} = \dfrac{-5}{0}$; slope does not exist

14. Slope does not exist

15. $y = \frac{1}{2}x$

x	y
0	0
2	1
-2	-1

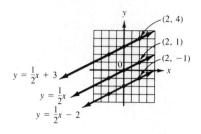

$y = \frac{1}{2}x + 3$

x	y
0	3
2	4
-2	2

a. $m = \dfrac{1 - 0}{2 - 0} = \frac{1}{2}$

$y = \frac{1}{2}x - 2$

x	y
0	-2
2	-1
-2	-3

b. $m = \dfrac{4 - 3}{2 - 0} = \frac{1}{2}$

c. $m = \dfrac{-1 - (-2)}{2 - 0} = \frac{1}{2}$

16.

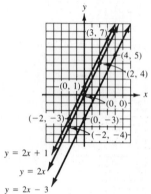

a. $m = 2$
b. $m = 2$
c. $m = 2$

$y = 2x + 1$
$y = 2x$
$y = 2x - 3$

17. $y = -3x$

x	y
0	0
2	-6
-1	3

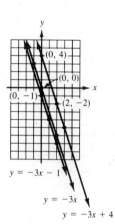

$y = -3x + 4$

x	y
0	4
2	-2
3	-5

$y = -3x - 1$

x	y
0	-1
2	-7
1	-4

a. $m = \dfrac{-6 - 0}{2 - 0} = -3$

b. $m = \dfrac{-2 - 4}{2 - 0} = -3$

c. $m = \dfrac{-7 - (-1)}{2 - 0} = -3$

18.

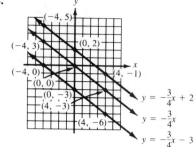

a. $m = -\frac{3}{4}$
b. $m = -\frac{3}{4}$
c. $m = -\frac{3}{4}$

$y = -\frac{3}{4}x + 2$
$y = -\frac{3}{4}x$
$y = -\frac{3}{4}x - 3$

19. $y = -\frac{1}{3}x + 2$

x	y
0	2
3	1
6	0

$y = 3x + 2$

x	y
0	2
1	5
-1	-1

$y = 3x + 2$ $y = \frac{1}{3}x + 2$

a. $m = \dfrac{2 - 1}{0 - 3} = \dfrac{1}{-3} = -\frac{1}{3}$

b. $m = \dfrac{2 - 5}{0 - 1} = \dfrac{-3}{-1} = 3$

20. $y = -2x - 3$

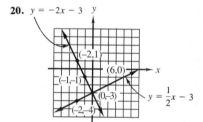

$y = \frac{1}{2}x - 3$

a. $m = \frac{1}{2}$
b. $m = -2$

Exercises 9.5 (page 465)

1. $3x = 2y - 4$
$3x - 2y + 4 = 0$

2. $2x - 3y - 7 = 0$

3. $y = -\frac{3}{4}x - 2$
$4y = -3x - 8$
$3x + 4y + 8 = 0$

4. $3x + 5y + 20 = 0$

5. $2(3x + y) = 5(x - y) + 4$
$6x + 2y = 5x - 5y + 4$
$x + 7y - 4 = 0$

6. $4x - 9y + 5 = 0$

7. $y - y_1 = m(x - x_1)$
$y - 4 = \frac{1}{2}(x - 3)$
$2y - 8 = x - 3$
$0 = x - 2y + 5$ or $x - 2y + 5 = 0$

8. $x - 3y + 13 = 0$

9. $y - y_1 = m(x - x_1)$
$y - (-2) = -\frac{2}{3}[x - (-1)]$
$3(y + 2) = -2(x + 1)$
$3y + 6 = -2x - 2$
$2x + 3y + 8 = 0$

10. $5x + 4y + 22 = 0$

11. $y = mx + b$
$y = \frac{3}{4}x - 3$
$4y = 3x - 12$
$3x - 4y - 12 = 0$

12. $2x - 7y - 14 = 0$

13.
$$y = mx + b$$
$$y = -\tfrac{2}{5}x + \tfrac{1}{2} \quad \text{LCD} = 10$$
$$10y = -4x + 5$$
$$4x + 10y - 5 = 0$$

14. $20x + 12y - 9 = 0$

15. $m = \dfrac{y_2 - y_1}{x_2 - x_1} = \dfrac{4 - (-1)}{2 - 4} = -\dfrac{5}{2}$

16. $3x + 2y - 11 = 0$

Using (2, 4):
$$y - y_1 = m(x - x_1)$$
$$y - 4 = -\tfrac{5}{2}(x - 2)$$
$$2y - 8 = -5x + 10$$
$$5x + 2y - 18 = 0$$

17. $m = \dfrac{4 - 0}{3 - 0} = \dfrac{4}{3}$
$$y - y_1 = m(x - x_1)$$
$$y - 0 = \tfrac{4}{3}(x - 0)$$
$$3y = 4x$$
$$4x - 3y = 0$$

18. $5x - 2y = 0$

19a. $4y = -3x - 12 \Rightarrow y = -\tfrac{3}{4}x - 3$
b. $m = -\tfrac{3}{4}$ **c.** y-intercept = -3
d. $y - (-4) = -\tfrac{3}{4}(x - 2)$, or $y + 4 = -\tfrac{3}{4}(x - 2)$
e. $y - 5 = \tfrac{4}{3}[x - (-3)]$, or $y - 5 = \tfrac{4}{3}(x + 3)$

20a. $y = -\tfrac{2}{5}x - 2$ **b.** $m = -\tfrac{2}{5}$ **c.** y-intercept = -2
d. $y + 4 = -\tfrac{2}{5}(x - 2)$ **e.** $y - 5 = \tfrac{5}{2}(x + 3)$

21a. $3y = -2x + 9 \Rightarrow y = -\tfrac{2}{3}x + 3$
b. $m = -\tfrac{2}{3}$ **c.** y-intercept = 3
d. $y - (-4) = -\tfrac{2}{3}(x - 2)$, or $y + 4 = -\tfrac{2}{3}(x - 2)$
e. $y - 5 = \tfrac{3}{2}[x - (-3)]$, or $y - 5 = \tfrac{3}{2}(x + 3)$

22a. $y = -\tfrac{2}{5}x + 3$ **b.** $m = -\tfrac{2}{5}$ **c.** y-intercept = 3
d. $y + 4 = -\tfrac{2}{5}(x - 2)$ **e.** $y - 5 = \tfrac{5}{2}(x + 3)$

23a. $3y = 5x - 30 \Rightarrow y = \tfrac{5}{3}x - 10$
b. $m = \tfrac{5}{3}$ **c.** y-intercept = -10
d. $y - (-4) = \tfrac{5}{3}(x - 2)$, or $y + 4 = \tfrac{5}{3}(x - 2)$
e. $y - 5 = -\tfrac{3}{5}[x - (-3)]$, or $y - 5 = -\tfrac{3}{5}(x + 3)$

24a. $y = \tfrac{4}{7}x - 4$ **b.** $m = \tfrac{4}{7}$ **c.** y-intercept = -4
d. $y + 4 = \tfrac{4}{7}(x - 2)$ **e.** $y - 5 = -\tfrac{7}{4}(x + 3)$

25a. $2y = -x + 6 \Rightarrow y = -\tfrac{1}{2}x + 3$
b. $m = -\tfrac{1}{2}$ **c.** y-intercept = 3
d. $y - (-4) = -\tfrac{1}{2}(x - 2)$, or $y + 4 = -\tfrac{1}{2}(x - 2)$
e. $y - 5 = \tfrac{2}{1}[x - (-3)]$, or $y - 5 = 2(x + 3)$

26a. $y = -\tfrac{1}{5}x + 1$ **b.** $m = -\tfrac{1}{5}$ **c.** y-intercept = 1
d. $y + 4 = -\tfrac{1}{5}(x - 2)$ **e.** $y - 5 = 5(x + 3)$

27. Because the y-coordinate of the given point is 5, the equation is $y = 5$ or $y - 5 = 0$.

28. $y = -3$ or $y + 3 = 0$

29. Because the x-coordinate of the given point is -3, the equation is $x = -3$ or $x + 3 = 0$.

30. $x = -2$ or $x + 2 = 0$

Exercises 9.6 (page 471)

1. $y = x^2$
$$y = (-2)^2 = 4$$
$$y = (-1)^2 = 1$$
$$y = 0^2 = 0$$
$$y = 1^2 = 1$$
$$y = 2^2 = 4$$
The x- and y-intercepts are (0, 0).

x	y
-2	4
-1	1
0	0
1	1
2	4

2.

x	y
-3	$2\tfrac{1}{4}$
-2	1
-1	$\tfrac{1}{4}$
0	0
1	$\tfrac{1}{4}$
2	1
3	$2\tfrac{1}{4}$

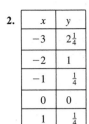

The x- and y-intercepts are (0, 0).

3. $y = x^2 - 2x$
$$y = (-2)^2 - 2(-2)$$
$$= 4 + 4 = 8$$
$$y = (-1)^2 - 2(-1)$$
$$= 1 + 2 = 3$$
$$y = 0^2 - 2(0) = 0$$
$$y = 1^2 - 2(1) = -1$$
$$y = 2^2 - 2(2) = 0$$
$$y = 3^2 - 2(3) = 3$$
$$y = 4^2 - 2(4) = 8$$

x	y
-2	8
-1	3
0	0
1	-1
2	0
3	3
4	8

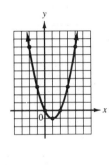

The x-intercepts are (0, 0) and (2, 0).
The y-intercept is (0, 0).

4.

x	y
-2	-10
-1	-4
0	0
1	2
2	2
3	0
4	-4

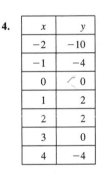

The x-intercepts are (0, 0) and (3, 0).
The y-intercept is (0, 0).

5. $x^2 + 2x - 2$ can't be factored; we can't find the x-intercepts.
If $x = 0$, $y = 0^2 + 2(0) - 2 = -2$; the y-intercept is -2.
$$y = x^2 + 2x - 2$$
$$y = (-4)^2 + 2(-4) - 2 = 6$$
$$y = (-3)^2 + 2(-3) - 2 = 1$$
$$y = (-2)^2 + 2(-2) - 2 = -2$$
$$y = (-1)^2 + 2(-1) - 2 = -3$$
$$y = (1)^2 + 2(1) - 2 = 1$$
$$y = (2)^2 + 2(2) - 2 = 6$$

x	y
-4	6
-3	1
-2	-2
-1	-3
0	-2
1	1
2	6

y-intercept →

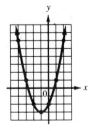

6. We can't find the x-intercepts. The y-intercept is 2.

x	y
-4	10
-3	5
-2	2
-1	1
0	2
1	5
2	10

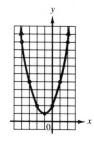

7. If $y = 0$, $0 = 2x - x^2 = x(2 - x)$; the x-intercepts are 0 and 2.
If $x = 0$, $y = 2(0) - 0^2 = 0$; the y-intercept is 0.

$y = 2x - x^2$
$y = 2(-2) - (-2)^2$
  $= -4 - 4 = -8$
$y = 2(-1) - (-1)^2$
  $= -2 - 1 = -3$
$y = 2(1) - (1)^2$
  $= 2 - 1 = 1$
$y = 2(3) - (3)^2$
  $= 6 - 9 = -3$
$y = 2(4) - (4)^2$
  $= 8 - 16 = -8$

x	y
-2	-8
-1	-3
0	0
1	1
2	0
3	-3
4	-8

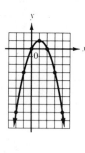

8. The x-intercepts are 0 and -2; the y-intercept is 0.

x	y
-3	3
-2	0
-1	-1
0	0
1	3

9. If $x = 0$, $y = 0^3 = 0$; the y-intercept is 0.

$y = x^3$
$y = (-2)^3 = -8$
$y = (-1)^3 = -1$
$y = 1^3 = 1$
$y = 2^3 = 8$

x	y
-2	-8
-1	-1
0	0
1	1
2	8

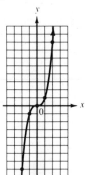

10. The y-intercept is 4.

x	y
-2	2
-1	6
0	4
1	2
2	6

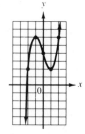

Exercises 9.7 (page 479)

(Check points will usually not be shown.)

1. $x + 2y < 4$
Boundary line is $x + 2y = 4$.

x	y
0	2
4	0

Boundary line is dashed because equality sign is not included with $<$.
Half-plane includes the origin because $(0, 0)$ makes $x + 2y < 4$ true.

2.

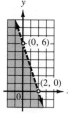

3. $2x - 3y > 6$
Boundary line is $2x - 3y = 6$.

x	y
0	-2
3	0

Half-plane does not include the origin because $2(0) - 3(0) \not> 6$
$0 \not> 6$

4.

5. $x \geq -2$. All points to the right of and including the line $x = -2$

6.

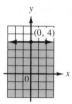

7. Boundary is $y = 4$, a horizontal line.
The correct half-plane includes the origin.

8.

9. $3y - 4x \geq 12$
Boundary line is $3y - 4x = 12$.

x	y
0	4
-3	0

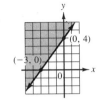

The correct half-plane does not include the origin.

10.

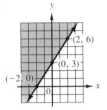

11. Boundary line is $3x + 2y = -6$.

x	y
0	-3
-2	0
-4	3

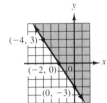

The correct half-plane includes the origin.

12.

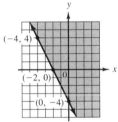

13. $3x - 4y \geq 10$
Boundary line is $3x - 4y = 10$.

x	y
0	$-2\frac{1}{2}$
$3\frac{1}{3}$	0

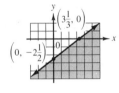

The correct half-plane does not include the origin.

14.

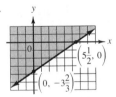

15. $2x + 2 + 3 \leq 6x - 3$
$2x - 6x \leq -2 - 3 - 3$
$-4x \leq -8$
$4x \geq 8$
$x \geq 2$

Boundary is $x = 2$, a vertical line. The correct half-plane does not include the origin.

16.

17. $8y + 12 - 1 \leq 10y - 5$

$$
\begin{array}{r}
8y + 11 \leq 10y - 5 \\
-8y + 5 \quad -8y + 5 \\
\hline
16 \leq 2y \\
8 \leq y \\
y \geq 8
\end{array}
$$

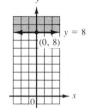

Boundary is $y = 8$, a horizontal line.
The correct half-plane does not include the origin.

18.

19. LCD $= 15$

$$15\left(\frac{x}{5}\right) + 15\left(-\frac{y}{3}\right) > 15(1)$$

$$3x - 5y > 15$$

Boundary line is $3x - 5y = 15$.
The correct half-plane does not include the origin.

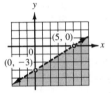

x	y
0	-3
5	0

20.

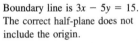

21. LCD $= 20$

$$(20)\left(\frac{y}{4}\right) - (20)\left(\frac{x}{5}\right) > 1(20)$$

$$5y - 4x > 20$$

Boundary is $5y - 4x = 20$.

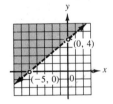

x	y
0	4
-5	0

The correct half-plane does not include the origin.

22.

23. LCD = 6

$$\cancel{6}^{2}\left(\frac{2x+y}{\cancel{3}_{1}}\right) + \cancel{6}^{3}\left(\frac{-(x-y)}{\cancel{2}_{1}}\right) \geq \cancel{6}^{1}\left(\frac{5}{\cancel{6}_{1}}\right)$$

$$4x + 2y - 3x + 3y \geq 5$$
$$x + 5y \geq 5$$

Boundary line is $x + 5y = 5$.
The correct half-plane does not include the origin.

x	y
0	1
5	0

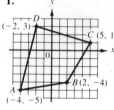

24.

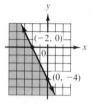

Review Exercises 9.8 (page 481)

1.

2a. y-coordinate = 4
b. abscissa = −2
c. x-coordinate = −2
d.

3. $x - y = 5$

x	y
0	−5
5	0
3	−2

4.

5. $x = -2$ is a vertical line.

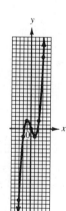

6.

7. $4y - 5x = 20$

x	y
0	5
−4	0
−2	$2\frac{1}{2}$

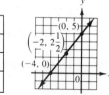

8.

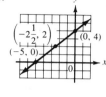

9. $x + 2y = 0$

$$x = -2y$$

x	y
0	0
−4	2
−2	1

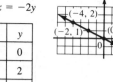

10. The slope does not exist.

11. $m = \dfrac{y_2 - y_1}{x_2 - x_1} = \dfrac{5 - (-6)}{-3 - 2} = \dfrac{11}{-5} = -\dfrac{11}{5}$

12. $-\dfrac{9}{5}$

13.
$$y - y_1 = m(x - x_1)$$
$$y - (-5) = -\tfrac{2}{3}(x - 3)$$
$$3(y + 5) = -2(x - 3)$$
$$3y + 15 = -2x + 6$$
$$2x + 3y + 9 = 0$$

14. $x + 3y - 27 = 0$

15. $m = \dfrac{y_2 - y_1}{x_2 - x_1} = \dfrac{-2 - 4}{1 - (-3)} = \dfrac{-6}{4} = -\dfrac{3}{2}$

$$y - y_1 = m(x - x_1)$$
$$y - (-2) = -\tfrac{3}{2}(x - 1)$$
$$2(y + 2) = -3(x - 1)$$
$$2y + 4 = -3x + 3$$
$$3x + 2y + 1 = 0$$

16. $y = 5$ or $y - 5 = 0$

17. $x = -2$ (The x-coordinate of *every* point on the line is −2.)
$x + 2 = 0$

18.

x	y
−2	10
−1	4
0	0
1	−2
2	−2
3	0
4	4

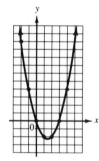

19. $y = x^3 - 3x$
$$y = (-3)^3 - (3)(-3)$$
$$= -27 + 9 = -18$$
$$y = (-2)^3 - 3(-2)$$
$$= -8 + 6 = -2$$
$$y = (-1)^3 - 3(-1)$$
$$= -1 + 3 = 2$$
$$y = 0^3 - 3(0) = 0$$
$$y = 1^3 - 3(1) = -2$$
$$y = 2^3 - 3(2) = 2$$
$$y = 3^3 - 3(3) = 18$$

x	y
−3	−18
−2	−2
−1	2
0	0
1	−2
2	2
3	18

20.

x	y
-4	-18
-3	-2
-2	2
-1	0
0	-2
1	2
2	18

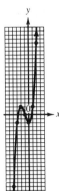

21. Boundary line is $3x - 7y = 21$.

x	y
0	-3
7	0
2	$-2\frac{1}{7}$

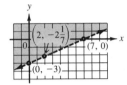

$(0, 0)$ makes the inequality true. Therefore, the correct half-plane includes the origin.

22.

23. The boundary is $y = 3$, which is a horizontal line. $(0, 0)$ makes the inequality false. Therefore, the correct half-plane does not include the origin.

24.

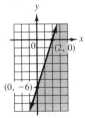

25. LCD $= 12$

$$(12)\left(\frac{y}{3}\right) - (12)\left(\frac{x}{4}\right) \le 1(12)$$

$$4y - 3x \le 12$$

Boundary is $4y - 3x = 12$.

x	y
0	3
-4	0
-2	$1\frac{1}{2}$

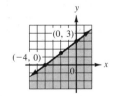

The correct half-plane includes the origin.

Chapter 9 Diagnostic Test (page 487)

Following each problem number is the textbook section number (in parentheses) where that kind of problem is discussed.

1. (9.1) a.

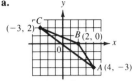

    b. $D = (4, 5)$; $E = (-4, 1)$; $F = (-2, -6)$; $G = (5, -4)$

2. (9.2) a. $(0, 0)$ is not a solution, because $5(0) - 3(0) = 0 \ne 15$.
    b. $(3, 0)$ is a solution, because $5(3) - 3(0) = 15$.
    c. $(0, 3)$ is not a solution, because $5(0) - 3(3) = -9 \ne 15$.
    d. $(0, -5)$ is a solution, because $5(0) - 3(-5) = 15$.
    e. $(6, 5)$ is a solution, because $5(6) - 3(5) = 15$.

3. (9.3) a. Since $y = -2$ is a horizontal line, the y-coordinate of every point on the line is -2.

    b. $3x + y = 3$

    Set $x = 0$. Then $y = 3$.
    Set $y = 0$. Then $x = 1$.
    Checkpoint: Set $x = 2$.
    Then $y = -3$.

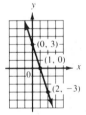

4. (9.3) a. $2x - y = 4$

    Set $x = 0$. Then $y = -4$.
    Set $y = 0$. Then $x = 2$.
    Checkpoint: Set $x = 4$.
    Then $y = 4$.

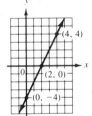

    b. $4x + 9y = 18$

    Set $x = 0$. Then $y = 2$.
    Set $y = 0$. Then $x = 4\frac{1}{2}$.
    Checkpoint: Set $x = 2$.
    Then $y = 1\frac{1}{9}$.

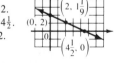

5. (9.4) a. $m = \dfrac{y_2 - y_1}{x_2 - x_1} = \dfrac{1 - 2}{-3 - 5} = \dfrac{-1}{-8} = \dfrac{1}{8}$

    b. $m = \dfrac{y_2 - y_1}{x_2 - x_1} = \dfrac{0 - 2}{-3 - [-3]} = \dfrac{-2}{0}$

    The slope does not exist.

    c. $m = \dfrac{y_2 - y_1}{x_2 - x_1} = \dfrac{-4 - [-4]}{0 - 2} = \dfrac{0}{-2} = 0$

6. (9.5) a. $x = -3$ (The x-coordinate of every point on the line is -3.) $x + 3 = 0$

    b. $y = 4$ (The y-coordinate of every point on the line is 4.) $y - 4 = 0$

7. (9.5) Using the slope-intercept form, we have

$$y = mx + b$$
$$y = 5x + 3$$
$$-5x + y - 3 = 0$$
$$5x - y + 3 = 0$$

8. (9.4 and 9.5) **a.** $m = \dfrac{y_2 - y_1}{x_2 - x_1} = \dfrac{-4 - 2}{1 - [-3]} = \dfrac{-6}{4} = -\dfrac{3}{2}$

b. Using the point-slope form, we have

$$y - y_1 = m(x - x_1)$$
$$y - 2 = -\tfrac{3}{2}(x - [-3]) \quad \text{LCD} = 2$$

$$2(y - 2) = \overset{1}{\cancel{2}}\left(-\frac{3}{\underset{1}{\cancel{2}}}\right)(x + 3)$$

$$2y - 4 = -3(x + 3)$$
$$2y - 4 = -3x - 9$$
$$3x + 2y + 5 = 0$$

9. (9.6)

x	y	
-3	6	$y = x^2 - x - 6$
-3	6	$y = (-3)^2 - (-3) - 6 = 6$
-2	0	$y = (-2)^2 - (-2) - 6 = 0$
-1	-4	$y = (-1)^2 - (-1) - 6 = -4$
0	-6	$y = (0)^2 - (0) - 6 = -6$
1	-6	$y = (1)^2 - (1) - 6 = -6$
2	-4	$y = (2)^2 - (2) - 6 = -4$
3	0	$y = (3)^2 - (3) - 6 = 0$
4	6	$y = (4)^2 - (4) - 6 = 6$

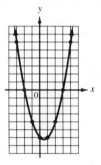

10. (9.7) Boundary line is $3x - 2y = 6$.

x	y
0	-3
2	0
4	3

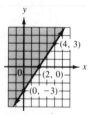

$(0, 0)$ makes the inequality true. Therefore, the half-plane includes the origin.

Cumulative Review Exercises: Chapters 1–9 (page 488)

1. $A = \dfrac{h}{2}(B + b)$

$$A = \tfrac{5}{2}(11 + 7)$$
$$A = \frac{5}{\cancel{2}}(\overset{9}{\cancel{18}}) = 45$$

2. $\dfrac{9b^2}{49a^6}$

3.
$$3z - 5 \overline{)\,6z^2 - z - 7} \quad 2z + 3 \text{ R } 8$$
$$\underline{6z^2 - 10z}$$
$$9z - 7$$
$$\underline{9z - 15}$$
$$8$$

4. $\dfrac{2x^2 + 1}{(x - 2)(x + 1)}$

5. $\dfrac{b + 1}{a(a + b)} \cdot \dfrac{a(a + b)(a - b)}{(b + 1)(b - 2)} = \dfrac{a - b}{b - 2}$

6. $\dfrac{xy - 2}{y}$

7. LCD $= (x - 6)(x - 1)(2x + 3)$

$$\frac{3(2x + 3)}{(x - 6)(x - 1)(2x + 3)} - \frac{4(x - 6)}{(2x + 3)(x - 1)(x - 6)}$$

$$= \frac{6x + 9 - (4x - 24)}{(x - 6)(x - 1)(2x + 3)}$$

$$= \frac{6x + 9 - 4x + 24}{(x - 6)(x - 1)(2x + 3)}$$

$$= \frac{2x + 33}{(x - 6)(x - 1)(2x + 3)}$$

8. $\tfrac{8}{7}$, or $1\tfrac{1}{7}$

9. LCD $= 18$

$$18\left(\frac{2x}{9}\right) - 18\left(\frac{11}{3}\right) = \frac{5x}{6}(18)$$

$$\begin{array}{rcr} 4x - 66 = & & 15x \\ -4x & & -4x \\ \hline -66 = & & 11x \end{array}$$

$$x = -6$$

The solution is -6.

10. $k = \dfrac{-2h^2}{2hC - 5}$ or $k = \dfrac{2h^2}{5 - 2hC}$

11. $y = 3x - 2$

Set $x = 0$. Then $y = -2$.
Set $y = 0$. Then $x = \tfrac{2}{3}$.
Checkpoint: Set $x = 2$. Then $y = 4$.

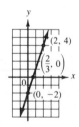

12.

x	y
-1	5
0	0
1	-3
2	-4
3	-3
4	0
5	5

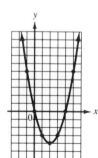

13. $m = \dfrac{y_2 - y_1}{x_2 - x_1} = \dfrac{-4 - 1}{2 - 7} = \dfrac{-5}{-5} = 1.$

$$y - y_1 = m(x - x_1)$$
$$y - 1 = 1(x - 7)$$
$$y - 1 = x - 7$$
$$-x + y + 6 = 0$$
$$x - y - 6 = 0$$

14. Two answers: One number is $-\frac{1}{2}$ and the other is $3\frac{1}{2}$, or one number is $\frac{1}{2}$ and the other is $-3\frac{1}{2}$.

15. Let $\quad x =$ number of pounds of Delicious apples
$50 - x =$ number of pounds of Jonathan apples

$$98x + 85(50 - x) = 4{,}458$$
$$98x + 4{,}250 - 85x = 4{,}458$$
$$13x = 208$$
$$x = 16$$
$$50 - x = 34$$

The mixture contains 16 lb of Delicious apples and 34 lb of Jonathan apples.

16. Altitude is 8 cm, base is 11 cm

17. Let $x =$ the number.

$\dfrac{1}{x} =$ its reciprocal

$$x + \frac{1}{x} = \frac{53}{14} \quad \text{LCD} = 14x$$

$$(14x)x + (14x)\left(\frac{1}{x}\right) = \left(\frac{53}{14}\right)(14x)$$

$$14x^2 + 14 = 53x$$
$$14x^2 - 53x + 14 = 0$$
$$(2x - 7)(7x - 2) = 0$$

$2x - 7 = 0$	$7x - 2 = 0$
$2x = 7$	$7x = 2$
$x = \frac{7}{2}$	$x = \frac{2}{7}$
$\frac{1}{x} = \frac{2}{7}$	$\frac{1}{x} = \frac{7}{2}$

Therefore, there are two answers: The number is $\frac{2}{7}$ and its reciprocal is $\frac{7}{2}$, or the number is $\frac{7}{2}$ and its reciprocal is $\frac{2}{7}$.

18. Two answers: The integers are -2 and -1, or the integers are 3 and 4.

19. Let $x =$ number of cc of water. (There will be $[10 + x]$ cc in the mixture.)

There is 0% of disinfectant in water

$$10(0.17) + x(0) = (10 + x)(0.002)$$
$$10(0.17) = (10 + x)(0.002)$$
$$10(170) = (10 + x)(2)$$
$$1{,}700 = 20 + 2x$$
$$1{,}680 = 2x$$
$$840 = x$$
$$x = 840$$

Therefore, 840 cc of water must be added.

20. 55 words per minute

21. Let $\quad x =$ cost per lens for cheaper lenses
$x + 23 =$ cost per lens for better lenses

$$\frac{\text{Total cost of}}{\text{cheap lenses}} = \frac{\text{Total cost of}}{\text{better lenses}}$$

$$18x = 15(x + 23)$$
$$18x = 15x + 345$$
$$3x = 345$$
$$x = 115$$
$$x + 23 = 138$$

Therefore, the cheaper lenses cost \$115 each and the more expensive ones cost \$138 each.

22. \$5.50, \$8.25

Exercises 10.1 (page 492)

1a. $(0, 8)$ is not a solution, because in the second equation $2(0) - (8) = 0 - 8 = -8 \neq 2$.

b. $(1, 0)$ is not a solution, because in the first equation, $3(1) + (0) = 3 + 0 = 3 \neq 8$.

c. $(2, 2)$ is a solution, because $3(2) + (2) = 6 + 2 = 8$ and $2(2) - (2) = 4 - 2 = 2$.

2a. Not a solution   **b.** A solution   **c.** Not a solution

3a. $(3, 2)$ is not a solution, because in the first equation $2(3) + 3(2) = 6 + 6 = 12 \neq 6$.

b. $(0, 2)$ is a solution, because $2(0) + 3(2) = 0 + 6 = 6$ and $4(0) + 6(2) = 0 + 12 = 12$.

c. $(3, 0)$ is a solution, because $2(3) + 3(0) = 6 + 0 = 6$ and $4(3) + 6(0) = 12 + 0 = 12$.

4a. A solution   **b.** A solution   **c.** A solution

5a. $(0, 4)$ is not a solution, because in the first equation, $(0) + (4) = 0 + 4 = 4 \neq 2$.

b. $(0, 2)$ is not a solution, because in the second equation, $(0) + (2) = 0 + 2 = 2 \neq 4$.

c. $(4, 0)$ is not a solution, because in the first equation $(4) + (0) = 4 + 0 = 4 \neq 2$.

6a. Not a solution   **b.** Not a solution   **c.** Not a solution

Exercises 10.2 (page 497)

1. (1) $\begin{cases} 2x + y = 6 \\ 2x - y = -2 \end{cases}$

(1) Intercepts $(0, 6)$, $(3, 0)$
(2) Intercepts $(0, 2)$, $(-1, 0)$

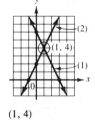

$(1, 4)$

2.

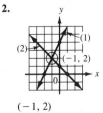

$(-1, 2)$

3. (1) $\begin{cases} x - 2y = -6 \\ 4x + 3y = 20 \end{cases}$

(1) Intercepts $(-6, 0)$, $(0, 3)$
(2) Intercepts $(5, 0)$, $\left(0, 6\frac{2}{3}\right)$

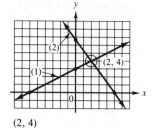

$(2, 4)$

4.

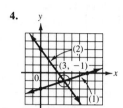

$(3, -1)$

5. (1) $\begin{cases} x + 2y = 0 \\ x - 2y = -2 \end{cases}$

(1)

x	y
0	0
2	-1

(2)

x	y
0	1
-2	0

$\left(-1, \frac{1}{2}\right)$

6.

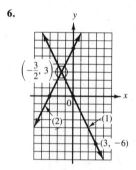

$\left(-\frac{3}{2}, 3\right)$

7. (1) $\begin{cases} 3x - 2y = -9 \\ x + y = 2 \end{cases}$

(1) Intercepts $(-3, 0)$, $\left(0, 4\frac{1}{2}\right)$
(2) Intercepts $(2, 0)$, $(0, 2)$

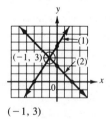

$(-1, 3)$

8.

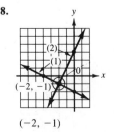

$(-2, -1)$

9. (1) $\begin{cases} 10x - 4y = 20 \\ 6y - 15x = -30 \end{cases}$

(1) Intercepts $(2, 0)$, $(0, -5)$
(2) Intercepts $(2, 0)$, $(0, -5)$
Since both lines have the same intercepts, they are the same line.

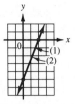

Dependent. Many solutions.

10.

Dependent. Many solutions.

11. (1) $\begin{cases} 2x - 4y = -2 \\ -3x + 6y = 12 \end{cases}$

(1)

x	y
-1	0
-3	-1
4	$\frac{5}{2}$

(2)

x	y
0	2
-4	0
3	$\frac{7}{2}$

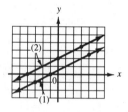

No solution (parallel lines)

12.

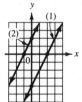

No solution (parallel lines)

Exercises 10.3 (page 505)

1. (1) $\begin{cases} 2x - y = -4 \\ x + y = -2 \end{cases}$

$\quad 3x \quad\quad = -6$

$\quad\quad x = -2$

Substituting -2 for x in the equation (2),

(2) $x + y = -2$
$\quad -2 + y = -2$
$\quad\quad y = 0$

Solution: $(-2, 0)$

2. $(2, 2)$

3. $\begin{array}{r} \begin{cases} x - 2y = 10 \\ -1] \ x + y = 4 \end{cases} \end{array} \Rightarrow \begin{array}{r} x - 2y = 10 \\ \underline{-x - y = -4} \\ -3y = 6 \\ y = -2 \end{array}$

Substituting -2 for y in the second equation,

$x + y = 4$
$x - 2 = 4$
$\quad x = 6$

Solution: $(6, -2)$

4. $(0, 1)$

5. (1) $\begin{cases} x - 3y = 6 \\ 4x + 3y = 9 \end{cases}$

$\quad 5x = 15$

$\qquad\quad x = 3$

Substituting 3 for x in equation (1),

(1) $x - 3y = 6$

$\ 3 - 3y = 6$

$\ -3y = 3$

$\qquad y = -1$

Solution: $(3, -1)$

6. $(1, 0)$

7. $\begin{array}{l} 2] \\ 1] \end{array}\begin{cases} 1x + 1y = 2 \\ 3x - 2y = -9 \end{cases} \Rightarrow \begin{array}{l} 2x + 2y = 4 \\ 3x - 2y = -9 \end{array}$

$ 5x = -5$

$ x = -1$

Substituting -1 for x in the first equation,

$x + y = 2$

$-1 + y = 2$

$y = 3$

Solution: $(-1, 3)$

8. $(-2, -1)$

9. $\begin{array}{l} 2] \\ 1] \end{array}\begin{cases} x + 2y = 0 \\ -2x + y = 0 \end{cases} \Rightarrow \begin{array}{l} 2x + 4y = 0 \\ -2x + y = 0 \end{array}$

$ 5y = 0$

$ y = 0$

Substituting 0 for y in the first equation,

$x + 2y = 0$

$x + 2(0) = 0$

$x = 0$

Solution: $(0, 0)$

10. Dependent
(many solutions)

11. $\begin{array}{l} 3] \\ -4] \end{array}\begin{cases} 4x + 3y = 2 \\ 3x + 5y = -4 \end{cases} \Rightarrow \begin{array}{l} 12x + 9y = 6 \\ -12x - 20y = 16 \end{array}$

$ -11y = 22$

$ y = -2$

Substituting -2 for y in the first equation,

$4x + 3y = 2$

$4x + 3(-2) = 2$

$4x - 6 = 2$

$4x = 8$

$x = 2$

Solution: $(2, -2)$

12. $(3, -2)$

13. $\begin{array}{l} 9] \ 3] \\ -6]-2] \end{array}\begin{cases} 6x - 10y = 6 \\ 9x - 15y = -4 \end{cases} \Rightarrow \begin{array}{l} 18x - 30y = 18 \\ -18x + 30y = 8 \end{array}$

$ 0 = 26 \quad \text{False}$

Inconsistent (no solution)

14. Inconsistent (no solution)

15. $\begin{array}{l} 2] \\ 1] \end{array}\begin{cases} 3x - 5y = -2 \\ -6x + 10y = 4 \end{cases} \Rightarrow \begin{array}{l} 6x - 10y = -4 \\ -6x + 10y = 4 \end{array}$

$ 0 = 0 \quad \text{True}$

Dependent (many solutions)

16. Dependent (many solutions)

Exercises 10.4 (page 510)

1. (1) $\begin{cases} 2x - 3y = 1 \\ x = y + 2 \end{cases}$

Substituting $y + 2$ for x in equation (1):

(1) $ 2x - 3y = 1$

$ 2(y + 2) - 3y = 1$

$ 2y + 4 - 3y = 1$

$ -y = -3$

$ y = 3$

Substituting 3 for y in equation (2):

(2) $x = y + 2$

$ x = 3 + 2 = 5.$

Solution: $(5, 3)$

2. $(2, 7)$

3. (1) $\begin{cases} 3x + 4y = 2 \\ y = x - 3 \end{cases}$

Substituting $x - 3$ for y in equation (1):

$3x + 4(x - 3) = 2$

$3x + 4x - 12 = 2$

$ 7x = 14$

$ x = 2$

Substituting 2 for x in equation (2):

$y = x - 3 = 2 - 3 = -1$

Solution: $(2, -1)$

4. $(1, 3)$

5. (1) $\begin{cases} 4x + y = 2 \\ 7x + 3y = 1 \end{cases} \Rightarrow y = 2 - 4x$

Substituting $2 - 4x$ for y in equation (2):

(2) $ 7x + 3y = 1$

$ 7x + 3(2 - 4x) = 1$

$ 7x + 6 - 12x = 1$

$ -5x = -5$

$ x = 1$

Substituting 1 for x in $y = 2 - 4x$:

$y = 2 - 4(1) = -2$

Solution: $(1, -2)$

6. $(3, -2)$

7. (1) $\begin{cases} 4x - y = 3 \\ 8x - 2y = 6 \end{cases} \Rightarrow y = 4x - 3$

Substituting $4x - 3$ for y in equation (2):

(2) $ 8x - 2y = 6$

$ 8x - 2(4x - 3) = 6$

$ 8x - 8x + 6 = 6$

$ 6 = 6 \quad \text{True}$

Dependent (many solutions)

8. Dependent (many solutions)

9. (1) $\begin{cases} x + 3 = 0 \\ 3x - 2y = 6 \end{cases} \Rightarrow x = -3$

Substituting -3 for x in equation (2):

(2) $ 3x - 2y = 6$

$ 3(-3) - 2y = 6$

$ -9 - 2y = 6$

$ -2y = 15$

$ y = \dfrac{15}{-2} = -7\frac{1}{2}$

Solution: $\left(-3, -7\frac{1}{2}\right)$

10. $\left(-\frac{3}{5}, 4\right)$

11. (1) $\begin{cases} 8x + 4y = 7 \\ 3x + 6y = 6 \end{cases} \Rightarrow 3x = 6 - 6y$

$$x = \frac{6 - 6y}{3} = \frac{\overset{1}{\cancel{3}}(2 - 2y)}{\underset{1}{\cancel{3}}} = 2 - 2y$$

Substituting $2 - 2y$ for x in equation (1):

(1) $8x + 4y = 7$
$8(2 - 2y) + 4y = 7$
$16 - 16y + 4y = 7$
$-12y = -9 \Rightarrow y = \frac{3}{4}$

Substituting $\frac{3}{4}$ for y in $x = 2 - 2y$:

$$x = 2 - 2\left(\frac{3}{4}\right) = \frac{1}{2}$$

Solution: $\left(\frac{1}{2}, \frac{3}{4}\right)$

12. $\left(\frac{3}{5}, \frac{1}{4}\right)$

13. (1) $\begin{cases} 3x - 2y = 8 \\ 2y - 3x = 4 \end{cases} \Rightarrow 2y = 3x + 4$

$$y = \frac{3x + 4}{2}$$

Substituting $\frac{3x + 4}{2}$ for y in equation (1):

(1) $3x - 2y = 8$

$$3x - \overset{1}{\cancel{2}}\left(\frac{3x + 4}{\underset{1}{\cancel{2}}}\right) = 8$$

$3x - 3x - 4 = 8$
$-4 = 8$ False

Inconsistent (no solution)

14. Inconsistent (no solution)

15. (1) $\begin{cases} 8x + 5y = 2 \\ 7x + 4y = 1 \end{cases} \Rightarrow 4y = 1 - 7x$

$$y = \frac{1 - 7x}{4}$$

Substituting $\frac{1 - 7x}{4}$ for y in equation (1):

$$8x + 5\left(\frac{1 - 7x}{4}\right) = 2$$

$$\frac{4}{1} \cdot \frac{8x}{1} + \frac{\overset{1}{\cancel{4}}}{1} \cdot \frac{5}{1}\left(\frac{1 - 7x}{\underset{1}{\cancel{4}}}\right) = \frac{4}{1} \cdot \frac{2}{1}$$

$32x + 5 - 35x = 8$
$-3x = 3$

$$x = \frac{3}{-3} = -1$$

Substituting -1 for x in $y = \frac{1 - 7x}{4}$:

$$y = \frac{1 - 7(-1)}{4} = \frac{8}{4} = 2$$

Solution: $(-1, 2)$

16. $(4, 1)$

17. (1) $\begin{cases} 4x + 4y = 3 \\ 6x + 12y = -6 \end{cases} \Rightarrow 4x = 3 - 4y$

$$x = \frac{3 - 4y}{4}$$

18. $\left(2\frac{1}{2}, -2\frac{1}{3}\right)$

Substituting $\frac{3 - 4y}{4}$ for x in equation (2):

(2) $6x + 12y = -6$

$$\overset{3}{\cancel{6}}\left(\frac{3 - 4y}{\underset{2}{\cancel{4}}}\right) + 12y = -6 \quad \text{LCD} = 2$$

$$\frac{\overset{1}{\cancel{2}}}{1} \cdot 3\left(\frac{3 - 4y}{\underset{1}{\cancel{2}}}\right) + 2(12y) = 2(-6)$$

$9 - 12y + 24y = -12$
$12y = -21$

$$y = -\frac{21}{12} = -\frac{7}{4} = -1\frac{3}{4}$$

Substituting $-\frac{7}{4}$ for y in $x = \frac{3 - 4y}{4}$:

$$x = \frac{3 - 4\left(-\frac{7}{4}\right)}{4} = \frac{3 + 7}{4} = \frac{10}{4} = 2\frac{1}{2}$$

Solution: $\left(2\frac{1}{2}, -1\frac{3}{4}\right)$

Exercises 10.5 (page 515)

(The checks will not be shown.)

1. Let x = one number
y = other number

(1) $\begin{cases} x + y = 80 \\ x - y = 12 \end{cases}$
$\ \ 2x = 92$
$\ \ x = 46$

2. 47 and 43

Substituting 46 for x in equation (1):

$46 + y = 80$
$y = 34$

One number is 46 and the other is 34.

3. Let x = measure of first angle
y = measure of second angle

(1) $\begin{cases} x + y = 90 \\ x - y = 60 \end{cases}$
$\ \ 2x = 150$
$\ \ x = 75$

4. $106°$ and $74°$

Substituting 75 for x in equation (1):

$75 + y = 90$
$y = 15$

The angles measure $75°$ and $15°$.

5. Let x = smaller number
y = larger number

(1) $\begin{cases} 2x + 3y = 34 \\ 5x - 2y = 9 \end{cases}$

6. 7 and 4

2] $2x + 3y = 34 \Rightarrow 4x + 6y = 68$
3] $5x - 2y = 9 \Rightarrow 15x - 6y = 27$
$\ 19x = 95$
$x = 5$

Substituting 5 for x in equation (1),

(1) $2x + 3y = 34$
$2(5) + 3y = 34$
$10 + 3y = 34$
$3y = 24$
$y = 8$

The smaller number is 5 and the larger one is 8.

7. Let x = number of pounds of apple chunks
   y = number of pounds of granola

(1) $\begin{cases} x + y = 6 \\ 4.24x + 2.10y = 16.88 \end{cases}$
(2)

$\begin{aligned} -210] \quad x + y &= 6 \Rightarrow -210x - 210y = -1{,}260 \\ 100] \quad 4.24x + 2.10y &= 16.88 \Rightarrow \underline{424x + 210y = 1{,}688} \\ 214x &= 428 \\ x &= 2 \end{aligned}$

Substituting 2 for x in equation (1), $2 + y = 6 \Rightarrow y = 4$

Jason bought 2 lb of apple chunks and 4 lb of granola.

8. 43 lb Grade A and 57 lb Grade B

9. Let x = number of 18¢ stamps
   y = number of 22¢ stamps

(1) $\begin{cases} x + y = 22 \\ 0.18x + 0.22y = 4.48 \end{cases}$
(2)

$\begin{aligned} -18] \quad x + y &= 22 \Rightarrow -18x - 18y = -396 \\ 100] \quad 0.18x + 0.22y &= 4.48 \Rightarrow \underline{18x + 22y = 448} \\ 4y &= 52 \\ y &= 13 \end{aligned}$

Substituting 13 for y in equation (1), $x + 13 = 22 \Rightarrow x = 9$

Don bought thirteen 22¢ stamps and nine 18¢ stamps.

10. Forty-four 22¢ stamps and six 45¢ stamps

11. Let w = width
       ℓ = length
    $2w + 2\ell$ = perimeter

(1) $\begin{cases} 2w + 2\ell = 19 \\ \ell = w + 1.5 \end{cases}$
(2)

Substituting $w + 1.5$ for ℓ in equation (1), we get

$\begin{aligned} 2w + 2\ell &= 19 \\ 2w + 2(w + 1.5) &= 19 \\ 2w + 2w + 3 &= 19 \\ 4w &= 16 \\ w &= 4 \end{aligned}$

Substituting 4 for w in equation (2), we get

$\begin{aligned} \ell &= w + 1.5 \\ &= (4) + 1.5 = 5.5 \end{aligned}$

The width is 4 ft and the length is 5.5 ft, or 5 ft 6 in.

12. Width = 5 ft; length = 7 ft 6 in.

13. Let n = numerator
    d = denominator

(1) $\dfrac{n}{d} = \dfrac{2}{3}$ and (2) $\dfrac{n+4}{d-2} = \dfrac{6}{7}$

$\begin{aligned} 2d &= 3n \\ d &= \dfrac{3n}{2} \end{aligned}$ $\begin{aligned} 6(d-2) &= 7(n+4) \\ 6d - 12 &= 7n + 28 \\ \text{(3)}\quad 6d &= 7n + 40 \end{aligned}$

Substituting $\dfrac{3n}{2}$ for d in equation (3) $\Rightarrow 6\left(\dfrac{3n}{2}\right) = 7n + 40$:

$\begin{aligned} 9n &= 7n + 40 \\ 2n &= 40 \\ n &= 20 \end{aligned}$

Substituting 20 for n in $d = \dfrac{3n}{2}$:

$$d = \dfrac{3(20)}{2} = 30$$

Therefore, the original fraction is $\dfrac{20}{30}$.

14. $\dfrac{36}{48}$

15. Let x = speed of boat in still water (in mph)
   y = speed of current (in mph)

Note: We multiply both sides of (1) by $\frac{1}{7}$ and both sides of (2) by $\frac{1}{9}$.

(1) $\begin{cases} 7(x + y) = 252 \\ 9(x - y) = 252 \end{cases} \Rightarrow \begin{aligned} x + y &= 36 \\ x - y &= 28 \end{aligned}$ $\Big\rangle$ Add
(2)

$\begin{aligned} 2x &= 64 \\ x &= 32 \end{aligned}$

Substituting 32 for x in $x + y = 36$:

$$32 + y = 36 \Rightarrow y = 4$$

The speed of the boat in still water is 32 mph and the speed of the current is 4 mph.

16. The speed of the plane in still air is 120 mph. The speed of the wind is 30 mph.

17. Let t = cost of tie in cents
    p = cost of pin in cents

(1) $\begin{cases} t + p = 110 \\ t = 100 + p \end{cases}$
(2)

Substitute $100 + p$ for t in equation (1):

(1) $\begin{aligned} t + p &= 110 \\ 100 + p + p &= 110 \\ 2p &= 10 \\ p &= 5 \end{aligned}$

Substituting 5 for p in equation (2):

$$t = 100 + p = 100 + 5 = 105$$

The pin costs 5¢, and the tie costs $1.05.

18. 7 on upper branch; 5 on lower branch

Review Exercises 10.6 (page 518)

1. (1) $\begin{cases} x + y = 6 \\ x - y = 4 \end{cases}$
   (2)

   (1) Intercepts (6, 0), (0, 6): checkpoint (3, 3)
   (2) Intercepts (4, 0), (0, −4); checkpoint (3, −1)

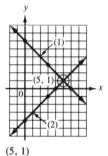

(5, 1)

2.

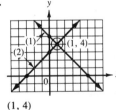

(1, 4)

3. $3x - y = 6$ $6x - 2y = 12$

x	y
0	−6
2	0
4	6

x	y
0	−6
2	0
4	6

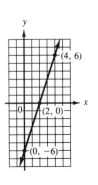

The system is dependent (many solutions).

4. (3, 2)

5. $\begin{array}{r} 3]\{4x - 8y = 4\} \Rightarrow \quad 12x - 24y = \quad 12 \\ -4]\{3x - 6y = 3\} \Rightarrow \underline{-12x + 24y = -12} \\ 0 = \quad 0 \quad \text{True} \end{array}$

Dependent (many solutions)

6. Inconsistent (no solution)

7. (1) $\begin{cases} x = y + 2 \\ (2) \ 4x - 5y = 3 \end{cases}$ **8.** (4, 9)

Substitute $y + 2$ for x in (2):

(2) $4x - 5y = 3$
    $4(y + 2) - 5y = 3$
    $4y + 8 - 5y = 3$
        $-y = -5 \Rightarrow y = 5$

Substitute $y = 5$ in $x = y + 2$:

            $x = 5 + 2 = 7$

Solution: (7, 5)

9. $\begin{cases} x + y = 4 \\ 2x - y = 2 \end{cases} \Rightarrow y = 4 - x$ **10.** (−3, −2)

Substituting $4 - x$ for y in the second equation:

$2x - (4 - x) = 2$
$2x - 4 + x = 2$
    $3x = 6$
    $x = 2$

Substituting 2 for x in $y = 4 - x$:

        $y = 4 - 2 = 2$

Solution: (2, 2)

11. Let x = larger number   **12.** $\frac{24}{30}$
    y = smaller number

(1) $\begin{cases} x + y = \quad 84 \\ (2) \ x - y = \quad 22 \end{cases}$ Add
    $2x \quad = 106$
    $x = \quad 53$

Substituting 53 for x in equation (1):

(1) $x + y = 84$
    $53 + y = 84$
        $y = 31$

The larger number is 53 and the smaller is 31.

13. Let x = number of boxes of paper for laser printer
    y = number of boxes of paper for copy machine

(1) $-425] \quad x + \quad y = 20 \Rightarrow -425x - 425y = -8,500$
(2) $100] \ 7.50x + 4.25y = 107.75 \Rightarrow \underline{750x + 425y = \quad 10,775}$
        $325x \quad = \quad 2,275$
            $x = 7$

Substituting 7 for x in equation (1):

$7 + y = 20 \Rightarrow y = 13$

The manager purchased 7 boxes of paper for the laser printer and 13 boxes of paper for the copy machine.

14. The speed of the boat in still water is 27 mph. The speed of the current is 6 mph.

Chapter 10 Diagnostic Test (page 521)

Following each problem is the textbook section number (in parentheses) where that kind of problem is discussed.

The checks will not be shown for Problems 1–7.

1. (10.2) (1) $\begin{cases} 2x + 3y = 6 \\ (2) \ x + 3y = 0 \end{cases}$

$2x + 3y = 6$

x	y
0	2
3	0
−3	4

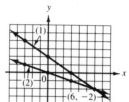

$x + 3y = 0$

x	y
0	0
−3	1
3	−1

Solution: (6, −2)

2. (10.3) (1) $\begin{cases} x + 3y = 5 \\ (2) \ -x + 4y = 2 \end{cases}$ Add
        $7y = 7$
        $y = 1$

Substituting 1 for y in equation (1): $x + 3(1) = 5 \Rightarrow x = 2$

Solution: (2, 1)

3. (10.3) $\begin{array}{r} 2]\{4x + 3y = \quad 5\} \Rightarrow \quad 8x + 6y = 10 \\ 3]\{5x - 2y = 12\} \Rightarrow \underline{15x - 6y = 36} \\ 23x \quad = 46 \\ x = \quad 2 \end{array}$

Substituting 2 for x in the first equation: $4(2) + 3y = 5 \Rightarrow y = -1$

Solution: (2, −1)

4. (10.4) (1) $\begin{cases} 2x - 5y = -20 \\ (2) \quad x = y - 7 \end{cases}$

$2(y - 7) - 5y = -20$
$2y - 14 - 5y = -20$
$-3y - 14 = -20$
    $-3y = -6$
    $y = 2$

Substituting 2 for y in equation (2): $x = (2) - 7 = -5$

Solution: (−5, 2)

5. (10.3) $\begin{array}{r} 5]\{3x + 4y = -1\} \Rightarrow 15x + 20y = \quad -5 \\ 4]\{2x - 5y = -16\} \Rightarrow \underline{8x - 20y = -64} \\ 23x \quad = -69 \\ x = -3 \end{array}$

Substituting −3 for x in the first equation:
$3(-3) + 4y = -1 \Rightarrow y = 2$

Solution: (−3, 2)

6. (10.3) $\begin{array}{r} -2]\{5x + \quad y = 10\} \Rightarrow -10x - 2y = -20 \\ 1]\{10x + 2y = \quad 9\} \Rightarrow \underline{10x + 2y = \quad 9} \\ 0 = -11 \quad \text{False} \end{array}$

Inconsistent (no solution)

7. (10.3) $\begin{array}{r} 10]5]\{ \quad 6x - \quad 9y = \quad 3\} \Rightarrow \quad 30x - 45y = \quad 15 \\ 6]3]\{-10x + 15y = -5\} \Rightarrow \underline{-30x + 45y = -15} \\ 0 = 0 \quad \text{True} \end{array}$

Dependent (many solutions)

8. (10.5) Let x = one number
y = other number

(1) $\begin{cases} x + y = 13 \\ x - y = 37 \end{cases}$ Add

$2x \quad = 50$

$x = 25$

Substituting 25 for x in equation (1):

$25 + y = 13 \Rightarrow y = -12$

Check: $25 + (-12) = 13$
$25 - (-12) = 25 + 12 = 37$

The numbers are 25 and -12.

9. (10.5) Let x = speed of plane in still air (in mph)
y = speed of wind (in mph)

Then $x + y$ = speed with wind
and $x - y$ = speed against wind

Note: We will have small numbers to work with if we multiply both sides of (1) by $\frac{1}{11}$ and both sides of (2) by $\frac{1}{7.5}$.

(1) $\begin{cases} 11(x - y) = 1,650 \\ 7.5(x + y) = 1,650 \end{cases}$

$\frac{1}{11}]$ $11(x - y) = 1,650 \Rightarrow x - y = 150$ (3)

$\frac{1}{7.5}]$ $7.5(x + y) = 1,650 \Rightarrow x + y = 220$ (4)

$2x \quad = 370$

$x = 185$

Substituting 185 for x in equation (4): $185 + y = 220 \Rightarrow y = 35$

Check: The speed of the plane against the wind is 185 mph $-$

35 mph = 150 mph, and $150\frac{mi}{hr}(11 \text{ hr}) = 1,650$ mi. The speed of

the plane with the wind is 185 mph + 35 mph = 220 mph, and

$220\frac{mi}{hr}(7.5 \text{ hr}) = 1,650$ mi.

The speed of the plane in still air is 185 mph, and the speed of the wind is 35 mph.

10. (10.5) Let x = number of cm in length
y = number of cm in width

(1) $\begin{cases} x = y + 3 \\ 2x + 2y = 98 \end{cases}$

Substituting $y + 3$ for x in equation (2):

$2(y + 3) + 2y = 98$

$2y + 6 + 2y = 98$

$4y + 6 = 98$

$4y = 92$

$y = 23$

Substituting 23 for y in (1): $x = 23 + 3 = 26$

Check: The length of the rectangle is 3 cm more than its width. Its perimeter is 2(23 cm) + 2(26 cm) = 46 cm + 52 cm = 98 cm.

The length is 26 cm and the width is 23 cm.

Cumulative Review Exercises: Chapters 1–10 (page 522)

1. $\dfrac{2 + 5x}{6x - 1} = \dfrac{3}{4}$

$4(2 + 5x) = 3(6x - 1)$

$8 + 20x = 18x - 3$

$2x = -11$

$x = -\frac{11}{2}$ or $-5\frac{1}{2}$

The solution is $-5\frac{1}{2}$.

2. -5

3. LCD = $4a(2 - 3a)$

$\dfrac{1 + a}{2 - 3a} \cdot \dfrac{4a}{4a} + \dfrac{2 + a}{4a} \cdot \dfrac{2 - 3a}{2 - 3a} = \dfrac{4a + 4a^2}{4a(2 - 3a)} + \dfrac{4 - 4a - 3a^2}{4a(2 - 3a)}$

$= \dfrac{4a + 4a^2 + 4 - 4a - 3a^2}{4a(2 - 3a)} = \dfrac{a^2 + 4}{4a(2 - 3a)}$

4.

5. $4y = x^2 \Rightarrow y = \frac{1}{4}x^2$

x	y
-4	4
-2	1
-1	$\frac{1}{4}$
0	0
1	$\frac{1}{4}$
2	1
4	4

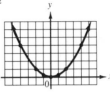

6. $x \le 5$

7. $3x - 4y > 12$

Boundary line is $3x - 4y = 12$.

x	y
0	-3
4	0

The correct half-plane does not include the origin because

$3(0) - 4(0) \not> 12$

$0 \not> 12$

8. $3x + 7y - 5 = 0$

9. (1) $\begin{cases} 2x - 3y = 3 \\ -2x + 3y = 5 \end{cases}$

$0 = 8$ False

Inconsistent (no solution)

10. Dependent (many solutions)

11. $3]\begin{cases} 1x + 5y = 11 \\ 3x + 4y = 11 \end{cases} \Rightarrow$ $\begin{array}{r} 3x + 15y = 33 \\ -3x - 4y = -11 \end{array}$

$11y = 22$

$y = 2$

Substituting 2 for y in the first equation:

$x + 5y = 11$

$x + 5(2) = 11$

$x + 10 = 11$

$x = 1$

Solution: (1, 2)

12. $\left(-\frac{7}{2}, \frac{9}{2}\right)$ or $\left(-3\frac{1}{2}, 4\frac{1}{2}\right)$

13. Not factorable   **14.** Not factorable

15. Let x = cost of each cheaper camera.
   $x + 48$ = cost of each more expensive camera

   $18x = 10(x + 48)$ The two costs are equal

   $18x = 10x + 480$

   $\underline{-10x \quad -10x}$

   $8x = \qquad 480$

   $x = 60$

   $x + 48 = 108$

   Check: $18(\$60) = \$1,080$ and $10(\$108) = \$1,080$

   Therefore, the cheaper cameras cost $60 each, and the more expensive cameras would have cost $108 each.

16. $15,500 at 14% and $11,800 at 21%

17. Let x = one number
   y = other number

   (1) $\left\{ \begin{array}{l} x + y = 6 \\ x - y = 40 \end{array} \right\}$ Add

   $2x \quad = 46$

   $x = 23$

   Substituting 23 for x in equation (1): $23 + y = 6$

   $y = -17$

   Check: $23 + (-17) = 6$ and $23 - (-17) = 23 + 17 = 40$

   One number is 23 and the other is -17.

18. The width is 12 m; the length is 17 m.

Exercises 11.1 (page 525)

1. 17   2. 23   3. $4xy^2$ 4. $9a^2b$ 5. $x + 1$

6. $m - n$ 7. 9   8. 100   9. 15   10. 29   11. $3a^2$

12. $7y^2$ 13. $x + y$ 14. $u + v$

Exercises 11.2A (page 528)

1. $\sqrt{25}\sqrt{x^2} = 5x$ 2. $10y$ 3. $\sqrt{81}\sqrt{s^4} = 9s^2$ 4. $8t^3$

5. $\sqrt{4}\sqrt{x^2} = 2x$ 6. $3y$ 7. $\sqrt{16}\sqrt{z^4} = 4z^2$ 8. $5b^3$

9. $\sqrt{49 \cdot 2} = \sqrt{49}\sqrt{2} = 7\sqrt{2}$ 10. $2\sqrt{5}$

11. $\sqrt{9 \cdot 2} = \sqrt{9}\sqrt{2} = 3\sqrt{2}$ 12. $3\sqrt{5}$

13. $\sqrt{4 \cdot 2} = \sqrt{4}\sqrt{2} = 2\sqrt{2}$ 14. $4\sqrt{2}$

15. $\sqrt{x^2 \cdot x} = \sqrt{x^2}\sqrt{x} = x\sqrt{x}$ 16. $y^2\sqrt{y}$

17. $\sqrt{m^6 \cdot m} = m^3\sqrt{m}$ 18. $n^4\sqrt{n}$

19. $\sqrt{4 \cdot 6} = \sqrt{4}\sqrt{6} = 2\sqrt{6}$ 20. $3\sqrt{6}$

21. $\sqrt{2 \cdot 16x^2y^2y} = \sqrt{2}\sqrt{16}\sqrt{x^2}\sqrt{y^2}\sqrt{y} = 4xy\sqrt{2y}$

22. $7b\sqrt{2ab}$

23. $\sqrt{2 \cdot 4s^4t^4t} = \sqrt{2}\sqrt{4}\sqrt{s^4}\sqrt{t^4}\sqrt{t} = 2s^2t^2\sqrt{2t}$ 24. $2xy^3\sqrt{11x}$

25. $\sqrt{2 \cdot 9a^2b^2b} = \sqrt{2}\sqrt{9}\sqrt{a^2}\sqrt{b^2}\sqrt{b} = 3ab\sqrt{2b}$ 26. $3c^2d\sqrt{3c}$

27. $\sqrt{4 \cdot 10x^6y^2} = \sqrt{4}\sqrt{10}\sqrt{x^6}\sqrt{y^2} = 2x^3y\sqrt{10}$ 28. $3a^3b^4\sqrt{15}$

29. $\sqrt{4 \cdot 15h^4hk^4} = \sqrt{4}\sqrt{15}\sqrt{h^4}\sqrt{h}\sqrt{k^4} = 2h^2k^2\sqrt{15h}$

30. $3m^3n^3\sqrt{10m}$

31. First express 280 in prime factored form: $280 = 2^2 \cdot 2 \cdot 5 \cdot 7$
   $\sqrt{280y^5z^6} = \sqrt{2^2 \cdot 70 \cdot y^4 \cdot y \cdot z^6} = 2y^2z^3\sqrt{70y}$

32. $3g^4h^4\sqrt{30h}$ 33. $\sqrt{100 \cdot 5 \cdot x^6 \cdot x \cdot y^8 \cdot y} = 10x^3y^4\sqrt{5xy}$

34. $6x^5y^6\sqrt{6xy}$

Exercises 11.2B (page 531)

1. $\dfrac{\sqrt{9}}{\sqrt{25}} = \dfrac{3}{5}$ 2. $\dfrac{6}{7}$ 3. $\dfrac{\sqrt{16}}{\sqrt{25}} = \dfrac{4}{5}$ 4. $\dfrac{9}{10}$

5. $\dfrac{\sqrt{y^4}}{\sqrt{x^2}} = \dfrac{y^2}{x}$ 6. $\dfrac{x^3}{y^4}$ 7. $\dfrac{\sqrt{4x^2}}{\sqrt{9}} = \dfrac{2x}{3}$ 8. $\dfrac{4a}{7}$

9. $\sqrt{\dfrac{1}{4k^2}} = \dfrac{\sqrt{1}}{\sqrt{4k^2}} = \dfrac{1}{2k}$ 10. $\dfrac{m}{3}$ 11. $\sqrt{\dfrac{4x^2}{y^2}} = \dfrac{\sqrt{4x^2}}{\sqrt{y^2}} = \dfrac{2x}{y}$

12. $\dfrac{x^2}{3y}$ 13. $\sqrt{\dfrac{1 \cdot 5}{5 \cdot 5}} = \dfrac{\sqrt{5}}{\sqrt{5^2}} = \dfrac{\sqrt{5}}{5}$ 14. $\dfrac{\sqrt{3}}{3}$

15. $\sqrt{\dfrac{7 \cdot 13}{13 \cdot 13}} = \dfrac{\sqrt{91}}{\sqrt{13^2}} = \dfrac{\sqrt{91}}{13}$ 16. $\dfrac{\sqrt{85}}{17}$

17. $\sqrt{\dfrac{1 \cdot 2}{18 \cdot 2}} = \dfrac{\sqrt{2}}{\sqrt{36}} = \dfrac{\sqrt{2}}{6}$ 18. $\dfrac{\sqrt{2}}{10}$

19. $\sqrt{\dfrac{7 \cdot 3}{27 \cdot 3}} = \dfrac{\sqrt{21}}{\sqrt{81}} = \dfrac{\sqrt{21}}{9}$ 20. $\dfrac{\sqrt{10}}{4}$

21. $\dfrac{3}{\sqrt{7}} \cdot \dfrac{\sqrt{7}}{\sqrt{7}} = \dfrac{3\sqrt{7}}{7}$ 22. $\dfrac{2\sqrt{3}}{3}$

23. $\dfrac{10}{\sqrt{5}} \cdot \dfrac{\sqrt{5}}{\sqrt{5}} = \dfrac{\overset{2}{\cancel{10}}\sqrt{5}}{\underset{1}{\cancel{5}}} = 2\sqrt{5}$

24. $7\sqrt{2}$ 25. $\sqrt{\dfrac{m^2(3)}{3(3)}} = \dfrac{m\sqrt{3}}{3}$ 26. $\dfrac{k\sqrt{5}}{5}$

27. $\dfrac{\sqrt{x^4xz^4}}{\sqrt{36y^2}} = \dfrac{x^2z^2\sqrt{x}}{6y}$ 28. $\dfrac{ac^3\sqrt{a}}{5b}$

29. $\sqrt{\dfrac{3a^2}{4b^2}} = \dfrac{a\sqrt{3}}{2b}$ 30. $\dfrac{v\sqrt{5}}{2}$

31. $\sqrt{\dfrac{b^2c^4d}{16d^3d}} = \dfrac{bc^2\sqrt{d}}{4d^2}$ 32. $\dfrac{h^2k^4\sqrt{p}}{7p^3}$

33. $\sqrt{\dfrac{4m^2n}{n^2}} = \dfrac{2m\sqrt{n}}{n}$ 34. $\dfrac{3y\sqrt{x}}{x}$

Exercises 11.3A (page 534)

1. 3   2. 7   3. 4   4. 9

5. $\sqrt{36} = 6$ 6. 8   7. $\sqrt{9x^2} = 3x$

8. $5y$ 9. $\sqrt{5 \cdot 10 \cdot 2} = \sqrt{100} = 10$ 10. 12

11. $\sqrt{4 \cdot 2 \cdot 2 \cdot 9 \cdot 4 \cdot 5} = \sqrt{4}\sqrt{2} \cdot 2\sqrt{9}\sqrt{4}\sqrt{5}$
   $= 2 \cdot 2 \cdot 3 \cdot 2\sqrt{5} = 24\sqrt{5}$

12. $60\sqrt{7}$ 13. $\sqrt{5 \cdot 5 \cdot 2 \cdot x \cdot x^2} = 5 \cdot x\sqrt{2 \cdot x} = 5x\sqrt{2x}$

14. $11y\sqrt{3y}$

15. $\sqrt{3 \cdot 2 \cdot 6 \cdot 5 \cdot x \cdot x^2 \cdot y^2} = 6 \cdot x \cdot y\sqrt{5x} = 6xy\sqrt{5x}$

16. $14ab\sqrt{5ab}$ 17. $\sqrt{100a^2b^2b} = 10ab\sqrt{b}$ 18. $9xy\sqrt{x}$

19. $\sqrt{2a \cdot 6 \cdot 3a} = \sqrt{36a^2} = 6a$ 20. $4h^2$

21. $(5 \cdot 2)\sqrt{2x \cdot 8x^3 \cdot 3x^5} = 10\sqrt{48x^9} = 10\sqrt{3 \cdot 16 \cdot x^8 \cdot x}$
   $= 10 \cdot 4 \cdot x^4\sqrt{3x} = 40x^4\sqrt{3x}$

22. $72M^3\sqrt{2M}$

23. $(8 \cdot 3)\sqrt{5x(15x^3)(6x)} = 24\sqrt{450x^5} = 24\sqrt{225 \cdot 2 \cdot x^4 \cdot x}$
   $= 24 \cdot 15 \cdot x^2 \cdot \sqrt{2x} = 360x^2\sqrt{2x}$

24. $450a^3\sqrt{3}$

Exercises 11.3B (page 536)

1. $\sqrt{\dfrac{20}{5}} = \sqrt{4} = 2$ **2.** $\dfrac{1}{2}$ **3.** $\sqrt{\dfrac{32}{2}} = \sqrt{16} = 4$ **4.** 7

5. $\dfrac{\sqrt{4}}{\sqrt{5}} = \dfrac{2}{\sqrt{5}} \cdot \dfrac{\sqrt{5}}{\sqrt{5}} = \dfrac{2\sqrt{5}}{5}$ **6.** $\dfrac{3\sqrt{7}}{7}$ **7.** $\sqrt{\dfrac{15x}{5x}} = \sqrt{3}$

8. $\sqrt{6}$ **9.** $\sqrt{\dfrac{75a^3}{5a}} = \sqrt{15a^2} = a\sqrt{15}$ **10.** $b^2\sqrt{3}$

11. $\sqrt{\dfrac{35s^4}{10s}} = \sqrt{\dfrac{7s^3}{2}} = \dfrac{\sqrt{7s^3}}{\sqrt{2}} \cdot \dfrac{\sqrt{2}}{\sqrt{2}} = \dfrac{s\sqrt{14s}}{2}$ **12.** $\dfrac{\sqrt{14t}}{2}$

13. $\sqrt{\dfrac{8xy^3}{12x^2y}} = \sqrt{\dfrac{2y^2}{3x}} = \dfrac{\sqrt{2y^2}}{\sqrt{3x}} = \dfrac{\sqrt{2y^2}}{\sqrt{3x}} \cdot \dfrac{\sqrt{3x}}{\sqrt{3x}} = \dfrac{y\sqrt{6x}}{3x}$

14. $\dfrac{a\sqrt{6b}}{3b}$ **15.** $\sqrt{\dfrac{72x^3y^2}{2xy^2}} = \sqrt{36x^2} = 6x$ **16.** $3y$

17. $\sqrt{\dfrac{x^4y}{5y}} = \dfrac{\sqrt{x^4}}{\sqrt{5}} \cdot \dfrac{\sqrt{5}}{\sqrt{5}} = \dfrac{x^2\sqrt{5}}{5}$ **18.** $\dfrac{m^3\sqrt{3}}{3}$

19. $\dfrac{4}{3}\sqrt{\dfrac{\overset{9}{45}m^3}{\underset{2}{10}m}} = \dfrac{4}{3} \dfrac{\sqrt{9m^2}}{\sqrt{2}} = \dfrac{4 \cdot \overset{1}{3}m}{\underset{1}{3}\sqrt{2}} \cdot \dfrac{\sqrt{2}}{\sqrt{2}} = \dfrac{\overset{2}{4}m\sqrt{2}}{\underset{1}{2}} = 2m\sqrt{2}$

20. $4x\sqrt{6x}$ **21.** $\dfrac{5}{7}\sqrt{\dfrac{35a^5}{5a}} = \dfrac{5}{7}\sqrt{7a^4} = \dfrac{5}{7}a^2\sqrt{7}$ **22.** $4x^3\sqrt{2}$

Exercises 11.4 (page 539)

1. $(2 + 5)\sqrt{3} = 7\sqrt{3}$ **2.** $7\sqrt{2}$ **3.** $(3 - 1)\sqrt{x} = 2\sqrt{x}$

4. $4\sqrt{a}$ **5.** $\left(\dfrac{3}{2} - \dfrac{1}{2}\right)\sqrt{2} = \dfrac{2}{2}\sqrt{2} = \sqrt{2}$ **6.** $\sqrt{3}$

7. $40\sqrt{5} + \sqrt{5} = (40 + 1)\sqrt{5} = 41\sqrt{5}$ **8.** $13\sqrt{7}$

9. $5 + \sqrt{5}$ **10.** $4 + \sqrt{6}$

11. $5\sqrt{2} + 5\sqrt{2} = (5 + 5)\sqrt{2} = 10\sqrt{2}$ **12.** $8\sqrt{2}$

13. $2\sqrt{17} - \sqrt{17} = (2 - 1)\sqrt{17} = \sqrt{17}$ **14.** $\sqrt{13}$

15. $3\sqrt{5} - 2\sqrt{7}$ **16.** $3\sqrt{10} - 3\sqrt{6}$

17. $2\sqrt{3} + \sqrt{4 \cdot 3} = 2\sqrt{3} + 2\sqrt{3} = 4\sqrt{3}$ **18.** $5\sqrt{2}$

19. $2\sqrt{25 \cdot 2} - \sqrt{16 \cdot 2} = 2 \cdot 5\sqrt{2} - 4\sqrt{2} = 10\sqrt{2} - 4\sqrt{2}$
$= 6\sqrt{2}$

20. $3\sqrt{6}$

21. $3\sqrt{16 \cdot 2} - \sqrt{4 \cdot 2} = 3 \cdot 4\sqrt{2} - 2\sqrt{2} = 12\sqrt{2} - 2\sqrt{2}$
$= 10\sqrt{2}$

22. $6\sqrt{3}$

23. $\sqrt{\dfrac{1 \cdot 2}{2 \cdot 2}} + \sqrt{4 \cdot 2} = \dfrac{1}{2}\sqrt{2} + 2\sqrt{2} = \left(\dfrac{1}{2} + 2\right)\sqrt{2} = \dfrac{5\sqrt{2}}{2}$

24. $\dfrac{7\sqrt{3}}{3}$

25. $\sqrt{4 \cdot 6} - \sqrt{\dfrac{2 \cdot 3}{3 \cdot 3}} = 2\sqrt{6} - \dfrac{\sqrt{6}}{3} = \left(2 - \dfrac{1}{3}\right)\sqrt{6} = \dfrac{5\sqrt{6}}{3}$

26. $\dfrac{13\sqrt{5}}{5}$

27. $10\sqrt{\dfrac{3 \cdot 5}{5 \cdot 5}} + \sqrt{4 \cdot 15} = \dfrac{10\sqrt{15}}{5} + 2\sqrt{15} = 2\sqrt{15} + 2\sqrt{15}$
$= 4\sqrt{15}$

28. $6\sqrt{3}$ **29.** $\dfrac{\sqrt{25}}{\sqrt{2}} - \dfrac{3}{\sqrt{2}} = \dfrac{5 - 3}{\sqrt{2}} = \dfrac{2}{\sqrt{2}} \cdot \dfrac{\sqrt{2}}{\sqrt{2}} = \dfrac{2\sqrt{2}}{2} = \sqrt{2}$

30. $\dfrac{7\sqrt{5}}{10}$

31. $3\sqrt{\dfrac{1 \cdot 6}{6 \cdot 6}} + \sqrt{4 \cdot 3} - 5\sqrt{\dfrac{3 \cdot 2}{2 \cdot 2}} = \dfrac{\overset{1}{3}}{1} \cdot \dfrac{\sqrt{6}}{\underset{2}{6}} + \sqrt{4}\sqrt{3} - \dfrac{5}{1} \cdot \dfrac{\sqrt{6}}{2}$
$= \dfrac{1}{2}\sqrt{6} + 2\sqrt{3} - \dfrac{5}{2}\sqrt{6} = \left(\dfrac{1}{2} - \dfrac{5}{2}\right)\sqrt{6} + 2\sqrt{3} = -2\sqrt{6} + 2\sqrt{3}$

32. $\sqrt{10} + 2\sqrt{5}$

33. $\approx \dfrac{1}{3}(2.646) + 2(1.732) = 0.882 + 3.464 = 4.346$

34. 7.906 if tables are used; 7.907 if a calculator is used

Exercises 11.5 (page 543)

1. $\sqrt{2}(\sqrt{2}) + \sqrt{2}(1) = 2 + \sqrt{2}$ **2.** $3 + \sqrt{3}$

3. $\sqrt{3}(2\sqrt{3}) + \sqrt{3}(1) = 6 + \sqrt{3}$ **4.** $15 + \sqrt{5}$

5. $\sqrt{x}(\sqrt{x}) + \sqrt{x}(-3) = x - 3\sqrt{x}$ **6.** $4\sqrt{y} - y$

7. $\sqrt{7}\sqrt{7} + 5\sqrt{7} + 2(3) = 7 + 5\sqrt{7} + 6 = 13 + 5\sqrt{7}$

8. $11 + 6\sqrt{3}$

9. $\sqrt{8}\sqrt{8} - 3\sqrt{2}\sqrt{8} + 2\sqrt{8}\sqrt{5} - 3\sqrt{2} \cdot 2\sqrt{5}$
$= 8 - 3\sqrt{16} + 2\sqrt{40} - 6\sqrt{10}$
$= 8 - 3(4) + 2\sqrt{4 \cdot 10} - 6\sqrt{10}$
$= 8 - 12 + 2 \cdot 2\sqrt{10} - 6\sqrt{10} = -4 + 4\sqrt{10} - 6\sqrt{10}$
$= -4 - 2\sqrt{10}$

10. 0

11. $(\sqrt{2} + \sqrt{14})^2 = (\sqrt{2})^2 + 2\sqrt{2}\sqrt{14} + (\sqrt{14})^2$
$= 2 + 2\sqrt{2^2 \cdot 7} + 14 = 16 + 2 \cdot 2\sqrt{7}$
$= 16 + 4\sqrt{7}$

12. $16 + 2\sqrt{55}$

13. $(\sqrt{2x} + 3)^2 = (\sqrt{2x})^2 + 2\sqrt{2x}(3) + (3)^2$
$= 2x + 6\sqrt{2x} + 9$

14. $7x + 8\sqrt{7x} + 16$

15. $\dfrac{\sqrt{8}}{\sqrt{2}} + \dfrac{\sqrt{18}}{\sqrt{2}} = \sqrt{\dfrac{8}{2}} + \sqrt{\dfrac{18}{2}} = \sqrt{4} + \sqrt{9} = 2 + 3 = 5$

16. 5

17. $\dfrac{\sqrt{20}}{\sqrt{5}} + \dfrac{5\sqrt{10}}{\sqrt{5}} = \sqrt{\dfrac{20}{5}} + 5\sqrt{\dfrac{10}{5}} = \sqrt{4} + 5\sqrt{2} = 2 + 5\sqrt{2}$

18. $2 + \dfrac{\sqrt{21}}{3}$ **19a.** $2 - \sqrt{3}$ **b.** $2\sqrt{5} + 7$

20a. $3\sqrt{2} + 5$ **b.** $\sqrt{7} - 4$

21. $\dfrac{3}{(\sqrt{2} - 1)} \dfrac{(\sqrt{2} + 1)}{(\sqrt{2} + 1)} = \dfrac{3(\sqrt{2} + 1)}{(\sqrt{2})^2 - 1^2} = \dfrac{3(\sqrt{2} + 1)}{2 - 1} = 3(\sqrt{2} + 1)$

22. $5(\sqrt{2} + 1)$

23. $\dfrac{6}{(\sqrt{3} - \sqrt{2})} \dfrac{(\sqrt{3} + \sqrt{2})}{(\sqrt{3} + \sqrt{2})} = \dfrac{6(\sqrt{3} + \sqrt{2})}{(\sqrt{3})^2 - (\sqrt{2})^2}$
$= \dfrac{6(\sqrt{3} + \sqrt{2})}{3 - 2} = 6(\sqrt{3} + \sqrt{2})$

24. $-8(\sqrt{2} + \sqrt{3})$

25. $\dfrac{6}{(\sqrt{5} + \sqrt{2})} \dfrac{(\sqrt{5} - \sqrt{2})}{(\sqrt{5} - \sqrt{2})} = \dfrac{6(\sqrt{5} - \sqrt{2})}{5 - 2} = \dfrac{\overset{2}{6}(\sqrt{5} - \sqrt{2})}{\underset{1}{3}}$
$= 2(\sqrt{5} - \sqrt{2})$

26. $2(\sqrt{7} - \sqrt{5})$

27. $\dfrac{(x - 4)}{(\sqrt{x} + 2)} \dfrac{(\sqrt{x} - 2)}{(\sqrt{x} - 2)} = \dfrac{\overset{1}{(x - 4)}(\sqrt{x} - 2)}{\underset{1}{(x - 4)}} = \sqrt{x} - 2$ **28.** $\sqrt{y} + 3$

29. $\approx \dfrac{1}{3}(2.646) - 2(1.732) = 0.882 - 3.464 = -2.582 \approx -2.58$

30. ≈ 6.79 **31.** $\approx \dfrac{3 + 2(3.317)}{6} = \dfrac{3 + 6.634}{6} = \dfrac{9.634}{6} \approx 1.61$

32. ≈ -0.61

Exercises 11.6 (page 549)

1. $\sqrt{x} = 5$ *Check:* $\sqrt{x} \overset{?}{=} 5$ **2.** 16
$(\sqrt{x})^2 = 5^2$ $\sqrt{25} \overset{?}{=} 5$
$x = 25$ $5 = 5$

The solution is 25.

3. $\sqrt{2x} = 4$ *Check* $\sqrt{2x} = 4$ **4.** 12
$(\sqrt{2x})^2 = 4^2$ $\sqrt{2(8)} \overset{?}{=} 4$
$2x = 16$ $\sqrt{16} \overset{?}{=} 4$
$x = 8$ $4 = 4$

The solution is 8.

5. $\sqrt{x - 3} = 2$ *Check:* $\sqrt{x - 3} = 2$ **6.** 32
$(\sqrt{x - 3})^2 = 2^2$ $\sqrt{7 - 3} \overset{?}{=} 2$
$x - 3 = 4$ $\sqrt{4} \overset{?}{=} 2$
$x = 7$ $2 = 2$

The solution is 7.

7. $(\sqrt{2x + 1})^2 = 9^2$ *Check:* $\sqrt{2x + 1} = 9$ **8.** 4
$2x + 1 = 81$ $\sqrt{2 \cdot 40 + 1} \overset{?}{=} 9$
$2x = 80$ $\sqrt{81} \overset{?}{=} 9$
$x = 40$ $9 = 9$

The solution is 40.

9. $(\sqrt{3x + 1})^2 = 5^2$ *Check:* $\sqrt{3x + 1} = 5$ **10.** 4
$3x + 1 = 25$ $\sqrt{3 \cdot 8 + 1} \overset{?}{=} 5$
$3x = 24$ $\sqrt{25} \overset{?}{=} 5$
$x = 8$ $5 = 5$

The solution is 8.

11. $(\sqrt{x + 1})^2 = (\sqrt{2x - 7})^2$ *Check:* $\sqrt{x + 1} = \sqrt{2x - 7}$
$x + 1 = 2x - 7$ $\sqrt{8 + 1} \overset{?}{=} \sqrt{2 \cdot 8 - 7}$
$8 = x$ $\sqrt{9} \overset{?}{=} \sqrt{9}$
$x = 8$ $3 = 3$

The solution is 8.

12. 3   **13.** $(\sqrt{3x - 2})^2 = (x)^2$ **14.** 2, 3
$3x - 2 = x^2$
$0 = x^2 - 3x + 2$
$0 = (x - 2)(x - 1)$

$x - 2 = 0$ | $x - 1 = 0$
$x = 2$ | $x = 1$

Check for x = 2: | *Check for x = 1:*
$\sqrt{3x - 2} = x$ | $\sqrt{3x - 2} = x$
$\sqrt{3(2) - 2} \overset{?}{=} 2$ | $\sqrt{3(1) - 2} \overset{?}{=} 1$
$\sqrt{4} \overset{?}{=} 2$ | $\sqrt{1} \overset{?}{=} 1$
$2 = 2$ | $1 = 1$

The solutions are 2 and 1.

15. $(\sqrt{4x - 1})^2 = (2x)^2$ **16.** $\frac{1}{3}$
$4x - 1 = 4x^2$
$0 = 4x^2 - 4x + 1$
$0 = (2x - 1)(2x - 1)$
$2x - 1 = 0$
$2x = 1$
$x = \frac{1}{2}$
Check: $\sqrt{4x - 1} = 2x$
$\sqrt{4\left(\frac{1}{2}\right) - 1} \overset{?}{=} 2\left(\frac{1}{2}\right)$
$\sqrt{2 - 1} \overset{?}{=} 1$
$1 = 1$

The solution is $\frac{1}{2}$.

17. $\sqrt{x - 3} = x - 5$ **18.** 5
$(\sqrt{x - 3})^2 = (x - 5)^2$
$x - 3 = x^2 - 10x + 25$
$0 = x^2 - 11x + 28$
$0 = (x - 4)(x - 7)$

$x - 4 = 0$ | $x - 7 = 0$
$x = 4$ | $x = 7$

Check for x = 4: | *Check for x = 7:*
$\sqrt{x - 3} + 5 = x$ | $\sqrt{x - 3} + 5 = x$
$\sqrt{4 - 3} + 5 \overset{?}{=} 4$ | $\sqrt{7 - 3} + 5 \overset{?}{=} 7$
$\sqrt{1} + 5 \overset{?}{=} 4$ | $\sqrt{4} + 5 \overset{?}{=} 7$
$1 + 5 \overset{?}{=} 4$ | $2 + 5 \overset{?}{=} 7$
$6 \neq 4$ | $7 = 7$

4 does not check. The only solution is 7.

19. $(\sqrt{2x + 2})^2 = (1 + \sqrt{x + 2})^2$ Squaring both sides   **20.** 5
$2x + 2 = 1 + 2\sqrt{x + 2} + (x + 2)$

$2x + 2 = 3 + 2\sqrt{x + 2} + x$
$\underline{-x - 3 \qquad -3 \qquad\qquad - x}$
$x - 1 = \qquad 2\sqrt{x + 2}$
$(x - 1)^2 = (2\sqrt{x + 2})^2$ Squaring both sides again
$x^2 - 2x + 1 = 4(x + 2)$
$x^2 - 2x + 1 = 4x + 8$
$\underline{-4x - 8 \qquad -4x - 8}$
$x^2 - 6x - 7 = 0$
$(x - 7)(x + 1) = 0$
$x - 7 = 0$ | $x + 1 = 0$
$x = 7$ | $x = -1$

Check for x = 7: | *Check for x = -1:*
$\sqrt{2x + 2} = 1 + \sqrt{x + 2}$ | $\sqrt{2x + 2} = 1 + \sqrt{x + 2}$
$\sqrt{2(7) + 2} \overset{?}{=} 1 + \sqrt{(7) + 2}$ | $\sqrt{2(-1) + 2} \overset{?}{=} 1 + \sqrt{(-1) + 2}$
$\sqrt{16} \overset{?}{=} 1 + \sqrt{9}$ | $\sqrt{0} \overset{?}{=} 1 + \sqrt{1}$
$4 \overset{?}{=} 1 + 3$ | $0 \overset{?}{=} 1 + 1$
$4 = 4$ | $0 \neq 2$

-1 does not check. The only solution is 7.

21. $(\sqrt{3x - 5})^2 = (\sqrt{x + 6} - 1)^2$ Squaring both sides
$3x - 5 = (x + 6) - 2\sqrt{x + 6} + 1$

$3x - 5 = x + 7 - 2\sqrt{x + 6}$
$\underline{-x - 7 \qquad -x - 7}$
$2x - 12 = \qquad -2\sqrt{x + 6}$
$(2x - 12)^2 = (-2\sqrt{x + 6})^2$ Squaring both sides again
$4x^2 - 48x + 144 = 4(x + 6)$
$4x^2 - 48x + 144 = 4x + 24$
$\underline{- 4x - 24 \qquad -4x - 24}$
$4x^2 - 52x + 120 = 0$
$4(x^2 - 13 + 30) = 0$
$4(x - 3)(x - 10) = 0$
$4 \neq 0$ | $x - 3 = 0$ | $x - 10 = 0$
   $x = 3$ | $x = 10$

Check for x = 3: | *Check for x = 10:*
$\sqrt{3x - 5} = \sqrt{x + 6} - 1$ | $\sqrt{3x - 5} = \sqrt{x + 6} - 1$
$\sqrt{3(3) - 5} \overset{?}{=} \sqrt{(3) + 6} - 1$ | $\sqrt{3(10) - 5} \overset{?}{=} \sqrt{(10) + 6} - 1$
$\sqrt{4} \overset{?}{=} \sqrt{9} - 1$ | $\sqrt{25} \overset{?}{=} \sqrt{16} - 1$
$2 \overset{?}{=} 3 - 1$ | $5 \overset{?}{=} 4 - 1$
$2 = 2$ | $5 \neq 3$

10 does not check. The only solution is 3.

22. 2

23. $I = \sqrt{\dfrac{P}{R}}; \quad R = 9, I = 7$ **24.** 100

$$7 = \sqrt{\dfrac{P}{9}}$$

$$(7)^2 = \left(\sqrt{\dfrac{P}{9}}\right)^2$$

$$49 = \dfrac{P}{9}$$

$$441 = P$$

The amount of power consumed is 441 watts.

25. $I = \sqrt{\dfrac{P}{R}}; \quad I = 5, P = 500$ **26.** 24

$$5 = \sqrt{\dfrac{500}{R}}$$

$$(5)^2 = \left(\sqrt{\dfrac{500}{R}}\right)^2$$

$$25 = \dfrac{500}{R}$$

$$25R = 500$$

$$R = 20$$

The resistance is 20 ohms.

27. $\sigma = \sqrt{npq}; \ \sigma = 3, q = \frac{3}{4}, n = 48$ **28.** $\frac{1}{5}$

$$3 = \sqrt{48\left(\frac{3}{4}\right)p}$$

$$3 = \sqrt{36p}$$

$$(3)^2 = (\sqrt{36p})^2 \quad \text{Squaring both sides}$$

$$9 = 36p$$

$$p = \tfrac{1}{4}$$

The probability of success is $\frac{1}{4}$.

29. $c = \sqrt{a^2 + b^2}; \quad c = 5, b = 3$ **30.** 12

$$5 = \sqrt{a^2 + 3^2}$$

$$(5)^2 = (\sqrt{a^2 + 9})^2$$

$$25 = a^2 + 9$$

$$0 = a^2 - 16$$

$$0 = (a + 4)(a - 4)$$

$$a + 4 = 0 \qquad \bigm| \qquad a - 4 = 0$$

$$a = -4 \qquad \qquad a = 4$$

↑ — Reject: a length cannot be negative

The length of one side is 4.

Review Exercises 11.7 (page 552)

1a. $9x$ **b.** $\frac{1}{2}$ **c.** $x - 5$ **2.** 10   **3.** 3   **4.** $a + b$

5. $\sqrt{x^2 \cdot x} = \sqrt{x^2}\sqrt{x} = x\sqrt{x}$ **6.** $6a^2b$

7. $\sqrt{a^2 \cdot a \cdot b^2 \cdot b} = \sqrt{a^2}\sqrt{a}\sqrt{b^2}\sqrt{b} = ab\sqrt{ab}$ **8.** 11

9. $\sqrt{64} = 8$ **10.** $5\sqrt{2}$

11. $\sqrt{2^2 \cdot 3^2 \cdot 5} = \sqrt{2^2}\sqrt{3^2}\sqrt{5} = 2 \cdot 3\sqrt{5} = 6\sqrt{5}$

12. $3\sqrt{2}$ **13.** $3\sqrt{2} - 2\sqrt{2} = (3 - 2)\sqrt{2} = \sqrt{2}$

14. $\sqrt{3}$ **15.** $\dfrac{1 \cdot \sqrt{5}}{\sqrt{5} \cdot \sqrt{5}} = \dfrac{\sqrt{5}}{5}$ **16.** $\dfrac{\sqrt{7}}{7}$

17. $\dfrac{6 \cdot \sqrt{3}}{\sqrt{3} \cdot \sqrt{3}} = \dfrac{\overset{2}{\cancel{6}}\sqrt{3}}{\underset{1}{\cancel{3}}} = 2\sqrt{3}$ **18.** $\dfrac{3\sqrt{2}}{2}$

19. $\sqrt{4 \cdot 2} - \sqrt{\dfrac{1 \cdot 2}{2 \cdot 2}} = 2\sqrt{2} - \dfrac{\sqrt{2}}{2} = 2\sqrt{2} - \tfrac{1}{2}\sqrt{2} = \dfrac{3\sqrt{2}}{2}$

20. $\dfrac{8\sqrt{6}}{3}$

21. $\sqrt{2}(\sqrt{8}) + \sqrt{2}(\sqrt{18}) = \sqrt{16} + \sqrt{36} = 4 + 6 = 10$ **22.** 23

23. $(2\sqrt{3})^2 + 2(2\sqrt{3})(1) + (1)^2 = 12 + 4\sqrt{3} + 1 = 13 + 4\sqrt{3}$

24. $19 - 6\sqrt{2}$

25. $\dfrac{8 \cdot (\sqrt{3} + 2)}{(\sqrt{3} - 2)(\sqrt{3} + 2)} = \dfrac{8(\sqrt{3} + 2)}{3 - 4} = -8(\sqrt{3} + 2)$

26. $6(\sqrt{5} - 2)$

27. $\sqrt{x} = 4$ *Check:* $\sqrt{x} = 4$ **28.** 12

$(\sqrt{x})^2 = 4^2$ $\sqrt{16} \overset{?}{=} 4$

$x = 16$ $4 = 4$

The solution is 16.

29. $(\sqrt{2x - 1})^2 = 5^2$ *Check:* $\sqrt{2x - 1} = 5$ **30.** 3

$2x - 1 = 25$ $\sqrt{2(13) - 1} \overset{?}{=} 5$

$2x = 26$ $\sqrt{25} \overset{?}{=} 5$

$x = 13$ $5 = 5$

The solution is 13.

31. $(\sqrt{7x - 6})^2 = (x)^2$ *Check* $\sqrt{7x - 6} = x$

$7x - 6 = x^2$ *for $x = 6$:* $\sqrt{7(6) - 6} \overset{?}{=} 6$

$0 = x^2 - 7x + 6$ $\sqrt{36} \overset{?}{=} 6$

$0 = (x - 6)(x - 1)$ $6 = 6$

$x - 6 = 0 \quad \bigm| \quad x - 1 = 0$ *for $x = 1$:* $\sqrt{7(1) - 6} \overset{?}{=} 1$

$x = 6 \qquad \quad \ x = 1$ $\sqrt{1} \overset{?}{=} 1$

                                       $1 = 1$

The solutions are 6 and 1.

32. 1, 8

33. $(\sqrt{2x + 7})^2 = (\sqrt{x} + 2)^2$ *Check* $\sqrt{2x + 7} = \sqrt{x} + 2$

$2x + 7 = x + 4\sqrt{x} + 4$ *for $x = 9$:* $\sqrt{2(9) + 7} \overset{?}{=} \sqrt{9} + 2$

$x + 3 = 4\sqrt{x}$ $\sqrt{25} \overset{?}{=} 3 + 2$

$(x + 3)^2 = (4\sqrt{x})^2$ $5 = 5$

$x^2 + 6x + 9 = 16x$ *for $x = 1$:* $\sqrt{2(1) + 7} \overset{?}{=} \sqrt{1} + 2$

$x^2 - 10x + 9 = 0$ $\sqrt{9} \overset{?}{=} 1 + 2$

$(x - 9)(x - 1) = 0$ $3 = 3$

$x - 9 = 0 \quad \bigm| \quad x - 1 = 0$

$x = 9 \qquad \quad \ x = 1$

The solutions are 9 and 1.

34. 1

35. $c = \sqrt{a^2 + b^2}; \quad c = 13, b = 12$

$$13 = \sqrt{a^2 + 12^2}$$

$$(13)^2 = (\sqrt{a^2 + 144})^2$$

$$169 = a^2 + 144$$

$$0 = a^2 - 25$$

$$0 = (a + 5)(a - 5)$$

$$a + 5 = 0 \qquad \bigm| \qquad a - 5 = 0$$

$$a = -5 \qquad \qquad a = 5$$

↑ — Reject: a length cannot be negative

The length of one side is 5.

Chapter 11 Diagnostic Test (page 557)

Following each problem number is the textbook section number (in parentheses) where that kind of problem is discussed.

1. (11.1)   **a.** $3x^3y$ **b.** $2x + 5$

2. (11.2) $\sqrt{16z^2} = \sqrt{16}\sqrt{z^2} = 4z$

3. (11.2) $\sqrt{50} = \sqrt{25 \cdot 2} = \sqrt{25}\sqrt{2} = 5\sqrt{2}$

4. (11.2) $\sqrt{x^6y^5} = \sqrt{x^6 \cdot y^4 \cdot y} = \sqrt{x^6}\sqrt{y^4}\sqrt{y} = x^3y^2\sqrt{y}$

5. (11.2) $\sqrt{\dfrac{3}{14}} = \sqrt{\dfrac{3 \cdot 14}{14 \cdot 14}} = \dfrac{\sqrt{42}}{14}$

6. (11.2) $\sqrt{\dfrac{48}{3n^2}} = \sqrt{\dfrac{16}{n^2}} = \dfrac{4}{n}$

7. (11.2) $\sqrt{60} = \sqrt{4 \cdot 15} = \sqrt{4}\sqrt{15} = 2\sqrt{15}$

8. (11.3) $\sqrt{13}\sqrt{13} = 13$

9. (11.3) $\sqrt{3}\sqrt{75y^2} = \sqrt{225y^2} = \sqrt{225}\sqrt{y^2} = 15y$

10. (11.5) $\sqrt{5}(2\sqrt{5} - 3) = \sqrt{5}(2\sqrt{5}) - \sqrt{5}(3)$
$= 2 \cdot \sqrt{5 \cdot 5} - 3\sqrt{5}$
$= 2 \cdot 5 - 3\sqrt{5} = 10 - 3\sqrt{5}$

11. (11.5) $(\sqrt{13} + \sqrt{3})(\sqrt{13} - \sqrt{3}) = (\sqrt{13})^2 - (\sqrt{3})^2$
$= 13 - 3 = 10$

12. (11.5) $(4\sqrt{x} + 3)(3\sqrt{x} + 4)$
$= (4\sqrt{x})(3\sqrt{x}) + 4(4\sqrt{x}) + 3(3\sqrt{x}) + (3)(4)$
$= 12\sqrt{x \cdot x} + \underline{16\sqrt{x} + 9\sqrt{x}} + 12$
$= 12x + 25\sqrt{x} + 12$

13. (11.5) $(8 + \sqrt{10})^2$
$= (8)^2 + 2(8)(\sqrt{10}) + (\sqrt{10})^2$
$= 64 + 16\sqrt{10} + 10$
$= 74 + 16\sqrt{10}$

14. (11.3) $\dfrac{\sqrt{5}}{\sqrt{80}} = \sqrt{\dfrac{5}{80}} = \sqrt{\dfrac{1}{16}} = \dfrac{1}{4}$

15. (11.3) $\dfrac{\sqrt{3}}{\sqrt{6}} = \sqrt{\dfrac{3}{6}} = \sqrt{\dfrac{1}{2}} = \sqrt{\dfrac{1 \cdot 2}{2 \cdot 2}} = \dfrac{\sqrt{2}}{2}$

16. (11.5) $\dfrac{\sqrt{18} - \sqrt{36}}{\sqrt{3}} = \dfrac{\sqrt{18}}{\sqrt{3}} - \dfrac{\sqrt{36}}{\sqrt{3}} = \sqrt{\dfrac{18}{3}} - \sqrt{\dfrac{36}{3}}$
$= \sqrt{6} - \sqrt{12} = \sqrt{6} - 2\sqrt{3}$

17. (11.3) $\dfrac{\sqrt{x^4 y}}{\sqrt{3y}} = \sqrt{\dfrac{x^4 y}{3y}} = \sqrt{\dfrac{x^4 (3)}{3(3)}}$
$= \dfrac{\sqrt{x^4}\sqrt{3}}{\sqrt{9}} = \dfrac{x^2\sqrt{3}}{3}$ or $\dfrac{x^2}{3}\sqrt{3}$

18. (11.5) $\dfrac{6}{1 + \sqrt{5}} = \dfrac{6}{(1 + \sqrt{5})} \cdot \dfrac{(1 - \sqrt{5})}{(1 - \sqrt{5})} = \dfrac{6(1 - \sqrt{5})}{1^2 - (\sqrt{5})^2}$
$= \dfrac{6(1 - \sqrt{5})}{1 - 5} = \dfrac{\overset{3}{\cancel{6}}(1 - \sqrt{5})}{\underset{2}{\cancel{-4}}} = -\dfrac{3}{2}(1 - \sqrt{5})$

19. (11.4) $8\sqrt{t} - \sqrt{t} = (8 - 1)\sqrt{t} = 7\sqrt{t}$

20. (11.4) $7\sqrt{3} - \sqrt{27} = 7\sqrt{3} - 3\sqrt{3} = (7 - 3)\sqrt{3} = 4\sqrt{3}$

21. (11.4) $\sqrt{54} + 4\sqrt{24} = 3\sqrt{6} + 4 \cdot 2\sqrt{6}$
$= 3\sqrt{6} + 8\sqrt{6} = 11\sqrt{6}$

22. (11.4) $\sqrt{45} + 6\sqrt{\dfrac{4}{5}} = 3\sqrt{5} + 6\sqrt{\dfrac{4 \cdot 5}{5 \cdot 5}} = 3\sqrt{5} + 6\left(\dfrac{2}{5}\right)\sqrt{5}$
$= \left(3 + \dfrac{12}{5}\right)\sqrt{5} = \left(\dfrac{15}{5} + \dfrac{12}{5}\right)\sqrt{5} = \left(\dfrac{27}{5}\right)\sqrt{5}$ or $\dfrac{27\sqrt{5}}{5}$

23. (11.6) $\sqrt{8x} = 12$ *Check:* $\sqrt{8x} = 12$
$(\sqrt{8x})^2 = (12)^2$ $\sqrt{8(18)} \overset{?}{=} 12$
$8x = 144$ $\sqrt{144} \overset{?}{=} 12$
$x = 18$ $12 = 12$

The solution is 18.

24. (11.6) $\sqrt{5x + 11} = 6$ *Check:* $\sqrt{5x + 11} = 6$
$(\sqrt{5x + 11})^2 = 6^2$ $\sqrt{5(5) + 11} \overset{?}{=} 6$
$5x + 11 = 36$ $\sqrt{36} \overset{?}{=} 6$
$5x = 25$ $6 = 6$
$x = 5$

The solution is 5.

25. (11.6)

$\sqrt{6x - 5} = x$
$(\sqrt{6x - 5})^2 = x^2$
$6x - 5 = x^2$
$0 = x^2 - 6x + 5$
$0 = (x - 5)(x - 1)$

$x - 5 = 0$ | $x - 1 = 0$
$x = 5$ | $x = 1$

Check for x = 5:   *Check for x = 1:*

$\sqrt{6x - 5} = x$ $\sqrt{6x - 5} = x$
$\sqrt{6(5) - 5} \overset{?}{=} 5$ $\sqrt{6(1) - 5} \overset{?}{=} 1$
$\sqrt{25} \overset{?}{=} 5$ $\sqrt{1} \overset{?}{=} 1$
$5 = 5$ $1 = 1$

The solutions are 5 and 1.

Cumulative Review Exercises: Chapters 1–11 (page 558)

1.  $\left(\dfrac{36x^4y^{-3}}{12y^{-1}}\right)^{-2} = \left(\dfrac{5x^4y^{-2}}{2}\right)^{-2} = \dfrac{5^{-2}x^{-8}y^4}{2^{-2}} = \dfrac{2^2 y^4}{5^2 x^8} = \dfrac{4y^4}{25x^8}$

2. $4(3 + h)(3 - h)$

3. $\dfrac{(x - 1)(x + 2)}{(x - 1)(x + 1)} \cdot \dfrac{x(x - 4)}{(x + 2)(x - 4)} = \dfrac{x}{x + 1}$

4. $x \le 4$

(number line shown, arrow from -7 to 7 with point at 4)

$-7 \ -6 \ -5 \ -4 \ -3 \ -2 \ -1 \ 0 \ 1 \ 2 \ 3 \ 4 \ 5 \ 6 \ 7$

5. LCD = 6

$\dfrac{\overset{3}{\cancel{6}}}{1} \cdot \dfrac{2x - 3}{\underset{1}{\cancel{2}}} = \dfrac{\overset{1}{\cancel{6}}}{1} \cdot \dfrac{5x + 4}{\cancel{6}} - \dfrac{\overset{2}{\cancel{6}}}{1} \cdot \dfrac{5}{\underset{1}{\cancel{3}}}$
$3(2x - 3) = 1(5x + 4) - 2(5)$
$6x - 9 = 5x + 4 - 10$
$x = 3$

The solution is 3.

6. $\dfrac{2x^2 - 7x - 10}{(x + 2)(x - 1)}$

7. (1) $\begin{cases} 4x - 3y = -9 \\ 3y + x = -6 \end{cases}$
(1) Intercepts: $\left(-2\frac{1}{4}, 0\right)$, $(0, 3)$
(2) Intercepts: $(-6, 0)$, $(0, -2)$

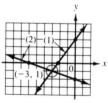

Solution: $(-3, -1)$

8. $(6, -5)$

9. (1) $\begin{cases} 2x - 5y = 2 \\ 3x - 7y = 2 \end{cases} \Rightarrow 2x = 5y + 2 \Rightarrow x = \dfrac{5y + 2}{2}$

Substitute $\dfrac{5y + 2}{2}$ for x in (2):

(2) $3x - 7y = 2$
$3\left(\dfrac{5y + 2}{2}\right) - 7y = 2$ LCD = 2
$\dfrac{\overset{1}{\cancel{2}}}{1} \cdot \dfrac{3}{1}\left(\dfrac{5y + 2}{\underset{1}{\cancel{2}}}\right) - \dfrac{2}{1} \cdot \dfrac{7y}{1} = \dfrac{2}{1} \cdot \dfrac{2}{1}$
$15y + 6 - 14y = 4$
$y = -2$

Substitute $y = -2$ in $x = \dfrac{5y + 2}{2}$;

$x = \dfrac{5(-2) + 2}{2} = \dfrac{-10 + 2}{2} = \dfrac{-8}{2} = -4$

Solution: $(-4, -2)$

10. $-\frac{7}{4}$ **11.** $2x$ **12.** $25 + 2\sqrt{154}$

13. $(\sqrt{5})(3) + (\sqrt{5})(\sqrt{80}) = 3\sqrt{5} + \sqrt{400} = 3\sqrt{5} + 20$

14. $3\sqrt{7}$

15. $\dfrac{8}{(\sqrt{6} - 2)(\sqrt{6} + 2)} \cdot \dfrac{(\sqrt{6} + 2)}{} = \dfrac{8(\sqrt{6} + 2)}{(\sqrt{6})^2 - 2^2} = \dfrac{8(\sqrt{6} + 2)}{6 - 4} = \dfrac{\overset{4}{\cancel{8}}(\sqrt{6} + 2)}{\underset{1}{\cancel{2}}}$

$\qquad = 4(\sqrt{6} + 2)$

16. $\frac{14}{3}$ or $4\frac{2}{3}$

(In Exercises 17–20, the checks will not be shown.)

17. Let $x =$ one number   **18.** 12 mi
$\qquad y =$ other number

$\begin{aligned}(1) \quad & \left\{x + y = \tfrac{1}{2}\right\} \\ (2) \quad & \left\{\underline{x - y = 7\tfrac{1}{2}}\right\}\end{aligned}$ $\Big\rangle$ Add

$\qquad\qquad 2x = 8$

$\qquad\qquad\ x = 4$

Substitute 4 for x in (1):

$4 + y = \tfrac{1}{2} \Rightarrow y = \tfrac{1}{2} - 4 = -3\tfrac{1}{2}$

One number is 4 and the other is $-3\tfrac{1}{2}$.

19. Let $\qquad x =$ number of dimes   **20.** 33, 35, and 37
$\qquad 32 - x =$ number of nickels

$10x + 5(32 - x) = 245$
$10x + 160 - 5x = 245$

$\begin{aligned} 5x + 160 &= \quad 245 \\ \underline{-160} &\quad \underline{-160} \\ 5x \quad &= \quad 85 \\ x &= 17 \\ 32 - x &= 15 \end{aligned}$

Sylvia has 17 dimes and 15 nickels.

Exercises 12.1 (page 562)

1. $2x^2 - 5x - 3 = 0$ $\begin{cases} a = 2 \\ b = -5 \\ c = -3 \end{cases}$

2. $3x^2 + 2x - 4 = 0$ $\begin{cases} a = 3 \\ b = 2 \\ c = -4 \end{cases}$

3. $6x^2 - 1x + 0 = 0$ $\begin{cases} a = 6 \\ b = -1 \\ c = 0 \end{cases}$

4. $3x^2 - 2x + 0 = 0$ $\begin{cases} a = 3 \\ b = -2 \\ c = 0 \end{cases}$

5. LCD $= 2$

$\dfrac{\overset{1}{\cancel{2}}}{1} \cdot \dfrac{3x}{\underset{1}{\cancel{2}}} + 2 \cdot 5 = 2 \cdot x^2$ $\begin{cases} a = 2 \\ b = -3 \\ c = -10 \end{cases}$

$\qquad 3x + 10 = 2x^2$
$2x^2 - 3x - 10 = 0$

6. $3x^2 + 2x - 12 = 0$ $\begin{cases} a = 3 \\ b = 2 \\ c = -12 \end{cases}$

7. LCD $= 12$

$12(x^2) + \dfrac{\overset{3}{\cancel{12}}}{1}\left(\dfrac{-5x}{\cancel{4}}\right) + \dfrac{\overset{4}{\cancel{12}}}{1}\left(\dfrac{2}{\cancel{3}}\right) = 12 \cdot 0$ $\begin{cases} a = 12 \\ b = -15 \\ c = 8 \end{cases}$

$\qquad\qquad 12x^2 - 15x + 8 = 0$

8. $30x^2 + 9x - 5 = 0$ $\begin{cases} a = 30 \\ b = 9 \\ c = -5 \end{cases}$

9. $\begin{aligned} x^2 - 3x &= 4 \\ 1x^2 - 3x - 4 &= 0 \end{aligned}$ $\begin{cases} a = 1 \\ b = -3 \\ c = -4 \end{cases}$

10. $2x^2 + 2x - 12 = 0$ $\begin{cases} a = 2 \\ b = 2 \\ c = -12 \end{cases}$

11. $\begin{aligned} 3x^2 + 3x &= x^2 + 3x + 2 \\ 2x^2 + 0x - 2 &= 0 \end{aligned}$ $\begin{cases} a = 2 \\ b = 0 \\ c = -2 \end{cases}$

12. $3x^2 - 2x + 3 = 0$ $\begin{cases} a = 3 \\ b = -2 \\ c = 3 \end{cases}$

Exercises 12.2 (page 566)

(The checks will not be shown.)

1. $(x + 3)(x - 2) = 0$ **2.** 3, -2

$\begin{aligned} x + 3 &= 0 & \Big| \quad x - 2 &= 0 \\ x &= -3 & x &= 2 \end{aligned}$

The solutions are -3 and 2.

3. $x^2 + x - 12 = 0$ **4.** 3, -5
$\ \ (x + 4)(x - 3) = 0$

$\begin{aligned} x + 4 &= 0 & \Big| \quad x - 3 &= 0 \\ x &= -4 & x &= 3 \end{aligned}$

The solutions are -4 and 3.

5. $\ \ 2x^2 - x - 1 = 0$ **6.** $\frac{1}{2}$, -1
$\ \ (2x + 1)(x - 1) = 0$

$\begin{aligned} 2x + 1 &= 0 & \Big| \quad x - 1 &= 0 \\ x &= -\tfrac{1}{2} & x &= 1 \end{aligned}$

The solutions are $-\tfrac{1}{2}$ and 1.

7. $\qquad \dfrac{x}{8} = \dfrac{2}{x}$ A proportion   **8.** 6, -6

$\qquad\qquad x^2 = 16$ Cross-multiplying
$\qquad\ x^2 - 16 = 0$
$(x - 4)(x + 4) = 0$

$\begin{aligned} x - 4 &= 0 & \Big| \quad x + 4 &= 0 \\ x &= 4 & x &= -4 \end{aligned}$

The solutions are 4 and -4.

9. $\qquad \dfrac{x + 2}{3} = \dfrac{-1}{x - 2}$ A proportion   **10.** 1, -1

$(x + 2)(x - 2) = 3(-1)$ Cross-multiplying
$\qquad\quad x^2 - 4 = -3$
$\qquad\quad x^2 - 1 = 0$
$(x + 1)(x - 1) = 0$

$\begin{aligned} x + 1 &= 0 & \Big| \quad x - 1 &= 0 \\ x &= -1 & x &= 1 \end{aligned}$

The solutions are -1 and 1.

11. $(x + 1)(x + 8) = 0$ **12.** -1, -6

$\begin{aligned} x + 1 &= 0 & \Big| \quad x + 8 &= 0 \\ x &= -1 & x &= -8 \end{aligned}$

The solutions are -1 and -8.

13. $\ \ 2x(x + 2) = 0$ **14.** 0, 3

$\begin{aligned} 2x &= 0 & \Big| \quad x + 2 &= 0 \\ x &= 0 & x &= -2 \end{aligned}$

The solutions are 0 and -2.

15. $x^2 - x - 2 = 0$ **16.** $3, -2$
$(x - 2)(x + 1) = 0$

$x - 2 = 0$	$x + 1 = 0$
$x = 2$	$x = -1$

The solutions are 2 and -1.

17. LCD $= 2x$ **18.** $1, 6$

$$\frac{\overset{1}{\cancel{2x}}}{1} \cdot \frac{x}{\underset{1}{\cancel{2}}} + \frac{\overset{1}{\cancel{2x}}}{1} \cdot \frac{2}{\underset{1}{\cancel{4}}} = \frac{\overset{1}{\cancel{2x}}}{1} \cdot \frac{5}{\underset{1}{\cancel{2}}}$$

$$x^2 + 4 = 5x$$
$$x^2 - 5x + 4 = 0$$
$$(x - 1)(x - 4) = 0$$

$x - 1 = 0$	$x - 4 = 0$
$x = 1$	$x = 4$

The solutions are 1 and 4.

19. LCD $= 4(x + 1)$ **20.** $2, 3$

$$\frac{\overset{1}{\cancel{4(x + 1)}}}{1} \cdot \frac{x - 1}{\underset{1}{\cancel{4}}} + \frac{\overset{1}{\cancel{4(x + 1)}}}{1} \cdot \frac{6}{\underset{1}{\cancel{(x + 1)}}} = 4(x + 1)2$$

$$(x + 1)(x - 1) + 24 = 8(x + 1)$$
$$x^2 - 1 + 24 = 8x + 8$$
$$x^2 - 8x + 15 = 0$$
$$(x - 3)(x - 5) = 0$$

$x - 3 = 0$	$x - 5 = 0$
$x = 3$	$x = 5$

The solutions are 3 and 5.

21. $\dfrac{2x^2}{1} = \dfrac{2 - x}{3}$ A proportion   **22.** $1\frac{1}{2}, \frac{1}{4}$

$$3(2x^2) = 1(2 - x) \quad \text{Cross-multiplying}$$
$$6x^2 = 2 - x$$
$$6x^2 + x - 2 = 0$$
$$(2x - 1)(3x + 2) = 0$$

$2x - 1 = 0$	$3x + 2 = 0$
$2x = 1$	$3x = -2$
$x = \frac{1}{2}$	$x = -\frac{2}{3}$

The solutions are $\frac{1}{2}$ and $-\frac{2}{3}$.

23. Let $w =$ width
$w + 5 =$ length
Area $= w(w + 5)$

$$w(w + 5) = 24$$
$$w^2 + 5w = 24$$
$$w^2 + 5w - 24 = 0$$
$$(w + 8)(w - 3) = 0$$

$w + 8 = 0$	$w - 3 = 0$
$w = -8$	$w = 3$
Reject; a length cannot be negative.	$w + 5 = 3 + 5 = 8$

The width is 3 and the length is 8.

24. Width $= 7$; length $= 11$

25. Let $x =$ first even integer
$x + 2 =$ second even intger

$$x(x + 2) + 4 = 84$$
$$x^2 + 2x + 4 = 84$$
$$x^2 + 2x - 80 = 0$$
$$(x + 10)(x - 8) = 0$$

$x + 10 = 0$	$x - 8 = 0$
$x = -10$	$x = 8$
$x + 2 = -8$	$x + 2 = 10$

Two answers: The integers are -10 and -8, or the integers are 8 and 10.

26. Two answers: The integers are 6 and 7, or the integers are -7 and -6.

27. Let $x =$ rate from LA to Mexico
$x + 20 =$ rate from Mexico to LA

$$r \cdot t = d. \text{ Therefore, } t = \frac{d}{r}.$$

$$\text{Time from LA to Mexico} = \frac{120}{x}$$

$$\text{Time from Mexico to LA} = \frac{120}{x + 20}$$

Total time is 5 hr

$$\frac{120}{x} + \frac{120}{x + 20} = 5 \quad \text{LCD} = x(x + 20)$$

$$\overset{1}{\cancel{x}}(x + 20)\frac{120}{\cancel{x}} + x(\overset{1}{\cancel{x + 20}})\frac{120}{\cancel{x + 20}} = 5 \cdot x(x + 20)$$

$$120x + 2{,}400 + 120x = 5x^2 + 100x$$
$$240x + 2{,}400 = 5x^2 + 100x$$
$$0 = 5x^2 - 140x - 2{,}400$$
$$0 = 5(x^2 - 28x - 480)$$
$$0 = 5(x + 12)(x - 40)$$

$5 \neq 0$	$x + 12 = 0$	$x - 40 = 0$
	Reject $x = -12$	$x = 40$

Therefore, average speed to Mexico is 40 mph.

28. 45 mph

Exercises 12.3 (page 571)

1. $8x^2 - 4x = 0$ **2.** $0, \frac{1}{3}$
$4x(2x - 1) = 0$

$4x = 0$	$2x - 1 = 0$
$x = 0$	$2x = 1$
	$x = \frac{1}{2}$

The solutions are 0 and $\frac{1}{2}$.

3. $x^2 = 9$ **4.** $6, -6$
$x = \pm\sqrt{9}$
$x = \pm 3$

The solutions are 3 and -3.

5. $x(x - 4) = 0$ **6.** $0, 16$

$x = 0$	$x - 4 = 0$
	$x = 4$

The solutions are 0 and 4.

7. $x^2 = -16$
$x = \pm\sqrt{-16}$ The solutions are not real numbers.

8. The solutions are not real numbers.

9. $x^2 = 4$

$x = \pm\sqrt{4}$

$x = \pm 2$

The solutions are 2 and -2.

10. ± 4

11. $x^2 = \frac{4}{5}$

$x = \pm\sqrt{\frac{4}{5}}$

$x = \pm\dfrac{2}{\sqrt{5}} = \pm\dfrac{2}{\sqrt{5}}\cdot\dfrac{\sqrt{5}}{\sqrt{5}}$

$x = \pm\dfrac{2\sqrt{5}}{5}$

The solutions are $\dfrac{2\sqrt{5}}{5}$ and $-\dfrac{2\sqrt{5}}{5}$.

12. $\pm\dfrac{5\sqrt{3}}{3}$

13. $2x^2 = 8$

$x^2 = 4$

$x = \pm\sqrt{4}$

$x = \pm 2$

The solutions are 2 and -2.

14. ± 3

15. $x^2 + 3x = 3x - 4$

$x^2 = -4$

$x = \pm\sqrt{-4}$ The solutions are not real numbers.

16. The solutions are not real numbers.

17. $2x + 6 = 6 + x^2 + 2x$

$x^2 = 0$

$x = \pm\sqrt{0}$

$x = 0$

The solution is 0.

18. 0

19. $6x^2 - 8x = 6 - 8x$

$6x^2 = 6$

$x^2 = 1$

$x = \pm\sqrt{1}$

$x = \pm 1$

The solutions are 1 and -1.

20. $\pm\frac{1}{5}\sqrt{30}$

Exercises 12.4 (page 576)

(Checks will not be shown.)

1. $3x^2 - 1x - 2 = 0 \quad \begin{cases} a = 3 \\ b = -1 \\ c = -2 \end{cases}$

$x = \dfrac{-(-1) \pm \sqrt{(-1)^2 - 4(3)(-2)}}{2(3)}$

$= \dfrac{1 \pm \sqrt{1 + 24}}{6} = \dfrac{1 \pm \sqrt{25}}{6}$

$x = \dfrac{1 \pm 5}{6} = \begin{cases} \dfrac{1+5}{6} = \dfrac{6}{6} = 1 \\ \dfrac{1-5}{6} = \dfrac{-4}{6} = -\dfrac{2}{3} \end{cases}$

The solutions are 1 and $-\frac{2}{3}$.

2. $\frac{1}{2}, -2$

3. $1x^2 - 4x + 1 = 0 \quad \begin{cases} a = 1 \\ b = -4 \\ c = 1 \end{cases}$

$x = \dfrac{-(-4) \pm \sqrt{(-4)^2 - 4(1)(1)}}{2(1)} = \dfrac{4 \pm \sqrt{16 - 4}}{2(1)} = \dfrac{4 \pm \sqrt{12}}{2}$

$= \dfrac{4 \pm 2\sqrt{3}}{2} = \dfrac{2(2 \pm \sqrt{3})}{2} = 2 \pm \sqrt{3}$

The solutions are $2 + \sqrt{3}$ and $2 - \sqrt{3}$.

4. $2 \pm \sqrt{5}$

5. LCD $= 2x$

$\dfrac{2x}{1}\cdot\dfrac{x}{2} + \dfrac{2x}{1}\cdot\dfrac{2}{x} = \dfrac{2x}{1}\cdot\dfrac{5}{2}$

$x^2 + 4 = 5x$

$1x^2 - 5x + 4 = 0 \quad \begin{cases} a = 1 \\ b = -5 \\ c = 4 \end{cases}$

$x = \dfrac{-(-5) \pm \sqrt{(-5)^2 - 4(1)(4)}}{2(1)} = \dfrac{5 \pm \sqrt{25 - 16}}{2} = \dfrac{5 \pm \sqrt{9}}{2}$

$x = \dfrac{5 \pm 3}{2} = \begin{cases} \dfrac{5+3}{2} = \dfrac{8}{2} = 4 \\ \dfrac{5-3}{2} = \dfrac{2}{2} = 1 \end{cases}$

The solutions are 4 and 1.

6. 1, 6

7. $2x^2 - 8x + 5 = 0 \quad \begin{cases} a = 2 \\ b = -8 \\ c = 5 \end{cases}$

$x = \dfrac{-(-8) \pm \sqrt{(-8)^2 - 4(2)(5)}}{2(2)}$

$= \dfrac{8 \pm \sqrt{64 - 40}}{4} = \dfrac{8 \pm \sqrt{24}}{4} = \dfrac{8 \pm 2\sqrt{6}}{4}$

$= \dfrac{\overset{1}{2}(4 \pm \sqrt{6})}{\underset{2}{4}} = \dfrac{4 \pm \sqrt{6}}{2}$

The solutions are $\dfrac{4 + \sqrt{6}}{2}$ and $\dfrac{4 - \sqrt{6}}{2}$.

8. $\dfrac{3 \pm \sqrt{3}}{3}$

9. LCD $= x(x - 1)$

$\dfrac{\overset{1}{x(x-1)}}{1}\cdot\dfrac{1}{\underset{1}{x}} + \dfrac{x(\overset{1}{x-1})}{1}\cdot\dfrac{x}{\underset{1}{x-1}} = \dfrac{x(x-1)}{1}\cdot\dfrac{3}{1}$

$(x - 1) + x^2 = 3x(x - 1)$

$x - 1 + x^2 = 3x^2 - 3x \quad \begin{cases} a = 2 \\ b = -4 \\ c = 1 \end{cases}$

$0 = 2x^2 - 4x + 1$

$x = \dfrac{-(-4) \pm \sqrt{(-4)^2 - 4(2)(1)}}{2(2)}$

$= \dfrac{4 \pm \sqrt{16 - 8}}{4} = \dfrac{4 \pm \sqrt{8}}{4}$

$= \dfrac{4 \pm 2\sqrt{2}}{4} = \dfrac{2(2 \pm \sqrt{2})}{4} = \dfrac{2 \pm \sqrt{2}}{2}$

The solutions are $\dfrac{2 + \sqrt{2}}{2}$ and $\dfrac{2 - \sqrt{2}}{2}$.

10. $\dfrac{-1 \pm \sqrt{11}}{5}$

11. $x^2 + x + 1 = 0 \quad \begin{cases} a = 1 \\ b = 1 \\ c = 1 \end{cases}$

$x = \dfrac{-(1) \pm \sqrt{(1)^2 - 4(1)(1)}}{2(1)} = \dfrac{-1 \pm \sqrt{1 - 4}}{2}$

$= \dfrac{-1 \pm \sqrt{-3}}{2}$

The solutions are not real numbers.

12. The solutions are not real numbers.

13. $4x^2 + 4x - 1 = 0$ $\begin{cases} a = 4 \\ b = 4 \\ c = -1 \end{cases}$

$x = \dfrac{-(4) \pm \sqrt{(4)^2 - 4(4)(-1)}}{2(4)} = \dfrac{-4 \pm \sqrt{16 + 16}}{8}$

$= \dfrac{-4 \pm \sqrt{32}}{8} = \dfrac{-4 \pm 4\sqrt{2}}{8} = \dfrac{4(-1 \pm \sqrt{2})}{8} = \dfrac{-1 \pm \sqrt{2}}{2}$

The solutions are $\dfrac{-1 + \sqrt{2}}{2}$ and $\dfrac{-1 - \sqrt{2}}{2}$.

14. $\dfrac{1 \pm \sqrt{3}}{3}$

15. $3x^2 + 2x + 1 = 0$ $\begin{cases} a = 3 \\ b = 2 \\ c = 1 \end{cases}$

$x = \dfrac{-(2) \pm \sqrt{(2)^2 - 4(3)(1)}}{2(3)}$

$= \dfrac{-2 \pm \sqrt{4 - 12}}{6} = \dfrac{-2 \pm \sqrt{-8}}{6}$

The solutions are not real numbers.

16. The solutions are not real numbers.

17. Let x = number

$\dfrac{1}{x}$ = its reciprocal

$x - \dfrac{1}{x} = \dfrac{2}{3}$ LCD = $3x$

$(3x)x - (3x)\left(\dfrac{1}{x}\right) = \left(\dfrac{2}{3}\right)(3x)$

$3x^2 - 3 = 2x$

$3x^2 - 2x - 3 = 0$ $\begin{cases} a = 3 \\ b = -2 \\ c = -3 \end{cases}$

$x = \dfrac{-(-2) \pm \sqrt{(-2)^2 - 4(3)(-3)}}{2(3)}$

$= \dfrac{2 \pm \sqrt{4 + 36}}{6}$

$= \dfrac{2 \pm \sqrt{40}}{6} = \dfrac{2 \pm 2\sqrt{10}}{6} = \dfrac{2(1 \pm \sqrt{10})}{6} = \dfrac{1 \pm \sqrt{10}}{3}$

Two answers: The number is $\dfrac{1 + \sqrt{10}}{3}$ or the number is $\dfrac{1 - \sqrt{10}}{3}$.

18. Two answers: The number is $\dfrac{-5 + 3\sqrt{17}}{8}$ or the number is $\dfrac{-5 - 3\sqrt{17}}{8}$.

19. $4x^2 + 4x - 1 = 0$ $\begin{cases} a = 4 \\ b = 4 \\ c = -1 \end{cases}$

$x = \dfrac{-(4) \pm \sqrt{(4)^2 - 4(4)(-1)}}{2(4)} = \dfrac{-4 \pm \sqrt{16 + 16}}{8}$

$= \dfrac{-4 \pm \sqrt{16 \cdot 2}}{8} = \dfrac{-4 \pm 4\sqrt{2}}{8} = \dfrac{4(-1 \pm \sqrt{2})}{8} = \dfrac{-1 \pm \sqrt{2}}{2}$

$x \approx \dfrac{-1 \pm 1.414}{2} = \begin{cases} \dfrac{0.414}{2} \approx 0.21 \\ \dfrac{-2.414}{2} \approx -1.21 \end{cases}$

One solution is $\dfrac{-1 + \sqrt{2}}{2}$ (≈ 0.21) and the other is $\dfrac{-1 - \sqrt{2}}{2}$ (≈ -1.21).

20. One solution is $\dfrac{1 + \sqrt{3}}{3} \approx 0.91$ and the other is $\dfrac{1 - \sqrt{3}}{3} \approx -0.24$.

21. Let w = width

$w + 2$ = length

Area = $(w + 2)w = 2$

$w^2 + 2w = 2$

$1w^2 + 2w - 2 = 0$ $\begin{cases} a = 1 \\ b = 2 \\ c = -2 \end{cases}$

$w = \dfrac{-(2) \pm \sqrt{(2)^2 - 4(1)(-2)}}{2(1)} = \dfrac{-2 \pm \sqrt{4 + 8}}{2}$

$= \dfrac{-2 \pm \sqrt{12}}{2} = \dfrac{-2 \pm 2\sqrt{3}}{2} = \dfrac{\overset{1}{2}(-1 \pm \sqrt{3})}{\underset{1}{2}} = -1 \pm \sqrt{3}$

$w = \begin{cases} -1 + \sqrt{3} \approx -1 + 1.732 \approx 0.73 \\ -1 - \sqrt{3} \approx -1 - 1.732 \approx -2.73 \end{cases}$

Reject $w = -1 - \sqrt{3}$; a length cannot be negative.

If $w = -1 + \sqrt{3}$; $w + 2 = (-1 + \sqrt{3}) + 2 = 1 + \sqrt{3} \approx 2.73$.

The width is $-1 + \sqrt{3}$ (≈ 0.73) and the length is $1 + \sqrt{3}$ (≈ 2.73).

22. Width $= -2 + \sqrt{10} \approx 1.16$; length $= 2 + \sqrt{10} \approx 5.16$

23. Let x = length of a side of square

Area = x^2
Perimeter = $4x$

$4x = x^2 - 6$

$x^2 - 4x - 6 = 0$ $\begin{cases} a = 1 \\ b = -4 \\ c = -6 \end{cases}$

$x = \dfrac{-(-4) \pm \sqrt{(-4)^2 - 4(1)(-6)}}{2(1)}$

$= \dfrac{4 \pm \sqrt{16 + 24}}{2} = \dfrac{4 \pm \sqrt{40}}{2}$

$= \dfrac{4 \pm 2\sqrt{10}}{2} = \dfrac{2(2 \pm \sqrt{10})}{2} = 2 \pm \sqrt{10}$

$x = \begin{cases} 2 + \sqrt{10} \approx 2 + 3.162 \approx 5.16 \\ 2 - \sqrt{10} \approx 2 - 3.162 \approx -1.16 \end{cases}$

Reject $2 - \sqrt{10}$; a length cannot be negative.
The length of a side is $2 + \sqrt{10}$ (≈ 5.16).

24. The length of a side is $2 + \sqrt{6} \approx 4.45$.

Exercises 12.5 (page 580)

1. $x^2 = 16^2 + 12^2$ **2.** 15
$x^2 = 256 + 144$
$x^2 = 400$
$x = \sqrt{400} = 20$

3. $x^2 = 6^2 + 2^2$ **4.** $2\sqrt{5}$
$x^2 = 36 + 4$
$x^2 = 40$
$x = \sqrt{40} = \sqrt{4 \cdot 10} = 2\sqrt{10}$

5. $x^2 = (\sqrt{5})^2 + 2^2$ **6.** 4   **7.** $5^2 = x^2 + 3^2$ **8.** 5
$x^2 = 5 + 4$ $25 = x^2 + 9$
$x^2 = 9$ $x^2 = 16$
$x = \sqrt{9} = 3$ $x = \sqrt{16}$
                                        $x = 4$

9. $(3\sqrt{3})^2 = 3^2 + x^2$ **10.** $2\sqrt{2}$
$9 \cdot 3 = 9 + x^2$
$27 = 9 + x^2$
$x^2 = 18$
$x = \sqrt{18} = \sqrt{9 \cdot 2} = 3\sqrt{2}$

704

11. $x^2 = 10^2 + 6^2$
$x^2 = 100 + 36$
$x^2 = 136$
$x = \sqrt{136} = \sqrt{4 \cdot 34} = 2\sqrt{34}$

12. $4\sqrt{13}$

13. $(x + 1)^2 + (\sqrt{20})^2 = (x + 3)^2$
$x^2 + 2x + 1 + 20 = x^2 + 6x + 9$
$12 = 4x$
$x = 3$

14. 7

15. Let w = width
$(24)^2 + w^2 = (25)^2$
$576 + w^2 = 625$
$w^2 = 49$
$w = \sqrt{49}$
$w = 7$

The width is 7 cm

16. 9 in.

17. Let x = length of one leg (in meters)
$2x - 4$ = length of other leg (in meters)
$10^2 = (2x - 4)^2 + x^2$
$100 = 4x^2 - 16x + 16 + x^2$
$0 = 5x^2 - 16x - 84$
$0 = (5x + 14)(x - 6)$

$5x + 14 = 0 \qquad\qquad x - 6 = 0$
$x = -\frac{14}{5} \qquad\qquad\quad x = 6$
Reject $\qquad\qquad\qquad 2x - 4 = 12 - 4 = 8$

One leg is 6 m long and the other is 8 m long.

18. Width = 9 yd; length = 12 yd

19. Let x = length of diagonal (in inches).
$x^2 = (\sqrt{6})^2 + (\sqrt{6})^2$
$x^2 = 6 + 6$
$x^2 = 12$
$x = \sqrt{12}$
$x = 2\sqrt{3}$

The diagonal is $2\sqrt{3}$ in. long.

20. $2\sqrt{7}$ m

21. Let x = side.
$(4\sqrt{2})^2 = x^2 + x^2$
$32 = 2x^2$
$16 = x^2$
$\sqrt{16} = x$
$x = 4$

A side is 4 cm long.

22. 5 in.

23. Let x = length of diagonal (in cm).
$(x)^2 = (41.6)^2 + (41.6)^2$
$x^2 = 1{,}730.56 + 1{,}730.56$
$x^2 = 3{,}461.12$
$x = \sqrt{3{,}461.12}$
$x \approx 58.83$

The diagonal is about 58.83 cm long.

24. ≈ 2.66 m

Review Exercises 12.6 (page 584)

(The checks will not be shown.)

1. $x^2 + x - 6 = 0$
$(x + 3)(x - 2) = 0$
$x + 3 = 0 \qquad\qquad x - 2 = 0$
$x = -3 \qquad\qquad\quad x = 2$

The solutions are -3 and 2.

2. 0, 49

3. $x^2 - 2x - 4 = 0 \quad \begin{cases} a = 1 \\ b = -2 \\ c = -4 \end{cases}$

$x = \dfrac{-(-2) \pm \sqrt{(-2)^2 - 4(1)(-4)}}{2(1)}$

$= \dfrac{2 \pm \sqrt{4 + 16}}{2} = \dfrac{2 \pm \sqrt{20}}{2}$

$= \dfrac{2 \pm 2\sqrt{5}}{2} = \dfrac{2(1 \pm \sqrt{5})}{2} = 1 \pm \sqrt{5}$

The solutions are $1 + \sqrt{5}$ and $1 - \sqrt{5}$.

4. $2 \pm \sqrt{3}$

5. $x^2 - 5x = 0$
$x(x - 5) = 0$
$x = 0 \qquad\quad\; x - 5 = 0$
$\qquad\qquad\qquad x = 5$

The solutions are 0 and 5.

6. $\pm\frac{5}{6}$

7. LCD = $3(x - 2)$
$\dfrac{3(x-2)}{1} \cdot \dfrac{(x+2)}{3} = \dfrac{3(x-2)}{1} \cdot \dfrac{1}{(x-2)} + \dfrac{3(x-2)}{1} \cdot \dfrac{2}{3}$
$(x - 2)(x + 2) = 3 + 2(x - 2)$
$x^2 - 4 = 3 + 2x - 4$
$x^2 - 2x - 3 = 0$
$(x - 3)(x + 1) = 0$
$x - 3 = 0 \qquad\qquad x + 1 = 0$
$x = 3 \qquad\qquad\quad x = -1$

The solutions are 3 and -1.

8. The solutions are not real numbers.

9. $5x + 10 = x^2 + 5x$
$x^2 = 10$
$x = \pm\sqrt{10}$

The solutions are $\sqrt{10}$ and $-\sqrt{10}$.

10. $\pm 2\sqrt{3}$

11. LCD = $x(x + 1)$
$\dfrac{x(x+1)}{1} \cdot \dfrac{2}{x} + \dfrac{x(x+1)}{1} \cdot \dfrac{x}{(x+1)} = \dfrac{x(x+1)}{1} \cdot \dfrac{5}{1}$
$2(x + 1) + x^2 = 5x(x + 1)$
$2x + 2 + x^2 = 5x^2 + 5x$
$4x^2 + 3x - 2 = 0 \quad \begin{cases} a = 4 \\ b = 3 \\ c = -2 \end{cases}$

$x = \dfrac{-(3) \pm \sqrt{(3)^2 - 4(4)(-2)}}{2(4)} = \dfrac{-3 \pm \sqrt{9 + 32}}{8} = \dfrac{-3 \pm \sqrt{41}}{8}$

The solutions are $\dfrac{-3 + \sqrt{41}}{8}$ and $\dfrac{-3 - \sqrt{41}}{8}$.

12. $\dfrac{-1 \pm \sqrt{73}}{6}$

13. $3x^2 + 2x + 1 = 0 \quad \begin{cases} a = 3 \\ b = 2 \\ c = 1 \end{cases}$

$x = \dfrac{-(2) \pm \sqrt{(2)^2 - 4(3)(1)}}{2(3)}$

$= \dfrac{-2 \pm \sqrt{4 - 12}}{6} = \dfrac{-2 \pm \sqrt{-8}}{6}$

The solutions are not real numbers.

14. $\pm\dfrac{3\sqrt{5}}{5}$

15. $15^2 = 12^2 + x^2$
$225 = 144 + x^2$
$81 = x^2$
$\sqrt{81} = x$
$x = 9$

16. 12 in.

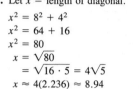

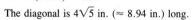

17. Let x = length of diagonal (in meters).   **18.** $8\sqrt{2}$ yd

$$x^2 = (\sqrt{10})^2 + (\sqrt{10})^2$$
$$x^2 = 10 + 10$$
$$x^2 = 20$$
$$x = \sqrt{20}$$
$$\quad = 2\sqrt{5}$$

The diagonal is $2\sqrt{5}$ m long.

19. Let x = length of diagonal.

$$x^2 = 8^2 + 4^2$$
$$x^2 = 64 + 16$$
$$x^2 = 80$$
$$x = \sqrt{80}$$
$$\quad = \sqrt{16 \cdot 5} = 4\sqrt{5}$$
$$x \approx 4(2.236) \approx 8.94$$

The diagonal is $4\sqrt{5}$ in. (≈ 8.94 in.) long.

20. The width is $(-3 + \sqrt{15})$ ft ≈ 0.87 ft, and the length is $(3 + \sqrt{15})$ ft ≈ 6.87 ft.

Chapter 12 Diagnostic Test (page 589)

Following each problem number is the textbook section number (in parentheses) where that kind of problem is discussed.

1. (12.1)   **a.** $5 + 3x^2 = 7x$

$$3x^2 - 7x + 5 = 0 \quad \begin{cases} a = 3 \\ b = -7 \\ c = 5 \end{cases}$$

b. $3(x + 5) = 3(4 - x^2)$
$$3x + 15 = 12 - 3x^2$$

$$3x^2 + 3x + 3 = 0 \quad \begin{cases} a = 3 \\ b = 3 \\ c = 3 \end{cases}$$

2. (12.4) $x^2 = 8x + 15$

$$x^2 - 8x - 15 = 0 \quad \begin{cases} a = 1 \\ b = -8 \\ c = -15 \end{cases}$$

$$x = \frac{-(-8) \pm \sqrt{(-8)^2 - 4(1)(-15)}}{2(1)}$$

$$= \frac{8 \pm \sqrt{64 + 60}}{2} = \frac{8 \pm \sqrt{124}}{2}$$

$$= \frac{8 \pm 2\sqrt{31}}{2} = \frac{2(4 \pm \sqrt{31})}{2} = 4 \pm \sqrt{31}$$

The solutions are $4 + \sqrt{31}$ and $4 - \sqrt{31}$.

3. (12.3) $5x^2 = 17x$
$$5x^2 - 17x = 0$$
$$x(5x - 17) = 0$$

$x = 0$ | $5x - 17 = 0$
$$5x = 17$$
$$x = \tfrac{17}{5}$$

The solutions are 0 and $\tfrac{17}{5}$.

4. (12.4) $x^2 + 2x + 6 = 0 \quad \begin{cases} a = 1 \\ b = 2 \\ c = 6 \end{cases}$

$$x = \frac{-(2) \pm \sqrt{(2)^2 - 4(1)(6)}}{2(1)} = \frac{-2 \pm \sqrt{4 - 24}}{2}$$

$$= \frac{-2 \pm \sqrt{-20}}{2} \quad \text{The solutions are not real numbers.}$$

5. (12.3) $6x^2 = 54$
$$x^2 = \tfrac{54}{6} = 9$$
$$x = \pm 3$$

The solutions are 3 and -3.

6. (12.4) $3(x + 1) = x^2$
$$3x + 3 = x^2$$

$$0 = x^2 - 3x - 3 \quad \begin{cases} a = 1 \\ b = -3 \\ c = -3 \end{cases}$$

$$x = \frac{-(-3) \pm \sqrt{(-3)^2 - 4(1)(-3)}}{2(1)}$$

$$= \frac{3 \pm \sqrt{9 + 12}}{2} = \frac{3 \pm \sqrt{21}}{2}$$

The solutions are $\dfrac{3 + \sqrt{21}}{2}$ and $\dfrac{3 - \sqrt{21}}{2}$.

7. (12.3) $x(x + 3) = x$
$$x^2 + 3x = x$$
$$x^2 + 2x = 0$$
$$x(x + 2) = 0$$

$x = 0$ | $x + 2 = 0$
$$x = -2$$

The solutions are 0 and -2.

8. (12.5) $(\sqrt{11})^2 + x^2 = (4)^2$
$$11 + x^2 = 16$$
$$x^2 = 5$$
$$x = \sqrt{5}$$

9. (12.5)   Let x = length of diagonal.
$$x^2 = (7)^2 + (7)^2$$
$$x^2 = 49 + 49$$
$$x^2 = 98$$
$$x = \sqrt{98} = 7\sqrt{2}$$

The diagonal is $7\sqrt{2}$ m long.

10. (12.5)   Let x = width of rectangle
$x + 6$ = length of rectangle
$$x(x + 6) = 78$$

$$x^2 + 6x - 78 = 0 \quad \begin{cases} a = 1 \\ b = 6 \\ c = -78 \end{cases}$$

$$x = \frac{-(6) \pm \sqrt{(6)^2 - 4(1)(-78)}}{2(1)} = \frac{-6 \pm \sqrt{36 + 312}}{2}$$

$$= \frac{-6 \pm \sqrt{348}}{2} = \frac{-6 \pm 2\sqrt{87}}{2} = \frac{2(-3 \pm \sqrt{87})}{2} = -3 \pm \sqrt{87}$$

We reject $-3 - \sqrt{87}$ because a length cannot be negative. The width is $(-3 + \sqrt{87})$ ft and the length is $[(-3 + \sqrt{87}) + 6]$ ft $= (3 + \sqrt{87})$ ft.

Cumulative Review Exercises: Chapters 1–12 (page 590)

1. No   **2.** No   **3.** 0   **4.** $\dfrac{4s^4t^6}{9}$ **5.** Not defined

6. $9 - 2\sqrt{3}$ **7.**

$$\begin{array}{r} 4x^2 + 3x - 2 \\ 5x - 3 \\ \hline -12x^2 - 9x + 6 \\ 20x^3 + 15x^2 - 10x \\ \hline 20x^3 + 3x^2 - 19x + 6 \end{array}$$

8. $26 - 2\sqrt{105}$ **9.** $\dfrac{x(\cancel{x+3})}{(2x+5)(\cancel{x+1})} \cdot \dfrac{(\cancel{x+1})(\cancel{x-3})}{(\cancel{x+3})(\cancel{x-3})} = \dfrac{x}{2x+5}$

10. $x - 11 - \dfrac{38}{x-3}$

11. $\dfrac{(x+4)(\cancel{x-2})}{1} \cdot \dfrac{x(\cancel{x+2})}{(\cancel{x+2})(\cancel{x-2})} = x(x+4)$ **12.** $\dfrac{6x}{x^2-9}$

13. $\dfrac{x^2+3x-2}{(x+2)(x+3)} - \dfrac{x-1}{(2x+1)(x+3)}$

$= \dfrac{(x^2+3x-2)(2x+1)}{(x+2)(x+3)(2x+1)} - \dfrac{(x-1)(x+2)}{(2x+1)(x+3)(x+2)}$

$= \dfrac{2x^3+7x^2-x-2}{(x+2)(x+3)(2x+1)} - \dfrac{x^2+x-2}{(2x+1)(x+3)(x+2)}$

$= \dfrac{(2x^3+7x^2-x-2)-(x^2+x-2)}{(x+2)(x+3)(2x+1)}$

$= \dfrac{2x^3+7x^2-x-2-x^2-x+2}{(x+2)(x+3)(2x+1)} = \dfrac{2x^3+6x^2-2x}{(x+2)(x+3)(2x+1)}$

14. $\dfrac{27\sqrt{2}}{2}$

(The checks will usually not be shown.)

15. $7 - 1(2x+3) = 3x - 2 - 1(4x-3)$
$7 - 2x - 3 = 3x - 2 - 4x + 3$

$4 - 2x = -x + 1$
$\underline{-1 + 2x} \quad \underline{2x - 1}$
$3 \quad = \quad x$
$x = 3$

The solution is 3.

16. $\left(\dfrac{29}{2}, -6\right)$ or $\left(14\tfrac{1}{2}, -6\right)$

17. (1) $2]$ $\begin{cases} 3x + 4y = 3 \\ 2x - 5y = 25 \end{cases}$ $\Rightarrow$ $\begin{aligned} 6x + 8y &= 6 \\ \underline{-6x + 15y} &= \underline{-75} \\ 23y &= -69 \\ y &= -3 \end{aligned}$ **18.** $-\dfrac{2}{3}$

Substituting -3 for y in (1):

(1) $3x + 4y = 3$
$3x + 4(-3) = 3$
$3x - 12 = 3$
$3x = 15$
$x = 5$

Solution: $(5, -3)$

19. $(\sqrt{5x+3})^2 = (10x)^2$
$5x + 3 = 100x^2$
$0 = 100x^2 - 5x - 3$
$0 = (20x + 3)(5x - 1)$

$20x + 3 = 0 \qquad \Big| \qquad 5x - 1 = 0$
$20x = -3 \qquad \Big| \qquad 5x = 1$
$x = -\dfrac{3}{20} \quad \Big| \quad x = \dfrac{1}{5}$

Check for $x = \tfrac{1}{5}$: $\qquad$ *Check for $x = -\tfrac{3}{20}$:*

$\sqrt{5\left(\tfrac{1}{5}\right) + 3} = 10\left(\tfrac{1}{5}\right) \qquad \sqrt{5\left(-\tfrac{3}{20}\right) + 3} = 10\left(-\tfrac{3}{20}\right)$

$\sqrt{1+3} \overset{?}{=} 2 \qquad\qquad \sqrt{-\tfrac{3}{4} + 3} \overset{?}{=} -\tfrac{3}{2}$

$\sqrt{4} = 2$ True $\qquad\qquad \sqrt{\tfrac{9}{4}} = -\tfrac{3}{2}$ False

The only solution is $\tfrac{1}{5}$.

20. $-\dfrac{3}{7}$ **21.** $12x - 3 \le 7x - 13$ **22.** $-7, 2$
$\qquad\qquad\qquad \underline{-7x + 3} \quad \underline{-7x + 3}$
$\qquad\qquad\qquad 5x \quad \le \quad -10$
$\qquad\qquad\qquad\qquad x \le -2$

23. $4x^2 - 12x + 3 = 0$ $\begin{cases} a = 4 \\ b = -12 \\ c = 3 \end{cases}$ **24.** $2, -1$

$x = \dfrac{-(-12) \pm \sqrt{(-12)^2 - 4(4)(3)}}{2(4)}$

$= \dfrac{12 \pm \sqrt{144 - 48}}{8}$

$= \dfrac{12 \pm \sqrt{96}}{8} = \dfrac{12 \pm 4\sqrt{6}}{8} = \dfrac{4(3 \pm \sqrt{6})}{8} = \dfrac{3 \pm \sqrt{6}}{2}$

The solutions are $\dfrac{3 + \sqrt{6}}{2}$ and $\dfrac{3 - \sqrt{6}}{2}$.

25. $7x^2 = 1$ **26.** $6 + 2\sqrt{2}$
$\quad x^2 = \tfrac{1}{7}$

$x = \pm\sqrt{\dfrac{1}{7}} = \pm\sqrt{\dfrac{1 \cdot 7}{7 \cdot 7}} = \pm\dfrac{\sqrt{7}}{7}$

The solutions are $\dfrac{\sqrt{7}}{7}$ and $-\dfrac{\sqrt{7}}{7}$.

27. $3(x^2 - 4x - 5) = 3(x - 5)(x + 1)$

28. $5(x+3)(x-3)$ **29.** Not factorable   **30.** $(7x+1)(x-2)$

In Exercises 31–34, the checks will not be shown.

31. Let x = number of inches corresponding to 12 ft
$\qquad y$ = number of inches corresponding to 18 ft

$\dfrac{\tfrac{1}{4} \text{ in.}}{1 \text{ ft}} = \dfrac{x \text{ in.}}{12 \text{ ft}} \qquad\qquad \dfrac{\tfrac{1}{4} \text{ in.}}{1 \text{ ft}} = \dfrac{y \text{ in.}}{18 \text{ ft}}$

$\qquad x = \tfrac{1}{4}(12) = 3 \qquad\qquad y = \tfrac{1}{4}(18) = \tfrac{9}{2}$

The drawing will be 3 in. by $\tfrac{9}{2}$ in.

32. Two answers: The integers are -3 and -2, or the integers are 4 and 5.

33. Let $\qquad x$ = number of pounds of stew meat
$\qquad\qquad 5 - x$ = number of pounds of back ribs

$2.10x + 1.20(5 - x) = 7.35$
$210x + 120(5 - x) = 735$
$210x + 600 - 120x = 735$
$90x = 135$
$x = \tfrac{135}{90} = \tfrac{3}{2}$ or $1\tfrac{1}{2}$
$5 - x = 5 - \tfrac{3}{2} = \tfrac{10}{2} - \tfrac{3}{2} = \tfrac{7}{2}$ or $3\tfrac{1}{2}$

Therefore, there are $1\tfrac{1}{2}$ lb of stew meat and $3\tfrac{1}{2}$ lb of back ribs.

34. The shorter leg is 5 cm and the longer one is 12 cm.

Appendix A Exercises A.1 (page 596)

1. Yes, because it is a collection of objects or things   **2.** Yes

3. Yes, because they have exactly the same members. Writing a member more than once tells you no more than when you write it once—namely, that that element is a member of the set.

4. $\{0, 1, 2\}$

5. $\{\ \}$, since there are no digits greater than 9

6. $\{0, 1, 2\}$

7. $\{10, 11, 12, \ldots\}$. This is the way we show the set of whole numbers greater than 9. It is an infinite set.

8. $\{5\}$

9. $\{\ \}$, since there are no whole numbers greater than 4 and at the same time less than 5

10. $\{0, 1, 2, 3\}$

11. 2, a, 3. Because of the way the set is written, we know that these are its elements.

12. 0, 1, 2, 3, 4, 5, 6, 7, 8, 9

13a. $n(\{1, 1, 3, 5, 5, 5\}) = n(\{1, 3, 5\}) = 3$. This set has three elements: 1, 3, and 5.

b. $n(\{0\}) = 1$. This set has one element: 0.

c. $n(\{a, b, g, x\}) = 4$. This set has four elements: a, b, g, and x.

d. $n(\{0, 1, 2, 3, 4, 5, 6, 7\}) = 8$. You can count its elements and see that there are eight of them.

e. $n(\varnothing) = 0$. The empty set has no elements.

14. $\varnothing$ has no elements; $\{0\}$ has one element, namely 0. Since the sets do not have exactly the same elements, they are not equal.

15a. The set of digits is finite because when we count its elements the counting comes to an end. In this case the counting ends at 10.

b. The set of whole numbers is infinite because when we attempt to count its elements the counting never comes to an end.

c. Finite. The counting ends at 7.

d. Finite. If we started counting the books in the ELAC library, we could eventually finish counting them.

16a. True **b.** True **c.** False **d.** False

17a. True **b.** False **c.** False **d.** True

18a. False **b.** True **c.** True **d.** False

Exercises A.2 (page 598)

1a. $\{3, 5\}$ is a proper subset of M because each of its elements, 3 and 5, is an element of M; and M has at least one element, such as 2, that is not an element of $\{3, 5\}$.

b. $\{0, 1, 7\}$ is not a subset of M because elements 0 and 7 are not elements of M.

c. $\varnothing$ is a proper subset of M because $\varnothing$ is a proper subset of every set except itself.

d. $\{2, 4, 1, 3, 5\}$ is an improper subset of M because each element of C is an element of M, and M has no element that is not an element of C.

2a. Proper **b.** Improper **c.** Proper **d.** Not a subset of P

3. $\{R, G, Y\}$ All the subsets with three elements
$\{R, G\}, \{R, Y\}, \{G, Y\}$ All the subsets with two elements
$\{R\}, \{G\}, \{Y\}$ All the subsets with one element
$\{ \ \}$ All the subsets with no elements

4. $\{\square, \triangle\}, \{\square\}, \{\triangle\}, \{ \ \}$

Exercises A.3 (page 600)

1a. $\{1, 5, 7\} \cup \{2, 4\} = \{1, 5, 7, 2, 4\}$
$\{1, 5, 7\} \cap \{2, 4\} = \{ \ \}$

b. $\{a, b\} \cup \{x, y, z, a\} = \{a, b, x, y, z\}$
$\{a, b\} \cap \{x, y, z, a\} = \{a\}$

c. $\{ \ \} \cup \{k, 2\} = \{k, 2\}$
$\{ \ \} \cap \{k, 2\} = \{ \ \}$

d. {river, boat} $\cup$ {boat, streams, down} = {river, boat, streams, down}
{river, boat} $\cap$ {boat, streams, down} = {boat}

2a. $\{1, 3, 5, 7, 2, 4, 6\}$ **b.** $\{1, 2, 3, 4, 5, 6, 7\}$ **c.** $\{ \ \}$
d. $\{6\}$

3. C and D are disjoint because they have no member in common; that is, $C \cap D = \varnothing$. All other pairs of sets have at least one member in common.

4a. $P = \{c, d, k\}$ **b.** $Q = \{k, j, f\}$ **c.** $P \cup Q = \{c, d, k, j, f\}$
d. $P \cap Q = \{k\}$ **e.** $U = \{a, e, g, h, c, d, k, j, f\}$

5a. $X \cap Y = \{5, 11\}$ because these are the only elements in both X and Y.

b. $Y \cap X = \{5, 11\}$ for the same reason as (a).

c. Yes. $X \cap Y = Y \cap X$ because they have exactly the same elements.

6a. $K \cap L = \{4, b\}$ **b.** $n(K \cap L) = 2$

c. $L \cup M = \{m, 4, 6, b, n, 7, t\}$ **d.** $n(L \cup M) = 7$

7.
$A \cup B$

8.
$P \cap Q$

9. $R \cap S$ because the shaded area is in both R and S. **10.** $Y \cup Z$

Appendix B Exercises B.1 (page 608)

1. $\dfrac{\overset{1}{\cancel{6}}}{\underset{3}{\cancel{9}}} = \dfrac{2}{3}$ **2.** $\dfrac{3}{5}$ **3.** $\dfrac{49}{24}$ Already in lowest terms

4. $\dfrac{7}{8}$ **5.** $\dfrac{5}{2} = 2\overset{\text{2 R 1}}{\overline{)5}} = 2\dfrac{1}{2}$ **6.** $1\dfrac{3}{8}$

7. $\dfrac{47}{25} = 25\overset{\text{1 R 22}}{\overline{)47}} = 1\dfrac{22}{25}$ **8.** $2\dfrac{4}{11}$
$\phantom{\dfrac{47}{25} = 25} \underline{25}$
$\phantom{\dfrac{47}{25} = 25)} 22$

9. $3\dfrac{7}{8} = \dfrac{3 \cdot 8 + 7}{8} = \dfrac{24 + 7}{8} = \dfrac{31}{8}$

10. $\dfrac{25}{9}$ **11.** $2\dfrac{5}{6} = \dfrac{2 \cdot 6 + 5}{6} = \dfrac{12 + 5}{6} = \dfrac{17}{6}$ **12.** $\dfrac{43}{5}$

13. LCD = 12 $\quad \dfrac{2}{3} = \dfrac{8}{12}$ **14.** $\dfrac{1}{2}$
$\phantom{\text{13. LCD = 12}\quad} +\dfrac{1}{4} = \dfrac{3}{12}$
$\phantom{\text{13. LCD = 12}\quad +} \overline{ \dfrac{11}{12}}$

15. LCD = 10 $\quad \dfrac{7}{10} = \dfrac{7}{10}$ **16.** $\dfrac{7}{20}$
$\phantom{\text{15. LCD = 10}\quad} -\dfrac{2}{5} = -\dfrac{4}{10}$
$\phantom{\text{15. LCD = 10}\quad -} \overline{ \dfrac{3}{10}}$

17. $\dfrac{\overset{1}{\cancel{3}}}{\underset{4}{\cancel{16}}} \cdot \dfrac{\overset{5}{\cancel{20}}}{\underset{3}{\cancel{9}}} = \dfrac{5}{12}$ **18.** $\dfrac{9}{28}$ **19.** $\dfrac{4}{3} \div \dfrac{8}{9} = \dfrac{\overset{1}{\cancel{4}}}{\underset{1}{\cancel{3}}} \cdot \dfrac{\overset{3}{\cancel{9}}}{\underset{2}{\cancel{8}}} = \dfrac{3}{2}$ **20.** 4

21. LCD = 15 $\quad 2\dfrac{2}{3} = 2\dfrac{10}{15}$ **22.** $7\dfrac{3}{4}$
$\phantom{\text{21. LCD = 15}\quad} +1\dfrac{3}{5} = 1\dfrac{9}{15}$
$\phantom{\text{21. LCD = 15}\quad +} \overline{ 3\dfrac{19}{15} = 4\dfrac{4}{15}}$

23. LCD = 10 $\quad 5\dfrac{4}{5} = 5\dfrac{8}{10}$ **24.** $2\dfrac{1}{6}$
$\phantom{\text{23. LCD = 10}\quad} -3\dfrac{7}{10} = -3\dfrac{7}{10}$
$\phantom{\text{23. LCD = 10}\quad -} \overline{ 2\dfrac{1}{10}}$

25. $4\frac{1}{5} \cdot 2\frac{1}{7} = \frac{\overset{3}{\cancel{21}}}{\underset{1}{\cancel{5}}} \cdot \frac{\overset{3}{\cancel{15}}}{\underset{1}{\cancel{7}}} = 9$ **26.** $10\frac{1}{2}$

27. $2\frac{2}{5} \div 1\frac{1}{15} = \frac{12}{5} \div \frac{16}{15} = \frac{\overset{3}{\cancel{12}}}{\cancel{5}} \cdot \frac{\overset{3}{\cancel{15}}}{\underset{4}{\cancel{16}}} = \frac{9}{4} = 2\frac{1}{4}$

28. $\frac{2}{3}$ **29.** $\dfrac{\frac{5}{8}}{\frac{5}{6}} = \frac{5}{8} \div \frac{5}{6} = \frac{\overset{1}{\cancel{5}}}{\cancel{8}} \cdot \frac{\overset{3}{\cancel{6}}}{\underset{1}{\cancel{5}}} = \frac{3}{4}$ **30.** $\frac{1}{2}$

31. 34.5
1.74
18.
0.016
54.256

32. 94.745 **33.** 356.40
34.67
321.73

34. 336.56 **35.** $100 \times 7.45 = 745$ **36.** 3,540

37. $\dfrac{46.8}{100} = 0.468$ **38.** 0.0895

39.
```
     9 4.7 8  (2 decimal places)
       7 0.9  (1 decimal place)
   ─────────
     8 5 3 0 2
   6 6 3 4 6
   ───────────
   6 7 1 9.9 0 2 ≈ 6,719.9
```

40. 2,798.5

41.
```
           0.8 2 8 ≈ 0.83
   7.2 5 )6.0 0 7 0 0
           5 8 0 0
          ─────────
           2 0 7 0
           1 4 5 0
          ─────────
             6 2 0 0
             5 8 0 0
            ─────────
               4 0 0
```

42. 1.71

43. $5\frac{3}{4} = \frac{23}{4} = $
```
       5.7 5
   4)2 3ᶾ0ᶾ0
```

44. 4.6 **45.** $0.65 = \dfrac{\overset{13}{\cancel{65}}}{\underset{20}{\cancel{100}}} = \frac{13}{20}$ **46.** $\frac{1}{4}$

47. $5.9 = 5 + .9 = 5 + \frac{9}{10} = 5\frac{9}{10}$ **48.** $4\frac{3}{10}$

49. $4.70 = 470\%$

50. 0.1% **51.** $0.25\,8 = 25.8\%$ **52.** 1,200%

53. $03.5\% = 0.035$ **54.** 0.0002 **55.** $1\,57\% = 1.57$

56. 0.178 **57.** $\frac{1}{8} = 0.12\,5 = 12.5\%$ **58.** 60%

59. $\frac{7}{12} \approx 0.58\,33 = 58.33\%$ **60.** 14.29%

61. $35\% = \frac{35}{100} = \frac{7}{20}$ **62.** $\frac{4}{5}$ **63.** $12\% = \frac{12}{100} = \frac{3}{25}$

64. $\frac{9}{50}$

Appendix C Exercises C.1 (page 615)

1. The set is a function, because no two ordered pairs have the same first element but different second elements. The domain is $\{4, 2, 3, 0\}$.

2. The set is a function. The domain is $\{6, -2, 4, 0\}$.

3. The set is not a function, because the pairs $(2, -5)$ and $(2, 0)$ have the same first element but different second elements.

4. The set is not a function.

5. The set is a function, because no two ordered pairs have the same first element but different second elements. The domain is $\{-8, 3, 6, 9\}$.

6. The set is a function. The domain is $\{-3, -1, 0, 3\}$.

7. The domain is the set of all real numbers except 12. (12 is an excluded value.)

8. The domain is the set of all real numbers except 2.

9. The domain is the set of all real numbers.

10. The domain is the set of all real numbers.

11. The radicand must be ≥ 0.
$$\begin{array}{r} x - 5 \geq 0 \\ \underline{+\,5 \quad +5} \\ x \geq 5 \end{array}$$
The domain is the set of all real numbers greater than or equal to 5.

12. The domain is the set of all real numbers greater than or equal to 10.

13. There is a restriction: $x \geq 0$. Therefore, the domain is the set of all real numbers greater than or equal to 0.

14. The domain is the set of all real numbers greater than or equal to -3.

15. (b) is the graph of a function; (a) and (c) are not graphs of functions.

16. (c) is the graph of a function; (a) and (b) are not graphs of functions.

Exercises C.2 (page 617)

1. $f(x) = (x + 2)^2$
   a. $f(0) = ([0] + 2)^2 = (2)^2 = 4$
   b. $f(2) = ([2] + 2)^2 = (4)^2 = 16$
   c. $f(-3) = ([-3] + 2)^2 = (-1)^2 = 1$
   d. $f(1) = ([1] + 2)^2 = (3)^2 = 9$

2a. 9 b. 25 c. 1 d. 49

3. $g(x) = x^2 + 4x + 4$
   a. $g(0) = (0)^2 + 4(0) + 4 = 0 + 0 + 4 = 4$
   b. $g(2) = (2)^2 + 4(2) + 4 = 4 + 8 + 4 = 16$
   c. $g(-3) = (-3)^2 + 4(-3) + 4 = 9 - 12 + 4 = 1$
   d. $g(1) = (1)^2 + 4(1) + 4 = 1 + 4 + 4 = 9$

4a. 9 b. 25 c. 1 d. 49

5. $F(x) = x^2 + 4$
   a. $F(0) = (0)^2 + 4 = 0 + 4 = 4$
   b. $F(2) = (2)^2 + 4 = 4 + 4 = 8$
   c. $F(-3) = (-3)^2 + 4 = 9 + 4 = 13$
   d. $F(1) = (1)^2 + 4 = 1 + 4 = 5$

6a. 9 b. 13 c. 25 d. 25

7. $f(x) = 3x^2 + x - 1$
   a. $f(0) = 3(0)^2 + (0) - 1 = 0 + 0 - 1 = -1$
   b. $f(-2) = 3(-2)^2 + (-2) - 1 = 12 - 2 - 1 = 9$
   c. $f(5) = 3(5)^2 + (5) - 1 = 75 + 5 - 1 = 79$
   d. $f(1) = 3(1)^2 + (1) - 1 = 3 + 1 - 1 = 3$

8a. -5 b. 4 c. 39 d. 0

9. $f(x) = x^3 - 1$
   a. $f(0) = (0)^3 - 1 = 0 - 1 = -1$
   b. $f(-1) = (-1)^3 - 1 = -1 - 1 = -2$
   c. $f(2) = (2)^3 - 1 = 8 - 1 = 7$
   d. $f(1) = (1)^3 - 1 = 1 - 1 = 0$

10a. 1 b. -7 c. 2 d. 0

11. $g(t) = \dfrac{4t^2 + 5t - 1}{t + 2}$

    a. $g(0) = \dfrac{4(0)^2 + 5(0) - 1}{(0) + 2} = \dfrac{0 + 0 - 1}{2} = -\dfrac{1}{2}$

    b. $g(-1) = \dfrac{4(-1)^2 + 5(-1) - 1}{(-1) + 2} = \dfrac{4 - 5 - 1}{1} = -2$

    c. $g(2) = \dfrac{4(2)^2 + 5(2) - 1}{(2) + 2} = \dfrac{16 + 10 - 1}{4} = \dfrac{25}{4} \text{ or } 6\dfrac{1}{4}$

    d. $g(4) = \dfrac{4(4)^2 + 5(4) - 1}{(4) + 2} = \dfrac{64 + 20 - 1}{6} = \dfrac{83}{6} \text{ or } 13\dfrac{5}{6}$

12a. $\dfrac{1}{3}$ **b.** -1 **c.** -1 **d.** 1

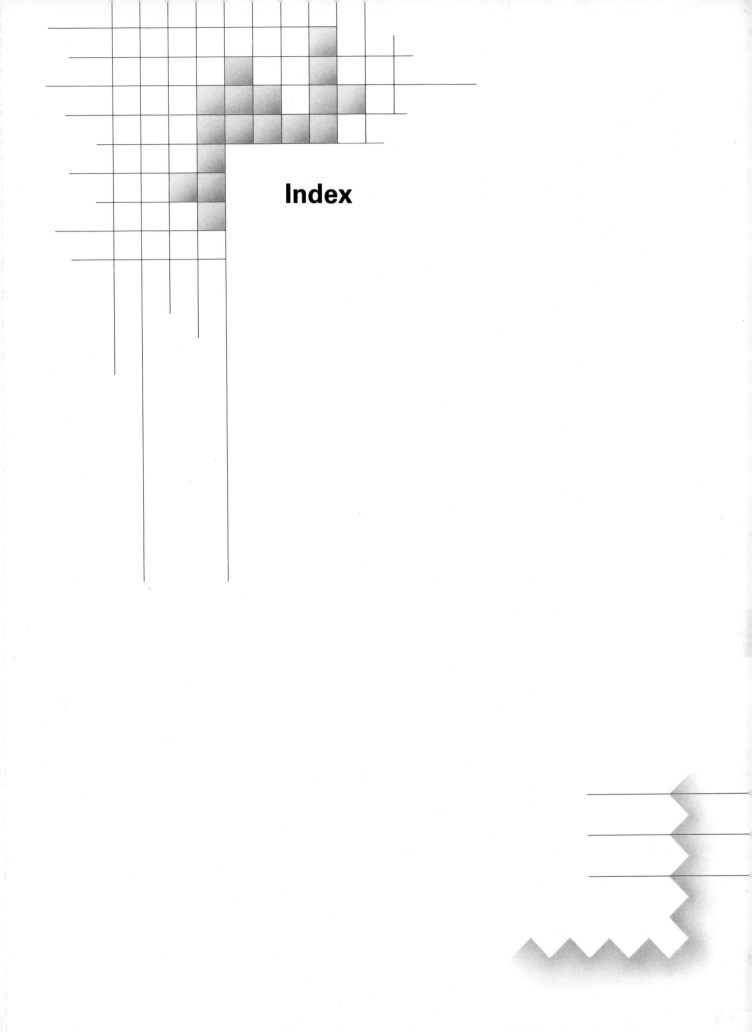

Index

Formula	The formula for finding:	Page Number
$A = lw$	The area of a rectangle, where A is the area, l is the length, and w is the width.	162
$P = 2l + 2w$	The perimeter of a rectangle, where P is the perimeter, l is the length, and w is the width.	162
$A = s^2$	The area of a square, where A is the area and s is the length of a side.	162
$P = 4s$	The perimeter of a square, where P is the perimeter and s is the length of a side.	162
$A = bh$	The area of a parallelogram, where A is the area, b is the base, and h is the altitude.	162
$A = \frac{1}{2}bh$	The area of a triangle, where A is the area, b is the base, and h is the altitude.	99
$P = a + b + c$	The perimeter of a triangle, where P is the perimeter and a, b, and c are the lengths of the sides.	162
$A = \pi r^2$	The area of a circle, where A is the area, r is the radius, and $\pi \approx 3.14$.	102
$C = 2\pi r$	The circumference of a circle, where C is the circumference, r is the radius, and $\pi \approx 3.14$.	162
$V = lwh$	The volume of a rectangular solid, where V is the volume, l is the length, w is the width, and h is the height.	162
$S = 2(lw + lh + wh)$	The surface area of a rectangular solid, where S is the surface area, l is the length, w is the width, and h is the height.	162
$V = s^3$	The volume of a cube, where V is the volume and s is the length of an edge.	162
$S = 6s^2$	The surface area of a cube, where S is the surface area and s is the length of an edge.	162
$V = \frac{4}{3}\pi r^3$	The volume of a sphere, where V is the volume, r is the radius, and $\pi \approx 3.14$.	103
$S = 4\pi r^2$	The surface area of a sphere, where S is the surface area, r is the radius, and $\pi \approx 3.14$.	162
$V = \pi r^2 h$	The volume of a right circular cylinder, where V is the volume, r is the radius, h is the altitude (or height), and $\pi \approx 3.14$.	163
$V = \frac{1}{3}\pi r^2 h$	The volume of a right circular cone, where V is the volume, r is the radius of the base, h is the altitude, and $\pi \approx 3.14$.	111
$I = prt$	Simple interest, where I is the interest, p is the principal, r is the rate, and t is the time.	102
$A = P(1 + rt)$	The amount of money in a savings account, where A is the amount, P is the amount originally invested, r is the annual interest rate, and t is the number of years.	99
$A = P(1 + i)^n$	Compound interest, where A is the compounded amount, P is the principal, i is the interest rate per period, and n is the number of periods.	103
$V = C - Crt$	The value of a depreciated item, where V is the present value, C is the original cost, r is the rate of depreciation, and t is the time.	102
$d = rt$	The distance traveled, where d is the distance, r is the rate of speed, and t is the time.	197
$s = \frac{1}{2}gt^2$	The distance a freely falling object falls, where s is the distance, g is the force due to gravity, and t is the time.	100
$T = \pi\sqrt{\dfrac{L}{g}}$	The time for a single swing of a pendulum, where T is the time, L is the length, g is the force due to gravity, and $\pi \approx 3.14$.	100
$C = \frac{5}{9}(F - 32)$	Degrees Celsius when degrees Fahrenheit are given.	100
$F = \frac{9}{5}C + 32$	Degrees Fahrenheit when degrees Celsius are given.	102
$S = \dfrac{a(1 - r^n)}{1 - r}$	The sum of a geometric series, where S is the sum, a is the first term, n is the number of terms, and r is the common ratio.	101
$I = \dfrac{E}{R}$	The current, where I is the current, E is the electromotive force, and R is the resistance.	101
$\sigma = \sqrt{npq}$	The standard deviation of a binomial distribution, where σ is the standard deviation, n is the number of trials, p is the probability of success, and q is the probability of failure.	102
$C = \dfrac{a}{a + 12} \cdot A$	A child's dosage of medicine, where C is the child's dosage, a is the age of the child, and A is the adult dosage.	103